Graduate Texts in Mathematics 276

Graduate Texts in Mathematics

Graduate Texts in Mathematics bridge the gap between passive study and creative understanding, offering graduate-level introductions to advanced topics in mathematics. The volumes are carefully written as teaching aids and highlight characteristic features of the theory. Although these books are frequently used as textbooks in graduate courses, they are also suitable for individual study.

More information about this series at http://www.springer.com/series/136

Manfred Einsiedler • Thomas Ward

Functional Analysis, Spectral Theory, and Applications

 Springer

Manfred Einsiedler
ETH Zürich
Zürich, Switzerland

Thomas Ward
School of Mathematics
University of Leeds
Leeds, UK

ISSN 0072-5285   ISSN 2197-5612   (electronic)
Graduate Texts in Mathematics
ISBN 978-3-319-86423-5   ISBN 978-3-319-58540-6   (eBook)
DOI 10.1007/978-3-319-58540-6

Mathematics Subject Classification (2010): 46-01, 47-
37A99, 47A60   01, 11N05, 20F69, 22B05, 35J25, 35P10, 35P20,

This Springer imprint is published by Springer Nature
The registered company is Springer International Publishing AG
The registered company address is: Gewerbestrasse 11, 6330 Cham, Switzerland

Preface

Believe us, we also asked ourselves what could be the rationale for 'Yet another book on functional analysis'.[1] Little indeed can justify this beyond our own enjoyment of the beauty and power of the topics introduced here.

Functional analysis might be described as a part of mathematics where analysis, topology, measure theory, linear algebra, and algebra come together to create a rich and fascinating theory. The applications of this theory are then equally spread throughout mathematics (and beyond).

We follow some fairly conventional journeys, and have of course been influenced by other books, most notably that of Lax [59]. While developing the theory we include reminders of the various areas that we build on (in the appendices and throughout the text) but we also reach some fairly advanced and diverse applications of the material usually called functional analysis that often do not find their place in a course on that topic.

The assembled material probably cannot be covered in a year-long course, but has grown out of several such introductory courses taught at the Eidgenössische Technische Hochschule Zürich by the first named author, with a slightly different emphasis on each occasion. Both the student and (especially) the lecturer should be brave enough to jump over topics and pick the material of most interest, but we hope that the student will eventually be sufficiently interested to find out what happens in the material that was not covered initially. The motivation for the topics discussed may by found in Chapter 1.

NOTATION AND CONVENTIONS

The symbols $\mathbb{N} = \{1, 2, \dots\}$, $\mathbb{N}_0 = \mathbb{N} \cup \{0\}$, and $\mathbb{Z}$ denote the natural numbers, non-negative integers and integers; $\mathbb{Q}$, $\mathbb{R}$, $\mathbb{C}$ denote the rational numbers, real numbers and complex numbers. The real and imaginary parts of a complex number are denoted by $x = \Re(x + iy)$ and $y = \Im(x + iy)$.

For functions f, g defined on a set X we write $f = \mathrm{O}(g)$ or $f \ll g$ if there is a constant $A > 0$ with $\|f(x)\| \leqslant A\|g(x)\|$ for all $x \in X$. When the implied constant A depends on a set of parameters $\mathcal{A}$, we write $f = \mathrm{O}_{\mathcal{A}}(g)$ or $f \ll_{\mathcal{A}} g$

(but we may also forget the index if the set of parameters will not vary at all in the discussion). A sequence $a_1, a_2, \ldots$ in any space will be denoted (a_n) (or $(a_n)_n$ if we wish to emphasize the index variable of the sequence). For two $\mathbb{C}$-valued functions f, g defined on $X \smallsetminus \{x_0\}$ for a topological space X containing x_0 we write $f = \mathrm{o}(g)$ as $x \to x_0$ if $\lim_{x \to x_0} \frac{f(x)}{g(x)} = 0$. This definition includes the case of sequences by letting $X = \mathbb{N} \cup \{\infty\}$ and $x_0 = \infty$ with the topology of the one-point compactification. Additional specific notation introduced throughout the text is collected in an index of notation on p. 600.

Prerequisites

We will assume throughout that the reader is familiar with linear algebra and quite frequently that she is also familiar with finite-dimensional real analysis and complex analysis in one variable. Further background and conventions in topology and measure theory are collected in two appendices, but let us note that throughout compact and locally compact spaces are implicitly assumed to be Hausdorff.

Organisation

There are 402 exercises in the text, 221 of these with hints in an appendix, all of which contribute to the reader's understanding of the material. A small number are essential to the development (of the ideas in the section or of later theories); these are denoted 'Essential Exercise' to highlight their significance.

We indicate the dependencies between the various chapters in the *Leitfaden* overleaf and in the guide to the chapters that follows it.

Acknowledgements

We are thankful for various discussions with Menny Aka, Uri Bader, Michael Björklund, Marc Burger, Elon Lindenstrauss, Shahar Mozes, René Rühr, Akshay Venkatesh, and Benjamin Weiss on some of the topics presented here. We also thank Emmanuel Kowalski for making available his notes on spectral theory and allowing us to raid them. We are grateful to several people for their comments on drafts of sections, including Menny Aka, Manuel Cavegn, Rex Cheung, Anthony Flatters, Maxim Gerspach, Tommaso Goldhirsch, Thomas Hille, Guido Lob, Manuel Lüthi, Clemens Macho, Alex Maier, Andrea Riva, René Rühr, Lukas Ruosch, Georg Schildbach, Samuel Stark, Andreas Wieser, Philipp Wirth, and Gao Yunting. Special thanks are due to Roland Prohaska, who proofread the whole volume in four months. Needless to say, despite these many helpful eyes, some typographical and other errors will remain — these are of course solely the responsibility of the authors.

The second named author also thanks Grete for her repeated hospitality which significantly aided this book's completion, and thanks Saskia and Toby for doing their utmost to prevent it.

Manfred Einsiedler, Zürich
Thomas Ward, Leeds
2nd April 2017

Leitfaden

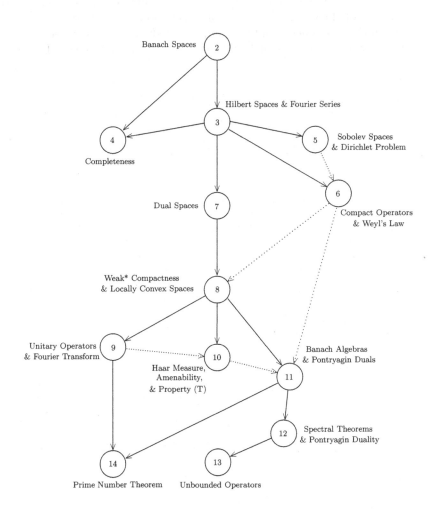

GUIDE TO CHAPTERS

Chapter 1 is mostly motivational in character and can be skipped for the theoretical discussions later.

Chapter 4 has a somewhat odd role in this volume. On the one hand it presents quite central theorems for functional analysis that also influence many of the definitions later in the volume, but on the other hand, by chance, the theorems are not crucial for our later discussions.

The dotted arrows in the *Leitfaden* indicate partial dependencies. Chapter 6 consists of two parts; the discussion of compact groups depends only on Chapter 3 while the material on Laplace eigenfunctions also builds on material from Chapter 5. The discussion of the adjoint operator and its properties in Chapter 6 is crucial for the spectral theory in Chapters 11, 12, and 13. Moreover, one section in Chapter 8 builds on and finishes our discussion of Sobolev spaces in Chapters 5 and 6. Finally, some of Chapter 11 needs the discussion of Haar measures in the first section of Chapter 10.

With these comments and the *Leitfaden* it should be easy to design many different courses of different lengths focused around the topic of Functional Analysis.

Contents

1   **Motivation** 1
    1.1   From Even and Odd Functions to Group Representations.... 1
    1.2   Partial Differential Equations and the Laplace Operator 5
          1.2.1   The Heat Equation 7
          1.2.2   The Wave Equation 10
          1.2.3   The Mantegna Fresco 11
    1.3   What is Spectral Theory? 12
    1.4   The Prime Number Theorem 13
    1.5   Further Topics 14

2   **Norms and Banach Spaces** 15
    2.1   Norms and Semi-Norms 15
          2.1.1   Normed Vector Spaces 16
          2.1.2   Semi-Norms and Quotient Norms................. 21
          2.1.3   Isometries are Affine 23
          2.1.4   A Comment on Notation 26
    2.2   Banach Spaces 26
          2.2.1   Proofs of Completeness 29
          2.2.2   The Completion of a Normed Vector Space 36
          2.2.3   Non-Compactness of the Unit Ball................ 38
    2.3   The Space of Continuous Functions 39
          2.3.1   The Arzela–Ascoli Theorem 40
          2.3.2   The Stone–Weierstrass Theorem................. 42
          2.3.3   Equidistribution of a Sequence 48
          2.3.4   Continuous Functions in L^p Spaces 51
    2.4   Bounded Operators and Functionals 55
          2.4.1   The Norm of Continuous Functionals on $C_0(X)$ 60
          2.4.2   Banach Algebras 61
    2.5   Ordinary Differential Equations 62
          2.5.1   The Volterra Equation 63
          2.5.2   The Sturm–Liouville Equation 66

   2.6   Further Topics 70

3   Hilbert Spaces, Fourier Series, Unitary Representations .. 71
   3.1   Hilbert Spaces 71
        3.1.1   Definitions and Elementary Properties 71
        3.1.2   Convex Sets in Uniformly Convex Spaces 75
        3.1.3   An Application to Measure Theory 83
   3.2   Orthonormal Bases and Gram–Schmidt 86
        3.2.1   The Non-Separable Case 90
   3.3   Fourier Series on Compact Abelian Groups 91
   3.4   Fourier Series on $\mathbb{T}^d$ 95
        3.4.1   Convolution on the Torus 97
        3.4.2   Dirichlet and Fejér Kernels 99
        3.4.3   Differentiability and Fourier Series................. 104
   3.5   Group Actions and Representations..................... 106
        3.5.1   Group Actions and Unitary Representations 107
        3.5.2   Unitary Representations of Compact Abelian Groups . 110
        3.5.3   The Strong (Riemann) Integral..................... 111
        3.5.4   The Weak (Lebesgue) Integral 113
        3.5.5   Proof of the Weight Decomposition 115
        3.5.6   Convolution 118
   3.6   Further Topics 120

4   Uniform Boundedness and the Open Mapping Theorem .. 121
   4.1   Uniform Boundedness................................... 121
        4.1.1   Uniform Boundedness and Fourier Series 123
   4.2   The Open Mapping and Closed Graph Theorems 126
        4.2.1   Baire Category................................... 126
        4.2.2   Proof of the Open Mapping Theorem 128
        4.2.3   Consequences: Bounded Inverses and Closed Graphs .. 130
   4.3   Further Topics 133

5   Sobolev Spaces and Dirichlet's Boundary Problem........ 135
   5.1   Sobolev Spaces and Embedding on the Torus 135
        5.1.1 L^2 Sobolev Spaces on $\mathbb{T}^d$ 135
        5.1.2   The Sobolev Embedding Theorem on $\mathbb{T}^d$ 138
   5.2   Sobolev Spaces on Open Sets 140
        5.2.1   Examples 144
        5.2.2   Restriction Operators and Traces..................... 146
        5.2.3   Sobolev Embedding in the Interior................. 149
   5.3   Dirichlet's Boundary Value Problem and Elliptic Regularity . 152
        5.3.1   The Semi-Inner Product........................... 153
        5.3.2   Elliptic Regularity for the Laplace Operator 155
        5.3.3   Dirichlet's Boundary Value Problem 160
   5.4   Further Topics 165

6   Compact Self-Adjoint Operators, Laplace Eigenfunctions . 167
   6.1   Compact Operators 168
      6.1.1   Integral Operators are Often Compact 170
   6.2   Spectral Theory of Self-Adjoint Compact Operators 174
      6.2.1   The Adjoint Operator 175
      6.2.2   The Spectral Theorem 176
      6.2.3   Proof of the Spectral Theorem 178
      6.2.4   Variational Characterization of Eigenvalues 181
   6.3   Trace-Class Operators.................................. 183
   6.4   Eigenfunctions for the Laplace Operator.................. 196
      6.4.1   Right Inverse and Compactness on the Torus 197
      6.4.2   A Self-Adjoint Compact Right Inverse 198
      6.4.3   Eigenfunctions on a Drum 199
      6.4.4   Weyl's Law 201
   6.5   Further Topics 208

7   Dual Spaces 209
   7.1   The Hahn–Banach Theorem and its Consequences 209
      7.1.1   The Hahn–Banach Lemma and Theorem 209
      7.1.2   Consequences of the Hahn–Banach Theorem 212
      7.1.3   The Bidual 213
      7.1.4   An Application of the Spanning Criterion 214
   7.2   Banach Limits, Amenable Groups, Banach–Tarski 217
      7.2.1   Banach Limits 217
      7.2.2   Amenable Groups 218
      7.2.3   The Banach–Tarski Paradox 223
   7.3   The Duals of $L^p_\mu(X)$ 227
      7.3.1   The Dual of $L^1_\mu(X)$ 228
      7.3.2   The Dual of $L^p_\mu(X)$ for $p > 1$ 230
      7.3.3   Riesz–Thorin Interpolation 233
   7.4   Riesz Representation: The Dual of $C(X)$ 239
      7.4.1   Uniqueness 240
      7.4.2   Totally Disconnected Compact Spaces 240
      7.4.3   Compact Spaces.................................. 243
      7.4.4   Locally Compact σ-Compact Metric Spaces 246
      7.4.5   Continuous Linear Functionals on $C_0(X)$ 248
   7.5   Further Topics 252

8   Locally Convex Vector Spaces 253
   8.1   Weak Topologies and the Banach–Alaoglu Theorem........ 253
      8.1.1   Weak* Compactness of the Unit Ball 256
      8.1.2   More Properties of the Weak and Weak* Topologies .. 257
      8.1.3   Analytic Functions and the Weak Topology 260
   8.2   Applications of Weak* Compactness 261
      8.2.1   Equidistribution.................................. 262

        8.2.2   Elliptic Regularity for the Laplace Operator 270
        8.2.3   Elliptic Regularity at the Boundary................. 278
    8.3   Topologies on the space of bounded operators 290
    8.4   Locally Convex Vector Spaces........................... 292
    8.5   Distributions as Generalized Functions 296
    8.6   Convex Sets 298
        8.6.1   Extreme Points and the Krein–Milman Theorem 301
        8.6.2   Choquet's Theorem.............................. 304
    8.7   Further Topics 311

9   **Unitary Operators and Flows, Fourier Transform** 313
    9.1   Spectral Theory of Unitary Operators.................... 313
        9.1.1   Herglotz's Theorem for Positive-Definite Sequences ... 314
        9.1.2   Cyclic Representations and the Spectral Theorem 316
        9.1.3   Spectral Measures 320
        9.1.4   Functional Calculus for Unitary Operators.......... 323
        9.1.5   An Application of Spectral Theory to Dynamics...... 326
    9.2   The Fourier Transform 329
        9.2.1   The Fourier Transform on $L^1(\mathbb{R}^d)$ 331
        9.2.2   The Fourier Transform on $L^2(\mathbb{R}^d)$ 337
        9.2.3   The Fourier Transform, Smoothness, Schwartz Space.. 340
        9.2.4   The Uncertainty Principle 342
    9.3   Spectral Theory of Unitary Flows 344
        9.3.1   Positive-Definite Functions; Cyclic Representations ... 344
        9.3.2   The Case $G = \mathbb{R}^d$ 346
        9.3.3   Stone's Theorem 350
    9.4   Further Topics 352

10 **Locally Compact Groups, Amenability, Property (T)** 353
    10.1 Haar Measure.. 353
    10.2 Amenable Groups 361
        10.2.1 Definitions and Main Theorem 362
        10.2.2 Proof of Theorem 10.15 363
        10.2.3 A More Uniform Følner Set....................... 371
        10.2.4 Further Equivalences and Properties 373
    10.3 Property (T)... 375
        10.3.1 Definitions and First Properties 375
        10.3.2 Main Theorems 377
        10.3.3 Proof of Kažan's Property (T), Connected Case 378
        10.3.4 Proof of Kažan's Property (T), Discrete Case 384
        10.3.5 Iwasawa Decomposition and Geometry of Numbers ... 392
    10.4 Highly Connected Networks: Expanders 400
        10.4.1 Constructing an Explicit Expander Family.......... 406
    10.5 Further Topics 408

11 Banach Algebras and the Spectrum...................... 409
  11.1 The Spectrum and Spectral Radius 409
    11.1.1 The Geometric Series and its Consequences 411
    11.1.2 Using Cauchy Integration 413
  11.2 C^*-algebras.. 417
  11.3 Commutative Banach Algebras and their Gelfand Duals 418
    11.3.1 Commutative Unital Banach Algebras 419
    11.3.2 Commutative Banach Algebras without a Unit 421
    11.3.3 The Gelfand Transform 422
    11.3.4 The Gelfand Transform for Commutative C^*-algebras . 423
  11.4 Locally Compact Abelian Groups........................ 425
    11.4.1 The Pontryagin Dual 428
  11.5 Further Topics 431

12 Spectral Theory and Functional Calculus 433
  12.1 Definitions and Basic Lemmas 433
    12.1.1 Decomposing the Spectrum 433
    12.1.2 The Numerical Range............................ 436
    12.1.3 The Essential Spectrum 437
  12.2 The Spectrum of a Tree 437
    12.2.1 The Correct Upper Bound for the Summing Operator . 439
    12.2.2 The Spectrum of S 441
    12.2.3 No Eigenvectors on the Tree 442
  12.3 Main Goals: The Spectral Theorem and Functional Calculus . 443
  12.4 Self-Adjoint Operators 446
    12.4.1 Continuous Functional Calculus 447
    12.4.2 Corollaries to the Continuous Functional Calculus 451
    12.4.3 Spectral Measures 453
    12.4.4 The Spectral Theorem for Self-Adjoint Operators..... 454
    12.4.5 Consequences for Unitary Representations 458
  12.5 Commuting Normal Operators 459
  12.6 Spectral Measures and the Measurable Functional Calculus .. 461
    12.6.1 Non-Diagonal Spectral Measures 461
    12.6.2 The Measurable Functional Calculus 462
  12.7 Projection-Valued Measures 468
  12.8 Locally Compact Abelian Groups and Pontryagin Duality ... 473
    12.8.1 The Spectral Theorem for Unitary Representations ... 474
    12.8.2 Characters Separate Points 477
    12.8.3 The Plancherel Formula........................... 478
    12.8.4 Pontryagin Duality 483
  12.9 Further Topics 485

13   Self-Adjoint and Symmetric Operators 487
     13.1 Examples and Definitions................................ 487
     13.2 Operators of the Form T^*T 491
     13.3 Self-Adjoint Operators 495
     13.4 Symmetric Operators 498
            13.4.1 The Friedrichs Extension 499
            13.4.2 Cayley Transform and Deficiency Indices 500
     13.5 Further Topics 502

14   The Prime Number Theorem 503
     14.1 Two Reformulations 503
     14.2 The Selberg Symmetry Formula and Banach Algebra Norm .. 507
            14.2.1 Dirichlet Convolution and Möbius Inversion......... 507
            14.2.2 The Selberg Symmetry Formula 509
            14.2.3 Measure-Theoretic Reformulation 514
            14.2.4 A Density Function and the Continuity Bound 517
            14.2.5 Mertens' Theorem 518
            14.2.6 Completing the Proof 520
     14.3 Non-Trivial Spectrum of the Banach Algebra............. 523
     14.4 Trivial Spectrum of the Banach Algebra 524
     14.5 Primes in Arithmetic Progressions 526
            14.5.1 Non-Vanishing of Dirichlet L-function at 1 529

Appendix A: Set Theory and Topology 537
     A.1 Set Theory and the Axiom of Choice..................... 537
     A.2 Basic Definitions in Topology 538
     A.3 Inducing Topologies 541
     A.4 Compact Sets and Tychonoff's Theorem.................. 545
     A.5 Normal Spaces 547

Appendix B: Measure Theory 551
     B.1 Basic Definitions and Measurability...................... 551
     B.2 Properties of the Integral 554
     B.3 The p-Norm 556
     B.4 Near-Continuity of Measurable Functions 558
     B.5 Signed Measures 561

Hints for Selected Problems 563

Notes 589

References 593
     Notation 598
     General Index 600

Chapter 1
Motivation

†We start by discussing some seemingly disparate topics that are all intimately linked to notions from functional analysis. Some of the topics have been important motivations for the development of the theory that came to be called functional analysis in the first place. We hope that the variety of topics discussed in Sections 1.1–1.4 and those mentioned in Section 1.5 will help to give some insight into the central role of functional analysis in mathematics.

1.1 From Even and Odd Functions to Group Representations

We recall the following elementary notions of symmetry and anti-symmetry for functions. A function $f : \mathbb{R} \to \mathbb{R}$ is said to be *even* if $f(-x) = f(x)$ for all $x \in \mathbb{R}$, and *odd* if $f(-x) = -f(x)$ for all $x \in \mathbb{R}$. Every function $f : \mathbb{R} \to \mathbb{R}$ can be split into an even and an odd component, since

$$f(x) = \underbrace{\tfrac{f(x)+f(-x)}{2}}_{\text{the even part}} + \underbrace{\tfrac{f(x)-f(-x)}{2}}_{\text{the odd part}} . \tag{1.1}$$

Exercise 1.1. Is the decomposition of a function into odd and even parts in (1.1) unique? That is, if $f = e + o$ with e even and o odd, is $e(x) = \tfrac{f(x)+f(-x)}{2}$?

As one might guess, behind the definition of even and odd functions and the decomposition in (1.1), is the group $\mathbb{Z}/2\mathbb{Z} = \{0, 1\}$ acting on $\mathbb{R}$ via the map $x \mapsto (-1)^{\ell}x$ for $\ell \in \mathbb{Z}/2\mathbb{Z}$. Here we are using $\ell \in \mathbb{Z}$ as a shorthand for the coset $\ell + 2\mathbb{Z} \in \mathbb{Z}/2\mathbb{Z}$.

---

† Chapter 1 is atypical for this book. The reader may, and the lecturer should, skip this chapter or return to it later, as convenient.

© Springer International Publishing AG 2017
M. Einsiedler, T. Ward, *Functional Analysis, Spectral Theory, and Applications*,
Graduate Texts in Mathematics 276, DOI 10.1007/978-3-319-58540-6_1

In order to generalize this observation, recall that an *action* of a group G on a set X is a map $G \times X \to X$, written $(g, x) \mapsto g \cdot x$, with the properties $g \cdot (h \cdot x) = (gh) \cdot x$ for all $g, h \in G$ and $x \in X$, and $e \cdot x = x$ for all $x \in X$, where $e \in G$ is the identity element. We will sometimes write $G \curvearrowright X$ for an action of G on X.

Having associated (in an informal way for the moment) the decomposition of a function into odd and even parts with the action of the group $\mathbb{Z}/2\mathbb{Z}$ on $\mathbb{R}$, the notion of group action suggests many generalizations of the decomposition. For this we note that an action of a group G on a space X gives rise to a linear action on the space of functions on X by the formula $(g, f) \mapsto f^g$ where $f^g(x) = f(g^{-1} \cdot x)$ for all $x \in X$, $g \in G$, and functions f on X.

Exercise 1.2. Show that $(\ell_1, \ell_2) \cdot (x_1, x_2) = \big((-1)^{\ell_1} x_1, (-1)^{\ell_2} x_2 \big)$ for $\ell_1, \ell_2 \in \mathbb{Z}/2\mathbb{Z}$ and $(x_1, x_2) \in \mathbb{R}^2$ defines an action of $(\mathbb{Z}/2\mathbb{Z})^2$ on $\mathbb{R}^2$. Show that every real- or complex-valued function on $\mathbb{R}^2$ can be decomposed uniquely into a sum of four functions that are even in both variables; even in x_1 and odd in x_2; odd in x_1 and even in x_2; and odd in both variables.

In order to define another action on $\mathbb{R}^2$ we write

$$k(\phi) = \begin{pmatrix} \cos \phi & -\sin \phi \\ \sin \phi & \cos \phi \end{pmatrix}$$

for the matrix of anti-clockwise rotation on $\mathbb{R}^2$ by the angle $\phi \in \mathbb{R}$. We also let $\mathbb{T} = \mathbb{R}/\mathbb{Z}$ be the one-dimensional circle group or 1-torus, and define the action of $\phi \in \mathbb{T}$ on $\mathbb{R}^2$ by the rotation by $k(2\pi\phi)$. Once again we will sometimes write t as a shorthand for the coset $t + \mathbb{Z} \in \mathbb{R}/\mathbb{Z}$; in particular the interval $[0, 1)$ may be identified with $\mathbb{T}$ using addition modulo 1. We note that we can also make $\mathbb{T}$ into a *topological group* by declaring cosets to be close if their representatives can be chosen to be close in $\mathbb{R}$ or equivalently by using the metric $\mathrm{d}(t_1 + \mathbb{Z}, t_2 + \mathbb{Z}) = \min_{n \in \mathbb{Z}} |t_1 - t_2 + n|$ for $t_1, t_2 \in \mathbb{R}$. In studying any situation with rotational symmetry on $\mathbb{R}^2$ one is naturally led to this action. What is the corresponding decomposition of functions for this action?

Clearly one distinguished class of functions is given by the functions invariant under rotation. That is, functions satisfying $f(v) = f(k(2\pi\phi)v)$ for all $\phi \in \mathbb{T}$. The graph of such a function is the surface obtained by rotating a graph of a real- or complex-valued function on $[0, \infty)$ about the z-axis.

Let us postpone the answers to the above questions and instead consider a finite analogue to the problem. Fix some integer $q \geqslant 1$ and define the group $G = \mathbb{Z}/q\mathbb{Z}$, which acts on $\mathbb{R}^2$ by letting $\ell + q\mathbb{Z} \in G$ act by rotation by the angle $\frac{2\pi\ell}{q}$ using $k(\frac{2\pi\ell}{q})$. To simplify the notation set $K = k(\frac{2\pi}{q})$ and write the action as $G \times \mathbb{R}^2 \ni (\ell + q\mathbb{Z}, v) \mapsto K^\ell v$ (which is well-defined since $K^q = I$). This simplification in the notation helps to clarify the underlying structure, and reflects one of the themes of functional analysis: thinking of progressively more complicated objects (numbers, vectors, functions, operators) as 'points' in a larger space allows the structures to be seen more clearly.

We say that a complex-valued function on $\mathbb{R}^2$ has *weight* n for this action if $f(K^\ell v) = e^{2\pi i n\ell/q} f(v)$ for every $v \in \mathbb{R}^2$ and $\ell + q\mathbb{Z} \in G$ (or equivalently only for $\ell = 1$). We now generalize the formula (1.1) to the case of the finite rotation group considered here. In fact, using the shorthand $\zeta = e^{2\pi i/q}$ we define for a complex-valued function f on $\mathbb{R}^2$ the functions

$$f_n(v) = \frac{1}{q} \sum_{\ell=0}^{q-1} \zeta^{-n\ell} f(K^\ell v) \tag{1.2}$$

for every $v \in \mathbb{R}^2$ and $n = 0, \ldots, q-1$. Since

$$f_n(Kv) = \frac{1}{q} \sum_{\ell=0}^{q-1} \zeta^{-n\ell} f(K^{\ell+1}v) = \frac{1}{q} \sum_{m=1}^{q} \zeta^{-n(m-1)} f(K^{\ell'}v) = \zeta^n f_n(v)$$

for every $v \in \mathbb{R}^2$, we see that f_n has weight n. By the geometric series formula, we see that $\sum_{n=0}^{q-1} \zeta^{-n\ell}$ equals q for $\ell = 0$ and 0 for $\ell = 1, \ldots, q-1$, so

$$\sum_{n=0}^{q-1} f_n(v) = \frac{1}{q} \sum_{n,\ell=0}^{q-1} \zeta^{-n\ell} f(K^\ell v) = \frac{1}{q} \sum_{\ell=0}^{q-1} \left(\sum_{n=0}^{q-1} \zeta^{-n\ell} \right) f(K^\ell v) = f(v)$$

for every $v \in \mathbb{R}^2$. Therefore, f can be written as a finite sum of functions of weight $n = 0, \ldots, q-1$.

Let us return now to the case of the group $\mathrm{SO}_2(\mathbb{R})$ of all rotations $k(2\pi t)$ for $t \in \mathbb{T}$. To guess what the classes of functions should be, we note that all of the symmetries of functions considered above can be phrased naturally in terms of the possible continuous group homomorphisms of the acting group to the group $\mathbb{S}^1 = \{ z \in \mathbb{C} \mid |z| = 1 \}$. It is easy to show (see Exercise 1.3 for the third, non-trivial, statement) that

(1) any homomorphism $\mathbb{Z}/2\mathbb{Z} \to \mathbb{S}^1$ has the form $\ell \mapsto (\pm 1)^\ell$,
(2) any homomorphism $\mathbb{Z}/q\mathbb{Z} \to \mathbb{S}^1$ has the form $\ell \mapsto e^{2\pi i n\ell/q} = \zeta^{n\ell}$ for some $n \in \{0, \ldots, q-1\}$, and
(3) any continuous homomorphism $\mathbb{T} \to \mathbb{S}^1$ has the form $\phi \mapsto \chi_n(\phi) = e^{2\pi i n\phi}$ for some $n \in \mathbb{Z}$, so there are infinitely many such homomorphisms, and they are naturally parameterized by the integers.

For any topological group G, we call the continuous homomorphisms from G to $\mathbb{S}^1$ the *(unitary) characters* of G.

Exercise 1.3. Show that any character of $\mathbb{T}$ has the form claimed in (3) above.

Notice that each of the characters in (1) and (2) corresponds to exactly one type of function in the decompositions of functions discussed above. Generalizing this correspondence, we turn to (3) and say that a complex-valued function $f : \mathbb{R}^2 \to \mathbb{C}$ has *weight* n (is of *type* n) if it satisfies $f(k(2\pi\phi)v) = \chi_n(\phi)f(v)$ for all $\phi \in \mathbb{T}$ and $v \in \mathbb{R}^2$.

One might now guess — and we will see in Chapter 3 that this is indeed the case — that any reasonable function $f : \mathbb{R}^2 \to \mathbb{C}$ can be written as a linear combination

$$f = \sum_{n \in \mathbb{Z}} f_n \tag{1.3}$$

where f_n has weight n. However, in contrast to (1.1) this is an infinite sum, so we are no longer talking about a purely algebraic phenomenon. The decomposition (1.3) — its existence and its properties — lies both in algebra and in analysis. We therefore have to become concerned both with the algebraic structure and with questions of convergence. Depending on the notion of convergence used, the class of reasonable functions turns out to vary. These classes of reasonable functions are in fact important examples of *Banach spaces*, which will be defined in Chapter 2.

The discussion above on decompositions into sums of functions of different weights will later be part of the treatment of *Fourier analysis* (see Chapter 3). For this we will initially study the mathematically simpler situation of the action of $\mathbb{T}$ on $\mathbb{T}$ by translation, $(x, y) \mapsto x + y$ for $x, y \in \mathbb{T}$. Adjusting the definitions above appropriately, we say that a function $f : \mathbb{T} \to \mathbb{C}$ has weight $n \in \mathbb{Z}$ if and only if f is a multiple of χ_n itself. We therefore seek, for a reasonable function $f : \mathbb{T} \to \mathbb{C}$, constants c_n for $n \in \mathbb{Z}$ with

$$f = \sum_{n \in \mathbb{Z}} c_n \chi_n. \tag{1.4}$$

The right-hand side of (1.4) is called the *Fourier series* of f. We will see later that it is relatively straightforward (at least in the abstract sense) to find the *Fourier coefficients* c_n via the identity

$$c_n = \int_{\mathbb{T}} f(x) \overline{\chi_n(x)} \, \mathrm{d}x$$

for all $n \in \mathbb{Z}$.

Fourier series arise naturally in many day-to-day applications. A string or a wind instrument playing a note is producing a periodic pressure wave with a certain frequency. The tone humans hear usually corresponds to this frequency, which is called the fundamental in music theory. There are also higher frequencies, usually integer multiples of the fundamental frequency, appearing in the wave. These frequences are called harmonics and the ratio between the Fourier coefficients of the harmonics and the Fourier coefficient of the fundamental make the distinctive sound of different instruments (for example, the flute and the clarinet) when playing the same fundamental note.

Returning to our discussion of symmetries for functions on $\mathbb{R}^2$ we will show similarly that for a reasonable function $f : \mathbb{R}^2 \to \mathbb{C}$ the function

$$f_n(v) = \int_{\mathbb{T}} \overline{\chi_n(\phi)} f(k(2\pi\phi)v) \, \mathrm{d}\phi$$

for $n \in \mathbb{Z}$ has weight n (compare this with (1.2)), and that (1.3) holds.

Exercise 1.4. Show that if the function $f_n(v) = \int_{\mathbb{T}} \overline{\chi_n(\phi)} f(k(2\pi\phi)v) \, \mathrm{d}\phi$ is well-defined (say the integral exists for almost every $v \in \mathbb{R}^2$), then it has weight n.

To summarize, we will introduce classes of functions (which will be examples of Banach spaces), and determine whether for functions in these classes the Fourier series (1.4) or the weight decomposition (1.3) converges, and in what sense the convergence does or does not happen.

For functions $f : \mathbb{R}^3 \to \mathbb{C}$ one can generalize the discussion above in many different ways, by considering the actions of various different groups as follows:

- $\mathbb{Z}/2\mathbb{Z}$, giving the familiar generalization of even and odd functions.
- $\mathbb{T} \cong \mathrm{SO}_2(\mathbb{R})$ acting by rotations in the x, y-plane about the z-axis. This gives a generalization of our discussion of functions $\mathbb{R}^2 \to \mathbb{C}$, and we will be able to treat this case in a similar way to the two-dimensional case.
- $\mathrm{SO}_3(\mathbb{R})$, the full group of orientation-preserving rotations of $\mathbb{R}^3$.

The last case in this list is more difficult to analyze than any of the cases discussed above. The additional complications arise since $\mathrm{SO}_3(\mathbb{R})$ is not abelian. In fact, the group $\mathrm{SO}_3(\mathbb{R})$ is simple, and as a result there are no non-trivial continuous homomorphisms $\mathrm{SO}_3(\mathbb{R}) \to \mathbb{S}^1$, so this cannot be used to define classes of functions in the same way. The case of $\mathrm{SO}_3(\mathbb{R})$ requires the theory of harmonic analysis and unitary representations of compact groups. We will not reach these important topics here, but will lay the ground for them and refer to the treatment in Folland [32, Ch. 5] or [26].

Although we have used actions of geometric origin to motivate the discussion above, the decompositions described hold more generally for general linear actions of finite abelian groups on vector spaces (often called *group representations*) and also for *unitary representations* of compact abelian groups (with $\mathbb{T}$ being the main example of a compact abelian group) on *Hilbert spaces*. Hilbert spaces are Banach spaces that are equipped with an inner product, and will be introduced in Chapter 3 where we will also discuss unitary representations for the first time.

1.2 Partial Differential Equations and the Laplace Operator

There is no need to motivate the study of differential equations, as they are of central importance across all sciences concerned with measurable quantities that change with respect to other variables of the system studied. Even the simplest ordinary differential equations can lead directly to the study of integral operators, which may be analyzed using tools from functional analysis. The reader familiar with the theorem of Picard and Lindelöf on existence and uniqueness of solutions to certain initial value problems and its proof will not be surprised by this connection. We refer to Section 2.4 for more on this.

However, here we would like to discuss two particular partial differential equations. As we will see later, the mathematical background needed for this, most of which comes from functional analysis, is much more interesting (meaning difficult) than that needed for ordinary differential equations. One of the objectives of this book is to make the informal discussion in this section more formal and rigorous. We will cover this topic in Chapters 5 and 6 (apart from a technical point, which we resolve in Section 8.2).

In both of the partial differential equations that we will discuss, we will need to express the difference between the value of a function at a point and its values in a neighbourhood of the point. To make the resulting equations more amenable for study one uses an infinitesimal version of this difference, which brings into the picture[2] the Laplace operator Δ (also sometimes denoted by ∇^2) defined by

$$\Delta f = \frac{\partial^2 f}{\partial x_1^2} + \cdots + \frac{\partial^2 f}{\partial x_d^2} \tag{1.5}$$

for a smooth function $f : \mathbb{R}^d \to \mathbb{R}$ because of the following simple observation.

Proposition 1.5 (Laplace and neighbourhood averages). *Let $U \subseteq \mathbb{R}^d$ be an open set, and suppose that $f : U \to \mathbb{R}$ is a C^2 function. Then*

$$\lim_{r \to 0} \frac{1}{r^2 \operatorname{vol}(B_r(x))} \int_{B_r(x)} \big(f(y) - f(x)\big) \, \mathrm{d}y = c\Delta f(x)$$

for any $x \in U$, where $\mathrm{d}y$ denotes integration with respect to the Lebesgue measure on the r-ball $B_r = \{y \in \mathbb{R}^d \mid \|y\| < r\} \subseteq \mathbb{R}^d$ and $c = \frac{1}{2(d+2)}$.

PROOF. Suppose for simplicity of notation that $x = 0$, and apply Taylor approximation to obtain

$$f(y) = f(0) + f'(0)y + \frac{1}{2} \sum_{i,j=1}^{d} \frac{\partial^2 f}{\partial x_i \partial x_j}(0)y_i y_j + \mathrm{o}\big(\|y\|^2\big),$$

as $y \to 0$, where $f'(0)$ is the total derivative of f at 0, and we use the notation $\mathrm{o}(\cdot)$ from p. vi. Now in the integral over the r-ball B_r the linear terms cancel out due to the symmetry of the ball. The same argument applies to the mixed quadratic terms. Thus we are left with

$$\int_{B_r} f(y) \, \mathrm{d}y = \operatorname{vol}(B_r)f(0) + \frac{1}{2} \sum_{i=1}^{d} \frac{\partial^2 f}{\partial x_i^2}(0) \int_{B_r} y_i^2 \, \mathrm{d}y + \operatorname{vol}(B_r)\mathrm{o}\big(r^2\big). \tag{1.6}$$

Next notice that $\int_{B_r} y_i^2 \, \mathrm{d}y = \int_{B_r} y_j^2 \, \mathrm{d}y$ for all $1 \leqslant i, j \leqslant d$ and

$$\int_{B_r} \|y\|^2 \, dy = \int_{B_1} r^2 \|z\|^2 r^d \, dz$$

using the substitution $y = rz$. It follows that

$$\int_{B_r} y_i^2 \, dy = \frac{1}{d} \sum_{j=1}^d \int_{B_r} y_j^2 \, dy = \frac{1}{d} \int_{B_r} \|y\|^2 \, dy = \frac{r^{d+2}}{d} \underbrace{\int_{B_1} \|z\|^2 \, dz}_{=:C}.$$

Combining this with (1.6) gives

$$\frac{1}{r^2} \frac{1}{\mathrm{vol}(B_r)} \int_{B_r} (f(y) - f(0)) \, dy = \frac{1}{r^2 \, \mathrm{vol}(B_r)} \frac{1}{2} \Delta f(0) \frac{r^{d+2}}{d} C + o(1)$$

$$= \underbrace{\frac{C}{2d \, \mathrm{vol}(B_1)}}_{=c} \Delta f(0) + o(1).$$

For completeness, we calculate the value of c using d-dimensional spherical coordinates. Every point $z \in \mathbb{R}^d$ is of the form $z = rv$ for some $r \geqslant 0$ and

$$v \in \mathbb{S}^{d-1} = \{w \in \mathbb{R}^d \mid \|w\| = 1\}.$$

Using this substitution we have

$$\mathrm{vol}(B_1) = \int_{\mathbb{S}^{d-1}} \int_0^1 r^{d-1} \, dr \, dv = \frac{1}{d} \, \mathrm{vol}(\mathbb{S}^{d-1}),$$

where the integration with respect to v uses the $(d-1)$-dimensional volume measure on the sphere $\mathbb{S}^{d-1}$. Similarly,

$$C = \int_{B_1} \|z\|^2 \, dz = \int_{\mathbb{S}^{d-1}} \int_0^1 r^{d+1} \, dr \, dv = \frac{1}{d+2} \, \mathrm{vol}(\mathbb{S}^{d-1}).$$

Thus

$$c = \frac{C}{2d \, \mathrm{vol}(B_1)} = \frac{\frac{1}{d+2} \mathrm{vol}(\mathbb{S}^{d-1})}{2d \frac{1}{d} \mathrm{vol}(\mathbb{S}^{d-1})} = \frac{1}{2(d+2)}.$$

$\square$

1.2.1 The Heat Equation

The *heat equation* describes how temperatures in a region $U \subseteq \mathbb{R}^d$ (representing a physical medium) evolve given an initial temperature distribution and some prescribed behaviour of the heat at the boundary ∂U. Inside the medium we expect the flow of heat to be proportional to the difference between the temperature at each point and the temperature in a neighbourhood of the point. If we write $u(x,t)$ for the temperature of the medium at the point x at the time t, then this suggests a relationship

$$\frac{\partial u}{\partial t} = \underbrace{\text{constant}}_{>0} \Delta_x u, \tag{1.7}$$

where

$$\Delta_x u = \Delta u = \frac{\partial^2 u}{\partial x_1^2} + \cdots + \frac{\partial^2 u}{\partial x_d^2}$$

is the Laplace operator with respect to the space variables $x_1, \ldots, x_d$ only. Equation (1.7) is called the *heat equation.* If we take the physical interpretation of this equation for granted, then we can use it to give heuristic explanations of some of the mathematical phenomena that arise.

Suppose first that we prescribe a time-independent temperature distribution at the boundary ∂U of the medium U, and then wait until the system has settled into thermal equilibrium. Experience (that is, physical intuition) suggests that in the long run (as time goes to infinity) the temperature distribution inside U will reach a stable (time-independent) configuration. That is, for any prescribed boundary value $b : \partial U \to \mathbb{R}$ we expect the heat equation on U to have a time-independent solution. More formally, we expect there to be a function $u : U \to \mathbb{R}$ satisfying

$$\begin{cases} \Delta u = 0 \\ u|_{\partial U} = b. \end{cases} \tag{1.8}$$

The boundary value problem (1.8) is the *Dirichlet boundary value problem,* the partial differential equation $\Delta u = 0$ is called the *Laplace equation,* and its solutions are called *harmonic functions.* Proving what the physical intuition suggests, namely that the Dirichlet boundary value problem does indeed have a (smooth) solution, will take us into the theory of Sobolev spaces. We will prove the existence of smooth solutions for the Dirichlet boundary value problem in Chapter 5 (and Section 8.2).

Leaving the Dirichlet problem to one side for now, we continue with the heat equation. Motivated by the methods of linear ordinary differential equations and their initial value problems, we would like to know how we can find other solutions to the partial differential equation while ignoring the boundary values. A simple kind of solution to seek would be those with separated variables, that is solutions of the form

$$u(x,t) = F(x)G(t)$$

with $x \in U \subseteq \mathbb{R}^d$ and $t \in \mathbb{R}$. The heat equation would then imply that

$$F(x)G'(t) = \frac{\partial u}{\partial t} = c(\Delta F(x))G(t)$$

and so (we may as well choose all physical constants to make $c = 1$) the quotient

$$\frac{G'(t)}{G(t)} = \frac{\Delta F(x)}{F(x)}$$

is independent of x and of t, and therefore is a constant (as this is not really a proof, we will not worry about the division by a quantity that may vanish). In summary, $u(x, t) = F(x)G(t)$ solves the equation

$$\frac{\partial u}{\partial t} = \Delta_x u$$

if $G(t) = e^{\lambda t}$ and $\Delta F = \lambda F$ for some constant λ, which one can quickly check (rigorously). Ignoring for the moment the values of F on the boundary ∂U, it is easy to find functions with $\Delta F = \lambda F$ for any $\lambda \in \mathbb{R}$ by using suitable exponential and trigonometric functions. However, these simple-minded solutions turn out not to be particularly useful. Only those special functions $F : \overline{U} \to \mathbb{R}$ with

$$\begin{cases} \Delta F = \lambda F \text{ inside } U \\ F|_{\partial U} = 0 \end{cases}$$

turn out to be useful in the general case. However, it is not clear that such functions even exist, nor for which values of λ they may exist.

Suppose now that the following non-trivial result — the existence of a *basis of eigenfunctions* — (which we will be able to prove in many special cases in Chapter 6) is known for the region $U \subseteq \mathbb{R}^d$.

Claim. Every sufficiently nice function $f : U \to \mathbb{R}$ can be decomposed into a sum $f = \sum_n F_n$ of functions $F_n : \overline{U} \to \mathbb{R}$ satisfying

$$\begin{cases} \Delta F_n = \lambda_n F_n \text{ for some } \lambda_n < 0 \\ F_n|_{\partial U} = 0. \end{cases}$$

We may then solve the partial differential equation

$$\frac{\partial u}{\partial t} = \Delta_x u$$

with boundary values

$$\begin{cases} u|_{\partial U \times \{t\}} = 0 \text{ for all } t \\ u|_{U \times \{0\}} = f \end{cases}$$

using the *principle of superposition* to obtain the general solution

$$u(x, t) = \sum_n F_n(x) e^{\lambda_n t}. \tag{1.9}$$

Since $\lambda_n < 0$ for each $n \geqslant 1$, the series (1.9) converges to 0 as $t \to \infty$ if it is absolutely convergent, in accordance with our physical intuition, since the boundary condition states that the temperature is kept at 0 at the boundary of the region for all $t > 0$. We conclude by mentioning that the claim above will follow from the study of the spectral theory of an operator, but the definition of the operator involved will be somewhat indirect.

1.2.2 The Wave Equation

The *wave equation* describes how an elastic membrane moves. We let $u(x,t)$ be the vertical position of the membrane at time t above the point with coordinate x. As the membrane has mass (and hence inertia) our assumption is that the vertical acceleration — a second derivative of position with respect to time t — of the membrane at time t above x will be proportional to the difference between the position of the membrane at that point and at nearby points. Hence we call

$$\frac{\partial^2 u}{\partial t^2} = c \Delta_x u \tag{1.10}$$

the wave equation. As in the case of the heat equation, we may as well choose physical units to arrange that $c = 1$.

Once more we may argue from physical intuition that the Dirichlet boundary problem for the wave equation always has a solution. Consider a wire loop above the boundary ∂U (notice that even at this vague level we are imposing some smoothness: our physical image of a wire loop may be very distorted but will certainly be piecewise smooth) and imagine a soap film whose edge is the wire. Then, after some initial oscillations,[†] we expect the soap film to stabilize, giving a solution to the Dirichlet boundary value problem in (1.8) defined by the shape of the wire.

In this context, what is the meaning of eigenfunctions of the Laplace operator that vanish on the boundary? To see this, imagine a drum whose skin has the shape U so that the vibrating membrane is fixed along the boundary ∂U, which is simply a flat loop. Suppose now that $F : \overline{U} \to \mathbb{R}$ satisfies

$$\begin{cases} \Delta F = \lambda F \text{ in } U \\ F|_{\partial U} = 0 \end{cases}$$

for some $\lambda < 0$, then we see that $u(x,t) = F(x)\cos(\sqrt{-\lambda}t)$ satisfies

$$\frac{\partial^2}{\partial t^2} u(x,t) = F(x) \underbrace{(-(\sqrt{-\lambda})^2)}_{=\lambda} \cos(\sqrt{-\lambda}t)$$

$$= \lambda F(x) \cos(\sqrt{-\lambda}t) = \Delta_x(F \cos(\sqrt{-\lambda}t))$$

and hence solves the wave equation. In other words, if we start the drum at time $t = 0$ with the prescribed shape given by the function F, then the

---

[†] In the real world there would also be a friction term, and the model for this is a modified wave equation (which we will not discuss further).

drum will produce a pure tone[†] of frequency $\frac{2\pi}{\sqrt{-\lambda}}$. This also sheds some light on which values of λ appear in the claim on p. 9 – namely those that correspond to the pure frequencies that the drum can produce. For a one-dimensional drum (that is, a string) these frequencies are easy to understand (see also Exercise 1.7). However, even for two-dimensional drums the precise eigenvalues remain mysterious. We will nonetheless be able to count these shape-specific eigenvalues asymptotically using the method of Weyl from 1911 (see Section 6.4).

Exercise 1.6. Assume that U satisfies the basis of eigenfunctions claim from p. 9, and that the Dirichlet boundary value problem always has a solution on U.
(a) Combine the above discussions to produce a general procedure to solve the boundary value problem (no rigorous proof is expected, but find the places that lack rigour)

$$\begin{cases} \dfrac{\partial u}{\partial t} = \Delta u \text{ in } U \times [0, \infty); \\ u|_{\partial U \times \{t\}} = b; u|_{U \times \{0\}} = f. \end{cases}$$

(b) Repeat (a) for the wave equation.

Exercise 1.7. For a circular string vibrating in one dimension — the wave equation over $\mathbb{T}$ — the basis of eigenfunctions claim is precisely the claim that every nice function can be represented by its Fourier series. Assuming that this holds, show the basis of eigenfunctions claim for the domain $U = (0, 1) \subseteq \mathbb{R}$. This relates to the wave equation for the clamped vibrating string on $[0, 1]$, that is, to the boundary conditions $y(0) = y(1) = 0$. (In fact the eigenfunctions are given by $x \mapsto \sin(\pi n x)$ with $n = 1, 2 \ldots$; no rigorous proof is expected, but explore the connection.)

1.2.3 The Mantegna Fresco

An illustration of how some of the ideas discussed above link together will be seen in Section 6.4.3, where we discuss eigenfunctions of the Laplacian on a disk. Here the circular symmetry is exploited as in Section 1.1, and the eigenfunctions may be used to decompose functions.

A remarkable application of these ideas was made to the problem of reconstructing a bombed fresco by Andrea Mantegna in a church in Padua. The damage resulted in the fresco being broken into approximately $88,000$ small pieces which needed to be reassembled using a black and white photograph; we refer to Fornasier and Toniolo [35] for the detailed description of how circular harmonics were used to render the computation required practicable. The partially reconstructed coloured image was then used to build a coloured image of the entire fresco.

[†] This preferred frequency for certain physical objects is part of the phenomena of *resonance,* and the design of large structures like buildings or bridges tries to prevent resonances that may lead to reinforcement of oscillations by wind, for example.

1.3 What is Spectral Theory?

As we will see later, the topics considered in Sections 1.1 and 1.2 are connected to *spectral theory*.

The goal of spectral theory, at its broadest, might be described as an attempt to 'classify' all linear operators. We will restrict our attention to Hilbert spaces, which is natural for two reasons. Firstly, it is much easier than the general case of operators on Banach spaces. Secondly, many of the most important applications belong to this simpler setting of operators on Hilbert spaces.

In finite-dimensional linear algebra the classification problem for linear operators is successfully solved by the theory of eigenvalues, eigenspaces, minimal and characteristic polynomials, which leads to a canonical normal form (the Jordan normal form) for any linear operator $\mathbb{C}^n \to \mathbb{C}^n$ for $n \geqslant 1$.

We will not be able to get such a general theory if the Hilbert space $\mathcal{H}$ is infinite-dimensional, but it turns out that many operators of great interest have properties which, in the finite-dimensional case, ensure an even simpler description. They may belong to any of the special classes of operators defined on a Hilbert space by means of the *adjoint* operation $T \mapsto T^*$: self-adjoint operators, unitary operators, or normal operators. For these, if $\dim \mathcal{H} = n$ and we work over $\mathbb{C}$, then there is an orthonormal basis $(e_1, \ldots, e_n)$ of *eigenvectors* of T with corresponding eigenvalues $(\lambda_1, \ldots, \lambda_n)$ so that

$$T\left(\sum_{j=1}^n \alpha_j e_j\right) = \sum_{j=1}^n \alpha_j \lambda_j e_j. \tag{1.11}$$

In other words, the map $\phi(\sum_{j=1}^n \alpha_j e_j) = (\alpha_1, \ldots, \alpha_n)$ is an isometry from $\mathcal{H}$ to $\mathbb{C}^n$ and we may rephrase (1.11) to become

$$T_1 = \phi \circ T \circ \phi^{-1} \tag{1.12}$$

where T_1 is the diagonal map defined by $T_1 : \mathbb{C}^n \ni (\alpha_i) \mapsto (\alpha_i \lambda_i) \in \mathbb{C}^n$. This is obvious, but gives a slightly different view of the classification problem. For any finite-dimensional Hilbert space $\mathcal{H}$, and normal operator T, we have found a *model space and operator* $(\mathbb{C}^n, T_1)$, such that $(\mathcal{H}, T)$ is *equivalent* to $(\mathbb{C}^n, T_1)$ in the sense of (1.12).

The theory we will describe in Chapters 9, 12 and 13 will be a generalization of this type of normal form reduction. This succeeds because the model spaces and operators are indeed simple: they are of the type $L^2_\mu(X)$ for some measure space (X, μ), and the operators are multiplication operators $M_g : f \mapsto gf$ for a suitable function $g : X \to \mathbb{C}$.

1.4 The Prime Number Theorem

Part of the inherent beauty of mathematics comes from the interplay between simple problems and the sophisticated theories that are sometimes required to solve these problems. The natural numbers are among the simplest mathematical objects, but number theory tends to use techniques from much of mathematics to study basic properties of $\mathbb{N}$. Additively $\mathbb{N}$ is quite simple, but multiplicatively $\mathbb{N}$ is much more complex as it is generated by the prime numbers $2, 3, 5, \ldots$. Perhaps because of mathematics' omnipresence across all of the sciences, its absolute (internal) truth, and the pre-eminent role played by the natural numbers, Gauss is alleged to have said "mathematics is the queen of the sciences and number theory is the queen of mathematics".

For modern number theory functional analysis is one of many essential tools. While we will not be able to really justify this statement without devoting a significant proportion of this volume to number theory, we do attempt a partial justification by giving a proof of the prime number theorem in Chapter 14. Prime numbers have been a source of inspiration for mathematicians certainly since Euclid proved (approx. 300 BCE) that there are infinitely many prime numbers. One of many mysteries concerning the prime numbers is their distribution or location within the natural numbers.

Exercise 1.8. Writing $p_1, p_2, p_3, \ldots$ for the primes $2, 3, 5, \ldots$ and $\pi(x) = |\{n \mid p_n \leqslant x\}|$ for the number of primes less than or equal to x, recall Euclid's argument using the fact that $p_1 p_2 \cdots p_n + 1$ has a prime divisor not in $\{p_1, \ldots, p_n\}$ to show that there are infinitely many primes. Use this to show that $p_n < 2^{2^n}$ and deduce that

$$\log(\log(x)) < \pi(x) \leqslant x \tag{1.13}$$

for all large x.

The property of being a prime is in some sense 'deterministic' but, measured appropriately, their appearance in $\mathbb{N}$ seems to mimic randomness. Leaving finer questions of this nature to one side, one may ask if Euclid's proof that there are infinitely many primes can be improved to a more effective statement concerning the growth rate of $\pi(x)$. That is, can (1.13) be improved, and if so, to what extent? Attempting to answer this question has been the source of many developments in number theory and other fields.

For instance, Euler had studied properties of what we now call the Riemann zeta function $\zeta(z) = \sum_{n \geqslant 1} \frac{1}{n^z}$ for $z \in \mathbb{R}$, obtaining in 1737 a result that says (in a manner of speaking) $\sum_{n=1}^{x} \frac{1}{p_n}$ is approximately the logarithm of $\sum_{n=1}^{x} \frac{1}{n}$, equivalently that $\sum_{n=1}^{x} \frac{1}{p_n}$ is approximately $\log \log x$, presaging an important result of Mertens from 1874. We will prove a weaker form of Mertens' theorem in Chapter 14 (Theorem 14.15). Euler's paper introduced several seminal ideas into number theory and arguably is the founding work of analytic number theory.

Based on tables of primes Gauss (in 1792 or 1793), Legendre (in 1797 or 1798) and Dirichlet (in 1838) made conjectures about the asymptotic growth of π; the latter two both suggesting that $\frac{\log x}{x}\pi(x) \to 1$ as $x \to \infty$. Chebyshev was the first to really establish the correct order of growth for π, proving in 1848 that if the sequence $\left(\frac{\log x}{x}\pi(x)\right)_{x \geqslant 1}$ converges at all as $x \to \infty$, then it must converge to 1, and proving in 1851 that there are constants $A, B > 0$ with $A\frac{x}{\log x} \leqslant \pi(x) \leqslant B\frac{x}{\log x}$ for all x.

Riemann's memoir of 1859 introduced methods of complex analysis into the subject, building on Euler's use of real analysis by allowing complex variables and relating analytic properties of the zeta function and its zeros to arithmetic properties of the primes. Finally, in 1886 Hadamard and de la Vallée-Poussin independently used complex analysis on the Riemann zeta function to prove the prime number theorem, $\frac{\log x}{x}\pi(x) \longrightarrow 1$ as $x \to \infty$.

We will present a proof of the prime number theorem (Theorem 14.1) in Chapter 14, closely following Tao's blog [103], which builds on a key step of Selberg's elementary proof of the prime number theorem from 1949 in [96] and Mertens' theorem. In combining these ingredients to prove the prime number theorem we will use the language and several tools developed in this volume. Moreover, we will close our discussion with an application of the spectral theory of finite abelian groups (as in the discussion from Section 1.1) proving Dirichlet's theorem from 1837 on primes in arithmetic progressions and the prime number theorem for primes in arithmetic progressions.

1.5 Further Topics

We list here two more topics that we will be able to discuss.

- Suppose that $f : \mathbb{R}^2 \to \mathbb{R}$ is a continuous function and the partial derivatives $\frac{\partial^k}{\partial x_1^k}f, \frac{\partial^k}{\partial x_2^k}f$ exist and are continuous for all $k \geqslant 1$. Then f is smooth (see Exercises 3.69 and 5.18).
- Another application of functional analysis that we wish to discuss concerns the construction of sparse but highly connected graphs called *expanders*, which are an important concept in graph theory and computer science, see Section 10.4.

Chapter 2
Norms and Banach Spaces

In this chapter we start the more formal treatment of functional analysis, giving the fundamental definitions and introducing some of the basic examples and their properties. We also discuss some theorems and constructions that may be considered part of topology or measure theory, to put them into the context of the theory developed here.

2.1 Norms and Semi-Norms

Throughout this book we will be working with real or complex vector spaces $(V, +, \cdot)$ (here $+$ is vector addition, and $\cdot$ scalar multiplication). We will call the elements of the field simply scalars if we want to avoid making the distinction between the real and complex case. For instance, in the fundamental definitions to come in this section, we treat the real and complex cases simultaneously.

We will assume familiarity with the following concepts from linear algebra: vector spaces, subspaces, quotient spaces, dimension (which may be infinite), linear maps, image and kernel of linear maps. The notion of a basis of a vector space will only be used to distinguish finite-dimensional vector spaces from infinite-dimensional ones. We will not usually try to describe the vector spaces that arise in functional analysis, or the linear maps between them, in terms of bases. An exception will arise in the study of Hilbert spaces (see Section 3.1) and in the study of certain (important but nonetheless special) operators on them (see Section 6.2).

Also recall that a subset $K \subseteq V$ of a vector space is said to be *convex* if for $k_1, k_2 \in K$ and $t \in [0, 1]$ we have that the *convex combination* $(1-t)k_1 + tk_2$ also belongs to K.

© Springer International Publishing AG 2017
M. Einsiedler, T. Ward, *Functional Analysis, Spectral Theory, and Applications*,
Graduate Texts in Mathematics 276, DOI 10.1007/978-3-319-58540-6_2

2.1.1 Normed Vector Spaces

Definition 2.1. Let V be a real or complex vector space. A map $\|\cdot\| : V \to \mathbb{R}$ is called a *norm* if it has the following properties

- (strict positivity) $\|v\| \geq 0$ for any $v \in V$, and $\|v\| = 0$ if and only if $v = 0$;
- (homogeneity) $\|\alpha v\| = |\alpha| \|v\|$ for all $v \in V$ and scalars α; and
- (triangle inequality) $\|v + w\| \leq \|v\| + \|w\|$ for all $v, w \in V$.

If $\|\cdot\|$ is a norm on V, then $(V, \|\cdot\|)$ is called a *normed vector space*.

It is easy to give examples of normed vector spaces, and we list a few standard examples here (more will appear throughout the text).

Example 2.2. The following are examples of normed real vector spaces, in which we write $v = (v_1, \ldots, v_d)^{\mathrm{t}}$ for elements of $\mathbb{R}^d$.

(1) $\mathbb{R}^d$ with the Euclidean norm $\|v\| = \|v\|_2 = \sqrt{|v_1|^2 + \cdots + |v_d|^2}$.
(2) $\mathbb{R}^d$ with $\|v\| = \|v\|_\infty = \max_{1 \leq i \leq d} |v_i|$.
(3) $\mathbb{R}^d$ with $\|v\| = \|v\|_1 = |v_1| + \cdots + |v_d|$.
(4) $\mathbb{R}^d$ with the norm defined by $\|v\|_B = \inf\{\alpha > 0 \mid \frac{1}{\alpha} v \in B\}$, where B is a non-empty, open, centrally symmetric (that is, with $B = -B$), convex, bounded (with respect to the Euclidean norm) subset of $\mathbb{R}^d$.
(5) Let X be any topological space (for example, a metric space; see Appendix A), and let $C_b(X) = \{f : X \to \mathbb{R} \mid f \text{ is continuous and bounded}\}$ with the *uniform* or *supremum* norm

$$\|f\| = \|f\|_\infty = \sup_{x \in X} |f(x)|.$$

Notice that if X is compact, then $C_b(X)$ coincides with $C(X)$, the space of continuous functions $X \to \mathbb{R}$. We note that our definition of compactness (see Definition A.18) contains the assumption that X is Hausdorff.
(6) A special case of (5) makes $C([0,1])$, and so also the subspace

$$C^1([0,1]) = \{f : [0,1] \to \mathbb{R} \mid f \text{ has a continuous derivative on } [0,1]\},$$

into a normed vector space. A different norm on $C^1([0,1])$ may be obtained by setting $\|f\|_{C^1([0,1])} = \max\{\|f\|_\infty, \|f'\|_\infty\}$.
(7) Finally, consider the vector space of real polynomials

$$\mathbb{R}[x] = \left\{ f = \sum_{k=0}^{N} c_f(k) x^k \mid N \in \mathbb{N}, c_f(k) \in \mathbb{R} \right\}$$

on which we can define any of the following norms (thinking of $f \in \mathbb{R}[x]$ really as the finitely supported but infinite vector

$$(c_f(0), c_f(1), \ldots, c_f(N), c_f(N+1), \ldots) \in \mathbb{R}^{\mathbb{N}_0}$$

of its coefficients with $c_f(k) = 0$ for $k > N$):

(a) $\|f\|_1 = \displaystyle\sum_{k=0}^{\infty} |c_f(k)|,$

(b) $\|f\|_2 = \left(\displaystyle\sum_{k=0}^{\infty} |c_f(k)|^2\right)^{1/2}$, or

(c) $\|f\|_\infty = \max\limits_{k \geqslant 0} |c_f(k)|.$

We could also think of polynomials as defining continuous functions on $[0,1]$, thus embedding $\mathbb{R}[x] \subseteq C^1([0,1]) \subseteq C([0,1])$, so that the norm $\|\cdot\|_{C^1([0,1])}$ or $\|\cdot\|_\infty$ may also be used.

The examples in Example 2.2 all generalize in the obvious way to form normed complex vector spaces, with the exception of (4), where additional requirements on the set B are required (see Exercise 2.3).

Exercise 2.3. (a) Verify that Example 2.2(1), (2), (3), (5), (6), and (7) define normed vector spaces over $\mathbb{R}$ or $\mathbb{C}$.
(b) Show that Example 2.2(4) defines a real normed vector space.
(c) Show that for a complex normed vector space $(V, \|\cdot\|)$ the open unit ball

$$B = B_1^V = \{v \in V \mid \|v\| < 1\}$$

has the property that $\alpha B = B$ for any $\alpha \in \mathbb{C}$ with $|\alpha| = 1$.
(d) Show that if $B \subseteq \mathbb{C}^d$ is non-empty, open, convex, bounded, and satisfies $\alpha B = B$ for any $\alpha \in \mathbb{C}$ with $|\alpha| = 1$, then there exists a norm on $\mathbb{C}^d$ whose open unit ball is B.

Throughout the text we will use notions from topology (see Appendix A for a summary).

Lemma 2.4 (Associated metric). *Suppose that $(V, \|\cdot\|)$ is a normed vector space. Then for every $v, w \in V$ we have*

$$\big|\|v\| - \|w\|\big| \leqslant \|v - w\|. \tag{2.1}$$

Moreover, writing $\mathsf{d}(v, w) = \|v - w\|$ for $v, w \in V$ defines a metric d on V such that the norm function $\|\cdot\| : V \to \mathbb{R}$ is continuous with respect to the topology induced by the metric d.

PROOF. For any $v, w \in V$, we have $\|v\| = \|v - w + w\| \leqslant \|v - w\| + \|w\|$ and similarly $\|w\| = \|w - v + v\| \leqslant \|v - w\| + \|v\|$, by Definition 2.1 (the triangle inequality and homogeneity), and the two inequalities together give (2.1).

To see that d is a metric we need to check the following defining properties of a metric:

- (strict positivity) that $\mathsf{d}(v, w) \geqslant 0$ for all $v, w \in V$ and $\mathsf{d}(v, w) = 0$ if and only if $v = w$ is clear by strict positivity in Definition 2.1.
- (symmetry) $\mathsf{d}(v, w) = \mathsf{d}(w, v)$ for all $v, w \in V$ follows by applying the homogeneity in Definition 2.1 with the choice $\alpha = -1$.

- (triangle inequality) Finally, we have

$$\mathsf{d}(u,w) = \|u - w\| = \|u - v + v - w\| \leqslant \|u - v\| + \|v - w\| = \mathsf{d}(u,v) + \mathsf{d}(v,w).$$

The norm is continuous at $v \in V$ if for every $\varepsilon > 0$ there exists some $\delta > 0$ such that $\mathsf{d}(u,v) < \delta$ implies $\big|\|u\| - \|v\|\big| < \varepsilon$. By (2.1), we may choose $\delta = \varepsilon$ to see this.   □

Notice that the triangle inequality makes addition continuous. Indeed, if we write

$$B_\varepsilon^{\|\cdot\|}(v) = \{w \in V \mid \|w - v\| < \varepsilon\}$$

for the ball of radius ε around $v \in V$, then we have

$$B_{\varepsilon/2}^{\|\cdot\|}(v_1) + B_{\varepsilon/2}^{\|\cdot\|}(v_2) \subseteq B_\varepsilon^{\|\cdot\|}(v_1 + v_2) \tag{2.2}$$

for every $\varepsilon > 0$. This means that $(v,w) \mapsto v + w$ is continuous at (v_1, v_2) and, since $v_1, v_2 \in V$ were arbitrary, shows that addition is continuous.

Scalar multiplication is also continuous. To see this fix a scalar α and a vector $v \in V$, and notice that

$$\beta w - \alpha v = (\beta - \alpha)w - \alpha(v - w).$$

So if $\varepsilon \in (0,1)$, $|\beta - \alpha| < \frac{\varepsilon}{\|v\|+1}$ for a scalar β and $\|w - v\| < \varepsilon$ for a vector $w \in V$, then $\|w\| < \|v\| + 1$ and hence

$$\|\beta w - \alpha v\| < \varepsilon(1 + |\alpha|). \tag{2.3}$$

This gives continuity of scalar multiplication at (α, v).

We now turn to the sense in which the topology induced by a norm determines the norm.

Lemma 2.5 (Equivalence of norms). *Two norms $\|\cdot\|$ and $\|\cdot\|'$ on the same vector space induce the same topology if and only if there exists a (Lipschitz) constant $c \geqslant 1$ such that*

$$\tfrac{1}{c}\|v\|' \leqslant \|v\| \leqslant c\|v\|' \tag{2.4}$$

for all $v \in V$. In this case we call the norms equivalent.

PROOF. If (2.4) holds, then the standard neighbourhoods of $v \in V$,

$$B_\varepsilon^{\|\cdot\|'}(v) = \{w \in V \mid \|w - v\|' < \varepsilon\}$$

and

$$B_\varepsilon^{\|\cdot\|}(v) = \{w \in V \mid \|w - v\| < \varepsilon\}$$

with respect to the two norms satisfy

$$B_{\frac{1}{c}\varepsilon}^{\|\cdot\|'}(v) \subseteq B_{\varepsilon}^{\|\cdot\|}(v) \subseteq B_{c\varepsilon}^{\|\cdot\|'}(v).$$

This implies that the topologies have the same notion of neighbourhood, and so are identical.

Suppose now that the two topologies are the same, so that $B_1^{\|\cdot\|}$ is a neighbourhood of 0 in this topology. Then there must be some $\varepsilon > 0$ with

$$B_{\varepsilon}^{\|\cdot\|'} \subseteq B_1^{\|\cdot\|}.$$

Equivalently, $\|v\|' < \varepsilon$ implies that $\|v\| < 1$. For any $v \in V \smallsetminus \{0\}$, if $w = \frac{\varepsilon}{2\|v\|'}v$ then

$$\|w\|' = \frac{\varepsilon}{2\|v\|'}\|v\|' < \varepsilon$$

and so

$$\frac{\varepsilon}{2\|v\|'}\|v\| = \|w\| < 1.$$

This implies that $\|v\| \leqslant \frac{2}{\varepsilon}\|v\|'$ for all $v \in V$, giving the second inequality in (2.4). Reversing the roles of $\|\cdot\|$ and $\|\cdot\|'$ gives the first inequality, and choosing c to be the larger of the two choices produced for c gives the lemma. $\quad\square$

The phenomenon seen in the proof of Lemma 2.5, where a property on all of V is determined by the local behaviour at 0, is something that will occur frequently. For $\mathbb{R}^d$ the notion of equivalence of norms has the following property.

Proposition 2.6 (Equivalence in finite dimensions). *Any two norms on $\mathbb{R}^d$ are equivalent, for any $d \geqslant 1$.*

As we will see in the proof, this is related to the compactness of the closed unit ball in $\mathbb{R}^d$.

PROOF OF PROPOSITION 2.6. Let $\|\cdot\|_1$ be the norm on $\mathbb{R}^d$ from Example 2.2(3), and let $\|\cdot\|'$ be an arbitrary norm on $\mathbb{R}^d$. It is enough to show that these two norms are equivalent. Write $e_1, \ldots, e_d$ for the standard basis of $\mathbb{R}^d$, and let $M = \max_{1 \leqslant i \leqslant d} \|e_i\|'$. Then

$$\|v\|' = \left\|\sum_{i=1}^{d} v_i e_i\right\|' \leqslant \sum_{i=1}^{d} |v_i| \|e_i\|' \leqslant M\|v\|_1, \tag{2.5}$$

where we have used the triangle inequality generalized by induction to finite sums and homogeneity of the norm. This gives one of the inequalities in (2.4). USING COMPACTNESS: To obtain the reverse inequality, notice first that

$$S_1 = \{v \in \mathbb{R}^d \mid \|v\|_1 = 1\}$$

is closed and bounded in the standard topology of $\mathbb{R}^d$ (which may either be seen as a consequence of the bounds $\frac{1}{d}\|\cdot\|_1 \leqslant \|\cdot\|_2 \leqslant \|\cdot\|_1$ or directly using $\| \cdot \|_1$), and so is compact by the Heine–Borel theorem. Also, (2.5) shows that $v \mapsto \|v\|'$ is a continuous function in the standard topology: (2.1) for $\| \cdot \|'$ and (2.5) together give $\big| \|v\|' - \|w\|' \big| \leqslant M\|v - w\|_1$, giving the continuity claimed just as in the end of the proof of Lemma 2.4 with $\delta = \frac{\varepsilon}{M}$. Together this implies that

$$m = \min_{v \in S_1} \|v\|' = \|v_0\|'$$

is attained for some $v_0 \in S_1$. By definition of S_1 we have $v_0 \neq 0$, so that $m > 0$ by the property of the norm $\| \cdot \|'$. Therefore, $v \in V \smallsetminus \{0\}$ implies that

$$\left\| \frac{v}{\|v\|_1} \right\|' \geqslant m,$$

or $v \in V$ implies that $\|v\|' \geqslant m\|v\|_1$, as required. $\square$

This might suggest that the equivalence of norms is a widespread phenomenon. However, once we leave the setting of finite-dimensional normed spaces, we will quickly see that a given normed space may have many inequivalent norms.

Exercise 2.7. Show that the norms $\|f\|_\infty$ and $\|f\|_{C^1([0,1])}$ for $f \in C^1([0,1])$ from Example 2.2(5)–(6) are not equivalent. Show that the norm $\|f\|_{C^1([0,1])}$ and the norm defined by $\|f\|_0 = |f(0)| + \|f'\|_\infty$ for $f \in C^1([0,1])$ are equivalent.

Exercise 2.8. Show that no two of the norms on $\mathbb{R}[x]$ from Example 2.2(7) are equivalent. However, some of the pairs of norms do satisfy an inequality of the form $\|f\| \leqslant c\|f\|'$ for some fixed $c > 0$ and any $f \in \mathbb{R}[x]$. Find those that do and identify the smallest relevant constant c in each case.

Exercise 2.9. Let V, W be normed vector spaces. Show that $V \times W$ with its canonical inherited vector space structure can be made into a normed vector space using either of the norms

$$\|(v, w)\|_p = \big(\|v\|_V^p + \|w\|_W^p \big)^{1/p}$$

for some $p \in [1, \infty)$, or

$$\|(v, w)\|_\infty = \max\{\|v\|_V, \|w\|_W\}.$$

Show that all of these norms are equivalent, and that they induce the product topology.

The next exercise also shows why we are careful in setting up the theory of normed spaces instead of just declaring that everything is a generalization of the finite-dimensional theory.

Exercise 2.10. We define the space $\ell^1(\mathbb{N}) = \{(x_n) \mid \sum_{n=1}^\infty |x_n| < \infty\}$ to be the space of all absolutely summable sequences.

- Show that $\|x\|_1 = \sum_{n=1}^\infty |x_n|$ for $x \in \ell^1(\mathbb{N})$ defines a norm, and that the subspace $c_c(\mathbb{N}) = \{x \mid x_n = 0 \text{ for all large enough } n\} \subseteq \ell^1(\mathbb{N})$ is dense.
- Let $V = \ell^1(\mathbb{N}) \times \ell^1(\mathbb{N})$ (with any of the equivalent norms from Exercise 2.9) and define the subspaces $V_1 = \ell^1(\mathbb{N}) \times \{0\}$ and

$$V_2 = \{(x, y) \in V \mid ny_n = x_n \text{ for all } n \in \mathbb{N}\}.$$

Show that V_1, V_2 are closed subspaces of V, but that

$$V_1 + V_2 = \{v_1 + v_2 \mid v_1 \in V_1, v_2 \in V_2\}$$

is not closed.

2.1.2 Semi-Norms and Quotient Norms

†The following weakening of Definition 2.1 is often useful.

Definition 2.11. A non-negative function $\| \cdot \| : V \to \mathbb{R}_{\geqslant 0}$ on a vector space V is called a *semi-norm* (or a *pseudo-norm*) if $\| \cdot \|$ satisfies the homogeneity property and the triangle inequality of a norm.

Thus a semi-norm is allowed to have a non-trivial subset (which will be a subspace, see below) on which it vanishes. A semi-norm gives rise to a pseudo-metric, which in turn gives rise to a topology on V. The resulting topology is Hausdorff if and only if the original semi-norm is a norm in the usual sense. Indeed, if $v \in V$ has $\|v\| = 0$, then v will belong to every neighbourhood of 0 in the topology defined by $\| \cdot \|$.

Example 2.12. Let $(X, \mathcal{B}, \mu)$ be a measure space (see Appendix B), and define

$$\mathscr{L}_\mu^1(X) = \{f : X \to \mathbb{R} \mid f \text{ is measurable and Lebesgue integrable w.r.t. } \mu\}.$$

On this space we can define a semi-norm

$$\|f\|_1 = \int_X |f| \, \mathrm{d}\mu,$$

and this is not a norm (unless the measure space (X, μ) has special properties; see Exercise 2.13).

Exercise 2.13. Characterize those measure spaces $(X, \mathcal{B}, \mu)$ on which the semi-norm from Example 2.12 on the space $\mathscr{L}_\mu^1(X)$ of Lebesgue integrable functions is a norm.

A reader familiar with measure theory may misread Example 2.12, so we should emphasize that $\mathscr{L}_\mu^1(X)$ denotes the space that contains genuine functions defined at each point of X. The usual solution to the problems created by the many functions on which the semi-norm vanishes is to define an equivalence class of a function f to consist of all functions that differ from f on a null set. This generalizes to a construction that allows any semi-norm on a vector space to be modified to give a norm (on a related vector space). To describe this construction, we first prove a simple lemma that starts to connect the algebraic properties of spaces equipped with a semi-norm to their topological properties.

---

† The construction in this section is satisfying and useful at times but, with the exception of Definition 2.11, is not critical for later developments.

Lemma 2.14 (Kernel of a semi-norm). *The* kernel

$$V_0 = \{v \in V \mid \|v\| = 0\}$$

of a semi-norm on a vector space is a closed subspace in the topology induced by the semi-norm.

PROOF. For $v, w \in V_0$ and any scalar α we have

$$0 \leqslant \|\alpha v + w\| \leqslant |\alpha|\|v\| + \|w\| = 0,$$

so V_0 is a subspace.

By the argument used in Lemma 2.4, we see that the semi-norm $\|\cdot\|$ is continuous with respect to the induced topology. It follows that the pre-image $V_0 = (\|\cdot\|)^{-1}(\{0\})$ is also closed.   □

Returning to Example 2.12, recall that for $f \in \mathscr{L}^1_\mu(X)$, $\|f\|_1 = 0$ is equivalent to the statement that $f = 0$ almost everywhere with respect to μ. Thus the usual equivalence class of a function f is precisely the coset $f + V_0$ defined by f with respect to the kernel $V_0 \subseteq \mathscr{L}^1_\mu(X)$ of the semi-norm. We define, as is standard, the quotient space

$$L^1_\mu(X) = \mathscr{L}^1_\mu(X)/V_0,$$

and note that the semi-norm $\|\cdot\|_1$ on $\mathscr{L}^1_\mu(X)$ gives rise to a norm, also denoted $\|\cdot\|_1$, on $L^1_\mu(X)$. For an introduction to the function spaces $L^p_\mu(X)$ for $p \in [1, \infty)$ we refer to Appendix B.3. Where the measure is clear from the context or has a standard choice (for example, the Lebesgue measure on $[0, 1]$), it is omitted from the notation. This construction is a special case of the following.

Lemma 2.15 (Quotient norm). *For any vector space V equipped with a semi-norm $\|\cdot\|$, and any closed subspace $W \subseteq V$, the expression*

$$\|v + W\|_{V/W} = \inf_{w \in W} \|v + w\|$$

for $v \in V$ defines a norm on the quotient space $V/W = \{v + W \mid v \in V\}$. For the kernel $W = V_0$ we have $\|v + V_0\|_{V/V_0} = \|v\|$ for $v \in V$.

PROOF. This is simply a matter of chasing the definitions through the statements. Let $v_1, v_2 \in V$ and $\varepsilon > 0$ be given. Then there exist $w_1, w_2 \in W$ with

$$\|v_i + w_i\| \leqslant \|v_i + W\|_{V/W} + \varepsilon$$

for $i = 1, 2$. Hence

$$\|v_1 + v_2 + W\|_{V/W} \leqslant \|v_1 + v_2 + w_1 + w_2\|$$
$$\leqslant \|v_1 + w_1\| + \|v_2 + w_2\|$$
$$\leqslant \|v_1 + W\|_{V/W} + \|v_2 + W\|_{V/W} + 2\varepsilon,$$

and so the triangle inequality holds for $\|\cdot\|_{V/W}$. Similarly, for any scalar α,

$$\|\alpha v_1 + \alpha w_1\| = |\alpha| \|v_1 + w_1\| \leqslant |\alpha| \big(\|v_1 + W\|_{V/W} + \varepsilon\big),$$

which gives

$$\|\alpha v_1 + W\|_{V/W} \leqslant |\alpha| \|v_1 + W\|_{V/W}.$$

If $\alpha = 0$ then this is clearly an equality, and if $\alpha \neq 0$ then we may apply the above to αv_1 and the scalar α^{-1} to give

$$\|v_1 + W\|_{V/W} \leqslant |\alpha|^{-1} \|\alpha v_1 + W\|_{V/W}.$$

However, this is the remaining half of the homogeneity property.

It remains to check that $\|\cdot\|_{V/W}$ is indeed a norm and not simply a semi-norm. Assume therefore that

$$\|v + W\|_{V/W} = 0.$$

Then for every $\varepsilon > 0$ there exists some $w \in W$ with $\|v - w\| < \varepsilon$. However, this shows that v belongs to the closure $\overline{W}$ of W. By assumption $\overline{W} = W$ is closed, so that $v \in W$ and $v + W = W$ is the zero element in the quotient space V/W.

For $W = V_0$ we have $\|v + w\| = \|v + w\| + \|-w\| \geqslant \|v\|$ for every $v \in V$ and $w \in V_0$, which gives the final claim. $\square$

Notice that we cannot expect the infimum in Lemma 2.15 to be a minimum in general (see, for example, Exercise 2.16).

Exercise 2.16. Let $(C([-1,1]), \|\cdot\|_\infty)$ be the normed vector space defined as in Example 2.2(5). Define

$$W = \left\{ f \in C([-1,1]) \mid \int_{-1}^{0} f(x)\,dx = \int_{0}^{1} f(x)\,dx = 0 \right\}.$$

Show that W is a closed subspace. Now let $f(x) = x$, calculate $\|f\|_{C([-1,1])/W}$, and show that the infimum is not achieved.

2.1.3 Isometries are Affine

[†]The following strengthening of the triangle inequality has interesting consequences.

[†] The results in Section 2.1.3 are interesting, but will not be needed later.

Definition 2.17. A norm $\| \cdot \|$ on a vector space V is *strictly sub-additive* if we have strict inequality in the triangle inequality except in the case when the two vectors are real non-negative multiples, more precisely

$$\|v + w\| < \|v\| + \|w\|$$

for all $v, w \in V$ unless $v, w \in \{t(v + w) \mid t \geqslant 0\}$.

Exercise 2.18. A normed space $(V, \| \cdot \|)$ is *strictly convex* if the closed unit ball is a strictly convex set, or equivalently if a line segment with end points $x \neq y$ in the unit sphere $\{v \in V \mid \|v\| = 1\}$ only intersects the unit sphere at its end points. Show that the closed unit ball in a normed linear space is strictly convex if and only if the norm is strictly sub-additive.

A map $f : V \to W$ between normed spaces is an *isometry* if

$$\|f(v) - f(v_0)\|_W = \|v - v_0\|_V$$

for all $v, v_0 \in V$.

Exercise 2.19. Show that the supremum norm $\| \cdot \|_\infty$ on $\mathbb{R}^2$ is not strictly convex. Give an example to show that an isometry between normed spaces need not be affine, by considering maps of the form $x \mapsto (x, f(x))$ from $(\mathbb{R}, |\cdot|)$ to $(\mathbb{R}^2, \|\cdot\|_\infty)$ for a suitably chosen function f.

Theorem 2.20 (Mazur–Ulam[3]). *Let V and W be normed linear spaces over $\mathbb{R}$, and let $M : V \to W$ be a function. Assume that either*

- *M is a surjective isometry, or*
- *M is an isometry and the norm on W is strictly sub-additive.*

Then M is affine, that is $M(v) = M_{\text{linear}}(v) + M(0)$ where $M_{\text{linear}} : V \to W$ is a linear isometry.

PROOF. Clearly the map $v \mapsto M(v) - M(0)$ is an isometry if M is an isometry, so we may assume that $M(0) = 0$ without loss of generality, and need to show in this case that M is linear.

MIDPOINT-PRESERVING MAPS: We claim first that if M preserves mid-points in the sense that

$$M\left(\tfrac{v_1 + v_2}{2}\right) = \tfrac{M(v_1) + M(v_2)}{2} \tag{2.6}$$

for all $v_1, v_2 \in V$ and satisfies $M(0) = 0$, then M is linear. To see this, pick v in V and apply (2.6) to the pairs v and 0, then to $\frac{1}{2}v$ and 0, and inductively to $\frac{1}{2^k}v$ and 0 to prove that $M(\frac{1}{2^k}v) = \frac{1}{2^k}M(v)$ for all $k \in \mathbb{N}$ and $v \in V$. Next apply (2.6) to $2v$ and 0, and inductively to $(\ell + 1)v$ and $(\ell - 1)v$ to prove that $M(\ell v) = \ell M(v)$ for all $\ell \in \mathbb{N}$ and $v \in V$. Finally, apply (2.6) to v and $-v$ to see that $M(-v) = -M(v)$ for all $v \in V$. This gives

$$M\left(\tfrac{k}{2^n}v\right) = \tfrac{k}{2^n}M(v)$$

for any $k \in \mathbb{Z}$, $n \in \mathbb{N}$ and $v \in V$, and so by continuity $M(av) = aM(v)$ for all $a \in \mathbb{R}$ and $v \in V$. With (2.6), this also gives $M(v_1 + v_2) = M(v_1) + M(v_2)$ for all $v_1, v_2 \in V$.

SUBADDITIVE NORMS: The case of the theorem under the sub-additivity hypothesis is now easily obtained. Suppose that M is an isometry and the norm on W is strictly sub-additive. Let $v_1, v_2 \in V$ have mid-point $z = \frac{v_1 + v_2}{2}$, so that

$$\|v_1 - z\| = \|z - v_2\| = \tfrac{1}{2}\|v_1 - v_2\|,$$

and hence (since M is an isometry)

$$\|M(v_1) - M(z)\| = \|M(z) - M(v_2)\| = \tfrac{1}{2}\|M(v_1) - M(v_2)\|.$$

Moreover, $M(v_1) - M(v_2) = (M(v_1) - M(z)) + (M(z) - M(v_2))$. Thus if the norm on W is strictly sub-additive then $(M(v_1) - M(z))$ and $(M(z) - M(v_2))$ must be real non-negative multiples of each other by strict sub-additivity, but as they have the same norm this forces them to be be equal. Solving this equation for $M(z)$ gives (2.6).

SURJECTIVE ISOMETRIES: Consider now the case where M is assumed to be a surjective isometry (and hence also bijective). We define for $z \in V$ the *reflection in z* to be the map $\psi_z : V \to V$ defined by $\psi_z(v) = 2z - v$. It is easy to check that ψ_z^2 is the identity, so ψ_z has inverse ψ_z and is a bijective isometry. Note that

$$\|\psi_z(v) - z\| = \|v - z\|, \tag{2.7}$$

and

$$\|\psi_z(v) - v\| = 2\|v - z\| \tag{2.8}$$

for all $v \in V$, which also implies that z itself is the only point fixed of ψ_z.

Now fix $v_1, v_2 \in V$ and write $z = \frac{v_1 + v_2}{2}$ as before for the midpoint. Let $\mathscr{B}$ be the group of all bijective isometries $V \to V$ that fix v_1 and v_2, and define

$$\lambda = \sup\{\|g(z) - z\| \mid g \in \mathscr{B}\}.$$

Since any $g \in \mathscr{B}$ is an isometry satisfying $g(v_1) = v_1$ we have

$$\|g(z) - z\| \leqslant \|g(z) - g(v_1)\| + \|v_1 - z\| = 2\|v_1 - z\|$$

and hence $\lambda < \infty$. We also have

$$
\begin{aligned}
2\|g(z) - z\| &= \|\psi_z g(z) - g(z)\| && \text{(by (2.8) for } v = g(z)) \\
&= \|g^{-1}\psi_z g(z) - z\| && \text{(since } g \text{ is an isometry)} \\
&= \|\psi_z g^{-1}\psi_z g(z) - z\| && \text{(by (2.7) for } v = g^{-1}\psi_z g(z))
\end{aligned}
$$

for any $g \in \mathscr{B}$. Furthermore, note that $\psi_z(v_1) = v_2$ and $\psi_z(v_2) = v_1$. It follows that $g \in \mathscr{B}$ implies $g' = \psi_z g^{-1} \psi_z g \in \mathscr{B}$ and so $\|g'(z) - z\| \leqslant \lambda$. Combining this with the above gives

$$2\|g(z) - z\| = \|g'(z) - z\| \leqslant \lambda,$$

for all $g \in \mathscr{B}$, and hence by definition of $\lambda \in [0, \infty)$ also $2\lambda \leqslant \lambda$. This forces λ to be 0, and therefore $g(z) = z$ for all $g \in \mathscr{B}$.

Now let $M : V \to W$ be a bijective isometry, and let $z' = \frac{M(v_1) + M(v_2)}{2}$. Then $h = \psi_z M^{-1} \psi_{z'} M \in \mathscr{B}$, so $h(z) = z$ and therefore $\psi_{z'} M(z) = M(z)$. On the other hand, the only point fixed by $\psi_{z'}$ is z' itself, so $M(z) = z'$ and M preserves mid-points as required. □

Exercise 2.21. Show that the vertex set of a graph consisting of vertices v_1, v_2, v_3, v_c and three edges connecting one central vertex v_c to the remaining three vertices v_1, v_2, v_3, endowed with the combinatorial distance given by $\mathsf{d}(v_j, v_k) = 2\delta_{jk}$ for $j, k \in \{1, 2, 3\}$ and $\mathsf{d}(v_j, v_c) = 1$ for $j = 1, 2, 3$ admits no isometric embedding into any Banach space with a strictly sub-additive norm.

2.1.4 A Comment on Notation

On several occasions in this section we considered different norms on the same vector space. This will happen less frequently in the theoretical parts of the text, and most of the time the normed vector space $(V, \| \cdot \|)$ will be equipped with a particular norm. Where we are dealing with a single norm, we will write

$$B_r = B_r^V = B_r^V = \{w \in V \mid \|w\| < r\}$$

for the open ball of radius r around 0, and

$$B_r(v) = B_r^V(v) = \{w \in V \mid \|w - v\| < r\} = B_r^V + v$$

for the open ball of radius r around $v \in V$.

We will also frequently write $\| \cdot \|_V$ for the natural norm on V. For example, in Example 2.2(6) we may write $\|f\|_{C^1([0,1])}$ for the natural norm of a function $f \in C^1([0, 1])$, but may also write $\|f\|_{C([0,1])} = \|f\|_\infty$ for the supremum norm of $f \in C^1([0, 1])$ thought of as an element of the large space $C([0, 1])$. At this point it is reasonable to ask what makes a norm be naturally associated to a given space, and this is partially explained in the next section.

2.2 Banach Spaces

We start by recalling a basic definition from analysis on metric spaces.

Definition 2.22. A sequence (x_n) in a metric space (X, d) is said to be a *Cauchy sequence* if for any $\varepsilon > 0$ there is an $N = N(\varepsilon)$ such that

$$\mathsf{d}(x_m, x_n) < \varepsilon$$

for any $m, n \geqslant N$. The metric space is called *complete* if every Cauchy sequence converges to an element of X.

From a purely logical point of view, 'converges to an element of X' should be written 'converges', but we wish to emphasize here that the limit of the sequence belongs to X (and not to some strictly larger space that contains X).

The notion of Cauchy sequences gives rise to one of the fundamental types of normed spaces in functional analysis.

Definition 2.23. A normed vector space $(V, \|\cdot\|)$ is a *Banach space* if V is complete with respect to (the metric induced by) the norm $\|\cdot\|$.

Once again there are many familiar examples of Banach spaces. As we will see there is often an almost canonical choice of norm $\|\cdot\|_V$ which makes a linear space V into a Banach space $(V, \|\cdot\|_V)$. It is clear that this property of a norm does not define it uniquely. In fact, any equivalent norm would induce the same topology, the same notion of Cauchy sequence, and therefore also make V into a Banach space.

Example 2.24. We start with a small number of examples, and postpone the proof that these are indeed Banach spaces to Section 2.2.1.

(1) The Euclidean space $\mathbb{R}^d$ with any of the norms from Example 2.2(1)–(4) from Section 2.1.1 forms a Banach space.
(2) Let X be any set. Then $B(X) = \{f : X \to \mathbb{R} \mid f \text{ is bounded}\}$, equipped with the norm
$$\|f\|_\infty = \sup_{x \in X} |f(x)|,$$
is a Banach space. Convergence of a sequence of functions in this space is also called *uniform convergence*. If $X = \mathbb{N}$, then one often writes
$$\ell^\infty = \ell^\infty(\mathbb{N}) = B(\mathbb{N}).$$

(3) Let X be a topological space. Then
$$C_b(X) = \{f \in B(X) \mid f \text{ is continuous}\}$$
is a closed subspace of $B(X)$ and so is also a Banach space. Notice that if X is compact then $C_b(X) = C(X)$.
(4) Let X be a locally compact topological space (that is, a Hausdorff topological space in which every point has a compact neighbourhood, see also Definition A.21). Then

$$C_0(X) = \{f \in C_b(X) \mid \lim_{x \to \infty} f(x) = 0\}$$

is a closed subspace of $C_b(X)$ and hence a Banach space. The notion of the limit of $f(x)$ as $x \to \infty$ used here is defined as follows: $\lim_{x \to \infty} f(x) = A$ if and only if for every $\varepsilon > 0$ there exists some compact set $K \subseteq X$ with $|f(x) - A| < \varepsilon$ for all $x \in X \setminus K$. If $X = \mathbb{N}$ (with the discrete topology), one often writes $c_0 = c_0(\mathbb{N}) = C_0(\mathbb{N})$ for this subspace of $\ell^\infty(\mathbb{N})$.

(5) The space $C^1([0,1])$ of continuously differentiable functions on $[0,1]$ with the norm

$$\|f\|_{C^1([0,1])} = \max\{\|f\|_\infty, \|f'\|_\infty\}$$

is a Banach space.

(6) Let $U \subseteq \mathbb{R}^d$ be non-empty and open, and fix $k \geqslant 1$. Then the space $C_b^k(U)$ of functions $U \to \mathbb{R}$ for which all partial derivatives up to order k exist and are continuous and bounded on U, equipped with the norm

$$\|f\|_{C_b^k(U)} = \max_{\|\alpha\|_1 \leqslant k} \|\partial_\alpha f\|_\infty,$$

is a Banach space, where ∂_α for $\alpha \in \mathbb{N}_0^d$ stands for the partial differential operator defined by

$$\partial_\alpha f = \frac{\partial^{\|\alpha\|_1}}{\partial x_1^{\alpha_1} \cdots \partial x_d^{\alpha_d}} f$$

of degree $\|\alpha\|_1 = \alpha_1 + \cdots + \alpha_d$. In the case $\alpha = e_j$ we will call this the jth partial derivative and write $\partial_j f = \partial_{e_j} f$.

(7) Fix $p \in [1, \infty)$ and let $(X, \mathcal{B}, \mu)$ be a measure space. Then

$$\|f\|_p = \left(\int_X |f|^p \, d\mu \right)^{1/p}$$

defines a semi-norm on the vector space

$$\mathscr{L}_\mu^p(X) = \{f : X \to \mathbb{R} \mid f \text{ is measurable and } \|f\|_p < \infty\}.$$

The associated space of equivalence classes, equal to the quotient

$$L_\mu^p(X) = \mathscr{L}_\mu^p(X)/V_0$$

by the kernel V_0 of the semi-norm $\|\cdot\|_p$, is a Banach space. We will write $L^p(X, \mu)$ and $L^p(\mu)$ in place of $L_\mu^p(X)$ when the space is clear, and in particular where it is useful to avoid multiple levels of subscript. Important special cases of this construction include the following:

(a) $(X, \mathcal{B}, \mu) = (\Omega, \mathcal{B}, m)$ where Ω is a Borel subset of $\mathbb{R}^d$, $\mathcal{B}$ is the Borel σ-algebra, and m is d-dimensional Lebesgue measure on Ω.

(b) $(X, \mathcal{B}, \mu) = (\mathbb{N}, \mathbb{P}(\mathbb{N}), \lambda_{\text{count}})$, where λ_{count} denotes the counting measure, which is defined on any subset of $\mathbb{N}$. In this case we will write

$$\ell^p = \ell^p(\mathbb{N}) = L^p_{\lambda_{\text{count}}}(\mathbb{N}).$$

(8) The analogue of (7) with $p = \infty$ is constructed slightly differently. As before, let $(X, \mathcal{B}, \mu)$ be a measure space. Then

$$\mathscr{L}^\infty(X) = \{f : X \to \mathbb{R} \mid f \text{ is measurable, } f \in B(X)\}$$

is already a Banach space with respect to $\|f\|_\infty$. However, one also defines

$$L^\infty_\mu(X) = \mathscr{L}^\infty(X)/W_\mu(X),$$

where

$$W_\mu(X) = \{f \in \mathscr{L}^\infty(X) \mid f = 0 \ \mu\text{-almost everywhere}\}$$

and $L^\infty_\mu(X)$ is equipped with the *essential supremum norm* defined by

$$\|f\|_{\text{esssup}} = \text{esssup}_{x \in X} |f(x)| = \inf\{\alpha > 0 \mid \mu(\{x \mid |f(x)| > \alpha\}) = 0\}. \tag{2.9}$$

We will generally follow the convention that the essential supremum norm of f is also denoted for simplicity by $\|f\|_\infty$. All of these ℓ^p and L^p spaces also have natural complex-valued analogues.

As is customary, we will quickly stop being too careful about the distinction between an element of $\mathscr{L}^\infty(X)$ and the equivalence class defined by it in $L^\infty_\mu(X)$. For example, $|f|(x) = |f(x)|$ for all $x \in X$ really depends on $f \in \mathscr{L}^\infty(X)$ and not just on the equivalence class, but (as we will see later in the proof of completeness) the norm defined in (2.9) is independent of the representative chosen for a given equivalence class.

Exercise 2.25. Show that a product of two normed vector spaces $V \times W$ is complete with respect to one of the norms from Exercise 2.9 if and only if both V and W are complete with respect to their own norms. Thus the product of two Banach spaces is a Banach space.

2.2.1 Proofs of Completeness

In this subsection we will explain why the examples from Example 2.24 are indeed Banach spaces. Depending on the background of the reader, parts of this section may be skipped. In each case it is proving completeness that really takes up what effort is required. The following principle will be used several times.

Essential Exercise 2.26. Let (X, d) be a metric space.
(a) Show that if Y is a subset of X that is complete (with respect to the restriction of the metric on X to Y), then Y is a closed subset of X.
(b) If X is complete, show that $Y \subseteq X$ is complete if and only if Y is closed.

EXAMPLE 2.24(1). If two norms are equivalent then they define the same notion of convergence and of Cauchy sequence. Thus it is enough to consider $\mathbb{R}^d$ with the norm $\|\cdot\|_\infty$ by Proposition 2.6. Now a Cauchy sequence (v_n) in $\mathbb{R}^d$ has the property that each component sequence $(v_n^{(i)})$ for a fixed i, where $v_n = (v_n^{(1)}, \ldots, v_n^{(d)})^t$ for all n, is itself a Cauchy sequence in $\mathbb{R}$. Since $\mathbb{R}$ is complete, there exists a limit $v^{(i)} = \lim_{n\to\infty} v_n^{(i)}$ for each i. These limits together define a vector $v = (v^{(1)}, \ldots, v^{(d)})^t$ and it is easy to see that v is the limit of (v_n) in $\mathbb{R}^d$. □

EXAMPLE 2.24(2). Let X be any set and let (f_n) be a Cauchy sequence in $B(X)$ with respect to $\|\cdot\|_\infty$. Then for any fixed $x \in X$ the sequence $(f_n(x))$ is a Cauchy sequence in $\mathbb{R}$, which therefore has a limit $f(x)$. This defines a function $f : X \to \mathbb{R}$. We need to show that $f \in B(X)$ and $f_n \to f$ as $n \to \infty$ with respect to $\|\cdot\|_\infty$. Since (f_n) is Cauchy, for any $\varepsilon > 0$ there is some $N(\varepsilon)$ with $\|f_m - f_n\|_\infty < \varepsilon$ for all $m, n \geqslant N(\varepsilon)$, and so $|f_m(x) - f_n(x)| < \varepsilon$ for any $x \in X$ and $m, n \geqslant N(\varepsilon)$. Now let $m \to \infty$ to see that $|f(x) - f_n(x)| \leqslant \varepsilon$ for all $n \geqslant N(\varepsilon)$. Setting $\varepsilon = 1$ and $n = N(1)$ gives $|f(x)| \leqslant 1 + \|f_{N(1)}\|_\infty$ for any $x \in X$, showing that $f \in B(X)$. For any $\varepsilon > 0$, we obtain $\|f - f_n\|_\infty \leqslant \varepsilon$ for all $n \geqslant N(\varepsilon)$ and hence that $f = \lim_{n\to\infty} f_n \in B(X)$, as required. If $|X|$ has cardinality d then this example reduces to the previous one. □

EXAMPLE 2.24(3). By definition, $C_b(X)$ is a subspace of $B(X)$, and we use the same norm on both spaces. Thus, if (f_n) is a Cauchy sequence in $C_b(X)$ then, by (2), there exists a limit $f = \lim_{n\to\infty} f_n \in B(X)$. It remains to show that $f \in C_b(X)$ — that is, to show that $C_b(X)$ is a closed subspace of $B(X)$. This is a familiar argument from real analysis. Given any $\varepsilon > 0$ there exists some n with $\|f_n - f\|_\infty < \varepsilon$. Since $f_n \in C_b(X)$ is continuous at x, there is a neighbourhood $U \subseteq X$ of x with $|f_n(y) - f_n(x)| < \varepsilon$ for all $y \in U$. Therefore,

$$|f(y) - f(x)| \leqslant \underbrace{|f(y) - f_n(y)|}_{\leqslant \|f - f_n\|_\infty < \varepsilon} + \underbrace{|f_n(y) - f_n(x)|}_{< \varepsilon} + \underbrace{|f_n(x) - f(x)|}_{\leqslant \|f - f_n\|_\infty < \varepsilon} < 3\varepsilon$$

for all $y \in U$. As the existence of such a neighbourhood holds for all $\varepsilon > 0$ and $x \in X$, we see that $f \in C_b(X)$ as required. □

EXAMPLE 2.24(4). Once again $C_0(X) \subseteq C_b(X)$ and we use the same norm. So if (f_n) is a Cauchy sequence in $C_0(X)$, then $f = \lim_{n\to\infty} f_n \in C_b(X)$ exists by (3). We only need to show that $f \in C_0(X)$. For this, let $\varepsilon > 0$ and choose $n \in \mathbb{N}$ with $\|f_n - f\|_\infty < \varepsilon$. Since $f_n \in C_0(X)$, there exists some compact set $K \subseteq X$ with $|f_n(x)| < \varepsilon$ for all $x \in X \setminus K$. Thus

$$|f(x)| \leqslant |f(x) - f_n(x)| + |f_n(x)| < 2\varepsilon$$

for all $x \in X \setminus K$. This implies that $f \in C_0(X)$ as required. □

EXAMPLE 2.24(5). Let (f_n) be a Cauchy sequence in $C^1([0,1])$ with respect to the norm $\|f\|_{C^1([0,1])} = \max\{\|f\|_\infty, \|f'\|_\infty\}$. Then each f_n lies in $C([0,1])$, and (f_n) is a Cauchy sequence with respect to the norm $\|\cdot\|_\infty$. Thus (3) applies and shows that f_n converges uniformly to some $f \in C([0,1])$. The same argument applies to the sequence (f_n') of derivatives, showing that (f_n') converges uniformly to some $g \in C([0,1])$. All that remains is to verify that

$$f' = g. \tag{2.10}$$

In order to show (2.10), it is convenient to rephrase the statement as an integral equation. Since f_n is continuously differentiable, we certainly have

$$f_n(x) = f_n(0) + \int_0^x f_n'(t)\,\mathrm{d}t. \tag{2.11}$$

Now note that

$$\left| \int_0^x f_n'(t)\,\mathrm{d}t - \int_0^x g(t)\,\mathrm{d}t \right| \leqslant \|f_n' - g\|_\infty$$

converges to 0 as $n \to \infty$. Since $f_n(x) \to f(x)$ and $f_n(0) \to f(0)$ as $n \to \infty$, we see that (2.11) implies

$$f(x) = f(0) + \int_0^x g(t)\,\mathrm{d}t.$$

By continuity of g, this is equivalent to (2.10), and now it is clear that

$$\|f_n - f\|_{C^1([0,1])} = \max\{\|f_n - f\|_\infty, \|f_n' - g\|_\infty\} \longrightarrow 0$$

as $n \to \infty$, as required. $\qquad\qquad\qquad\qquad\qquad\qquad\qquad\qquad\square$

EXAMPLE 2.24(6). Let $U \subseteq \mathbb{R}^d$ be an open subset, and let (f_n) be a Cauchy sequence in $C_b^k(U)$ with respect to the norm $\|\cdot\|_{C_b^k(U)}$. Just as in the argument for (5), we know from (3) that for any $\alpha \in \mathbb{N}_0^d$ with $\alpha_1 + \cdots + \alpha_d \leqslant k$, the sequence $(\partial_\alpha f_n)$ in $C_b(U)$ has a uniform limit $g_\alpha \in C_b(U)$. All that remains is to show that

$$g_\alpha = \partial_\alpha g_0. \tag{2.12}$$

Suppose therefore that $x \in U$ and $i \in \{1, \ldots, d\}$. It is enough (by induction) to show that

$$\partial_{e_i} g_\alpha = g_{\alpha + e_i} \tag{2.13}$$

for $\alpha_1 + \cdots + \alpha_d < k$. For the function f_n we have, for sufficiently small h, that

$$\partial_\alpha f_n(x + he_i) = \partial_\alpha f_n(x) + \int_0^h \partial_{\alpha + e_i} f_n(x + te_i)\,\mathrm{d}t,$$

so letting $n \to \infty$ and using the known uniform convergence just as in (5) gives

$$g_\alpha(x + he_i) = g_\alpha(x) + \int_0^h g_{\alpha + e_i}(x + te_i)\, dt,$$

which implies (2.13) and hence (2.12) by induction on $\alpha_1 + \cdots + \alpha_d$. As in (5) it now follows that $f_n \to g_0$ with respect to $\|\cdot\|_{C_b^k(U)}$ as $n \to \infty$.   □

Exercise 2.27. Generalize Example 2.24(6) to give a Banach space over $\mathbb{C}$ in two different ways as follows.
(a) Let $U \subseteq \mathbb{R}^d$ be open and consider $\mathbb{C}$-valued bounded differentiable functions with bounded continuous derivative (here there is little difference from the real case).
(b) Let $U \subseteq \mathbb{C}$ (or in $\mathbb{C}^d$) be open, and consider the space of bounded complex differentiable functions with bounded derivative.

For Examples 2.24(7) and (8) regarding integrable functions and bounded measurable functions, we will use two lemmas that we formulate more generally.

The usual definitions of convergence and absolute convergence of series extend easily to normed vector spaces as follows. A series $\sum_{n=1}^\infty v_n$ *converges* if the sequence of partial sums $(s_N)_{N \geqslant 1}$ converges, where $s_N = \sum_{n=1}^N v_n$ for all $N \geqslant 1$, and *converges absolutely* if the real-valued series $\sum_{n=1}^\infty \|v_n\|$ converges.

Lemma 2.28 (Absolute convergence). *A normed vector space* $(V, \|\cdot\|)$ *is a Banach space if and only if any absolutely convergent series in V is convergent.*

PROOF. If V is a Banach space and a series $\sum_{n=1}^\infty v_n$ is absolutely convergent, which means that $\sum_{n=1}^\infty \|v_n\| < \infty$, then the sequence of partial sums (s_n) defined by $s_n = \sum_{k=1}^n v_k$ is a Cauchy sequence, since for $m > n$ we have

$$\|s_m - s_n\| = \left\| \sum_{k=n+1}^m v_k \right\| \leqslant \sum_{k=n+1}^m \|v_k\|,$$

and the last sum can be made arbitrarily small by requiring n to be sufficiently large.

Assume now for the converse that $(V, \|\cdot\|)$ is a normed vector space in which every absolutely convergent series is convergent, and let (v_n) be a Cauchy sequence in V. In order to render the Cauchy property more uniform, we choose a subsequence of (v_n) as follows. For each $k \geqslant 1$ there exists some N_k such that

$$\|v_m - v_n\| < \frac{1}{2^k}$$

for all $m, n \geqslant N_k$. Using these numbers we define inductively an increasing sequence (n_k) by $n_1 = N_1$ and $n_k = \max\{n_{k-1} + 1, N_k\}$ for $k \geqslant 2$. The corresponding subsequence $(v_{n_k})_{k \geqslant 1}$ satisfies $\|v_{n_{k+1}} - v_{n_k}\| < \frac{1}{2^k}$. Now define

$$w_k = v_{n_{k+1}} - v_{n_k}$$

for all $k \geqslant 1$, so that $\sum_{k=1}^{\infty} \|w_k\| < \sum_{k=1}^{\infty} \frac{1}{2^k} = 1$ converges, and hence the infinite sum $\sum_{k=1}^{\infty} w_k = w \in V$ converges by our assumption on the normed space $(V, \|\cdot\|)$. For the ℓth partial sum of this series we obtain

$$\sum_{k=1}^{\ell} w_k = v_{n_{\ell+1}} - v_{n_1},$$

and so the subsequence (v_{n_k}) satisfies $v = \lim_{k \to \infty} v_{n_k} = w + v_{n_1}$. Finally, we use the fact that any Cauchy sequence with a convergent subsequence must converge; we quickly recall the argument. For any $\varepsilon > 0$ choose N with $\|v_m - v_n\| < \varepsilon$ for $m, n \geqslant N$ and choose K with $\|v_{n_k} - v\| < \varepsilon$ for $k \geqslant K$. Then if $k \geqslant K$ has $n_k \geqslant N$ we have $\|v_m - v\| \leqslant \|v_m - v_{n_k}\| + \|v_{n_k} - v\| < 2\varepsilon$ for all $m \geqslant N$, showing that the sequence converges. $\qquad\square$

Lemma 2.29 (Quotients of Banach spaces). *If $(V, \|\cdot\|)$ is a Banach space and $W \subseteq V$ is a closed subspace then $(V/W, \|\cdot\|_{V/W})$ is a Banach space.*

PROOF. Assume that (v_n) is a sequence with

$$\sum_{n=1}^{\infty} \|v_n + W\|_{V/W} < \infty.$$

Since Banach spaces can be characterized by absolute convergence (see Lemma 2.28), it suffices to show that

$$\sum_{n=1}^{\infty} (v_n + W)$$

exists. For this, choose for each $n \geqslant 1$ some $w_n \in W$ with

$$\|v_n + w_n\| \leqslant \|v_n + W\|_{V/W} + \frac{1}{2^n}.$$

Then $\sum_{n=1}^{\infty} \|v_n + w_n\| < \infty$, so the limit

$$v = \sum_{n=1}^{\infty} (v_n + w_n)$$

exists in V by Lemma 2.28. Also note that the canonical map $\pi : V \to V/W$ (defined by $\pi(v) = v + W$ for all $v \in V$) is continuous since $\|v - v_0\| < \varepsilon$ implies $\|(v + W) - (v_0 + W)\|_{V/W} < \varepsilon$ for all $v, v_0 \in V$ and $\varepsilon > 0$. This implies that

$$\sum_{n=1}^{\infty} (v_n + W) = v + W$$

converges.   □

We refer to Appendix B for basic properties of $\| \cdot \|_p$ on $L^p_\mu(X)$, and in particular for the triangle inequality. Moreover, in the proof below and in the remainder of the book we will need the monotone convergence and the dominated convergence theorems (Theorems B.7 and B.8, respectively).

EXAMPLE 2.24(7). Let (f_n) be a sequence in $L^p_\mu(X)$ with

$$M = \sum_{n=1}^\infty \|f_n\|_p < \infty.$$

By Lemma 2.28 it is enough to show that $\sum_{n=1}^\infty f_n$ converges in $L^p_\mu(X)$. For this, define a sequence of functions (g_n) by

$$g_n(x) = \sum_{k=1}^n |f_k(x)|.$$

Clearly $g_n(x) \nearrow g(x)$ for some measurable function $g : X \to [0, \infty]$. Note that

$$\int |g_n|^p \, \mathrm{d}\mu = \|g_n\|_p^p \leqslant \left(\sum_{k=1}^n \|f_k\|_p\right)^p \leqslant M^p$$

by the triangle inequality for $\| \cdot \|_p$. By monotone convergence, this implies that

$$\|g\|_p^p = \lim_{n \to \infty} \|g_n\|_p^p \leqslant M^p,$$

and so $g(x) < \infty$ for μ-almost every $x \in X$. Therefore,

$$f(x) = \sum_{n=1}^\infty f_n(x)$$

exists for μ-almost every $x \in X$, and hence defines a measurable function

$$f : X \to \mathbb{R}.$$

Strictly speaking we have only defined f on the complement of a null set, but we simplify the notation by ignoring this distinction here.

Since we also have $|f(x)| \leqslant g(x)$ for all x, we have $f \in L^p_\mu(X)$. It remains to show that

$$\left\| \sum_{k=1}^n f_k - f \right\|_p \longrightarrow 0 \tag{2.14}$$

as $n \to \infty$. For this, notice first that $|\sum_{k=1}^n f_k - f|^p \leqslant (2g)^p$ and by definition of f we also have $|\sum_{k=1}^n f_k - f|^p \longrightarrow 0$ as $n \to \infty$ and almost everywhere, so that we may apply dominated convergence to the sequence of integrals defined by

$$\left\|\sum_{k=1}^{n} f_k - f\right\|_p^p = \int_X \left|\sum_{k=1}^{n} f_k - f\right|^p \mathrm{d}\mu.$$

From this we obtain (2.14), as required. $\qquad\square$

EXAMPLE 2.24(8). Since a pointwise limit (and *a fortiori* a uniform limit) of a sequence of measurable functions is a measurable function, the subspace $\mathscr{L}^\infty(X) \subseteq B(X)$ is closed. Therefore, by part (2) of the example we see that $\mathscr{L}^\infty(X)$ is a Banach space with respect to the norm $\|\cdot\|_\infty$.

Now let $W_\mu = \{f \in \mathscr{L}^\infty \mid f = 0 \ \mu\text{-almost everywhere}\}$. Clearly W_μ is closed, since if $f_n \in W_\mu$ for all $n \geqslant 1$ and $f_n \to f$ uniformly, then

$$\{x \in X \mid f(x) \neq 0\} \subseteq \bigcup_{n \geqslant 1} \{x \in X \mid f_n(x) \neq 0\}$$

is a μ-null set. Therefore

$$L_\mu^\infty = \mathscr{L}^\infty / W_\mu$$

is a Banach space with respect to the quotient norm $\|\cdot\|_{\mathscr{L}^\infty / W_\mu}$. It remains to show that

$$\|f\|_{\mathscr{L}^\infty / W_\mu} = \inf_{g \in W_\mu} \|f + g\|_\infty$$

coincides, as claimed, with the essential supremum norm

$$\|f\|_{\mathrm{esssup}} = \inf\{\alpha > 0 \mid \mu\left(\{x \in X \mid |f(x)| > \alpha\}\right) = 0\}$$

as given in Example 2.24(8). For this, assume first that $\alpha > \|f\|_{\mathrm{esssup}}$ so that $N_\alpha = \{x \in X \mid |f(x)| > \alpha\}$ is a μ-null set, and hence $g_\alpha = -f \mathbb{1}_{N_\alpha} \in W_\mu$. It follows that

$$\|f\|_{\mathscr{L}^\infty / W_\mu} \leqslant \|f + g_\alpha\|_\infty \leqslant \alpha.$$

Since this holds for any $\alpha > \|f\|_{\mathrm{esssup}}$ it follows that

$$\|f\|_{\mathscr{L}^\infty / W_\mu} \leqslant \|f\|_{\mathrm{esssup}}.$$

If, on the other hand, $\alpha > \|f\|_{\mathscr{L}^\infty / W_\mu}$, then there exists some $g \in W_\mu$ with

$$\|f + g\|_\infty < \alpha,$$

and so

$$\{x \in X \mid |f(x)| > \alpha\} \subseteq \{x \in X \mid g(x) \neq 0\}$$

is a null set. Varying α once more, we see that $\|f\|_{\mathrm{esssup}} \leqslant \|f\|_{\mathscr{L}^\infty / W_\mu}$. $\qquad\square$

Exercise 2.30. Show that in the definition of $\|\cdot\|_{\mathrm{esssup}}$ and of $\|\cdot\|_{\mathscr{L}^\infty / W_\mu}$ (from the proof that Example 2.24(8) is a Banach space on p. 35) the infima are actually minima and hence that for $f \in L_\mu^\infty(X)$ we have $|f(x)| \leqslant \|f\|_{\mathrm{esssup}}$ μ-almost everywhere.

Finally, let us recall two facts from real analysis:

- If a series of real numbers is absolutely convergent, then the sum is independent of any rearrangement of the series. The latter property is also called *unconditional convergence.*
- If a convergent series of real numbers is not absolutely convergent, then the series may be rearranged to obtain any value in $\mathbb{R} \cup \{\pm\infty\}$ for the sum of the rearranged series. We say that a series is *conditionally convergent* if its convergence properties depend on the order of its elements.

In finite-dimensional spaces unconditional convergence is equivalent to absolute convergence, and in infinite dimensions an absolutely convergent series is unconditionally convergent (see the exercise below). However, any infinite-dimensional Banach space contains an unconditionally convergent series that is not absolutely convergent[4] (see Corollary 3.42 for a particular case).

Exercise 2.31. Show that in a normed vector space any absolutely convergent series is unconditionally convergent.

2.2.2 The Completion of a Normed Vector Space

Even though we have seen several examples of Banach spaces above, there are many natural normed vector spaces that are not Banach spaces. For example, $\mathbb{R}[x]$ is not a Banach space with respect to any of the five norms discussed in Example 2.2(7) (see also Exercise 2.62). As a result it is useful to know that any normed vector space has a completion (whose uniqueness properties we will discuss in Corollary 2.60).

Theorem 2.32 (Existence of a Completion). *Let $(V, \|\cdot\|)$ be a normed vector space. Then there exists a Banach space $(B, \|\cdot\|)$ which contains V as a dense subspace, and the indicated norm on B restricts to the original norm on the image of V in B.*

PROOF[†]. Let $W = \{(v_n) \in V^{\mathbb{N}} \mid (v_n) \text{ is a Cauchy sequence}\}$. It is straightforward to check that W is a vector space. We also define the semi-norm

$$\|(v_n)\|' = \lim_{n \to \infty} \|v_n\|,$$

which is well-defined as $(\|v_n\|)$ is a Cauchy sequence in $\mathbb{R}$, since (v_n) is a Cauchy sequence in V (due to (2.1)). The kernel of this semi-norm is the space

$$W_0 = \{(v_n) \mid v_n \to 0 \text{ as } n \to \infty\}$$

of null sequences (that is, sequences converging to 0) in V. We define

---

[†] The proof in this section can be skipped, as many natural normed vector spaces are already Banach spaces, and we will be able to give another shorter construction in Chapter 7 on p. 214.

$$B = W/W_0$$

and

$$\|b\|_B = \lim_{n \to \infty} \|v_n\|$$

where $b = (v_n) + W_0$.

It follows from our discussion concerning quotient norms (see Lemma 2.15) that $(B, \| \cdot \|_B)$ is a normed vector space. Moreover, B contains an isometric copy of V (that is, there is an isometry $V \to B$), since an element $v \in V$ can be identified with the equivalence class of the constant sequence

$$\phi(v) = (v, v, \dots) + W_0,$$

with the norm of this coset being

$$\|\phi(v)\|_B = \lim_{n \to \infty} \|v\| = \|v\|$$

by definition.

We claim that (the image of) V is dense in B. Given an equivalence class $b = (v_1, v_2, \dots) + W_0 \in B$ of a Cauchy sequence (v_n), for every $\varepsilon > 0$ there exists some N with $\|v_m - v_n\| < \varepsilon$ for $m, n \geq N$. Then

$$\|(v_1, v_2, \dots) + W_0 - \phi(v_N)\|_B = \lim_{n \to \infty} \|v_n - v_N\| \leq \varepsilon.$$

Using this for any $\varepsilon > 0$ shows that the image of V is dense in B.

It remains to show that B is complete with respect to $\| \cdot \|_B$. For this, assume that $(b_n)_{n \geq 1}$ is a Cauchy sequence in B. Since the image of V is dense in B we can find a sequence (v_n) of vectors in V with

$$\|b_n - \phi(v_n)\|_B < \frac{1}{n}$$

for each $n \in \mathbb{N}$. Then for every $\varepsilon > 0$ there exists some $N(\varepsilon)$ with

$$\|b_m - b_n\|_B < \varepsilon$$

and $\frac{1}{m}, \frac{1}{n} < \varepsilon$ for $m, n \geq N(\varepsilon)$, so that

$$\|v_m - v_n\| \leq \underbrace{\|\phi(v_m) - b_m\|_B}_{<\frac{1}{m}<\varepsilon} + \underbrace{\|b_m - b_n\|_B}_{<\varepsilon} + \underbrace{\|b_n - \phi(v_n)\|_B}_{<\frac{1}{n}<\varepsilon} < 3\varepsilon$$

for $m, n \geq N(\varepsilon)$. Therefore, (v_n) is a Cauchy sequence in V. We define

$$b = (v_1, v_2, \dots) + W_0 \in B \tag{2.15}$$

and see that

$$\|b - b_m\|_B \leqslant \|b - \phi(v_m)\|_B + \|\phi(v_m) - b_m\|_B < \lim_{n \to \infty} \underbrace{\|v_n - v_m\|}_{< 3\varepsilon} + \frac{1}{m} < 4\varepsilon$$

for $m \geqslant N(\varepsilon)$. Thus $b \in B$ defined by (2.15) is the limit of (b_n) and so B is a Banach space. $\square$

Exercise 2.33. Let $C_c(\mathbb{R})$ be the vector space of continuous functions $f : \mathbb{R} \to \mathbb{R}$ with

$$\mathrm{Supp}(f) = \overline{\{x \in \mathbb{R} \mid f(x) \neq 0\}}$$

compact, with the norm $\| \cdot \|_\infty$. Show that this space is not complete, and find a Banach space containing $C_c(\mathbb{R})$ as a dense subspace so that the induced norm obtained by restriction is $\| \cdot \|_\infty$. Can you do the same for the norm $\|f\|_\Psi = \|f\Psi\|_\infty$, where $\Psi : \mathbb{R} \to \mathbb{R}_{>0}$ is a fixed continuous function (for example, $\Psi(x) = e^{x^2}$)?

Exercise 2.34. Generalize Theorem 2.32 to metric spaces as follows. If (X, d) is a metric space, then a *completion* of (X, d) is a pair consisting of a complete metric space (X^*, d^*) and an isometry $\phi : X \to X^*$ with the property that $\phi(X)$ is dense in X^*. Prove that any metric space has a completion.

2.2.3 Non-Compactness of the Unit Ball

Many properties of finite-dimensional vector spaces are consequences of the Heine–Borel theorem, which implies that the closed unit ball in a finite-dimensional vector space is compact. Correspondingly, many interesting problems in infinite-dimensional Banach spaces are related to the opposite phenomenon.

Proposition 2.35 (Non-compactness of the unit ball). *The closed unit ball $\overline{B_1^V}$ in an infinite-dimensional normed space V is not compact.*

PROOF[†]. It is enough to construct a sequence (v_n) in V with

$$\|v_n\| \leqslant 1 \tag{2.16}$$

for all $n \geqslant 1$ and with

$$\|v_m - v_n\| \geqslant \tfrac{1}{2} \tag{2.17}$$

for all $m \neq n$ (for then such a sequence has no Cauchy subsequence, and therefore no convergent subsequence).

Choose $v_1 \in V$ with norm 1 (this is always possible by homogeneity). Suppose that we have already found $v_1, \dots, v_k \in V$ with (2.16) and (2.17) for $1 \leqslant n \leqslant k$ and $1 \leqslant m \neq n \leqslant k$. The subspace $W = \langle v_1, \dots, v_k \rangle$ spanned by the vectors $v_1, \dots, v_k$ is finite-dimensional, and therefore complete with respect to the induced norm (see Proposition 2.6). Thus W is a closed subspace,

---

[†] The proof in this subsection could be skipped: the result is negative and will only be used in the more concrete setting of Hilbert spaces, where the proof is a simple exercise.

and so we may consider the quotient norm $\|\cdot\|_{V/W}$ on the normed vector space V/W as in Lemma 2.15. Since V is not finite-dimensional, the quotient space V/W is non-trivial, and so we may choose some $v \in V$ with $v + W \neq 0$. Thus

$$d = \|v + W\|_{V/W} > 0.$$

It follows that there exists some $w \in W$ with $\|v + w\| \leqslant 2d$. Define

$$v_{k+1} = \frac{1}{\|v + w\|}(v + w),$$

so that $\|v_{k+1}\| = 1$. Also, for $1 \leqslant n \leqslant k$, we have $v_n \in W$ and so

$$\|v_{k+1} - v_n\| \geqslant \|v_{k+1} + W\|_{V/W} = \frac{1}{\|v + w\|}\|v + W\|_{V/W} \geqslant \frac{d}{2d} = \frac{1}{2}$$

as required. Thus by induction we obtain a sequence (v_n) with the claimed properties, and hence the proposition. $\qquad\square$

Given the negative statement in Proposition 2.35, a natural question to ask is how to characterize compact subsets of a Banach space. This depends on the space concerned (see Exercise 2.36). A vague principle is that one tries to extract topological and geometrical properties of finite subsets of the Banach space, and then compact subsets are sometimes characterized by suitable uniform versions of those properties. We will also illustrate this in the next section, where we will prove the Arzela–Ascoli theorem.

Exercise 2.36. Characterize the compact subsets of the following Banach spaces.
(a) The space c_0 of null sequences (that is, sequences (x_n) of scalars with $|x_n| \to 0$ as $n \to \infty$) with the norm $\|(x_n)\|_\infty = \sup_{n \geqslant 1} |x_n| = \max_{n \geqslant 1} |x_n|$.
(b) The space ℓ^p of p-summable sequences of scalars with $p \in [1, \infty)$. That is,

$$\ell^p = \left\{ (x_n) \mid \sum_{n=1}^{\infty} |x_n|^p < \infty \right\}$$

with the p-norm $\|(x_n)\|_p = \left(\sum_{n=1}^{\infty} |x_n|^p \right)^{1/p}$.

2.3 The Space of Continuous Functions

To illustrate the failure of compactness of the closed unit ball in a Banach space, we now discuss the Banach space of continuous functions $C(X)$ on a compact metric space (X, d). A subset $K \subseteq C(X)$ is said to be *equicontinuous* if for every $\varepsilon > 0$ there is a $\delta > 0$ such that

$$\mathsf{d}(x, y) < \delta \implies |f(x) - f(y)| < \varepsilon$$

for all $x, y \in X$ and $f \in K$. The key uniformity here is that a single δ may be used for all the functions $f \in K$.

2.3.1 The Arzela–Ascoli Theorem

Essential Exercise 2.37. Recall that a function $f : X \to \mathbb{R}$ on a metric space (X, d) is *uniformly continuous* if for any $\varepsilon > 0$ there is some $\delta > 0$ for which

$$\mathsf{d}(x, y) < \delta \implies |f(x) - f(y)| < \varepsilon$$

for all $x, y \in X$. Show that any continuous function on a compact metric space is uniformly continuous and hence that any finite set of continuous functions is equicontinuous.

Theorem 2.38 (Arzela–Ascoli). *Let (X, d) be a compact metric space, and let $C(X)$ be the Banach space of continuous (real- or complex-valued) functions on X with the supremum norm. A subset $K \subseteq C(X)$ is compact if and only if K is closed, bounded, and equicontinuous.*

PROOF. Suppose that $K \subseteq C(X)$ is compact, so it is closed and bounded. We will now show that it is also equicontinuous. Fix $\varepsilon > 0$. Then we may find finitely many functions $f_1, \ldots, f_n \in K$ such that

$$K \subseteq \bigcup_{i=1}^{n} B_\varepsilon(f_i) \tag{2.18}$$

by compactness, since $\{B_\varepsilon(f) \mid f \in K\}$ is an open cover of K. Each f_i is continuous and, since X is compact, each f_i is also uniformly continuous by Exercise 2.37. Since the family $\{f_i\}$ is finite, we can conclude that there is a $\delta > 0$ with

$$\mathsf{d}(x, y) < \delta \implies |f_i(x) - f_i(y)| < \varepsilon \tag{2.19}$$

for $i = 1, \ldots, n$. We now combine (2.18) and (2.19) for the given $\varepsilon > 0$. Fix some $f \in K$. By (2.18), there exists some i with $\|f - f_i\|_\infty < \varepsilon$. If $x, y \in X$ and $\mathsf{d}(x, y) < \delta$, then

$$|f(x) - f(y)| \leqslant \underbrace{|f(x) - f_i(x)|}_{<\varepsilon} + \underbrace{|f_i(x) - f_i(y)|}_{<\varepsilon \text{ by (2.19)}} + \underbrace{|f_i(y) - f(y)|}_{<\varepsilon} < 3\varepsilon,$$

showing equicontinuity.

PROVING COMPACTNESS: Now suppose that $K \subseteq C(X)$ is closed, bounded, and equicontinuous. To show that K is compact, let (f_n) be an arbitrary sequence in K. It will be enough to exhibit a Cauchy subsequence of (f_n), since by Example 2.24(3) such a subsequence will converge in $C(X)$, and by our assumption that K is closed the limit will be in K.

First notice that X contains a dense countable subset D since X is a compact metric space (a topological space is *separable* if it contains a dense countable subset). The argument given here shows that a compact metric space is separable. In fact, $X \subseteq \bigcup_{x \in X} B^X_{1/m}(x)$, which implies that

$$X \subseteq \bigcup_{y \in D_m} B^X_{1/m}(y) \tag{2.20}$$

for some finite subset $D_m \subseteq X$, so that the set $D = \bigcup_{m \geqslant 1} D_m$ is countable and dense.

Next notice that by our assumption on K there is some M with

$$\|f_n\|_\infty \leqslant M$$

for every $n \geqslant 1$. Let us write $I_M = [-M, M] \subseteq \mathbb{R}$ or $I_M = \overline{B^{\mathbb{C}}_M}$ depending on whether the field of scalars is $\mathbb{R}$ or $\mathbb{C}$. Then by Tychonoff's theorem (Theorem A.20 and Lemma A.17) I^D_M is a compact metric space with respect to the product topology. Define $\phi_n \in I^D_M$ by $\phi_n = f_n|_D$. By compactness of I^D_M there exists a subsequence (ϕ_{n_k}) which converges in I^D_M. This convergence is precisely the statement that

$$f_{n_k}(y) \to \phi(y)$$

as $k \to \infty$ for all $y \in D$ and some function $\phi : D \to I_M$. Note, however, that at this point no uniformity of the convergence is known.

We now upgrade the argument above to give the desired statement that (f_{n_k}) is a Cauchy sequence. Fix $\varepsilon > 0$. Then there exists some $\delta > 0$ with

$$d(x, y) < \delta \implies |f_{n_k}(x) - f_{n_k}(y)| < \varepsilon \tag{2.21}$$

for all $k \geqslant 1$ (this is possible by equicontinuity of K). Now choose some $m \in \mathbb{N}$ with $\frac{1}{m} \leqslant \delta$. Since

$$f_{n_k}(y) \longrightarrow \phi(y)$$

as $k \to \infty$ for $y \in D_m$, each of the sequences $(f_{n_k}(y))_k$ in I_M (with k varying) is a Cauchy sequence. Since m is fixed, there are only finitely many sequences concerned, so there exists some $N(\varepsilon)$ such that $k, \ell \geqslant N(\varepsilon)$ implies that

$$|f_{n_k}(y) - f_{n_\ell}(y)| < \varepsilon \tag{2.22}$$

for all $y \in D_m$. Now we combine (2.21) and (2.22) as follows. Given $x \in X$, by (2.20) there is some $y \in D_m$ with $d(x, y) < \frac{1}{m} \leqslant \delta$. For $k, \ell \geqslant N(\varepsilon)$ this implies

$$|f_{n_k}(x) - f_{n_\ell}(x)| \leqslant \underbrace{|f_{n_k}(x) - f_{n_k}(y)|}_{<\varepsilon \text{ by } (2.21)}$$

$$+ \underbrace{|f_{n_k}(y) - f_{n_\ell}(y)|}_{<\varepsilon \text{ by } (2.22)} + \underbrace{|f_{n_\ell}(y) - f_{n_\ell}(x)|}_{<\varepsilon \text{ by } (2.21)} < 3\varepsilon.$$

Thus $\|f_{n_k} - f_{n_\ell}\| < 3\varepsilon$ for all $k, \ell \geqslant N(\varepsilon)$, showing that the subsequence is Cauchy as required. $\qquad\square$

Exercise 2.39. (a) Prove the Arzela–Ascoli theorem for any compact space (that is, without assuming that the space is a metric space). To do this, define a subset K of $C(X)$ to be equicontinuous if for every $\varepsilon > 0$ and every $x \in X$ there exists a neighbourhood U of x with $|f(y) - f(x)| < \varepsilon$ for all $f \in K$.
(b) Extend the Arzela–Ascoli theorem to the space $C_0(X)$ of continuous functions vanishing at infinity with the uniform norm $\|f\|_\infty = \sup_{x \in X} |f(x)|$, where X is a locally compact metric (or just locally compact) space.

2.3.2 The Stone–Weierstrass Theorem

Let X be a compact topological space. We now prove a useful criterion for a subset of functions to be dense in $C(X)$. However, for this we will need to distinguish between the space $C_{\mathbb{R}}(X)$ of real-valued, and the space $C_{\mathbb{C}}(X)$ of complex-valued, continuous functions on X.

Theorem 2.40 (Stone–Weierstrass). *Let $(X, \mathcal{T})$ be a compact topological space.*

(a) *Suppose that $\mathcal{A} \subseteq C_{\mathbb{R}}(X)$ is a collection of functions that satisfy the following properties:*

- *(Algebra) $\mathcal{A}$ is a sub-algebra, meaning that $\mathcal{A}$ is a linear subspace of $C_{\mathbb{R}}(X)$ and, for any $f, g \in \mathcal{A}$, the pointwise product fg also belongs to $\mathcal{A}$;*
- *(Constants) the constant function $\mathbb{1}$ lies in $\mathcal{A}$;*
- *(Separation) the algebra $\mathcal{A}$ separates points: for any $x, y \in X$ with x not equal to y, there is some function $f \in \mathcal{A}$ with $f(x) \neq f(y)$.*

Then $\mathcal{A}$ is dense in $C_{\mathbb{R}}(X)$ with respect to $\| \cdot \|_\infty$.
(b) *Suppose that $\mathcal{A} \subseteq C_{\mathbb{C}}(X)$ satisfies all of the properties in (a) and, in addition, has*

- *(Complex conjugation) $\mathcal{A}$ is closed under conjugation, meaning that if f is in $\mathcal{A}$ then $\overline{f}$ is in $\mathcal{A}$.*

Then $\mathcal{A}$ is dense in $C_{\mathbb{C}}(X)$ with respect to $\| \cdot \|_\infty$.

Example 2.41. The algebra $\mathcal{A} = \mathbb{R}[t]$ (or $\mathbb{C}[t]$) of polynomials on a compact interval $X \subseteq \mathbb{R}$ satisfies Theorem 2.40. This recovers the classical *Weierstrass approximation theorem*.[5]

Uniform approximation in the complex setting is more complicated.[6] Let us note that closure under complex conjugation is necessary for the conclusion of the theorem to hold due to the following obstacle. If $D \subseteq \mathbb{C}$ is a compact set with non-empty interior (a closed disk, for example), then the limit of any sequence of holomorphic functions uniformly converging on D is also holomorphic in the interior D^o, so the continuous function $z \mapsto \bar{z}$ cannot be in the closure of the point-separating algebra $\mathbb{C}[z]$ in $C_{\mathbb{C}}(D)$.

We will start the proof of Theorem 2.40 with the following lemma, and will write $\overline{\mathcal{A}}$ for the closure of $\mathcal{A}$ with respect to $\| \cdot \|_\infty$ (notice that we also use $\bar{}$ to denote complex conjugation, but it should always be clear from the context whether closure or conjugation is meant).

Lemma 2.42. *Let $\mathcal{A}$ be a sub-algebra of $C_{\mathbb{R}}(X)$ containing the constant functions. Then $\overline{\mathcal{A}}$ is also a sub-algebra, and*

$$|f|, \max\{f, g\}, \min\{f, g\} \in \overline{\mathcal{A}}$$

for any $f, g \in \overline{\mathcal{A}}$.

PROOF. It is easy to check that the algebra operations are continuous with respect to $\| \cdot \|_\infty$. (See (2.2)–(2.3) for the vector space operations and generalize the argument to include the product operation for functions; see also Section 2.4.2 for a more general discussion containing this case.) Therefore $\overline{\mathcal{A}}$ is also an algebra. Recall that

$$\sqrt{1 + u} = (1 + u)^{1/2} = \sum_{n=0}^{\infty} \binom{1/2}{n} u^n$$

is a power series with radius of convergence 1 (this can be shown using real analysis by using the standard estimates for the Taylor approximation of $\sqrt{1 + u}$. Using complex analysis this follows from the fact that $\sqrt{1 + u}$ is holomorphic for $|u| < 1$.) In particular, we have

$$\sum_{n=0}^{\infty} \left|\binom{1/2}{n}\right| (1 - \varepsilon)^n < \infty.$$

Studying the coefficients more closely gives $\sum_{n=0}^{\infty} |\binom{1/2}{n}| < \infty$, and the reader who knows this may set $\varepsilon = 0$ and simplify the argument accordingly, but we will not use this. Suppose that $f \in \overline{\mathcal{A}}$, $M = \|f\|_\infty$, and $\varepsilon > 0$. Then the function

$$g_\varepsilon = \frac{1}{M^2 + \varepsilon}(f^2 + \varepsilon)$$

is in $\overline{\mathcal{A}}$ and takes on values in $[\varepsilon/(M^2 + \varepsilon), 1]$, and so

$$\sum_{n=0}^{\infty} \left|\binom{1/2}{n}\right| \|g_\varepsilon - 1\|_\infty^n < \infty,$$

which implies that

$$\sqrt{g_\varepsilon} = (1 + (g_\varepsilon - 1))^{1/2} = \sum_{n=0}^{\infty} \binom{1/2}{n} (g_\varepsilon - 1)^n$$

converges with respect to $\| \cdot \|_\infty$ by Example 2.24(3) and Lemma 2.28. We deduce that $\sqrt{f^2 + \varepsilon} \in \overline{\mathcal{A}}$. Now

$$0 \leqslant \sqrt{f^2 + \varepsilon} - |f| = \sqrt{f^2 + \varepsilon} - \sqrt{f^2} = \frac{f^2 + \varepsilon - f^2}{\sqrt{f^2 + \varepsilon} + \sqrt{f^2}} \leqslant \frac{\varepsilon}{\sqrt{\varepsilon}}.$$

In particular, $\left\| |f| - \sqrt{f^2 + \varepsilon} \right\|_\infty \leqslant \sqrt{\varepsilon}$, and so the fact that $\sqrt{f^2 + \varepsilon} \in \overline{\mathcal{A}}$ for all $\varepsilon > 0$ implies that $|f| \in \overline{\mathcal{A}}$. The identities

$$\max\{f, g\} = \tfrac{1}{2}(f + g) + \tfrac{1}{2}|f - g|$$

and

$$\min\{f, g\} = \tfrac{1}{2}(f + g) - \tfrac{1}{2}|f - g|$$

give the other parts of the lemma. $\square$

PROOF OF THEOREM 2.40. We start with the case of an algebra $\mathcal{A} \subseteq C_\mathbb{R}(X)$. Notice that by Lemma 2.42 the algebra $\overline{\mathcal{A}}$ is closed under taking finitely many maxima or minima: if $f_1, \ldots, f_n \in \overline{\mathcal{A}}$ then

$$\max\{f_1, \ldots, f_n\}, \min\{f_1, \ldots, f_n\} \in \overline{\mathcal{A}}.$$

We will use this property for a given $f \in C_\mathbb{R}(X)$ and $\varepsilon > 0$ to find a function $f_\varepsilon \in \overline{\mathcal{A}}$ with $\|f - f_\varepsilon\|_\infty < \varepsilon$. This then implies that $\overline{\mathcal{A}} = C_\mathbb{R}(X)$. The construction has three steps.

FIRST STEP: CORRECT VALUE AT TWO POINTS. Let $x_0, x \in X$ be (not necessarily distinct) points. Then there exists some $h_{x_0,x} \in \mathcal{A}$ with

$$\left. \begin{array}{l} h_{x_0,x}(x_0) = f(x_0) \\ h_{x_0,x}(x) = f(x). \end{array} \right\} \tag{2.23}$$

Indeed, if $x_0 = x$ then we simply take $h_{x_0,x} = f(x_0)\mathbb{1} \in \mathcal{A}$. If $x \in X \backslash \{x_0\}$ we know that $\mathcal{A}$ contains a function $\widetilde{h} \in \mathcal{A}$ with $\widetilde{h}(x) \neq \widetilde{h}(x_0)$ since the algebra separates points. In this case, we may find a linear combination $h_{x_0,x}$ of $\widetilde{h} \in \mathcal{A}$ and the constant function $\mathbb{1} \in \mathcal{A}$ with the desired property.

SECOND STEP: CORRECT VALUE AT ONE POINT, NOWHERE MUCH SMALLER. Let $x_0 \in X$. As our next step we claim that there exists a function $g_{x_0} \in \overline{\mathcal{A}}$ with

$$\left. \begin{array}{l} g_{x_0}(x_0) = f(x_0) \\ g_{x_0}(y) > f(y) - \varepsilon \end{array} \right\} \tag{2.24}$$

for all $y \in X$. That is, g_{x_0} is chosen to have the correct value at x_0 for the objective of approximating f, and to be not much smaller than f at every other point, as illustrated in Figure 2.1.

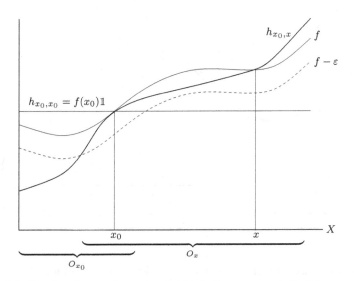

Fig. 2.1: The function g_{x_0} is constructed by finding $x_1, \ldots, x_n$ (in this case, x_0 and x) with the property that $g_{x_0} = \max\{h_{x_0,x_1}, \ldots, h_{x_0,x_n}\} > f - \varepsilon$.

We will construct g_{x_0} as a maximum after finding a finite subcover for the following open cover of X. For any $x \in X$ (including x_0) there exists an open neighbourhood O_x of x with

$$y \in O_x \implies h_{x_0,x}(y) > f(y) - \varepsilon, \qquad (2.25)$$

where $h_{x_0,x} \in \mathcal{A}$ is as in (2.23). This defines an open cover $\{O_x \mid x \in X\}$ of X. By compactness there exists some finite subcover

$$X = O_{x_1} \cup \cdots \cup O_{x_n}. \qquad (2.26)$$

We define

$$g_{x_0} = \max\{h_{x_0,x_1}, \ldots, h_{x_0,x_n}\} \in \overline{\mathcal{A}},$$

and notice that g_{x_0} satisfies

$$g_{x_0}(x_0) = \max\{f(x_0), \ldots, f(x_0)\} = f(x_0)$$

by (2.23), and by (2.26) for every $y \in X$ there is some $i \in \{1, \ldots, n\}$ for which $y \in O_{x_i}$, and hence

$$g_{x_0}(y) \geqslant h_{x_0,x_i}(y) > f(y) - \varepsilon$$

by (2.25).

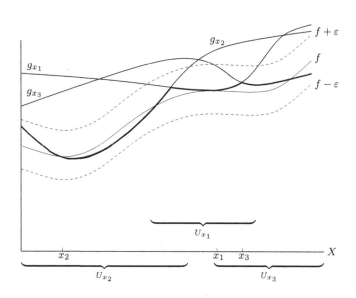

Fig. 2.2: The function $f_\varepsilon = \min\{g_{x_1}, \ldots, g_{x_m}\}$ is constructed with $\|f - f_\varepsilon\|_\infty < \varepsilon$.

THIRD STEP: NOWHERE MUCH SMALLER, NOWHERE MUCH BIGGER. The claim (2.24) above takes care of one half of the need to find an approximation to f within $\overline{\mathcal{A}}$. For every $x \in X$ we found some $g_x \in \overline{\mathcal{A}}$ that is nowhere much smaller than f, and is equal to f at x. We now vary the point x, and essentially repeat the argument to find an ε-approximation to f within $\overline{\mathcal{A}}$. Indeed, for every $x \in X$ there is an open neighbourhood U_x for which

$$y \in U_x \implies g_x(y) < f(y) + \varepsilon. \qquad (2.27)$$

By allowing $x \in X$ to vary this gives an open cover $\{U_x \mid x \in X\}$ of X, and once again by compactness there is a finite subcover

$$X = U_{x_1} \cup \cdots \cup U_{x_m}. \qquad (2.28)$$

We define

$$f_\varepsilon = \min\{g_{x_1}, \ldots, g_{x_m}\} \in \overline{\mathcal{A}},$$

and claim that $\|f - f_\varepsilon\|_\infty \leqslant \varepsilon$, as illustrated in Figure 2.2. For every $y \in X$ we have

$$g_{x_i}(y) > f(y) - \varepsilon$$

by the property of g_{x_i} in (2.24), and so $f_\varepsilon(y) > f(y) - \varepsilon$. By (2.28) every $y \in X$ lies in some U_{x_i} and so (2.27) implies that

$$f_\varepsilon(y) \leqslant g_{x_i}(y) < f(y) + \varepsilon.$$

Since $f_\varepsilon \in \overline{\mathcal{A}}$ and $\varepsilon > 0$ was arbitrary, we deduce that $f \in \overline{\mathcal{A}}$.

COMPLEX CASE. In the case of a complex sub-algebra $\mathcal{A} \subseteq C_{\mathbb{C}}(X)$ that is closed under conjugation we may consider

$$\mathcal{A}_{\mathbb{R}} = \mathcal{A} \cap C_{\mathbb{R}}(X).$$

This is again a sub-algebra that separates points if $\mathcal{A}$ separates points. Indeed, if $x, y \in X$ with $x \neq y$ then there is (by the assumption on $\mathcal{A}$) some $f \in \mathcal{A}$ with $f(x) \neq f(y)$. Let $u = \Re(f)$ and $v = \Im(f)$, so that

$$u = \frac{f + \overline{f}}{2}, v = \frac{f - \overline{f}}{2\mathrm{i}} \in \mathcal{A}_{\mathbb{R}}$$

by our assumption on $\mathcal{A}$. Thus $\mathcal{A}_{\mathbb{R}}$ also contains a function that separates x and y. By the real case, $\mathcal{A}_{\mathbb{R}}$ is dense in $C_{\mathbb{R}}(X)$, so by splitting an arbitrary function in $C_{\mathbb{C}}(X)$ into real and imaginary parts and approximating each of these with elements of $\mathcal{A}_{\mathbb{R}} \subseteq \mathcal{A}$ the theorem is proved. $\qquad\square$

Exercise 2.43. (a) Let X be as in Theorem 2.40. Show that the second requirement (Constants) on $\mathcal{A} \subseteq C(X)$ could also be replaced by the requirement

- (Nowhere vanishing) for every $x \in X$ there is a function $f \in \mathcal{A}$ with $f(x) \neq 0$.

(b) Let X be a locally compact space. Extend the Stone–Weierstrass theorem to $C_0(X)$ by considering a sub-algebra $\mathcal{A} \subseteq C_0(X)$ that separates points, is closed under conjugation, and vanishes nowhere.

Exercise 2.44. Define, for every infinite compact subset $K \subseteq \mathbb{R}$,

$$\|p\|_K = \|p|_K\|_{C(K)} = \sup_{x \in K} |p(x)|.$$

Show that $\| \cdot \|_K$ and $\| \cdot \|_L$ are inequivalent norms on $\mathbb{R}[x]$ if $K \neq L$ are two different infinite compact subsets.

Exercise 2.45. Let X and Y be two compact spaces. Prove that the linear hull of all functions of the form $(x, y) \in X \times Y \to f(x)g(y)$ for $f \in C(X)$ and $g \in C(Y)$ is dense in $C(X \times Y)$.

The following is also an easy consequence of the Stone–Weierstrass theorem.

Lemma 2.46 (Separability). *Let (X, d) be a compact metric space. Then the Banach space $C(X)$ is separable with respect to the topology induced by the supremum norm.*

PROOF. The space X is separable (this may be seen, for example, from the proof of Theorem 2.38) so we may choose a countable dense set $\{x_n \mid n \in \mathbb{N}\}$ in X. We now define $f_n(x) = \mathsf{d}(x, x_n)$ for all $x \in X$ and $n \geqslant 1$, and claim that these functions *separate points* in X. That is, if $x \neq y$ then there exists some n for which $f_n(x) \neq f_n(y)$. To see this, notice that by density there is some n with $\mathsf{d}(x, x_n) = f_n(x) < \frac{1}{2}\mathsf{d}(x, y)$, which implies that

$$f_n(y) = \mathsf{d}(y, x_n) \geqslant \mathsf{d}(y, x) - \mathsf{d}(x, x_n) > \tfrac{1}{2}\mathsf{d}(x, y).$$

Now let $\mathscr{A}_{\mathbb{Q}} = \mathbb{Q}[f_0 = 1, f_1, f_2, \dots]$ be the $\mathbb{Q}$-algebra generated by the functions $f_1, f_2, \dots$ together with the constant function $f_0 = 1$. Clearly $\mathscr{A}_{\mathbb{Q}}$ is countable, and the closure of $\mathscr{A}_{\mathbb{Q}}$ contains the algebra $\mathscr{A} = \mathbb{R}[f_0, f_1, f_2, \dots]$. Since $\mathscr{A}$ is an algebra that separates points, it is dense in $C_{\mathbb{R}}(X)$ (and $\mathscr{A} + \mathrm{i}\mathscr{A}$ is dense in $C_{\mathbb{C}}(X)$) by the Stone–Weierstrass theorem (Theorem 2.40).   □

2.3.3 Equidistribution of a Sequence

[†]As an application of the discussion above, and in particular of the Stone–Weierstrass theorem, we now describe the notion of equidistribution. A sequence $(x_n)_n$ of elements of a metric space X is *dense* if for every $x \in X$ there is a subsequence $(x_{n_k})_k$ that converges to x. A much finer property is given by equidistribution, which roughly speaking corresponds to the sequence spending the right proportion of time in any given part of the space. In this section we will define and discuss this notion carefully for $X = [0, 1]$.

A sequence $(x_n)_{n \geqslant 1}$ of points in $[0, 1]$ is said to be *equidistributed* or *uniformly distributed* if any one of the following equivalent conditions is satisfied:

(1) $\dfrac{1}{K}|\{k \in \mathbb{N} \mid 1 \leqslant k \leqslant K, x_k \in [a, b]\}| \to b - a$ as $K \to \infty$ for $0 \leqslant a < b \leqslant 1$.

(2) $\dfrac{1}{K} \sum\limits_{k=1}^{K} f(x_k) \longrightarrow \displaystyle\int_0^1 f(x)\, \mathrm{d}x$ as $K \to \infty$ for any $f \in C([0, 1])$ (that is, any continuous function).

(3) $\dfrac{1}{K} \sum\limits_{k=1}^{K} f(x_k) \longrightarrow \displaystyle\int_0^1 f(x)\, \mathrm{d}x$ as $K \to \infty$ for any $f \in \mathscr{R}([0, 1])$ (that is, any Riemann-integrable function).

(4) $\dfrac{1}{K} \sum\limits_{k=1}^{K} \chi_n(x_k) \longrightarrow \displaystyle\int_0^1 \chi_n(x)\, \mathrm{d}x = \begin{cases} 0 & \text{if } n \neq 0 \\ 1 & \text{if } n = 0 \end{cases}$ as $K \to \infty$ for any n in $\mathbb{Z}$, where $\chi_n(x) = \mathrm{e}^{2\pi i n x}$ for all $x \in [0, 1]$.

We will now sketch some of the implications between these equivalent statements (see Exercise 2.48) and will return to the topic of equidistribution in Chapter 8 from a more general point of view.

ALMOST A PROOF OF (4) $\implies$ (2). Consider the algebra of *trigonometric polynomials*

[†] The results of this section will not be needed in this form later, so may be skipped.

$$\mathcal{A} = \left\{ \sum_{n=-N}^{N} c_n \chi_n \mid c_n \in \mathbb{C}, N \in \mathbb{N} \right\}.$$

Using the complex version of the Stone–Weierstrass theorem (Theorem 2.40), it follows that $\mathcal{A}$ is dense in $C(\mathbb{T})$ with respect to the uniform metric (see Exercise 2.47 and Proposition 3.65), where $\mathbb{T} = \mathbb{R}/\mathbb{Z}$. This means that given $f \in C(\mathbb{T})$ and $\varepsilon > 0$, there is some $g \in \mathcal{A}$ with

$$\|f - g\|_\infty = \sup_{x \in \mathbb{T}} |f(x) - g(x)| < \varepsilon,$$

which implies that

$$\left| \int_0^1 f(x) \, dx - \int_0^1 g(x) \, dx \right| < \varepsilon$$

and

$$\left| \frac{1}{K} \sum_{k=1}^{K} f(x_k) - \frac{1}{K} \sum_{k=1}^{K} g(x_k) \right| < \varepsilon$$

for any $K \geqslant 1$. If K is sufficiently large then, by assumption,

$$\left| \frac{1}{K} \sum_{k=1}^{K} g(x_k) - \int_0^1 g(x) \, dx \right| < \varepsilon.$$

It follows that

$$\left| \frac{1}{K} \sum_{k=1}^{K} f(x_k) - \int_0^1 f(x) \, dx \right| < 3\varepsilon,$$

which is not quite the claim in (2) since $C(\mathbb{T})$ and $C([0,1])$ differ slightly. Indeed, any function $f : \mathbb{T} \to \mathbb{C}$ gives rise to a function $\overline{f} : \mathbb{R} \to \mathbb{C}$ via the diagram

which we can restrict to $[0,1]$, defining an element $g \in C([0,1])$. If $f : \mathbb{T} \to \mathbb{C}$ is continuous then so is g, but g will always satisfy $g(0) = g(1)$. On the other hand, if $g \in C([0,1])$ is a function satisfying $g(0) = g(1)$ then one can define a continuous function $f : \mathbb{T} \to \mathbb{C}$ by $f(t + \mathbb{Z}) = g(t)$ for $t \in [0,1]$, and obtain the result for such g.

The extension to general continuous functions on $[0,1]$ can be handled by the same method as in the proof that (2) implies (1) below, where we will only assume (2) for all $f \in C(\mathbb{T})$. $\qquad \square$

Exercise 2.47. Show that $\mathcal{A}$ as in the previous proof does indeed satisfy all the assumptions of Theorem 2.40,

PROOF OF (2) $\implies$ (1). Suppose first that $0 < a < b < 1$ and write $\mathbb{1}_{[a,b]}$ for the characteristic function of the interval $[a, b]$. Fix $\varepsilon > 0$ and choose continuous functions $f_-, f_+ : [0, 1] \to \mathbb{R}$ with

(a) $0 \leqslant f_-(x) \leqslant \mathbb{1}_{[a,b]}(x) \leqslant f_+(x) \leqslant 1$ for all $x \in [0, 1]$,

(b) $\displaystyle\int_0^1 (f_+ - f_-)\, \mathrm{d}x < \varepsilon$, and

(c) $f_+(0) = f_+(1) = f_-(0) = f_-(1) = 0$.

For example, the functions f_+ and f_- could be chosen to be piecewise linear, as illustrated in Figure 2.3. In this case the shaded region can easily be chosen to have total area bounded above by ε, as required in (b). By (c), the functions f_- and f_+ also define continuous functions on $\mathbb{T}$.

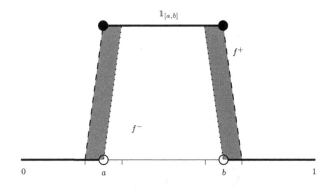

Fig. 2.3: The function $\mathbb{1}_{[a,b]}$ and the approximations f_- (drawn using dots) and f_+ (using dashes).

Since

$$\frac{1}{K}\sum_{k=1}^{K} f_-(x_k) \leqslant \frac{1}{K}\sum_{k=1}^{K} \mathbb{1}_{[a,b]}(x_k) \leqslant \frac{1}{K}\sum_{k=1}^{K} f_+(x_k)$$

for all $K \geqslant 1$, and the left-hand side converges to $\int_0^1 f_-(x)\, \mathrm{d}x$ while the right-hand side converges to $\int_0^1 f_+(x)\, \mathrm{d}x$ as $K \to \infty$, we obtain

$$(b-a)-\varepsilon \leqslant \liminf_{K\to\infty} \frac{1}{K}\sum_{k=1}^{K} \mathbb{1}_{[a,b]}(x_k) \leqslant \limsup_{K\to\infty} \frac{1}{K}\sum_{k=1}^{K} \mathbb{1}_{[a,b]}(x_k) \leqslant (b-a)+\varepsilon,$$

which implies the claim in (1) for $0 < a < b < 1$. The formula in (1) holds trivially if $f \equiv 1$, so we also get

$$\frac{1}{K}\sum_{k=1}^{K} \left(\mathbb{1}_{[0,a)}(x_k) + \mathbb{1}_{(b,1]}(x_k) \right) \longrightarrow 1 - (b - a)$$

as $K \to \infty$ by taking the difference. Suppose now that $0 = a < b < 1$. Then, for any sufficiently small $\varepsilon > 0$, we have

$$f_- = \mathbb{1}_{[\varepsilon,b]} \leqslant \mathbb{1}_{[0,b]} \leqslant \mathbb{1}_{[0,b+\varepsilon)} + \mathbb{1}_{(1-\varepsilon,1]} = f_+$$

and

$$\int_0^1 (f_+ - f_-)\,\mathrm{d}x < 3\varepsilon,$$

and the formula in (1) already holds for f_- and f_+. As before, this implies the claim for $\mathbb{1}_{[0,b]}$. The case of $0 < a < b = 1$ is similar. $\qquad\square$

Exercise 2.48. Prove the remaining implications to show that the four characterizations of equidistribution at the start of this section are indeed equivalent.

Example 2.49. A simple example of an equidistributed sequence may be obtained as follows. Fix $\alpha \in \mathbb{R} \setminus \mathbb{Q}$ and define $x_k = \{k\alpha\} \in [0,1)$ for $k \in \mathbb{N}$, where $\{t\}$ denotes the fractional part of the real number t. To see that this defines an equidistributed sequence, the characterization in (4) is the most convenient to use. For $n = 0$, we have $\chi_n \equiv 1$ and so $\frac{1}{K}\sum_{k=1}^K \chi_0(x_k) = 1$ for all K. If $n \neq 0$, then

$$\frac{1}{K}\sum_{k=1}^K \underbrace{\mathrm{e}^{2\pi i n k\alpha}}_{\neq 1} = \frac{1}{K}\sum_{k=1}^K \left(\mathrm{e}^{2\pi i n\alpha}\right)^k = \frac{\mathrm{e}^{2\pi i n(K+1)\alpha} - \mathrm{e}^{2\pi i n\alpha}}{K(\mathrm{e}^{2\pi i n\alpha} - 1)} \longrightarrow 0$$

as $K \to \infty$.

An amusing consequence of this example is a special case of Benford's law [7].

Exercise 2.50. Use the equidistribution from Example 2.49 to show the following. Write ℓ_n for the leading digit of 2^n written in decimal (so the sequence (ℓ_n) begins $(2, 4, 8, 1, 3, \dots)$). Then

$$\frac{1}{K}\,|\{k \in \mathbb{N} \mid 1 \leqslant k \leqslant K, \ell_k = 1\}| \longrightarrow \log_{10} 2$$

as $K \to \infty$.

2.3.4 Continuous Functions in L^p Spaces

Another important feature of continuous functions (of compact support) is that they form a dense subset of the L^p spaces. We say that a measure on a topological space is *locally finite* if each point of X has an open neighbourhood of finite measure.

Proposition 2.51 (Density of $C_c(X)$ in $L_\mu^p(X)$). *Let X be a locally compact σ-compact metric space equipped with a locally finite measure μ on the Borel σ-algebra $\mathcal{B}(X)$. Then, for any $p \in [1,\infty)$, $C_c(X)$ is dense in $L_\mu^p(X)$.*

PROOF[†]. We split the proof of the proposition into several steps.

COMPACT CASE. We assume first that X is a compact metric space, that μ is a finite Borel measure on X, and start with some preparatory observations. Fix $p \in [1, \infty)$ and let $f \in \mathscr{L}_\mu^p(X)$. Then

$$f = \Re f + i \Im f,$$

and it is enough to show that each of $\Re f$ and $\Im f$ can be approximated in L_μ^p by elements of $C_c(X)$. We may therefore assume that f is real-valued, and by writing $f = f^+ - f^-$ we may also assume that it takes values in $[0, \infty)$. Now notice that such a real-valued, non-negative function f is the pointwise limit of the simple functions

$$f_n(x) = \min\left\{n, \tfrac{1}{2^n}\lfloor 2^n f(x)\rfloor\right\} \nearrow f(x)$$

as $n \to \infty$, which implies that

$$\|f_n - f\|_p = \left(\int_X |f_n - f|^p \, \mathrm{d}\mu\right)^{1/p} \longrightarrow 0$$

as $n \to \infty$, by dominated convergence.

Thus it is sufficient to show that any simple function $f = \sum_{i=1}^N a_i \mathbb{1}_{B_i}$ (where $a_i \in \mathbb{R}$ and $B_i \in \mathcal{B}(X)$ have $\mu(B_i) < \infty$ for $i = 1, \ldots, N$) can be approximated by elements of $C(X)$. This in turn will follow if we can show that the characteristic function of any Borel set can be approximated by elements of $C(X)$ in the $\|\cdot\|_p$ norm.

DEFINING A σ-ALGEBRA. Having made these initial reductions, we can now turn to the heart of the argument (still assuming that X is compact). We define the family

$$\mathcal{A} = \left\{ B \in \mathcal{B} \mid \mathbb{1}_B \in \overline{C(X)}^{\|\cdot\|_p} \right\}$$

of all Borel sets whose characteristic function can be approximated by elements of $C(X)$ (the notation indicates that the closure is taken within $L_\mu^p(X)$ and with respect to the norm $\|\cdot\|_p$).

We claim that $\mathcal{A} = \mathcal{B}$, and will prove this by showing that

- $\mathcal{A}$ contains any open subset of X, and
- $\mathcal{A}$ is a σ-algebra.

OPEN SUBSETS. Let $O \subseteq X$ be open. Define the closed set $A = X \setminus O$ and the distance function

$$\mathrm{d}(x, A) = \inf_{y \in A} \mathrm{d}(x, y). \tag{2.29}$$

[†] The reader may be familiar with this result for the Lebesgue measure (for example), and this case is sufficient for much of the material that will follow. Thus she may skip the general proof and return to it at a later stage if needed.

This distance function satisfies

$$|\mathsf{d}(x_1, A) - \mathsf{d}(x_2, A)| \leqslant \mathsf{d}(x_1, x_2) \tag{2.30}$$

for all $x_1, x_2 \in X$. Indeed, given $x_1, x_2 \in X$ and $\varepsilon > 0$, there exists some $y \in A$ for which $\mathsf{d}(x_2, y) \leqslant \mathsf{d}(x_2, A) + \varepsilon$, and so

$$\mathsf{d}(x_1, A) \leqslant \mathsf{d}(x_1, y) \leqslant \mathsf{d}(x_1, x_2) + \mathsf{d}(x_2, y) \leqslant \mathsf{d}(x_1, x_2) + \mathsf{d}(x_2, A) + \varepsilon,$$

which implies that $\mathsf{d}(x_1, A) \leqslant \mathsf{d}(x_1, x_2) + \mathsf{d}(x_2, A)$, and hence (2.30) by the symmetry between x_1 and x_2. This shows the continuity of $x \mapsto \mathsf{d}(x, A)$ and so it follows that the function defined by $f_n(x) = \min\{1, n\mathsf{d}(x, A)\}$ lies in $C(X)$. Moreover, if $x \in A = X \setminus O$ then $f_n(x) = 0 = \mathbb{1}_O(x)$, while if $x \in O$ then $\mathsf{d}(x, A) > 0$ and $f_n(x) \nearrow 1 = \mathbb{1}_O(x)$. Thus $f_n \nearrow \mathbb{1}_O$ as $n \to \infty$ on X, and so

$$\|f_n - \mathbb{1}_O\|_p = \left(\int_O |f_n - \mathbb{1}_O|^p \, \mathrm{d}\mu \right)^{1/p} \longrightarrow 0$$

as $n \to \infty$ by dominated convergence. This shows that $O \in \mathcal{A}$, by definition.

COMPLEMENTS. Suppose that $A \in \mathcal{A}$. Then there exists a sequence of functions (f_n) in $C(X)$ with $\|f_n - \mathbb{1}_A\|_p \to 0$ as $n \to \infty$. Using the sequence $\mathbb{1} - f_n \in C(X)$, we see that $X \setminus A \in \mathcal{A}$.

FINITE INTERSECTIONS. Suppose that $A, B \in \mathcal{A}$. Then there exist sequences of functions $(f_n), (g_n)$ in $C(X)$ with $\|f_n - \mathbb{1}_A\|_p \to 0$ and $\|g_n - \mathbb{1}_B\|_p \to 0$ as $n \to \infty$. We may assume that f_n and g_n take on values in $[0, 1]$, for if not we can replace f_n by

$$\widetilde{f_n} = \max\{0, \min\{1, f_n\}\}$$

(which will approximate $\mathbb{1}_A$ equally well or better), and similarly g_n by $\widetilde{g_n}$. Then $f_n g_n \in C(X)$ and

$$f_n g_n - \mathbb{1}_{A \cap B} = f_n g_n - \mathbb{1}_A \mathbb{1}_B = (f_n - \mathbb{1}_A) g_n + \mathbb{1}_A (g_n - \mathbb{1}_B),$$

which implies

$$\|f_n g_n - \mathbb{1}_{A \cap B}\|_p \leqslant \|f_n - \mathbb{1}_A\|_p \|g_n\|_\infty + \|g_n - \mathbb{1}_B\|_p \longrightarrow 0$$

as $n \to \infty$. This shows that $A \cap B \in \mathcal{A}$ as desired.

COUNTABLE UNIONS. Let $A, B \in \mathcal{A}$. Then $X \setminus A, X \setminus B \in \mathcal{A}$ and hence

$$B \cup A = X \setminus \left((X \setminus A) \cap (X \setminus B) \right) \in \mathcal{A}$$

by the two steps above. This extends to finite unions by induction.

Now suppose that $A_1, A_2, \ldots$ all lie in $\mathcal{A}$, and fix $\varepsilon > 0$. Then there exists an ℓ such that

$$\mu \left(\bigcup_{k=1}^{\infty} A_k \setminus \bigcup_{k=1}^{\ell} A_k \right) < \varepsilon^p.$$

Thus

$$\left\| \mathbb{1}_{\bigcup_{k=1}^{\infty} A_k} - \mathbb{1}_{\bigcup_{k=1}^{\ell} A_k} \right\|_p = \left(\mu\left(\bigcup_{k=1}^{\infty} A_k \smallsetminus \bigcup_{k=1}^{\ell} A_k \right) \right)^{1/p} < \varepsilon.$$

However, since $\bigcup_{k=1}^{\ell} A_k \in \mathcal{A}$ for any $\ell \geqslant 1$, we already know that there exists an $f \in C(X)$ with

$$\left\| f - \mathbb{1}_{\bigcup_{k=1}^{\ell} A_k} \right\|_p < \varepsilon,$$

and so

$$\left\| f - \mathbb{1}_{\bigcup_{k=1}^{\infty} A_k} \right\|_p < 2\varepsilon.$$

Since $\varepsilon > 0$ was arbitrary, we deduce that $\bigcup_{k=1}^{\infty} A_k \in \mathcal{A}$.

CONCLUDING THE COMPACT CASE. By the arguments above, $\mathcal{A}$ is a σ-algebra containing all the open subsets of X. By definition, $\mathcal{A} \subseteq \mathcal{B}$ and so $\mathcal{A} = \mathcal{B}$ by definition of the Borel σ-algebra $\mathcal{B}$. As explained above, this implies that every simple function, and so also every function, in $L_\mu^p(X)$ can be approximated by continuous functions.

EXTENDING TO THE LOCALLY COMPACT CASE. Let us now extend the above to the general case where X is locally compact σ-compact metric and μ is locally finite. By Lemma A.22 we find a sequence (X_m) of compact subsets of X with $X_m \subseteq X_{m+1}^o$ for all $m \geqslant 1$, and with $X = \bigcup_{m=1}^{\infty} X_m$.

Given some $f \in L_\mu^p(X)$ we first note that the sequence $f_m = f \mathbb{1}_{X_m}$ converges to f with respect to $\| \cdot \|_p$ as $m \to \infty$ (by dominated convergence). Given some $\varepsilon > 0$ we choose m such that $\|f - f_m\|_p < \varepsilon$.

Next we apply the compact case above and hence we find some $g \in C(X_m)$ with $\|g - f_m\|_{L^p(X_m,\mu)} < \varepsilon$. Applying Tietze's extension theorem (Proposition A.29) we can extend g to an element $g \in C_c(X_{m+1}^o) \subseteq C_c(X)$. Using again the distance function in (2.29) with $A = X_m$ we define the sequence

$$g_n(x) = \bigl(1 - \min\{1, n\mathsf{d}(x, X_m)\}\bigr) g(x) \in C_c(X).$$

For $x \in X_m$ we have $g_n(x) = g(x)$ and for $x \notin X_m$ we have $|g_n(x)| \searrow 0$ as $n \to \infty$. Now notice that

$$\|g_n - f_m\|_p^p = \int_{X_{m+1}} |g_n - f_m|^p \, \mathrm{d}\mu = \|g - f_m\|_{L^p(X_m,\mu)}^p + \int_{X_{m+1} \smallsetminus X_m} |g_n|^p \, \mathrm{d}\mu,$$

where the first expression on the right is less than ε^p by construction of g and the second expression converges to 0 by dominated convergence as $n \to \infty$. Therefore, there exists some $n \geqslant 1$ such that $\|g_n - f_m\|_p < 2\varepsilon$. Combining this with the choice of f_m above, we obtain $\|f - g_n\|_p < 3\varepsilon$ and $g_n \in C_c(X)$, as desired. $\square$

2.4 Bounded Operators and Functionals

Just as in linear algebra, linear maps are of fundamental importance in functional analysis. However, in infinite-dimensional normed vector spaces continuity of linear maps is not guaranteed.

Lemma 2.52 (Continuity and boundedness). *Let* $L : V \to W$ *be a linear map between the two normed vector spaces* $(V, \|\cdot\|_V)$ *and* $(W, \|\cdot\|_W)$. *Then* L *is continuous if and only if the* operator norm

$$\|L\| = \|L\|_{\mathrm{op}} = \sup_{\substack{v \in V \\ \|v\|_V \leqslant 1}} \|Lv\|_W$$

is finite.

Definition 2.53. A continuous linear map $L : V \to W$ between normed vector spaces is called a *bounded linear operator*. We denote the space of all bounded operators from V to W by $\mathrm{B}(V, W)$. For brevity we write $\mathrm{B}(V)$ for $\mathrm{B}(V, V)$. If $W = \mathbb{R}$ (or $W = \mathbb{C}$ if the field of scalars is $\mathbb{C}$) then we also write V^* for $\mathrm{B}(V, \mathbb{R})$ (respectively $\mathrm{B}(V, \mathbb{C})$) the *dual space* of V, and elements of the dual space are called *linear functionals*.

Lemma 2.54 (Space of operators). *Let* $(V, \|\cdot\|_V)$ *and* $(W, \|\cdot\|_W)$ *be normed vector spaces. Then the space* $\mathrm{B}(V, W)$ *of bounded linear maps from* V *to* W *is also a normed vector space with addition and scalar multiplication defined pointwise as in any space of functions, and with the operator norm from Lemma 2.52. If* W *is a Banach space, then so is* $\mathrm{B}(V, W)$, *and in particular* V^* *is always a Banach space.*

PROOF OF LEMMA 2.52. The case $L = 0$ is trivial, so we may assume that L is not 0. Suppose that $\|L\|_{\mathrm{op}} < \infty$. Then for any $v_0 \in V$ we have

$$L\big(v_0 + B^V_{\varepsilon/\|L\|_{\mathrm{op}}}\big) \subseteq L(v_0) + B^W_\varepsilon$$

since $v \in B^V_{\varepsilon/\|L\|_{\mathrm{op}}} \setminus \{0\}$ implies that

$$\|Lv\|_W = \underbrace{\|v\|_V}_{<\varepsilon/\|L\|_{\mathrm{op}}} \underbrace{\big\|L\big(\|v\|_V^{-1}v\big)\big\|_W}_{\leqslant \|L\|_{\mathrm{op}}} < \varepsilon.$$

Hence $\|L\|_{\mathrm{op}} < \infty$ implies that L is continuous.

Suppose now that L is continuous. Then there exists some $\delta > 0$ such that

$$L\big(B^V_\delta\big) \subseteq B^W_1.$$

In particular, $\|v\|_V \leqslant 1$ implies that $\|L(\frac{\delta}{2}v)\|_W \leqslant 1$, and $\|Lv\|_W \leqslant \frac{2}{\delta}$. As this holds for all v with $\|v\|_V \leqslant 1$, we deduce that $\|L\|_{\mathrm{op}} \leqslant \frac{2}{\delta} < \infty$. $\square$

As the next exercise shows, the notion of boundedness for an operator makes a clear distinction between integration and differentiation of real-valued functions.

Exercise 2.55. Show that the operator $I : C([0,1]) \to C([0,1])$ defined as the integral

$$I(f)(x) = \int_0^x f(t)\, dt$$

is continuous. Use this to shorten the argument in the proof for Example 2.24(5) on p. 30. Show also that the operator $D : C^1([0,1]) \to C([0,1])$ defined as the derivative $D(f) = f'$ is not continuous if we use the norm $\|\cdot\|_\infty$ on both spaces.

Notice that the definition of the operator norm immediately gives the general inequality

$$\|Lv\|_W \leqslant \|L\|_{\mathrm{op}}\|v\|_V,$$

for all $v \in V$, and the operator norm may be characterized as being the smallest number C with the property that

$$\|Lv\|_W \leqslant C\|v\|_V, \tag{2.31}$$

for all $v \in V$. We will use both these statements frequently in the sequel without comment.

Essential Exercise 2.56. Prove that the operator norm of a bounded operator $L : V \to W$ between two normed vector spaces is the smallest constant $C \geqslant 0$ such that (2.31) holds for all $v \in V$.

PROOF OF LEMMA 2.54. As indicated in the lemma, for $L_1, L_2 \in B(V, W)$ and a scalar α we define $\alpha L_1 + L_2$ by $(\alpha L_1 + L_2)(v) = \alpha L_1(v) + L_2(v)$ for all $v \in V$. This is clearly another linear map. In order to bound its operator norm, let $v \in V$ with $\|v\|_V \leqslant 1$. Then

$$\begin{aligned}
\|(\alpha L_1 + L_2)(v)\|_W &= \|\alpha L_1(v) + L_2(v)\|_W \\
&\leqslant |\alpha|\|L_1(v)\|_W + \|L_2(v)\|_W \\
&\leqslant |\alpha|\|L_1\|_{\mathrm{op}} + \|L_2\|_{\mathrm{op}},
\end{aligned}$$

and so $\|\alpha L_1 + L_2\|_{\mathrm{op}} \leqslant |\alpha|\|L_1\|_{\mathrm{op}} + \|L_2\|_{\mathrm{op}}$. That is, the operator norm satisfies the triangle inequality and one half of the homogeneity property. The reverse inequality for homogeneity of the operator norm follows easily by considering the case $\alpha = 0$ and $\alpha \neq 0$ separately, as in the proof of Lemma 2.15. Strict positivity is clear, so we have shown that $B(V, W)$ is a normed vector space with the operator norm.

Now suppose that W is a Banach space and that (L_n) is a Cauchy sequence in $B(V, W)$. We claim that

$$L(v) = \lim_{n \to \infty} L_n(v)$$

defines an element L of $B(V,W)$ which is the limit of the sequence with respect to the operator norm. To see that $L(v)$ is well-defined it is enough to check that $(L_n(v))$ is a Cauchy sequence, which follows at once from the bound

$$\|L_m(v) - L_n(v)\|_W = \|(L_m - L_n)(v)\|_W \leqslant \|L_m - L_n\|_{\mathrm{op}}\|v\|_V,$$

which (for fixed v) may be made as small as we please for m, n large by the Cauchy property for the sequence (L_n).

To see that the limit L is a bounded operator one has to show that it is linear (which we leave as an exercise) and that it is bounded. For the latter, assume that $v \in V$ has $\|v\|_V \leqslant 1$ and choose $N(\varepsilon)$ as in the Cauchy property for (L_n) with $\|L_m - L_n\|_{\mathrm{op}} \leqslant \varepsilon$ for $m, n \geqslant N(\varepsilon)$. Continuity of the norm now gives

$$\|L_n(v) - L(v)\|_W = \lim_{m \to \infty} \|L_n(v) - L_m(v)\|_W \leqslant \varepsilon$$

for $n \geqslant N(\varepsilon)$. Taking the supremum over v with $\|v\|_V \leqslant 1$, we get

$$\|L - L_n\|_{\mathrm{op}} \leqslant \varepsilon,$$

so L is bounded with $\|L\|_{\mathrm{op}} \leqslant \|L_n\|_{\mathrm{op}} + \varepsilon$, and as $\varepsilon > 0$ is arbitrary we also see that $L_n \to L$ as $n \to \infty$ with respect to $\|\cdot\|_{\mathrm{op}}$. $\qquad\square$

A word about notation: Where the spaces concerned are clear, or where we wish to emphasize certain aspects of the spaces, we will for brevity often use $\|\cdot\|$ or $\|\cdot\|_X$ to mean the appropriate norm in that situation. Thus, for example, depending on context, the symbols $\|L\|$, $\|L\|_{\mathrm{op}}$, and $\|L\|_{B(V,W)}$ all mean the same thing. A good exercise for the reader is to ensure that they can identify the norms in each case.

Lemma 2.57 (Sub-multiplicativity of operator norms). *Let V, W, Z be three normed vector spaces, and let $R : V \to W$ and $S : W \to Z$ be bounded operators. Then $S \circ R : V \to Z$ is also a bounded operator, and*

$$\|S \circ R\| \leqslant \|S\|\|R\|.$$

In particular, if $L : V \to V$ is a bounded operator then $\|L^n\| \leqslant \|L\|^n$ for all $n \geqslant 1$.

PROOF. We have $\|S \circ R(v)\| \leqslant \|S\|\|R(v)\| \leqslant \|S\|\|R\|\|v\| \leqslant \|S\|\|R\|$ for any $v \in V$ with $\|v\| \leqslant 1$. $\qquad\square$

Exercise 2.58. Compute the operator norm of the continuous map $f \longmapsto f$ when viewed:
(a) as a map from the Banach space $C^1([0,1])$ to $C([0,1])$ (and where the former is equipped with the norm $\|f\|_{C^1([0,1])} = \max\{\|f\|_\infty, \|f'\|_\infty\}$ for $f \in C^1([0,1])$); and
(b) as a map $C([0,1]) \to L^1_m([0,1])$, where m denotes Lebesgue measure on $[0,1]$.
(c) Compute the operator norm of the composition of the maps from (a) and from (b).
(d) Now restrict the maps in (a), (b) and (c) to the subspace of functions f with $f(0) = 0$, and compute the operator norms again.

The following result is both quite easy and extremely useful for the theory to come.

Proposition 2.59 (Unique extension to completion). *Let V be a normed vector space, let $V_0 \subseteq V$ be a dense subspace, and assume that $L_0 : V_0 \to W$ is a bounded operator into a Banach space W. Then L_0 has a unique bounded extension $L : V \to W$, that is a bounded linear map $L : V \to W$ which satisfies $L|_{V_0} = L_0$. Moreover, $\|L\|_{B(V,W)} = \|L_0\|_{B(V_0,W)}$.*

We implicitly assume here that a subspace $V_0 \subseteq V$ is equipped with the restriction of the norm on V to V_0. This is important to remember in applications where the subspace may have other natural norms defined on it.

PROOF OF PROPOSITION 2.59. For any $v \in V$ there is a sequence (v_n) in V_0 with $v_n \to v$ as $n \to \infty$. In particular, this implies that (v_n) is a Cauchy sequence in V_0, and since

$$L_0 : V_0 \to W$$

is bounded (and so Lipschitz), it follows that $(L_0(v_n))$ is a Cauchy sequence in W. If (v'_n) is another sequence in V_0 with $v'_n \to v$ as $n \to \infty$ then it is clear that $v_n - v'_n \to 0$ as $n \to \infty$ and so

$$L_0(v_n) - L_0(v'_n) \longrightarrow 0$$

as $n \to \infty$ since L_0 is bounded (and so continuous at 0). Thus it makes sense to define an operator L on V by

$$L(v) = \lim_{n \to \infty} L_0(v_n) \in W,$$

because W is a Banach space. Notice that by density and the desired continuity of the extension, this is the only possible definition of a bounded operator that extends L_0. One can quickly check that L is a linear map from V to W. Moreover, if $v \in V$ and (v_n) is a sequence in V_0 with $v_n \to v$ as $n \to \infty$, then

$$\|L(v)\| = \lim_{n \to \infty} \|L_0(v_n)\| \leqslant \|L_0\| \underbrace{\lim_{n \to \infty} \|v_n\|}_{=\|v\|},$$

showing that L is bounded, with $\|L\| \leqslant \|L_0\|$. On the other hand $L|_{V_0} = L_0$, so $\|L\| \geqslant \|L_0\|$. □

Corollary 2.60. *Any two completions B_1 and B_2 of a given normed vector space V are isometrically isomorphic.*

Here a *completion* of a normed vector space V is a Banach space B containing an isometric dense copy of V, just as in the construction in Section 2.2.2.

PROOF OF COROLLARY 2.60. Suppose that $\phi_1 : V \to B_1$ and $\phi_2 : V \to B_2$ are isometric embeddings associated to the two completions, as illustrated in Figure 2.4.

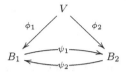

Fig. 2.4: The two given completions ϕ_1, ϕ_2 and the maps ψ_1, ψ_2 to be constructed.

Since ϕ_1 and ϕ_2 are isometries, the map

$$\phi_2 \circ \phi_1^{-1} : \phi_1(V) \longrightarrow \phi_2(V) \subseteq B_2$$
$$\phi_1(v) \longmapsto \phi_2(v)$$

is a well-defined bounded operator defined on a dense subset $\phi_1(V) \subseteq B_1$. By Proposition 2.59 there is an extension $\psi_1 : B_1 \to B_2$ with norm

$$\|\psi_1\| = \|\phi_2 \circ \phi_1^{-1}\| = 1.$$

Similarly there exists an extension $\psi_2 : B_2 \to B_1$ which extends $\phi_1 \circ \phi_2^{-1}$ and which also has norm $\|\psi_2\| = 1$. It follows that $\psi_2 \circ \psi_1$ and $\psi_1 \circ \psi_2$ are extensions of the identity map on $\phi_1(V)$ and on $\phi_2(V)$ respectively. By uniqueness of the extension in Proposition 2.59 we must have $\psi_2 \circ \psi_1 = I_{B_1}$ and $\psi_1 \circ \psi_2 = I_{B_2}$. We also see that $\|b\| = \|\psi_2(\psi_1(b))\| \leqslant \|\psi_1(b)\| \leqslant \|b\|$ for any $b \in B_1$, so that ψ_1 is an isometry from B_1 to B_2 with ψ_2 its inverse. $\square$

Exercise 2.61. Let $D = \{z \in \mathbb{C} \mid |z| < 1\} \subseteq \mathbb{C}$ be the open unit disk, and parameterize the circle of radius $r \in (0,1)$ by the map $\gamma_r : [0,1] \to \mathbb{C}$ defined by $\gamma_r(t) = re^{2\pi it}$. Let V be the space of functions $f \in C(\overline{D})$ holomorphic on D, and fix $p \in [1, \infty)$.
(a) Equip V with the norm

$$\|f\|_{H^p(D)} = \sup_{r \in (0,1)} \left(\int_0^1 |f(\gamma_r(t))|^p \, dt \right)^{1/p}.$$

Show that the linear map $E_z : f \longmapsto f(z)$ is continuous with respect to $\| \cdot \|_{H^p(D)}$ for all $z \in D$. Also show that if $O \subseteq D$ is open with compact closure $\overline{O} \subseteq D$, then

$$V \ni f \longmapsto f|_{\overline{O}} \in C(\overline{O})$$

is a bounded operator with respect to $\|\cdot\|_{H^p(D)}$ and $\|\cdot\|_\infty$ on $C(\overline{O})$. In particular, conclude that there exists a canonical injective map from the completion $H^p(D)$ of V, known as a *Hardy space*, into the space of holomorphic functions on D.
(b) Equip V with the norm

$$\|f\|_{A^p(D)} = \|f\|_{L^p(D)},$$

and repeat the problems from (a) to obtain the[7] *Bergman space* $A^p(D)$.

Exercise 2.62. For each of the five norms on $\mathbb{R}[x]$ given in Example 2.2(7), find a Banach space containing $\mathbb{R}[x]$ for which the induced norm obtained by restriction coincides with the given norm on $\mathbb{R}[x]$.

2.4.1 The Norm of Continuous Functionals on $C_0(X)$

Let X be a locally compact metric space, and let μ be a finite Borel measure. Then

$$\mu : f \longmapsto \int f \, d\mu$$

is a continuous functional on $C_0(X)$. Indeed,

$$\left| \int f \, d\mu \right| \leqslant \int |f| \, d\mu \leqslant \mu(X) \|f\|_\infty$$

shows the continuity by Lemma 2.52. More generally, if μ is a Borel measure on X and $g \in L^1_\mu(X)$ then

$$g \, d\mu : C_0(X) \ni f \longmapsto \int fg \, d\mu \tag{2.32}$$

is also a continuous functional on $C_0(X)$. Again this is easy to see since

$$\left| \int fg \, d\mu \right| \leqslant \int |f| |g| \, d\mu \leqslant \|f\|_\infty \|g\|_{L^1_\mu}. \tag{2.33}$$

In fact, a more precise statement holds, but this takes a little more work.

Lemma 2.63 (Operator norm of integration). *Suppose that μ is a Borel measure on a locally compact σ-compact metric space X and g is a function in $L^1_\mu(X)$. Then the norm of the functional on $C_0(X)$ defined in (2.32) is precisely $\|g\|_{L^1_\mu}$.*

PROOF[†]. Let

$$h(x) = \arg(\overline{g(x)}) = \begin{cases} \overline{\dfrac{g(x)}{|g(x)|}} & \text{if } g(x) \neq 0, \\ 0 & \text{if } g(x) = 0. \end{cases}$$

Clearly $h \in L^\infty_\mu(X)$ and

$$\int hg \, d\mu = \int |g| \, d\mu = \|g\|_{L^1_\mu}.$$

We wish to approximate h by continuous functions. Fix $\varepsilon > 0$. By Lusin's theorem (Theorem B.17) applied to the finite measure ν defined by $d\nu = |g| \, d\mu$

[†] The result of Section 2.4.1 will initially only be used in more concrete settings. The reader may therefore skip the proof and return to it later if needed.

there exists a compact set $K \subseteq X$ such that the restriction $h|_K$ of h to K is continuous, and $\int_{X \smallsetminus K} |g| \, d\mu < \varepsilon$. By Tietze's extension theorem (Proposition A.29) the restriction $h|_K$ can be extended to a continuous function $f_\varepsilon \in C_c(X)$ of compact support. We may assume that $\|f_\varepsilon\|_\infty \leqslant 1$, because if this is not the case we may replace f_ε by the continuous function

$$
\begin{cases}
f_\varepsilon(x) & \text{if } |f_\varepsilon(x)| \leqslant 1, \\
\frac{f_\varepsilon(x)}{|f_\varepsilon(x)|} & \text{if } |f_\varepsilon(x)| \geqslant 1.
\end{cases}
$$

Thus

$$
\|g \, d\mu\|_{\mathrm{op}} \geqslant \left| \int_X f_\varepsilon g \, d\mu \right| \geqslant \left| \int_K f_\varepsilon g \, d\mu \right| - \left| \int_{X \smallsetminus K} f_\varepsilon g \, d\mu \right|
$$
$$
\geqslant \int_K |g| \, d\mu - \int_{X \smallsetminus K} |g| \, d\mu \geqslant \|g\|_{L^1_\mu} - 2\varepsilon.
$$

Since $\varepsilon > 0$ was arbitrary, this shows that

$$
\|g\|_{L^1_\mu} \leqslant \|g \, d\mu\|_{\mathrm{op}},
$$

and the reverse inequality follows from (2.33). $\qquad\square$

Exercise 2.64. Let X be a compact metric space, μ a Borel measure on X, and g a function in $L^1_\mu(X)$. Give and prove a precise criterion in terms of properties of g for the existence of a function $f \in C(X)$ with $\|f\|_\infty \leqslant 1$ such that $|\int fg \, d\mu| = \|g\|_1$.

2.4.2 Banach Algebras

In many situations it makes sense to multiply elements of a normed vector space with each other. Recall that an *algebra* is a vector space and simultaneously a ring in such a way that the two structures are compatible: addition in the vector space and addition in the ring are the same, the scalar multiplication and the ring multiplication satisfy $(\alpha x)y = x(\alpha y) = \alpha(xy)$ for all scalars α and elements $x, y \in A$.

Definition 2.65. Let $\mathcal{A}$ be a Banach space, and assume there is a multiplication operation $(x, y) \mapsto xy$ from $\mathcal{A} \times \mathcal{A}$ to $\mathcal{A}$ such that addition and multiplication make $\mathcal{A}$ into an algebra, with the sub-multiplicativity property that

$$
\|xy\| \leqslant \|x\| \|y\|
$$

for all $x, y \in A$. Then $\mathcal{A}$ is called a *Banach algebra*. Elements a, b of an algebra are said to commute if $ab = ba$, and the algebra is said to be *commutative* if any $a, b \in A$ commute.

Recall that a ring or an algebra does not need to have a unit; if a nontrivial ring $\mathcal{A}$ has a unit $1_{\mathcal{A}}$ satisfying $1_{\mathcal{A}}a = a1_{\mathcal{A}} = a$ for all $a \in \mathcal{A}$ then it is called *unital*.

The additional axiom on the norm makes the product operation continuous by the following argument. Fix $\varepsilon \in (0,1)$ and $x, y \in \mathcal{A}$. Then $\|x' - x\| < \varepsilon < 1$ and $\|y' - y\| < \varepsilon$ together imply that

$$\|x'y' - xy\| \leqslant \|x'(y' - y)\| + \|(x' - x)y\|$$
$$\leqslant (\|x'\| + \|y\|)\,\varepsilon \leqslant (\|x\| + 1 + \|y\|)\,\varepsilon. \qquad (2.34)$$

Since $\varepsilon \in (0,1)$ was arbitrary, this shows the continuity of the product map at $(x, y) \in \mathcal{A} \times \mathcal{A}$.

Example 2.66. (1) The continuous functions $C(X)$ on a compact topological space X with the supremum norm form a Banach algebra with respect to the pointwise multiplication operation $(fg)(x) = f(x)g(x)$ for all $x \in X$. Notice that the constant function 1 is a unit in this ring.

(2) Let X be a non-compact topological space. Then $C_0(X)$ is a Banach algebra with respect to the supremum norm and pointwise multiplication as in (1) above, but it does not have a unit.

(3) If V is any Banach space, then $B(V) = B(V, V)$ is a Banach algebra with respect to composition. The sub-multiplicativity property of the operator norm in Definition 2.65 is precisely the content of Lemma 2.57. The algebra has a unit, namely the identity map $I(v) = v$ for all $v \in V$.

(4) A special case of (3) above is the case $V = \mathbb{R}^n$. By choosing a basis for $\mathbb{R}^n$ we may identify $B(\mathbb{R}^n)$ with the space of $n \times n$ real matrices.

In a Banach algebra with unit, we can apply many well-known functions to its elements and obtain new elements of the Banach algebra. For example, if a is any element of a unital Banach algebra, then we may define

$$\exp a = \sum_{n=0}^{\infty} \frac{a^n}{n!},$$

where $a^0 = 1_{\mathcal{A}}$ is the unit in $\mathcal{A}$. The series defines an element of $\mathcal{A}$ by Lemma 2.28. We will return to the topic of Banach algebras in Chapter 11.

2.5 Ordinary Differential Equations

We want to briefly indicate how even the simplest differential equations can lead directly to the study of integral operators, which may be analyzed using tools introduced above (and in Chapter 6).

Consider first the differential equation

$$f''(x) + f(x) = g(x) \tag{2.35}$$

with the initial values

$$f(0) = 1, f'(0) = 0. \tag{2.36}$$

Let us recall briefly the familiar approach to solving such an equation. First one finds all solutions to the homogeneous equation

$$f''(x) + f(x) = 0,$$

giving

$$f(x) = A \sin x + B \cos x \tag{2.37}$$

for constants A and B. Then one moves on to the problem of finding one particular solution f_p to the equation

$$f_p''(x) + f_p(x) = g(x), \tag{2.38}$$

ignoring the initial values, which may be done by a sophisticated guess if g is sufficiently simple, or by using the method of variation of parameters (that is, treating A and B as functions of x rather than constants). Finally, taking the sum of f from (2.37) and a solution to (2.38), one chooses the constants A and B in the solution to the homogeneous equation to satisfy the initial values. Rather than going through this in detail, we claim that the function

$$f(x) = \cos(x) + \int_0^x \sin(x - t)g(t)\,dt \tag{2.39}$$

is a solution to the initial value problem. This is easily checked by a calculation: $f(0) = 1$ clearly, and

$$f'(x) = -\sin x + \sin(x - x)g(x) + \int_0^x \cos(x - t)g(t)\,dt,$$

so $f'(0) = 0$. Finally,

$$f''(x) = -\cos x + \cos(x - x)g(x) - \int_0^x \sin(x - t)g(t)\,dt$$
$$= -f(x) + g(x),$$

as required. To summarize, we have shown that (2.35)–(2.36) together are equivalent to (2.39).

2.5.1 The Volterra Equation

If the original differential equation in (2.35) is changed slightly, to take the form

$$f''(x) + f(x) = \sigma(x)f(x), \tag{2.40}$$

with the same initial values $f(0) = 1$ and $f'(0) = 0$, then the discussion above does not solve the equation. Nonetheless, the ideas are still useful, since it transforms the equation into the integral equation

$$f(x) = \cos(x) + \int_0^x \sin(x-t) \underbrace{\sigma(t)f(t)}_{g(t)} \, dt. \tag{2.41}$$

Now define $k(x,t) = \sin(x-t)\sigma(t)$ so that (2.41) takes the form

$$f = u + K(f), \tag{2.42}$$

where $u(x) = \cos x$ and

$$K(f)(x) = \int_0^x k(x,t)f(t) \, dt.$$

Due to its inventors and its nature K is called a *Hilbert–Schmidt integral operator*. The function k will be referred to as the *kernel* of the integral operator.

Solving the perturbed equation (2.40) with initial values turns out to be straightforward at the level of abstraction aimed at in functional analysis. We can rewrite the equation (2.42) as a Volterra equation

$$(I - K)f = u$$

where I is the identity map. The solution f is then given by applying the inverse operator $(I - K)^{-1}$ to u, which we may calculate (in this particular case) using an operator form of the geometric series (in this context the geometric series is usually called a *von Neumann series*),

$$(I - K)^{-1} = \sum_{n=0}^{\infty} K^n,$$

and hence

$$f = \sum_{n=0}^{\infty} K^n u.$$

The heuristic above is made formal in the following lemma.

Lemma 2.67. *Suppose that $k \in C([0,1]^2)$. Then*

$$K(f)(x) = \int_0^x k(x,t)f(t) \, dt$$

defines a bounded linear operator $K : C([0,1]) \to C([0,1])$ *with* $\|K\| \leqslant \|k\|_\infty$, *and more generally with* $\|K^n\| \leqslant \|k\|_\infty^n / n!$ *for* $n \geqslant 1$. *In particular, the geometric series*

$$(I - K)^{-1} = \sum_{n=0}^\infty K^n$$

converges in $\mathrm{B}\big(C([0,1])\big)$. *It follows that the integral equation* $(I - K)f = u$ *has a unique solution for any* $u \in C([0,1])$. *For* $u(x) = \cos x$, $\sigma \in C([0,1])$ *and* $k(x,t) = \sin(x-t)\sigma(t)$ *with* $x, t \in [0,1]$, *this solution belongs to* $C^2([0,1])$, *and solves the initial value problem*

$$\begin{cases} f'' + f = \sigma f \\ f(0) = 1, \, f'(0) = 0 \end{cases} \tag{2.43}$$

on $[0,1]$.

PROOF. As k is uniformly continuous, it is easy to check that $K(f) \in C([0,1])$ for every $f \in C([0,1])$. Indeed, if $\varepsilon > 0$ then there exists some $\delta > 0$ for which

$$|x_1 - x_2| < \delta \implies |k(x_1, t) - k(x_2, t)| < \varepsilon$$

for all $t \in [0,1]$. Multiplying by $f(t)$ and integrating from 0 to x shows that

$$|x_1 - x_2| < \delta \implies |K(f)(x_1) - K(f)(x_2)| < \|f\|_\infty \varepsilon.$$

Also, K is linear, and

$$\|Kf\|_\infty \leqslant \sup_{x \in [0,1]} \left| \int_0^x k(x,t) f(t) \, \mathrm{d}t \right| \leqslant \|k\|_\infty \|f\|_\infty,$$

so K defines a bounded linear operator with $\|K\| \leqslant \|k\|_\infty$.

To prove the estimate on $\|K^n\|$ we need to be a bit more careful. For every $x \in [0,1]$ we have

$$|K(f)(x)| \leqslant \int_0^x |k(x,t) f(t)| \, \mathrm{d}t$$
$$\leqslant x \|k\|_\infty \|f\|_\infty.$$

Suppose we have already shown for $x \in [0,1]$ that

$$|K^n(f)(x)| \leqslant \frac{x^n}{n!} \|k\|_\infty^n \|f\|_\infty. \tag{2.44}$$

Then

$$\left|K^{n+1}(f)(x)\right| \leqslant \int_0^x |k(x,t)|\,|K^n(f)(t)|\;\mathrm{d}x$$

$$\leqslant \int_0^x \frac{t^n}{n!}\|k\|_\infty^{n+1}\|f\|_\infty\,\mathrm{d}x = \frac{x^{n+1}}{(n+1)!}\|k\|_\infty^{n+1}\|f\|_\infty$$

for all $x \in [0,1]$. By induction on n, it follows that (2.44) holds for all $n \geqslant 1$. Hence $\|K^n\| \leqslant \|k\|_\infty^n/n!$ for all $n \geqslant 1$, as claimed.

By Lemma 2.54, $\mathrm{B}\bigl(C([0,1])\bigr)$ is a Banach space. It follows by Lemma 2.28 that the absolutely convergent series $\sum_{n=0}^\infty K^n$ also converges in $\mathrm{B}\bigl(C([0,1])\bigr)$. However,

$$(I-K)\left(\sum_{n=0}^\infty K^n\right) = \left(\sum_{n=0}^\infty K^n\right)(I-K) = \sum_{n=0}^\infty K^n - \sum_{n=1}^\infty K^n = I,$$

so the sum $\sum_{n=0}^\infty K^n$ is the inverse of $I-K$ and, for any $u \in C([0,1])$, the equation $(I-K)f = u$ has the unique solution $f = (I-K)^{-1}u$. In the case $u(x) = \cos x$, $k(x,t) = \sin(x-t)\sigma(t)$ for $x,t \in [0,1]$ and $\sigma \in C([0,1])$, the calculation after (2.39) shows that the solution f belongs to $C^2([0,1])$ and solves (2.40) with the initial values $f(0) = 1$ and $f'(0) = 0$. $\qquad\square$

2.5.2 The Sturm–Liouville Equation

We now make two more small changes to the initial value problem (2.35) and (2.36). Fix a parameter $\lambda > 0$ and consider instead the *Sturm–Liouville equation*

$$f'' + \lambda^2 f = g, \tag{2.45}$$

with the *boundary conditions*

$$f(0) = f(1) = 0.$$

These boundary conditions (made at the two end points of $[0,1]$) replace the initial value conditions in (2.36), and that change has a surprisingly deep impact on the resulting equation.

As we recall below the space of functions satisfying (2.45) is, if non-empty, a two-dimensional affine subspace of functions, so that the additional boundary conditions might lead to a unique solution.

We may proceed just as before. The functions of the form

$$f(x) = A\cos(\lambda x) + B\sin(\lambda x)$$

give all solutions to the homogeneous differential equation $f'' + \lambda^2 f = 0$. Next one needs to find a particular solution f_p to

$$f_p'' + \lambda^2 f_p = g$$

(ignoring the boundary conditions). After this, one would use the solutions to the homogeneous differential equation to satisfy the boundary conditions. Explicitly, given f_p we can calculate the vector

$$\begin{pmatrix} f_p(0) \\ f_p(1) \end{pmatrix} \tag{2.46}$$

and try to express it as a linear combination of the two vectors

$$\begin{pmatrix} \cos(\lambda 0) \\ \cos(\lambda 1) \end{pmatrix} = \begin{pmatrix} 1 \\ \cos \lambda \end{pmatrix}$$

and

$$\begin{pmatrix} \sin(\lambda 0) \\ \sin(\lambda 1) \end{pmatrix} = \begin{pmatrix} 0 \\ \sin \lambda \end{pmatrix}.$$

If

$$\det \begin{pmatrix} 1 & 0 \\ \cos \lambda & \sin \lambda \end{pmatrix} = \sin \lambda$$

is non-zero, then this is always possible and we find a unique solution to the boundary value problem. However, if $\lambda \in \pi\mathbb{Z}$ then $\sin \lambda = 0$ and we may be unlucky with the value of the vector (2.46): if the vectors

$$\begin{pmatrix} f_p(0) \\ f_p(1) \end{pmatrix}, \begin{pmatrix} 1 \\ \cos \lambda \end{pmatrix}$$

are linearly independent, then there will not be a solution to the boundary value problem.

This obstruction to being able to find a solution to the boundary value problem may be phrased in terms of another integral operator.

Lemma 2.68. *Define the continuous* Green function *on* $[0,1]^2$ *by*

$$G(s,t) = \begin{cases} s(t-1) & for\ 0 \leqslant s \leqslant t \leqslant 1; \\ t(s-1) & for\ 0 \leqslant t \leqslant s \leqslant 1. \end{cases}$$

Then for $f, h \in C([0,1])$ *the conditions*

$$\begin{rcases} f(0) = f(1) = 0 \\ f \in C^2([0,1])\ and\ f'' = h \end{rcases} \tag{2.47}$$

are equivalent to the operator equation $f = Kh$, *where* K *is the operator defined by*

$$K(h)(s) = \int_0^1 G(s,t)h(t)\,dt. \tag{2.48}$$

PROOF. Assume first that $f = Kh$. Then

$$f(0) = \int_0^1 \underbrace{G(0,t)}_{=0} h(t)\, dt = 0,$$

and $f(1) = 0$ for the same reason. Moreover,

$$f(s) = \int_0^s t(s-1)h(t)\, dt + \int_s^1 s(t-1)h(t)\, dt,$$

$$f'(s) = \underline{s(s-1)h(s)} + \int_0^s th(t)\, dt$$

$$- \underline{s(s-1)h(s)} + \int_s^1 (t-1)h(t)\, dt,$$

and

$$f''(s) = sh(s) - (s-1)h(s) = h(s),$$

so f is a solution of the boundary value problem (2.47).

To see the converse, notice that the boundary value problem has a solution (by the argument above). However, our previous discussion of the boundary value problem associated to the Sturm–Liouville equation (2.45) (which needs to be modified for the case $\lambda = 0$) shows that in this case the solution is unique. Thus the equivalence of (2.47) and $f = Kh$ is established.   □

Exercise 2.69. Modify the argument for the Sturm–Liouville equation for the case $\lambda = 0$, and show that the solution is always unique.

In particular, the fact that $s_n(x) = \sin(\pi n x)$ for any $n \in \mathbb{N}$ satisfies

$$\left.\begin{array}{r} s_n(0) = s_n(1) = 0 \\ s_n'' = -(\pi n)^2 s_n \end{array}\right\}$$

implies by Lemma 2.68 that

$$s_n = -(\pi n)^2 K(s_n).$$

In other words, the values

$$\mu_n = -(\pi n)^{-2}$$

for $n = 1, 2, \ldots$ are *eigenvalues* of the integral operator K (actually these are all the eigenvalues of K; see Exercise 2.70).

Thus we can rephrase our earlier observation regarding the equivalent formulations

$$\left.\begin{array}{r} f'' + \lambda^2 f = g \\ f(0) = f(1) = 0 \end{array}\right\} \Longleftrightarrow f = K(-\lambda^2 f + g) \Longleftrightarrow \left(I + \lambda^2 K\right) f = K(g)$$

by saying that this differential equation always has a unique solution for any g unless $\lambda = \pi n$ corresponds to one of the eigenvalues $\mu_n = -(\pi n)^{-2} = -\lambda^{-2}$ of K.

In Chapter 6 we will start the discussion of eigenvalues of operators, and as discussed in Section 1.3 it is easier for this to restrict to the case of operators on Hilbert spaces. As we will show in Chapter 6 the operator K also makes sense in $L^2([0,1])$ and is completely diagonalizable on that space.

Exercise 2.70 (A special case of elliptic regularity). Suppose that the function f in $L^2([0,1])$ satisfies $Kf = \lambda f$ for some λ in $\mathbb{R}\smallsetminus\{0\}$, where K is the operator (2.48) discussed in connection with the Sturm–Liouville problem. Show that f must be smooth on $(0,1)$, and deduce that f and λ must satisfy the conditions found above.

Exercise 2.71. In this exercise we generalize the connection between the Sturm–Liouville boundary value problem and integral operators. Let $a < b$ be real numbers, and assume that $p \in C^1([a,b])$ and $q \in C([a,b])$ are real-valued functions with $p > 0$ and $q > 0$. We define the second-order differential operator

$$L(f) = (pf')' + qf.$$

Also let $\alpha_1, \alpha_2, \beta_1, \beta_2 \in \mathbb{R}$ and define the boundary conditions

$$B_1(f) = \alpha_1 f(a) + \alpha_2 f'(a) = 0,$$
$$B_2(f) = \beta_1 f(b) + \beta_2 f'(b) = 0.$$

Assume that f_1 and f_2 are *fundamental solutions*[†] of the differential equation

$$L(f) = 0$$

such that we also have

$$B_1(f_1) = B_2(f_2) = 0,$$

but $B_1(f_2) \neq 0$ and $B_2(f_1) \neq 0$. Show that

$$p(f_1 f_2' - f_1' f_2) = c$$

is a constant. Using this, define an associated Green function

$$G(s,t) = \begin{cases} \frac{1}{c} f_1(s) f_2(t) & \text{for } a \leqslant s \leqslant t \leqslant b, \\ \frac{1}{c} f_1(t) f_2(s) & \text{for } a \leqslant t \leqslant s \leqslant b, \end{cases}$$

and show that for $h \in C([a,b])$ the boundary-value problem

$$\left. \begin{array}{c} B_1(f) = B_2(f) = 0 \\ L(f) = h \end{array} \right\}$$

is equivalent to the equation

$$f(s) = K(h)(s) = \int_a^b G(s,t) h(t) \, dt.$$

Calculate G explicitly for the equation $L(f) = f''$, $B_1(f) = f(a)$ and $B_2(f) = f'(b)$.

[†] That is, the functions f_1, f_2 form a basis of the vector space of all solutions.

2.6 Further Topics

The material in this chapter represents the basic language and some of the main examples of functional analysis. Let us mention briefly some directions in which the theory continues.

- In Chapter 3 and the following chapters we will start to see why we insisted on completeness in the definition of Banach spaces.
- We have seen the definition of dual spaces, but have not yet found a description of any dual space. This will be corrected in the next chapter and more generally in Chapter 7, where we will describe the dual spaces of many of the Banach spaces that we discussed here.
- How can one construct a *generalized limit notion* that assigns to every bounded sequence a limit, and still has many of the expected properties? One such property is linearity (but notice, for example, that lim sup is not a linear function on the space of bounded sequences). Another such property is translation-invariance with respect to the underlying group (for a sequence in the normal sense, this group would be $\mathbb{Z}$). After we construct this so-called Banach limit, we ask which groups have similar notions of generalized limits. We will discuss these topics in Sections 7.2 and 10.2.
- Clearly there is some hidden notion of convergence of *measures* to the Lebesgue measure in Section 2.3.3. In order to formulate this precisely, we will need to define an appropriate topology on a space of measures. This topology will be called the weak* topology (read as 'weak star' topology; see Chapter 8), and as we will show the space of probability measures on a compact metric space is itself a compact metric space in this topology. This result helps to provide a coherent setting for many equidistribution results.
- Some natural spaces (examples include $C_c(X)$ and $C^\infty([0,1])$) do not fit into the framework of Banach spaces, but do fit into the more general context of locally convex spaces. These will be introduced in Chapter 8.
- Convexity will also turn out to be fundamental for many discussions in functional analysis. One of the goals in Chapter 8 will be to analyze how the extreme points of a convex compact set determine the set.
- Banach algebras will be discussed in greater detail in Chapter 11, which lays the foundations for the more advanced spectral theory in Chapters 12 and 13.

The reader is advised to continue with the next chapter (or at least the first three or four sections of it), after which she may select parts of the text.

Chapter 3
Hilbert Spaces, Fourier Series, and Unitary Representations

In this chapter we define Hilbert spaces as a special case of Banach spaces, pick up some of the informal claims from Section 1.1, and prove them. In particular, we will introduce Fourier series in two different settings: the first abstract, and the second being the — at least soon to be — familiar setting of the torus. In Section 3.5 we discuss the spectral theory of compact abelian groups as well as two notions of integrals for Banach space-valued functions.

3.1 Hilbert Spaces

The notion of a Hilbert space is a fundamental idea in functional analysis. We will see in this section that a Hilbert space is a Banach space of a special sort, and the additional structure entailed by the extra hypothesis turns out to be highly significant.

3.1.1 Definitions and Elementary Properties

Definition 3.1. An *inner product space* or a *pre-Hilbert space* is a vector space over $\mathbb{R}$ (or $\mathbb{C}$) with an *inner product* $\langle \cdot, \cdot \rangle : V \times V \to \mathbb{R}$ (or $\mathbb{C}$) with the following properties:

- (Strict positivity) $\langle v, v \rangle > 0$ for all $v \in V \smallsetminus \{0\}$;
- ((Conjugate-)Symmetry) $\langle v, w \rangle = \overline{\langle w, v \rangle}$ for all $v, w \in V$; and
- (Linearity) for any fixed $w \in V$ the map $v \longmapsto \langle v, w \rangle$ is linear.

If the first property of strict positivity is replaced by the weaker axiom of

- (Positivity) $\langle v, v \rangle \geqslant 0$ for all $v \in V$,

then we call $\langle \cdot, \cdot \rangle$ a *semi-inner product*.

© Springer International Publishing AG 2017
M. Einsiedler, T. Ward, *Functional Analysis, Spectral Theory, and Applications*,
Graduate Texts in Mathematics 276, DOI 10.1007/978-3-319-58540-6_3

Notice that over $\mathbb{R}$ a consequence is linearity of the map $w \mapsto \langle v, w \rangle$ in the second variable for fixed v, so that $\langle \cdot, \cdot \rangle$ is bilinear. In the complex case, we have *semi-linearity* in the second argument, that is

$$\langle v, \alpha_1 w_1 + \alpha_2 w_2 \rangle = \overline{\alpha_1} \langle v, w_1 \rangle + \overline{\alpha_1} \langle v, w_2 \rangle$$

for any $v, w_1, w_2 \in \mathcal{H}$ and $\alpha_1, \alpha_2 \in \mathbb{C}$. A map L from a complex vector space V to $\mathbb{C}$ is *semi-linear* ($\frac{1}{2}$-linear) if $L(\alpha_1 v_1 + \alpha_2 v_2) = \overline{\alpha_1} L(v_1) + \overline{\alpha_2} L(v_2)$ for all vectors $v_1, v_2 \in V$ and scalars $\alpha_1, \alpha_2 \in \mathbb{C}$, and a map $B : V \times V \to \mathbb{C}$ is sesqui-linear ($1\frac{1}{2}$-linear) if the map $v \in V \mapsto B(v, w)$ is linear for any $w \in V$ and the map $w \in V \mapsto B(v, w)$ is semi-linear for any $v \in V$. Thus the inner product $\langle \cdot, \cdot \rangle$ is *sesqui-linear*.

In an inner product space, we will see shortly that defining

$$\|v\| = \sqrt{\langle v, v \rangle} \tag{3.1}$$

gives a norm on V.

Proposition 3.2 (Cauchy–Schwarz). *Let $(V, \langle \cdot, \cdot \rangle)$ be an inner product space. Then we have the Cauchy–Schwarz inequality,*

$$|\langle v, w \rangle| \leqslant \|v\| \|w\| \tag{3.2}$$

for all $v, w \in V$, where equality holds if and only if v and w are linearly dependent. Moreover, the function $\| \cdot \|$ defined in (3.1) is a norm on V, so that every inner product space is also a normed vector space.

If $\langle \cdot, \cdot \rangle$ is only assumed to be a semi-inner product on V, then the induced function $\|v\| = \sqrt{\langle v, v \rangle}$ for $v \in V$ is a semi-norm and the inequality in (3.2) also holds in that case.

Definition 3.3. A *Hilbert space* is an inner product space $(\mathcal{H}, \langle \cdot, \cdot \rangle)$ which is complete with respect to the norm $\| \cdot \|$ induced by the inner product as in (3.1).

PROOF OF PROPOSITION 3.2. To see that $\| \cdot \| : V \to \mathbb{R}_{\geqslant 0}$ from (3.1) defines a norm, we need to check the following properties:

- strict positivity of $\| \cdot \|$, which follows at once from the strict positivity property of the inner product, and
- homogeneity of $\|\cdot\|$, which follows from linearity and (conjugate-)symmetry, since

$$\|\alpha v\|^2 = \langle \alpha v, \alpha v \rangle = |\alpha|^2 \|v\|^2.$$

For the proof of the triangle inequality we will need the Cauchy–Schwarz inequality (3.2), which we will prove now. We note that the latter holds trivially if $w = 0$. So assume that $w \neq 0$. By definition, we have

$$0 \leqslant \|v + tw\|^2 = \langle v + tw, v + tw \rangle$$
$$= \langle v, v \rangle + \langle tw, v \rangle + \langle v, tw \rangle + \langle tw, tw \rangle$$
$$= \|v\|^2 + t\langle w, v \rangle + \overline{t\langle w, v \rangle} + |t|^2\|w\|^2$$
$$= \|v\|^2 + 2\Re(t\overline{\langle v, w \rangle}) + |t|^2\|w\|^2 \qquad (3.3)$$

for any scalar t by linearity and (conjugate-)symmetry of the inner product. We set

$$t = -\frac{\langle v, w \rangle}{\|w\|^2}.$$

Then the inequality (3.3) becomes

$$0 \leqslant \|v\|^2 - 2\frac{|\langle v, w \rangle|^2}{\|w\|^2} + \frac{|\langle v, w \rangle|^2}{\|w\|^4}\|w\|^2,$$

or

$$0 \leqslant \|v\|^2\|w\|^2 - |\langle v, w \rangle|^2,$$

giving (3.2).

Reading the manipulations above in the reverse direction we see that equality in (3.2) gives

$$\|v + tw\|^2 = \langle v + tw, v + tw \rangle = 0,$$

which forces $v + tw = 0$ by the positivity property. If, on the other hand, v and w are linearly dependent with $v = \alpha w$ for some scalar α, then

$$|\langle v, w \rangle| = |\langle \alpha w, w \rangle| = |\alpha|\|w\|^2 = \|v\|\|w\|$$

by the homogeneity property of $\|\cdot\|$.

It remains to show the

- triangle inequality, which may be seen as follows:

$$\|v + w\|^2 = \langle v + w, v + w \rangle = \|v\|^2 + 2\Re\langle v, w \rangle + \|w\|^2$$
$$\leqslant \|v\|^2 + 2\|v\|\|w\| + \|w\|^2 = (\|v\| + \|w\|)^2$$

by the Cauchy–Schwarz inequality.

Analyzing the proof above shows that the strict positivity of $\langle \cdot, \cdot \rangle$ was only needed to obtain strict positivity of the induced norm and in the proof of the equality case of the Cauchy–Schwarz inequality. $\qquad \square$

As we have seen, the triangle inequality for the norm needs the Cauchy–Schwarz inequality. Another reason for the importance of this fundamental inequality is that it gives us continuity of the inner product, which we will use frequently.

Essential Exercise 3.4. Show that an inner product on an inner product space is jointly continuous with respect to the induced norm: if $v_n \to v$ and $w_n \to w$ as $n \to \infty$, then $\langle v_n, w_n \rangle \to \langle v, w \rangle$ as $n \to \infty$.

We record a few elementary properties of inner product spaces.

- The *parallelogram identity*,

$$\|v + w\|^2 + \|v - w\|^2 = 2\|v\|^2 + 2\|w\|^2 \tag{3.4}$$

  for all $v, w \in V$.
- The relationship with *linear functionals:* for fixed $w \in V$ the map ϕ_w defined by $\phi_w(v) = \langle v, w \rangle$ is a linear functional with norm $\|\phi_w\| = \|w\|$.
- The relationship with *geometry*: the vector $\frac{\langle v, w \rangle}{\|w\|^2} w$ (appearing as tw in the proof of Proposition 3.2 above) is the orthogonal projection of v onto the subspace spanned by w. Moreover, if $\langle v, w \rangle = 0$ then we recover Pythagoras' theorem in the form $\|v + w\|^2 = \|v\|^2 + \|w\|^2$.

These are easy to check. For the first, expand the left-hand side to obtain

$$\|v + w\|^2 + \|v - w\|^2 = \langle v + w, v + w \rangle + \langle v - w, v - w \rangle$$
$$= \|v\|^2 + 2\Re \langle v, w \rangle + \|w\|^2 + \|v\|^2 - 2\Re \langle v, w \rangle + \|w\|^2.$$

The second claim is a consequence of the linearity of the inner product, the Cauchy–Schwarz inequality and the definition of the operator norm. The two final claims follow by expanding $\langle v - \frac{\langle v, w \rangle}{\|w\|^2} w, w \rangle$, respectively the square norm $\|v + w\|^2 = \langle v + w, v + w \rangle$.

Exercise 3.5. (a) Show that any real inner product space satisfies the *polarization identity*

$$\langle x, y \rangle = \tfrac{1}{4} \left(\|x + y\|^2 - \|x - y\|^2 \right)$$

which expresses the inner product in terms of the norm.
(b) Show that the parallelogram identity (3.4) *characterizes* the real inner product spaces among the real normed spaces in the following sense. If a real normed vector space satisfies the parallelogram identity, then an inner product can be defined in such a way that the norm arises from the inner product.
(c) Generalize the polarization identity to complex inner product spaces, and show the complex analogue of (b).

Example 3.6. We have already seen several Hilbert spaces without making explicit the underlying inner product.

(1) $\mathbb{R}^d$ (or $\mathbb{C}^d$) with $\langle v, w \rangle = \sum_{i=1}^d v_i \overline{w_i}$, (also written $v \cdot w$) giving the 2-norm

$$\|v\|_2 = \left(\sum_{i=1}^d |v_i|^2 \right)^{1/2}.$$

(2) $\ell^2 = \ell^2(\mathbb{N})$, the space of square-summable sequences of scalars, with inner product

$$\langle v, w \rangle = \sum_{i=1}^{\infty} v_i \overline{w_i}$$

and the 2-norm

$$\|v\|_2 = \left(\sum_{i=1}^{\infty} |v_i|^2 \right)^{1/2}.$$

Equivalently $\ell^2(\mathbb{N}) = L^2_{\lambda_{\text{count}}}(\mathbb{N})$, where λ_{count} is the counting measure on $\mathbb{N}$.

(3) $L^2_\mu(X)$ for a measure space $(X, \mathcal{B}, \mu)$ with the inner product

$$\langle f, g \rangle = \int f \overline{g} \, d\mu, \tag{3.5}$$

giving the 2-norm

$$\|f\|_2 = \left(\int |f|^2 \, d\mu \right)^{1/2}.$$

Notice that in Example 3.6(2) and (3), the spaces are themselves defined as the set of sequences or functions with finite 2-norm. We recall how this implies that the inner product is well-defined and note that (3) contains (2) as a special case.

Lemma 3.7. *If $(X, \mathcal{B}, \mu)$ is a measure space and $f, g \in L^2_\mu(X)$, then the right-hand side of* (3.5) *is well-defined.*

PROOF. Since $(|f| - |g|) = |f|^2 - 2|fg| + |g|^2 \geqslant 0$ we have

$$2 \int |fg| \, d\mu \leqslant \int |f|^2 \, d\mu + \int |g|^2 \, d\mu < \infty,$$

which proves that $f\overline{g}$ is integrable. $\qquad\square$

Definition 3.8. Let V, W be two Hilbert spaces and $M : V \to W$ a linear map. If M is both an isometry and a bijection, then M is called a *unitary operator*.

Essential Exercise 3.9. Show that a bijective linear operator $M : V \to W$ is unitary if and only if $\langle Mv_1, Mv_2 \rangle_W = \langle v_1, v_2 \rangle_V$ for all $v_1, v_2 \in V$.

3.1.2 Convex Sets in Uniformly Convex Spaces

From the equality case of the Cauchy–Schwarz inequality, which is itself used in the proof of the triangle inequality, it follows quickly that a norm in an inner product space is strictly sub-additive (see Definition 2.17). Thus the Mazur–Ulam theorem concerning isometries (Theorem 2.20) applies in particular to Hilbert spaces.

While the emphasis in this section is on Hilbert spaces, we will isolate a more abstract convexity property which is precisely what is needed for several proofs in this section.

Exercise 3.10. (a) Show that the norm in a Hilbert space is strictly sub-additive (see Definition 2.17).
(b) Show that the norm in a uniformly convex vector space (as defined below) is strictly sub-additive.

Definition 3.11. A normed vector space $(V, \| \cdot \|)$ is called *uniformly convex* if

$$\|x\|, \|y\| \leqslant 1 \implies \left\| \frac{x+y}{2} \right\| \leqslant 1 - \eta(\|x-y\|),$$

for all $x, y \in V$ where $\eta : [0, 2] \to [0, 1]$ is a monotonically increasing function with $\eta(r) > 0$ for all $r > 0$.

Fig. 3.1: If x and y are not close to each other, then the mid-point is uniformly closer to zero (independent of the choice of x and y).

Heuristically, we can think of Definition 3.11 as having the following geometrical meaning, illustrated in Figure 3.1. If vectors x and y have norm (length) one, then their mid-point $\frac{x+y}{2}$ has significantly smaller norm unless x and y are very close together. This accords closely with the geometrical intuition from finite-dimensional spaces with Euclidean distance.

Lemma 3.12. *A Hilbert space* $(\mathcal{H}, \langle \cdot, \cdot \rangle)$ *is uniformly convex.*

PROOF. For $x, y \in \mathcal{H}$ with $\|x\|, \|y\| \leqslant 1$ we have

$$\left\| \frac{x+y}{2} \right\| = \sqrt{\tfrac{1}{2}\|x\|^2 + \tfrac{1}{2}\|y\|^2 - \tfrac{1}{4}\|x-y\|^2}$$

$$\leqslant \sqrt{1 - \tfrac{1}{4}\|x-y\|^2} = 1 - \eta(\|x-y\|)$$

by the parallelogram identity, with $\eta(r) = 1 - \sqrt{1 - \tfrac{1}{4}r^2}$. $\qquad\square$

The following theorem, whose conclusion is illustrated in Figure 3.2, will have many important consequences for the study of Hilbert spaces.

Theorem 3.13 (Unique approximation within a closed convex set).
Let $(V, \|\cdot\|)$ be a Banach space with a uniformly convex norm, let $K \subseteq V$ be a non-empty closed convex subset, and assume that $v_0 \in V$. Then there exists a unique element $w \in K$ that is closest to v_0 in the sense that w is the only element of K with $\|w - v_0\| = \inf_{k \in K} \|k - v_0\|$.

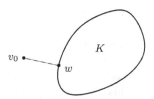

Fig. 3.2: The unique closest element of K to v_0.

The following exercise shows that the unique existence of the best approximation is by no means guaranteed.

Exercise 3.14. (1) Let $K \subseteq V$ be a non-empty compact subset of a normed vector space $(V, \|\cdot\|)$ or let V be finite-dimensional and K closed. Show the existence of a best approximation of any $v_0 \in V$ within K.
(2) Let $V = \mathbb{R}^2$ equipped with the norm $\|\cdot\|_\infty$ and let K be the closed unit ball. Find a point $v_0 \in V$ that has more than one best approximation within K. Describe the points that have exactly one best approximation within K.
(3) Let $V = \ell_{\mathbb{R}}^1(\mathbb{N})$ equipped with the norm $\|\cdot\|_1$. Let

$$K = \left\{ (x_n) \in \ell_{\mathbb{R}}^1(\mathbb{N}) \mid x_n \geqslant 0 \text{ and } \sum_{n=1}^{\infty} a_n x_n = 1 \right\},$$

where (a_n) is a fixed sequence in $(0, 1)$ with $\lim_{n \to \infty} a_n = 1$. Show that K is closed and convex. Let $v_0 = 0$ and show that there is no best approximation of v_0 within K. Conclude in particular from this that the closed unit ball in $\ell_{\mathbb{R}}^1(\mathbb{N})$ is not compact and that $\ell_{\mathbb{R}}^1(\mathbb{N})$ is not uniformly convex.

Exercise 3.15. [8]Let $(X, \mathcal{B}, \mu)$ be a measure space with $L_\mu^p(X)$, for $p \in [1, \infty]$, the associated function spaces.
(a) Show that $a^p + b^p \leqslant (a^2 + b^2)^{p/2}$ for any $p \in [2, \infty)$ and $a, b > 0$.
(b) Show that $\left\| \frac{f+g}{2} \right\|_p^p + \left\| \frac{f-g}{2} \right\|_p^p \leqslant \frac{1}{2} \|f\|_p^p + \frac{1}{2} \|g\|_p^p$ for any $p \in [2, \infty)$ and f, g in $L_\mu^p(X)$.
(c) Deduce from (b) that $L_\mu^p(X)$ is uniformly convex for $p \in [2, \infty)$.
(d) Show that $L_\mu^1(X)$ and $L_\mu^\infty(X)$ are in general not uniformly convex.

PROOF OF THEOREM 3.13. By translating both the set K and the point v_0 by $-v_0$, we may assume without loss of generality that $v_0 = 0$. We define

$$s = \inf_{k \in K} \|k - v_0\| = \inf_{k \in K} \|k\|.$$

If $s = 0$, then we must have $0 \in K$ since K is closed, and the only choice is then $w = v_0 = 0$ (the uniqueness of w is a consequence of the strict positivity of the norm). So assume that $s > 0$. By multiplying by the scalar $\frac{1}{s}$ we may also assume without loss of generality that $s = 1$. Notice that once we have found a point $w \in K$ with norm 1, then its uniqueness is an immediate consequence of the uniform convexity: if $w_1, w_2 \in K$ have $\|w_1\| = \|w_2\| = 1$, then $\frac{w_1 + w_2}{2} \in K$ because K is convex. Also, $\|\frac{w_1 + w_2}{2}\| = 1$ by the triangle inequality and since $s = 1$. By uniform convexity this implies that $w_1 = w_2$.

EXISTENCE: Turning to the existence, let us first give the idea of the proof. Choose a sequence (k_n) in K with $\|k_n\| \to 1$ as $n \to \infty$. Then the midpoints $\frac{k_n + k_m}{2}$ also lie in K, since K is convex, and thus the mid-point must have norm greater than or equal to 1, since $s = 1$. Therefore k_n and k_m must be close together by uniform convexity, so (k_n) is a Cauchy sequence. Since V is complete and K is closed, this will give a point $w \in K$ with $\|w\| = 1 = s$ as required.

To make this more precise, it is easier to apply uniform convexity to the normalized vectors

$$x_n = \frac{1}{s_n} k_n,$$

where $s_n = \|k_n\|$. The mid-point of x_n and x_m can now be expressed as

$$\frac{x_m + x_n}{2} = \frac{1}{2s_m} k_m + \frac{1}{2s_n} k_n = \left(\frac{1}{2s_m} + \frac{1}{2s_n}\right)(ak_m + bk_n)$$

with

$$a = \frac{\frac{1}{2s_m}}{\frac{1}{2s_m} + \frac{1}{2s_n}} \geqslant 0,$$

$$b = \frac{\frac{1}{2s_n}}{\frac{1}{2s_m} + \frac{1}{2s_n}} \geqslant 0,$$

and $a + b = 1$. Therefore $ak_m + bk_n \in K$ by convexity, and so

$$\left\|\frac{x_m + x_n}{2}\right\| = \left(\frac{1}{2s_m} + \frac{1}{2s_n}\right)\|ak_m + bk_n\| \geqslant \frac{1}{2s_m} + \frac{1}{2s_n}.$$

Let η be as in Definition 3.11 and fix $\varepsilon > 0$. Choose $N = N(\varepsilon)$ large enough to ensure that $m \geqslant N$ implies that

$$\frac{1}{s_m} > 1 - \eta(\varepsilon).$$

Then $m, n \geqslant N$ implies that

$$\frac{1}{2s_m} + \frac{1}{2s_n} > 1 - \eta(\varepsilon),$$

which together with the definition of uniform convexity gives

$$1 - \eta\big(\|x_m - x_n\|\big) \geqslant \big\|\tfrac{x_m+x_n}{2}\big\| > 1 - \eta(\varepsilon).$$

By monotonicity of the function η this implies that

$$\|x_m - x_n\| < \varepsilon$$

for all $m, n \geqslant N$, showing that (x_n) is a Cauchy sequence. As V is assumed to be complete, we deduce that (x_n) converges to some $x \in V$ with $\|x\| = 1$. Since $s_n \to 1$ and $k_n = s_n x_n$ as $n \to \infty$ it follows that $\lim_{n\to\infty} k_n = x$. As K is closed the limit x belongs to K and by construction is an (hence the unique) element in K closest to $v_0 = 0$. $\qquad\square$

Definition 3.16. Let $\mathcal{H}$ be a Hilbert space, and $A \subseteq \mathcal{H}$ any subset. Then the *orthogonal complement* of A is defined to be

$$A^\perp = \{h \in \mathcal{H} \mid \langle h, a \rangle = 0 \text{ for all } a \in A\}.$$

Corollary 3.17 (Orthogonal decomposition). *Let $\mathcal{H}$ be a Hilbert space, and let $Y \subseteq \mathcal{H}$ be a closed subspace. Then $Y^\perp$ is a closed subspace with*

$$\mathcal{H} = Y \oplus Y^\perp,$$

meaning that every $h \in \mathcal{H}$ can be written in the form $h = y + z$ with $y \in Y$ and $z \in Y^\perp$, and y and z are unique with these properties. Moreover, we have $Y = \big(Y^\perp\big)^\perp$ and

$$\|h\|^2 = \|y\|^2 + \|z\|^2 \tag{3.6}$$

if $h = y + z$ with $y \in Y$ and $z \in Y^\perp$.

In a two-dimensional real vector space, (3.6) is familiar as Pythagoras' theorem.

PROOF OF COROLLARY 3.17. As $h \mapsto \langle h, y \rangle$ is a (continuous linear) functional for each $y \in Y$, the set $Y^\perp$ is an intersection of closed subspaces and hence is a closed subspace. Using strict positivity of the inner product, it is easy to see that $Y \cap Y^\perp = \{0\}$; from this the uniqueness of the decomposition $h = y + z$ with $y \in Y$ and $z \in Y^\perp$ follows at once.

So it remains to show the existence of this decomposition. Fix $h \in \mathcal{H}$, and apply Theorem 3.13 (which we may as a Hilbert space is uniformly convex by Lemma 3.12) with the closed convex set $K = Y$ to find a point $y \in Y$ that is closest to h. Let $z = h - y$, so that for any $v \in Y$ and any scalar t we have

$$\|z\|^2 \leqslant \|h - \underbrace{(tv + y)}_{\in Y}\|^2 = \|z - tv\|^2 = \|z\|^2 - 2\Re\big(t\langle v, z\rangle\big) + |t|^2\|v\|^2.$$

In particular, for $t \in \mathbb{R}$ the function $t \mapsto \|h - (tv + y)\|^2$ is a quadratic polynomial with its minimum at $t = 0$. Taking the derivative at $t = 0$ gives $\Re(\langle v, z \rangle) = 0$ for all $v \in Y$. Similarly, restricting to $t = is$ with $s \in \mathbb{R}$ and taking the derivative once more it follows that $\Im\langle v, z \rangle = 0$ for all $v \in Y$. Thus $z \in Y^\perp$, and hence

$$\|h\|^2 = \langle h, h \rangle = \langle y + z, y + z \rangle = \|y\|^2 + \|z\|^2,$$

showing (3.6).

It is clear from the definitions that $Y \subseteq (Y^\perp)^\perp$. If $v \in (Y^\perp)^\perp$ then we may write $v = y + z$ for some $y \in Y$ and $z \in Y^\perp$ by the first part of the proof. However, $0 = \langle v, z \rangle = \|z\|^2$ implies that $v = y$ and so $Y = (Y^\perp)^\perp$. $\square$

An immediate consequence of Corollary 3.17 is the following.

Corollary 3.18 (Orthogonal projection). *For a closed subspace Y of a Hilbert space $\mathcal{H}$, the* orthogonal projection *onto Y, defined by*

$$P_Y : \mathcal{H} \longrightarrow Y$$
$$h \longmapsto y$$

where y is the unique element of Y with $h - y \in Y^\perp$, is a bounded linear operator with $\|P_Y\| \leqslant 1$ (and with $\|P_Y\| = 1$ unless $Y = \{0\}$) satisfying (and characterized by) $\langle h, y \rangle = \langle P_Y h, y \rangle$ for all $h \in \mathcal{H}$ and $y \in Y$. Moreover, if $\mathcal{H} = Y \oplus Y^\perp$, then the orthogonal decomposition from Corollary 3.17 is given by

$$h = P_Y h + P_{Y^\perp} h.$$

Recall that we write $V^* = B(V, \mathbb{R})$ or $B(V, \mathbb{C})$ for the dual space of a normed vector space V, equipped with the operator norm. The following gives our first classification of a dual space, and is crucial for the further development of the theory as well as of its applications.

Corollary 3.19 (Fréchet–Riesz representation). *For a Hilbert space $\mathcal{H}$ the map sending $h \in \mathcal{H}$ to $\phi(h) \in \mathcal{H}^*$ defined by $\phi(h)(x) = \langle x, h \rangle$ is a linear (resp. semi-linear in the complex case) isometric isomorphism between $\mathcal{H}$ and its dual space $\mathcal{H}^*$.*

PROOF. By the axioms of the inner product, we know that ϕ is (semi-)linear. By the Cauchy–Schwarz inequality and since $\phi(h)(h) = \|h\|^2$ we also know that ϕ is isometric. It remains to show that ϕ is onto. Let

$$\ell : \mathcal{H} \to \mathbb{R} \text{ (or } \mathbb{C})$$

be a linear functional. Then $Y = \ker(\ell)$ is a closed linear subspace of $\mathcal{H}$ (since ℓ is continuous). If $Y = \mathcal{H}$ then $\ell = 0$ and so $\phi(0) = \ell$. So suppose that $Y \neq \mathcal{H}$, in which case we can choose[†] a non-zero element $z \in Y^{\perp}$.

We claim that

$$\ell = \phi\left(\overline{\frac{\ell(z)}{\|z\|^2}}z\right).$$

Indeed, if $x \in \mathcal{H}$ then $\ell(z)x - \ell(x)z \in \ker(\ell) = Y$ and so $\langle \ell(z)x - \ell(x)z, z\rangle = 0$ by choice of z. In other words, we have shown that $\ell(z)\langle x, z\rangle = \ell(x)\|z\|^2$, which is equivalent to $\ell(x) = \left\langle x, \frac{\overline{\ell(z)}}{\|z\|^2}z\right\rangle$ for $x \in \mathcal{H}$, as claimed. $\square$

The following exercise shows that completeness is essential in Corollaries 3.17 and 3.19.

Exercise 3.20. Consider the space ℓ_c^2 of sequences with finite support with the ℓ^2 inner product, and the subspace $V = \{x \in \ell_c^2 \mid \sum_{n\geqslant 1} \frac{x_n}{n} = 0\}$. Show that V is closed, and that its orthogonal complement in ℓ_c^2 is empty. Deduce that the bounded linear functional sending $(x_n) \in \ell_c^2$ to $\sum_{n\geqslant 1} \frac{x_n}{n} \in \mathbb{C}$ cannot be represented as $x \mapsto \langle x, y\rangle$ for any $y \in \ell_c^2$.

Exercise 3.21. The following is known as the *Lax–Milgram lemma*.
(a) Suppose that $\mathcal{H}$ is a Hilbert space, and suppose that $B : \mathcal{H} \times \mathcal{H} \longrightarrow \mathbb{R}$ (or $\mathbb{C}$) is bilinear (or sesqui-linear in the complex case). Finally, assume that B is bounded in the sense that there is some $M > 0$ with

$$|B(x, y)| \leqslant M\|x\|\|y\|$$

for all $x, y \in \mathcal{H}$. Show that there exists a unique linear operator $T : \mathcal{H} \to \mathcal{H}$ with

$$B(x, y) = \langle Tx, y\rangle$$

for which $\|T\|_{\mathrm{op}} \leqslant M$.
(b) Assume in addition that B is *coercive*, meaning that there exists some $c > 0$ such that $|B(x, x)| \geqslant c\|x\|^2$ for all $x \in \mathcal{H}$. Show in this case that the operator T from (a) has an inverse, and that $\|T^{-1}\|_{\mathrm{op}} \leqslant \frac{1}{c}$.

Exercise 3.22. Use Corollary 3.19 to show that if $\mathcal{H}$ is a Hilbert space, then $\mathcal{H}^*$ is also a Hilbert space, and exhibit a natural isometric isomorphism between $\mathcal{H}$ and $\mathcal{H}^{**}$.

Essential Exercise 3.23. Show that the completion of an inner product space is a Hilbert space.

Exercise 3.24. Recall the definition of the Hardy space $H^2(D)$ (or the definition of the Bergman space $A^2(D)$) for $D = B_1^{\mathbb{C}}$ from Exercise 2.61. Show that for every $a \in D$ there is a function k_a in $H^2(D)$ (respectively $k_a \in A^2(D)$) with

$$f(a) = \langle f, k_a\rangle_{H^2(D)}$$

or

$$f(a) = \langle f, k_a\rangle_{A^2(D)}$$

respectively. The function $D \times D \ni (a, w) \mapsto k_a(w)$ is called a *reproducing kernel*. Determine k_a explicitly in both cases.

[†] The space $Y^{\perp}$ is one-dimensional, since $\ell|_{Y^{\perp}}$ has trivial kernel. Hence z is really uniquely determined up to a scalar multiple.

We recall (and extend) the definition of linear hull as follows.

Definition 3.25. Let $(V, \| \cdot \|)$ be a normed vector space, and let $S \subseteq V$ be a subset. The *linear hull* of S, written $\langle S \rangle$, is the smallest subspace of V containing S. Thus $\langle S \rangle$ consists of all linear combinations $\sum_{s \in F} c_s s$ for $F \subseteq S$ finite and scalars c_s. The *closed linear hull* $\overline{\langle S \rangle}$ is the smallest closed subspace of V containing S — it is the closure of the linear hull.

Corollary 3.26 (Characterization of the closed linear hull). *Let $\mathcal{H}$ be a Hilbert space and $S \subseteq \mathcal{H}$ a subset. Then $\overline{\langle S \rangle} = \left(S^{\perp} \right)^{\perp}$.*

PROOF. Let $Y = \overline{\langle S \rangle}$ be the closed linear hull. By orthogonal decomposition of Hilbert spaces (see Corollary 3.17), $Y = \left(Y^{\perp} \right)^{\perp}$. We claim that $Y^{\perp} = S^{\perp}$, which together with the last statement gives the corollary. To see the claim, notice that for any $x \in \mathcal{H}$ we have

$$x \in S^{\perp} \iff \langle x, y \rangle = 0 \text{ for } y \in \langle S \rangle$$
$$\iff \langle x, y \rangle = 0 \text{ for } y \in \overline{\langle S \rangle}$$

by (semi-)linearity in the second argument and continuity of the inner product. $\quad\square$

We have seen that a closed subspace Y in a Hilbert space has a closed orthogonal complement $Y^{\perp}$, allowing the Hilbert space to be written as the direct sum $Y \oplus Y^{\perp}$. The next two exercises explore the same question for Banach spaces, where no such simple conclusion can be drawn.

Exercise 3.27. Let $(V, \| \cdot \|)$ be a normed space. Two subspaces V_1, V_2 of V are said to be *algebraically complemented* if $V_1 + V_2 = V$ and $V_1 \cap V_2 = \{0\}$. In that case we define linear maps $\pi : V \to V/V_2$ by $v \mapsto v + V_2$, $\phi : V_1 \times V_2 \to V$ by $(v_1, v_2) \mapsto v_1 + v_2$, and $P : V \to V_1$ by $v_1 + v_2 \mapsto v_1$ where $v \in V, v_1 \in V_1$ and $v_2 \in V_2$. We may call P the projection of V onto V_1 along V_2. Show that the following are equivalent:

(1) V_1 and V_2 are closed subspaces of V and the map $\pi|_{V_1}$ is a homeomorphism (where V/V_2 is equipped with the quotient norm from Lemma 2.15);
(2) the map ϕ is a homeomorphism (where $V_1 \times V_2$ is equipped with any of the norms from Exercise 2.9);
(3) P is a bounded operator.

If any of these equivalent conditions hold, then the subspaces are called *topologically complemented*.

A closed subspace W of a normed space V is called *complemented* if there is a closed subspace W' with the property that W and W' are topologically complemented. The next exercise shows that a closed subspace is not necessarily complemented.[9]

Exercise 3.28. Prove that the closed subspace $c_0 = \{x = (x_n) \in \ell^{\infty} \mid \lim_{n \to \infty} x_n = 0\}$ is not complemented in ℓ^{∞} as follows.

(1) Show that $(\ell^\infty)^*$ contains a countable subset A with the property that if $x \in \ell^\infty$ has $a(x) = 0$ for all $a \in A$ then $x = 0$, and deduce that the same holds for any space isomorphic to a closed subspace of ℓ^∞.

Using the following steps, show that $V = \ell^\infty/c_0$ does not have the property in (1) and hence that c_0 cannot be complemented by Exercise 3.27.

(2) Use an enumeration of $\mathbb{Q}$ to construct for each $i \in I = \mathbb{R} \setminus \mathbb{Q}$ a sequence

$$x^{(i)} = (x_n^{(i)}) \in \ell^\infty$$

with values in $\{0, 1\}$ and with infinite support

$$\mathrm{Supp}(x^{(i)}) = \{n \in \mathbb{N} \mid x_n^{(i)} = 1\}$$

in such a way that $\mathrm{Supp}(x^{(i)}) \cap \mathrm{Supp}(x^{(j)})$ is finite for all $i \neq j$.

(3) Show that for any finite non-empty subset $J \subseteq I$ and any list of numbers $(b_i)_{i \in J}$ with $|b_i| = 1$ for all $i \in J$ we have

$$\left\| \sum_{i \in J} b_i x^{(i)} \right\|_{\ell^\infty/c_0} = 1.$$

(4) Deduce from (3) that for any continuous linear functional $f \in V^*$ and $n \in \mathbb{N}$ the set $\{i \in I \mid |f(x^{(i)})| > \frac{1}{n}\}$ is finite. Conclude that for any countable subset A of V^* there is some $i \in I$ with the property that $a(x^{(i)}) = 0$ for all $a \in A$.

3.1.3 An Application to Measure Theory

We will show in this section how the results from Section 3.1.2 can be used in measure theory. Before stating the main result, we recall some definitions. A measure ν is *absolutely continuous* with respect to another measure μ, written $\nu \ll \mu$, if any measurable set N with $\mu(N) = 0$ must also satisfy $\nu(N) = 0$. Two measures μ and ν are *singular* with respect to each other, written $\nu \perp \mu$, if there exist disjoint measurable sets $X_\mu, X_\nu \subseteq X$ with $X = X_\mu \sqcup X_\nu$ and with $\nu(X_\mu) = 0 = \mu(X_\nu)$. Finally, recall that a measure μ is *σ-finite* if there is a decomposition of X into measurable sets,

$$X = \bigsqcup_{i=1}^{\infty} X_i,$$

with $\mu(X_i) < \infty$ for all $i \geqslant 1$.

Proposition 3.29 (Lebesgue decomposition, Radon–Nikodym derivative). *Let μ and ν be two σ-finite measures on a measurable space $(X, \mathcal{B})$. Then ν can be decomposed as $\nu = \nu_{\mathrm{abs}} + \nu_{\mathrm{sing}}$ into the sum of two σ-finite measures with $\nu_{\mathrm{abs}} \ll \mu$ and $\nu_{\mathrm{sing}} \perp \mu$. Moreover, there exists a (μ-almost everywhere uniquely determined) measurable function $f \geqslant 0$, called the Radon–Nikodym derivative and often denoted by $f = \frac{\mathrm{d}\nu_{\mathrm{abs}}}{\mathrm{d}\mu}$, such that*

$$\nu_{\mathrm{abs}}(B) = \int_B f \, \mathrm{d}\mu$$

for any measurable $B \subseteq X$.

PROOF. Suppose that μ and ν are both finite measures (the general case can be reduced to this case by using the assumption that μ and ν are both σ-finite; see Exercise 3.31).

We define a new measure $m = \mu + \nu$ and will work with the real Hilbert space $\mathcal{H} = L_m^2(X)$. On this Hilbert space we define a linear functional ϕ by

$$\phi(g) = \int g \, \mathrm{d}\nu$$

for $g \in \mathcal{H} = L_m^2(X)$. Note that equivalence of functions modulo m implies equivalence of functions modulo ν, and moreover that for $g \in \mathcal{H}$ we have

$$\int |g| \, \mathrm{d}\nu \leqslant \int |g| \, \mathrm{d}m \leqslant \|g\|_{L_m^2} \|\mathbb{1}\|_{L_m^2},$$

where we have used the fact that $m = \mu + \nu$, that μ is a positive measure, and the Cauchy–Schwarz inequality on $\mathcal{H}$. Therefore, $\phi(g)$ is well-defined for $g \in \mathcal{H}$ and satisfies $|\phi(g)| \leqslant \|g\|_{L_m^2} \|\mathbb{1}\|_{L_m^2}$. By Fréchet–Riesz representation (Corollary 3.19) there is some $k \in \mathcal{H} = L_m^2$ such that

$$\int g \, \mathrm{d}\nu = \phi(g) = \int g k \, \mathrm{d}m \qquad (3.7)$$

for all $g \in L_m^2(X)$. We claim that k takes values in $[0, 1]$ almost surely with respect to m. Indeed, for any $B \in \mathcal{B}$ we have $0 \leqslant \nu(B) \leqslant m(B)$, so (using the fact that $g = \mathbb{1}_B$)

$$0 \leqslant \int_B k \, \mathrm{d}m \leqslant m(B).$$

Using the choices $B = \{x \in X \mid k(x) < 0\}$ and $B = \{x \in X \mid k(x) > 1\}$ implies the claim that k takes values in $[0, 1]$, m almost surely. Modifying k if necessary we will assume that k only takes values in $[0, 1]$.

Since $m = \mu + \nu$, we can reformulate (3.7) as

$$\int g(1 - k) \, \mathrm{d}\nu = \int g k \, \mathrm{d}\mu. \qquad (3.8)$$

This holds by construction for all simple functions g, and hence for all non-negative measurable functions by monotone convergence. Now define

$$X_{\mathrm{sing}} = \{x \in X \mid k(x) = 1\},$$
$$X_\mu = X \smallsetminus X_{\mathrm{sing}} = \{x \in X \mid k(x) < 1\},$$

and $\nu_{\text{sing}} = \nu|_{X_{\text{sing}}}$. By definition, $\nu_{\text{sing}}(X_\mu) = 0$, and by equation (3.8) applied with $g = \mathbb{1}_{X_{\text{sing}}}$ we also have $\mu(X_{\text{sing}}) = 0$. Therefore $\nu_{\text{sing}} \perp \mu$. We also define $\nu_{\text{abs}} = \nu|_{X_\mu}$ so that $\nu = \nu_{\text{sing}} + \nu_{\text{abs}}$. Finally, define the function

$$f = \begin{cases} \frac{k}{1-k} & \text{on } X_\mu, \\ 0 & \text{on } X_{\text{sing}}. \end{cases}$$

For any measurable $g \geqslant 0$ we may now apply (3.8) to obtain

$$\int_X g \, d\nu_{\text{abs}} = \int_{X_\mu} \tfrac{g}{1-k}(1-k) \, d\nu = \int_{X_\mu} \tfrac{g}{1-k} k \, d\mu = \int_X gf \, d\mu,$$

which shows that $f = \frac{d\nu_{\text{abs}}}{d\mu}$ is a Radon–Nikodym derivative and also shows that $\nu_{\text{abs}} \ll \mu$.

If f_1, f_2 are both Radon–Nikodym derivatives of ν_{abs} with respect to μ, then

$$\int_B f_1 \, d\mu = \nu_{\text{abs}}(B) = \int_B f_2 \, d\mu$$

for all measurable $B \subseteq X$, which implies that $f_1 = f_2$ μ-almost surely by considering $B = \{x \in X \mid f_1(x) < f_2(x)\}$ and $B = \{x \in X \mid f_1(x) > f_2(x)\}$. $\quad\square$

Exercise 3.30. For every $n \geqslant 1$, let $(X_n, \mathcal{B}_n, \mu_n)$ be a measure space.
(a) Define $X = \bigsqcup_{n=1}^\infty X_n$. Show that $\mathcal{B} = \{B \subseteq X \mid B \cap X_n \in \mathcal{B}_n \text{ for every } n \geqslant 1\}$ defines a σ-algebra on X.
(b) For every $B \in \mathcal{B}$ as in (a) define $\mu(B) = \sum_{n=1}^\infty \mu_n(B \cap X_n)$ and show that μ is a measure on $\mathcal{B}$.

Essential Exercise 3.31. Complete the proof of Proposition 3.29 in the σ-finite case.

Exercise 3.32. Use Lemma 2.63 and Proposition 3.29 to calculate the norm of the functional

$$C_c(X) \ni f \longmapsto \int f \, d\mu - \int f \, d\nu$$

for two finite Borel measures μ and ν on a locally compact metric space X (which may or may not be mutually singular).

Exercise 3.33. Let $(X, \mathcal{B})$ be a measurable space and denote the space of signed measures on X (as defined in Section B.5) by $\mathcal{M}(X)$.
(a) Given a signed measure $d\nu = g \, d\mu$ with a finite measure μ and $g \in L^1_\mu(X)$, define $\|\nu\|$ to be $\|g\|_{L^1(\mu)}$. Show that this yields a well-defined norm on $\mathcal{M}(X)$.
(b) Show that $\mathcal{M}(X)$ is a Banach space with respect to this norm.

The notion of conditional expectation with respect to a sub-σ-algebra is a powerful tool, and finds wide applications in probability (see Loéve [64], [65]) and ergodic theory (see [27], for example). Roughly speaking it is an 'orthogonal projection' defined on L^1, and we invite the reader to construct it in this manner in the following exercise.

Exercise 3.34. Let $(X, \mathcal{B}, \mu)$ be a probability space, and let $\mathcal{A} \subseteq \mathcal{B}$ be a sub-σ-algebra.
(a) Show that there exists a bounded operator, called the *conditional expectation*,

$$E_\mu(\,\cdot\,|\,\mathcal{A}) : L^1_\mu(X, \mathcal{B}) \longrightarrow L^1_\mu(X, \mathcal{A})$$
$$f \longmapsto E_\mu(f\,|\,\mathcal{A})$$

such that

$$\int_A f \, \mathrm{d}\mu = \int_A E_\mu(f\,|\,\mathcal{A}) \, \mathrm{d}\mu \tag{3.9}$$

for all $A \in \mathcal{A}$.
(b) Show that (3.9) uniquely characterizes $E_\mu(f\,|\,\mathcal{A}) \in L^1_\mu(X, \mathcal{A})$ as an equivalence class.
(c) Show that $f \in L^1_\mu(X, \mathcal{B})$ and $g \in L^\infty_\mu(X, \mathcal{A})$ implies that $E_\mu(fg\,|\,\mathcal{A}) = gE_\mu(f\,|\,\mathcal{A})$.
(d) Show that $\|E_\mu(f\,|\,\mathcal{A})\|_1 \leqslant \|f\|_1$ for $f \in L^1_\mu(X, \mathcal{B})$.

3.2 Orthonormal Bases and Gram–Schmidt

Definition 3.35. A finite or countable list (x_n) of vectors in an inner product space $(V, \langle \cdot, \cdot \rangle)$ is called *orthonormal* if

$$\langle x_m, x_n \rangle = \delta_{mn} = \begin{cases} 1 & \text{if } m = n, \\ 0 & \text{if } m \neq n \end{cases}$$

for all $m, n \geqslant 1$. In other words, we require that all the vectors have length one, and are mutually orthogonal.

As one might expect, this notion is fundamental for Hilbert spaces, and gives rise to the following satisfying abstract result, which as we will see lays the ground for Fourier analysis.

Proposition 3.36 (The closed linear hull of an orthonormal list).
Let $\mathcal{H}$ be a Hilbert space. Then the closed linear hull of an orthonormal list (x_n) is given by

$$\overline{\langle \{x_n\} \rangle} = \left\{ \sum_n a_n x_n \mid \text{the sum converges in } \mathcal{H} \right\},$$

where the sum $v = \sum_n a_n x_n$ converges in $\mathcal{H}$ if and only if $\sum_n |a_n|^2 < \infty$. In that case we also have $\|v\| = \left(\sum_n |a_n|^2 \right)^{1/2}$ and $\langle v, x_m \rangle = a_m$ for $m \geqslant 1$. Hence the linear map ϕ that sends the sequence (a_n) with $\sum_n |a_n|^2 < \infty$ to $\sum_n a_n x_n \in \overline{\langle \{x_n\} \rangle}$ is a unitary isomorphism of Hilbert spaces.

We note that the series $\sum_n a_n x_n$ need not be absolutely convergent, since $\ell^2(\mathbb{N}) \supsetneq \ell^1(\mathbb{N})$.

PROOF OF PROPOSITION 3.36. Suppose first that $(x_1, \ldots, x_N)$ is a finite orthonormal list. Then we may define a map ϕ from $\mathbb{K}^N$ (with $\mathbb{K}$ being the field of scalars $\mathbb{R}$ or $\mathbb{C}$) to $\mathcal{H}$ by setting $\phi((a_n)) = \sum_{n=1}^{N} a_n x_n$. Using the assumption of orthonormality it follows that

$$\|\phi((a_n))\|^2 = \sum_{m,n} \langle a_m x_m, a_n x_n \rangle = \sum_n |a_n|^2 = \|(a_n)\|_2^2 \qquad (3.10)$$

and

$$\langle \phi((a_n)), x_j \rangle = \sum_n \langle a_n x_n, x_j \rangle = a_j \qquad (3.11)$$

for $j = 1, \ldots, N$. This proves the statements in the case of a finite list.

Now suppose that the list is infinite. In this case we define the space

$$c_c(\mathbb{N}) = \big\{ (a_n) \in \ell^2(\mathbb{N}) \mid a_n = 0 \text{ for all but finitely many } n \big\}$$

and the linear map $\phi : c_c(\mathbb{N}) \to \mathcal{H}$ by

$$(a_n) \longmapsto \sum_{n=1}^{\infty} a_n x_n,$$

where the sum is actually finite by definition of the space $c_c(\mathbb{N})$, so that properties (3.10) and (3.11) clearly still hold in this case. We note that $\phi(c_c(\mathbb{N}))$ is the linear hull of the set $\{x_n\}$.

Now notice that $c_c(\mathbb{N}) \subseteq \ell^2(\mathbb{N})$ is dense. Indeed, if $a = (a_n) \in \ell^2(\mathbb{N})$ and we define

$$a_n^{(N)} = \begin{cases} a_n & \text{if } n \leqslant N; \\ 0 & \text{if } n > N \end{cases}$$

for $N \in \mathbb{N}$, then

$$\left\| (a_n^{(N)}) - (a_n) \right\|_2^2 = \sum_{n=N+1}^{\infty} |a_n|^2 \longrightarrow 0$$

as $N \to \infty$. By Proposition 2.59 there is therefore a unique extension of ϕ to $\ell^2(\mathbb{N})$, defined by

$$\phi((a_n)) = \lim_{N \to \infty} \phi((a_n^{(N)})). \qquad (3.12)$$

By continuity of the norm and the inner product, properties (3.10) and (3.11) extend to all of $\ell^2(\mathbb{N})$. By (3.10), ϕ is an isometry from $\ell^2(\mathbb{N})$ onto its image, so the image is complete and therefore closed in $\mathcal{H}$. Since $\phi(c_c(\mathbb{N})) = \langle\{x_n\}\rangle$, it follows that $\phi(\ell^2(\mathbb{N})) = \overline{\langle\{x_n\}\rangle}$ is the closed linear hull. Finally (3.12) shows that the series $\sum_{n=1}^{\infty} a_n x_n$ converges if $(a_n) \in \ell^2(\mathbb{N})$, and if $\sum_{n=1}^{\infty} a_n x_n$ converges, then (3.10) (applied to the partial sums) also implies that $\sum_{n=1}^{\infty} |a_n|^2$ converges. $\qquad\qquad\square$

The argument in the proof above can also be used for orthogonal subspaces as in the following exercise.

Essential Exercise 3.37. (a) Let $(\mathcal{H}_n)$ be a finite or countable list of Hilbert spaces. Then we define the *direct Hilbert space sum*

$$\bigoplus_n \mathcal{H}_n = \left\{ (v_n) \mid v_n \in \mathcal{H}_n \text{ and } \sum_n \|v_n\|^2 < \infty \right\}$$

to consist of all 'square summable sequences'. Show that

$$\langle (v_n), (w_n) \rangle_\oplus = \sum_n \langle v_n, w_n \rangle$$

defines an inner product on the direct sum, making it into a Hilbert space. (b) Let $\mathcal{H}$ be a Hilbert space and $(\mathcal{H}_n)$ a finite or countable list of mutually orthogonal closed subspaces of $\mathcal{H}$. Show that there is a canonical isometric isomorphism

$$\phi : \bigoplus_n \mathcal{H}_n \longrightarrow \overline{\left\langle \bigcup_n \mathcal{H}_n \right\rangle}$$

analogous to Proposition 3.36, and describe the inverse map of ϕ using orthogonal projections.

Definition 3.38. A list of orthonormal vectors in a Hilbert space $\mathcal{H}$ is said to be *complete* (or to be an *orthonormal basis*) if its closed linear hull is $\mathcal{H}$.

We note that strictly speaking this notion of a *Schauder basis* of an infinite-dimensional Hilbert space does not coincide with the notion of a basis in the sense of linear algebra as we allow (in contrast to the standard definition) infinite converging sums to represent arbitrary vectors as linear combinations of the basis vectors. We invite the reader to compare our discussion here with the proof of the existence of a basis for an infinite-dimensional vector space (relying on the axiom of choice and often called a *Hamel basis*), and hope that the reader agrees with us that the notion of an orthonormal basis of a Hilbert space is much more natural than the notion of a Hamel basis in our context. More importantly, the notion of an orthonormal basis will prove to be much more useful in the following discussions.

Theorem 3.39 (Gram–Schmidt). *Every separable Hilbert space $\mathcal{H}$ has an orthonormal basis. If $\mathcal{H}$ is n-dimensional, then $\mathcal{H}$ is isomorphic to $\mathbb{R}^n$ or $\mathbb{C}^n$. If $\mathcal{H}$ is not finite-dimensional, then $\mathcal{H}$ is isomorphic to $\ell^2(\mathbb{N})$.*

Here isomorphic means isomorphic as Hilbert spaces, so there is a linear bijection between the spaces that preserves the inner product. The proof of Theorem 3.39 is simply an interpretation of the familiar Gram–Schmidt orthonormalization procedure.

PROOF OF THEOREM 3.39. Let $\{y_1, y_2, \dots\} \subseteq \mathcal{H}$ be a dense countable subset. We are going to use the vectors $\{y_n\}$ to construct an orthonormal list of vectors which has the same linear hull. This is built up from the simple geometrical observation that if a vector v does not lie in the linear span of a finite set of vectors, then something from the linear span may be added to v to produce a non-zero vector orthogonal to the linear span.

We may assume that $y_1 \neq 0$ and define $x_1 = \frac{y_1}{\|y_1\|}$. Suppose now that we have already constructed orthonormal vectors $x_1, \ldots, x_n$ by using the vectors $y_1, \ldots, y_k$ with $k \geqslant n$ in such a way that

$$V_n = \langle x_1, \ldots, x_n \rangle = \langle y_1, \ldots, y_k \rangle.$$

If $y_{k+1} \in V_n$ we simply increase k but not n. If $y_{k+1} \notin V_n$ we decompose y_{k+1} into a sum $v+w$ with $v \in V_n$ and $w \in V_n^\perp$ using the orthogonal decomposition in Hilbert spaces (see Corollary 3.17). Since $w \neq 0$, we may define $x_{n+1} = \frac{w}{\|w\|}$ and obtain an extended list of orthonormal vectors satisfying

$$V_{n+1} = \langle x_1, \ldots, x_{n+1} \rangle = \langle y_1, \ldots, y_{k+1} \rangle.$$

Continuing this construction, we see that either $\mathcal{H}$ is n-dimensional for some $n \geqslant 0$ and we have produced an orthonormal basis $\{x_1, \ldots, x_n\}$ for $\mathcal{H}$, or that $\mathcal{H}$ is infinite-dimensional, and we construct an infinite list $x_1, x_2, \ldots$ of orthonormal vectors in $\mathcal{H}$. In this case the linear hull of $\{x_1, x_2, \ldots\}$ contains the original dense set $\{y_1, y_2, \ldots\}$ and so the closed linear hull must be all of $\mathcal{H}$, showing that $\{x_n\}$ is an orthonormal basis of $\mathcal{H}$. The remaining statement that $\mathcal{H}$ is isomorphic to $\ell^2(\mathbb{N})$ follows from Proposition 3.36. $\square$

Exercise 3.40. Give a direct proof that the closed unit ball in an infinite-dimensional Hilbert space is not compact by using the material from this section.

Exercise 3.41. Recall the Hardy and Bergman spaces $H^2(D)$ and $A^2(D)$ on the unit disk $D = B_1^{\mathbb{C}}$ from Exercise 2.61. Describe the spaces $H^2(D)$ and $A^2(D)$ in terms of the sequence of Taylor coefficients (a_n) of the Taylor expansion $f(z) = \sum_{n \geqslant 0} a_n z^n$ of elements of the space.

As we have noted above, a convergent sum obtained from an orthonormal list may not always be absolutely convergent. However, as we will show now, it is always unconditionally convergent in the sense that the order of summation is irrelevant.

Corollary 3.42. *Let (x_n) be a countable orthonormal list in a Hilbert space $\mathcal{H}$ and let $(a_n) \in \ell^2(\mathbb{N})$. Then $\sum_{n=1}^\infty a_n x_n$ converges unconditionally, meaning that for any permutation $\jmath : \mathbb{N} \to \mathbb{N}$ we have $\sum_{m=1}^\infty a_{\jmath(m)} x_{\jmath(m)} = \sum_{n=1}^\infty a_n x_n$. In particular, it makes sense to speak of a countable orthonormal basis even if we do not specify an enumeration of the basis.*

PROOF. Let (x_n), (a_n), and $\jmath : \mathbb{N} \to \mathbb{N}$ be as in the corollary. By Proposition 3.36 the series $v = \sum_{n=1}^\infty a_n x_n$ converges, and since

$$\sum_{m=1}^\infty |a_{\jmath(m)}|^2 = \sum_{n=1}^\infty |a_n|^2$$

the same applies to $w = \sum_{m=1}^{\infty} a_{j(m)} x_{j(m)}$. Also by Proposition 3.36 we have $\langle v, x_n \rangle = a_n$ and $\langle w, x_{j(m)} \rangle = a_{j(m)}$ for all $m, n \in \mathbb{N}$. As j is a permutation we see that $\langle v - w, x_n \rangle = 0$ for all $n \in \mathbb{N}$. As $v - w$ belongs to the closed linear hull of $\{x_n\}$, it follows that $v = w$, as claimed.

Suppose now $B \subseteq \mathcal{H}$ is a countable set consisting of mutually orthogonal unit vectors with dense linear hull. Then we may choose an enumeration $B = \{x_n \mid n \in \mathbb{N}\}$ and obtain an orthonormal basis in the sense of Definition 3.38. By the above the properties of the orthonormal basis and also the coordinates $\langle v, x \rangle$ of $v \in \mathcal{H}$ associated to a given element of $x \in B$ remain unchanged if a different enumeration is being used. $\square$

3.2.1 The Non-Separable Case

While the motivation generated by natural examples and the notational convenience of thinking of countable collections as sequences incline one strongly to the separable case, there is no reason to restrict attention completely to separable Hilbert spaces.

Example 3.43. Let I be a set, equipped with the discrete topology and the counting measure λ_{count} defined on the σ-algebra $\mathbb{P}(I)$ of all subsets of I. Then $\ell^2(I) = L^2(I, \mathbb{P}(I), \lambda_{\text{count}})$ is a Hilbert space, and it comprises all functions $a : I \to \mathbb{R}$ (or $\mathbb{C}$) for which the support $\text{Supp}(a) = \{i \in I \mid a_i \neq 0\}$ is finite or countable, and for which $\sum_{i \in I} |a_i|^2 = \sum_{i \in \text{Supp}(a)} |a_i|^2 < \infty$.

Theorem 3.44 (Non-separable Gram–Schmidt). *Let $\mathcal{H}$ be a Hilbert space that is not separable. Then there is an orthonormal basis consisting of elements x_i for every i in an uncountable index set I. Moreover, we have an isomorphism $\mathcal{H} \cong \ell^2(I)$, where the isomorphism between $\mathcal{H}$ and $\ell^2(I)$ is given by*

$$\ell^2(I) \ni a \longmapsto \sum_{i \in \text{Supp}(a)} a_i x_i$$

and the sum on the right is countable and convergent.

PROOF. We will construct a maximal orthonormal set of vectors by using Zorn's lemma (see Appendix A.1). Define a partially ordered[†] set

$$\mathscr{F} = \{(I, x_\bullet) \mid \text{the function } x_\bullet : I \to \mathcal{H} \text{ has orthonormal image}\},$$

with partial order defined by $(I, x_\bullet) \preccurlyeq (J, y_\bullet)$ if $I \subseteq J$ and $x_\bullet = y_\bullet|_I$. In this partially ordered set every totally ordered subset (or chain) has an upper bound, which can be found by simply taking the union of the index sets and the natural extension of the partially defined functions to the union.

[†] In order to ensure that this definition does indeed define a set, we could add the requirement that I is a subset of $\mathcal{H}$, and let $x_\bullet$ be the identity.

It follows that there exists a maximal element $(I, x.)$ of this partially ordered set by Zorn's lemma. Using this, define an isometry $\phi : \ell^2(I) \to \mathcal{H}$ by

$$a \longmapsto \sum_{i \in \mathrm{Supp}(a)} a_i x_i$$

first on the subset of all elements $a \in \ell^2(I)$ with $|\mathrm{Supp}(a)| < \infty$, and then, by applying the automatic extension to the closure (Proposition 2.59), on all of $\ell^2(I)$. This again defines an isomorphism from $\ell^2(I)$ to the complete, and hence closed, subspace $Y = \phi(\ell^2(I)) \subseteq \mathcal{H}$. We claim that $Y = \mathcal{H}$, for otherwise there would exist some $x \in Y^{\perp}$ of norm one by the orthogonal decomposition of Hilbert spaces (Corollary 3.17), and using this element x we can define a new element of $\mathscr{F}$ which is strictly bigger than the maximal element $(I, x.)$ in the partial order. This contradiction shows the claim, and hence proves the theorem. $\square$

3.3 Fourier Series on Compact Abelian Groups

Definition 3.45. A *topological group* is a group G that carries a topology with respect to which the maps $(g, h) \mapsto gh$ and $g \mapsto g^{-1}$ are continuous as maps $G \times G \to G$ and $G \to G$ respectively. A *compact* (σ-*compact, locally compact,* and so on) group is a topological group for which the topological space is compact (σ-compact, locally compact, and so on). We similarly extend other topological and algebraic properties to topological groups. For example, a metric compact abelian group is a compact metric topological space with an abelian group structure satisfying the continuity conditions above.

Below we will be largely concerned with specific metric abelian groups, in which the circle or 1-torus

$$\mathbb{T} = \mathbb{R}/\mathbb{Z} \cong \mathbb{S}^1$$

with its metric inherited from the usual metric on $\mathbb{R}$ (see the footnote on p. 2) and the d-torus

$$\mathbb{T}^d = \mathbb{R}^d/\mathbb{Z}^d \cong \left(\mathbb{S}^1\right)^d$$

are the main examples (which will also be discussed in Section 3.4 from a slightly different, more concrete, point of view). The notation $\mathbb{T}$ will be used for the additive circle and $\mathbb{S}^1 = \{z \in \mathbb{C} \mid |z| = 1\}$ for the multiplicative circle. Here compact and abelian are necessary assumptions for the type of result we prove, and dropping either of these two assumptions changes the theory significantly. However, we assume metrizability mainly for convenience, because it gives separability of $C(G)$ by Lemma 2.45.

We will use the following facts about compact abelian groups as 'black boxes' (that is, we will not need to know how they are proved at this stage).

However, we will also see in some examples below that these are often easy
to prove if the group is given concretely.

Theorem (Existence of Haar measure[†(10)]). Every locally compact σ-
compact metric group G has a left Haar measure m_G, satisfying (and, up to
positive multiples, characterized by) the properties:

- $m_G(K) < \infty$ for any compact set $K \subseteq G$;
- $m_G(O) > 0$ for any non-empty open set $O \subseteq G$; and
- $m_G(gB) = m_G(B)$ for all measurable $B \subseteq G$ and $g \in G$.

We will usually be dealing with σ-compact metrizable groups, which simplifies the measure theory needed, but the existence of Haar measure only
requires the group to be locally compact. For $G = \mathbb{T}^d$, which as a measurable space can be identified with $[0,1)^d$, the Haar measure is simply the d-
dimensional Lebesgue measure restricted to $[0,1)^d$.

Exercise 3.46. Show that the Lebesgue measure on $[0,1)^d$ considered as a measure on $\mathbb{T}^d$
satisfies all the properties of the Haar measure.

Knowing the defining third property $m_G(x + B) = m_G(B)$ for all Borel
subsets $B \subseteq G$, the formula

$$\int_G f(x)\,\mathrm{d}m(x) = \int_G f(g+x)\,\mathrm{d}m(x) \tag{3.13}$$

follows immediately for simple functions and then by monotone convergence
also for positive integrable functions, and then by taking differences for all
integrable functions.

Before stating the next fact, we recall that a *(unitary) character* on a
topological group G is a continuous homomorphism

$$\chi : G \longrightarrow \mathbb{S}^1 = \{z \in \mathbb{C} \mid |z| = 1\}. \tag{3.14}$$

The *trivial character* is the character defined by $\chi(g) = 1$ for all $g \in G$. A
collection $\mathscr{F}$ of functions on G (and, in particular, a collection of characters)
is said to *separate points* if for any $g, g' \in G$ with $g \neq g'$ there is some $f \in \mathscr{F}$
with $f(g) \neq f(g')$.

Theorem (Completeness of characters[‡]). On every locally compact σ-
compact metric abelian group G there are enough characters to separate
points.

For $G = \mathbb{T}$ this is trivial, because the single character χ defined by

$$\chi(x + \mathbb{Z}) = \mathrm{e}^{2\pi i x}$$

[†] This will be proved in Section 10.1.

[‡] This will be established in Section 12.8, and holds more generally for locally compact
abelian groups.

for $x \in \mathbb{R}$ already separates points since it is an isomorphism between $\mathbb{T}$ and $\mathbb{S}^1$. For $G = \mathbb{T}^d$ the characters $\chi_1, \ldots, \chi_d$, where

$$\chi_j(x + \mathbb{Z}^d) = e^{2\pi i x_j}$$

for $x = (x_1, \ldots, x_d)^t \in \mathbb{R}^d$, separate points since if $x \neq y$ we must have some $j \in \{1, \ldots, d\}$ with $x_j \neq y_j$, and then $\chi_j(x) \neq \chi_j(y)$.

In some discussions about characters we will parameterize the collection of all characters using some index set. For example, we will see shortly that the characters on $\mathbb{T}^d$ are parameterized by elements $n \in \mathbb{Z}^d$ in a natural way if we define for $n \in \mathbb{Z}^d$ the character χ_n on $\mathbb{T}^d$ by

$$\chi_n(x + \mathbb{Z}^d) = e^{2\pi i n \cdot x}$$

where $n \cdot x$ denotes the usual inner product $\mathbb{R}^d$. We will write $x \in \mathbb{T}^d$ as a shorthand for the element $x + \mathbb{Z}^d \in \mathbb{T}^d$, and whenever convenient we identify $x \in \mathbb{T}^d$ with $x \in [0,1)^d$.

Assuming the existence of a Haar measure and the completeness of characters as above for a compact metric abelian group G, we will now describe the theory of Fourier series on G. This will give a complete description of $L^2(G) = L^2_{m_G}(G)$ where m_G is the Haar measure on G. For convenience we normalize m_G to satisfy $m_G(G) = 1$.

Theorem 3.47 (Fourier series). *Assume that a metric compact abelian group G has a Haar measure and satisfies completeness of characters. Then the set of characters is finite or countably infinite and forms an orthonormal basis of $L^2(G)$. That is, the set of characters is an orthonormal set and any $f \in L^2(G)$ may be written as*

$$f = \sum_\chi a_\chi \chi,$$

where the sum, which runs over all the characters of G, is convergent[11] *with respect to $\| \cdot \|_2$, the equality is meant as elements of $L^2(G)$, the coefficients are given by $a_\chi = \langle f, \chi \rangle$, and they satisfy*

$$\sum_\chi |a_\chi|^2 = \|f\|_2^2.$$

The final equality is a form of *Parseval's theorem* or *Parseval's formula*.

PROOF OF THEOREM 3.47. Let χ be a non-trivial character on G, so that there is some element $g \in G$ with $\chi(g) \neq 1$. Since χ is continuous and by assumption $m(G) = 1$, the function χ is integrable, and

$$\int_G \chi(x) \, \mathrm{d}m(x) = \int_G \chi(g + x) \, \mathrm{d}m(x) = \chi(g) \int_G \chi(x) \, \mathrm{d}m(x). \qquad (3.15)$$

In fact, in (3.15) we used the defining invariance property of the Haar measure extended to integrals as in (3.13) and the fact that a character is in particular a homomorphism. However, we have chosen g with $\chi(g) \neq 1$ so (3.15) gives

$$\int_G \chi \, dm = 0.$$

Now let χ_1, χ_2 be any characters, and write $\chi = \chi_1 \overline{\chi_2}$. Then χ is also a character, and since $\overline{\chi_2(g)} = \chi_2(g)^{-1}$, we see that χ is trivial if and only if $\chi_1 = \chi_2$. Therefore the calculation above gives

$$\langle \chi_1, \chi_2 \rangle = \int_G \chi_1 \overline{\chi_2} \, dm = \delta_{\chi_1, \chi_2} = \begin{cases} m(G) = 1 & \text{if } \chi_1 = \chi_2; \\ 0 & \text{if } \chi_1 \neq \chi_2, \end{cases}$$

so the characters form an orthonormal set (and this is a consequence of the properties of the Haar measure).

To show that there are only countably many characters on G, notice that by orthonormality of the set of characters, the L^2 distance between any two distinct characters is $\sqrt{2}$. By Lemma 2.46, $C(G)$ is separable with respect to the $\| \cdot \|_\infty$ norm. This extends to $L^2(G)$ with respect to $\| \cdot \|_2$ since the bound

$$\|f\|_2^2 = \int_G \underbrace{|f(g)|^2}_{\leqslant \|f\|_\infty^2} \, dm(g) \leqslant \|f\|_\infty^2$$

shows that the embedding $C(G) \to L^2(G)$, which we know has dense image by Proposition 2.51, is continuous. It follows that there can be only countably many distinct characters, since an uncountable collection would give rise to an uncountable collection of disjoint open balls of radius $\frac{1}{2}\sqrt{2}$, contradicting separability.

In order to show completeness we will use the completeness of characters from p. 92. Define the complex linear hull $\mathscr{A} = \langle \chi \mid \chi \text{ a character on } G \rangle$, and notice that $\mathscr{A}$ is an algebra since the product of two characters is another character. Also notice that $\mathscr{A}$ is closed under conjugation, since

$$\overline{\chi}(g) = \overline{\chi(g)} = \chi(g)^{-1} = \chi(-g)$$

for $g \in G$ defines another character if χ is a character. Since by completeness of characters the algebra $\mathscr{A}$ separates points in G, the Stone–Weierstrass theorem now implies that $\mathscr{A}$ is dense in $C(G)$ with respect to $\| \cdot \|_\infty$. However, by the continuity of the embedding from $C(G)$ to $L^2(G)$ the closed linear hull of $\mathscr{A}$ in $L^2(G)$ contains $C(G)$ and so by Proposition 2.51 must be all of $L^2(G)$. Now the theorem follows from the description of the closed linear hull of an orthonormal list in Theorem 3.39. $\square$

Exercise 3.48. Let $G \subseteq \mathbb{T}^d$ be a closed subgroup. Show that any character χ on G is the restriction of a character of the form χ_n for some $n \in \mathbb{Z}^d$ (by using the arguments from the proof of Theorem 3.47).

Exercise 3.49. Find all the characters on $G = \mathbb{Z}/q\mathbb{Z}$ and prove Theorem 3.47 directly for this case.

Exercise 3.50 (Uncertainty principle on finite groups). Generalize Exercise 3.49 to a finite abelian group G, and write $\widehat{G}$ for the group of characters on G.
(a) Show that $\widehat{G}$ is a group under the operation of pointwise multiplication, and that we have $|\widehat{G}| = |G|$.
(b) Define the discrete Fourier transform of $f : G \to \mathbb{C}$ to be the function $\widehat{f} : \widehat{G} \to \mathbb{C}$ defined by $\widehat{f}(\chi) = \frac{1}{|G|} \sum_{g \in G} f(g) \overline{\chi(g)}$. Show that $\sum_{\chi \in \widehat{G}} |\widehat{f}(\chi)|^2 = \frac{1}{|G|} \sum_{g \in G} |f(g)|^2$ (Parseval's formula).
(c) Prove that $\|\widehat{f}\|_\infty = \max_{\chi \in \widehat{G}} |\widehat{f}(\chi)| \leqslant \frac{1}{|G|} \sum_{g \in G} |f(g)| = \|f\|_1$.
(d) Use the Cauchy–Schwarz inequality, Parseval's formula, and the inequality from (c) to deduce the following uncertainty principle[12]: $|\operatorname{Supp} f| \cdot |\operatorname{Supp} \widehat{f}| \geqslant |G|$ for $f \in L^2(G) \smallsetminus \{0\}$.

Exercise 3.51. (a) Find all the characters on $G = (\mathbb{Z}/N\mathbb{Z})^{\mathbb{N}}$ (endowed with the product topology). Show the existence of a Haar measure and the completeness of characters from p. 92 for this case.
(b) Now set $N = 2$ and notice that $G = (\mathbb{Z}/2\mathbb{Z})^{\mathbb{N}}$ is, as a measure space, isomorphic to $(0, 1)$ with the Lebesgue measure (by using the binary expansion of real numbers). Interpret the characters of G as maps on $(0, 1)$ to obtain the orthonormal basis known as the *Walsh system*.

Exercise 3.52. Let $p \in \mathbb{N}$ be a prime number. Describe all the characters on the compact group of p-adic integers $G = \mathbb{Z}_p$, defined by

$$G = \varprojlim_{n \to \infty} \mathbb{Z}/(p^n\mathbb{Z}) = \left\{ (z_n) \in \prod_{n=1}^{\infty} \mathbb{Z}/(p^n\mathbb{Z}) \mid z_n \equiv z_{n+1} \bmod p^n\mathbb{Z} \right\}.$$

Show the existence of a Haar measure and the completeness of characters from p. 92 for this case.

The following exercise shows how large the class of metric compact abelian groups really is.

Exercise 3.53. Let Γ be a countable abelian group and use it to define

$$G = \{(z_\gamma) \in \mathbb{T}^\Gamma \mid z_{\gamma_1 + \gamma_2} = z_{\gamma_1} + z_{\gamma_2} \text{ for all } \gamma_1, \gamma_2 \in \Gamma\}.$$

(a) Show that G is a metric compact abelian group in the induced topology from the product topology on $\mathbb{T}^\Gamma$.
(b) Use the theorem on completeness of the characters from p. 92 to show that the group of characters on G is isomorphic to Γ.

3.4 Fourier Series on $\mathbb{T}^d$

The discussion in Section 3.3 applies in particular to the torus $G = \mathbb{T}^d$, giving the basic theory of Fourier series of L^2 functions there, but this case is so important that we will treat it in greater detail here. Along the way, we will give a proof for Fourier series on the torus that will be independent of the theorem regarding Fourier series on general groups (Theorem 3.47).

For this section, we will define a character on $\mathbb{T}^d$ to be a function of the form

$$\chi_n(x) = \mathrm{e}^{2\pi \mathrm{i} n \cdot x} = \mathrm{e}^{2\pi \mathrm{i}(n_1 x_1 + \cdots + n_d x_d)}$$

for all $x \in \mathbb{T}^d$, for some $n \in \mathbb{Z}^d$. We will see in Corollary 3.67 that these are indeed all the characters on $\mathbb{T}^d$ in the sense of Section 3.3 (also see Exercise 3.48). We note that $\overline{\chi_n(x)} = \chi_{-n}(x) = \chi_n(-x)$ for $n \in \mathbb{Z}^d$ and $x \in \mathbb{T}^d$.

A *trigonometric polynomial* is a finite linear combination

$$p = \sum_{n \in F} a_n \chi_n$$

of characters, where $F \subseteq \mathbb{Z}^d$ is a finite set and $a_n \in \mathbb{C}$ for all $n \in F$. As in the proof of Theorem 3.47, one can use the complex version of the Stone–Weierstrass theorem to show that every continuous function can be approximated by a trigonometric polynomial. In this section we will give another proof of this using *convolution*.

Theorem 3.54 (Fourier series on the torus). *The characters χ_n with n in $\mathbb{Z}^d$ form an orthonormal basis for $L^2(\mathbb{T}^d)$, so that every $f \in L^2(\mathbb{T}^d)$ is given by an L^2 convergent Fourier series*

$$f = \sum_{n \in \mathbb{Z}^d} a_n \chi_n, \tag{3.16}$$

where the a_n are the Fourier coefficients *defined by*

$$a_n = a_n(f) = \langle f, \chi_n \rangle = \int_{\mathbb{T}^d} f(t) \chi_n(-t) \, \mathrm{d}t$$

for $n \in \mathbb{Z}^d$. Moreover,

$$\|f\|_2^2 = \sum_{n \in \mathbb{Z}^d} |a_n|^2. \tag{3.17}$$

Exercise 3.55. (a) Phrase Theorem 3.47 for $d = 1$ using the function $\chi_0 = 1$ and the functions $x \mapsto \cos(2\pi n x)$ and $x \mapsto \sin(2\pi n x)$ for $n \geqslant 1$.
(b) For every $n \geqslant 1$ choose d_n such that $f_n(x) = d_n \sin(\pi n x)$ has norm one in $L^2((0, 1))$. Show that $(f_n)_{n \geqslant 1}$ forms an orthonormal basis of $L^2((0, 1))$. Notice that each f_n satisfies the boundary conditions $f_n(0) = f_n(1) = 0$, which are called the Dirichlet boundary conditions.
(c) For every $n \geqslant 0$ choose d_n such that $g_n(x) = d_n \cos(\pi n x)$ has norm one in $L^2((0, 1))$. Show that $(g_n)_{n \geqslant 0}$ forms an orthonormal basis of $L^2((0, 1))$. Note that every g_n satisfies $g_n'(0) = g_n'(1) = 0$, which are called the Neumann boundary conditions.
(d) Find an orthonormal basis $(h_n)_{n \geqslant 1}$ of $L^2((0, 1))$ that consists of smooth functions satisfying the mixed boundary conditions $h_n(0) = h_n'(1) = 0$ for all $n \geqslant 1$.

Exercise 3.56. (a) Rephrase Theorem 3.47 for $\mathbb{T}^d$ for real-valued functions using sine and cosine functions.
(b) Find an orthonormal basis of $L^2((0,1)^d)$ satisfying the Dirichlet boundary conditions (that is, the basis should consist of smooth functions that vanish on the boundary of $[0,1]^d$).

The relation (3.17) is Parseval's formula, and it may be viewed as an infinite-dimensional form of Pythagoras' theorem. We will see later (see Theorem 4.9 in Section 4.1.1) that it is too much to ask for the Fourier series of a continuous function to converge uniformly, or even pointwise. However, some additional smoothness assumptions do imply uniform convergence of the Fourier series, and this will be the starting point for our excursion into the theory of Sobolev spaces in Chapter 5 and Section 6.4.

Theorem 3.57 (Differentiability and Fourier series). *Suppose that f is a function in $C^k(\mathbb{T}^d)$ for some $k \geqslant 1$. Let $\alpha = (\alpha_1, \ldots, \alpha_d) \in \mathbb{N}_0^d$ be a multi-index with $\|\alpha\|_1 \leqslant k$. Then the Fourier coefficient $a_n(\partial_\alpha f)$ of $\partial_\alpha f$ is given by*

$$a_n(\partial_\alpha f) = (2\pi i n_1)^{\alpha_1} \cdots (2\pi i n_d)^{\alpha_d} a_n(f). \qquad (3.18)$$

If $k > d/2$, then the Fourier series on the right-hand side of (3.16) converges absolutely, and

$$\|f\|_\infty \leqslant \sum_{n \in \mathbb{Z}^d} |a_n(f)| \ll_d \sqrt{\|f\|_2^2 + \|\partial_{e_1}^k f\|_2^2 + \cdots + \|\partial_{e_d}^k f\|_2^2}.$$

The reader may wonder in what sense the absolute convergence is meant, and the answer is in all of them: With respect to $\|\cdot\|_2$, pointwise at every point, and with respect to $\|\cdot\|_\infty$.

In order to prove these results (independently from the previous section) we will need to discuss convolution.

3.4.1 Convolution on the Torus

Definition 3.58 (Convolution). Fix $p, q \in [1, \infty]$ satisfying $\frac{1}{p} + \frac{1}{q} = 1$, let f be an element of $L^p(\mathbb{T}^d)$ and g an element of $L^q(\mathbb{T}^d)$. Then the *convolution* of f and g is the function $f * g$ defined by

$$f * g(x) = \int_{\mathbb{T}^d} f(t) g(x - t) \, \mathrm{d}t.$$

A pair of numbers p, q related as in Definition 3.58 are called *Hölder conjugate* or *conjugate exponents* due to Hölder's inequality $\|fg\|_1 \leqslant \|f\|_p \|g\|_q$ for $f \in L^p(\mathbb{T}^d)$ and $g \in L^q(\mathbb{T}^d)$ (see Theorem B.15). This implies in particular that the integral defining $f * g(x)$ exists for all $x \in \mathbb{T}^d$.

Lemma 3.59. *Let $p, q \in [1, \infty]$ be Hölder conjugate numbers, let $f \in L^p(\mathbb{T}^d)$ and $g \in L^q(\mathbb{T}^d)$. Then*

(1) $f * g = g * f$;

(2) $f * \chi_n = \left(\int f(t) \overline{\chi_n}(t) \, dt \right) \chi_n$; and

(3) $\langle \chi_m, \chi_n \rangle = \delta_{m,n}$.

PROOF. The first formula follows by a simple substitution (see Exercise 3.46):

$$f * g(x) = \int_{\mathbb{T}^d} f(t) g(\underbrace{x - t}_{=u}) \, dt = \int_{\mathbb{T}^d} f(x - u) g(u) \, du = g * f(x).$$

The second formula follows from the definition, since

$$f * \chi_n(x) = \int_{\mathbb{T}^d} f(t) \chi_n(x - t) \, dt$$

$$= \int_{\mathbb{T}^d} f(t) \chi_n(x) \chi_n(-t) \, dt = \int_{\mathbb{T}^d} f(t) \overline{\chi_n(t)} \, dt \, \chi_n(x).$$

For the last identity (which is a general property of characters, as we have seen in Section 3.3), note that

$$\chi_m(t) \overline{\chi_n(t)} = \chi_{m-n}(t) = e^{2\pi i ((m_1 - n_1)t_1 + \cdots + (m_d - n_d)t_d)},$$

and integrate this character over $\mathbb{T}^d$ to obtain the result. □

Lemma 3.60 (Continuity). *Let $p, q \in [1, \infty]$ be Hölder conjugate numbers. If $f \in L^p(\mathbb{T}^d)$ and $p < \infty$ then the shifted function $f^x \in L^p(\mathbb{T}^d)$ defined by $f^x(t) = f(t - x)$ depends continuously on $x \in \mathbb{T}^d$ in the $\| \cdot \|_p$ norm. Moreover, if $f \in L^p(\mathbb{T}^d)$ and $g \in L^q(\mathbb{T}^d)$, then $f * g \in C(\mathbb{T}^d)$.*

PROOF. For $1 \leqslant p < \infty$, $C(\mathbb{T}^d)$ is dense in $L^p(\mathbb{T}^d)$ by Proposition 2.51. Now fix $f \in L^p(\mathbb{T}^d)$, $\varepsilon > 0$ and choose $F \in C(\mathbb{T}^d)$ with $\|f - F\|_p < \varepsilon$. Then by uniform continuity of F there exists some $\delta > 0$ for which

$$d(x, y) < \delta \implies \|F^x - F^y\|_\infty < \varepsilon \implies \|F^x - F^y\|_p < \varepsilon.$$

Since shifting functions preserves their integrals and their p-norms, we deduce that $d(x, y) < \delta$ implies that

$$\|f^x - f^y\|_p \leqslant \|f^x - F^x\|_p + \|F^x - F^y\|_p + \|F^y - f^y\|_p < 3\varepsilon,$$

showing the continuity of the map $\mathbb{T}^d \ni x \mapsto f^x \in L^p(\mathbb{T}^d)$.

Now suppose that $f \in L^p(\mathbb{T}^d)$ and $g \in L^q(\mathbb{T}^d)$. By Lemma 3.59(1) we may switch f and g if necessary and assume $q < \infty$. Fix $\varepsilon > 0$ and choose $\delta > 0$ so that $\|g^x - g^y\|_q < \varepsilon$ holds whenever $d(x, y) < \delta$. Using the fact that the function g essentially appears in the shifted form g^x in the definition of $f * g(x)$, we now obtain

$$|f * g(x) - f * g(y)| = \left| \int_{\mathbb{T}^d} f(t) \left(g(x - t) - g(y - t) \right) \mathrm{d}t \right|$$

$$\leqslant \int_{\mathbb{T}^d} |f(t)||g(x - t) - g(y - t)| \, \mathrm{d}t$$

$$\leqslant \|f\|_p \|g^x - g^y\|_q \leqslant \varepsilon \|f\|_p$$

by the Hölder inequality, whenever $\mathrm{d}(x, y) < \delta$ (strictly speaking the function $(\tilde{g})^x(t) = g(x - t)$ with $\tilde{g}(t) = g(-t)$ appears in the definition of $f * g(x)$, but using $\|\tilde{g}^x - \tilde{g}^y\|_p = \|g^x - g^y\|_p$ this does not make much of a difference). As $\varepsilon > 0$ was arbitrary we see that $f * g$ is continuous. $\qquad \square$

3.4.2 Dirichlet and Fejér Kernels

Let us assume first that $d = 1$. By Lemma 3.59(2) the nth term in the Fourier series of f is given by $a_n(f)\chi_n = f * \chi_n$ with $a_n(f) = \langle f, \chi_n \rangle$ for every $n \in \mathbb{Z}$. Thus the partial sums of the Fourier series satisfy

$$\sum_{n=-N}^{N} a_n(f)\chi_n = f * \left(\sum_{n=-N}^{N} \chi_n \right).$$

This observation motivates the following definition.

Definition 3.61. The Nth *Dirichlet kernel* is the function $D_N \in C(\mathbb{T})$ defined by

$$D_N = \sum_{n=-N}^{N} \chi_n.$$

The 8th Dirichlet kernel is illustrated in Figure 3.3.

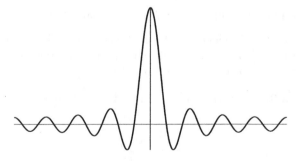

Fig. 3.3: The 8th Dirichlet kernel on the interval $[-\frac{1}{2}, \frac{1}{2}]$.

Fig. 3.4: The 8th Fejér kernel on the interval $[-\frac{1}{2}, \frac{1}{2}]$.

Lemma 3.62 (Dirichlet kernel). *The Dirichlet kernel D_N is real-valued and can also be expressed in the form*

$$D_N(x) = \begin{cases} 2N + 1 & \text{if } x = 0 \in \mathbb{T}, \\ \frac{e^{2\pi i(N+1)x} - e^{-2\pi i Nx}}{e^{2\pi i x} - 1} = \frac{\sin((N+\frac{1}{2})2\pi x)}{\sin(\pi x)} & \text{if } x \neq 0, \end{cases}$$

and satisfies

$$\int_{\mathbb{T}} D_N(x) \, dx = 1.$$

PROOF. The case $x = 0$ and the integral calculation follow immediately from the definitions. To check the formula for $x \neq 0$ we notice that the Dirichlet kernel is a geometric series and use the relation $\sin \phi = \frac{e^{i\phi} - e^{-i\phi}}{2i}$:

$$D_N(x) = \sum_{n=-N}^{N} \left(e^{2\pi i x}\right)^n = e^{-2\pi i Nx} \left(1 + \cdots + \left(e^{2\pi i x}\right)^{2N}\right)$$

$$= e^{-2\pi i Nx} \frac{\left(e^{2\pi i x}\right)^{2N+1} - 1}{e^{2\pi i x} - 1} = \frac{e^{2\pi i(N+1)x} - e^{-2\pi i Nx}}{e^{2\pi i x} - 1}$$

$$= \frac{e^{2\pi i(N+\frac{1}{2})x} - e^{-2\pi i(N+\frac{1}{2})x}}{e^{\pi i x} - e^{-\pi i x}} = \frac{\sin\left((N + \frac{1}{2})2\pi x\right)}{\sin(\pi x)}.$$

The latter formula also implies that D_N is real-valued. $\qquad\square$

By the above, the Dirichlet kernel is real-valued but takes on both positive and negative values. By averaging we obtain another kernel that only takes on positive values, which will be crucial later.

Definition 3.63. The Mth Fejér kernel is the function $F_M \in C(\mathbb{T})$ defined by

$$F_M = \frac{1}{M} \sum_{m=0}^{M-1} D_m.$$

The 8th Fejér kernel is shown in Figure 3.4.

Lemma 3.64 (Fejér kernel). *The Mth Fejér kernel is given by*

$$F_M(x) = \begin{cases} M & \text{if } x = 0, \\ \frac{1}{M}\left(\frac{\sin(M\pi x)}{\sin(\pi x)}\right)^2 & \text{if } x \neq 0 \end{cases}$$

and satisfies the following properties:

- $F_M(x) \geqslant 0$ *for all* $x \in \mathbb{T}$;
- $\int_{\mathbb{T}} F_M(x)\, dx = 1$;
- $F_M(x) \to 0$ *as* $M \to \infty$ *uniformly on every set of the form* $[\delta, 1 - \delta]$ *for* $\delta > 0$.

PROOF. We first verify the formula claimed for F_M. If $x = 0$, then

$$F_M(0) = \frac{1}{M}\sum_{m=0}^{M-1}(2m+1) = M.$$

For $x \neq 0$ we use Lemma 3.62 and obtain

$$\begin{aligned}
F_M(x) &= \frac{1}{M}\sum_{m=0}^{M-1}\frac{e^{2\pi i(m+1)x} - e^{-2\pi imx}}{e^{2\pi ix} - 1} \\
&= \frac{1}{M}\frac{1}{(e^{2\pi ix} - 1)}\left(e^{2\pi ix}\sum_{m=0}^{M-1}e^{2\pi imx} - \sum_{m=0}^{M-1}e^{-2\pi imx}\right) \\
&= \frac{1}{M}\frac{e^{\pi ix}}{(e^{2\pi ix} - 1)}\left(e^{\pi ix}\frac{e^{2\pi iMx} - 1}{e^{2\pi ix} - 1} - e^{-\pi ix}\frac{e^{-2\pi iMx} - 1}{e^{-2\pi ix} - 1}\right) \\
&= \frac{1}{M}\frac{e^{\pi ix}}{(e^{2\pi ix} - 1)}\left(\frac{e^{\pi ix}}{e^{2\pi ix} - 1}\left(e^{2\pi iMx} - 1\right)\right. \\
&\qquad\qquad\qquad\qquad \left. + \frac{e^{-\pi ix}}{1 - e^{-2\pi ix}}\left(e^{-2\pi iMx} - 1\right)\right) \\
&= \frac{1}{M}\frac{1}{(e^{\pi ix} - e^{-\pi ix})^2}\left(e^{2\pi iMx} - 2 + e^{-2\pi iMx}\right) \\
&= \frac{1}{M}\frac{(e^{\pi iMx} - e^{-\pi iMx})^2}{(e^{\pi ix} - e^{-\pi ix})^2} = \frac{1}{M}\frac{\sin^2(M\pi x)}{\sin^2(\pi x)}.
\end{aligned}$$

Now it is clear that $F_M(x) \geqslant 0$.

Since $\int_{\mathbb{T}} D_m(x)\, dx = 1$ for all $m \geqslant 0$ we also have $\int_{\mathbb{T}} F_M(x)\, dx = 1$. Finally, for $x \in [\delta, 1 - \delta]$ we have $\sin(\pi x) \geqslant \frac{\pi}{2}\delta$ and hence

$$F_M(x) \leqslant \frac{1}{M}\left(\frac{2}{\pi\delta}\right)^2 \longrightarrow 0$$

uniformly for $x \in [\delta, 1 - \delta]$ as $M \to \infty$. $\qquad\square$

Proposition 3.65 (Density of trigonometric polynomials). *For a continuous function f on $\mathbb{T}$ we have $f * F_M \to f$ as $M \to \infty$ with respect to $\|\cdot\|_\infty$. In particular, trigonometric polynomials are dense in $C(\mathbb{T})$.*

This behaviour of the sequence of functions (F_M) with respect to convolution is also described by saying that the sequence (F_M) is an *approximate identity*. As we will see in the proof, this property holds for any sequence of functions that satisfy the last three properties of the Fejér kernel in Lemma 3.64.

PROOF OF PROPOSITION 3.65. Let $f \in C(\mathbb{T})$ and fix $\varepsilon > 0$. Then there exists some $\delta \in (0, \frac{1}{2})$ for which $\mathrm{d}(x, y) < \delta \implies |f(x) - f(y)| < \varepsilon$. Now we estimate the difference $f * F_M(x) - f(x)$ as follows. By commutativity of convolution and the facts that $F_M \geqslant 0$ and $\int_{\mathbb{T}} F_M(t)\,\mathrm{d}t = 1$ we have

$$
\begin{aligned}
|f * F_M(x) - f(x)| &= |F_M * f(x) - f(x)| \\
&= \left| \int_{\mathbb{T}} F_M(t) f(x - t)\,\mathrm{d}t - \int_{\mathbb{T}} F_M(t) f(x)\,\mathrm{d}t \right| \\
&\leqslant \int_{\mathbb{T}} F_M(t) |f(x - t) - f(x)|\,\mathrm{d}t.
\end{aligned}
$$

Now we split the range of integration into the interval $[\delta, 1 - \delta]$ and its complement:

$$
|f * F_M(x) - f(x)| \leqslant \int_{\delta}^{1-\delta} F_M(t) \underbrace{|f(x - t) - f(x)|}_{\leqslant 2\|f\|_\infty}\,\mathrm{d}t
$$
$$
+ \int_{-\delta}^{\delta} F_M(t) \underbrace{|f(x - t) - f(x)|}_{< \varepsilon}\,\mathrm{d}t.
$$

The first integral goes to zero as $M \to \infty$ since $F_M \to 0$ uniformly as $M \to \infty$ on $[\delta, 1 - \delta]$. Hence for large enough M, the first integral is smaller than ε. The second integral is bounded above by ε as $F_M \geqslant 0$ and

$$
\int_{\mathbb{T}} F_M(t)\,\mathrm{d}t = 1.
$$

As δ is independent of x, the same is true for M, and we see that $f * F_M$ converges uniformly to f.

To see the final statement of the proposition note that $f * F_M$ is a trigonometric polynomial (by linearity and Lemma 3.59(2)).   □

Exercise 3.66. Analyze where the above proof fails if we replace the Fejér kernel by the Dirichlet kernel.

PROOF OF THEOREM 3.54. We start with the case $d = 1$. Proposition 3.65 shows that the linear hull $\mathscr{A}$ of the characters (that is, the space of trigonometric polynomials) is dense in $C(\mathbb{T})$ with respect to $\|\cdot\|_\infty$. Therefore the same holds with respect to $\|\cdot\|_2$ in $L^2(\mathbb{T})$. By Lemma 3.59, the characters on $\mathbb{T}$ form an orthonormal set. Thus the description of the closed linear hull of the characters follows from Proposition 3.36 and proves the theorem for $d = 1$.

The case of $d \geqslant 2$ is similar once we have shown that the space of trigonometric polynomials is dense in the continuous functions. For this, notice first that

$$\widetilde{F_M}(x_1, \ldots, x_d) = F_M(x_1) \cdots F_M(x_d)$$

is a trigonometric polynomial satisfying

- $\widetilde{F_M} \geqslant 0$,
- $\int_{\mathbb{T}^d} \widetilde{F_M}(x) \, dx = 1$, and
- $\int_{[-\delta,\delta]^d} \widetilde{F_M}(x) \, dx = \left(\int_{-\delta}^{\delta} F_M(t) \, dt \right)^d > 1 - \varepsilon$

for $\varepsilon, \delta > 0$ and large enough M (how large depending on ε and δ). Next notice that $f * \widetilde{F_M}$ is a trigonometric polynomial for any $f \in C(\mathbb{T}^d)$. The argument is now similar to the case $d = 1$: we again show that the sequence $(\widetilde{F_M})$ is an approximate identity. Given $f \in C(\mathbb{T}^d)$ and $\varepsilon > 0$ we can choose $\delta > 0$ such that $|f(x - t) - f(x)| < \varepsilon$ for $x \in \mathbb{T}^d$ and $t \in [-\delta, \delta]^d$. This implies that

$$\left| f * \widetilde{F_M}(x) - f(x) \right| \leqslant \int_{\mathbb{T}^d} |f(x - t) - f(x)| \widetilde{F_M}(t) \, dt$$

$$\leqslant \int_{[-\delta,\delta]^d} \underbrace{|f(x - t) - f(x)|}_{<\varepsilon} \widetilde{F_M}(t) \, dt$$

$$+ \int_{\mathbb{T}^d \smallsetminus [-\delta,\delta]^d} 2\|f\|_\infty \widetilde{F_M}(t) \, dt$$

$$< \varepsilon + 2\|f\|_\infty \varepsilon,$$

as required. As in the case $d = 1$ this implies that the set of characters forms an orthonormal basis, and the theorem follows. $\qquad\square$

Let us briefly describe how the definition of a character in Section 3.3 relates to the characters χ_n for $n \in \mathbb{Z}^d$ that we used here (see also Exercise 3.48).

Corollary 3.67 (Description of characters). *Every character of* $\mathbb{T}^d$ *in the sense of the definition given on p. 92 is of the form* $\chi = \chi_n$ *for some* $n \in \mathbb{Z}^d$.

PROOF. Let $\chi : \mathbb{T} \to \mathbb{S}^1$ be a continuous homomorphism. By Proposition 3.65, χ can be approximated uniformly by a trigonometric polynomial f. If χ does not appear in this trigonometric polynomial, then f is orthogonal to χ (by the argument on p. 94 based on (3.15)) and so cannot be close to χ. However, by definition of the Fejér kernel, the characters appearing in f are those of the form χ_n for $n \in \mathbb{Z}$.

Now suppose that $d > 1$ and $\chi : \mathbb{T}^d \to \mathbb{S}^1$ is a continuous homomorphism. Writing $x = (x_1, \ldots, x_d)$ we have

$$\chi(x) = \underbrace{\chi(x_1, 0, 0, \ldots, 0)}_{\chi^{(1)}(x_1)} \underbrace{\chi(0, x_2, 0, \ldots, 0)}_{\chi^{(2)}(x_2)} \cdots \underbrace{\chi(0, 0, \ldots, x_d)}_{\chi^{(d)}(x_d)},$$

where each $\chi^{(i)} : \mathbb{T} \to \mathbb{S}^1$ is a continuous homomorphism for $1 \leqslant i \leqslant d$. By the argument above we have $\chi^{(i)} = \chi_{n_i}$ for some $n_i \in \mathbb{Z}$, and so the result follows.   □

Exercise 3.68 (Polynomial approximate identities on $\mathbb{R}$). Give another proof of the Weierstrass approximation theorem (Example 2.41) using the *Landau kernel* defined by $L_n(x) = \ell_n(1 - x^2)^n$ for $x \in \mathbb{R}$, where $\ell_n > 0$ is chosen to make $\int_{-1}^{1} L_n(x)\,dx = 1$.
(a) Given $f \in C([0, 1])$, extend f continuously to $\mathbb{R}$ with support in $[-c, 1 + c]$ for some small $c > 0$ (the choice of which will matter in (b)). Show that

$$L_n * f(x) = \int_{\mathbb{R}} L_n(t) f(x - t)\,dt$$

is a polynomial.
(b) State, prove, and use an appropriate approximate identity property of the sequence (L_n) to show that $L_n * f(x)$ converges uniformly on $[0, 1]$ to f as $n \to \infty$.

3.4.3 Differentiability and Fourier Series

We now turn to the interplay between Fourier series and differentiation, with the goal of proving Theorem 3.57. As we will see, this relationship will be a simple but important consequence of integration by parts.

Suppose that $f \in C^1(\mathbb{T}^d)$ and $j \in \{1, \ldots, d\}$. Notice that

$$
\begin{aligned}
\int_{\mathbb{T}} \partial_j f(x) \overline{\chi_n(x)}\,dx_j &= f(x) \overline{\chi_n(x)} \, \Big|_{x_j=0}^{x_j=1} - \int_{\mathbb{T}} f(x) \partial_j \overline{\chi_n(x)}\,dx_j \\
&= - \int_{\mathbb{T}} f(x) \partial_j \overline{\chi_n(x)}\,dx_j \\
&= 2\pi i n_j \int_{\mathbb{T}} f(x) \overline{\chi_n(x)}\,dx_j
\end{aligned}
\tag{3.19}
$$

by integration by parts and periodicity, since

$$(\partial_j \chi_n)(x) = \partial_j e^{2\pi i (n_1 x_1 + \cdots + n_d x_d)} = 2\pi i n_j e^{2\pi i (n_1 x_1 + \cdots + n_d x_d)}.$$

Integrating over the remaining variables, we see that the Fourier coefficients of f and of the partial derivative $\partial_j f$ satisfy the relation

$$a_n(\partial_j f) = \int_{\mathbb{T}^d} \partial_j f(x)\overline{\chi_n(x)}\, dx = 2\pi i n_j \int_{\mathbb{T}^d} f(x)\overline{\chi_n(x)}\, dx \qquad \text{(by (3.19))}$$

$$= 2\pi i n_j a_n(f). \qquad (3.20)$$

PROOF OF THEOREM 3.57. The formula (3.18) follows from (3.20) by induction on k. To prove the last claim of the theorem, we will show that

$$\sum_{n \in \mathbb{Z}^d} |a_n(f)| \ll_d \sqrt{\|f\|_2^2 + \|\partial_1^k f\|_2^2 + \cdots + \|\partial_d^k f\|_2^2} \qquad (3.21)$$

for $f \in C^k(\mathbb{T}^d)$ and $k > \frac{d}{2}$. Assuming (3.21) for the moment, we see that

$$\sum_{n \in \mathbb{Z}^d} a_n(f)\chi_n$$

is an absolutely convergent series with respect to $\|\cdot\|_\infty$, and so converges to some limit F in $C(\mathbb{T}^d)$ with respect to $\|\cdot\|_\infty$. However, since $\|\cdot\|_2 \leqslant \|\cdot\|_\infty$ the same function F is also a limit with respect to $\|\cdot\|_2$. Hence by Fourier series on the torus (Theorem 3.54) we have $F = f$ first as an identity in $L^2(\mathbb{T}^d)$ (and hence almost everywhere), but as both functions are continuous, also in $C(\mathbb{T}^d)$ (and hence everywhere).

To prove (3.21), we start by expressing the right-hand side in terms of the Fourier coefficient $a_n = a_n(f)$. By Parseval's theorem in (3.17) applied to $\partial_{e_j}^k f$ we have

$$\|\partial_{e_j}^k f\|_2^2 = \sum_{n \in \mathbb{Z}^d} |a_n(\partial_{e_j}^k f)|^2 = \sum_{n \in \mathbb{Z}^d} (2\pi n_j)^{2k} |a_n(f)|^2,$$

where we have used (3.18) in the last step. Therefore we can simplify the sum under the square root in (3.21) to give the estimate

$$\|f\|_2^2 + \|\partial_{e_1}^k f\|_2^2 + \cdots + \|\partial_{e_d}^k f\|_2^2 = \sum_{n \in \mathbb{Z}^d} \left(1 + (2\pi)^{2k} \sum_{j=1}^d n_j^{2k}\right) |a_n(f)|^2$$

$$\gg_d \sum_{n \in \mathbb{Z}^d} \left(1 + \|n\|_2^{2k}\right) |a_n(f)|^2$$

since $\|n\|_2 \leqslant \sqrt{d}\max_{1 \leqslant j \leqslant d} |n_j|$ and hence $\|n\|_2^{2k} \ll_d \sum_{j=1}^d |n_j|^{2k}$. We claim that

$$\left(\left(1 + \|n\|_2^{2k}\right)^{-1/2}\right)_{n \in \mathbb{Z}^d} \in \ell^2(\mathbb{Z}^d) \qquad (3.22)$$

for $k > \frac{d}{2}$. From this claim the inequality (3.21) follows quickly by the Cauchy–Schwarz inequality:

$$\sum_{n \in \mathbb{Z}^d} |a_n(f)| = \sum_{n \in \mathbb{Z}^d} \left(1 + \|n\|_2^{2k}\right)^{-1/2} \left(1 + \|n\|_2^{2k}\right)^{1/2} |a_n(f)|$$

$$\leqslant \left\| \left(\left(1 + \|n\|_2^{2k}\right)^{-1/2}\right)_{n \in \mathbb{Z}^d} \right\|_{\ell^2(\mathbb{Z}^d)} \sqrt{\sum_{n \in \mathbb{Z}^d} \left(1 + \|n\|_2^{2k}\right) |a_n(f)|^2}$$

$$\ll_d \sqrt{\|f\|_2^2 + \|\partial_{e_1}^k f\|_2^2 + \cdots + \|\partial_{e_d}^k f\|_2^2}.$$

To verify the claim that

$$\sum_{n \in \mathbb{Z}^d} \frac{1}{1 + \|n\|_2^{2k}} < \infty$$

we split up the sum. Firstly, by running through the possibilities of the signs of the n_j, it is sufficient to show convergence for $n_1, \ldots, n_d \geqslant 0$. Secondly, using the symmetry of the summands with respect to permutation of the variables, we may restrict the sum to those $n \in \mathbb{Z}^d$ for which $n_2, n_3, \ldots, n_d \leqslant n_1$, and we may also assume that $n_1 \geqslant 1$. Now $1 + \|n\|_2^{2k} \geqslant n_1^{2k}$, so

$$\sum_{n \in \mathbb{Z}^d} \frac{1}{1 + \|n\|_2^{2k}} \ll_d \sum_{n_1=1}^{\infty} \sum_{n_2, \ldots, n_d=0}^{n_1} \frac{1}{n_1^{2k}}$$

$$= \sum_{n_1=1}^{\infty} \frac{(n_1+1)^{d-1}}{n_1^{2k}} \ll_d \sum_{n_1=1}^{\infty} \frac{1}{n_1^{2k+1-d}},$$

and the last sum converges if $2k > d$. This implies the claim above, the inequality (3.21), and hence the theorem. $\qquad \square$

The above is already sufficient to answer another claim from Section 1.5. We state a special case of the inheritance of smoothness below, and return again to this topic in Chapter 5.

Exercise 3.69. Let f be a real-valued function defined on an open subset $U \subseteq \mathbb{R}^2$. Suppose that f is continuous and that $\partial_1^4 f$, $\partial_2^4 f$ exist and are continuous. Show that $\partial_1 \partial_2 f$ exists and is continuous.

3.5 Group Actions and Representations

We are going to describe in this section in particular how functions on $\mathbb{R}^2$ can be decomposed into functions that have special rotational symmetries as alluded to in Section 1.1. The same argument will also apply to symmetries with respect to rotations about the z-axis in $\mathbb{R}^3$ (but not to all rotations in $\mathbb{R}^3$), and to many other situations. A convenient framework that incorporates all of these examples and much more is the following set-up.

3.5.1 Group Actions and Unitary Representations

Definition 3.70. Let G be a topological group, and let X be a topological space. A *continuous group action* of G on X is a continuous map

$$\cdot : G \times X \longrightarrow X$$
$$(g, x) \longmapsto g{\cdot}x$$

with $g{\cdot}(h{\cdot}x) = (gh){\cdot}x$ for $g, h \in G$ and $x \in X$, and $e{\cdot}x = x$ for all $x \in X$, where $e \in G$ is the identity element.

In this definition we have used multiplicative notation for the group operation in G, but as usual if G is abelian we will often use additive notation.

Definition 3.71. Let G be a topological group with a continuous action on a topological space X, and let μ be a measure on the Borel sets of X. Then we say that the G-action is *measure-preserving* (or that the measure is G-*invariant*) if $\mu(g{\cdot}B) = \mu(B)$ for all $g \in G$ and any Borel set $B \subseteq X$.

It is straightforward (see Exercise 3.72) to see that a measure-preserving action also preserves integration with respect to μ in the sense that

$$\int_X f^g \, d\mu = \int_X f(g^{-1}{\cdot}x) \, d\mu(x) = \int_X f(x) \, d\mu(x) = \int_X f \, d\mu, \qquad (3.23)$$

for all integrable functions f and $g \in G$, where we define $f^g(x) = f(g^{-1}{\cdot}x)$ for all $x \in X$ (the inverse in the definition of f^g is only necessary if G is non-abelian). In particular, if $f_1, f_2 \in L^2_\mu(X)$ then $\|f_1^g\|_2 = \|f_1\|_2$ and, more generally

$$\langle f_1^g, f_2^g \rangle = \langle f_1, f_2 \rangle . \qquad (3.24)$$

We can associate to the action of g on X an operator π_g on $L^2_\mu(X)$ defined by

$$\pi_g f = f^g = f \circ g^{-1}.$$

Using (3.24) and the relation $\pi_g \pi_{g^{-1}} = \pi_{g^{-1}} \pi_g = I$, we see that π_g is unitary (see Definition 3.8 and Exercise 3.9) for all $g \in G$.

Essential Exercise 3.72. Prove that a group action is measure-preserving if and only if (3.23) holds for all integrable f.

Definition 3.73. A *unitary representation* of a topological group G on a Hilbert space $\mathcal{H}$ is a map $\pi : G \to B(\mathcal{H})$, written as π_g (or $\pi(g)$, $g{\cdot}v$, or v^g for $v \in \mathcal{H}$), such that

- (Identity) $\pi_e = I$, the identity operator on $\mathcal{H}$;
- (Composition) $\pi_{g_1} \circ \pi_{g_2} = \pi_{g_1 g_2}$ for $g_1, g_2 \in G$;
- (Unitary) $\pi_g : \mathcal{H} \to \mathcal{H}$ is unitary for every $g \in G$; and

- (Continuity) for any given $v \in \mathcal{H}$, the map $g \mapsto \pi_g v \in \mathcal{H}$ is continuous.

We note that the first three properties together state that π is a homomorphism into the group of unitary operators.

Lemma 3.74 (Continuity). *Let G be a locally compact metric group G acting continuously on a locally compact σ-compact metric space X. Suppose that the action is measure-preserving with respect to a locally finite*[†] *measure μ on the Borel sets of X. Then the induced action of G on $\mathcal{H} = L_\mu^2(X)$ is a unitary representation of G on $\mathcal{H}$. More generally, for any $p \in [1, \infty)$ and $f \in L_\mu^p(X)$ we have that $G \ni g \mapsto f^g \in L_\mu^p(X)$ is continuous with respect to $\| \cdot \|_p$ and satisfies $\|f^g\|_p = \|f\|_p$.*

Lemma 3.74 is an important motivation for the study of unitary representations in general. We will return to this topic in several settings.

PROOF OF LEMMA 3.74. By Exercise 3.72 (see the hints on p. 563), π_g defined by $\pi_g f(x) = f(g^{-1} \cdot x)$ for $x \in X$ and $f \in L_\mu^p(X)$ satisfies $\|\pi_g f\|_p = \|f\|_p$ for every $g \in G$. The first properties of a unitary representation hold trivially. For the second, let f be an element of $L_\mu^p(X)$ and $g_1, g_2 \in G$. Applying the definition twice we obtain

$$\pi_{g_1}(\pi_{g_2} f)(x) = \pi_{g_2} f(g_1^{-1} \cdot x) = f(g_2^{-1} \cdot (g_1^{-1} \cdot x))$$
$$= f((g_1 g_2)^{-1} \cdot x) = \pi_{g_1 g_2} f(x)$$

for $x \in X$, giving the second defining property of a unitary representation.

For the continuity we essentially have to repeat the argument from the proof of Lemma 3.60 regarding convolutions on the torus. So let $p \in [1, \infty)$ and $f \in L_\mu^p(X)$ and fix $\varepsilon > 0$. By Proposition 2.51, there exists some function $F \in C_c(X)$ with

$$\|f - F\|_p < \varepsilon. \tag{3.25}$$

Let U be a compact neighbourhood of $e \in G$, so that

$$K = U \cdot \operatorname{Supp} F \subseteq X$$

is compact and hence has finite measure. The map

$$(g, x) \longmapsto F(g^{-1} \cdot x)$$

is continuous and so is uniformly continuous on the set $U \times K$. Hence there exists a $\delta > 0$ for which

$$\mathsf{d}(g, e) < \delta \implies g \in U \text{ and } \left| F(g^{-1} \cdot y) - F(y) \right| < \varepsilon / \sqrt[p]{\mu(K)}$$

for all $y \in K$. Note also that $F(g^{-1} \cdot y) \neq 0$ for $y \in X$ and $g \in U$ implies $g^{-1} \cdot y \in \operatorname{Supp} F$ and hence $y \in K$. Thus

---

[†] That is, a measure μ with the property that at every point there is an open neighbourhood of the point with finite μ-measure.

$$\|\pi_g F - F\|_p^p = \int_X \left| F(g^{-1} \cdot x) - F(x) \right|^p \, d\mu(x)$$

$$= \int_K \underbrace{\left| F(g^{-1} \cdot x) - F(x) \right|^p \, d\mu(x)}_{< \varepsilon^p / \mu(K)} < \varepsilon^p$$

for all g with $\mathsf{d}(g, e) < \delta$. Together with (3.25), this gives

$$\|\pi_g f - f\|_p < 3\varepsilon$$

for all $g \in U$ with $\mathsf{d}(g, e) < \delta$. Therefore, $G \ni g \mapsto \pi_g f$ is continuous at $g = e$. However, this gives the general case since

$$\|\pi_g f - \pi_{g_0} f\|_p = \|\pi_{g_0^{-1} g} f - f\|_p < 3\varepsilon$$

if we assume that g is sufficiently close to g_0 so that $\mathsf{d}(g_0^{-1} g, e) < \delta$. As $\varepsilon > 0$ and $g_0 \in G$ were arbitrary, continuity of $g \mapsto \pi_g f$ follows. $\square$

We explain now that any continuous group action and invariant measure gives rise to a convolution, which will play a critical role in the following discussions.

Lemma 3.75. *Let G be a locally compact σ-compact metric group with a left Haar measure m_G, X a locally compact σ-compact metric space with a continuous group action $\cdot$ of G on X, and μ a locally finite G-invariant measure on X. Let $\phi \in L^1(G)$, $p \in [1, \infty)$ and $f \in L_\mu^p(X)$. Then the integral*

$$\phi * f(x) = \int_G \phi(g) f(g^{-1} \cdot x) \, dm_G(g)$$

exists for μ-almost every $x \in X$, and $\|\phi * f\|_p \leqslant \|\phi\|_1 \|f\|_p$.

We note that the inequality also shows that $\phi * f \in L_\mu^p(X)$ only depends on the equivalence classes of $f \in L_\mu^p(X)$ and $\phi \in L^1(G)$.

PROOF OF LEMMA 3.75. Let $(Y, \mathcal{B}, \nu)$ be a probability space. Recall that the map $[0, \infty) \ni t \mapsto t^p$ is convex, which implies that $\left(\int f \, d\nu \right)^p \leqslant \int f^p \, d\nu$ for any non-negative simple function f on Y. Using monotone convergence we obtain the same for any non-negative measurable f on Y (that is, a special case of Jensen's inequality).

Consider now G, ϕ, X, and f as in the lemma, but assume first that ϕ is a non-negative measurable function with $\int \phi \, dm_G = 1$, and $f \geqslant 0$. We apply Jensen's inequality to the function $g \mapsto f(g^{-1} \cdot x)$ and the probability measure $\phi \, dm_G$ on G to obtain

$$\left(\int_G f(g^{-1} \cdot x) \phi(g) \, dm_G(g) \right)^p \leqslant \int_G f(g^{-1} \cdot x)^p \phi(g) \, dm_G(g).$$

Fubini's theorem shows that $\phi * f(x) = \int_G f(g^{-1} \cdot x) \phi(g) \, dm_G(g)$ depends measurably on $x \in X$. Integrating over X with respect to μ gives

$$\int_X (\phi * f)^p \, d\mu \leqslant \int_X \int_G f(g^{-1} \cdot x)^p \phi(g) \, dm_G(g) \, d\mu(x)$$

$$= \int_G \|f\|_p^p \phi(g) \, dm_G(g) = \|f\|_p^p,$$

where we also applied Fubini's theorem to the function

$$G \times X \ni (g, x) \longmapsto f(g^{-1} \cdot x)^p \phi(g)$$

and used the fact that π_g preserves the p-norm (see Lemma 3.74). Taking the pth root, we obtain $\|\phi * f\|_p \leqslant \|f\|_p$ in the case considered.

If $f \in L_\mu^p(X)$ and $\phi \in L^1(G)$, then we can apply the above to $\tilde{f} = |f|$ and $\tilde{\phi} = \|\phi\|_1^{-1} |\phi|$. Since

$$|\phi * f|(x) = \left| \int_G \phi(g) f(g^{-1} \cdot x) \, dm_G(g) \right|$$

$$\leqslant \|\phi\|_1 \int_G \tilde{\phi}(g) \tilde{f}(g^{-1} \cdot x) \, dm_G(g) = \|\phi\|_1 \tilde{\phi} * \tilde{f}(x),$$

the lemma follows from the previous case. $\qquad\square$

Essential Exercise 3.76. Prove Lemma 3.75 in the case $p = \infty$.

We will sometimes consider direct sums of representations as in the next exercise.

Exercise 3.77. Let G be a topological group, and let π_n be a unitary representation of G on the Hilbert space $\mathcal{H}_n$ for $n \geqslant 1$. Define $\mathcal{H}_\oplus = \bigoplus_n \mathcal{H}_n$ as in Exercise 3.37. Show that $\pi_{\oplus,g}(v_n) = (\pi_{n,g} v_n)$ for $g \in G$ and $(v_n) \in \mathcal{H}_\oplus$ defines a unitary representation of G.

3.5.2 Unitary Representations of Compact Abelian Groups

We now describe the decomposition of elements in a complex Hilbert space into elements of special types with respect to a unitary representation of a compact abelian group.

Definition 3.78. Let G be a topological group and let π be a unitary representation of G on the complex Hilbert space $\mathcal{H}$. Let $\chi : G \to \mathbb{S}^1$ be a character on G. Then $v \in \mathcal{H}$ is of *type* χ or has *weight* χ (type or weight $n \in \mathbb{Z}^d$ if $\chi = \chi_n$ in the case $G = \mathbb{T}^d$) if $\pi_g v = \chi(g) v$ for all $g \in G$. We also define the *weight space* $\mathcal{H}_\chi = \{v \in \mathcal{H} \mid v \text{ has weight } \chi\}$.

We note that $\mathcal{H}_\chi = \bigcap_{g \in G} \ker(\pi_g - \chi(g)I)$ is in fact just a common eigenspace if one considers all operators π_g for $g \in G$ simultaneously, and in particular is closed. Here χ gives us the various eigenvalues $\chi(g)$ as $g \in G$ (and hence the operator π_g) varies.

Lemma 3.79 (Orthogonality). *Let G, π, and $\mathcal{H}$ be as in Definition 3.78, then $\mathcal{H}_\chi \perp \mathcal{H}_\eta$ for any two characters $\chi \neq \eta$ of G.*

PROOF. Let $v \in \mathcal{H}_\chi$, $w \in \mathcal{H}_\eta$ and $g \in G$. Then

$$\chi(g)\langle v, w \rangle = \langle \chi(g)v, w \rangle = \langle \pi_g v, w \rangle = \langle v, \pi_{-g} w \rangle = \langle v, \eta(-g)w \rangle = \eta(g)\langle v, w \rangle.$$

However, for some $g \in G$ we have $\chi(g) \neq \eta(g)$ and so $\langle v, w \rangle = 0$. $\square$

The following result gives as a special case the decomposition of functions on $\mathbb{R}^2$ into components of different weights for $\mathrm{SO}_2(\mathbb{R})$.

Theorem 3.80 (Spectral theorem of compact abelian groups). *Let G be a compact metric abelian group and let π be a unitary representation on a complex Hilbert space $\mathcal{H}$. Then the weight spaces $\mathcal{H}_\chi$ are closed, mutually orthogonal, subspaces of $\mathcal{H}$ and*

$$\mathcal{H} = \bigoplus_\chi \mathcal{H}_\chi,$$

where the sum is over all characters χ of G. More concretely, every $v \in \mathcal{H}$ can be written as a convergent sum $\sum_\chi v_\chi$ in $\mathcal{H}$, where

$$v_\chi = \overline{\chi} *_\pi v = \int_G \overline{\chi}(g)\pi_g v \, dm_G(g) \tag{3.26}$$

has weight χ.

We need to explain the meaning of the convolution by χ in (3.26) by extending the notion of Riemann or Lebesgue integration to functions taking values in a Hilbert space (or even a Banach space). In Lemma 3.75 we have already seen one possible interpretation of the convolution in the case where the Hilbert space is $L^2_\mu(X)$ and the unitary representation is induced from a measure-preserving action. Although this is an interesting case, there are other ways to come by a unitary representation. Hence we will discuss in the following two subsections two more general definitions of such integrals.

3.5.3 The Strong (Riemann) Integral

One way to interpret $\overline{\chi} *_\pi v$ is to generalize the Riemann integral to this context, giving rise to a special case of the *Bochner* or *strong* integral. This approach applies to any unitary representation as in Theorem 3.80 (and more generally). Recall that for a fixed $v \in \mathcal{H}$ we have assumed in that theorem that $\pi_g v \in \mathcal{H}$ depends continuously on $g \in G$, and since the map $g \mapsto \chi(g)$ is continuous we also see that $\overline{\chi(g)}\pi_g v \in \mathcal{H}$ depends continuously on $g \in G$. Extracting the essential assumptions from this application we obtain the assumptions of the next statement.

Proposition 3.81 (Strong integration). *Let (X, d) be a compact metric space and let μ be a finite Borel measure on X. Let V be a Banach space, and let $f : X \to V$ be a continuous map. Then we can define Riemann sums of the form*

$$R(f, \xi) = \sum_{P \in \xi} m_G(P) f(x_P) \in V,$$

where $\xi = \{P_1, \ldots, P_k\}$ is a partition of X into finitely many non-empty measurable sets and $x_P \in P$ is a point chosen arbitrarily in each partition element $P \in \xi$. Suppose that

$$\lim_{n \to \infty} \max_{P \in \xi_n}(\mathrm{diam}(P)) = 0 \tag{3.27}$$

along a sequence of partitions (ξ_n). Then the sequence of associated Riemann sums $(R(f, \xi_n))$ forms a Cauchy sequence in V. The limit is a well-defined Riemann integral $_R\!\int f_X \, \mathrm{d}\mu \in V$ that is independent of the choice of the partition and sample points.

PROOF. Since f is continuous on a compact metric space, it is also uniformly continuous. Fix some $\varepsilon > 0$ and let $\delta > 0$ be such that $\mathrm{d}(x, y) < \delta$ implies $\|f(x) - f(y)\| < \varepsilon$. Suppose $\xi = \{P_1, \ldots, P_k\}$ and $\zeta = \{Q_1, \ldots, Q_\ell\}$ are two partitions such that $\mathrm{diam}(P_i) < \delta$ and $\mathrm{diam}(Q_j) < \delta$ for all i, j. Let

$$\eta = \xi \vee \zeta = \{P_i \cap Q_j \mid P_i \cap Q_j \neq \varnothing, i = 1, \ldots, k \text{ and } j = 1, \ldots, \ell\}.$$

Fix some choice of sample points for ξ, ζ, and η and note that $x_{P_i} \in P_i$ and $z_{P_i \cap Q_j} \in P_i \cap Q_j \subseteq P_i$ implies $\mathrm{d}(x_{P_i}, z_{P_i \cap Q_j}) < \delta$ for every i and j. This gives

$$\|R(f, \xi) - R(f, \eta)\| = \left\| \sum_{i=1}^{k} \sum_{j}{}' (f(x_{P_i}) - f(z_{P_i \cap Q_j})) \mu(P_i \cap Q_j) \right\|$$

$$\leqslant \sum_{i=1}^{k} \sum_{j}{}' \|f(x_{P_i}) - f(z_{P_i \cap Q_j})\| \mu(P_i \cap Q_j)$$

$$\leqslant \sum_{i=1}^{k} \sum_{j}{}' \varepsilon \mu(P_i \cap Q_j) = \varepsilon \mu(X),$$

where we write $\sum_{j}'$ for the sum over those $j \in \{1, \ldots, \ell\}$ with $P_i \cap Q_j \neq \varnothing$. The same holds for the Riemann sums $R(f, \zeta)$ and $R(f, \eta)$, which implies that

$$\|R(f,\xi) - R(f,\zeta)\| \leqslant 2\mu(X)\varepsilon.$$

This implies the lemma: If (ξ_n) is a sequence satisfying (3.27), then for every $\varepsilon > 0$ there exists some N such that $\xi = \xi_m, \zeta = \xi_n$ satisfy the above discussions whenever $m, n \geqslant N$. This implies that $(R(f, \xi_n))$ is a Cauchy sequence. If (ζ_n) is another such sequence, we may mix the two sequences of partitions into another sequence (by, for example, setting $\eta_{2n-1} = \xi_n$ and $\eta_{2n} = \zeta_n$ for all $n \in \mathbb{N}$) satisfying (3.27). Since the Riemann sums for this sequence also form a Cauchy sequence, we see that the limit is indeed independent of the choice of the sequence of partitions and the choice of the sample points. $\qquad\square$

Note that in the context of Theorem 3.80 we may set $X = G$, the measure $\mu = m_G$, $V = \mathcal{H}$, and $f(g) = \overline{\chi(g)}\pi_g(v)$ and use Proposition 3.81 to obtain a definition of $\overline{\chi} \underset{\pi}{*} v$.

Essential Exercise 3.82. Using the same notation and assumptions as in Proposition 3.81, show that $R\text{-}\!\int_X f \, d\mu$ depends linearly on f and satisfies

$$\left\| R\text{-}\!\int_X f \, d\mu \right\| \leqslant \int_X \|f\| \, d\mu.$$

3.5.4 The Weak (Lebesgue) Integral

In the following we will describe an integral for functions taking values in a Hilbert space, and this is also the approach that requires the least amount of structure for the function. On the other hand, this approach does not allow integration of functions taking values in any Banach space, but works similarly for any dual space (and so also for the class of reflexive Banach spaces to be defined later). This is a special case of the *weak*, *Pettis* or *Gelfand–Pettis* integral. In the general setting the properties of the Pettis integral involve subtle developments in measure theory; we refer to Talagrand [102] for the details.

Proposition 3.83 (Weak integration). *Let $(X, \mathcal{B}, \mu)$ be a measure space and let $\mathcal{H}$ be a Hilbert space. Let the function $f : X \to \mathcal{H}$ have the properties that $x \mapsto \|f(x)\|$ is measurable and integrable, and that for any $v \in \mathcal{H}$ the map $x \mapsto \langle v, f(x) \rangle$ is measurable. Then there exists a unique element of $\mathcal{H}$ denoted $\int_X f \, d\mu$ and called the* weak integral *of f with*

$$\left\langle v, \int_X f \, d\mu \right\rangle = \int_X \langle v, f(x) \rangle \, d\mu(x) \tag{3.28}$$

for all $v \in \mathcal{H}$. Moreover, $\int_X f \, d\mu$ depends linearly on f and satisfies

$$\left\| \int_X f \, \mathrm{d}\mu \right\| \leqslant \int_X \|f\| \, \mathrm{d}\mu.$$

PROOF. To see this we only have to show that the right-hand side of (3.28) defines a continuous functional on $\mathcal{H}$, for then the Fréchet–Riesz representation theorem (Corollary 3.19) implies the claimed existence and uniqueness.

By the Cauchy–Schwarz inequality we have

$$\left| \int_X \langle v, f(x) \rangle \, \mathrm{d}\mu(x) \right| \leqslant \int_X |\langle v, f(x) \rangle| \, \mathrm{d}\mu(x) \leqslant \|v\| \int_X \|f(x)\| \, \mathrm{d}\mu(x),$$

so the hypotheses show that the integral converges, and hence the map is well-defined. Moreover, for any scalar α and $v, w \in \mathcal{H}$ we have by linearity of the inner product that

$$\int_X \langle \alpha v + w, f(x) \rangle \, \mathrm{d}\mu(x) = \alpha \int_X \langle v, f(x) \rangle \, \mathrm{d}\mu(x) + \int_X \langle w, f(x) \rangle \, \mathrm{d}\mu(x),$$

showing linearity of the functional.

This shows that $\int_X f \, \mathrm{d}\mu \in \mathcal{H}$ exists, and that $\left\| \int_X f \, \mathrm{d}\mu \right\| \leqslant \int_X \|f\| \, \mathrm{d}\mu$. Using uniqueness, one can now check that this definition depends linearly on f. $\qquad\square$

Lemma 3.84. *Let (X, d) be a compact metric space and let μ be a finite Borel measure on X. Let $\mathcal{H}$ be a Hilbert space, and let $f : X \to H$ be a continuous map. Then $R\text{-}\int_X f \, \mathrm{d}\mu = \int_X f \, \mathrm{d}\mu$, that is, the strong and weak integrals agree.*

PROOF. Let (ξ_n) be a sequence of partitions as in Proposition 3.81 defining $R\text{-}\int_X f \, \mathrm{d}\mu$ as the limit of the Riemann sums $R(f, \xi_n)$. Let $w \in \mathcal{H}$ and notice that

$$\langle w, R(f, \xi_n) \rangle = \sum_{P \in \xi_n} \langle w, f(x_P) \rangle \, \mu(P) = \int_X F_n \, \mathrm{d}\mu,$$

where F_n is the simple function with values $F_n(x) = \langle w, f(x_P) \rangle$ for all $x \in P$ and $P \in \xi_n$. Note that $F_n(x) \to \langle w, f(x) \rangle$ as $n \to \infty$ by continuity of f and the assumption (3.27) on (ξ_n). Letting $n \to \infty$ we can apply the definition of the strong integral and dominated convergence to obtain

$$\left\langle w, R\text{-}\int_X f \, \mathrm{d}\mu \right\rangle = \int_X \langle w, f(x) \rangle \, \mathrm{d}\mu(x).$$

As this holds for any $w \in \mathcal{H}$, the lemma follows from the construction of the weak integral in Proposition 3.83. $\qquad\square$

In the context of unitary representations of a group G on a Hilbert space $\mathcal{H}$, we can use the notions of integration above to define convolution with measures or with L^1 functions as follows.

Definition 3.85. Let π be a unitary representation of a topological group G on a Hilbert space $\mathcal{H}$. Let μ be a finite measure on G. Then for $v \in \mathcal{H}$ we define the *convolution operator*

$$\mu \underset{\pi}{*} v = \int_G \pi_g v \, \mathrm{d}\mu(g).$$

If G has a left Haar measure m_G and $\phi \in L^1(G) = L^1_{m_G}(G)$, then for $v \in \mathcal{H}$ we define

$$\phi \underset{\pi}{*} v = \int_G \phi(g) \pi_g v \, \mathrm{d}m_G(g).$$

Essential Exercise 3.86. (a) Let π, G, $\mathcal{H}$, μ, and ϕ be as in Definition 3.85. Show that $\|\mu \underset{\pi}{*} v\| \leqslant \mu(G)\|v\|$ and $\|\phi \underset{\pi}{*} v\| \leqslant \|\phi\|_1 \|v\|$ for all $v \in \mathcal{H}$, so that $\mu \underset{\pi}{*} (\cdot)$ and $\phi \underset{\pi}{*} (\cdot)$ define two bounded operators on $\mathcal{H}$.
(b) Suppose that the unitary representation π on $\mathcal{H} = L^2_\mu(X)$ is induced by a measure-preserving action of G as in Lemma 3.75. Let ν be a finite measure on G, $\phi \in L^1(G)$ and $f \in \mathcal{H}$. Generalize Lemma 3.75 to also give a definition of $\nu * f$. Show that the pointwise defined functions $\nu * f$ and $\phi * f$ satisfy (3.28) (or equivalently that $\nu * f = \nu \underset{\pi}{*} f$ and $\phi * f = \phi \underset{\pi}{*} f$).

3.5.5 Proof of the Weight Decomposition

We are now ready to prove Theorem 3.80 which, apart from the generalized context, is a simple extension of the theorem regarding Fourier series on compact abelian groups (Theorem 3.47). We will use the assumptions of the theorem in this section without further remark.

Lemma 3.87 (Convolution with χ). *Let χ be a character of G. Then $\overline{\chi} \underset{\pi}{*} v$ has weight χ for any v in $\mathcal{H}$.*

PROOF. We need to prove that $\pi_g(\overline{\chi} \underset{\pi}{*} v) = \chi(g)(\overline{\chi} \underset{\pi}{*} v)$. Then

$$\langle w, \pi_g(\overline{\chi} \underset{\pi}{*} v) \rangle = \langle \pi_{-g} w, \overline{\chi} \underset{\pi}{*} v \rangle$$

$$= \int_G \langle \pi_{-g} w, \overline{\chi}(h) \pi_h v \rangle \, \mathrm{d}m_G(h)$$

$$= \int_G \langle w, \overline{\chi}(h) \pi_{g+h} v \rangle \, \mathrm{d}m_G(h)$$

$$= \int_G \langle w, \overline{\chi}(h' - g) \pi_{h'} v \rangle \, \mathrm{d}m_G(h')$$

$$= \int_G \langle w, \chi(g) \overline{\chi}(h') \pi_{h'} v \rangle \, \mathrm{d}m_G(h') = \langle w, \chi(g) \overline{\chi} \underset{\pi}{*} v \rangle$$

for $w \in \mathcal{H}$, which gives the lemma. $\qquad \square$

Lemma 3.88 (Convolution with χ). *If χ is a character and $v \in \mathcal{H}_\chi$ then we have $\overline{\chi} \underset{\pi}{*} v = v$. If $\eta \neq \chi$ are two different characters and $v \in \mathcal{H}_\eta$, then $\overline{\chi} \underset{\pi}{*} v = 0$.*

PROOF. If $v \in \mathcal{H}_\chi$ then

$$\overline{\chi} \underset{\pi}{*} v = \int_G \overline{\chi(g)} \pi_g v \, dm_G(g) = \int_G \overline{\chi(g)} \chi(g) v \, dm_G(g) = v,$$

since we assume that $m_G(G) = 1$ (strictly speaking, we should argue by taking the inner product with another vector, as in Proposition 3.83; here, and in similar cases below, we keep this implicit when convenient). Also for $v \in H_\eta$ we have

$$\overline{\chi} \underset{\pi}{*} v = \int_G \overline{\chi(h)} \pi_h v \, dm_G = \int_G \overline{\chi(h)} \eta(h) v \, dm_G = \left(\int_G \overline{\chi} \eta \, dm_G \right) v = 0.$$

$\square$

PROOF OF THEOREM 3.80. Let $\mathcal{H}'$ be the closed linear hull of $\mathcal{H}_\chi$ for all characters χ of G. By Lemma 3.79 the various weight spaces are mutually orthogonal, and (as already noted) closed, which gives us the following description of their closed linear hull

$$\mathcal{H}' = \bigoplus_\chi \mathcal{H}_\chi = \left\{ \sum_\chi v_\chi \mid v_\chi \in \mathcal{H}_\chi \text{ and } \sum_\chi \|v_\chi\|^2 < \infty \right\},$$

see Exercise 3.37. If $\mathcal{H}' = \mathcal{H}$ then the theorem follows from Exercise 3.86(a) (which specializes Proposition 3.83) and Lemma 3.88. In fact, (3.26) can then be shown as follows: if $v = \sum_\chi v_\chi$ is an element of $\mathcal{H}'$ with $v_\chi \in \mathcal{H}_\chi$ for every character χ of G, then continuity of convolution implies

$$\overline{\chi} \underset{\pi}{*} v = \overline{\chi} \underset{\pi}{*} \left(\sum_\chi v_\chi \right) = \overline{\chi} \underset{\pi}{*} v_\chi = v_\chi$$

for any character χ of G.

To see that $\mathcal{H}' = \mathcal{H}$ we show that any vector $v \in \mathcal{H}$ can be approximated by a vector in the linear hull of the spaces $\mathcal{H}_\chi$. Fix v and $\varepsilon > 0$. By continuity of the unitary representation there exists some $\delta > 0$ such that $\|\pi_g v - v\| < \varepsilon$ for $g \in B_\delta^G$. By Urysohn's lemma (Lemma A.27) there exists a function f in $C(G)$ with $f(0) > 0$, $f \geqslant 0$, and $f(g) = 0$ for all $g \in G \setminus B_\delta^G$.

Replacing f by a multiple of itself we may also suppose $\int_G f \, dm_G = 1$. Using this function together with the last part of the proof of Theorem 3.47 (or the more concrete argument from Section 3.4.2 in the case of the torus) we see that there is a trigonometric polynomial (that is, a finite linear combination of characters) F with the following properties:

- $F \geqslant 0$;

- $\displaystyle\int_G F \, dm_G = 1$; and

- $\displaystyle\int_{B_\delta^G} F \, dm_G > 1 - \varepsilon$.

The careful reader might notice that the approximation may not be real-valued nor have integral one. To deal with this issue we let F_1 be the first ε'-approximation and define F_2 to be the function $\frac{1}{2}(F_1 + \overline{F_1}) + \varepsilon'$, which is a positive $2\varepsilon'$-approximation since $f \geqslant 0$. Now define $F = (\int F_2 \, dm_G)^{-1} F_2$. Since $\int f \, dm_G = 1$, we have $|\int F_2 \, dm_G - 1| < 2\varepsilon'$ and so F is an ε-approximation if ε' is sufficiently small.

Then $F \underset{\pi}{*} v \in \mathcal{H}'$ is a finite linear combination of elements from weight spaces by Lemma 3.87. However, we also claim that

$$\| F \underset{\pi}{*} v - v \| \leqslant \varepsilon \, (1 + 2\|v\|). \tag{3.29}$$

To see this, let $w \in \mathcal{H}$, and then notice that

$$|\langle w, F \underset{\pi}{*} v - v \rangle| = \left| \left\langle w, \int_G F(g) \pi_g v \, dm_G(g) \right\rangle - \left\langle w, \left(\int_G F \, dm_G \right) v \right\rangle \right|$$

$$= \left| \int_G \langle w, F(g) \, (\pi_g v - v) \rangle \, dm_G(g) \right|$$

$$\leqslant \int_{B_\delta^G} \underbrace{|\langle w, \pi_g v - v \rangle|}_{\leqslant \|w\|\varepsilon} F(g) \, dm_G(g)$$

$$+ \int_{G \smallsetminus B_\delta^G} \underbrace{|\langle w, \pi_g v - v \rangle|}_{\leqslant 2\|w\|\|v\|} F(g) \, dm_G(g)$$

$$\leqslant \|w\|\varepsilon \, (1 + 2\|v\|),$$

which implies (3.29) since we may apply the inequality with $w = F \underset{\pi}{*} v - v$. $\qquad\square$

For the case of the continuous action of $\mathbb{T}$ on $\mathbb{R}^2$ defined by

$$\phi \cdot \begin{pmatrix} x_1 \\ x_2 \end{pmatrix} = k(2\pi\phi)x = \begin{pmatrix} \cos(2\pi\phi) & -\sin(2\pi\phi) \\ \sin(2\pi\phi) & \cos(2\pi\phi) \end{pmatrix} \begin{pmatrix} x_1 \\ x_2 \end{pmatrix},$$

for $\phi \in \mathbb{T}$ and all $x = (x_1, x_2)^{\mathrm{t}} \in \mathbb{R}^2$, the above immediately gives the following corollary (that we hinted at in Section 1.1).

Corollary 3.89. *Every function $f \in L^2(\mathbb{R}^2)$ can be written uniquely as a sum $\sum_{n \in \mathbb{Z}} f_n$ that converges with respect to $\| \cdot \|_2$, where*

$$f_n(x) = \int_{\mathbb{T}} \overline{\chi_n(\phi)} f(k(2\pi\phi)x) \, d\phi$$

and f_n has weight n with respect to the rotation action of $\mathbb{T}$ on $\mathbb{R}^2$ for all integers $n \in \mathbb{Z}$.

Exercise 3.90. Let $f \in C^\infty(\mathbb{R}^2) \cap L^2(\mathbb{R}^2)$ (or, more generally, $f \in C^\infty(\mathbb{R}^2)$). Show that the decomposition of f given by Corollary 3.89 converges uniformly on compact subsets of $\mathbb{R}^2$ to f. How much smoothness is needed to arrive at this uniform convergence?

3.5.6 Convolution

We conclude the chapter by generalizing the convolution considered in Section 3.4.1 to a more general context, combining it with the discussion regarding the convolution with respect to a unitary representation.

Proposition 3.91 (Convolutions). *Let G be a locally compact σ-compact metric group with a left Haar measure m_G. Define the convolution $f_1 * f_2$ of $f_1, f_2 \in L^1(G)$ by*

$$f_1 * f_2(g) = \int f_1(h) f_2(h^{-1}g) \, dm_G(h)$$

for all $g \in G$. The integral defining $f_1 * f_2(g)$ exists for m_G-almost every g in G, and defines an element $f_1 * f_2 \in L^1(G)$ with $\|f_1 * f_2\|_1 \leqslant \|f_1\|_1 \|f_2\|_1$. In other words, the convolution makes $L^1(G)$ into a separable Banach algebra. Suppose in addition π is a unitary representation of G on the Hilbert space $\mathcal{H}$. Then

$$f_1 \underset{\pi}{*} (f_2 \underset{\pi}{*} v) = (f_1 * f_2) \underset{\pi}{*} v,$$

where $\underset{\pi}{*}$ is defined in Definition 3.85. In other words, $\mathcal{H}$ is a module for the Banach algebra $L^1(G)$.

PROOF. Let G be as in the proposition. Then $G \ni h \longmapsto g \cdot h = gh$ for every $g \in G$ defines (by the definition of topological group) a continuous group action of G on $X = G$, called *the left action of G* (on G). By the definition of a left Haar measure, m_G is a G-invariant locally finite measure on G for the left action of G. Therefore, we may apply Lemma 3.75 to this action, giving $\|f_1 * f_2\|_1 \leqslant \|f_1\|_1 \|f_2\|_1$ for $f_1, f_2 \in L^1(G)$. Clearly the map $(f_1, f_2) \mapsto f_1 * f_2$ is bilinear in $f_1, f_2 \in L^1(G)$. It remains to show associativity for the operation. For this assume that $f_1, f_2, f_3 \in L^1(G)$, and then note that

$$(f_1 * f_2) * f_3(g) = \int_G (f_1 * f_2)(h) f_3(h^{-1}g) \, dm_G(h)$$

$$= \int_G \int_G f_1(k) f_2(\underbrace{k^{-1}h}_{\ell}) f_3(h^{-1}g) \, dm_G(k) \, dm_G(h).$$

Using Fubini and the substitution $\ell = k^{-1}h$ for a fixed k gives

$$(f_1 * f_2) * f_3(g) = \int_G f_1(k) \underbrace{\int_G f_2(\ell) f_3(\ell^{-1} k^{-1} g) \, dm_G(\ell)}_{(f_2 * f_3)(k^{-1}g)} \, dm_G(k)$$

$$= f_1 * (f_2 * f_3)(g),$$

as required. For the proof of separability of $L^1(G)$ we apply Lemma A.22 to find a sequence (K_n) of compact subsets $K_n \subseteq G$ with $K_n \subseteq K_{n+1}^o$ for all $n \geqslant 1$ and with $X = \bigcup_{n=1}^{\infty} K_n$. By Lemma 2.46, $C(K_n)$ is separable, which implies the same for the subset $C_c(K_n^o)$, for every $n \geqslant 1$. Since $m_G(K_n) < \infty$ the inclusion of $C_c(K_n^o)$ into $L^1(G)$ is continuous, and so the image of $C_c(K_n^o)$ in $L^1(G)$ is also separable. By density of $C_c(G) = \bigcup_{n=1}^{\infty} C_c(K_n^o)$ in $L^1(G)$ (Proposition 2.51) this proves separability of $L^1(G)$.

For the second part we suppose π is a unitary representation of G on the Hilbert space $\mathcal{H}$ and that $v, w \in \mathcal{H}$. Then

$$\langle w, f_1 \underset{\pi}{*} (f_2 \underset{\pi}{*} v) \rangle = \int \left\langle w, f_1(h) \pi_h \left(\int f_2(g) \pi_g v \, dm_G(g) \right) \right\rangle dm_G(h)$$

$$= \int \overline{f_1(h)} \left\langle \pi_h^{-1} w, \int f_2(g) \pi_g v \, dm_G(g) \right\rangle dm_G(h)$$

$$= \iint \overline{f_1(h)} \left\langle \pi_h^{-1} w, f_2(g) \pi_g v \right\rangle dm_G(g) \, dm_G(h)$$

$$= \iint \langle w, f_1(h) f_2(g) \pi_{hg} v \rangle \, dm_G(g) \, dm_G(h).$$

Using the substition $k = hg$ in the inner integral and exchanging the order of integration we obtain

$$\langle w, f_1 \underset{\pi}{*} (f_2 \underset{\pi}{*} v) \rangle = \iint \langle w, f_1(h) f_2(h^{-1} k) \pi_k v \rangle \, dm_G(k) \, dm_G(h)$$

$$= \int \langle w, f_1 * f_2(k) \pi_k v \rangle \, dm_G(k) = \langle w, (f_1 * f_2) \underset{\pi}{*} v \rangle.$$

By the definition of the weak integral in Proposition 3.83, this implies the proposition. $\square$

Essential Exercise 3.92. Suppose in addition to the assumptions of Proposition 3.91 that G is abelian and that the Haar measure is invariant under[†] $g \mapsto -g$. Show that $f_1 * f_2 = f_2 * f_1$ for any $f_1, f_2 \in L^1(G)$.

Exercise 3.93. Recall from Exercise 3.33 that the space of signed measures on G is a Banach space. Define a convolution for elements in the space of signed measures on G, and extend Proposition 3.91 to this case.

[†] The assumption that G is abelian actually implies this by the uniqueness properties of the Haar measure (see Proposition 10.2).

Exercise 3.94. Show that there is a continuous injective algebra homomorphism from the Banach algebra $\ell^1(\mathbb{Z})$ (which may be thought of as $L^1(\mathbb{Z})$ with respect to the counting measure on $\mathbb{Z}$ and convolution) to $C(\mathbb{T})$, where multiplication is pointwise multiplication.

3.6 Further Topics

- Hilbert spaces are at the heart of many developments. We will start to see this in the context of Sobolev spaces and the Laplace differential operator in Chapters 5 and 6.

- Another case where the Hilbert space splits into eigenspaces will be considered in Chapter 6.

- The spectral theory of a single unitary operator (equivalently, a unitary representation of the group $G = \mathbb{Z}$) is actually more delicate than the case considered above (see Exercise 6.1, for example), where we showed that the Hilbert space splits into a sum of generalized eigenspaces (the weight spaces). We will treat the case of a single unitary operator only in Chapter 9 (which will build on the material in Chapters 7 and 8).

- The topic of Fourier series on $\mathbb{T}^d$ leads naturally to the study of the Fourier integral on $\mathbb{R}^d$ (see Section 9.2). The concepts of Fourier series and Fourier integrals on $\mathbb{T}^d$ and $\mathbb{R}^d$, respectively, find a common generalization in the theory of Pontryagin duality (see Section 12.8).

- The case of unitary representations for compact abelian groups considered in this chapter was quite straightforward and is only the beginning of the important theory of unitary representations of locally compact groups. For locally compact abelian groups this is strongly related to Pontryagin duality; see Sections 11.4 and 12.8. For compact groups the main theorem in this direction is the Peter–Weyl theorem [85] (which is covered in Folland [32]). For many other groups that are neither abelian nor compact this topic is also important and can have many interesting surprises.

- One such surpise may be the so-called property (T) that was introduced by Každan in 1967 and has become important in many parts of mathematics since then. Building on the material in Chapter 9 we will study this notion in Section 10.3.

- We have seen in this chapter that the notion of a left Haar measure leads to many interesting concepts. For a concretely given group it is often not difficult to find its left Haar measure. In Chapter 10 (which relies on Chapter 7) we will prove the existence of the left Haar measure in general.

The reader may continue with Chapter 4, 5, 6, or 7 (with some of the material of Chapter 6 building on Chapter 5).

Chapter 4
Uniform Boundedness and the Open Mapping Theorem

In this chapter we present the main consequences of completeness for Banach spaces.

4.1 Uniform Boundedness

Our first result is the *principle of uniform boundedness* or the *Banach–Steinhaus theorem*.

Theorem 4.1 (Banach–Steinhaus). *Let X be a Banach space and let Y be a normed vector space. Let $\{T_\alpha \mid \alpha \in A\}$ be a family of bounded linear operators from X to Y. Suppose that for each $x \in X$, the set $\{T_\alpha x \mid \alpha \in A\}$ is a bounded subset of Y. Then the set $\{\|T_\alpha\| \mid \alpha \in A\}$ is bounded.*

The reader in a hurry may also first prove the Baire category theorem (Theorem 4.12) and derive Theorem 4.1 relatively quickly from it (see Exercise 4.16). We refrain from doing this here as it might help her to see the argument behind the Baire category theorem once here in the concrete application and once in the general case.

PROOF OF THEOREM 4.1. Assume first that there is an open ball $B_\varepsilon(x_0)$ on which

$$\{T_\alpha x \mid \alpha \in A\}$$

is *uniformly bounded:* that is, there is a constant K such that

$$\|x - x_0\| < \varepsilon \implies \|T_\alpha x\| \leqslant K \tag{4.1}$$

for all $\alpha \in A$. Then we claim that it is possible to find a bound on the family $\{\|T_\alpha\| \mid \alpha \in A\}$ of the norms of the operators. Indeed, for any $y \neq 0$ define

$$z = \frac{\varepsilon}{2\|y\|}y + x_0.$$

© Springer International Publishing AG 2017

M. Einsiedler, T. Ward, *Functional Analysis, Spectral Theory, and Applications*, Graduate Texts in Mathematics 276, DOI 10.1007/978-3-319-58540-6_4

Then $z \in B_\varepsilon(x_0)$ by construction, so (4.1) implies that $\|T_\alpha z\| \leqslant K$. Now by linearity of T_α the triangle inequality shows that

$$\frac{\varepsilon}{2\|y\|}\|T_\alpha y\| - \|T_\alpha x_0\| \leqslant \left\|\frac{\varepsilon}{2\|y\|}T_\alpha y + T_\alpha x_0\right\| = \|T_\alpha z\| \leqslant K,$$

which can be solved for $\|T_\alpha y\|$ to give

$$\|T_\alpha y\| \leqslant 2\frac{K + \|T_\alpha x_0\|}{\varepsilon}\|y\| \leqslant 2\frac{K + K'}{\varepsilon}\|y\|,$$

where $K' = \sup_\alpha \|T_\alpha x_0\| \leqslant K < \infty$. It follows that

$$\|T_\alpha\| \leqslant \frac{4K}{\varepsilon}$$

for every $\alpha \in A$, as required.

To finish the proof we have to show that there is a ball on which property (4.1) holds. This is proved by contradiction. Assume that there is no ball on which (4.1) holds. Fix an arbitrary open ball B_0. By assumption there is a point $x_1 \in B_0$ such that

$$\|T_{\alpha_1}x_1\| > 1$$

for some index $\alpha_1 \in A$. Since each T_α is continuous, there is a ball $B_{\varepsilon_1}(x_1)$ with $\|T_{\alpha_1}y\| \geqslant 1$ for all $y \in \overline{B_{\varepsilon_1}(x_1)}$. Assume without loss of generality that $\overline{B_{\varepsilon_1}(x_1)} \subseteq B_0$ and $\varepsilon_1 < 1$. By assumption, in this new ball the family $\{T_\alpha x \mid \alpha \in A\}$ is not bounded, so there is a point $x_2 \in B_{\varepsilon_1}(x_1)$ with

$$\|T_{\alpha_2}x_2\| > 2$$

for some index $\alpha_2 \in A$. We continue in the same way. By continuity of α_2 there is a ball $B_{\varepsilon_2}(x_2)$ with $\overline{B_{\varepsilon_2}(x_2)} \subseteq B_{\varepsilon_1}(x_1)$ and with $\|T_{\alpha_2}y\| \geqslant 2$ for all $y \in \overline{B_{\varepsilon_2}(x_2)}$. Assume without loss of generality that $\varepsilon_2 < \frac{1}{2}$.

Repeating this process produces points $x_1, x_2, \ldots$, indices $\alpha_1, \alpha_2, \ldots$, and positive numbers $\varepsilon_1, \varepsilon_2, \ldots$ such that $\overline{B_{\varepsilon_n}(x_n)} \subseteq B_{\varepsilon_{n-1}}(x_{n-1})$, $\varepsilon_n < \frac{1}{n}$, and

$$\|T_{\alpha_n}y\| \geqslant n \quad \text{for all } y \in \overline{B_{\varepsilon_n}(x_n)}$$

for all $n \geqslant 1$. Now the sequence (x_n) is clearly Cauchy (since $x_m \in B_{\varepsilon_n}(x_n)$ for all $m \geqslant n$, and so $\mathsf{d}(x_m, x_n) < \varepsilon_n < 1/n$), and therefore converges to some $z \in X$. By construction, $z \in \overline{B_{\varepsilon_n}(x_n)}$ and $\|T_{\alpha_n}z\| \geqslant n$ for all $n \geqslant 1$, which contradicts the hypothesis that the set $\{T_\alpha z \mid \alpha \in A\}$ is bounded. $\square$

Corresponding to the operator norm defined in Lemma 2.52 there is of course a notion of convergence in the space $\mathrm{B}(X, Y)$ of bounded linear operators from X to Y. A sequence (T_n) in $\mathrm{B}(X, Y)$ is *uniformly convergent* to $T \in \mathrm{B}(X, Y)$ if $\|T_n - T\| \to 0$ as $n \to \infty$ (so uniform convergence of a sequence of operators is simply convergence in the operator norm).

A different (and weaker, despite the name) notion of convergence for a sequence of operators is given by the following definition. We will discuss this and other notions of convergence again in Section 8.3.

Definition 4.2. A sequence (T_n) in $B(X, Y)$ is *strongly convergent* if for any $x \in X$ the sequence $(T_n x)$ converges in Y. If there is a $T \in B(X, Y)$ with $\lim_{n \to \infty} T_n x = Tx$ for all $x \in X$, then (T_n) is *strongly convergent* to T.

Corollary 4.3. *Let X be a Banach space, and Y any normed vector space. If a sequence (T_n) in $B(X, Y)$ is strongly convergent, then there exists an operator $T \in B(X, Y)$ such that (T_n) is strongly convergent to T.*

PROOF. For each $x \in X$ the sequence $(T_n x)$ is bounded since it is convergent. By the uniform boundedness principle (Theorem 4.1), there is a constant K such that $\|T_n\| \leqslant K$ for all n. Hence

$$\|T_n x\| \leqslant K \|x\| \quad \text{for all} \ \ x \in X. \tag{4.2}$$

We now define $T : X \to Y$ by $Tx = \lim_{n \to \infty} T_n x$ for all $x \in X$. It is clear that T is linear, and (4.2) shows that $\|Tx\| \leqslant K\|x\|$ for all $x \in X$, so T is bounded. The construction of T means that (T_n) converges strongly to T. $\square$

We note that the conclusions of Theorem 4.1 and of Corollary 4.3 crucially rely on the assumption that X is a Banach space (see also Exercise 4.6). (For Corollary 4.3 it is also crucial that we have restricted our attention to sequences.)

Exercise 4.4. Prove that uniform convergence implies strong convergence, and find an example of a sequence of bounded operators from a Banach space into a Banach space to show that strong convergence does not imply uniform convergence.

Exercise 4.5. Phrase the definition of a unitary representation of a metric group (Definition 3.73) using the notion of strong convergence for operators.

Exercise 4.6. Let $c_c(\mathbb{N}) \subseteq \ell^\infty(\mathbb{N})$ be the space of sequences with finite support equipped with the supremum norm. Define $T : c_c(\mathbb{N}) \to c_c(\mathbb{N})$ by

$$T(x_1, x_2, x_3, \dots) = (x_1, 2x_2, 3x_3, \dots)$$

for all $(x_1, x_2, x_2, \dots) \in c_c(\mathbb{N})$. Show that T is not bounded. Construct a sequence of bounded linear operators T_k on $c_c(\mathbb{N})$ with $T_k x \to Tx$ as $k \to \infty$ for all x in $c_c(\mathbb{N})$.

4.1.1 Uniform Boundedness and Fourier Series

This section gives an application of Theorem 4.1 to classical Fourier analysis on $\mathbb{T}$ (see Section 3.4 for the background). Recall that if $f \in C(\mathbb{T})$ then the Fourier coefficients of f are defined by $a_m = \langle f, \chi_m \rangle$ where $\chi_m(x) = e^{2\pi i m x}$, for $m \in \mathbb{Z}$ and $x \in \mathbb{T}$. The nth partial sum of the Fourier series is

$$s_n(x) = \sum_{m=-n}^{n} a_m e^{2\pi i m x}.$$

Recall that one of the basic goals of Fourier analysis is to clarify the relationship between the sequence of partial sums (s_n) and the function f. That is, to understand in what sense does the function s_n approximate f for large n (if it does at all). We now ask if the sequence of functions (s_n) converges uniformly or pointwise to f for $f \in C(\mathbb{T})$.

Recall from Definition 3.61 that the Dirichlet kernel D_n is defined by

$$D_n(x) = \sum_{k=-n}^{n} e^{2\pi i k x} = \frac{\sin((n+\frac{1}{2})2\pi x)}{\sin(\pi x)}$$

for $x \in \mathbb{T}$. By the discussion in Section 3.4.2 we have

$$s_n(0) = \int_{\mathbb{T}} f(x) D_n(-x)\,\mathrm{d}x = \int_{\mathbb{T}} f(x) D_n(x)\,\mathrm{d}x.$$

Lemma 4.7. *The linear functional $T_n : C(\mathbb{T}) \to \mathbb{R}$ defined by*

$$T_n f = \int_{\mathbb{T}} f(x) D_n(x)\,\mathrm{d}x$$

is bounded, with

$$\|T_n\| = \int_{\mathbb{T}} |D_n(x)|\,\mathrm{d}x.$$

This is a very special case of the general argument in Lemma 2.63, but we include it for the case at hand as this is easier to prove.

PROOF. For any function $f \in C(\mathbb{T})$ we have

$$|T_n f| \leqslant \int_{\mathbb{T}} |f(x)||D_n(x)|\,\mathrm{d}x \leqslant \|f\|_\infty \int_{\mathbb{T}} |D_n(x)|\,\mathrm{d}x,$$

so

$$\|T_n\| \leqslant \int_{\mathbb{T}} |D_n(x)|\,\mathrm{d}x.$$

Fix $\delta > 0$. Since D_n is analytic it can only have finitely many sign changes in $[0,1]$. Therefore, we may find a continuous (this could even be chosen to be piecewise-linear, for example) function f_n with $\|f_n\|_\infty \leqslant 1$ that differs from $\mathrm{sign}(D_n(x))$ only on a finite union of intervals whose total length is less than $\frac{1}{\|D_n\|_\infty}\delta$. The triangle inequality for integrals now gives

$$\left| \int_{\mathbb{T}} f_n(x) D_n(x)\,\mathrm{d}x \right| > \int_{\mathbb{T}} |D_n(x)|\,\mathrm{d}x - 2\delta,$$

which proves the lemma as $\delta > 0$ was arbitrary. $\qquad\square$

Lemma 4.8. *The Dirichlet kernel D_n from Definition 3.61 satisfies*

$$\int_{\mathbb{T}} |D_n(x)| \, dx = \int_{\mathbb{T}} \left| \frac{\sin((n + \frac{1}{2})2\pi x)}{\sin(\pi x)} \right| \, dx \longrightarrow \infty$$

as $n \to \infty$.

PROOF. Recall that $|\sin t| \leqslant |t|$ for all $t \in \mathbb{R}$. It follows that

$$\int_{\mathbb{T}} \left| \frac{\sin((n + \frac{1}{2})2\pi x)}{\sin(\pi x)} \right| \, dx \geqslant \int_0^1 \frac{1}{\pi x} |\sin((2n + 1)\pi x)| \, dx.$$

Now $|\sin t| \geqslant \frac{1}{2}$ for all $t \in \pi\mathbb{Z} + [\frac{\pi}{6}, \frac{5\pi}{6}]$. In particular, it follows that if

$$(2n + 1)\pi x \in \bigcup_{k=0}^{2n} [(k + \tfrac{1}{6})\pi, (k + \tfrac{5}{6})\pi]$$

then

$$|\sin((2n + 1)\pi x)| \geqslant \tfrac{1}{2}.$$

Together this gives

$$\int_{\mathbb{T}} \left| \frac{\sin((n + \frac{1}{2})2\pi x)}{\sin(\pi x)} \right| \, dx \geqslant \sum_{k=0}^{2n} \int_{(k + \frac{1}{6})\pi/(2n+1)}^{(k + \frac{5}{6})\pi/(2n+1)} \frac{1}{\pi(k + \frac{5}{6})/(2n + 1)} \frac{1}{2} \, dx$$

$$= \frac{1}{2\pi} \sum_{k=0}^{2n} \frac{2n+1}{k + \frac{5}{6}} \frac{\frac{4}{6}\pi}{2n+1} \longrightarrow \infty$$

as $n \to \infty$. $\qquad\qquad\qquad\qquad\qquad\qquad\qquad\qquad\qquad\qquad\qquad\qquad\qquad\qquad$ □

Theorem 4.9. *There exists a continuous function $f \in C(\mathbb{T})$ whose Fourier series diverges at $x = 0$.*

PROOF. As noted before Lemma 4.7, we have $T_n f = s_n(0)$ for all $f \in C(\mathbb{T})$. Moreover, for a fixed $f \in C(\mathbb{T})$, if the Fourier series of f converges at 0, then the family $\{T_n f \mid n \geqslant 1\}$ is bounded (since each element is just a partial sum of a convergent series). Thus if the Fourier series of f converges at 0 for all $f \in C(\mathbb{T})$, then for each $f \in C(\mathbb{T})$ the set $\{T_n f \mid n \geqslant 1\}$ is bounded. By Theorem 4.1, this implies that the set $\{\|T_n\| \mid n \geqslant 1\}$ is bounded, which contradicts Lemmas 4.7 and 4.8.

It follows that there must be some $f \in C(\mathbb{T})$ whose Fourier series does not converge at 0 (and in fact the partial sums must be unbounded). $\qquad\qquad$ □

In principle the proofs of Theorem 4.1 and Theorem 4.9 allow one to construct the function f as in Theorem 4.9 more concretely, at least as the limit of a Cauchy sequence of explicit continuous functions. Comparing Theorem 4.9 with the absolute convergence claim in Theorem 3.57 and the result regarding the Fejér kernel in Proposition 3.65, we see that this limit function is not continuously differentiable and that the Fourier series of f at 0 is an oscillating function with the property that the Césaro averages of the diverging sequence $(s_n(0))$ actually converge to $f(0)$.

4.2 The Open Mapping and Closed Graph Theorems

Recall that a continuous map has the property that the pre-image of any open set is open, but in general the image of an open set is not open. We now show that bounded linear maps between Banach spaces on the other hand have the following special property.

Theorem 4.10 (Open mapping theorem). *Let X and Y be Banach spaces, and let T be a bounded linear map from X onto Y. Then T maps open sets in X onto open sets in Y.*

The assumption that X maps *onto* Y is essential. Consider, for example, the projection $(x, y) \mapsto (x, 0)$ from $\mathbb{R}^2 \to \mathbb{R}^2$ to see this.

The proof of Theorem 4.10 uses the Baire category theorem,[13] which states that a complete non-empty metric space cannot be written as a countable union of nowhere dense subsets.

4.2.1 Baire Category

Definition 4.11. A subset $S \subseteq X$ of a metric space (X, d) is said to be *nowhere dense* if for every point $x \in \overline{S}$, and for every $\varepsilon > 0$, $B_\varepsilon(x) \cap (X \backslash \overline{S})$ is non-empty (equivalently, if $(\overline{S})^o = \varnothing$). A set is called *meagre* or *first category* if it is a countable union of nowhere dense sets.

We will think of a nowhere dense set as being small and want to extend this to meagre sets. The next result is needed as a justification of that interpretation.

Theorem 4.12 (Baire category theorem). *A complete non-empty metric space cannot be written as a countable union of nowhere dense sets. Indeed, the complement of a countable union of nowhere dense sets is dense.*

This is often described by saying that a complete metric space is of *second category*. As we will see, the method of proof is similar to the proof of Theorem 4.1 (see also Exercise 4.16).

We note that in a normed linear space the closed unit ball $\overline{B}_r(x)$ coincides with the closure $\overline{B_r(x)}$ of the open ball. However, in a metric space they may be entirely different: if any set X is given the discrete metric defined by $\mathsf{d}(x, y) = 1$ if $x \neq y$ and 0 if $x = y$, then $\overline{B}_1(x) = X$ while $\overline{B_1(x)} = \{x\}$ for any $x \in X$.

PROOF OF THEOREM 4.12. Let X be a complete non-empty metric space, and suppose that (X_j) is a sequence of nowhere dense subsets of X (that is, the sets $\overline{X_j}$ all have empty interior for $j = 1, 2, \ldots$). Fix an arbitrary ball $B_\varepsilon(x_0)$ with $\varepsilon > 0$ and $x_0 \in X$. Since $\overline{X_1}$ does not contain $B_\varepsilon(x_0)$, there must be a point x_1 in $B_\varepsilon(x_0)$ with $x_1 \notin \overline{X_1}$. It follows that there is some $r_1 > 0$ such that the closed ball $\overline{B}_{r_1}(x_1) = \{y \in X \mid d(y, x_1) \leqslant r_1\}$ satisfies $\overline{B}_{r_1}(x_1) \subseteq B_\varepsilon(x_0)$ and $\overline{B}_{r_1}(x_1) \cap \overline{X_1} = \varnothing$. Assume without loss of generality that $r_1 < 1$.

Similarly, there is some $x_2 \in X$ and $r_2 > 0$ such that $\overline{B}_{r_2}(x_2) \subseteq B_{r_1}(x_1)$, and $\overline{B}_{r_2}(x_2) \cap \overline{X_2} = \varnothing$, and without loss of generality $r_2 < \frac{1}{2}$. Notice that $\overline{B}_{r_2}(x_2) \cap \overline{X_1} = \varnothing$ since $\overline{B}_{r_2}(x_2) \subseteq B_{r_1}(x_1)$.

Inductively, we construct a sequence of decreasing closed balls $\overline{B}_{r_n}(x_n)$ such that $\overline{B}_{r_n}(x_n) \cap \overline{X_j} = \varnothing$ for $1 \leqslant j \leqslant n$, and $r_n \to 0$ as $n \to \infty$. It follows that (x_n) is a Cauchy sequence, and the limit z lies in the intersection of all the closed balls $\overline{B}_{r_n}(x_n)$, so $z \notin \overline{X_j}$ for all $j \geqslant 1$. This implies that

$$z \in B_\varepsilon(x_0) \smallsetminus \bigcup_{j \geqslant 1} \overline{X_j} \neq \varnothing,$$

which gives the result since $\varepsilon > 0$ and $x_0 \in X$ were arbitrary.   □

Exercise 4.13. Prove the Baire category theorem for compact topological spaces (that is, without the assumption that the space is metric).

By taking complements we can also phrase the Baire category theorem in terms of G_δ-sets.

Definition 4.14. A countable intersection of open sets in a topological space is called a G_δ-set.

Corollary 4.15 (Baire category theorem). *Let (X, d) be a complete metric space, and assume that $G_n \subseteq X$ is a dense G_δ-set for each $n \geqslant 1$. Then the intersection $\bigcap_{n=1}^\infty G_n$ is also a dense G_δ-set.*

PROOF. By assumption we can write each G_n in the form $G_n = \bigcap_{k=1}^\infty O_{n,k}$, where each $O_{n,k}$ is open and dense. It follows that

$$\bigcap_{n=1}^\infty G_n = \bigcap_{n=1}^\infty \bigcap_{k=1}^\infty O_{n,k}$$

is a G_δ-set, and that it is sufficient to consider the case where each $G_n = O_n$ is open and dense. In that case, $X_n = X \smallsetminus G_n$ is closed and so

$$B_\varepsilon(x) \cap (X \smallsetminus \overline{X_n}) = B_\varepsilon(x) \cap (X \smallsetminus X_n) = B_\varepsilon(x) \cap G_n \neq \varnothing$$

for any open ball $B_\varepsilon(x)$ since G_n is dense. Therefore, X_n is nowhere dense for each $n \geqslant 1$. By Theorem 4.12 the complement of $\bigcup_{n=1}^\infty X_n$ is dense, and this is precisely $\bigcap_{n=1}^\infty G_n$, by construction.   □

Exercise 4.16. Prove the Banach–Steinhaus theorem (Theorem 4.1) using the Baire category theorem (Theorem 4.12).

Let us mention that the notion of a dense G_δ-set is the topological version of being a 'large' set, while a set is measure-theoretically 'large' if its complement is a null set. Both notions of being large share similar features, and in particular a countable intersection of large sets in either sense is also large.[14] However, these two notions are quite different. Example 4.17 shows how to construct topologically large sets that are measure-theoretically small, and *vice-versa*.

Example 4.17. For every $\varepsilon > 0$ there exists an open set $O_\varepsilon \subseteq \mathbb{R}$ which contains $\mathbb{Q}$ and has Lebesgue measure less than ε. This may be found, for example, by listing the elements of $\mathbb{Q}$ as $\{x_1, x_2, \dots\}$ and setting

$$O_\varepsilon = \bigcup_{k \geq 1} B_{\varepsilon/2^{k+2}}(x_k).$$

Then $G = \bigcap_{n \geq 1} O_{1/n}$ is a dense G_δ and a null set, and its complement $\mathbb{R} \setminus G$ is meagre and of full measure.

The Baire category theorem can be used to show the existence of elements of a complete metric space with certain properties. If the set of elements of a complete space which do not satisfy the property can be obtained as a countable union of nowhere dense sets, there must be elements that satisfy the property (indeed, there exists a dense set of such elements).

Exercise 4.18. (a) Assume that X and Y are metric spaces. Show that for any $f : X \to Y$ the set $\{x \in X \mid f \text{ is continuous at } x\}$ is a G_δ-set.
(b) Show that the map $f : \mathbb{R} \to \mathbb{R}$ defined by

$$f(x) = \begin{cases} \frac{1}{q} & \text{if } x = \frac{p}{q} \in \mathbb{Q}; \\ 0 & \text{if } x \in \mathbb{R} \setminus \mathbb{Q} \end{cases}$$

is continuous at each irrational point and is not continuous at each rational, where we assume that $\frac{p}{q}$ is written in lowest terms and has $q \geq 1$.
(c) Use (a) to show that no function could have the reverse properties of the function in (b).

Exercise 4.19. Show that $\{f \in L^1((0,1)) \mid \||f|_{(a,b)}\|_\infty = \infty \text{ whenever } 0 \leq a < b \leq 1\}$ contains a dense G_δ-subset of $L^1((0,1))$.

Exercise 4.20. Show that the set of functions in $C([0,1])$ that are nowhere differentiable contains a dense G_δ-set.

4.2.2 Proof of the Open Mapping Theorem

Recall that we write B_r^X and B_r^Y for the open balls of radius r and centre 0 in X and Y, respectively.

Lemma 4.21. *Assume that X is a normed vector space, Y is a Banach space, and $T : X \to Y$ is a bounded, surjective linear operator. For any $\varepsilon > 0$, there is a $\delta > 0$ such that*

$$\overline{TB_\varepsilon^X} \supseteq B_\delta^Y. \qquad (4.3)$$

PROOF. Since

$$X = \bigcup_{n=1}^{\infty} nB_\varepsilon^X,$$

and T is onto, we have $Y = T(X) = \bigcup_{n=1}^{\infty} nTB_\varepsilon^X$. By the Baire category theorem (Theorem 4.12) applied to Y it follows that, for some n, the set $n\overline{TB_\varepsilon^X}$ contains some ball $B_r^Y(z)$ in Y. Then, by linearity, $\overline{TB_\varepsilon^X}$ must contain the ball $B_\delta^Y(y)$, where $y = \frac{1}{n}z$ and $\delta = \frac{1}{n}r$. We note that

$$B_\delta^Y(y) - B_\delta^Y(y) = \{y_1 - y_2 \mid y_1, y_2 \in B_\delta^Y(y)\} = B_{2\delta}^Y$$

and similarly $B_{2\varepsilon}^X = B_\varepsilon^X - B_\varepsilon^X$. Therefore,

$$B_{2\delta}^Y \subseteq \overline{TB_\varepsilon^X} - \overline{TB_\varepsilon^X} \subseteq \overline{TB_{2\varepsilon}^X}$$

and (4.3) follows. $\qquad \square$

The above lemma only used the fact that Y is a Banach space. Using the hypothesis that X is also a Banach space, we are able to prove the main step towards the theorem in the following lemma.

Lemma 4.22. *Let $T : X \to Y$ be as in Theorem 4.10. For any $\varepsilon > 0$ there is a $\delta > 0$ such that*

$$TB_\varepsilon^X \supseteq B_\delta^Y. \qquad (4.4)$$

PROOF. Choose a sequence (ε_n) with each $\varepsilon_n > 0$ and with $\sum_{n=1}^{\infty} \varepsilon_n < \varepsilon$. By Lemma 4.21 there is a sequence (δ_n) of positive numbers such that

$$\overline{TB_{\varepsilon_n}^X} \supseteq B_{\delta_n}^Y \qquad (4.5)$$

for all $n \geqslant 1$. Without loss of generality, assume that $\delta_n \to 0$ as $n \to \infty$. (Actually this holds unless Y is very special indeed.) Now let $\delta = \delta_1$.

Let y be any point in $B_\delta^Y = B_{\delta_1}^Y$. By (4.5) there is a point $x_1 \in B_{\varepsilon_1}^X$ such that Tx_1 is as close to y as we wish, say with $\|y - Tx_1\| < \delta_2$. Since

$$y - Tx_1 \in B_{\delta_2}^Y,$$

the inclusion (4.5) with $n = 2$ implies that there exists a point $x_2 \in B_{\varepsilon_2}^X$ such that $\|y - Tx_1 - Tx_2\| < \delta_3$. Continuing, we obtain a sequence (x_n) in X such that $\|x_n\| < \varepsilon_n$ for all n, and

$$\left\| y - T\left(\sum_{k=1}^{n} x_k \right) \right\| < \delta_{n+1}. \qquad (4.6)$$

This argument may be paraphrased as follows. At each stage we approximate the current element $y - T\left(\sum_{k=1}^{n-1} x_k\right)$ in Y up to an error δ_{n+1} that we know can be dealt with later. This pushes the problem along until it ultimately vanishes in the limit.

Since $\|x_n\| < \varepsilon_n$, the series $\sum_n x_n$ is absolutely convergent, so by Lemma 2.28 it is convergent; write $x = \sum_n x_n$. Then

$$\|x\| \leqslant \sum_{n=1}^{\infty} \|x_n\| \leqslant \sum_{n=1}^{\infty} \varepsilon_n < \varepsilon.$$

The map T is continuous, so (4.6) shows that $y = Tx$, since $\delta_n \to 0$ as $n \to \infty$.

That is, for any $y \in B_\delta^Y$ we have found a point $x \in B_\varepsilon^X$ such that $Tx = y$, proving (4.4).   □

PROOF OF THEOREM 4.10. Let $O \subseteq X$ be a non-empty open subset and let x be an element of O. Then there is a ball B_ε^X such that

$$x + B_\varepsilon^X \subseteq O.$$

By Lemma 4.22, $TB_\varepsilon^X \supseteq B_\delta^Y$ for some $\delta > 0$. Hence

$$T(O) \supseteq T(x + B_\varepsilon^X) = Tx + T(B_\varepsilon^X) \supseteq Tx + B_\delta^Y.$$

To summarize, we have shown that for every $x \in O$ the point Tx is in the interior of $T(O)$.   □

Exercise 4.23. Let X and Y be Banach spaces and $T : X \to Y$ a bounded operator. Show that the following three conditions are equivalent:

(a) T is surjective;
(b) T is open;
(c) there exists a dense subspace $Y' \subseteq Y$ and a constant $c > 0$ such that for any $y \in Y'$ there exists some $x \in X$ with $Tx = y$ and $\|x\|_X \leqslant c\|y\|_Y$.

4.2.3 Consequences: Bounded Inverses and Closed Graphs

As an application of Theorem 4.10, we establish a general property of inverse maps. As is standard, $T : X \to Y$ means that T is defined on all of X, but sometimes it is convenient (or necessary) to permit an operator T to only be defined on a domain D_T which is then a (possibly proper) subspace of X.

Definition 4.24. Let $T : X \to Y$ be an injective linear operator. Define the *inverse* of T, denoted T^{-1}, by requiring that $T^{-1}y = x$ if and only if $Tx = y$. Then the domain of T^{-1} is the linear subspace $TX \subseteq Y$, and T^{-1} is a linear operator on its domain.

Clearly $T^{-1}Tx = x$ for all $x \in X$, and $TT^{-1}y = y$ for all y in the domain of T^{-1}. We also say that T^{-1} is a *left inverse* of T.

Proposition 4.25 (Bounded inverse). *Let X and Y be Banach spaces, and let T be a bijective bounded operator from X to Y. Then T^{-1} is also a bounded operator.*

PROOF. Since T^{-1} is a linear operator, we only need to show it is continuous (which is equivalent to boundedness by Lemma 2.52). By Theorem 4.10, T maps open sets onto open sets. For the map T^{-1} this shows that the pre-image $(T^{-1})^{-1}(O) = T(O)$ of an open set $O \subseteq X$ is open in Y. Therefore, T^{-1} is continuous.   □

Corollary 4.26 (Equivalent norms). *If X is a Banach space with respect to two norms $\| \cdot \|^{(1)}$ and $\| \cdot \|^{(2)}$ and there is a constant K such that*

$$\|x\|^{(1)} \leqslant K\|x\|^{(2)},$$

for all $x \in X$, then the two norms are equivalent. That is, there is another constant $K' > 0$ with

$$\|x\|^{(2)} \leqslant K'\|x\|^{(1)}$$

for all $x \in X$.

PROOF. Consider the map $T : x \mapsto x$ from $(X, \| \cdot \|^{(2)})$ to $(X, \| \cdot \|^{(1)})$. By assumption, T is bounded, so by Proposition 4.25, T^{-1} is also bounded, giving the bound in the other direction.   □

Definition 4.27. Let T be a linear operator from a normed linear space X into a normed linear space Y, with domain a linear subspace $D_T \subseteq X$. The *graph* of T is the set

$$G_T = \{(x, Tx) \mid x \in D_T\} \subseteq X \times Y.$$

If G_T is a closed subspace of $X \times Y$ then T is a *closed operator.*

Notice as usual that this notion becomes trivial in finite dimensions in the following sense. If X and Y are finite-dimensional, then the graph of T is simply some linear subspace, which is automatically closed. Also it is easy to see that a continuous operator has a closed graph. The next theorem — the converse — is called the *closed graph theorem*. Notice that this converse is not a purely topological fact. For instance, the set consisting of the graph of the hyperbola $xy = 1$ and the origin is the closed graph of a discontinuous function $f : \mathbb{R} \to \mathbb{R}$.

Theorem 4.28 (Closed graph theorem). *Let X and Y be Banach spaces, and $T : X \to Y$ a linear operator with $D_T = X$. If T is closed, then it is continuous.*

PROOF. Fix the norm $\|(x, y)\| = \|x\|_X + \|y\|_Y$ on $X \times Y$. The graph G_T is, by hypothesis, a closed subspace of $X \times Y$, so G_T is itself a Banach space. Consider the projection $P : G_T \to X$ defined by $P(x, Tx) = x$. Then P is clearly bounded, linear, and bijective. It follows by Proposition 4.25 that P^{-1} is a bounded linear operator from X to G_T, so

$$\|(x, Tx)\| = \|P^{-1}x\| \leqslant K\|x\|_X$$

for all $x \in X$, for some constant K. It follows that $\|x\|_X + \|Tx\|_Y \leqslant K\|x\|_X$ for all $x \in X$, so T is bounded, and hence continuous by Lemma 2.52. $\qquad\square$

Exercise 4.29 (Hellinger–Toeplitz theorem). Suppose that $A : \mathcal{H} \to \mathcal{H}$ is a linear operator on a Hilbert space $\mathcal{H}$ that is self-adjoint in the sense that

$$\langle Ax, y \rangle = \langle x, Ay \rangle$$

for all $x, y \in \mathcal{H}$. Show that this implies A is bounded.

Corollary 4.30. *Let (X, μ) be a σ-finite measure space and $g : X \to \mathbb{C}$ a measurable function. If $T : f \mapsto gf$ maps $L^2_\mu(X)$ to $L^2_\mu(X)$, then $g \in L^\infty_\mu(X)$ and $\|T\| = \|g\|_\infty$.*

PROOF. Notice that the hypotheses in the statement do not require that the map is continuous, but simply ask that the range lies in $L^2_\mu(X)$. However, if (f_n, gf_n) has $f_n \to f$ and $gf_n \to \psi$ as $n \to \infty$ in $L^2_\mu(X)$, (that is, a sequence in the graph that converges to (f, ψ)), then we can extract a subsequence along which both convergences hold μ-almost everywhere. Along this subsequence gf_n converges almost everywhere to gf and to ψ, so that

$$gf = \psi \in L^2_\mu(X),$$

and hence (f, ψ) also lies in the graph of T. It follows that T is closed, and hence continuous by Theorem 4.28.

Knowing now that T is bounded, there is a constant $C = \|T\| \geqslant 0$ such that $\|gf\|_2 \leqslant C\|f\|_2$ for any $f \in L^2_\mu(X)$. Let

$$X_C = \{x \in X \mid |g(x)| > C\},$$

which we claim is a null set. Assuming the opposite, let $B \subseteq X_C$ be a measurable subset of positive finite measure and let $f = \mathbb{1}_B$ be its characteristic function. Then

$$C^2 \mu(B) < \int_B |g|^2 |f|^2 \, \mathrm{d}\mu = \|gf\|_2^2 \leqslant C^2 \|f\|_2^2 = C^2 \mu(B)$$

gives a contradiction, which implies that $\mu(X_C) = 0$. Hence $|g|$ is almost everywhere less than or equal to C, and in particular $g \in L^\infty_\mu(X)$. Moreover, $\|g\|_\infty \leqslant \|T\|$ and the opposite inequality follows directly from the definition. $\qquad\square$

Some important and natural operators are unbounded. An example is the derivative operator $D_0 : f \mapsto f'$ considered on $L^2((0,1))$, which is originally defined on the dense subset $\{f \in C^1((0,1)) \mid f, f' \in L^2((0,1))\}$. It is not continuous, but as we will see in Chapter 5, considering the closure of its graph gives a closed (but still unbounded) *weak differential operator* defined on a dense subspace of L^2. In particular, we see that closed operators provide a suitable framework for studying unbounded operators.

The closed graph theorem says that these generalized operators, namely closed operators, are usually only defined on a proper subset of the first Banach space unless they actually are bounded.

4.3 Further Topics

The above consequences of completeness are useful in several different areas. We indicate below a few applications and extensions of these results.

- As already mentioned, in the next two chapters we will study partial differential operators as examples of closed operators. We can do this without developing any theory on unbounded operators, by simply studying the graphs of these operators.
- As we have seen the Banach–Steinhaus theorem (Theorem 4.1) has interesting consequences for the notion of strong convergence (Corollary 4.3). We will discuss the corresponding topology again in Section 8.3.
- We will encounter multiplication operators as in Corollary 4.30 again in Chapter 9, 12, and 13
- The notion of closed operators is the starting point of the theory of *self-adjoint unbounded* operators, which we will study systematically in Chapter 13.

The reader may continue with Chapter 5, 6, or 7 (with some of the material of Chapter 6 building on Chapter 5).

Chapter 5
Sobolev Spaces and Dirichlet's Boundary Problem

Using the theory of Fourier series developed in Section 3.4, we will now develop the notion of Sobolev spaces and prove the Sobolev embedding theorem. Sobolev spaces combine familiar notions of smoothness (that is, differentiability) with bounds on L^p norms. We will set $p = 2$ and so will have all the tools of Hilbert spaces at our disposal, but the theory can be extended to all $p \geqslant 1$. The Sobolev embedding theorem and elliptic regularity for the Laplace operator will allow us to prove in Section 5.3 the existence of solutions to the Dirichlet boundary value problem introduced in Section 1.2.

5.1 Sobolev Spaces and Embedding on the Torus

In this section we are going to construct Hilbert spaces of functions on $\mathbb{T}^d$ depending on a parameter $k \in \mathbb{N}_0$. Unlike the equivalence classes $f \in L^2(\mathbb{T}^d)$, these functions may be continuous or even differentiable, depending on k (see Section 5.1.2).

5.1.1 L^2 Sobolev Spaces on $\mathbb{T}^d$

Definition 5.1. Let $k \geqslant 0$ be an integer. We (initially) define the (L^2) *Sobolev space* $H^k(\mathbb{T}^d)$ to be the closure of $C^\infty(\mathbb{T}^d)$ inside

$$V = \bigoplus_{\|\alpha\|_1 \leqslant k} L^2(\mathbb{T}^d),$$

where the direct sum runs over all multi-indices $\alpha \in \mathbb{N}_0^d$ with $\|\alpha\|_1 \leqslant k$ and a function $f \in C^\infty(\mathbb{T}^d)$ is identified with the tuple $\phi_k(f) = (\partial_\alpha f)_{\|\alpha\|_1 \leqslant k} \in V$.

In order to make this definition more palatable, we now describe some special cases.

© Springer International Publishing AG 2017
M. Einsiedler, T. Ward, *Functional Analysis, Spectral Theory, and Applications*,
Graduate Texts in Mathematics 276, DOI 10.1007/978-3-319-58540-6_5

(1) If $k = 0$, then there is only the multi-index $\alpha = 0$ and so $H^0(\mathbb{T}^d)$ is the closure of $C^\infty(\mathbb{T}^d)$ in $L^2(\mathbb{T}^d)$ with respect to $\| \cdot \|_2$. As we have seen, $C^\infty(\mathbb{T}^d)$ is dense in $C(\mathbb{T}^d)$ with respect to $\| \cdot \|_\infty$ (indeed, the trigonometric polynomials are already dense) and $C(\mathbb{T}^d)$ is dense in $L^2(\mathbb{T}^d)$ with respect to $\| \cdot \|_2$, so we obtain $H^0(\mathbb{T}^d) = L^2(\mathbb{T}^d)$. We will also write $\|f\|_{H^0} = \|f\|_2$ for $f \in H^0(\mathbb{T}^d)$.

(2) Now let $k = 1$, and in this case there are $d + 1$ multi-indices. Hence

$$H^1(\mathbb{T}^d) = \overline{\phi_1(C^\infty(\mathbb{T}^d))}$$

is the closure of $\phi_1(C^\infty(\mathbb{T}^d))$ in $\left(L^2(\mathbb{T}^d)\right)^{d+1}$, where we used the embedding $\phi_1 : f \mapsto (f, \partial_1 f, \ldots, \partial_d f) \in V$. So, by our definition, elements of $H^1(\mathbb{T}^d)$ are $(d + 1)$-tuples of functions on $\mathbb{T}^d$. In order to be able to think of these as single functions on $\mathbb{T}^d$ (which is how we will think of Sobolev spaces), notice that the last d terms of the $(d+1)$-tuple are uniquely determined by the first term. This is clear for $\phi_1(f)$ with $f \in C^\infty(\mathbb{T}^d)$, but also remains true in the closure $H^1(\mathbb{T}^d)$, as we show next.

Lemma 5.2 (Fourier series of weak derivatives). *Suppose that the vector $(f, f_1, \ldots, f_d)$ belongs to $H^1(\mathbb{T}^d)$ and the Fourier series of f is given by*

$$f = \sum_{n \in \mathbb{Z}^d} c_n \chi_n.$$

Then

$$f_j = \sum_{n \in \mathbb{Z}^d} 2\pi i n_j c_n \chi_n. \tag{5.1}$$

PROOF. For $f \in L^2(\mathbb{T}^d)$ and $n \in \mathbb{Z}^d$, write $a_n(f)$ for the nth Fourier coefficient. We start with the formula

$$a_n\left(\partial_j f\right) = \langle \partial_j f, \chi_n \rangle = 2\pi i n_j \langle f, \chi_n \rangle = 2\pi i n_j a_n(f)$$

for all $n \in \mathbb{Z}^d$ and all $f \in C^\infty(\mathbb{T}^d)$, see (3.18). Using continuity of the inner product and the definition of $H^1(\mathbb{T}^d)$, this formula automatically extends to all $(f, f_1, \ldots, f_d) \in H^1(\mathbb{T}^d)$. Expanding f_j into its Fourier series (see Theorem 3.54) gives the lemma.   □

The lemma now shows in full generality that the first component f of any element $(f, f_1, \ldots, f_d) \in H^1(\mathbb{T}^d)$ determines all the other components. Thus we can identify an element of $H^1(\mathbb{T}^d)$ with the associated element $f \in L^2(\mathbb{T}^d)$, and will write $f \in H^1(\mathbb{T}^d)$ and $\partial_j f = f_j \in L^2(\mathbb{T}^d)$ for $j = 1, \ldots, d$. We will also call the other components $\partial_j f$ *weak derivatives* (this will be further justified in Section 5.2), as these generalize the notion of partial derivative for smooth functions. However, the norm associated to $f \in H^1(\mathbb{T}^d)$ is

$$\|f\|_{H^1} = \sqrt{\|f\|_2^2 + \sum_{j=1}^{d} \|\partial_j f\|_2^2}.$$

Summarizing the above discussion for the case $k = 1$, we may interpret $H^1(\mathbb{T}^d)$ as the domain (and, in the original definition, as the graph) of the closed operator $\nabla : H^1(\mathbb{T}^d) \ni f \mapsto (\partial_{x_1} f, \dots, \partial_{x_d} f) \in (L^2(\mathbb{T}^d))^d$. This discussion generalizes to any $k \geqslant 1$ as follows.

Proposition 5.3 (Forgetting regularity, weak derivative). *Fix k and ℓ with $0 \leqslant \ell < k$. Then the identity map on $C^\infty(\mathbb{T}^d)$ uniquely extends to an injective continuous operator $\imath_{k,\ell} : H^k(\mathbb{T}^d) \longrightarrow H^\ell(\mathbb{T}^d)$ of norm one. Moreover, for every $j \in \{1, \dots, d\}$ the partial derivative extends uniquely to a continuous operator*

$$\partial_j : H^k(\mathbb{T}^d) \longrightarrow H^{k-1}(\mathbb{T}^d)$$

with norm less than or equal to one. Finally, (3.18) holds similarly for all f in $H^k(\mathbb{T}^d)$ and for all $\alpha \in \mathbb{N}_0^d \backslash \{(0, \dots, 0)\}$ with $\|\alpha\|_1 \leqslant k$.

PROOF. For the first claim consider the map

$$\pi_{k,\ell} : \bigoplus_{\|\alpha\|_1 \leqslant k} L^2(\mathbb{T}^d) \longrightarrow \bigoplus_{\|\alpha\|_1 \leqslant \ell} L^2(\mathbb{T}^d)$$

$$(f_\alpha)_{\|\alpha\|_1 \leqslant k} \longmapsto (f_\alpha)_{\|\alpha\|_1 \leqslant \ell}$$

and notice that $\pi_{k,\ell}(\phi_k(f)) = \phi_\ell(f)$ for all $f \in C^\infty(\mathbb{T}^d)$. Therefore the extended map $\imath_{k,\ell}$ is simply the restriction of this projection to $H^k(\mathbb{T}^d)$, and so has norm less than or equal to one. Using constant functions we see that the norm of $\imath_{k,\ell}$ is equal to one. Injectivity will follow from the last claim of the proposition.

For the second claim, regarding the operator $\partial_j : H^k(\mathbb{T}^d) \to H^{k-1}(\mathbb{T}^d)$, we modify the argument above as follows. Consider the projection map

$$\pi_j : \bigoplus_{\|\alpha\|_1 \leqslant k} L^2(\mathbb{T}^d) \longrightarrow \bigoplus_{\|\alpha\|_1 \leqslant k-1} L^2(\mathbb{T}^d)$$

$$(f_\alpha)_{\|\alpha\|_1 \leqslant k} \longmapsto (f_{\alpha+e_j})_{\|\alpha\|_1 \leqslant k-1}$$

which clearly has norm one. Figure 5.1 illustrates the difference between the projection $\pi_{k,\ell}$ and the projection π_j in a simple example. For $f \in C^\infty(\mathbb{T}^d)$ we see that $\pi_j(\phi_k(f)) = \phi_{k-1}(\partial_{e_j}(f))$, which (as above) shows that the restriction of π_j to $H^k(\mathbb{T}^d)$ is the desired operator $\partial_j : H^k(\mathbb{T}^d) \to H^{k-1}(\mathbb{T}^d)$.

The final claim of the proposition follows from the description of the Fourier series of the weak derivative in Lemma 5.2 for $k = 1$ and induction. $\square$

Now justified by Proposition 5.3, we identify an element $f = (f_\alpha)_{\|\alpha\|_1 \leqslant k}$ in $H^k(\mathbb{T}^d)$ with its first component f_0 in $L^2(\mathbb{T}^d)$. The other components are

Fig. 5.1: With $d = 2$, $\ell = 2$, $k = 3$ and $j = 1$ the projection defining $\imath_{k,\ell}$ corresponds to the set of multi-index α highlighted on the left. The projection defining ∂_j corresponds to the set highlighted on the right.

identified with the weak derivative, so we write

$$f = f_0 = \partial_0 f \in H^k(\mathbb{T}^d) \subseteq L^2(\mathbb{T}^d), \partial_\alpha f = f_\alpha \in L^2(\mathbb{T}^d)$$

for all multi-indices α with $\|\alpha\|_1 \leqslant k$. In this notation our norm becomes

$$\|f\|_{H^k} = \sqrt{\sum_{\|\alpha\|_1 \leqslant k} \|\partial_\alpha f\|_2^2}.$$

5.1.2 The Sobolev Embedding Theorem on $\mathbb{T}^d$

As we have seen in the discussion above, each of the spaces $H^k(\mathbb{T}^d)$ consists of certain L^2 functions on $\mathbb{T}^d$. For $k = 0$ we have $H^0(\mathbb{T}^d) = L^2(\mathbb{T}^d)$. A natural question for $k \geqslant 1$ is to ask which functions in $L^2(\mathbb{T}^d)$ lie in $H^k(\mathbb{T}^d)$. Using Fourier series we can give a formal answer to this, and this will have interesting and important consequences which will be discussed below. Another consequence of this lemma is that it makes it meaningful to define H^k for $k \in \mathbb{R}$ by using the convergence property in the lemma as a definition — we will not pursue this further.

Lemma 5.4 (Characterizing $H^k(\mathbb{T}^d)$ by the Fourier series). *Let $k \geqslant 0$ be an integer and let $f = \sum_{n \in \mathbb{Z}^d} c_n \chi_n \in L^2(\mathbb{T}^d)$. Then $f \in H^k(\mathbb{T}^d)$ if and only if*

$$\sum_{n \in \mathbb{Z}^d} |c_n|^2 \|n\|_2^{2k} < \infty. \tag{5.2}$$

PROOF. For $n = (n_1, \ldots, n_d) \in \mathbb{Z}^d$ and $\alpha = (\alpha_1, \ldots, \alpha_d) \in \mathbb{N}_0^d$ we write n^α for $(n_1^{\alpha_1}, \ldots, n_d^{\alpha_d})$. If $f = \sum_{n \in \mathbb{Z}^d} c_n \chi_n \in H^k(\mathbb{T}^d)$ then, by Proposition 5.3 and by Fourier series on the torus (Theorem 3.54),

$$(n^\alpha c_n)_{n \in \mathbb{Z}^d} \in \ell^2(\mathbb{Z}^d)$$

for all α with $\|\alpha\|_1 \leqslant k$. We apply this to $\alpha = ke_1, ke_2, \ldots, ke_d$ and see that

$$\sum_{j=1}^{d} \sum_{n \in \mathbb{Z}^d} n_j^{2k} |c_n|^2 < \infty.$$

Using the bound $\|n\|_2^{2k} \ll \sum_{j=1}^{d} n_j^{2k}$ for all $n \in \mathbb{Z}^d$, we get (5.2) as required.

Conversely, assume (5.2). Then for any $\alpha \in \mathbb{N}_0^d \setminus \{(0, \ldots, 0)\}$ with $\|\alpha\|_1 \leqslant k$ we have $|n^\alpha| \leqslant \|n\|_2^k$, and so

$$\sum_{n \in \mathbb{Z}^d} |(2\pi i n)^\alpha c_n|^2 < \infty.$$

The characterization of convergence for a series involving an orthonormal basis (Proposition 3.36) shows that every component of $\phi_k \left(\sum_{\|n\|_2 \leqslant N} c_n \chi_n \right)$ converges as $N \to \infty$, and so we deduce that the images of these partial sums converge in $H^k(\mathbb{T}^d)$. As the first component of the limit vector equals f, it follows that $f \in H^k(\mathbb{T}^d)$. $\qquad \square$

Exercise 5.5. Show that $H^1(\mathbb{T}^d)$ is a meagre subset of $L^2(\mathbb{T}^d)$.

The following theorem shows (in a more constructive manner than the previous exercise) how special the elements of the subset $H^k(\mathbb{T}^d)$ within $L^2(\mathbb{T}^d)$ become once k is sufficiently large. If $k > \frac{d}{2}$, then any element of $H^k(\mathbb{T}^d)$ agrees almost surely (and will be identified) with a continuous function. Increasing k further also gives some differentiability of this continuous function.

Theorem 5.6 (Sobolev embedding on the torus). *Let k and ℓ be non-negative integers with $k > \ell + \frac{d}{2}$. Then the inclusion map from $C^\infty(\mathbb{T}^d)$ to $C^\ell(\mathbb{T}^d)$ has a continuous extension to $H^k(\mathbb{T}^d)$. In particular, any function $f \in H^k(\mathbb{T}^d)$ has a uniquely defined continuous representative belonging to $C^\ell(\mathbb{T}^d)$ with $\|f\|_{C^\ell} \ll_d \|f\|_{H^k}$, where we also denote the continuous representative by f.*

The proof will show that most of the work has already been done.

PROOF OF THEOREM 5.6. Let us start with the case $\ell = 0$. In this case we already know that

$$\|f\|_\infty \ll_d \sqrt{\|f\|_2^2 + \|\partial_{e_1}^k f\|_2^2 + \cdots + \|\partial_{e_d}^k f\|_2^2}$$

for $f \in C^\infty(\mathbb{T}^d)$ by Theorem 3.57. However, the square root on the right-hand side is bounded above by $\|f\|_{H^k}$, which shows that the inclusion map

$$\imath : \left(C^\infty(\mathbb{T}^d), \| \cdot \|_{H^k} \right) \longrightarrow \left(C(\mathbb{T}^d), \| \cdot \|_\infty \right)$$

is a bounded operator. Since $C(\mathbb{T}^d)$ is a Banach space, this operator extends to $H^k(\mathbb{T}^d)$ by Proposition 2.59. We still need to argue that this extension really does select a continuous representative. For this, notice that the composition of the inclusion maps

$$C^\infty(\mathbb{T}^d) \xrightarrow{\imath} C(\mathbb{T}^d) \longrightarrow L^2(\mathbb{T}^d)$$

is the inclusion map $\phi_0 : C^\infty(\mathbb{T}^d) \to L^2(\mathbb{T}^d)$, whose unique extension to $H^k(\mathbb{T}^d)$ is $\imath_{k,0}$ (see Proposition 5.3). Hence the composition of the constructed extension $\imath : H^k(\mathbb{T}^d) \to C(\mathbb{T}^d)$ with the inclusion into $L^2(\mathbb{T}^d)$ coincides with $\imath_{k,0}$ and so $\imath(f) \in C(\mathbb{T}^d)$ is a representative of the equivalence class $\imath_{k,0}(f) \in L^2(\mathbb{T}^d)$ associated to $H^k(\mathbb{T}^d)$.

Now let $\ell \geqslant 1$ satisfy $\ell + \frac{d}{2} < k$. Note (for example, by using the argument from Example 2.24(6)) that $C^\ell(\mathbb{T}^d)$ is a Banach space with the norm

$$\|f\|_{C^\ell} = \max_{\|\gamma\|_1 \leqslant \ell} \|\partial_\gamma f\|_\infty.$$

We apply the above to $\partial_\gamma f$ and obtain from Proposition 5.3 that

$$\|\partial_\gamma f\|_\infty \ll_d \|\partial_\gamma f\|_{H^{k-\ell}} \leqslant \|f\|_{H^k}$$

for $f \in C^\infty(\mathbb{T}^d)$ and $\|\gamma\|_1 \leqslant \ell$. Therefore,

$$\|f\|_{C^\ell} \ll_d \|f\|_{H^k}$$

for $f \in C^\infty(\mathbb{T}^d)$, and the inclusion $C^\infty(\mathbb{T}^d) \to C^\ell(\mathbb{T}^d)$ once again gives rise to a bounded operator $\imath_\ell : H^k(\mathbb{T}^d) \to C^\ell(\mathbb{T}^d)$. Composing with the inclusion map from $C^\ell(\mathbb{T}^d)$ to $L^2(\mathbb{T}^d)$ we again see that $f \in H^k(\mathbb{T}^d)$ agrees almost everywhere with $\imath_\ell(f) \in C^\ell(\mathbb{T}^d)$. $\square$

5.2 Sobolev Spaces on Open Sets

Much of the discussion in Section 5.1.1 regarding Sobolev spaces on $\mathbb{T}^d$ can be quickly generalized to open subsets of $\mathbb{R}^d$. However, in Section 5.1.2 we frequently made use of Fourier series in the arguments, so we will go through the definitions and elementary properties once again, without appealing to Fourier series.

We will define spaces of functions on an open subset $U \subseteq \mathbb{R}^d$ that form Hilbert spaces, and in which the elements are once more continuous or even differentiable depending on a regularity parameter k (and the dimension d). We have a choice regarding the behaviour of the functions at the boundary ∂U of $U \subseteq \mathbb{R}^d$, giving rise to two different Sobolev spaces for every $k \geqslant 1$.

Definition 5.7. Let $d \geqslant 1$ and $k \geqslant 0$ be integers, and let $U \subseteq \mathbb{R}^d$ be an open subset. Then the *(L^2) Sobolev space* $H^k(U)$ is defined[†] to be the closure of

$$\{(\partial_\alpha f)_\alpha \mid f \in C^\infty(U), \partial_\alpha f \in L^2(U) \text{ for } \|\alpha\|_1 \leqslant k\} \tag{5.3}$$

inside $\bigoplus_{\|\alpha\|_1 \leqslant k} L^2(U)$, where as before we take the direct sum over all $\alpha \in \mathbb{N}_0^d$ with $\|\alpha\|_1 \leqslant k$.

Even though the closure $H^k(U)$ contains many new functions that are not in $C^\infty(U)$, those new elements

$$(f_\alpha)_{\|\alpha\|_1 \leqslant k} \in H^k(U) \tag{5.4}$$

still have some of the properties of the elements in the subspace (5.3) used to define $H^k(U)$. In fact, as we will show below, $f = f_0$ determines all the other components f_α of the vector (5.4), and these are derivatives of f in the following weaker sense (which, as we will see, turns integration by parts into the definition of a derivative).

Definition 5.8. Suppose that $\alpha \in \mathbb{N}_0^d$ and $f, g \in L^2(U)$. Then g is called a *weak α-partial derivative* (or an *α-partial derivative in the sense of distributions*) of f, written $g = \partial_\alpha f$, if

$$\int_U f \partial_\alpha \phi \, \mathrm{d}x = (-1)^{\|\alpha\|_1} \int_U g \phi \, \mathrm{d}x$$

for all $\phi \in C_c^\infty(U)$. In the case $\alpha = e_j$, we will call this the weak jth partial derivative and write $g = \partial_j f$.

We view the functions ϕ appearing in this definition (and similar instances) as 'test functions'.

Example 5.9. Let $U = (-1, 1) \subseteq \mathbb{R}$ and define the functions

$$f(x) = \begin{cases} x & \text{if } x \geqslant 0 \\ 0 & \text{if } x < 0 \end{cases} \quad \text{and} \quad g(x) = \begin{cases} 1 & \text{if } x \geqslant 0 \\ 0 & \text{if } x < 0 \end{cases}$$

for $x \in (-1, 1)$. Then f has weak e_1-partial derivative g. In fact, for ϕ in $C_c^\infty((-1, 1))$ we have

$$\int_{-1}^1 f \phi' \, \mathrm{d}x = \int_0^1 x \phi'(x) \, \mathrm{d}x = x\phi(x) \Big|_0^1 - \int_0^1 \phi \, \mathrm{d}x = 0 - \int_{-1}^1 g(x) \, \mathrm{d}x,$$

as required.

[†] In the literature another notation that is used is $W^{k,2}$. The more general case of $W^{k,p}$ is defined similarly using $L^p(U)$ instead of $L^2(U)$.

Lemma 5.10 (Weak derivatives). *Let $U \subseteq \mathbb{R}^d$ be open. A weak α-partial derivative of an $L^2(U)$ function f is uniquely determined as an element of $L^2(U)$ if it exists. If $(f_\alpha)_{\|\alpha\|_1 \leqslant k} \in H^k(U)$ then $f = f_0$ has $f_\alpha = \partial_\alpha f$ as a weak α-partial derivative for α with $\|\alpha\|_1 \leqslant k$. In particular, $f = f_0$ determines all the elements of the vector in $H^k(U)$.*

For convenience in this section we restrict attention at several points to real-valued functions. As a $\mathbb{C}$-valued function f on U belongs to $H^k(U)$ (or $H_0^k(U)$) for some $k \geqslant 0$ if and only if both $\Re(f)$ and $\Im(f)$ belong to that space, this is not a significant restriction.

PROOF. If g is a weak α-partial derivative of f then the inner product

$$\langle g, \phi \rangle = \int_U g\phi \, \mathrm{d}x = (-1)^{\|\alpha\|_1} \int f \partial_\alpha \phi \, \mathrm{d}x$$

is determined by f for all $\phi \in C_c^\infty(U)$. As $C_c^\infty(U)$ is dense in $L^2(U)$ (see Exercise 5.11 or Exercise 5.17(e)), we see that g is uniquely determined by f.

Let $f \in C^\infty(U)$ with $\partial_{e_j} f \in L^2(U)$ and $\phi \in C_c^\infty(U)$. Then for every $y \in \mathbb{R}^d$ the set $K_y = \mathrm{Supp}\, \phi \cap (y + \mathbb{R}e_j)$ is a compact subset of $U_y = U \cap (y + \mathbb{R}e_j)$ which is relatively open in $y + \mathbb{R}e_j$ for every $y \in \mathbb{R}^d$. Therefore U_y is a countable union of intervals and finitely many of these are sufficient to cover K_y. Using integration by parts for the integral along the jth coordinate x_j gives

$$\int_{U_y} f(x)\partial_{e_j}\phi(x) \, \mathrm{d}x_j = -\int_{U_y} \partial_{e_j} f(x)\phi(x) \, \mathrm{d}x_j$$

for any $y \in U$, where the boundary terms vanish since $\phi \in C_c^\infty(U)$. Integrating over the remaining variables shows that $\partial_{e_j} f$ is indeed also a weak e_j-partial derivative. By induction on $\|\alpha\|_1$, this implies that

$$\langle f_0, \partial_\alpha \phi \rangle = (-1)^{\|\alpha\|_1} \langle f_\alpha, \phi \rangle$$

first for all $(f_\alpha)_{\|\alpha\|_1 \leqslant k}$ in the subspace (5.3) inside $\bigoplus_{\|\alpha\|_1 \leqslant k} L^2(U)$, and then by continuity of the scalar product for all $(f_\alpha)_{\|\alpha\|_1 \leqslant k} \in H^k(U)$. This implies the lemma. $\qquad\square$

Essential Exercise 5.11. Let $U \subseteq \mathbb{R}^d$ be open. Show that $C_c^\infty(U) \subseteq L^p(U)$ is dense for any $p \in [1, \infty)$.

Definition 5.12 (Modifying Definition 5.7). Lemma 5.10 justifies the following notational convention. We identify the elements of $H^k(U)$ with functions $f \in L^2(U)$ and equip the space $H^k(U)$ with the norm

$$\|f\|_{H^k(U)} = \sqrt{\sum_{\|\alpha\|_1 \leqslant k} \|\partial_\alpha f\|_{L^2(U)}^2}.$$

Using this identification, the subspace (5.3) will from now on be referred to as $C^\infty(U) \cap H^k(U)$.

Proposition 5.13 (Forgetting regularity and the weak partial derivative). *For $k > \ell \geqslant 0$ there is a natural injection $\imath_{k,\ell} : H^k(U) \to H^\ell(U)$ of norm one, extending the identity on $C^\infty(U)$. For any multi-index α with $\|\alpha\|_1 \leqslant k$ there is a natural operator*

$$\partial_\alpha : H^k(U) \longrightarrow H^{k-\|\alpha\|_1}(U)$$
$$f \longmapsto \partial_\alpha f$$

of norm less than or equal to one, which extends ∂_α on $C^\infty(U) \cap H^k(U)$.

This may be proved along the same lines as Proposition 5.3.

We may obtain other Sobolev spaces — which will be subspaces of $H^k(U)$ — by requiring additional decay properties at the boundary ∂U.

Definition 5.14. We define $H_0^k(U) = \overline{C_c^\infty(U)} \subseteq H^k(U)$ to be the closure of all smooth compactly supported functions in $H^k(U)$.

We will see later that elements of $H_0^k(U)$ 'vanish in the square-mean norm sense' at ∂U if $k \geqslant 1$.

Let us add the following remark to Definition 5.7. We defined $H^k(U)$ to consist of those $f \in L^2(U)$ that have weak α-partial derivatives $\partial_\alpha f \in L^2(U)$ for all α with $\|\alpha\|_1 \leqslant k$ such that the vector

$$(\partial_\alpha f)_{\|\alpha\|_1 \leqslant k} \in \bigoplus_{\|\alpha\|_1 \leqslant k} L^2(U)$$

can be approximated by vectors corresponding to elements of

$$C^\infty(U) \cap H^k(U).$$

One may ask whether this approximation statement can be proved instead of assumed.

Exercise 5.15. Show that $f \in L^2(\mathbb{T}^d)$ belongs to $H^k(\mathbb{T}^d)$ if and only if there exists, for every $\alpha \in \mathbb{N}_0^d$ with $\|\alpha\|_1 \leqslant k$, a weak α-partial derivative $\partial_\alpha f \in L^2(\mathbb{T}^d)$. Here the weak partial derivative is defined in terms of smooth test functions $\phi \in C^\infty(\mathbb{T}^d)$.

The analogue of Exercise 5.15 also holds[15] for certain open subsets U of $\mathbb{R}^d$, but we will not use this possible alternative definition of the Sobolev spaces here (and will return to this question in Section 8.2.2). We note that due to the boundary of U this equivalence is a bit harder to prove. For this proof, but more importantly also for the material that follows in this chapter, we need some more background concerning smooth functions on $\mathbb{R}^d$, which we outline in the following series of exercises.

Exercise 5.16 (A smooth function on $\mathbb{R}$). Show that the function

$$\psi(t) = \begin{cases} e^{1/t} & \text{for } t < 0, \\ 0 & \text{for } t \geqslant 0 \end{cases}$$

is smooth on $\mathbb{R}$.

Essential Exercise 5.17 (Smooth approximate identities on $\mathbb{R}^d$). Define a real function $\jmath$ on $\mathbb{R}^d$ by

$$\jmath(x) = c\psi(\|x\|_2^2 - 1) = \begin{cases} ce^{1/\left(\|x\|_2^2 - 1\right)} & \text{for } \|x\|_2 < 1, \\ 0 & \text{for } \|x\|_2 \geqslant 1, \end{cases}$$

where $c > 0$ is chosen so that $\int_{\mathbb{R}^d} \jmath(x)\,dx = \int_{B_1} \jmath(x)\,dx = 1$. For $\varepsilon > 0$, also define $\jmath_\varepsilon(x) = \frac{1}{\varepsilon^d}\jmath\left(\frac{x}{\varepsilon}\right)$ for $x \in \mathbb{R}^d$. Show the following.

(a) $\jmath \in C_c^\infty(\mathbb{R}^d)$.
(b) The function f_ε defined by $f_\varepsilon(x) = f * \jmath_\varepsilon(x) = \int_{\mathbb{R}^d} f(y)\jmath_\varepsilon(x - y)\,dy$ converges uniformly to f as $\varepsilon \to 0$ on any compact subset of $\mathbb{R}^d$ for any $f \in C(\mathbb{R}^d)$.
(c) $f_\varepsilon \in C^\infty(\mathbb{R}^d)$ for any $f \in C(\mathbb{R}^d)$.
(d) $\operatorname{Supp} f_\varepsilon \subseteq \operatorname{Supp} f + \overline{B_\varepsilon}$.
(e) Generalize the above in the appropriate sense to any $f \in L^p(\mathbb{R}^d)$ and any $p \in [1, \infty)$. Derive the statement in Exercise 5.11 from this.

Exercise 5.18. Generalize and prove the first claim from Section 1.5 for functions defined on $\mathbb{T}^d$ or on an open subset of $\mathbb{R}^d$.

Exercise 5.19. Suppose $U \subseteq \mathbb{R}^d$ is open, bounded, and *star-shaped* with centre 0 in the sense that $\overline{U} \subseteq \lambda U$ for all $\lambda > 1$ (see, for example, Figure 5.2). Let $f, f_1, \ldots, f_d$ be in $L^2(U)$ and suppose f_j is the weak e_j-partial derivative of f for $j = 1, \ldots, d$. Show that $f \in H^1(U)$.

5.2.1 Examples

We illustrate the theory above with some simple examples, which will be justified below.

Example 5.20. Let $d = 1$, $U = (0, 1)$ and $k = 1$. Then every $f \in H^1((0, 1))$ has a continuous representative $\imath(f) \in C([0, 1])$ with

$$\|\imath(f)\|_\infty \leqslant 2\|f\|_{H^1}. \tag{5.5}$$

This is again an instance of the Sobolev embedding theorem, which we will prove for open subsets in Theorem 5.34 below. However, the instance discussed here can be proven quite directly and independently. Also see Exercise 5.22 and Exercise 5.23.

Example 5.21. Let $d = 2$, $U = B_{1/2} \setminus \{0\}$ and $k = 1$. Then the function f defined by $f(x) = \log|\log\|x\||$ lies in $H^1(U)$ and cannot be extended to an element of $C(\overline{U})$. Also see Exercise 5.25 and Exercise 5.26.

JUSTIFICATION OF EXAMPLE 5.20. For $x, y \in U$ and $f \in C^\infty(U) \cap H^1(U)$ we clearly have

$$f(y) = f(x) + \int_x^y f'(s)\,ds.$$

Notice that for fixed x and y the integral on the right is a continuous functional on $H^1(U)$, but for the terms $f(y)$ and $f(x)$ this is not clear yet. Now integrate over $x \in (0, 1)$ to get

$$
\begin{aligned}
f(y) &= \int_0^1 f(x)\,dx + \int_0^1 \int_x^y f'(s)\,ds\,dx \\
&= \int_0^1 f(x)\,dx + \int_0^1 \int_0^1 f'(s)\sigma(y, x, s)\,ds\,dx,
\end{aligned}
$$

where

$$
\sigma(y, x, s) = \begin{cases}
1 & \text{if } x < s < y, \\
-1 & \text{if } y < s < x, \text{ and} \\
0 & \text{if } s \text{ is not between } x \text{ and } y.
\end{cases}
$$

Applying Fubini's theorem we get

$$f(y) = \int_0^1 f(x)\,dx + \int_0^1 f'(s)k(y, s)\,ds, \qquad (5.6)$$

where

$$
k(y, s) = \int_0^1 \sigma(y, x, s)\,dx = \begin{cases}
s & \text{if } s < y, \\
s - 1 & \text{if } s > y.
\end{cases}
$$

Hence (5.6) expresses the value of f at $y \in U$ as the sum $\langle f, \mathbb{1}_U \rangle + \langle f', k(y, \cdot) \rangle$, which is clearly continuous on $H^1((0, 1))$, and since $\|k(y, \cdot)\|_{L^2} \leqslant 1$ we also have $|f(y)| \leqslant 2\|f\|_{H^1}$. Moreover, we may use (5.6) for $y = 0$ and $y = 1$ as a definition of $f(0)$ and $f(1)$, and then

$$
|f(y_1) - f(y_2)| = \left| \int_0^1 f'(s)\left(k(y_1, s) - k(y_2, s)\right)ds \right|
$$
$$
\leqslant \|f'\|_2 \|k(y_1, \cdot) - k(y_2, \cdot)\|_2 = \|f'\|_2 \sqrt{|y_1 - y_2|} \qquad (5.7)
$$

for all $y_1, y_2 \in [0, 1]$. It follows that any $f \in C^\infty(U) \cap H^1(U)$ extends to a continuous function satisfying (5.5). Applying automatic extension to the closure (Proposition 2.59) to the so described map from $C^\infty(U) \cap H^1(U)$ to $C(\overline{U})$, the claims in Example 5.20 follow.   □

Exercise 5.22. Let $U = (0,1)$ and let $f : \overline{U} \to \mathbb{C}$ be a function continuous on $\overline{U}$ and continuously differentiable at all but finitely many points of U. If we also have $f' \in L^2(U)$, show that $f \in H^1(U)$.

Exercise 5.23. Show that Example 5.20 can be generalized to the statement that any function $f \in H^k((0,1))$ (continuously extended) belongs to $C^{k-1}([0,1])$. Show also that $H_0^1((0,1))$ is mapped under the embedding from Example 5.20 into the space

$$\{f \in C([0,1]) \mid f(0) = f(1) = 0\}.$$

Exercise 5.24. Show that not every function in $H_0^1((0,1)) \cap H^2((0,1))$ is also in $H_0^2((0,1))$.

JUSTIFICATION OF EXAMPLE 5.21. It is easy to check that $f \in L^2(U)$. Since f is also smooth on U, we only have to check that $\partial_j f \in L^2(U)$. By the chain rule we have

$$\partial_j \left(\log |\log \|x\|| \right) = \frac{1}{\log \|x\|} \frac{1}{\|x\|} \frac{x_j}{\|x\|}.$$

Taking the square and integrating with respect to $\mathrm{d}x\,\mathrm{d}y = r\,\mathrm{d}\phi\,\mathrm{d}r$ we get

$$\|\partial_j f\|_{L^2(U)} \leqslant \sqrt{\int_0^{1/2} \int_0^{2\pi} \frac{1}{(r \log r)^2} r\,\mathrm{d}\phi\,\mathrm{d}r} \ll \sqrt{\int_0^{1/2} \frac{1}{r(\log r)^2}\,\mathrm{d}r} < \infty.$$

$\square$

Exercise 5.25. Extend Example 5.21 by showing that $f(x) = \log|\log \|x\||$ defines an element of $H^1(B_{1/2})$.

Exercise 5.26. Let $U = B_1^{\mathbb{R}^d}$, and let $f_\alpha(x) = \|x\|^\alpha$ for $x \in U$. For which values of α do we have $f_\alpha \in H^k(U)$?

5.2.2 Restriction Operators and Traces

Essential Exercise 5.27 (Open subsets). Let $V \subseteq U \subseteq \mathbb{R}^d$ be open subsets. Let $k \geqslant 0$.
(a) Show that the restriction $\cdot|_V : H^k(U) \to H^k(V)$ is a bounded operator.
(b) Show that the extension operator sending functions in $H_0^k(V)$ to $H_0^k(U)$, defined by extending the functions to be zero on $U \smallsetminus V$, is a bounded operator.
(c) Show that for $k \geqslant 1$ in general $H_0^k(U)|_V$ does not belong to $H_0^k(V)$. Show that one cannot define an extension operator from $H^k(V)$ to $H^k(U)$ by simply extending the functions to be zero on $U \smallsetminus V$.

In order to get a better geometric understanding of what it means for a function to belong to $H^1(U)$, we will now show that an element $f \in H^1(U)$, when restricted to any hyperplane, still belongs to L^2. Notice that since a hyperplane is a null set, any property claimed for the restriction to a hyperplane cannot be demanded — indeed does not even make sense — for a function $f \in L^2(U)$ (as in this case f is really an equivalence class of functions). For notational simplicity we start with the case $U = (0,1)^d$ and describe how Example 5.20 can be generalized to higher dimensions.

Example 5.28. Let $U = (0,1)^d$, and $S = (0,1)^{d-1}$ and write

$$S_y = S \times \{y\} \subseteq \overline{U}$$

for $y \in [0,1]$. For every $y \in [0,1]$ there is a natural *restriction operator* to $L^2(S_y)$, called the *trace* on S_y,

$$H^1(U) \ni f \longmapsto f\big|_{S_y} \in L^2(S_y),$$

which for $y \in (0,1)$ is the continuous extension of the restriction operator

$$C^\infty(U) \cap H^1(U) \ni f \longmapsto f\big|_{S_y}.$$

Moreover, if we identify the space $L^2(S_y)$ with $L^2(S)$ for all $y \in [0,1]$ (by simply identifying S_y with S via the projection $S_y \ni (x,y) \mapsto x \in S$), then we also have

$$\left\| f\big|_{S_{y_1}} - f\big|_{S_{y_2}} \right\|_{L^2(S)} \leqslant \|f\|_{H^1(U)} \sqrt{|y_1 - y_2|}.$$

Exercise 5.29. Prove the statements of Example 5.28 by the following steps:
(a) Fix some $f \in C^\infty(U) \cap H^1(U)$ and apply Fubini's theorem to see that the restriction of f to $\{x\} \times (0,1)$ belongs to $H^1((0,1))$ for almost every $x \in (0,1)^{d-1}$. Now apply Example 5.20 (or more precisely (5.6)) to show that

$$f(x,y) = \int_0^1 f(x,s)\,\mathrm{d}s + \int_0^y s\partial_2 f(x,s)\,\mathrm{d}s + \int_y^1 (s-1)\partial_2 f(x,s)\,\mathrm{d}s.$$

Notice that this also gives a definition for the trace in the cases $y = 0$ and $y = 1$. Use this to estimate the L^2 norm of the restriction of f to S_y.
(b) For the last statement show that for any $f \in C^\infty(U) \cap H^1(U)$,

$$|f(x,y_1) - f(x,y_2)| \leqslant \sqrt{|y_1 - y_2|} \cdot \|\partial_2 f\|_{L^2(\{x\} \times (0,1))}.$$

Exercise 5.30. Prove an extension of Example 5.28 for any bounded open set $U \subseteq \mathbb{R}^d$ and the image of $\phi\big([0,1]^{d-1} \times \{0\}\big)$ for a smooth map $\phi : [0,1]^d \to U$.

We now consider a general open set $U \subseteq \mathbb{R}^d$ and define the trace for elements of $H_0^1(U)$. For the statement that such functions vanish in the square-mean sense at ∂U we want to assume that U has a sufficiently regular boundary in the following sense (this may feel familiar after recalling the implicit function theorem).

Definition 5.31. Let $U \subseteq \mathbb{R}^d$ be an open set and $k \in \mathbb{N}_0 \cup \{\infty\}$. We say that U has a C^k-*smooth boundary* if for every $z^{(0)} \in \partial U$ there exists a neighbourhood $B_\varepsilon(z^{(0)})$, a rotated coordinate system (which we denote by $x_1, \ldots, x_{d-1}, y$) so that $z^{(0)}$ corresponds to $(x_1^{(0)}, \ldots, x_{d-1}^{(0)}, y^{(0)})$, and a function

$$\phi \in C^k\big(B_\varepsilon(x_1^{(0)}, \ldots, x_{d-1}^{(0)})\big)$$

such that $U \cap B_\varepsilon(z^{(0)}) = \{(x, y) \in B_\varepsilon(z^{(0)}) \mid y < \phi(x)\}$, as illustrated in Figure 5.2. If $k = \infty$ then we simply say that U has *smooth boundary*.

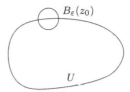

Fig. 5.2: A set with smooth boundary.

This includes examples like $U = B_r(x)$, but excludes $U = (0, 1)^d$ if $k \geqslant 1$ and $d \geqslant 2$. Notice that an open set with a C^k-smooth boundary need not be connected, simply connected, or bounded. Also note that the rotation within Definition 5.31 does not affect whether a function belongs to $H^k(U)$. In fact, since a rotation R preserves the H^k norm, a convergent sequence (f_n) in $C^\infty(U) \cap H^k(U)$ is mapped to another convergent sequence $(f_n \circ R)$ in $H^k(R^{-1}U)$.

Exercise 5.32. Let $U \subseteq \mathbb{R}^d$ be a bounded open set, and let Φ be a diffeomorphism (a rotation, for example) defined on a neighbourhood of $\overline{U}$. Let $k \geqslant 0$ be an integer. Show that $H^k(U) \ni f \mapsto f \circ \Phi \in H^k(\Phi^{-1}(U))$ is an isomorphism (and in the case of a rotation, an isometry) between $H^k(U)$ and $H^k(\Phi^{-1}(U))$.

Proposition 5.33 (Trace on graphs). *Let $U \subseteq \mathbb{R}^d$ be a bounded open set. We let $\varepsilon > 0$, denote the variables $(z_1, \ldots, z_d) \in \mathbb{R}^d$ by $(x_1, \ldots, x_{d-1}, y)$, and assume that*

$$\phi : \overline{B_\varepsilon^{\mathbb{R}^{d-1}}(x^{(0)})} \longrightarrow \mathbb{R}$$

is continuous. Then there is a natural restriction operator, called the trace *on the graph* $\mathrm{Graph}(\phi) = \{(x, \phi(x)) \mid x \in \overline{B_\varepsilon^{\mathbb{R}^{d-1}}(x^{(0)})}\}$ *of ϕ,*

$$H_0^1(U) \ni f \longmapsto f\big|_{\mathrm{Graph}(\phi)} \in L^2(\mathrm{Graph}(\phi))$$

which satisfies

$$\left\| f\big|_{\mathrm{Graph}(\phi)} \right\|_{L^2(\mathrm{Graph}(\phi))} \leqslant \sqrt{\delta_\phi} \|\partial_d f\|_{L^2(U)},$$

where δ_ϕ is chosen to have $(x, \phi(x) + \delta_\phi) \notin U$ for all $x \in B_\varepsilon(x^{(0)})$ (see Figure 5.3) and we use the measure $\mathrm{d}x_1 \cdots \mathrm{d}x_{d-1}$ on $\mathrm{Graph}(\phi)$.

Consider now the case of a bounded open set $U \subseteq \mathbb{R}^d$ with C^0-smooth boundary and the function $\phi_\delta(x) = \phi(x) - \delta$ with ϕ as in Definition 5.31. Then, by Proposition 5.33, the trace of an element $f \in H_0^1(U)$ on this translated portion of the boundary has L^2 norm of order $\mathrm{O}(\sqrt{\delta}\|f\|_{H^1(U)})$. This

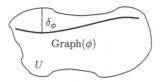

Fig. 5.3: The trace on the graph of ϕ.

explains the earlier claim that $f \in H_0^1(U)$ vanishes in the square-mean sense at ∂U.

We also note that if we set $\phi(x) = y_0$ to be constant we obtain that the L^2 norm of the restriction of f to $U \cap \mathbb{R}^{d-1} \times \{y_0\}$ is bounded by a multiple of $\|\partial_y f\|_2$. Integrating the square of this inequality over y_0 we obtain

$$\|f\|_2 \ll_U \|\partial_y f\|_2 \tag{5.8}$$

for all $f \in H_0^1(U)$, where the implicit constant depends on the bounded open set U.

PROOF OF PROPOSITION 5.33. Recall that we write x for the first $(d-1)$ coordinates and y for the last coordinate. For $f \in C_c^\infty(U)$ we have

$$f(x, \phi(x)) = - \int_{\phi(x)}^{\phi(x)+\delta_\phi} \partial_d f(x, s) \, ds \tag{5.9}$$

by our assumption on δ_ϕ. Taking the square and applying the Cauchy–Schwarz inequality gives

$$|f(x, \phi(x))|^2 = \left| \int_{\phi(x)}^{\phi(x)+\delta_\phi} \partial_d f(x, s) \, ds \right|^2 \leqslant \delta_\phi \int_{\phi(x)}^{\phi(x)+\delta_\phi} |\partial_d f(x, s)|^2 \, ds.$$

$$\tag{5.10}$$

Integrated over x this gives

$$\left\| f|_{\mathrm{Graph}(\phi)} \right\|_{L^2(\mathrm{Graph}(\phi))} \leqslant \sqrt{\delta_\phi} \, \|\partial_d f\|_{L^2(U)} \leqslant \sqrt{\delta_\phi} \|f\|_{H^1(U)}.$$

Using automatic extension to the closure (Proposition 2.59), this implies the proposition. $\qquad\square$

5.2.3 Sobolev Embedding in the Interior

We now extend the Sobolev embedding theorem (Theorem 5.6) to open subsets $U \subseteq \mathbb{R}^d$ (but, for now, leave open the question regarding the behaviour of f at ∂U).

Theorem 5.34 (Sobolev embedding on open subsets of $\mathbb{R}^d$). *Let U be an open subset of $\mathbb{R}^d$ and let $\ell \geqslant 0$ and $k > \frac{d}{2} + \ell$ be integers. Then any function in $H^k(U)$ (has a continuous representative that) also lies in $C^\ell(U)$.*

In the following exercise we simplify the geometry again and only look at the unit cube, but go further than in the theorem above.

Exercise 5.35. Let $U = (0,1)^d$ as in Example 5.28.
(a) Extend Example 5.28 by showing that $f \in H^k(U)$ implies that $f|_{S_y} \in H^{k-1}(S_y)$ for $y \in [0,1]$.
(b) Use (a) to prove the following weak version of the Sobolev embedding theorem. For any $\ell \geqslant 0$ there is a natural map from $H^{d+\ell}(U)$ to $C^\ell(U)$ that extends the identity on the functions in $C^\infty(U) \cap H^{d+\ell}(U)$.
(c) Extend the arguments from (b) to the boundary, by showing that there is a natural map from $H^{d+\ell}(U)$ to $C^\ell(\overline{U})$.
(d) Now improve the needed regularity in (c) in the following way: Show that there is a natural map from $H^k(U)$ to $C^\ell(S_0)$ if $k > \ell + 1 + \frac{d-1}{2}$ (by also applying the Sobolev embedding theorem on S_0).

The proof of Theorem 5.34 will be (apart from Exercise 5.18) the first example of a technique that we will use frequently: If a given statement is already known to hold on $\mathbb{T}^d$ (where we can use Fourier series to prove it), then one can sometimes obtain the same statement for open subsets of $\mathbb{R}^d$ by moving the functions or the problem to $\mathbb{T}^d$. For this the following lemma and the notation $\mathbb{T}^d_R = \mathbb{R}^d/(2R\mathbb{Z}^d)$ for $R > 0$ will be useful. We also define $H^k(\mathbb{T}^d_R)$ in the same way as we defined $H^k(\mathbb{T}^d)$ except that we will use the fundamental domain $[-R, R)^d$ and the restriction of the Lebesgue measure to it to define the L^2 norm and the derived Sobolev norms. Of course the theorems of the previous section also hold in that context (possibly with different multiplicative constants).

Lemma 5.36 (Transfering regularity). *Let $U \subseteq \mathbb{R}^d$ be open, $k \geqslant 1$, and $\chi \in C_c^\infty(V)$ for some open $V \subseteq U$. Then $M_\chi : H^k(U) \to H_0^k(V)$ defined by $M_\chi(f) = \chi f$ is a bounded operator. Let $R > 0$ and assume now that $U \subseteq B_R$. For a function f on U we define $P(f)$ on $\mathbb{T}^d_R$ by first extending f to $[-R, R)^d$ by setting it to be zero outside of U and then identifying $[-R, R)^d$ with $\mathbb{T}^d_R$. Then $P : H_0^k(U) \to H^k(\mathbb{T}^d_R)$ is a linear isometry. Finally, $f \in H^k(U)$, $\chi \in C_c^\infty(U)$ and $P(\chi f) \in H^\ell(\mathbb{T}^d_R)$ for some $\ell > k$ implies that $\chi f \in H_0^\ell(U)$.*

PROOF. For $f \in C^\infty(U) \cap H^k(U)$ we have $M_\chi(f) = \chi f \in C_c^\infty(V) \subseteq H_0^k(V)$ and the partial derivatives of χf are sums of products of partial derivatives of χ and partial derivatives of f of lower order (the full expansion is given by Leibniz' rule). Since $\chi \in C_c^\infty(V)$ this gives the estimate

$$\|\partial_\alpha(\chi f)\|_{L^2(V)} \ll \sup_{\|\beta\|_1 \leqslant \|\alpha\|_1} \|\partial_\beta \chi\|_\infty \sup_{\|\gamma\|_1 \leqslant \|\alpha\|_1} \|\partial_\gamma f\|_{L^2(U)}$$

for all α with $\|\alpha\|_1 \leqslant k$, which leads to $\|M_\chi(f)\|_{H^k(V)} \ll_\chi \|f\|_{H^k(U)}$. From this it follows that the operator M_χ is a bounded operator from $H^k(U)$

into $H_0^k(V)$ (and that the weak partial derivatives $\partial_\alpha(\chi f)$ for $f \in H^k(U)$ are obtained by the same Leibniz rule as the partial derivates $\partial_\alpha(\chi f)$ for f in $C^\infty(U)$.

For the second statement of the lemma notice first that

$$P(C_c^\infty(U)) \subseteq C^\infty(\mathbb{T}_R^d)$$

(which would not be true for $C^\infty(U) \cap H^k(U)$). Since the norm on $H^k(\mathbb{T}_R^d)$ is defined by integration over the Lebesgue measure on $(-R, R)^d$, and since

$$U \subseteq (-R, R)^d$$

we obtain $\|P(f)\|_{H^k(\mathbb{T}_R^d)} = \|f\|_{H^k(U)}$ for f in $C_c^\infty(U)$. Hence the automatic extension of P to an operator from $H_0^k(U)$ to $H^k(\mathbb{T}_R^d)$ (Proposition 2.59) exists and is an isometry.

Assume now that $P(\chi f)$ lies in $H^\ell(\mathbb{T}_R^d)$ for some $f \in H^k(U)$, $\ell > k$, and $\chi \in C_c^\infty(U)$. Let ψ in $C_c^\infty(U)$ be chosen with $\psi \equiv 1$ on a neighbourhood of $\mathrm{Supp}\,\chi$ and with $\mathrm{Supp}\,\psi \subseteq U$ (see Exercise 5.37). Since any $g \in C^\infty(\mathbb{T}_R^d)$ can be identified with a $(2R\mathbb{Z})^d$-periodic smooth function g on $\mathbb{R}^d$, the map

$$C^\infty(\mathbb{T}_R^d) \ni g \longmapsto \psi g \in C_c^\infty(U)$$

is well-defined. Arguing just as in the first part of the proof, we get that multiplication by ψ defines a bounded operator from $H^\ell(\mathbb{T}_R^d)$ to $H_0^\ell(U)$. Applying this map to $g = P(\chi f) \in H^\ell(\mathbb{T}_R^d)$ we get $\psi P(\chi f) = \chi f \in H_0^\ell(U)$. $\qquad\square$

The existence of functions in $C_c^\infty(U)$ that are equal to one on large subsets of U (as used in the above proof) will frequently be useful.

Essential Exercise 5.37 (Smooth approximate characteristic functions). Let $K \subseteq U$ be a compact subset of an open subset $U \subseteq \mathbb{R}^d$. Find a smooth function $\psi \in C_c^\infty(U)$ with $\psi|_K \equiv 1$.

PROOF OF THEOREM 5.34. Let $x_0 \in U$ and $\varepsilon > 0$ be such that

$$V = B_{2\varepsilon}^{\mathbb{R}^d}(x_0) \subseteq U.$$

Assume $k > \frac{d}{2} + \ell$. We fix some $\chi \in C_c^\infty(V)$ satisfying $\chi|_{B_\varepsilon(x_0)} \equiv 1$. Note that $\chi \in C_c^\infty(V) \subseteq C_c^\infty(B_R)$ for $R = \|x_0\| + 2\varepsilon$. The theorem follows from the existence of the following chain of operators

$$H^k(U) \xrightarrow{P \circ M_\chi} H^k(\mathbb{T}_R^d) \xrightarrow{\imath} C^\ell(\mathbb{T}_R^d) \xrightarrow{\cdot|_{B_\varepsilon(x_0)}} C_b^\ell(B_\varepsilon(x_0)) \longrightarrow L^2(B_\varepsilon(x_0)),$$

where $M_\chi : H^k(U) \to H_0^k(V)$ and $P : H_0^k(V) \to H^k(\mathbb{T}_R^d)$ are as in Lemma 5.36, $\imath$ is from the Sobolev embedding theorem for the torus (Theorem 5.6), and $\cdot|_{B_\varepsilon(x_0)}$ denotes the operator sending a function to its restriction to $B_\varepsilon(x_0)$. We also note that the composition of these operators applied to $f \in H^k(U)$ simply gives the restriction of f to $B_\varepsilon(x_0)$ (since this holds initially for $f \in C^\infty(U) \cap H^k(U)$). This gives the theorem (see also Exercise 5.38). □

Exercise 5.38 (Merging lemma for continuous functions). If we wish to be pedantic the above proof is not yet complete since we have only shown that for every point x there exists a neighbourhood $B_\varepsilon(x)$ such that we can find a C^ℓ-version of the restriction of f to the given neighbourhood. However, we actually claimed that there is a version of f on all of U which is in $C^\ell(U)$. To complete the proof, prove or recall the following statements:

(a) Suppose that U is covered by a family of open subsets B_τ for $\tau \in T$ and for every $\tau \in T$ we are given some $f_\tau \in C(B_\tau)$. Assume that $f_{\tau_1}|_{B_{\tau_1} \cap B_{\tau_2}} = f_{\tau_2}|_{B_{\tau_1} \cap B_{\tau_2}}$ for every τ_1, τ_2 in T. Then there exists some $f \in C(U)$ with $f_\tau = f|_{B_\tau}$ for all $\tau \in T$.

(b) Use the fact that U is σ-compact to construct a countable cover $B_n = B_{\varepsilon_n}(x_n)$ of U with $B_{2\varepsilon_n}(x_n) \subseteq U$.

(c) Complete the proof of Theorem 5.34 above, using (a) and (b).

Exercise 5.39. Let $U \subseteq \mathbb{R}^d$ be open and $K \subseteq U$ a compact subset. Show that

$$\|f\|_{K,\infty} \ll_{K,U} \|f\|_{H^k(U)}$$

for $f \in H^k(U)$ and $k > \frac{d}{2}$.

5.3 Dirichlet's Boundary Value Problem and Elliptic Regularity

In this section we will combine the discussion of Sobolev spaces from Section 5.2, the Fréchet–Riesz representation theorem (Corollary 3.19), and a simple orthogonality relation to solve the Dirichlet boundary value problem

$$\left. \begin{array}{r} \Delta u = 0 \\ u|_{\partial U} = b \end{array} \right\} \tag{5.11}$$

introduced and motivated in Section 1.2.1 for certain domains $U \subseteq \mathbb{R}^d$ and certain functions $b : \partial U \to \mathbb{R}$. Recall that a function g is said to be *harmonic* if $\Delta g = 0$, where

$$\Delta g = \frac{\partial^2 g}{\partial x_1^2} + \cdots + \frac{\partial^2 g}{\partial x_d^2} = \partial_1^2 g + \cdots + \partial_d^2 g$$

is the Laplacian of g.

We note that (with the exception of Lemma 5.48 and its proof) we restrict our attention in this section to real-valued functions. In the following we will also write $\langle \cdot, \cdot \rangle_{L^2(U)}$ to denote the inner product on $L^2(U)$, and similarly for other Hilbert spaces, to emphasize the difference between the various inner products used, especially for the semi-inner product $\langle \cdot, \cdot \rangle_1$ as introduced in the next lemma. Recall from Definition 3.1 that a semi-inner product satisfies positivity instead of strict positivity.

Lemma 5.40 (Orthogonality). *Let $U \subseteq \mathbb{R}^d$ be open, $\phi \in C_c^\infty(U)$, and assume that $g \in C^2(U) \cap H^1(U)$ is a harmonic function. Then ϕ and g are orthogonal with respect to the semi-inner product $\langle \cdot, \cdot \rangle_1$ defined by*

$$\langle u, v \rangle_1 = \sum_{j=1}^d \langle \partial_j u, \partial_j v \rangle_{L^2(U)} \tag{5.12}$$

for $u, v \in H^1(U)$.

Proof. Fix $j \in \{1, \ldots, d\}$ and $\phi \in C_c^\infty(U)$. Then as in Lemma 5.10 we can use integration by parts to obtain

$$\int_U \partial_j g \partial_j \phi \, dx = - \int_U (\partial_j^2 g) \phi \, dx$$

since the boundary terms vanish. Integrating over the remaining variables and summing over all $j = 1, \ldots, d$, we get $\langle g, \phi \rangle_1 = \langle -\Delta g, \phi \rangle_{L^2(U)} = 0$ by the assumption on g. $\qquad\square$

Motivated by Lemma 5.40, the approach is to decompose a function f in $C^1(\overline{U})$ as $f = g + v$, where $v \in H_0^1(U)$ and g is 'orthogonal to' $H_0^1(U)$ with respect to the semi-inner product $\langle \cdot, \cdot \rangle_1$. As harmonic functions have this orthogonality property by Lemma 5.40, there is some hope that g will be harmonic and indeed it will turn out to be. Morevoer, v will vanish at ∂U in the square-mean sense and so $f|_{\partial U} = g|_{\partial U}$ at least in the square-mean sense.

As we wish to use the semi-inner product from (5.12) in the definition of the orthogonal complement, we will have to discuss properties of this semi-inner product. We will then show that g is smooth and harmonic, and it is this step that relies on a general phenomenon called *elliptic regularity*, the Laplace operator being an example of an elliptic differential operator. We will show in Section 5.3.3, for $d = 2$, that g extends continuously to the boundary ∂U and agrees with $f|_{\partial U}$ there. Finally, we will discuss in Section 8.2.2 the behaviour at a smooth boundary in any dimension.

5.3.1 The Semi-Inner Product

Let $U \subseteq \mathbb{R}^d$ be an open bounded set.

Lemma 5.41 (Semi-inner product). *The semi-inner product $\langle \cdot, \cdot \rangle_1$ restricted to $H_0^1(U)$ is an inner product, and the norm defined by this inner product is equivalent to $\| \cdot \|_{H^1(U)}$. The semi-norm $\| \cdot \|_1$ induced by $\langle \cdot, \cdot \rangle_1$ on $C^\infty(U) \cap H^1(U)$ has as its kernel the subspace of all locally constant functions.*

Here the kernel is the subspace of all functions f with $\langle f, f \rangle_1 = \|f\|_1^2 = 0$. A function f on U is called *locally constant* if for every $x \in U$ there is a neighbourhood V of x such that $f|_V$ is constant. If U is connected, then any locally constant function is constant.

PROOF OF LEMMA 5.41. Let $f \in H_0^1(U)$. We have $\|f\|_{L^2(U)} \ll \|\partial_{x_d} f\|_{L^2(U)}$ by (5.8). Thus

$$\sqrt{\langle f, f \rangle_1} \leqslant \|f\|_{H^1(U)} = \sqrt{\|f\|_{L^2(U)}^2 + \sum_{j=1}^d \|\partial_j f\|_{L^2(U)}^2} \ll \sqrt{\langle f, f \rangle_1}$$

for $f \in H_0^1(U)$, proving the first statement in the proposition.

If $f \in C^\infty(U) \cap H^1(U)$ is locally constant then it is clear that $\langle f, f \rangle_1 = 0$. On the other hand, if $f \in C^\infty(U) \cap H^1(U)$ has $\langle f, f \rangle_1 = 0$, then $\partial_j f = 0$ almost everywhere and for all j, so f is locally constant. $\qquad\square$

We are now ready to exhibit the desired orthogonal decomposition.

Proposition 5.42 (Existence of weak solution). *Let $U \subseteq \mathbb{R}^d$ be an open bounded set with C^1-smooth boundary, and let $f \in C^1(\overline{U})$ (that is, f and all of its partial derivatives are continuous and extend continuously to $\overline{U}$). Then there exists some v in $H_0^1(U)$ such that $g = f - v \in H^1(U)$ is weakly harmonic in the sense that $\langle g, \Delta\phi \rangle_{L^2(U)} = 0$ for all $\phi \in C_c^\infty(U)$.*

As before with ∂ and $\boldsymbol{\partial}$, we will think of this statement as giving meaning to '$\Delta g = 0$' by writing $\langle \Delta g, \phi \rangle_{L^2(U)} = \langle g, \Delta\phi \rangle_{L^2(U)} = 0$ for all $\phi \in C_c^\infty(U)$. If $\Delta g = 0$, then we say that g is *weakly harmonic*.

PROOF OF PROPOSITION 5.42. We equip $C_c^\infty(U)$ with the inner product $\langle \cdot, \cdot \rangle_1$. By Lemma 5.41, $\|\cdot\|_1$ defines an inner product on $H_0^1(U)$, which makes $H_0^1(U)$ into a Hilbert space. Let $f \in C^1(\overline{U})$ be as in the statement of the proposition, and notice that

$$\ell(u) = \langle u, f \rangle_1 = \sum_{j=1}^d \langle \partial_j u, \partial_j f \rangle_{L^2(U)} \tag{5.13}$$

defines a linear functional on $H_0^1(U)$ since

$$|\ell(u)| \leqslant \sum_{j=1}^d |\langle \partial_j u, \partial_j f \rangle_{L^2(U)}| \leqslant \sum_{j=1}^d \|\partial_j u\|_{L^2(U)} \|\partial_j f\|_{L^2(U)} \ll_f \|u\|_{H_0^1(U)}$$

for all $u \in H_0^1(U)$. Applying the Fréchet–Riesz representation theorem (Corollary 3.19) for the Hilbert space $H_0^1(U)$ with the inner product $\langle \cdot, \cdot \rangle_1$ we find some $v \in H_0^1(U)$ with

$$\ell(\phi) = \langle \phi, v \rangle_1 \tag{5.14}$$

for all $\phi \in H_0^1(U)$. This implies for every $\phi \in C_c^\infty(U)$ that

$$\langle \Delta\phi, f - v\rangle_{L^2(U)} = \sum_{j=1}^{d} \langle \partial_j^2\phi, f - v\rangle_{L^2(U)} \qquad \text{(by definition of } \Delta\text{)}$$

$$= -\sum_{j=1}^{d} \langle \partial_j\phi, \partial_j f - \partial_j v\rangle_{L^2(U)} \qquad \text{(by Lemma 5.10)}$$

$$= -\ell(\phi) + \ell(\phi) = 0, \qquad \text{(by (5.13) and (5.14))}$$

completing the proof. □

5.3.2 Elliptic Regularity for the Laplace Operator

In this section we will upgrade the conclusion from the previous section to show that the weakly harmonic function g is actually smooth and harmonic. The principle at work here is much more general, and is called *elliptic regularity*. We will again rely on Fourier series in the argument, and this will only give the result in the interior of U and not at the boundary ∂U. For this reason, it is natural to start with functions that have little structure on ∂U, as in the following definition.

Definition 5.43. A measurable function f on U is called *locally L^p* for some $p \in [1, \infty]$ if $\mathbb{1}_K f \in L^p(U)$ for every compact set $K \subseteq U$. In this case we write $f \in L^p_{\text{loc}}(U)$. A measurable function f on U is called *locally H^k* for some $k \in \mathbb{N}_0$ if $\chi f \in H^k(U)$ for all $\chi \in C_c^\infty(U)$. In this case we write $f \in H^k_{\text{loc}}(U)$.

Notice that the characteristic function $\mathbb{1}_K$ localizes f and removes the values of f near the boundary ∂U; in the second case χ has the same effect but is chosen to be C^∞ so as not to disturb any of the smoothness properties of f. Clearly $L^p(U) \subseteq L^p_{\text{loc}}(U)$, and by Lemma 5.36 we also have $H^k(U) \subseteq H^k_{\text{loc}}(U)$.

Exercise 5.44. Let f be a measurable function on an open subset $U \subseteq \mathbb{R}^d$. Show that f lies in $H^k_{\text{loc}}(U)$ if and only if $f|_{K^\circ} \in H^k(K^\circ)$ for every compact $K \subseteq U$, if and only if $\chi f \in H^k_0(U)$ for all $\chi \in C_c^\infty(U)$ (we have not used this as our definition as we will need the functions χ as in Definition 5.43 in the proofs anyway).

Theorem 5.45 (Elliptic regularity for Δ inside open subsets of $\mathbb{R}^d$). *Suppose that $U \subseteq \mathbb{R}^d$ is open and bounded, and $g \in H^1_{\text{loc}}(U)$. Assume that $\Delta g \in H^k_{\text{loc}}(U)$ for $k \in \mathbb{N}_0$, in the sense that there exists some $u \in H^k_{\text{loc}}(U)$ with*

$$\langle \Delta g, \phi\rangle_{L^2(U)} = \langle g, \Delta\phi\rangle_{L^2(U)} = \langle u, \phi\rangle_{L^2(U)}$$

for all $\phi \in C_c^\infty(U)$. Then $g \in H^{k+2}_{\text{loc}}(U)$.

The assumption of boundedness is not important, but simplifies the discussion slightly and is sufficient for all our applications of the theorem. Similarly, the assumption on the regularity of g could be significantly weakened.

Roughly speaking, the theorem says that if Δg exists, then the Sobolev-regularity of g must be two more than that of Δg. In other words, any non-smoothness of g will be visible also in Δg, or there is no cancellation of singularities when Δg is calculated from g. This remarkable result has many striking consequences, a few of which we list here.

Corollary 5.46. *If $g \in H^1_{\mathrm{loc}}(U)$ has $\Delta g = u \in C^\infty(U)$ (or g is weakly harmonic in the sense that $\Delta g = 0$), then $g \in C^\infty(U)$ satisfies $\Delta g = u$ (respectively is harmonic).*

PROOF. Since $u \in H^k_{\mathrm{loc}}(U)$ for all $k \in \mathbb{N}_0$, Theorem 5.45 implies that

$$g \in H^{k+2}_{\mathrm{loc}}(U)$$

for all $k \geqslant 0$. Hence $\chi g \in H^{k+2}(U)$ for all $k \geqslant 0$ and all functions $\chi \in C^\infty_c(U)$.

By the Sobolev embedding theorem for open subsets (Theorem 5.34), this implies that $\chi g \in C^\infty(U)$ for all $\chi \in C^\infty_c(U)$. Choosing $\chi \in C^\infty_c(U)$ equal to 1 on a neighbourhood of a given $x \in U$ shows that $g \in C^\infty(U)$, since it is C^∞ in a neighbourhood of each point. Finally, integration by parts gives

$$\langle \Delta g, \phi \rangle = \langle g, \Delta \phi \rangle = \langle \Delta g, \phi \rangle = \langle u, \phi \rangle$$

for all $\phi \in C^\infty_c(U)$. By density of $C^\infty_c(U) \subseteq L^2(U)$ and continuity of Δg and u we see that $\Delta g = u$. $\qquad\square$

The following might be even more surprising and will be important in the next chapter.

Corollary 5.47. *If $g \in H^1_{\mathrm{loc}}(U)$ is a weak eigenfunction of Δ in the sense that there exists a $\lambda \in \mathbb{C}$ with $\langle \Delta g, \phi \rangle_{L^2(U)} = \langle g, \Delta \phi \rangle_{L^2(U)} = \lambda \langle g, \phi \rangle_{L^2(U)}$ for all $\phi \in C^\infty_c(U)$, then $g \in C^\infty(U)$ and $\Delta g = \lambda g$.*

PROOF. By assumption, $\Delta g = \lambda g \in H^1_{\mathrm{loc}}(U)$ and so by Theorem 5.45 we also have $g \in H^3_{\mathrm{loc}}(U)$. However, this shows that $\Delta g = \lambda g \in H^3_{\mathrm{loc}}(U)$ and Theorem 5.45 may be applied again to see that $g \in H^5_{\mathrm{loc}}(U)$, and so on. It follows that $g \in H^k_{\mathrm{loc}}(U)$ for all $k \geqslant 0$, and arguing as in the proof of Corollary 5.46 we see that $g \in C^\infty(U)$. $\qquad\square$

We will prove Theorem 5.45 in two steps: firstly we deal with the case of functions on $\mathbb{T}^d$ (which turns out to be easy because of Fourier series), and secondly we show how to transfer the theorem from $\mathbb{T}^d$ to open subsets of U. Morally the second step (the transfer) should be the easy step as we are discussing the 'Laplace operator' on both of these spaces. However, some care is necessary as Δ has different meanings on $\mathbb{T}^d$ and on U since the spaces of allowed test functions in the definition of Δ are $C^\infty(\mathbb{T}^d)$ and $C^\infty_c(U)$, respectively.

Lemma 5.48 (Elliptic regularity on $\mathbb{T}^d$). *Let $g \in L^2(\mathbb{T}^d)$, and assume that $\Delta g = u \in H^k(\mathbb{T}^d)$ so $\langle \Delta g, \phi \rangle_{L^2(\mathbb{T}^d)} = \langle g, \Delta \phi \rangle_{L^2(\mathbb{T}^d)} = \langle u, \phi \rangle_{L^2(\mathbb{T}^d)}$ for all $\phi \in C^\infty(\mathbb{T}^d)$. Then $g \in H^{k+2}(\mathbb{T}^d)$.*

PROOF. If $\Delta g = u$ as in the lemma, then u is uniquely determined by g. Indeed, if $g = \sum_{n \in \mathbb{Z}^d} c_n \chi_n$ is the Fourier series of g then

$$u = - \sum_{n \in \mathbb{Z}^d} c_n (2\pi)^2 \|n\|_2^2 \chi_n$$

is the Fourier series of u. This follows from Fourier series on the torus (Theorem 3.54) since the characters χ_n are eigenfunctions of the Laplace operator:

$$\Delta \chi_n = (2\pi i)^2 \|n\|_2^2 \chi_n = -(2\pi)^2 \|n\|_2^2 \chi_n$$

and so

$$\langle u, \chi_n \rangle_{L^2(\mathbb{T}^d)} = \langle g, \Delta \chi_n \rangle_{L^2(\mathbb{T}^d)} = -(2\pi)^2 \|n\|_2^2 \underbrace{\langle g, \chi_n \rangle_{L^2(\mathbb{T}^d)}}_{=c_n}.$$

By assumption $u \in H^k(\mathbb{T}^d)$, which shows that

$$\sum_{n \in \mathbb{Z}^d} \left| c_n (2\pi)^2 \|n\|_2^2 \right|^2 \|n\|_2^{2k} < \infty$$

by the characterization of $H^k(\mathbb{T}^d)$ in terms of the Fourier series in Lemma 5.4. However, this is equivalent to

$$\sum_{n \in \mathbb{Z}^d} |c_n|^2 \|n\|_2^{2(k+2)} < \infty,$$

which implies that $g \in H^{k+2}(\mathbb{T}^d)$, again by Lemma 5.4. $\qquad\square$

At first it is a bit hard to pinpoint where the non-cancellation of singularities in Theorem 5.45 really comes from. However, if one is determined to find a single step within the proof to blame for this, then it would be the fact that the eigenvalues of the character χ_n grow with the rate $\|n\|_2^2$. This would, for instance, not be true for the non-elliptic (hyperbolic) partial differential operator $D = \partial_1^2 - \partial_2^2$ corresponding to the wave equation in two dimensions — for this operator the eigenvalues on characters can cancel (that is, be zero or much smaller than $\|n\|_2^2$) and there are also many non-smooth solutions to $Dg = 0$.

Before extending Lemma 5.48 to open subsets (Theorem 5.45), we state how the localizing step in the definition of $H_{\text{loc}}^\ell(U)$ in which g is replaced by χg affects weak derivatives and the assumption regarding Δg. As the statements of the next two lemmas should be easy to believe and the proofs rely only on the definitions and are arguably a bit tedious, we postpone their proofs until after the proof of Theorem 5.45.

Lemma 5.49. *Let $U \subseteq \mathbb{R}^d$ be open and $g \in H^\ell_{\mathrm{loc}}(U)$ with $\ell \geqslant 1$. Then there exists a weak partial derivative $\boldsymbol{\partial}_j g \in H^{\ell-1}_{\mathrm{loc}}(U) \subseteq L^2_{\mathrm{loc}}(U)$ with*

$$\langle \boldsymbol{\partial}_j g, \phi \rangle_{L^2(U)} = - \langle g, \partial_j \phi \rangle_{L^2(U)}$$

for all $\phi \in C^\infty_c(U)$ and $j = 1, \ldots, d$.

Lemma 5.50. *Let $U \subseteq \mathbb{R}^d$ be open and bounded. If g lies in $H^\ell_{\mathrm{loc}}(U)$ for some $\ell \geqslant 1$, $\boldsymbol{\Delta} g = u \in H^k_{\mathrm{loc}}(U)$ with $k \geqslant 0$, and $\chi \in C^\infty_c(U)$, then*

$$\boldsymbol{\Delta}(\chi g) = \underbrace{\chi u}_{\in H^k} + \underbrace{(\boldsymbol{\Delta}\chi)g}_{\in H^\ell} + \underbrace{2\sum_{j=1}^d (\partial_j \chi)(\boldsymbol{\partial}_j g)}_{\in H^{\ell-1}} \in H^{\min\{k,\ell-1\}}(U).$$

Notice that if g happened to be smooth, then the formula in the lemma calculates $\boldsymbol{\Delta}(\chi g)$ since in that case

$$\boldsymbol{\Delta}(\chi g) = \sum_{j=1}^d \partial_j^2(\chi g) = \sum_{j=1}^d \partial_j \left(\chi(\partial_j g) + (\partial_j \chi)g \right)$$

$$= \sum_{j=1}^d \chi(\partial_j^2 g) + 2(\partial_j \chi)(\partial_j g) + (\partial_j^2 \chi)g.$$

PROOF OF THEOREM 5.45. Let $R > 0$ and $U \subseteq B_R$ be an open subset of $\mathbb{R}^d$. Suppose $k \in \mathbb{N}_0$ and that $g \in H^1_{\mathrm{loc}}(U)$ has $\boldsymbol{\Delta} g = u \in H^k_{\mathrm{loc}}(U)$ weakly. We wish to show that $g \in H^{k+2}_{\mathrm{loc}}(U)$, and will do so by showing that $g \in H^\ell_{\mathrm{loc}}(U)$ by induction on $\ell \in \{1, \ldots, k+2\}$. The case $\ell = 1$ is the assumption in the theorem. So suppose that $1 \leqslant \ell \leqslant k+1$, $g \in H^\ell_{\mathrm{loc}}(U)$, and fix $\chi \in C^\infty_c(U)$. Then $\min(k, \ell-1) = \ell-1$ and $\boldsymbol{\Delta}(\chi g) = u_1 \in H^{\ell-1}(U)$ weakly by Lemma 5.50. This means that

$$\langle u_1, \phi \rangle_{L^2(U)} = \langle \chi g, \Delta \phi \rangle_{L^2(U)} \tag{5.15}$$

for all $\phi \in C^\infty_c(U)$.

Using the fact that $\chi \in C^\infty_c(U)$ and $U \subseteq B_R$, we can now make a switch in this formula to $\mathbb{T}^d_R$ as follows. For this we will use Lemma 5.36 and its notation. Let $\psi \in C^\infty_c(U)$ be such that $\psi \equiv 1$ on $\mathrm{Supp}\,\chi$. Then $\psi\phi$ lies in $C^\infty_c(U)$ for any $\phi \in C^\infty(\mathbb{T}^d_R)$, where $\mathbb{T}^d_R = \mathbb{R}^d/(2R\mathbb{Z})^d$ and functions on $\mathbb{T}^d_R$ are identified with $(2R\mathbb{Z})^d$-periodic functions on $\mathbb{R}^d$. Applying (5.15) to $\psi\phi$ we get

$$\langle P(\psi u_1), \phi \rangle_{L^2(\mathbb{T}^d_R)} = \langle \psi u_1, \phi \rangle_{L^2(U)} = \langle u_1, \psi\phi \rangle_{L^2(U)} = \langle \chi g, \Delta(\psi\phi) \rangle_{L^2(U)}.$$

Since ψ is one and its derivatives are zero at any point of $\mathrm{Supp}\,\chi$, we can remove ψ on the right-hand side. One may wonder why ψ is introduced in the first place, since it is only brought in so that it can be removed again. The answer lies in the definition of Δ which depends crucially on a choice of test functions. In particular, Δ is defined differently on $\mathbb{T}^d$ and on U — we use ψ to bridge between these two definitions.

We now obtain

$$\langle P(\psi u_1), \phi \rangle_{L^2(\mathbb{T}_R^d)} = \langle \chi g, \Delta \phi \rangle_{L^2(U)} = \langle P(\chi g), \Delta \phi \rangle_{L^2(\mathbb{T}_R^d)}$$

for any $\phi \in C^\infty(\mathbb{T}_R^d)$. In fact, by definition of $H_{\mathrm{loc}}^k(U)$ we have $\chi g \in H^k(U)$. By Lemma 5.36 $\psi \chi g \in H_0^k(U)$, but $\psi \chi g = \chi g$ by our choice of ψ, so we deduce that $\chi g \in H_0^k(U)$ and $P(\chi g) \in H^k(\mathbb{T}_R^d)$ is defined by Lemma 5.36.

In other words, we have shown that $\Delta(P(\chi g)) = P(\psi u_1) \in H^{\ell-1}(\mathbb{T}_R^d)$ weakly by Lemma 5.36 (where elements of $C^\infty(\mathbb{T}_R^d)$ are used as the test functions). By Lemma 5.48 it follows that $P(\chi g) \in H^{\ell+1}(\mathbb{T}_R^d)$, which by the last claim of Lemma 5.36 allows us to pull the statement back to U and deduce that χg lies in $H^{\ell+1}(U)$. Since this holds for all $\chi \in C_c^\infty(U)$ we see that $g \in H_{\mathrm{loc}}^{\ell+1}(U)$. Repeating the argument and increasing ℓ each time, we eventually reach $\ell = k+1$ and then $g \in H_{\mathrm{loc}}^{\ell+1}(U) = H_{\mathrm{loc}}^{k+2}(U)$. $\square$

We now give the proof of the lemmas that were used in the theorem. In the remainder of the chapter we will only consider the inner product on $L^2(U)$ and hence will again simply write $\langle \cdot, \cdot \rangle$.

PROOF OF LEMMA 5.49. Fix some $j \in \{1, \ldots, d\}$. Let $V \subseteq U$ be an open subset with $\overline{V} \subseteq U$ compact, and choose $\chi \in C_c^\infty(U)$ with $\chi \equiv 1$ on V. Then $\chi g \in H^\ell(U)$ has a weak partial derivative along x_j, which we will denote by $g_{\chi,j} \in H^{\ell-1}(U)$. By definition, we now have for the inner products in $L^2(U)$ that

$$\langle g_{\chi,j}, \phi \rangle = - \langle \chi g, \partial_j \phi \rangle = - \langle g, \partial_j \phi \rangle$$

for all $\phi \in C_c^\infty(V) \subseteq C_c^\infty(U)$. This shows that $g_{\chi,j}|_V \in H^{\ell-1}(V)$ is the weak partial derivative of $g|_V$ along x_j and by the properties of weak derivatives (Lemma 5.10) is therefore uniquely determined by $g|_V$ (and independent of χ).

Now write $U = \bigcup_{n \geqslant 1} V_n$ for an increasing sequence of open subsets of U with compact closures within[†] U, and define g_j to be $g_{\chi_n,j}$ on V_n where χ_n and $g_{\chi_n,j}$ are the functions as above corresponding to the set $V = V_n$. By Lemma 5.10 the function g_j is well-defined almost everywhere.

Let $\phi \in C_c^\infty(U)$, then there exists some n with $\mathrm{Supp}\,\phi \subseteq V_n$, which gives $\langle g_j, \phi \rangle = \langle g_{\chi_n,j}, \phi \rangle = - \langle g, \partial_j \phi \rangle$. Moreover, $\phi g_j = \phi \chi_n g_{\chi_n,j} \in H^{\ell-1}(U)$ by Lemma 5.36. As these two facts hold for every ϕ in $C_c^\infty(U)$, we see that $g_j \in H_{\mathrm{loc}}^{\ell-1}(U)$ is a weak partial derivative of g along x_j. $\square$

[†] For example, $V_n = \{x \mid \mathrm{d}(x, \mathbb{R}^d \smallsetminus U) > \frac{1}{n}\} \cap B_n$.

PROOF OF LEMMA 5.50. By assumption $u \in H^k_{\text{loc}}(U)$ and so $\chi u \in H^k(U)$ by definition of $H^k_{\text{loc}}(U)$ (Definition 5.43). Similarly, $g \in H^\ell_{\text{loc}}(U)$ and so $(\Delta\chi)g$ lies in $H^\ell(U)$. Finally, by assumption, $g \in H^\ell_{\text{loc}}(U)$ with $\ell \geqslant 1$ and so $\partial_j g$ lies in $H^{\ell-1}_{\text{loc}}(U)$ by Lemma 5.49, which gives $(\partial_j\chi)(\partial_j g) \in H^{\ell-1}(U)$. Therefore,

$$\chi u + (\Delta\chi)g + 2\sum_{j=1}^{d}(\partial_j\chi)(\partial_j g) \in H^{\min\{k,\ell-1\}}(U)$$

and it remains to show that this function is equal to $\Delta(g\chi)$ weakly. For this, recall that $\Delta g = u$, let $\phi \in C^\infty_c(U)$, and calculate

$$\left\langle \chi u + (\Delta\chi)g + 2\sum_{j=1}^{d}(\partial_j\chi)(\partial_j g), \phi \right\rangle$$

$$= \langle u, \chi\phi \rangle + \langle g, (\Delta\chi)\phi \rangle + 2\sum_{j=1}^{d}\langle \partial_j g, (\partial_j\chi)\phi \rangle$$

$$= \langle g, \Delta(\chi\phi) \rangle + \langle g, (\Delta\chi)\phi \rangle - 2\sum_{j=1}^{d}\langle g, \partial_j((\partial_j\chi)\phi) \rangle$$

$$= \left\langle g, (\Delta\chi)\phi + 2\sum_{j=1}^{d}(\partial_j\chi)(\partial_j\phi) + \chi\Delta\phi \right.$$

$$\left. + (\Delta\chi)\phi - 2\sum_{j=1}^{d}(\partial_j^2\chi)\phi - 2\sum_{j=1}^{d}(\partial_j\chi)(\partial_j\phi) \right\rangle$$

$$= \langle g, \chi\Delta\phi \rangle = \langle \chi g, \Delta\phi \rangle .$$

As $\phi \in C^\infty_c(U)$ was arbitrary we see that

$$\Delta(\chi g) = \chi u + (\Delta\chi)g + 2\sum_{j=1}^{d}(\partial_j\chi)(\partial_j g)$$

weakly. $\square$

5.3.3 Dirichlet's Boundary Value Problem

For $k \geqslant 0$ and a bounded open set $U \subseteq \mathbb{R}^d$ the function space $C^k(\overline{U})$ consists of all continuous functions with $f|_U \in C^k(U)$ such that the partial derivatives extend continuously to the closure $\overline{U}$. If U has C^k-smooth boundary, then the function space $C^k(\partial U)$ is defined using the assumption that ∂U has the structure of a manifold; the local charts allow smoothness properties to be transported onto ∂U from a suitable subset of $\mathbb{R}^{d-1}$; an example of how this may be done appears in the proof of Theorem 5.51 below.

Theorem 5.51 (Dirichlet's boundary value problem). *Let $U \subseteq \mathbb{R}^d$ be open and bounded with C^1-smooth boundary and let $f \in C^1(\partial U)$. Then there exists a function $g \in H^1(U)$ such that $g|_U$ in $C^\infty(U)$ is harmonic and $g|_{\partial U}$ is equal to f in the square-mean sense. If $d = 2$ then g extends uniquely to an element in $C(\overline{U})$, and $g|_{\partial U} = f$.*

PROOF OF THE EXISTENCE OF A SOLUTION IN THE SQUARE-MEAN SENSE. Since U has C^1-smooth boundary, we can find an extension of f, again denoted by f, to a function in $C^1(\overline{U})$. We only sketch the argument. In an open neighbourhood of a point $z^{(0)}$ in ∂U as in Definition 5.31, such a function can, for example, be chosen to be independent of the y-coordinate. Using compactness we find finitely many open sets $V_1, \ldots, V_k$ covering ∂U and bounded C^1-functions f_j defined on V_j with $f_j(x) = f(x)$ for $x \in \partial U \cap V_j$ and $j = 1, \ldots, k$. Using a smooth partition $\psi_1, \ldots, \psi_k$ of unity (that is, functions $\psi_j \in C_c^\infty(V_j)$ for $j = 1, \ldots, k$ with $\sum_{j=1}^k \psi_j|_{\partial U} \equiv 1$; see Exercise 5.52) with the cover $V_1, \ldots, V_k, V_{k+1} = U$ we may define $f = \sum_{j=1}^k \psi_j f_j$ (where $\psi_j f_j(x) = 0$ for $x \notin V_j$ and $j = 1, \ldots, k$) and restrict it to $\overline{U}$.

Now apply Proposition 5.42 to f, giving functions v in $H_0^1(U)$ and g in $H^1(U)$ such that $f = g + v$ with g weakly harmonic in U. Now Corollary 5.46 implies that $g \in C^\infty(U)$, so that Δg (in the usual sense) is well-defined. By integration by parts (see Lemma 5.10), it follows that

$$\langle \Delta g, \phi \rangle = \langle g, \Delta \phi \rangle = \langle \Delta g, \phi \rangle = 0$$

for all $\phi \in C_c^\infty(U)$, showing that $\Delta g = 0$. That is, g is a harmonic function.

Moreover, v vanishes at ∂U in the square-mean sense (Proposition 5.33), so $g|_{\partial U} = f|_{\partial U}$ in $L^2(\partial U)$. $\qquad\square$

Essential Exercise 5.52 (Smooth partion of unity). Let $U \subseteq \mathbb{R}^d$ be open and bounded, let $V_1, \ldots, V_k$ be an open cover of $\overline{U}$ and let $V_0 = \mathbb{R}^d \setminus \overline{U}$. Show that there exist smooth functions $\psi_j \in C_c^\infty(V_j)$ for $j = 0, \ldots, k$ (always extended to $\mathbb{R}^d$ by setting $\psi_j(x) = 0$ for $x \in \mathbb{R}^d \setminus V_j$) such that $\sum_{j=0}^k \psi_j = 1$.

Using the averaging property of harmonic functions (Proposition 5.53) and Lemma 5.55 (which only works in two dimensions), we will upgrade the first part to give the second part of the theorem for $d = 2$.

Proposition 5.53 (Mean value principle). *Let $U \subseteq \mathbb{R}^d$ be open, and let $\phi \in C^\infty(U)$ be a harmonic function. Let $x_0 \in U$ and $r > 0$ be chosen with $\overline{B_r(x_0)} \subseteq U$. Then the value of the harmonic function at x_0 is equal to the average over the sphere of radius r around x_0, that is*

$$\phi(x_0) = \frac{1}{\sigma(r\mathbb{S}^{d-1})} \int_{r\mathbb{S}^{d-1}} \phi(x_0 + x) \, d\sigma(x),$$

where σ denotes the natural area measure on the sphere $r\mathbb{S}^{d-1}$.

PROOF. Without loss of generality $x_0 = 0$. The proof consists of applying the d-divergence theorem to the vector field $f(x) = \phi(x)\nabla v(x) - v(x)\nabla\phi(x)$, where $v : \mathbb{R}^d\backslash\{0\} \to \mathbb{R}$ is an auxiliary function. In fact, v is defined by

$$v(x) = \begin{cases} \log\|x\|_2 & \text{for } d = 2, \\ \dfrac{1}{\|x\|_2^{d-2}} & \text{for } d > 2. \end{cases}$$

A direct calculation shows that it satisfies

$$\nabla v = \begin{cases} \dfrac{1}{\|x\|_2}\dfrac{x}{\|x\|_2} & \text{for } x \neq 0 \text{ and } d = 2, \\ \dfrac{2-d}{\|x\|_2^{d-1}}\dfrac{x}{\|x\|_2} & \text{for } x \neq 0 \text{ and } d > 2, \end{cases}$$

and $\Delta v = \text{div}\,\nabla v = 0$ for $x \neq 0$. For $\varepsilon \in (0, r)$ the divergence theorem applied to f on the annulus $B_r\backslash B_\varepsilon \subseteq \mathbb{R}^d$ has the form

$$\int_{\partial B_r} f \cdot n\,\mathrm{d}\sigma - \int_{\partial B_\varepsilon} f \cdot n\,\mathrm{d}\sigma = \int_{B_r\backslash B_\varepsilon} \text{div}\,f\,\mathrm{d}x_1\cdots\mathrm{d}x_d,$$

where $n = \dfrac{x}{\|x\|_2}$ is the normalized outward normal vector to the sphere of radius $\|x\|_2$ at x and we also write σ for the area measure on ∂B_ε. For the right-hand side, we calculate for f as above

$$\text{div}\,f = \text{div}\,(\phi\nabla v - v\nabla\phi) = \nabla\phi \cdot \nabla v + \phi\Delta v - \nabla v \cdot \nabla\phi - v\Delta\phi = 0.$$

Therefore

$$\int_{\partial B_r} f \cdot n\,\mathrm{d}\sigma = \int_{\partial B_\varepsilon} f \cdot n\,\mathrm{d}\sigma. \tag{5.16}$$

For x with $\|x\|_2 = r$ we have

$$\begin{aligned} f(x) \cdot n &= (\phi(x)\nabla v(x) - v(x)\nabla\phi(x)) \cdot n \\ &= \phi(x)\frac{c_1}{\|x\|_2^{d-1}}\frac{x}{\|x\|_2} \cdot \frac{x}{\|x\|_2} - v(x)\nabla\phi(x) \cdot n \\ &= \phi(x)\frac{c_1}{r^{d-1}} - c_2\nabla\phi(x) \cdot n, \end{aligned}$$

where $c_1 = 1$ if $d = 2$ or $c_1 = 2 - d$ if $d > 2$ and $c_2 = v(x)$ which is a constant for $\|x\| = r$ since v depends only on $\|x\|_2$. Furthermore, we notice that

$$\int_{\partial B_r} \nabla\phi \cdot n\,\mathrm{d}\sigma = \int_{B_r} \underbrace{\text{div}\,\nabla\phi}_{=\Delta\phi=0}\,\mathrm{d}x_1\cdots\mathrm{d}x_d = 0.$$

Using this and the analogous formula for $\|x\| = \varepsilon$ allows us to write (5.16) as

$$\frac{1}{r^{d-1}} \int_{r\mathbb{S}^{d-1}} \phi \, \mathrm{d}\sigma = \frac{1}{\varepsilon^{d-1}} \int_{\varepsilon\mathbb{S}^{d-1}} \phi(x) \, \mathrm{d}\sigma.$$

Now divide by the area of $\mathbb{S}^{d-1}$ and notice that we get

$$\frac{1}{\sigma(r\mathbb{S}^{d-1})} \int_{r\mathbb{S}^{d-1}} \phi \, \mathrm{d}\sigma = \frac{1}{\sigma(\varepsilon\mathbb{S}^{d-1})} \int_{\varepsilon\mathbb{S}^{d-1}} \phi \, \mathrm{d}\sigma \longrightarrow \phi(0)$$

as $\varepsilon \to 0$, by continuity of ϕ. $\qquad\square$

Exercise 5.54. Use Proposition 5.53 to prove that any bounded harmonic function on $\mathbb{R}^d$ is constant.

Lemma 5.55 (Convergence on average). *Suppose that $U \subseteq \mathbb{R}^2$ is open and bounded with C^1-smooth boundary. Let $v \in H_0^1(U)$, which we extend to a function on $\mathbb{R}^2$ by setting it equal to 0 outside U. Then for any $z^{(0)} \in \partial U$ we have*

$$\frac{1}{\varepsilon^2} \int_{z^{(0)}+B_\varepsilon} |v| \, \mathrm{d}x \, \mathrm{d}y \longrightarrow 0 \tag{5.17}$$

as $\varepsilon \to 0$, uniformly for all $z^{(0)} \in \partial U$.

PROOF. To prove (5.17) for $z^{(0)} \in \partial U$ we use the assumption that U has C^1-smooth boundary and rotate the coordinate system so that $B_\delta(z^{(0)}) \cap U$ can be described as in Definition 5.31 (see also Exercise 5.32). By a further rotation and by shrinking δ if necessary we may also assume that $|\phi'(x)| < 1$ for all $x \in B_\delta(x_0)$ with the notation $z^{(0)} = (x_0, y_0)$ and $\phi \in C^1(B_\varepsilon(x_0))$ as in Definition 5.31. We claim that for $\varepsilon \in (0, \frac{1}{4}\delta)$, $x_1 \in B_\delta(x_0)$, and $y_1 = \phi(x_1)$ we have

$$\int_{x_1-\varepsilon}^{x_1+\varepsilon} \int_{y_1-\varepsilon}^{y_1+\varepsilon} |v| \, \mathrm{d}x \, \mathrm{d}y \leqslant 2\varepsilon^2 \sqrt{\int_{x_1-\varepsilon}^{x_1+\varepsilon} \int_{y_1-\varepsilon}^{y_1+\varepsilon} |\partial_2 v|^2 \, \mathrm{d}x \, \mathrm{d}y}, \tag{5.18}$$

first for $v \in C_c^\infty(U)$, which then extends by continuity to all $v \in H_0^1(U)$. We will call the domain of integration $[x_1 - \varepsilon, x_1 + \varepsilon] \times [y_1 - \varepsilon, y_1 + \varepsilon]$ the ε-*box around* (x_1, y_1).

To prove (5.18) for $v \in C_c^\infty(U)$ we note first that $(x, y_1 + \varepsilon) \notin U$ for all x with $|x - x_1| < \varepsilon$ (by the local description of U in terms of ϕ, the assumption that $|\phi'(x)| < 1$, and the mean value theorem). If $(x, y) \in U$ with $|x - x_1|, |y - y_1| < \varepsilon$ this gives

$$v(x, y) = -\int_y^{y_1+\varepsilon} \partial_2 v(x, s) \, \mathrm{d}s.$$

Now integrate the absolute value of $v(x, y)$ with respect to x and y to get

$$\int_{x_1-\varepsilon}^{x_1+\varepsilon} \int_{y_1-\varepsilon}^{y_1+\varepsilon} |v(x,y)| \, dy \, dx \leqslant \int_{x_1-\varepsilon}^{x_1+\varepsilon} \int_{y_1-\varepsilon}^{y_1+\varepsilon} \int_y^{y_1+\varepsilon} |\partial_2 v(x,s)| \, ds \, dy \, dx$$

$$= \int_{x_1-\varepsilon}^{x_1+\varepsilon} \int_{y_1-\varepsilon}^{y_1+\varepsilon} |\partial_2 v(x,s)| |s - y_1 + \varepsilon| \, ds \, dx,$$

where we simply switched the order of integration, used the identity

$$\mathbb{1}_{[y,y_1+\varepsilon]}(s) = \mathbb{1}_{[y_1-\varepsilon,s]}(y)$$

for $s, y \in [y_1 - \varepsilon, y_1 + \varepsilon]$, and evaluated the integral over y. Writing Q_ε for the ε-box $[x_1 - \varepsilon, x_1 + \varepsilon] \times [y_1 - \varepsilon, y_1 + \varepsilon]$ around (x_1, y_1) and applying Cauchy–Schwarz to the last integral we get

$$\iint_{Q_\varepsilon} |v(x,y)| \, dy \, dx \leqslant \sqrt{\iint_{Q_\varepsilon} |\partial_2 v|^2 \, dx \, dy} \sqrt{\iint_{Q_\varepsilon} |s - y_1 + \varepsilon|^2 \, ds \, dx}$$

$$\leqslant 2\sqrt{\iint_{Q_\varepsilon} |\partial_2 v|^2 \, dx \, dy} \sqrt{\varepsilon^4}$$

since $|s - y_1 + \varepsilon| \leqslant 2\varepsilon$ for all $s \in (y_1 - \varepsilon, y_1 + \varepsilon)$. This gives the estimate claimed.

It is clear that (5.18) implies (5.17) since $\partial_2 v \in L^2(U)$ and hence the L^2 norm of $\partial_2 v$ restricted to smaller and smaller subsets of neighbourhoods of $z^{(1)}$ converges to zero.

The uniformity claim in the lemma also follows from the discussion above. Indeed, we only assumed that $x_1 \in \left(x_0 - \frac{\delta}{4}, x_0 + \frac{\delta}{4}\right)$ and $\varepsilon \in \left(0, \frac{\delta}{4}\right)$.

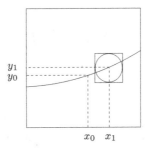

Fig. 5.4: The point $z^{(1)} = (x_1, y_1) \in \partial U$ and the ε-box Q_ε, containing the ε-ball.

We can make the L^2 norms of $\partial_2 v$ restricted to the ε-box around $z^{(1)}$ uniformly small for all $z^{(1)}$ as above by using the fact that

$$\int_{x_0-\frac{\delta}{2}}^{x_0+\frac{\delta}{2}} \int_{\phi(x_1)-2\varepsilon}^{\phi(x_1)} |\partial_2 v| \, dy \, dx \longrightarrow 0$$

as $\varepsilon \to 0$ by dominated convergence. Since the ε-ball around $z^{(1)}$ as in (5.17) is contained in the ε-box Q_ε around $z^{(1)}$ as in (5.18) we obtain (5.17) uniformly for $z^{(1)} = (x_1, \phi(x_1)) \in \partial U$ and $x_1 \in (x_0 - \frac{\delta}{2}, x_0 + \frac{\delta}{2})$. Using the fact that ∂U is compact we can find finitely many neighbourhoods of points $z^{(0)} \in \partial U$ as above, and the lemma follows. $\qquad\square$

Exercise 5.56. Describe what prevents the proof of Lemma 5.55 from extending to higher dimensions by trying to emulate the calculations involved.

PROOF OF POINTWISE BOUNDARY CONDITION IN THEOREM 5.51. Recall that we already established that f can be written as $f = g + v$ with $g \in H^1(U)$ harmonic and $v \in H_0^1(U)$. Let $z \in U$ be a point of distance ε from ∂U, and write

$$\overline{f}(z) = \frac{4}{\varepsilon^2 \pi} \int_{z+B_{\varepsilon/2}} f(w) \, dw$$

for the average of f over the ball of radius $\varepsilon/2$ with centre z (and similarly define $\overline{g}$ and $\overline{v}$). Then $\overline{g}(z) = g(z)$ by the mean value property in Proposition 5.53. By uniform continuity of f we also have $f(z) - \overline{f}(z) = o(1)$ as $\varepsilon \to 0$. Finally, by the convergence on average of $\overline{v}(z)$ in Lemma 5.55, $\overline{v}(z) = o(1)$ as $\varepsilon \to 0$. Thus

$$g(z) - f(z) = \overline{g}(z) - \overline{f}(z) - \overline{v}(z) + o(1)$$

for all $z \in U$ at distance ε from ∂U as $\varepsilon \to 0$, where the sum of the three functions on the right is equal to 0. This shows that g extends continuously to ∂U and agrees there with f, and so concludes the proof of the theorem. $\qquad\square$

We finish our discussion of the Dirichlet boundary value problem by outlining how to establish the uniqueness of solutions by the methods we have developed so far.

Exercise 5.57. Let $U \subseteq \mathbb{R}^d$ be open and bounded with smooth boundary. To avoid complications arising from the geometry of the boundary, you may also suppose that $U \subseteq \mathbb{R}^d$ is convex (for instance, U could be as in Figure 5.2 on p. 148). Let f be a function in $C^1(\partial U)$ and suppose that $g_1, g_2 \in C^\infty(U) \cap H^1(U)$ both solve the Dirichlet boundary value problem. Taking the difference $g = g_1 - g_2$ we obtain an element of $H^1(U)$ that vanishes at ∂U in the square-mean sense. Show that this implies $g \in H_0^1(U)$ and deduce that $g = 0$.

5.4 Further Topics

- We will continue our excursion into Sobolev spaces in Chapter 6, where we will prove the existence of an orthonormal basis consisting of eigenfunctions of the Laplace operator (see Section 6.4).

- In Section 8.2.2 we will return to the topic of elliptic regularity one more time and will present an argument that also gives the result at the boundary of U.

Chapter 6
Compact Self-Adjoint Operators and Laplace Eigenfunctions

There is no doubt that eigenvalues and eigenvectors are of fundamental importance in linear algebra and in its applications, both within and outside mathematics. In finite dimensions eigenvectors always exist over $\mathbb{C}$ because the corresponding eigenvalues arise as zeros of a polynomial. However, even in finite dimensions the eigenvectors are not guaranteed to give a basis of the space (because there may be non-trivial Jordan blocks). However, if the linear map is self-adjoint (over $\mathbb{R}$ or over $\mathbb{C}$; see Definition 6.22) or unitary (over $\mathbb{C}$) then there is an orthonormal basis consisting of eigenvectors. That is, in these cases the linear map can be diagonalized.

In infinite dimensions the inherent complications of linear algebra in finite dimensions have added to them entirely new phenomena, illustrated by the following exercise.

Exercise 6.1. (a) Let $\mathcal{H} = \ell^2(\mathbb{Z})$ and let $U : \mathcal{H} \to \mathcal{H}$ be the operator defined by by $U((x_n)_{n \in \mathbb{Z}}) = (x_{n+1})_{n \in \mathbb{Z}}$, which simply shifts the sequence by one step to the left. Show that the operator U has no eigenvectors.
(b) Let $\mathcal{H} = \ell^2(\mathbb{N})$ and define the operator $S : \mathcal{H} \to \mathcal{H}$ by $S((x_n)_{n \in \mathbb{N}}) = (x_{n+1})_{n \in \mathbb{N}}$, which again simply shifts the sequence one step to the left, but now 'forgets' the first entry of the sequence. Show that S has uncountably many different eigenvalues. In particular, deduce that there are too many eigenvalues to hope for a diagonalization (since the space $\mathcal{H}$ is separable, a 'diagonal' map would only have countably many eigenvalues).

There is an important class of operators for which some of the difficulties illustrated in Exercise 6.1 do not arise. These are the compact operators which will be defined in Section 6.1. In Section 6.2 we then prove that compact self-adjoint operators can be diagonalized using an orthonormal basis, and relate this to the Sturm–Liouville equation from Section 2.5.2.

Using this and results from Chapter 5, we will prove in Section 6.4 the existence of a basis of eigenfunctions of the Laplace operator claimed in Section 1.2 for a bounded domain $U \subseteq \mathbb{R}^d$. At first sight this is surprising, since the Laplace operator Δ is not even bounded on $L^2(U)$ (it is also not defined on all of $L^2(U)$, which is a related issue by the closed graph theorem (Theorem 4.28)), but we will find a compact operator defined on all of $L^2(U)$ whose eigenfunctions are precisely the eigenfunctions of Δ.

© Springer International Publishing AG 2017
M. Einsiedler, T. Ward, *Functional Analysis, Spectral Theory, and Applications*,
Graduate Texts in Mathematics 276, DOI 10.1007/978-3-319-58540-6_6

6.1 Compact Operators

Bounded linear operators with finite-dimensional image have the property that the image of a bounded set has compact closure. Requiring the latter property gives rise to a natural generalization of the class of operators with finite-dimensional image.

Definition 6.2. Let V and W be normed vector spaces, and let $L : V \to W$ be a linear operator. Then L is said to be a *compact operator* if the closure

$$\overline{L(B_1^V)} \subseteq W$$

of the image of the unit ball is compact in W. We will sometimes write $K(V, W)$ for the space of compact operators, and if $V = W$ we will write $K(V)$ for the space of compact operators from V to V.

We will see in Example 6.5 that $L(\overline{B_1^V})$ is in general not closed, even if L is a compact operator. Since compact sets are bounded, every compact operator is also bounded, but the converse does not hold. For example, the identity operator $V \to V$ on an infinite-dimensional normed vector space is not a compact operator by Proposition 2.35. As noted above, if $L : V \to W$ is a bounded operator and $L(V)$ is finite-dimensional, then L is a compact operator. We will see many more examples after we prove a few basic properties of compact operators.

Lemma 6.3 (Composition). *Let V_1, V_2, V_3 be normed vector spaces, and let $L_1 : V_1 \to V_2$ and $L_2 : V_2 \to V_3$ be bounded operators. If L_1 or L_2 is a compact operator, then so is $L_2 \circ L_1$.*

PROOF. Suppose that L_1 is compact. Then $L_2\big(L_1\big(B_1^{V_1}\big)\big) \subseteq L_2\big(\overline{L_1\big(B_1^{V_1}\big)}\big)$ and the latter is compact since $L_1\big(B_1^{V_1}\big)$ has compact closure and L_2 is continuous. It follows that $L_2\big(L_1\big(B_1^{V_1}\big)\big)$ is contained in a compact subset of V_3, so its closure is compact, and therefore $L_2 \circ L_1$ is a compact operator.

If L_2 is compact, then $L_2 \circ L_1\big(B_1^{V_1}\big) \subseteq \overline{L_2\big(\|L_1\|_{\mathrm{op}} B_1^{V_2}\big)} = \|L_1\|_{\mathrm{op}}\overline{L_2\big(B_1^{V_2}\big)}$, which is compact, and so $L_1 \circ L_2$ is again compact. $\square$

Exercise 6.4. Let V and W be two normed vector spaces. Show that

$$K(V, W) = \{L : V \to W \mid L \text{ is a compact operator}\}$$

is a linear subspace. Deduce that if $V = W$ is a Banach space, then $K(V) = K(V, V)$ is a two-sided ideal in the Banach algebra $B(V)$. That is, if $L \in K(V)$ and $A \in B(V)$, then $A \circ L$ and $L \circ A$ lie in $K(V)$.

Example 6.5. (a) The inclusion map

$$\imath : \big(C^1([0, 1]), \|\cdot\|_{C^1}\big) \longrightarrow \big(C([0, 1]), \|\cdot\|_\infty\big)$$

is a compact operator. This follows from the Arzela–Ascoli theorem (Theorem 2.38), since

$$\imath\big(B_1^{C^1([0,1])}\big) \subseteq \{f \in C([0,1]) \mid \|f\|_\infty \leqslant 1, f \text{ is 1-Lipschitz}\}.$$

This example shows that it is necessary to take the closure of the image and not just the image of the closed ball: for example, the function f defined by $f(x) = |x - \frac{1}{2}|$ belongs to $\overline{L(B_1^{C^1([0,1])})}$ but not to $L(B_1^{C^1([0,1])})$.
(b) For $f \in C([0,1])$ and $x \in [0,1]$, define

$$T(f)(x) = \int_0^x f(t)\,\mathrm{d}t.$$

Then $T : C([0,1]) \to C([0,1])$ is compact, since $T : C([0,1]) \to C^1([0,1])$ is bounded and the inclusion $C^1([0,1]) \to C([0,1])$ is compact by (a).

Exercise 6.6. For which $p \geqslant 1$ is the operator sending $f \in L^p([0,1])$ to the function in $C([0,1])$ defined by $x \mapsto \int_0^x f(t)\,\mathrm{d}t$ a compact operator?

Many of the compact operators that we will encounter have a similar flavour to the example above. They either map from a space of functions with more regularity properties (in this instance, differentiability) to a space of functions with fewer regularity properties (in this case, continuity), or are integral operators. The next lemma is a useful tool for proving compactness of bounded operators.

Lemma 6.7 (Uniform approximation). *Let V be a normed vector space, and let W be a Banach space. Suppose that (L_n) is a sequence of compact operators $V \to W$, and suppose that $L_n \to L \in \mathrm{B}(V,W)$ as $n \to \infty$ with respect to the operator norm. Then L is a compact operator as well.*

Lemma 6.7 improves the claim from Exercise 6.4 in that the two-sided ideal $\mathrm{K}(V)$ in $\mathrm{B}(V)$ is even closed for any Banach space V (see also Exercise 6.8).

PROOF OF LEMMA 6.7. Let $M = \overline{L(B_1^V)} \subseteq W$. Since W is assumed to be a Banach space, M is complete. It remains to show that M is totally bounded (see Section A.4 for the notion and for the equivalence to compactness). Let $\varepsilon > 0$ and choose L_n with $\|L_n - L\| < \varepsilon$. Since L_n is compact, we know that $\overline{L_n(B_1^V)}$ is compact and hence is totally bounded. It follows that there exist elements $w_1, \ldots, w_m \in \overline{L_n(B_1^V)}$ with

$$\overline{L_n(B_1^V)} \subseteq \bigcup_{i=1}^m B_\varepsilon^W(w_i).$$

For each w_i there exists some $v_i \in B_1^V$ with $\|w_i - L_n(v_i)\| < \varepsilon$.
If now $v \in B_1^V$, then for some $i \in \{1, \ldots, m\}$ we have

$$\|L_n(v) - L_n(v_i)\| < 2\varepsilon.$$

Now $\|L_n - L\| < \varepsilon$ and $\|v\|, \|v_i\| < 1$ so

$$\|L(v) - L(v_i)\| \leqslant \|L(v) - L_n(v)\| + \|L_n(v) - L_n(v_i)\| + \|L_n(v_i) - L(v_i)\| < 4\varepsilon.$$

It follows that

$$L\big(B_1^V\big) \subseteq \bigcup_{i=1}^m B_{4\varepsilon}^W (L(v_i)),$$

which implies that the points $L(v_i)$ for $i = 1, \ldots, m$ are 5ε-dense in the set $M = \overline{L\big(B_1^V\big)}$. As ε was arbitrary, M is therefore totally bounded, so M is a compact set and hence L is a compact operator. $\qquad \square$

Exercise 6.8. Continuing the discussion from Exercise 6.4, show that $B(V)/K(V)$ becomes a Banach algebra — the Calkin algebra — by defining $(A + K(V))(B + K(V))$ to be $AB + K(V)$ for all $A, B \in B(V)$ and using the quotient norm $\| \cdot \|_{B(V)/K(V)}$.

Exercise 6.9. In each of the following, justify your claim.
(a) Is the inclusion map $\imath_{k+1,k} : H^{k+1}(\mathbb{T}^d) \longrightarrow H^k(\mathbb{T}^d)$ from Proposition 5.3 a compact operator?
(b) Let $U \subseteq \mathbb{R}^d$ be an open set. Show that the inclusion $\imath_{k+1,k} : C_b^{k+1}(U) \longrightarrow C_b^k(U)$ is a compact operator if $\overline{U} \subseteq \mathbb{R}^d$ is compact. Show that for $U = \mathbb{R}$ and $k = 0$ (or for any $k \geqslant 0$), the inclusion map $\imath_{1,0}$ (or $\imath_{k+1,k}$) is not a compact operator.
(c) Is the inclusion map $C(\mathbb{T}^d) \to L^2(\mathbb{T}^d)$ a compact operator?

6.1.1 Integral Operators are Often Compact

We explore here briefly the realm of integral operators and show that many (but not all) are in fact compact operators.

Lemma 6.10 (Integral operators defined by continuous kernels). *Assume that (X, d_X) and (Y, d_Y) are compact metric spaces. Let μ be a finite Borel measure on X, and let k be a function in $C(X \times Y)$. Then the operator $K : L_\mu^2(X) \longrightarrow C(Y)$ defined by*

$$K(f)(y) = \int_X f(x) k(x, y) \, \mathrm{d}\mu(x)$$

is a compact operator.

PROOF. We first need to show that K is well-defined. To see this, notice that

$$\int_X |f(x)| |k(x, y)| \, \mathrm{d}\mu(x) \leqslant \|f\|_2 \|k(\cdot, y)\|_2 \leqslant \|f\|_2 \|k\|_\infty \mu(X)^{1/2},$$

where $k(\cdot, y)$ denotes the function on X obtained by fixing the coordinate y in Y. This shows that the integral defining $K(f)(y)$ is well-defined and that

$$|K(f)(y)| \leqslant \|k\|_\infty \mu(X)^{1/2} \|f\|_2. \tag{6.1}$$

We now must show that $K(f)$ is continuous, and in doing so we will obtain equicontinuity of the image of the unit ball, which together with (6.1) and the Arzela–Ascoli theorem will give the compactness of K. Since $X \times Y$ is compact, k is uniformly continuous, and so for any $\varepsilon > 0$ there is a $\delta > 0$ for which $d_Y(y_1, y_2) < \delta$ implies that $|k(x, y_1) - k(x, y_2)| < \varepsilon$ for all $x \in X$. Therefore

$$|K(f)(y_1) - K(f)(y_2)| \leqslant \int_X |f(x)||k(x, y_1) - k(x, y_2)| \, d\mu(x) \leqslant \varepsilon \mu(X)^{1/2} \|f\|_2$$

if $d_Y(y_1, y_2) < \delta$, by the same argument as above. Hence $K(f) \in C(Y)$ and the image of the unit ball $B_1^{L_\mu^2(X)}$ is an equicontinuous bounded family of functions. By the Arzela–Ascoli theorem (Theorem 2.38) the closure of $K(B_1^{L_\mu^2(X)})$ is a compact subset of $C(Y)$, and so K is a compact operator. $\qquad\square$

Proposition 6.11 (Hilbert–Schmidt [46]). *Let* $(X, \mathcal{B}_X, \mu)$ *and* $(Y, \mathcal{B}_Y, \nu)$ *be σ-finite measure spaces. Let* $k \in L_{\mu \times \nu}^2(X \times Y)$. *Then the Hilbert–Schmidt integral operator* $K : L_\mu^2(X) \to L_\nu^2(Y)$ *defined by*

$$K(f)(y) = \int_X f(x)k(x, y) \, d\mu(x)$$

for ν-almost every $y \in Y$ defines a compact operator.

Exercise 6.12. Assume in addition that X, Y are compact metric spaces and μ, ν are finite measures on the Borel σ-algebras of X and Y, respectively. Deduce Proposition 6.11 in this case as a corollary of Lemma 6.10.

PROOF OF PROPOSITION 6.11. Note first that

$$\int_X |f(x)k(x, y)| \, d\mu(x) \leqslant \|f\|_{L_\mu^2} \left(\int_X |k(x, y)|^2 \, d\mu(x) \right)^{1/2}. \tag{6.2}$$

Squaring and integrating over Y gives

$$\int_Y \left(\int_X |f(x)k(x, y)| \, d\mu(x) \right)^2 d\nu(y) \leqslant \|f\|_{L_\mu^2}^2 \|k\|_{L_{\mu \times \nu}^2}^2 < \infty$$

by Fubini's theorem. Thus (6.2) is finite almost everywhere, so $K(f)(y)$ is well-defined for ν-almost every y. The bound above also shows that

$$\|K(f)\|_{L_\nu^2} \leqslant \|f\|_{L_\mu^2} \|k\|_{L_{\mu \times \nu}^2}.$$

Hence $K : L_\mu^2(X) \to L_\nu^2(Y)$ is well-defined, clearly linear, and

$$\|K\|_{\mathrm{op}} \leqslant \|k\|_{L^2_{\mu \times \nu}}.$$

If k is a simple function of the form

$$k(x,y) = \sum_{i=1}^{n} c_i \mathbb{1}_{A_i \times B_i} \tag{6.3}$$

for some measurable sets $A_i \subseteq X$, $B_i \subseteq Y$ of finite measure and constants c_i in $\mathbb{C}$, then

$$K(f) = \sum_{i=1}^{n} c_i \left(\int_{A_i} f \, \mathrm{d}\mu \right) \mathbb{1}_{B_i}$$

is a bounded operator with finite-dimensional range and so is also compact. We wish to apply uniform approximation as in Lemma 6.7 to show that the compactness extends to all operators of the form described in the lemma. Since we already showed that the operator norm is bounded from above by the L^2 norm of the kernel k, we only have to show that any $k \in L^2_{\mu \times \nu}(X \times Y)$ can be written as the limit in $L^2_{\mu \times \nu}(X \times Y)$ of a sequence of functions (k_n) with each k_n of the form (6.3). Indeed, if K_n is the operator associated to k_n then $\|K - K_n\|_{\mathrm{op}} \leqslant \|k - k_n\|_2 \to 0$ as $n \to \infty$, which together with the previous discussion and Lemma 6.7 gives the compactness of K.

In order to show that $k \in L^2_{\mu \times \nu}$ can be obtained as the limit of simple functions as in (6.3), note first that simple functions are dense in $L^2_{\mu \times \nu}$. Hence it is sufficient to show that a characteristic function $\mathbb{1}_D$ for a measurable set $D \subseteq X \times Y$ of finite measure can be approximated by functions of the form $\sum_{i=1}^{n} \mathbb{1}_{A_i \times B_i}$, where

$$A_1 \times B_1, \ldots, A_m \times B_m$$

are all disjoint and have finite $\mu \times \nu$-measure. Let us write $X = \bigcup_{n=1}^{\infty} X_n$ and $Y = \bigcup_{n=1}^{\infty} Y_n$ with $X_1 \subseteq X_2 \subseteq \cdots$, $Y_1 \subseteq Y_2 \subseteq \cdots$ and with $\mu(X_n) < \infty$ and $\nu(Y_n) < \infty$ for all $n \geqslant 1$. Then

$$\mathcal{A} = \{ D \in \mathcal{B}_X \otimes \mathcal{B}_Y \mid \text{the claim above holds for } D \cap (X_n \times Y_n) \text{ for all } n \geqslant 1 \}$$

is a σ-algebra containing all rectangles $A \times B$ for $A \in \mathcal{B}_X$ and $B \in \mathcal{B}_Y$. It follows that $\mathcal{A} = \mathcal{B}_X \otimes \mathcal{B}_Y$. Finally, if $D \subseteq X \times Y$ has finite measure, then

$$\left\| \mathbb{1}_D - \mathbb{1}_{D \cap (X_n \times Y_n)} \right\|_{L^2_{\mu \times \nu}} \longrightarrow 0$$

as $n \to \infty$ by dominated convergence. Therefore the simple functions as in (6.3) are indeed dense, which gives the proposition. $\qquad \square$

Exercise 6.13. Prove that the collection $\mathcal{A}$ in the proof of Proposition 6.11 is a σ-algebra.

Exercise 6.14. Let $g \in L^2(\mathbb{T}^d)$. Show that $L^2(\mathbb{T}^d) \ni f \mapsto f * g \in C(\mathbb{T}^d)$ defines a compact operator from $(L^2(\mathbb{T}^d), \|\cdot\|_2)$ to $(C(\mathbb{T}^d), \|\cdot\|_\infty)$.

Not all integral operators are compact, as shown by the Holmgren operators.

Proposition 6.15 (Holmgren). *Let $(X, \mathcal{B}_X, \mu)$ and $(Y, \mathcal{B}_Y, \nu)$ be σ-finite measure spaces. Let $k : X \times Y \to \mathbb{R}$ be measurable on $X \times Y$, with*

$$\sup_{x \in X} \int_Y |k(x,y)| \, d\nu(y) < \infty$$

and

$$\sup_{y \in Y} \int_X |k(x,y)| \, d\mu(x) < \infty.$$

Then the integral operator K defined by

$$K(f)(y) = \int f(x) k(x,y) \, d\mu(x) \tag{6.4}$$

is a bounded operator $K : L^2_\mu \to L^2_\nu$. Moreover,

$$\|K\| \leqslant \left(\sup_{x \in X} \int_Y |k(x,y)| \, d\nu(y) \right)^{1/2} \left(\sup_{y \in Y} \int_X |k(x,y)| \, d\mu(x) \right)^{1/2} < \infty.$$

PROOF. The proof that the integral in (6.4) makes sense for ν-almost every y in Y, and defines an element in L^2_ν, is less straightforward than the proof of Proposition 6.11, and uses the Fréchet–Riesz representation theorem (Corollary 3.19). Suppose that $f \in L^2_\mu \backslash \{0\}$ and $g \in L^2_\nu \backslash \{0\}$, and consider the integral

$$I = \int_{X \times Y} |f(x) k(x,y) g(y)| \, d\mu \times \nu(x,y).$$

Notice that for any real numbers $a, b \geqslant 0$ and $c > 0$, we always have

$$ab \leqslant ab + \left(\sqrt{\tfrac{c}{2}} a - \sqrt{\tfrac{1}{2c}} b \right)^2 = \frac{ca^2}{2} + \frac{b^2}{2c}.$$

Applying this and Fubini's theorem to the definition of I with $a = |f(x)|$ and $b = |g(y)|$ gives

$$I \leqslant \iint_{X \times Y} |k(x,y)| \left(\tfrac{c}{2} |f(x)|^2 + \tfrac{1}{2c} |g(y)|^2 \right) \, d\mu(x) \, d\nu(y)$$

$$\leqslant \frac{c}{2} \int_X \int_Y |k(x,y)| d\nu(y) |f(x)|^2 d\mu(x) + \frac{1}{2c} \int_Y \int_X |k(x,y)| d\mu(x) |g(y)|^2 d\nu(y)$$

$$\leqslant \frac{c}{2} \|f\|^2_{L^2_\mu} \underbrace{\sup_{x \in X} \int_Y |k(x,y)| \, d\nu(y)}_{s_X} + \frac{1}{2c} \|g\|^2_{L^2_\nu} \underbrace{\sup_{y \in Y} \int_X |k(x,y)| \, d\mu(x)}_{s_Y}.$$

If s_X or s_Y is 0, then $k = 0$ $\mu \times \nu$-almost everywhere and the proposition holds trivially. If not, we optimize the parameter c by setting $c = \sqrt{\frac{s_Y}{s_X}} \frac{\|g\|_{L^2_\nu}}{\|f\|_{L^2_\mu}}$, and obtain

$$\int_{X \times Y} |f(x)k(x,y)g(y)| \, \mathrm{d}\mu \times \nu(x,y) \leqslant \sqrt{s_X s_Y} \|f\|_{L^2_\mu} \|g\|_{L^2_\nu}.$$

It follows that $(x,y) \mapsto f(x)k(x,y)g(y)$ is $\mu \times \nu$-integrable on $X \times Y$, and that

$$\phi : g \longmapsto \int_Y g(y) \int_X f(x)k(x,y) \, \mathrm{d}\mu(x) \, \mathrm{d}\nu(y) \tag{6.5}$$

is a continuous functional on L^2_ν with $\|\phi\| \leqslant \sqrt{s_X s_Y} \|f\|_{L^2_\mu}$. We conclude first that

$$K(f)(y) = \int_X f(x)k(x,y) \, \mathrm{d}\mu(x)$$

is well-defined ν-almost everywhere. Using the Fréchet–Riesz representation theorem (Corollary 3.19) the functional ϕ can be represented by taking the inner product with a function in L^2_ν again with norm bounded by $\sqrt{s_X s_Y} \|f\|_{L^2_\mu}$. Varying the element g in (6.5) we see that $K(f)$ must be this function, and we obtain $\|K(f)\|_{L^2_\nu} \leqslant \sqrt{s_X s_Y} \|f\|_{L^2_\mu}$. $\qquad\square$

The main difference between Hilbert–Schmidt integral operators and Holmgren integral operators is that the latter are not automatically compact.

Exercise 6.16. Let $X = Y = \mathbb{R}$ and $\mu = \nu = \lambda$, the Lebesgue measure on $\mathbb{R}$. Define

$$k(x,y) = \begin{cases} 1 & \text{for } \|x - y\| \leqslant 1, \\ 0 & \text{otherwise.} \end{cases}$$

Show that the corresponding Holmgren operator K as defined in Proposition 6.15 is not a compact operator on $L^2_\lambda(\mathbb{R})$.

6.2 Spectral Theory of Self-Adjoint Compact Operators

There is a general spectral theory of compact operators $L : V \to V$ on Banach spaces. However, as we will discuss later, our applications do not need that level of generality and the statement and proof for the simpler case of self-adjoint operators is significantly easier. For these reasons we will restrict to that case below and refer to Lax [59, Ch. 21] for the general result.

6.2.1 The Adjoint Operator

Let $\mathcal{H}_1, \mathcal{H}_2$ be Hilbert spaces, and let $A : \mathcal{H}_1 \to \mathcal{H}_2$ be a bounded operator. For any fixed $v_2 \in \mathcal{H}_2$ the map $\mathcal{H}_1 \ni v_1 \mapsto \langle Av_1, v_2 \rangle_{\mathcal{H}_2}$ is linear and bounded since $|\langle Av_1, v_2 \rangle| \leqslant \|Av_1\| \|v_2\| \leqslant \|A\|_{\mathrm{op}} \|v_2\| \|v_1\|$. Therefore, by the Fréchet–Riesz representation theorem (Corollary 3.19) applied to $\mathcal{H}_1$ there exists some uniquely determined element, which will be denoted $A^* v_2 \in \mathcal{H}_1$, with the properties that

$$\langle v_1, A^* v_2 \rangle_{\mathcal{H}_1} = \langle Av_1, v_2 \rangle_{\mathcal{H}_2} \tag{6.6}$$

for all $v_1 \in \mathcal{H}_1$, and

$$\|A^* v_2\| \leqslant \|A\|_{\mathrm{op}} \|v_2\|. \tag{6.7}$$

This defines a bounded operator $A^* : \mathcal{H}_2 \to \mathcal{H}_1$, called the *adjoint* of A. This map is indeed linear, since

$$\langle v_1, A^*(v_2 + \alpha v_2') \rangle = \langle Av_1, v_2 + \alpha v_2' \rangle = \langle Av_1, v_2 \rangle + \overline{\alpha} \langle Av_1, v_2' \rangle$$
$$= \langle v_1, A^* v_2 \rangle + \overline{\alpha} \langle v_1, A^* v_2' \rangle = \langle v_1, A^* v_2 + \alpha A^* v_2' \rangle$$

for $v_1 \in \mathcal{H}_1$, $v_2, v_2' \in \mathcal{H}_2$ and any scalar α. By (6.7) we have $\|A^*\|_{\mathrm{op}} \leqslant \|A\|_{\mathrm{op}}$, so A^* is bounded. Taking conjugates in (6.6) implies that $A^{**} = A$, so $\|A\|_{\mathrm{op}} = \|A^*\|_{\mathrm{op}}$.

Essential Exercise 6.17. (a) Show that the map $A \mapsto A^*$ is semi-linear. (b) Let $A : \mathcal{H}_1 \to \mathcal{H}_2$ and $B : \mathcal{H}_2 \to \mathcal{H}_3$ be bounded operators between Hilbert spaces. Show that $(BA)^* = A^* B^*$.

Exercise 6.18. Show that $\mathrm{im}(T)^{\perp} = \ker(T^*)$ and $\ker(T)^{\perp} = \overline{\mathrm{im}(T^*)}$ for a linear operator T between Hilbert spaces.

The adjoint operation allows us to give an alternate definition of unitarity.

Definition 6.19. An operator $U : \mathcal{H}_1 \to \mathcal{H}_2$ between two Hilbert spaces is *unitary* if $U^* U = I_{\mathcal{H}_1}$ and $UU^* = I_{\mathcal{H}_2}$, which we also write as $U^* = U^{-1}$.

Exercise 6.20. (a) Show that an operator $U : \mathcal{H}_1 \to \mathcal{H}_2$ is unitary in the sense of Definition 6.19 if and only if it is a bijective isometry (that is, a bijection with $\|Uv\|_{\mathcal{H}_2} = \|v\|_{\mathcal{H}_1}$ for all $v \in \mathcal{H}_1$). (b) Suppose that $U : \mathcal{H}_1 \to \mathcal{H}_2$ is an isometry. Show that $U^* U = I_{\mathcal{H}_1}$ and that UU^* is the orthogonal projection $P_{\mathrm{im}(U)}$ from $\mathcal{H}_2$ onto the closed subspace $\mathrm{im}(U) \subseteq \mathcal{H}_2$.

Exercise 6.21 (Von Neumann's mean ergodic theorem [78]). Let $U : \mathcal{H} \to \mathcal{H}$ be a unitary operator on a Hilbert space $\mathcal{H}$ and let $I = \{v \in \mathcal{H} \mid Uv = v\}$ be the subspace of invariant vectors.
(a) Show that I is closed and that $\{Uv - v \mid v \in \mathcal{H}\}$ is dense in $I^{\perp}$.
(b) Show that $\frac{1}{n} \sum_{j=0}^{n-1} U^n v \to P_I v$ as $n \to \infty$, where P_I is the orthogonal projection onto I.

Definition 6.22. A bounded operator $A : \mathcal{H} \to \mathcal{H}$ on a Hilbert space $\mathcal{H}$ is called *self-adjoint* if $A^* = A$.

The next exercise revisits the maps introduced in Exercise 6.1.

Exercise 6.23. (a) Define $U : \ell^2(\mathbb{Z}) \to \ell^2(\mathbb{Z})$ by $U\left((x_n)_{n \in \mathbb{Z}}\right) = (x_{n+1})_{n \in \mathbb{Z}}$. Show that the operator U is unitary.
(b) Define $S : \ell^2(\mathbb{N}) \to \ell^2(\mathbb{N})$ by $S\left((x_n)_{n \in \mathbb{N}}\right) = (x_{n+1})_{n \in \mathbb{N}}$. Show that $\|S\|_{\mathrm{op}} = 1$, but that S is not an isometry.
(c) Define $T : \ell^2(\mathbb{N}) \to \ell^2(\mathbb{N})$ by $T\left((x_n)\right) = (0, x_1, x_2, \dots)$, which shifts the sequence to the right and fills in the first entry of the new sequence with a 0. Show that $\|T\|_{\mathrm{op}} = 1$, that $T = S^*$ is an isometry, is not surjective, and has no eigenvectors.

Exercise 6.24 (Decomposition of isometries). Let $\mathcal{H}$ be a Hilbert space and $U : \mathcal{H} \to \mathcal{H}$ an isometry. Show that there exists an orthogonal decomposition $\mathcal{H} = \mathcal{H}_{\mathrm{shift}} \oplus \mathcal{H}_{\mathrm{unitary}}$ into two closed subspaces with the property that $\mathcal{H}_{\mathrm{shift}} = \bigoplus_{n \geqslant 0} U^n V$ for some closed subspace V, and $U|_{\mathcal{H}_{\mathrm{unitary}}} : \mathcal{H}_{\mathrm{unitary}} \to \mathcal{H}_{\mathrm{unitary}}$ is unitary.

The next exercise is not simply another example. It turns out to really be the basis of the powerful spectral theory of normal bounded operators as well as self-adjoint unbounded operators.

Essential Exercise 6.25. Let $(X, \mathcal{B}, \mu)$ be a measure space, $\mathcal{H} = L_\mu^2(X)$, let $g : X \to \mathbb{C}$ be a measurable function, and let M_g be the multiplication operator $M_g : f \mapsto gf$ for $f \in \mathcal{H}$.
(a) What properties of g ensure that $M_g : \mathcal{H} \to \mathcal{H}$ is well-defined and bounded? What is $\|M_g\|_{\mathrm{op}}$?
(b) When is M_g a bounded self-adjoint operator? That is, what property of g is equivalent to $\langle M_g f_1, f_2 \rangle = \langle f_1, M_g f_2 \rangle$ holding for all $f_1, f_2 \in \mathcal{H}$? What property of g is equivalent to M_g being unitary?
(c) When does M_g have $\lambda \in \mathbb{C}$ as an eigenvalue?
(d) Suppose that $X = \mathbb{R}$ and let $g(x) = x$, and assume that μ is an arbitrary finite compactly supported Borel measure on $\mathbb{R}$. Characterize in terms of μ the property that M_g can be diagonalized. That is, characterize the property that $\mathcal{H}$ has an orthonormal basis $\{e_n \mid n \in \mathbb{N}\}$ and a sequence of scalars (λ_n) such that $M_g\left(\sum_{n=1}^{\infty} x_n e_n\right) = \sum_{n=1}^{\infty} \lambda_n x_n e_n$ for every $(x_n) \in \ell^2(\mathbb{N})$.

Exercise 6.26. Let $\mathcal{H} = \mathbb{C}^n$ be a finite-dimensional Hilbert space with respect to the usual inner product. Show that the linear operator defined by a matrix $A = (a_{i,j})$ is self-adjoint if and only if A is equal to its own conjugate transpose (that is, $a_{i,j} = \overline{a_{ji}}$ for all i, j). Such matrices are also called *Hermitian*.

6.2.2 The Spectral Theorem

The spectral theorem presented here generalizes to an infinite-dimensional setting the familiar fact that a Hermitian matrix has real eigenvalues and can be diagonalized using a unitary matrix. We will assume separability of Hilbert spaces in this section in order to make use of an orthonormal basis that consists of a sequence. Properly formulated, the next result holds more generally, and in particular allows the kernel of A to be a non-separable space. Both the finite-dimensional and the inseparable case can easily be extracted from the proof we give.

Theorem 6.27 (Spectral theorem for compact self-adjoint operators). *Let $\mathcal{H}$ be a separable infinite-dimensional Hilbert space, and let A be a compact self-adjoint operator on $\mathcal{H}$. Then there exists a sequence of real eigenvalues (λ_n) with $\lambda_n \to 0$ as $n \to \infty$, and an orthonormal basis $\{v_n\}$ of eigenvectors with $Av_n = \lambda_n v_n$ for all $n \geqslant 1$.*

In other words, a compact self-adjoint operator is diagonalizable, each non-zero eigenvalue has finite multiplicity, and 0 is the only possible accumulation point of the set of eigenvalues. Given these properties — which will turn out to be extremely useful — it is worth asking if there are any such operators.

Clearly such operators exist in the following sense. If $\{e_n\}$ is an orthonormal basis of a Hilbert space $\mathcal{H}$, and (λ_n) is a sequence of real numbers with $\lambda_n \to 0$ as $n \to \infty$, then we may define an operator $A : \mathcal{H} \to \mathcal{H}$ by

$$A\left(\sum_{n=1}^{\infty} x_n e_n\right) = \sum_{n=1}^{\infty} \lambda_n x_n e_n$$

for any convergent series $\sum_{n=1}^{\infty} x_n e_n$. It may then be checked that A is compact and self-adjoint. Of course, Theorem 6.27 does not tell us anything we did not already know about such an operator.

A more interesting kind of example is found among the integral operators. Let $\mathcal{H} = L_\mu^2(X)$, where $(X, \mathcal{B}, \mu)$ is a σ-finite measure space, and suppose that $k \in L_{\mu \times \mu}^2(X \times X)$ satisfies $k(x, y) = \overline{k(y, x)}$ for $\mu \times \mu$-almost every point $(x, y) \in X \times X$. Then the operator K defined by

$$K(f)(y) = \int_X f(x) k(x, y) \, \mathrm{d}\mu(x)$$

is compact by Proposition 6.11, and is self-adjoint since

$$\langle f_1, K^*(f_2) \rangle = \langle K(f_1), f_2 \rangle = \int_X \int_X f_1(x) k(x, y) \, \mathrm{d}\mu(x) \overline{f_2(y)} \, \mathrm{d}\mu(y)$$

$$= \int_X f_1(x) \overline{\int_X f_2(y) \underbrace{\overline{k(x, y)}}_{=k(y,x)} \, \mathrm{d}\mu(y)} \, \mathrm{d}\mu(x) = \langle f_1, K(f_2) \rangle$$

for all $f_1, f_2 \in L_\mu^2(X)$ by Fubini's theorem. Hence Theorem 6.27 applies, but in this case it is *a priori* not at all clear how one could find the eigenvalues or eigenvectors for the operator.

Example 6.28. Notice that the integral operator from Section 2.5.2 defined by the kernel

$$G(s, t) = \begin{cases} s(t - 1) & \text{for } 0 \leqslant s \leqslant t \leqslant 1; \\ t(s - 1) & \text{for } 0 \leqslant t \leqslant s \leqslant 1 \end{cases}$$

satisfies the conditions above, and so the eigenfunctions found in Section 2.5.2 coincide with the eigenvectors which must exist by Theorem 6.27.

In fact as we saw in Section 3.4 (see Exercise 3.55(b) and its hint on p. 566) the functions $s_1, s_2, \ldots$ form an orthonormal basis of $L^2([0,1])$ which makes K a diagonalizable operator. These notions also explain the argument from Section 2.5.2 quite clearly: If $g = \sum_{n=1}^{\infty} d_n s_n$ and we are looking for $f = \sum_{n=1}^{\infty} c_n s_n$ with $(I + \lambda^2 K)f = g$, then $(1 + \lambda^2 \mu_n)c_n = d_n$ for all $n \in \mathbb{N}$, which can be solved for c_n unless $\lambda^2 = -\mu_n^{-1}$ and $d_n \neq 0$.

Exercise 6.29. Let K be the Hilbert–Schmidt integral operator on $L_\mu^2(X)$ defined by a kernel $k \in L_{\mu \times \mu}^2(X \times X)$ with $k(x,y) = \overline{k(y,x)}$ as above. Prove that the generalized Fredholm integral equation of the second kind $f = \lambda K(f) + \phi$ has a solution for any function $\phi \in L_\mu^2(X)$ if and only if $\lambda \lambda_n \neq 1$ for all n, where (λ_n) is the sequence of eigenvalues of K on $L_\mu^2(X)$.

We will see another class of compact self-adjoint operators in Section 6.4.

6.2.3 Proof of the Spectral Theorem

Lemma 6.30 (Invariance of orthogonal complement). *Let $A : \mathcal{H} \to \mathcal{H}$ be a bounded operator on a Hilbert space. If $V \subseteq \mathcal{H}$ is an A-invariant subspace (that is, a subspace with $A(V) \subseteq V$), then $V^\perp$ is A^*-invariant.*

PROOF. If $v' \in V^\perp$ and $v \in V$, then $\langle A^* v', v \rangle = \langle v', Av \rangle = 0$. As this holds for all $v \in V$, we must have $A^* v' \in V^\perp$. □

As we will see, Lemma 6.30 reduces the proof of Theorem 6.27 mostly to finding a single eigenvector e_1, as we can then apply the lemma to $V = \langle e_1 \rangle$ and $A = A^*$ to see that $V^\perp$ is A-invariant.

We now approach the central statement concerning the existence of an eigenvalue. Before doing this, it is useful to recall how one proves the complete diagonalizability of self-adoint operators on $\mathbb{R}^d$. By compactness, we may choose $e \in \mathbb{S}^{d-1} = \{v \in \mathbb{R}^d \mid \|v\|_2 = 1\}$ such that the quadratic form $\langle Ax, x \rangle$ achieves its maximum at $x = e$. Using Lagrange multipliers one can then check that e is an eigenvector of A. The vector e is then an eigenvector with eigenvalue $\lambda \in \mathbb{R}$ of absolute value $|\lambda| = \|A\|_{\mathrm{op}}$. This relies in an essential way on the compactness of the unit sphere $\mathbb{S}^{d-1}$, which as we know fails in infinite-dimensional Hilbert spaces, and it is here that the additional assumptions on A will become important.

Lemma 6.31 (The norm and the quadratic form). *Let $A : \mathcal{H} \to \mathcal{H}$ be a bounded self-adjoint operator on a Hilbert space. Then*

$$\|A\| = \sup_{\|x\| \leqslant 1} |\langle Ax, x \rangle|. \tag{6.8}$$

Notice that if A is self-adjoint, then $\langle Ax, x \rangle \in \mathbb{R}$ for all $x \in \mathcal{H}$, since

$$\overline{\langle Ax, x \rangle} = \langle x, Ax \rangle = \langle A^* x, x \rangle = \langle Ax, x \rangle.$$

PROOF OF LEMMA 6.31. Let us write

$$s(A) = \sup_{\|x\| \leqslant 1} |\langle Ax, x \rangle|$$

for the right-hand side of (6.8). Then, by the Cauchy–Schwarz inequality,

$$|\langle Ax, x \rangle| \leqslant \|Ax\| \|x\| \leqslant \|A\| \|x\|^2 \leqslant \|A\|$$

for all $x \in \mathcal{H}$ with $\|x\| \leqslant 1$. Hence $s(A) \leqslant \|A\|$.

The proof of the opposite inequality is slightly more involved. For $\lambda > 0$, we have

$$\left\langle A(\lambda x \pm \tfrac{1}{\lambda} Ax), \lambda x \pm \tfrac{1}{\lambda} Ax \right\rangle = \left\langle A(\lambda x), \lambda x \right\rangle + \left\langle A^2(\tfrac{1}{\lambda} x), A(\tfrac{1}{\lambda} x) \right\rangle \pm 2\|Ax\|^2.$$

Taking the difference of the two equations we see that

$$\begin{aligned}
4\|Ax\|^2 &= \left\langle A(\lambda x + \tfrac{1}{\lambda} Ax), \lambda x + \tfrac{1}{\lambda} Ax \right\rangle - \left\langle A(\lambda x - \tfrac{1}{\lambda} Ax), \lambda x - \tfrac{1}{\lambda} Ax \right\rangle \\
&\leqslant s(A) \left(\|\lambda x + \tfrac{1}{\lambda} Ax\|^2 + \|\lambda x - \tfrac{1}{\lambda} Ax\|^2 \right)
\end{aligned}$$

since the two inner products appearing are of the form $\langle Au, u \rangle$ and thus satisfy $|\langle Au, u \rangle| \leqslant s(A)\|u\|^2$. Now we apply the parallelogram identity (3.4) to obtain

$$4\|Ax\|^2 \leqslant 2s(A)\left(\lambda^2 \|x\|^2 + \tfrac{1}{\lambda^2}\|Ax\|^2\right).$$

Assuming that $\|Ax\| \neq 0$, we set $\lambda^2 = \frac{\|Ax\|}{\|x\|}$ and get

$$4\|Ax\|^2 \leqslant 2s(A)\left(\frac{\|Ax\|}{\|x\|}\|x\|^2 + \frac{\|x\|}{\|Ax\|}\|Ax\|^2\right) = 4s(A)\|Ax\|\|x\|,$$

and so $\|Ax\| \leqslant s(A)\|x\|$ for all $x \in \mathcal{H}$. This shows that $\|A\| \leqslant s(A)$. $\qquad \square$

We are now ready to prove the existence of an eigenvector.

Lemma 6.32 (Main step: finding the first eigenvector). *Let A be a compact self-adjoint operator on a non-trivial Hilbert space. Then either $\|A\|$ or $-\|A\|$ is an eigenvalue of A.*

PROOF. If $\|A\| = 0$ then $A = 0$ and there is nothing to prove, so we may assume that $\|A\| > 0$. By Lemma 6.31 there exists a scalar α with $|\alpha| = \|A\|$ and a sequence (x_n) in $\mathcal{H}$ with $\|x_n\| = 1$ for all $n \geqslant 1$ and with $\langle Ax_n, x_n \rangle \to \alpha$ as $n \to \infty$. As remarked before the proof of Lemma 6.31, $\langle Ax_n, x_n \rangle$ is real and so $\alpha \in \{\|A\|, -\|A\|\}$. Now notice that

$$\begin{aligned}
0 \leqslant \|Ax_n - \alpha x_n\|^2 &= \|Ax_n\|^2 - 2\Re\left(\alpha \langle Ax_n, x_n \rangle\right) + \alpha^2 \|x_n\|^2 \\
&= \|Ax_n\|^2 - 2\alpha \langle Ax_n, x_n \rangle + \alpha^2 \\
&\leqslant 2\|A\|^2 - 2\alpha \langle Ax_n, x_n \rangle \longrightarrow 2\|A\|^2 - 2\|A\|^2 = 0
\end{aligned}$$

as $n \to \infty$. In particular, this shows that (Ax_n) converges if and only if (αx_n) converges, and that the limits agree if this is the case. However, since $\|x_n\| = 1$ and A is a compact operator there exists a subsequence (x_{n_k}) for which Ax_{n_k} converges, say

$$Ax_{n_k} \longrightarrow \alpha x \tag{6.9}$$

as $k \to \infty$ for some $x \in \mathcal{H}$. Therefore, $\alpha x_{n_k} \to \alpha x$ as $k \to \infty$ as well, and hence $x_{n_k} \to x$ as $k \to \infty$. Since A is continuous, we deduce that $Ax_{n_k} \to Ax$ as $k \to \infty$. Together with (6.9) we have $Ax = \alpha x$, and since $\|x_{n_k}\| = 1$ for all $k \geqslant 1$ and $x_{n_k} \to x$ as $k \to \infty$ we also have $\|x\| = 1$ and hence $x \neq 0$. $\square$

Now we combine the arguments above to prove the spectral theorem for compact self-adjoint operators.

PROOF OF THEOREM 6.27. By assumption, $\mathcal{H}$ is an infinite-dimensional Hilbert space and $A : \mathcal{H} \to \mathcal{H}$ is a compact self-adjoint operator. By Lemma 6.32 there exists an eigenvector e_1 with eigenvalue $\lambda_1 \in \mathbb{R}$, and with $|\lambda_1| = \|A\|$. We may assume without loss of generality that $\|e_1\| = 1$.

Suppose now, for the purposes of an induction argument, that we have already found orthonormal eigenvectors $e_1, \ldots, e_n$ with corresponding eigenvalues $\lambda_1, \ldots, \lambda_n$. Let $V_n = \langle e_1, \ldots, e_n \rangle$ be the linear span of these vectors, and notice that $A(V_n) \subseteq V_n$ since they are eigenvectors for A. By Lemma 6.30 we have $A^*(V_n^\perp) \subseteq V_n^\perp$, but since $A^* = A$ this means that $A(V_n^\perp) \subseteq V_n^\perp$. Write

$$A_n = A|_{V_n^\perp} : V_n^\perp \longrightarrow V_n^\perp$$

for the restriction of A to $V_n^\perp$. Then A_n is a compact operator because A is compact, and is self-adjoint because A is self-adjoint.[†] Therefore, we may apply Lemma 6.32 again to the operator $A_n : V_n^\perp \to V_n^\perp$ to find another eigenvector e_{n+1} orthogonal to $e_1, \ldots, e_n$ with eigenvalue λ_{n+1} satisfying $|\lambda_{n+1}| = \|A_n\|$, and $\|e_{n+1}\| = 1$.

Repeating the argument, we find an orthonormal sequence (e_n) of eigenvectors with $Ae_n = \lambda_n e_n$ and $\lambda_n \in \mathbb{R}$. We need to show that $\lambda_n \to 0$ as $n \to \infty$. By construction we have

$$|\lambda_{n+1}| = \|A_n\| = \|A|_{V_n^\perp}\| \leqslant \|A|_{V_{n-1}^\perp}\| = \|A_{n-1}\| = |\lambda_n|,$$

so that by induction we have

$$|\lambda_1| \geqslant |\lambda_2| \geqslant \cdots . \tag{6.10}$$

If $\lambda_n \nrightarrow 0$ as $n \to \infty$, then there is some $\varepsilon > 0$ such that $|\lambda_n| > \varepsilon$ for all $n \geqslant 1$ by (6.10). This shows that $\varepsilon e_n = A\left(\frac{\varepsilon}{\lambda_n} e_n\right) \in A(B_1)$ for all $n \geqslant 1$, and since $e_n \perp e_m$ for $n \neq m$ we must have $\|\varepsilon e_n - \varepsilon e_m\| = \varepsilon\sqrt{2}$ for $n \neq m$. This shows that the sequence (εe_n) lies in $\overline{A(B_1)}$ (which is compact because A

[†] If $w_1, w_2 \in V_n^\perp$ then $\langle A_n^* w_1, w_2 \rangle = \langle w_1, A_n w_2 \rangle = \langle A^* w_1, w_2 \rangle = \langle Aw_1, w_2 \rangle$ and Aw_1 lies in $V_n^\perp$, so we have $A_n^* = A|_{V_n^\perp} = A_n$.

is a compact operator) but cannot have a convergent subsequence, which is a contradiction.

Also, since

$$|\lambda_n| = \|A|_{V_n^\perp}\| \geqslant \|A|_{V^\perp}\|,$$

where $V = \langle e_1, e_2, \dots \rangle$, we see that $A|_{V^\perp} = 0$.

Thus far we have not used the assumption that $\mathcal{H}$ is separable, and the statement at the end of the last paragraph is the general result. Assuming now that $\mathcal{H}$ is separable, we can choose an orthonormal basis of $V^\perp$ (which might be zero, in which case the theorem is already proved, or might be finite-dimensional). Listing this orthonormal basis of $V^\perp$ together with the basis of V already constructed proves the theorem. $\qquad\square$

Exercise 6.33. Let $\mathcal{H}$ be a separable Hilbert space, and let $A_1, A_2, \dots$ be a sequence of commuting self-adjoint bounded operators on $\mathcal{H}$. Using Theorem 6.27, state and prove a simultaneous spectral theorem for the sequence assuming either of the properties below:

(1) A_n is compact for all $n \geqslant 1$; or
(2) A_1 is compact and $\ker(A_1) = \{0\}$.

6.2.4 Variational Characterization of Eigenvalues

† In the following we let A be a compact self-adjoint operator on a separable infinite-dimensional Hilbert space $\mathcal{H}$ (or a Hermitian matrix in $\mathrm{Mat}_{n,n}(\mathbb{C})$). Applying Theorem 6.27 we find a (finite or countable) sequence of positive eigenvalues $\varphi_1(A) \geqslant \varphi_2(A) \geqslant \cdots > 0$ and a (finite or countable) sequence of negative eigenvalues $\nu_1(A) \leqslant \nu_2(A) \leqslant \cdots < 0$, with corresponding orthonormal eigenvectors $v_1, v_2, \dots$ and $w_1, w_2, \dots$, respectively, so that

$$A = \sum_j \varphi_j(A) v_j \otimes v_j^* + \sum_j \nu_j(A) w_j \otimes w_j^*,$$

where we define $v^*(w) = \langle w, v \rangle$ and $u \otimes v^*(w) = v^*(w)u$ for $u, v, w \in \mathcal{H}$. This decomposition is known as the *spectral resolution* of A.

In many situations it is useful to be able to say something about the eigenvalues of $A + B$ (even for Hermitian matrices A and B) in terms of the eigenvalues of A and of B, a fundamentally non-linear problem. The variational approach to finding eigenvalues dates back to Cauchy's interlacing theorem [17] (see Exercise 6.36). There are three elementary observations that can be made in this direction.

- Assuming that $\mathcal{H}$ is infinite-dimensional the spectral resolution shows that the *numerical range* $\{\langle Av, v \rangle \mid \|v\| = 1\}$ coincides with the real interval $[\nu_1, \varphi_1]$ unless there are no negative or positive eigenvalues. In

---

† The material of this subsection motivates some later arguments in this chapter but is strictly speaking not necessary.

the former case we obtain the numerical range $(0, \varphi_1]$ or $[0, \varphi_1]$ (and hence set $\nu_1 = 0$) and in the latter case we obtain $[\nu_1, 0)$ or $[\nu_1, 0]$ (and hence set $\varphi_1 = 0$). In particular,

$$\varphi_1(A) = \sup_{\|v\|=1} \langle Av, v \rangle, \tag{6.11}$$

where the supremum is achieved if $\varphi_1 > 0$, and

$$\nu_1(A) = \inf_{\|v\|=1} \langle Av, v \rangle, \tag{6.12}$$

which is again achieved if $\nu_1 < 0$. Thus $\varphi_1(A + B) \leqslant \varphi_1(A) + \varphi_1(B)$ and $\nu_1(A + B) \geqslant \nu_1(A) + \nu_1(B)$.

- In the case of a Hermitian matrix $A \in \mathrm{Mat}_{n,n}(\mathbb{C})$, we do not have to distinguish between positive and negative eigenvalues and may simply write $\lambda_1(A) \leqslant \cdots \leqslant \lambda_n(A)$ for its eigenvalues. Viewing the functions λ_j on the linear space of Hermitian matrices the above applies (again without giving 0 a special role) as well, and we see that λ_n is a convex function and λ_1 a concave one.
- Also note that the trace map $\mathrm{tr} : \mathrm{Mat}_{n,n}(\mathbb{C}) \longrightarrow \mathbb{C}$ is linear. Hence

$$\lambda_1(A+B) + \cdots + \lambda_n(A+B) = \lambda_1(A) + \cdots + \lambda_n(A) + \lambda_1(B) + \cdots + \lambda_n(B) \tag{6.13}$$

for any $A, B \in \mathrm{Mat}_{n,n}(\mathbb{C})$.

More detailed assertions about the relationships between the eigenvalues of Hermitian matrices are the subject of Horn's conjecture.[16] We will not go into the details of this, but state as exercises some special cases which can be proved with elementary methods and which are widely used in other parts of mathematics. The first of these is the min-max principle.

Exercise 6.34. Generalize the identities (6.11) and (6.12) by proving the *Courant–Fischer–Weyl theorem* or *min-max principle* for compact self-adjoint operators as follows. Fix some $k \geqslant 1$ and in the following let V vary over all k-dimensional subspaces of an infinite-dimensional separable Hilbert space $\mathcal{H}$. Show that

$$\inf_V \max_{v \in V, \|v\|=1} \langle Av, v \rangle = \nu_k(A),$$

where we set $\nu_k(A) = 0$ if there are fewer than k negative eigenvalues. Similarly

$$\sup_V \min_{v \in V, \|v\|=1} \langle Av, v \rangle = \varphi_k(A), \tag{6.14}$$

where we set $\varphi_k(A) = 0$ if there are fewer than k positive eigenvalues. Formulate and prove the result also for Hermitian matrices.

Exercise 6.35. Deduce from Exercise 6.34 the *Weyl monotonicity principle*[17] as follows. For compact self-adjoint operators A and B write $A \leqslant B$ if $\langle Av, v \rangle \leqslant \langle Bv, v \rangle$ for all v. Show that if $A \leqslant B$ then $\nu_j(A) \leqslant \nu_j(B)$ and $\varphi_j(A) \leqslant \varphi_j(B)$ (where we set $\nu_j = 0$ and $\varphi_j = 0$ if there are not sufficient eigenvalues of the necessary sign) for all j. Formulate and prove the result also for Hermitian matrices.

Exercise 6.36. Use Exercise 6.34 to prove *Cauchy's interlacing theorem* as follows. Let $A \in \mathrm{Mat}_{n,n}(\mathbb{C})$ be a Hermitian matrix. A matrix $B \in \mathrm{Mat}_{m,m}(\mathbb{C})$ with $m \leqslant n$ is called a *compression* of A if there is an orthogonal projection Q from $\mathbb{C}^n$ onto an m-dimensional subspace with $QAQ^* = B$. Show that $\lambda_j(A) \leqslant \lambda_j(B) \leqslant \lambda_{n-m+j}(A)$ for $1 \leqslant j \leqslant m$ in this case.

6.3 Trace-Class Operators

[†] The trace is undoubtedly one of the fundamental functions on the space of matrices. Recall that for any $n \geqslant 1$ the trace is defined by $\mathrm{tr}(A) = \sum_{k=1}^{n} A_{kk}$ for all $A = (A_{jk}) \in \mathrm{Mat}_{n,n}(\mathbb{C})$ and that it satisfies $\mathrm{tr}(AB) = \mathrm{tr}(BA)$ for A and B in $\mathrm{Mat}_{n,n}(\mathbb{C})$, so that

$$\mathrm{tr}(S^{-1}AS) = \mathrm{tr}(A) \tag{6.15}$$

for any $A \in \mathrm{Mat}_{n,n}(\mathbb{C})$ and $S \in \mathrm{GL}_n(\mathbb{C})$. The identity (6.15) means that the trace is well-defined on the space of linear maps of a finite-dimensional vector space (specifically, independent of the choice of basis). Using a Hilbert space structure on $\mathbb{C}^n$ and fixing an orthonormal basis $v_1, \ldots, v_n$ we note that $\langle Av_j, v_k \rangle$ is the coefficient of v_k when expressing Av_j in terms of the orthonormal basis for $j, k = 1, \ldots, n$. Hence

$$\mathrm{tr}(A) = \sum_{j=1}^{n} \langle Av_j, v_j \rangle.$$

It is desirable to extend the definition of the trace to operators on an infinite-dimensional Hilbert space $\mathcal{H}$. However, since $\mathrm{tr}(I_n) = n$ for the identity matrix $I_n \in \mathrm{Mat}_{n,n}(\mathbb{C})$ and all $n \geqslant 1$, it is clear that the trace cannot have a reasonable definition on all operators on $\mathcal{H}$, and in particular not on the identity. The following definition gives the natural domain of the trace functional.

Definition 6.37. Let $\mathcal{H}$ be a Hilbert space. A linear operator $A : \mathcal{H} \to \mathcal{H}$ is called *trace-class* if its *trace-class norm*

$$\|A\|_{\mathrm{tc}} = \sup_{(v_n),(w_n)} \sum_{n=1}^{N} |\langle Av_n, w_n \rangle|$$

is finite, where the supremum is taken over all integers $N \geqslant 0$ and over any two finite lists of orthonormal vectors $(v_1, \ldots, v_N)$ and $(w_1, \ldots, w_N)$ of the same length N.

[†] In this section we present an important class of compact operators. However, it is not needed for the further developments in this volume.

In the following we will assume that $\mathcal{H}$ is separable and complex (once again separability is only needed to simplify the notation and some steps in the proofs, but is not crucial for the results and as every real Hilbert space $\mathcal{H}$ has a complexification $\mathcal{H}_{\mathbb{C}} = \mathcal{H} \otimes_{\mathbb{R}} \mathbb{C} = \mathcal{H} \oplus i\mathcal{H}$ as in Exercise 6.51, the assumption that $\mathcal{H}$ is a complex vector space is also not a significant restriction). We list a few consequences of this definition for trace-class operators $A, B : \mathcal{H} \to \mathcal{H}$, which in particular justify our calling it a norm.

- If $v \in \mathcal{H}$ is a unit vector with $Av \neq 0$ then we may set $w = \frac{1}{\|Av\|} Av$, and apply the definition with $N = 1$, v and w to see that

$$\|Av\| = \langle Av, w \rangle \leqslant \|A\|_{\text{tc}},$$

  so $\|A\|_{\text{op}} \leqslant \|A\|_{\text{tc}}$ and hence a trace-class operator is bounded.
- If α is a scalar, then

$$\|\alpha A\|_{\text{tc}} = \sup_{(v_n),(w_n)} \sum_{n=1}^{N} |\langle \alpha A v_n, w_n \rangle| = |\alpha| \|A\|_{\text{tc}}.$$

- The triangle inequality follows, since

$$\|A + B\|_{\text{tc}} = \sup_{(v_n),(w_n)} \sum_{n=1}^{N} |\langle (A+B) v_n, w_n \rangle|$$

$$\leqslant \sup_{(v_n),(w_n)} \sum_{n=1}^{N} \left(|\langle A v_n, w_n \rangle| + |\langle B v_n, w_n \rangle| \right) \leqslant \|A\|_{\text{tc}} + \|B\|_{\text{tc}}.$$

  Thus the space $\text{TC}(\mathcal{H}) = \{ A \in \text{B}(\mathcal{H}) \mid \|A\|_{\text{tc}} < \infty \}$ of trace-class operators is a linear subspace of the space of bounded operators, and $\| \cdot \|_{\text{tc}}$ is a norm on $\text{TC}(\mathcal{H})$.
- Noting that a unitary operator maps an orthonormal list of vectors to an orthonormal list of vectors, we see that $\|AU\|_{\text{tc}} = \|UA\|_{\text{tc}} = \|A\|_{\text{tc}}$ for any unitary $U : \mathcal{H} \to \mathcal{H}$.
- By writing a bounded operator of $\mathcal{H}$ as a linear combination of four unitary operators (see Lemma 6.38 below) we see that if $A \in \text{TC}(\mathcal{H})$ and $B \in \text{B}(\mathcal{H})$ then $AB, BA \in \text{TC}(\mathcal{H})$.

Lemma 6.38 (Four unitary operators). *Any bounded operator on a separable complex Hilbert space may be written as a linear combination of four unitary operators.*

This will be shown using the spectral theory of self-adjoint operators in Section 12.4.2 (after Corollary 12.45) and will be used here as a black box.

Theorem 6.39 (Trace functional). *Let $\mathcal{H}$ be a separable complex Hilbert space. Then there exists a linear functional* $\text{tr} : \text{TC}(\mathcal{H}) \longrightarrow \mathbb{C}$ *with the following properties:*

(1) $|\operatorname{tr}(A)| \leqslant \|A\|_{\mathrm{tc}}$,
(2) $\operatorname{tr}(A) = \operatorname{tr}(U^{-1}AU)$, and
(3) $\operatorname{tr}(AB) = \operatorname{tr}(BA)$

for all $A \in \mathrm{TC}(\mathcal{H})$, $B \in \mathrm{B}(\mathcal{H})$ and unitary $U \in \mathrm{B}(\mathcal{H})$. Moreover,

$$(4)\ \operatorname{tr}(A) = \sum_{n=1}^{\infty} \langle Av_n, v_n \rangle$$

for any $A \in \mathrm{TC}(\mathcal{H})$ and orthonormal basis (v_n) of $\mathcal{H}$.

PROOF OF THEOREM 6.39: ASSUMING INDEPENDENCE. Using $w_n = v_n$ for all $n \geqslant 1$ in the definition of $\|A\|_{\mathrm{tc}}$ shows that

$$\sum_{n=1}^{\infty} |\langle Av_n, v_n \rangle| \leqslant \|A\|_{\mathrm{tc}},$$

so the right-hand side of (4) converges absolutely and gives a definition of the linear functional tr satisfying (1).

The difficult part of the theorem is to show that in (4) the right-hand side is independent of the choice of the orthonormal basis. Assuming this for now, property (2) follows at once as the operation sending A to $U^{-1}AU$ corresponds to choosing the orthonormal basis (Uv_n) in place of (v_n). In particular, $\operatorname{tr}(AU) = \operatorname{tr}(UA)$ for all $A \in \mathrm{TC}(\mathcal{H})$, and by applying Lemma 6.38 we can write every bounded operator B as a linear combination of four unitary operators and deduce by linearity of the trace that (3) holds as well. □

For the proof that the right-hand side of (4) is independent of the choice of the orthonormal basis two lemmas are needed.

Lemma 6.40 (Orthonormal approximations). *Let (v_n) be an orthonormal basis of a Hilbert space $\mathcal{H}$ and let $w_1, \ldots, w_m$ be orthonormal vectors. Then for every $\varepsilon > 0$ there exists some $N \geqslant 1$ and orthonormal vectors $w'_1, \ldots, w'_m \in \langle v_1, \ldots, v_N \rangle$ satisfying $\|w_j - w'_j\| < \varepsilon$ for $j = 1, \ldots, m$.*

PROOF. Let $\pi_N : \mathcal{H} \to \langle v_1, \ldots, v_N \rangle$ be the orthogonal projection. By the properties of an orthonormal basis (Proposition 3.36) we have $\pi_N(w) \to w$ as $N \to \infty$ for any $w \in \mathcal{H}$. Applying this projection to $w_1, \ldots, w_m$ gives

$$w''_j = \pi_N(w_j) \to w_j \tag{6.16}$$

as $N \to \infty$. For every N we now apply the Gram–Schmidt procedure to obtain

$$w_1' = c_1 w_1''$$
$$w_2' = c_2 \left(w_2'' - \langle w_2'', w_1' \rangle w_1' \right)$$
$$w_3' = c_3 \left(w_3'' - \langle w_3'', w_1' \rangle w_1' - \langle w_3'', w_2' \rangle w_2' \right)$$
$$\vdots$$
$$w_m' = c_m \left(w_m'' - \langle w_m'', w_1' \rangle w_1' - \cdots - \langle w_m'', w_{m-1}' \rangle w_{m-1}' \right),$$

where the constants $c_1, \dots, c_m > 0$ are chosen to normalize the vectors to have unit length. As $w_1, \dots, w_m$ are orthogonal and due to (6.16), a simple induction on $j = 1, \dots, m$ shows that c_j exists for all large enough N, that $c_j \to 1$ and also $w_j' \to w_j$ as $N \to \infty$. $\qquad\square$

Lemma 6.41 (Tail estimate). *Let $\mathcal{H}$ be a Hilbert space, let $A \in \mathrm{TC}(\mathcal{H})$ and let (v_n) be an orthonormal basis of $\mathcal{H}$. Then for every $\varepsilon > 0$ there exists some N such that for every extension $v_{N+1}', v_{N+2}', \dots$ of $v_1, \dots, v_N$ to an orthonormal basis of $\mathcal{H}$ we have*

$$\sum_{n=N+1}^{\infty} |\langle A v_n', v_n' \rangle| \leqslant \varepsilon.$$

PROOF. Fix some $\varepsilon > 0$ and some $N \geqslant 0$ and suppose the claim in the lemma does not hold for N. Then there exist orthonormal vectors $w_1, \dots, w_m$ in $\langle v_1, \dots, v_N \rangle^{\perp}$ such that

$$\sum_{k=1}^{m} |\langle A w_k, w_k \rangle| > \varepsilon. \tag{6.17}$$

Now apply Lemma 6.40 to the orthonormal vectors $w_1, \dots, w_m$ and the orthonormal basis $v_{N+1}, v_{N+2} \dots$ of the Hilbert space $\langle v_1, \dots, v_N \rangle^{\perp}$ to find some $N' > N$ large enough and a very good orthonormal approximation $w_1', \dots, w_m'$ to $w_1, \dots, w_m$ inside $\langle v_{N+1}, v_{N+2}, \dots, v_{N'} \rangle$. In particular, we may suppose that (6.17) also holds for $w_1', \dots, w_m'$.

We now apply the argument above infinitely often to achieve a contradiction to the hypothesis that $A \in \mathrm{TC}(\mathcal{H})$. Indeed, set $N_0 = 0$ to find some $N_1 > N_0$ and orthonormal vectors $w_{1,1}, \dots, w_{1,m_1}$ in $\langle v_1, \dots, v_{N_1} \rangle$ so that (6.17) also holds for $w_{1,1}, \dots, w_{1,m_1}$. Assuming we have already found $N_0 < N_1 < \cdots < N_\ell$ and orthonormal vectors $w_{j,1}, \dots, w_{j,m_j}$ in $\langle v_{N_{j-1}+1}, \dots, v_{N_j} \rangle$ with the same estimate for all $j = 1, \dots, \ell$, we may apply the same argument to find $w_{\ell+1,1}, \dots, w_{\ell+1,m_{\ell+1}}$ in $\langle v_{N_\ell+1}, \dots, v_{N_{\ell+1}} \rangle$ with the same properties. However, the bound

$$\ell\varepsilon < \sum_{j=1}^{\ell} \sum_{k=1}^{m_j} |\langle A w_{j,k}, w_{j,k} \rangle| \leqslant \|A\|_{\mathrm{tc}}$$

shows that the construction above has to stop, proving the lemma. $\qquad\square$

PROOF OF THEOREM 6.39: INDEPENDENCE. With these two lemmas we are now ready to prove that $\sum_{n=1}^{\infty} \langle Av_n, v_n \rangle$ is independent of the choice of the orthonormal basis (v_n) of $\mathcal{H}$ for a trace-class operator A. So let (w_n) be another orthonormal basis of $\mathcal{H}$, choose a positive ε, and choose N such that the conclusion of Lemma 6.41 holds for both bases (v_n) and (w_n). Let

$$V = \langle v_1, \ldots, v_N, w_1, \ldots, w_N \rangle,$$

and extend $v_1, \ldots, v_N$ with vectors $v'_{N+1}, \ldots, v'_M$ to an orthonormal basis of V. Similarly, we may find an orthonormal basis $w_1, \ldots, w_N, w'_{N+1}, \ldots, w'_M$ of V. Define a linear map $A_V : V \to V$ by sending $v \in V$ to $\pi_V(Av)$ where π_V is the orthogonal projection $\mathcal{H} \to V$. Note that $\langle A_V v, w \rangle = \langle Av, w \rangle$ for any two $v, w \in V$. By the tail estimate in Lemma 6.41 we have

$$\sum_{k=N+1}^{M} |\langle Av'_k, v'_k \rangle| \leqslant \varepsilon$$

and

$$\sum_{k=N+1}^{M} |\langle Aw'_k, w'_k \rangle| \leqslant \varepsilon.$$

Finally, note that

$$\sum_{k=1}^{N} \langle Av_k, v_k \rangle + \sum_{k=N+1}^{M} \langle Av'_k, v'_k \rangle = \sum_{k=1}^{N} \langle Aw_k, w_k \rangle + \sum_{k=N+1}^{M} \langle Aw'_k, w'_k \rangle$$

as both sides express the trace of the linear map A_V on the finite-dimensional space V. The choice of N now implies that $\sum_{k=1}^{\infty} \langle Av_k, v_k \rangle$ is within ε of the finite sum $\sum_{k=1}^{N} \langle Av_k, v_k \rangle$, which is within 2ε of $\sum_{k=1}^{N} \langle Aw_k, w_k \rangle$. The latter in turn is within ε of $\sum_{k=1}^{\infty} \langle Aw_k, w_k \rangle$ again by the choice of N. As $\varepsilon > 0$ was arbitrary the claimed independence follows. $\qquad\square$

As explained directly after the theorem, Theorem 6.39 follows from the independence and the black box Lemma 6.38.

Proposition 6.42 (Compactness). *Every trace-class operator on a complex Hilbert space $\mathcal{H}$ is compact.*

Before proving this, notice the following property of the trace-class norm. As the supremum is taken over $(v_n)_{n=1,\ldots,N}$ and $(w_n)_{n=1,\ldots,N}$ separately, we could multiply each w_n by an appropriate scalar α_n with $|\alpha_n| = 1$ to ensure that

$$\langle Av_n, \alpha_n w_n \rangle \geqslant 0.$$

Therefore we may also write

$$\|A\|_{\mathrm{tc}} = \sup_{(v_n),(w_n)} \left| \sum_{n=1}^{N} \langle Av_n, w_n \rangle \right|,$$

and if $A \in TC(\mathcal{H})$ then for every $\varepsilon > 0$ there exist finite orthonormal lists $(v_n)_{n=1,\dots,N}$ and $(w_n)_{n=1,\dots,N}$ with

$$\sum_{n=1}^{N} \langle Av_n, w_n \rangle > \|A\|_{\mathrm{tc}} - \varepsilon.$$

In the proof of Proposition 6.42 we will approximate a trace-class operator by operators with finite-dimensional range and then apply Lemma 6.7. In order to do this, it will be useful to understand the behaviour of the trace-class norm for matrices. For this we endow $\mathbb{C}^n$ and $\mathbb{C}^{n+1}$ with the standard inner product, and identify $\mathbb{C}^n$ with $\mathbb{C}^n \times \{0\} \subseteq \mathbb{C}^{n+1}$.

Lemma 6.43 (Trace-class norm for matrices). *Let us assume that a matrix $A_0 \in \mathrm{Mat}_{n,n}(\mathbb{C})$, the vectors $b, c \in \mathbb{C}^n$, and the scalar $d \in \mathbb{C}$ together define*

$$A_1 = \begin{pmatrix} A_0 & b \\ c^{\mathrm{t}} & d \end{pmatrix} \in \mathrm{Mat}_{n+1,n+1}(\mathbb{C})$$

and satisfy $\|A_0\|_{\mathrm{tc}} \geqslant (1 - \varepsilon^2)\|A_1\|_{\mathrm{tc}}$ for some $\varepsilon \in (0,1)$. Then

$$\left\| \begin{pmatrix} b \\ d \end{pmatrix} \right\| \leqslant \sqrt{5}\varepsilon\|A_1\|_{\mathrm{tc}}.$$

PROOF. From the fact that $e_{n+1} \perp \mathbb{C}^n$ (by the identification between $\mathbb{C}^n$ and $\mathbb{C}^n \times \{0\}$) and the definition of the trace-class norms, we have

$$\|A_0\|_{\mathrm{tc}} + |d| \leqslant \|A_1\|_{\mathrm{tc}}$$

and so

$$|d| \leqslant \|A_1\|_{\mathrm{tc}} - \|A_0\|_{\mathrm{tc}} \leqslant \varepsilon^2\|A_1\|_{\mathrm{tc}} \leqslant \varepsilon\|A_1\|_{\mathrm{tc}} \tag{6.18}$$

by the hypotheses. The lemma will follow from Pythagoras' theorem after we have shown the more delicate estimate

$$\|b\| \leqslant 2\varepsilon\|A_1\|_{\mathrm{tc}} \tag{6.19}$$

for the vector b.

To highlight the main point in the argument for (6.19) let us treat the case $n = 1$ first. In that case $A_0 = a, b, c, d \in \mathbb{C}$. If $\theta \in \mathbb{C}$ has $|\theta| = 1$ then

$$v = \begin{pmatrix} \sqrt{1 - \varepsilon^2} \\ \varepsilon\theta \end{pmatrix}$$

and

$$w = \begin{pmatrix} 1 \\ 0 \end{pmatrix}$$

are both unit vectors, and we may apply the definition of the trace-class norm $\|A_1\|_{\mathrm{tc}}$ to just these two vectors and obtain

$$|\langle A_1 v, w \rangle| = \left| a\sqrt{1 - \varepsilon^2} + b\varepsilon\theta \right| \leqslant \|A_1\|_{\mathrm{tc}}.$$

By choosing the argument of θ correctly this gives

$$|a|\sqrt{1 - \varepsilon^2} + |b|\varepsilon \leqslant \|A_1\|_{\mathrm{tc}}, \tag{6.20}$$

and the assumption of the lemma implies that

$$\|A_0\|_{\mathrm{tc}}\sqrt{1 - \varepsilon^2} \geqslant \left(1 - \varepsilon^2\right)^{3/2} \|A_1\|_{\mathrm{tc}} \geqslant \left(1 - \varepsilon^2\right)^2 \|A_1\|_{\mathrm{tc}}$$
$$\geqslant \left(1 - 2\varepsilon^2\right) \|A_1\|_{\mathrm{tc}}. \tag{6.21}$$

Combining (6.20) and (6.21) with $\|A_0\|_{\mathrm{tc}} = |a|$ gives

$$\left(1 - 2\varepsilon^2\right) \|A_1\|_{\mathrm{tc}} + |b|\varepsilon \leqslant \|A_1\|_{\mathrm{tc}},$$

which is equivalent to (6.19).

The idea of the proof of (6.19) in the general case is similar. However, since b is a vector for $n \geqslant 2$, some additional preparations are needed.

By compactness of the closed and bounded subset

$$\mathrm{U}_n(\mathbb{R}) = \{A \in \mathrm{Mat}_{n,n}(\mathbb{C}) \mid A^*A = I\} \subseteq \mathrm{Mat}_{n,n}(\mathbb{C}) \cong \mathbb{R}^{2n^2}$$

and the comment after the statement of Proposition 6.42 above, there exist orthonormal bases $v_1, \ldots, v_n \in \mathbb{C}^n$ and $w_1, \ldots, w_n \in \mathbb{C}^n$ with

$$\|A_0\|_{\mathrm{tc}} = \sum_{j=1}^{n} \langle A_0 v_j, w_j \rangle.$$

We define a unitary matrix $U \in \mathrm{Mat}_{n,n}(\mathbb{C})$ by requiring that $U^* v_j = w_j$ for all $j = 1, \ldots, n$, so that

$$\|A_0\|_{\mathrm{tc}} = \sum_{j=1}^{n} \langle A_0 v_j, U^* v_j \rangle = \sum_{j=1}^{n} \langle U A_0 v_j, v_j \rangle = \mathrm{tr}(U A_0). \tag{6.22}$$

Since we have now expressed $\|A_0\|_{\mathrm{tc}}$ as the trace of $U A_0$, independence of the trace on the choice of orthonormal basis implies that (6.22) holds for an arbitrary orthonormal basis $v_1', \ldots, v_n'$ of $\mathbb{C}^n$. In particular, from the equality and the definition of the trace-class norm we deduce that

$$\langle U A_0 v_j', v_j' \rangle \geqslant 0 \tag{6.23}$$

for any choice of orthonormal basis of $\mathbb{C}^n$ and $j = 1, \ldots, n$. Now let the orthonormal basis $v_1', \ldots, v_n'$ of $\mathbb{C}^n$ be chosen so that $Ub = \|b\|v_n'$. We extend U to a unitary operator on $\mathbb{C}^{n+1}$ by setting $Ue_{n+1} = e_{n+1}$, and note that

$$UA_1 e_{n+1} = U(b + de_{n+1}) = \|b\|v_n' + de_{n+1}$$

by definition of A_1 and the choice of orthonormal basis. We now consider the two orthonormal lists

$$v_1', \ldots, v_{n-1}', \sqrt{1 - \varepsilon^2}v_n' + \varepsilon e_{n+1}$$

and

$$U^{-1}v_1', \ldots, U^{-1}v_{n-1}', U^{-1}v_n'.$$

Using the definition of the trace-class norm $\|A_1\|_{\mathrm{tc}}$ we get

$$\|A_1\|_{\mathrm{tc}} \geqslant \sum_{j=1}^{n-1} \langle UA_1 v_j', v_j' \rangle + \sqrt{1 - \varepsilon^2} \langle UA_1 v_n', v_n' \rangle + \varepsilon\|b\| \langle v_n, v_n \rangle$$

$$\geqslant \sqrt{1 - \varepsilon^2}\|A_0\|_{\mathrm{tc}} + \varepsilon\|b\|,$$

where we have used (6.23) and (6.22) in the last step. This is the analogue to (6.20) with $|a|$ replaced by $\|A_0\|_{\mathrm{tc}}$. Together with (6.21) we obtain (6.19) in the general case. $\square$

PROOF OF PROPOSITION 6.42. Let A be a trace-class operator on $\mathcal{H}$. We will construct, for every $\varepsilon > 0$, an operator $A_\varepsilon : \mathcal{H} \to \mathcal{H}$ with finite-dimensional range and with

$$\|A - A_\varepsilon\|_{\mathrm{op}} \ll \varepsilon\|A\|_{\mathrm{tc}}. \tag{6.24}$$

This will imply the proposition by Lemma 6.7.

To construct A_ε we choose orthonormal vectors $v_1, \ldots, v_n$ and $w_1, \ldots, w_n$ in $\mathcal{H}$ with

$$\sum_{k=1}^{n} \langle Av_k, w_k \rangle \geqslant (1 - \varepsilon^2)\|A\|_{\mathrm{tc}}$$

as in the definition of $\|\cdot\|_{\mathrm{tc}}$ and using the comment after Proposition 6.42. We define A_ε by setting it equal to A on $\langle v_1, \ldots, v_n \rangle$ and to 0 on $\langle v_1, \ldots, v_n \rangle^\perp$. For any vector $v' \in \langle v_1, \ldots, v_n \rangle$ and $v'' \in \langle v_1, \ldots, v_n \rangle^\perp$ we have

$$(A - A_\varepsilon)(v' + v'') = Av'',$$

and we claim that

$$\|Av''\| \leqslant \sqrt{5}\varepsilon\|A\|_{\mathrm{tc}}\|v''\|, \tag{6.25}$$

which will imply (6.24). To prove (6.25) we may assume that

$$v'' = v_{n+1} \in \langle v_1, \ldots, v_n \rangle^\perp$$

is a unit vector. We write

$$Av_{n+1} = \sum_{j=1}^{n} b_j w_j + d w_{n+1}$$

for $b \in \mathbb{C}^n$, $d \in \mathbb{C}$, and $w_{n+1} \in \langle w_1, \ldots, w_n \rangle^{\perp}$ a unit vector.

We apply Lemma 6.43 to the matrix $A_1 \in \mathrm{Mat}_{n+1,n+1}(\mathbb{C})$ defined by

$$(A_1)_{jk} = \langle Av_k, w_j \rangle$$

for $j, k = 1, \ldots, n+1$. Then

$$A_1 = \begin{pmatrix} A_0 & b \\ c^{\mathrm{t}} & d \end{pmatrix} \in \mathrm{Mat}_{n+1,n+1}(\mathbb{C})$$

for some matrix $A_0 \in \mathrm{Mat}_{n,n}(\mathbb{C})$ and $c \in \mathbb{C}^n$. By choice of the orthonormal lists (v_n) and (w_n) we have

$$(1 - \varepsilon^2) \|A\|_{\mathrm{tc}} \leqslant \sum_{j=1}^{n} \langle Av_j, w_j \rangle = \mathrm{tr}\, A_0 \leqslant \|(A_0)\|_{\mathrm{tc}}. \tag{6.26}$$

Since a different choice of an orthonormal basis of $\mathbb{C}^{n+1}$ corresponds to a different choice of an orthonormal basis of $\langle v_1, \ldots, v_{n+1} \rangle$ or of $\langle w_1, \ldots, w_{n+1} \rangle$, it follows that $\|A_1\|_{\mathrm{tc}} \leqslant \|A\|_{\mathrm{tc}}$. Arguing a bit more carefully we may find, again by compactness, some unitary matrix $U \in \mathrm{Mat}_{n+1,n+1}(\mathbb{C})$ with

$$\|A_1\|_{\mathrm{tc}} = \mathrm{tr}(U^* A_1) = \sum_{j=1}^{n+1} \langle U^* A_1 e_j, e_j \rangle.$$

Since $\langle U^* A_1 e_j, e_j \rangle = \langle Av_j, \sum_{k=1}^{n+1} u_{kj} w_k \rangle$ and the vectors $\sum_{k=1}^{n+1} u_{kj} w_k \in \mathcal{H}$ are orthonormal for $j = 1, \ldots, n+1$, the inequality $\|A_1\|_{\mathrm{tc}} \leqslant \|A\|_{\mathrm{tc}}$ follows.

Combining the estimate $\|A_1\|_{\mathrm{tc}} \leqslant \|A\|_{\mathrm{tc}}$ with (6.26) and Lemma 6.43, we get (6.25) for any $v'' \in \langle v_1, \ldots, v_n \rangle^{\perp}$ with $\|v''\| = 1$.

It follows that A is the limit of A_ε defined as above as $\varepsilon \searrow 0$ (with respect to the operator norm), and so Lemma 6.7 implies the proposition. □

The results above regarding the trace and the trace-class are satisfying, but the concepts would not be important without non-trivial examples of trace-class operators. We next discuss the relationship with the class of self-adjoint (compact) operators, which gives us many examples.

We say that a self-adjoint operator A on a Hilbert space $\mathcal{H}$ is *positive* if $\langle Av, v \rangle \geqslant 0$ for all $v \in \mathcal{H}$.

Proposition 6.44. *Let $\mathcal{H}$ be a complex Hilbert space and A a bounded operator on $\mathcal{H}$. If A is self-adjoint and positive and (v_n) is an orthonormal basis of $\mathcal{H}$, then $\|A\|_{\mathrm{tc}} = \sum_{n=1}^{\infty} \langle Av_n, v_n \rangle$.*

In particular, $\mathrm{tr}(A) = \|A\|_{\mathrm{tc}}$, where in the case of a positive operator A with $\|A\|_{\mathrm{tc}} = \infty$ this extends our definition of the trace.

PROOF OF PROPOSITION 6.44. The inequality $\|A\|_{\mathrm{tc}} \geqslant \sum_{n=1}^{\infty} \langle Av_n, v_n \rangle$ follows directly from the definition of the trace-class norm. For the opposite inequality we may suppose that $S = \sum_{n=1}^{\infty} \langle Av_n, v_n \rangle$ is finite and let (x_k) and (y_k) be two orthonormal lists of length K as in the definition of the trace-class norm. We wish to show

$$\sum_{k=1}^{K} |\langle Ax_k, y_k \rangle| \leqslant S. \tag{6.27}$$

Using Lemma 6.40 we can find some $N \geqslant 1$ and orthonormal approximations of (x_k) and (y_k) within $V = \langle v_1, \dots, v_N \rangle$. Letting $N \to \infty$ later on, it suffices to show (6.27) for the approximations within V and we will use the same letters to denote the approximations. We extend the orthonormal lists (x_k) and (y_k) to orthonormal bases of V. Using the comment after Proposition 6.42 we may adjust the y_k once more and assume without loss of generality that $\langle Ax_k, y_k \rangle \geqslant 0$ for $k = 1, \dots, N$ without changing the value of the left-hand side in (6.27). We also define a unitary operator $U : V \to V$ satisfying $U^* x_k = y_k$ for $k = 1, \dots, N$. In other words, we wish to estimate

$$\sum_{k=1}^{K} \langle Ax_k, y_k \rangle \leqslant \sum_{k=1}^{N} \langle Ax_k, y_k \rangle = \sum_{k=1}^{N} \langle Ax_k, U^* x_k \rangle = \mathrm{tr}(UA_V), \tag{6.28}$$

where we let A_V be the positive self-adjoint operator $v \in V \mapsto \pi_V(Av)$ and $\pi_V : \mathcal{H} \to V$ is the orthogonal projection. As a trace (on the finite-dimensional space V) can be calculated in any basis we may also calculate $\mathrm{tr}(UA_V)$ using an orthonormal basis $v_1', \dots, v_N'$ of V consisting of eigenvectors of A_V. Let $\lambda_1, \dots, \lambda_N$ in $\mathbb{R}_{\geqslant 0}$ be the corresponding eigenvalues. Since we have $\mathrm{tr}(UA_V) \geqslant 0$ by (6.28), we obtain

$$\mathrm{tr}(UA_V) = \sum_{n=1}^{N} \langle UA_V v_j', v_j' \rangle = \left| \sum_{n=1}^{N} \lambda_n \langle U v_j', v_j' \rangle \right|$$

$$\leqslant \sum_{n=1}^{N} \lambda_n = \mathrm{tr}(A_V) = \sum_{n=1}^{N} \langle Av_n, v_n \rangle \leqslant S.$$

This, together with (6.28), implies (6.27), first for the approximations of (x_k) and (y_k) in V, and then using Lemma 6.40 and letting $N \to \infty$ as indicated earlier for any two lists of orthonormal vectors in $\mathcal{H}$. Hence $\|A\|_{\mathrm{tc}} \leqslant S$ and the proposition follows. $\qquad\square$

Exercise 6.45. Let A be a compact operator. Show that A has a polar decomposition of the form $A = QP$ where $\ker(A) = \ker(Q) = \ker(P)$, $Q|_{(\ker(A))^{\perp}}$ is an isometry, and P is positive, self-adjoint, and compact. Show that P is trace-class if and only if A is.

Corollary 6.46 (Lidskiĭ's theorem [61]). *Let $\mathcal{H}$ be a separable complex Hilbert space. A self-adjoint bounded operator A on $\mathcal{H}$ is trace-class if and only if it is compact and its eigenvalues λ_n (allowing repetitions as in Theorem 6.27) satisfy*

$$\sum_{n=1}^{\infty} |\lambda_n| = \|A\|_{\mathrm{tc}} < \infty.$$

If A is indeed trace-class, then

$$\mathrm{tr}(A) = \sum_{n=1}^{\infty} \lambda_n$$

and this sum converges absolutely.

PROOF. Assume first that A is self-adjoint and trace-class. Then Proposition 6.42 implies that A is compact. Using the orthonormal basis consisting of eigenvectors v_n with eigenvalues λ_n from Theorem 6.27 and the definition of the trace it follows that $\sum_{n=1}^{\infty} |\lambda_n| \leqslant \|A\|_{\mathrm{tc}} < \infty$ and $\mathrm{tr}(A) = \sum_{n=1}^{\infty} \lambda_n$.

Let now A be a compact self-adjoint operator and assume $\sum_{n=1}^{\infty} |\lambda_n| < \infty$, where λ_n are the eigenvalues of A. Again let v_n be an orthonormal basis of $\mathcal{H}$ consisting of eigenvectors for A (with corresponding eigenvalues λ_n). We let $\alpha_n \in \{\pm 1\}$ be chosen with $\lambda_n \alpha_n = |\lambda_n|$ for all $n \geqslant 1$. Define the positive self-adjoint operator P on $\mathcal{H}$ by setting $Pv_n = |\lambda_n| v_n$ and the unitary operator U on $\mathcal{H}$ by setting $Uv_n = \alpha_n v_n$ for all $n \geqslant 1$ and linearly extending both to $\mathcal{H}$, so that $A = UP$. By Proposition 6.44 we have $\|P\|_{\mathrm{tc}} = \sum_{n=1}^{\infty} |\lambda_n|$, and by the initial properties of the trace-class norm this gives

$$\|A\|_{\mathrm{tc}} = \|UP\|_{\mathrm{tc}} = \|P\|_{\mathrm{tc}} = \sum_{n=1}^{\infty} |\lambda_n| < \infty,$$

which implies the corollary. $\qquad\square$

Let us indicate how the trace appears frequently in applications, and how to calculate it in these special circumstances.

Exercise 6.47. Let $\ell > d$ and $k \in H^\ell(\mathbb{T}^d \times \mathbb{T}^d)$. Show that the Hilbert–Schmidt integral operator $K(f)(x) = \int_{\mathbb{T}^d} k(x,y) f(y)\, \mathrm{d}y$ on $L^2(\mathbb{T}^d)$ is trace-class and that the trace is given by the integral along the diagonal, that is $\mathrm{tr}(K) = \int_{\mathbb{T}^d} k(x,x)\, \mathrm{d}x$.

Proposition 6.48. *Let X be a compact metric space, let μ be a finite measure on X, and let $k \in C(X \times X)$ be a continuous kernel with the property that the associated Hilbert–Schmidt operator K defined by*

$$K(f)(x) = \int_X k(x,y) f(y)\, \mathrm{d}\mu(y)$$

is trace-class. Then

$$\mathrm{tr}(K) = \int_X k(x,x)\, \mathrm{d}\mu(x).$$

PROOF. Let (ξ_ℓ) be a sequence of finite measurable partitions of X that become finer in the sense that

$$\max_{P \in \xi_\ell} \operatorname{diam}(P) \longrightarrow 0 \qquad (6.29)$$

as $\ell \to \infty$, and assume that the sequence is *refining*, meaning that each element of ξ_ℓ is a union of elements of $\xi_{\ell+1}$ for $\ell \geqslant 1$. For $P \in \xi_\ell$ we also define the unit vector

$$w_P = \frac{1}{\sqrt{\mu(P)}} \mathbb{1}_P$$

and notice that $\{w_P \mid P \in \xi_\ell\}$ is an orthonormal basis of its linear hull W_ℓ for all $\ell \geqslant 1$ (if $\mu(P) = 0$ we do not associate a vector w_P to this partition element). Since (ξ_ℓ) is refining, we have $W_\ell \subseteq W_{\ell+1}$ for all $\ell \geqslant 1$.

Now define an orthonormal sequence (v_n) by starting with w_P for $P \in \xi_1$ (in some fixed order) and extending it via Gram–Schmidt first to an orthonormal basis $v_1, \ldots, v_{n(2)}$ of W_2, then to an orthonormal basis $v_1, \ldots, v_{n(3)}$ of W_3, and so on. It is clear that the assumption (6.29) implies that the characteristic function of every open set belongs to the closure of $\bigcup_{\ell \geqslant 1} W_\ell$. This in turn implies, as in the proof of Proposition 2.51, that

$$\overline{\bigcup_{\ell \geqslant 1} W_\ell} = L^2_\mu(X)$$

and hence that (v_n) is an orthonormal basis of $L^2_\mu(X)$. Therefore

$$\operatorname{tr}(K) = \sum_{n=1}^{\infty} \langle K v_n, v_n \rangle = \lim_{\ell \to \infty} \sum_{n=1}^{n(\ell)} \langle K v_n, v_n \rangle = \lim_{\ell \to \infty} \sum_{P \in \xi_\ell} \langle K w_P, w_P \rangle$$

since $\sum_{n=1}^{n_\ell} \langle K v_n, v_n \rangle = \operatorname{tr}(\pi_{W_\ell} K|_{W_\ell})$, where π_{W_ℓ} denotes the orthogonal projection onto W_ℓ, and this trace can also be computed in the orthonormal basis $\{w_P \mid P \in \xi_\ell\}$. Now we may use the definition of K to see that

$$\sum_{P \in \xi_\ell} \langle K w_P, w_P \rangle = \sum_{P \in \xi_\ell} \frac{1}{\mu(P)} \int_P K(\mathbb{1}_P) \, d\mu = \sum_{P \in \xi_\ell} \frac{1}{\mu(P)} \int_{P \times P} k(x, y) d\mu \times \mu(x, y).$$

Now fix $\varepsilon > 0$ and use uniform continuity of k to find an ℓ sufficiently large to ensure that

$$|k(x, y) - k(x, x)| < \varepsilon$$

whenever $x, y \in P$ for some $P \in \xi_\ell$. We may also suppose that ℓ is large enough to have

$$\left| \operatorname{tr}(K) - \sum_{P \in \xi_\ell} \langle K w_P, w_P \rangle \right| < \varepsilon.$$

Together we see that

$$\left| \operatorname{tr}(K) - \int k(x,x)\,\mathrm{d}\mu(x) \right|$$

$$\leqslant \varepsilon + \left| \sum_{P \in \xi_\ell} \frac{1}{\mu(P)} \int_{P \times P} k(x,y)\,\mathrm{d}\mu{\times}\mu(x,y) - \int_P k(x,x)\,\mathrm{d}\mu(x) \right|$$

$$\leqslant \varepsilon + \sum_{P \in \xi_\ell} \frac{1}{\mu(P)} \int_{P \times P} \left| \underbrace{(k(x,y) - k(x,x))}_{< \varepsilon} \,\mathrm{d}\mu{\times}\mu(x,y) \right| \leqslant (1 + \mu(X))\,\varepsilon.$$

As $\varepsilon > 0$ was arbitrary, the proposition follows.   □

Exercise 6.49. Generalize Proposition 6.48 by assuming that X is a σ-compact, locally compact, metric space, μ is locally finite, and $k \in C(X \times X) \cap L^2_{\mu \times \mu}(X \times X)$ defines a trace-class integral operator.

Exercise 6.50. Let $\mathcal{H}$ be a (separable) Hilbert space.
(a) Show that $\mathrm{TC}(\mathcal{H})$ is complete (and hence a Banach space) with respect to the trace-class norm.
(b) What is the closure of $\mathrm{TC}(\mathcal{H})$ with respect to the operator norm?

Exercise 6.51. Let $\mathcal{H}$ be a real Hilbert space.
(a) Define $\mathcal{H}_{\mathbb{C}} = \mathcal{H} \otimes_{\mathbb{R}} \mathbb{C} = \mathcal{H} \oplus \mathrm{i}\mathcal{H}$ and define

$$\langle a_1 + \mathrm{i}a_2, b_1 + \mathrm{i}b_2 \rangle_{\mathbb{C}} = \langle a_1, b_1 \rangle + \langle a_2, b_2 \rangle + \mathrm{i}(\langle a_2, b_1 \rangle - \langle a_1, b_2 \rangle).$$

Show that $\langle \cdot, \cdot \rangle_{\mathbb{C}}$ is a complex inner product making $\mathcal{H}_{\mathbb{C}}$ into a complex Hilbert space.
(b) We used Lemma 6.38 (concerning complex Hilbert spaces) twice in this section. Use (a) to show that the results of this section also hold in the case of a real Hilbert space.

Exercise 6.52. Let $(T, \mathcal{B}, \mu)$ be a probability space, $\mathcal{H}$ a Hilbert space, and

$$T \ni t \longmapsto A_t \in \mathrm{TC}(\mathcal{H})$$

a map such that $t \in T \mapsto \langle A_t v, w \rangle$ is measurable for every $v, w \in \mathcal{H}$. Also suppose that $\int_T \|A_t\|_{\mathrm{tc}}\,\mathrm{d}\mu(t) < \infty$. Show that $A = \int A_t\,\mathrm{d}\mu(t)$ (defined via $v \mapsto \int A_t v\,\mathrm{d}\mu(t)$ as in Section 3.5.4) is trace-class, and that $\operatorname{tr}(A) = \int \operatorname{tr}(A_t)\,\mathrm{d}\mu(t)$.

The next two exercises give a tool for showing that certain Hilbert–Schmidt integral operators (as in Proposition 6.48) are trace-class.

Exercise 6.53. Let $\mathcal{H}$ be a Hilbert space with an orthonormal basis (e_n). Define the *Hilbert–Schmidt norm*

$$\|A\|^2_{\mathrm{HS}} = \sum_{j,k} |\langle Ae_j, e_k \rangle|^2$$

and the space of Hilbert–Schmidt operators $\mathrm{HS}(\mathcal{H}) = \{ A \in \mathrm{B}(\mathcal{H}) \mid \|A\|_{\mathrm{HS}} < \infty \}$.
(a) Show that $A \in \mathrm{HS}(\mathcal{H})$ if and only if $A^* \in \mathrm{HS}(\mathcal{H})$, and that $\|A^*\|_{\mathrm{HS}} = \|A\|_{\mathrm{HS}}$ for all $A \in \mathrm{HS}(\mathcal{H})$.
(b) Show that the definition of the Hilbert–Schmidt norm is independent of the choice of orthonormal basis.
(c) Show that $\mathrm{HS}(\mathcal{H})$ forms a two-sided ideal in $\mathrm{B}(\mathcal{H})$. That is, for any $A \in \mathrm{HS}(\mathcal{H})$ and $B \in \mathrm{B}(\mathcal{H})$ we have $AB \in \mathrm{HS}(\mathcal{H})$ and $BA \in \mathrm{HS}(\mathcal{H})$.

(d) Find an inner product on $\mathrm{HS}(\mathcal{H})$ which induces the norm $\|\cdot\|_{\mathrm{HS}}$, and show that $\mathrm{HS}(\mathcal{H})$ is a Hilbert space with this inner product.
(e) Show that $\mathrm{HS}(\mathcal{H})$ is also a Banach algebra, meaning that $\|AB\|_{\mathrm{HS}} \leqslant \|A\|_{\mathrm{HS}}\|B\|_{\mathrm{HS}}$.
(f) Show that $\mathrm{HS}(\mathcal{H})$ is a closed subspace of $B(\mathcal{H})$ if and only if $\mathcal{H}$ is finite-dimensional.
(g) Show that every Hilbert–Schmidt operator is compact.
(h) Assume now that $\mathcal{H} = L^2((0,1))$. For every $k \in L^2((0,1)^2)$ we define the associated *Hilbert–Schmidt integral operator* as in Proposition 6.11. Show that the space of Hilbert–Schmidt integral operators corresponds exactly to $\mathrm{HS}(\mathcal{H})$. In particular, show that for any operator $A \in \mathrm{HS}(\mathcal{H})$ the corresponding kernel k_A is given by

$$k_A(x,y) = \sum_{i,j} \langle Ae_i, e_j \rangle \, e_i(x)\overline{e_j(y)}.$$

Exercise 6.54. (a) Show that if $A, B \in \mathrm{HS}(\mathcal{H})$ then AB is trace-class.
(b) If C is trace-class then there are operators $A, B \in \mathrm{HS}(\mathcal{H})$ with $C = AB$.

Exercise 6.55. [18] Let $\mathcal{H}$ be a Hilbert space with respect to the inner product $\langle \cdot, \cdot \rangle_{\mathcal{H}}$, and write $\|\cdot\|_{\mathcal{H}}$ for the induced norm on $\mathcal{H}$. Let $\langle \cdot, \cdot \rangle_0$ be a semi-inner product on $\mathcal{H}$, and write $\|\cdot\|_0$ for the induced semi-norm on $\mathcal{H}$. Assume that $\|\cdot\|_0 \leqslant \|\cdot\|_{\mathcal{H}}$.
(a) Show that there exists a unique positive bounded self-adjoint operator A such that

$$\langle v, w \rangle_0 = \langle Av, w \rangle_{\mathcal{H}}.$$

The *relative trace* of $\|\cdot\|_0$ with respect to $\|\cdot\|_{\mathcal{H}}$ is defined as the trace of A (which might be infinity).
(b) Let $k > \frac{d}{2}$, $\mathcal{H} = H^k(U)$ for some open subset $U \subseteq \mathbb{R}^d$, and $\langle f, g \rangle_0 = f(x)\overline{g(x)}$ for some fixed $x \in U$. Show that A as in (a) has finite trace (and so $\|\cdot\|_0$ has finite relative trace with respect to $\|\cdot\|_{\mathcal{H}}$).
(c) Let μ be a compactly supported measure on U. Combine (b) with Exercise 6.52 to show that the semi-norm $\|f\|_{L^2(\mu)} = \left(\int |f|^2 \, \mathrm{d}\mu\right)^{1/2}$ for $f \in H^k(U)$ has finite relative trace with respect to $\|\cdot\|_{\mathcal{H}}$.

6.4 Eigenfunctions for the Laplace Operator

We will prove in this section the claim from Section 1.2 that for any open bounded subset $U \subseteq \mathbb{R}^d$ there is a basis of $L^2(U)$ consisting of eigenfunctions of the Laplace operator such that these functions also vanish (in the square-mean sense) at the boundary of U.

In the proof we will first go back to the case of the d-dimensional torus, even though (or actually precisely because) we already have an orthonormal basis consisting of eigenfunctions of the Laplacian in this setting, namely the characters. In Section 6.4.2 we will define a right inverse of Δ defined on $L^2(U)$ for an open subset U of $\mathbb{R}^d$ — a setting in which we do not know the eigenfunctions of the Laplacian. Finally, we will ask in Section 6.4.4 about the growth rate of the eigenvalues and prove Weyl's law for Jordan measurable open domains. We start by stating the main theorem, which will be proved in Section 6.4.2.

Theorem 6.56 (Existence of basis of Laplace eigenfunctions). *Let U be an open bounded subset of $\mathbb{R}^d$. Then there exists an orthonormal basis $\{f_n\}$ of $L^2(U)$ of functions in $H_0^1(U)$ which are smooth in U and have $\Delta f_n = \lambda_n f_n$, with $\lambda_n < 0$ for all $n \geqslant 1$, and $\lambda_n \to -\infty$ as $n \to \infty$.*

6.4.1 Right Inverse and Compactness on the Torus

We already used the fact that the characters on $\mathbb{T}^d$ are eigenfunctions of the Laplace operator on $\mathbb{T}^d$ in the proof of elliptic regularity (Lemma 5.48). Obtaining a compact self-adjoint right inverse to Δ is quite easy on the torus $\mathbb{T}^d$.

Exercise 6.57. Define $L_0^2(\mathbb{T}^d) = \{f \in L^2(\mathbb{T}^d) \mid \int_{\mathbb{T}^d} f \, dx = 0\}$, and prove that there exists a compact self-adjoint operator $S : L_0^2(\mathbb{T}^d) \longrightarrow L_0^2(\mathbb{T}^d)$ with the property that $\Delta S f = f$ for all $f \in L_0^2(\mathbb{T}^d)$.

For the discussion on an open subset we will need the following lemma.

Lemma 6.58 (Compactness on the torus). *The operator*

$$\imath_{1,0} : H^1(\mathbb{T}^d) \longrightarrow H^0(\mathbb{T}^d) = L^2(\mathbb{T}^d)$$

is compact.

PROOF. For the proof we define

$$K = \{f \in L^2(\mathbb{T}^d) \mid \|f\|_2 \leqslant 1 \text{ and } \partial_j f \text{ exists with } \|\partial_j f\|_2 \leqslant 1 \text{ for } j = 1, \ldots, d\},$$

and note that $\imath_{1,0}\left(B_1^{H^1(\mathbb{T}^d)}\right) \subseteq K$. Hence it suffices to show that K is totally bounded. Let now $f = \sum_{n \in \mathbb{Z}^d} a_n \chi_n \in K$. The definition of $\partial_j f$ implies that $\langle \partial_j f, \chi_n \rangle = -\langle f, \partial_j \chi_n \rangle = 2\pi i n_j a_n$. Using $\|f\|_2 \leqslant 1$ and $\|\partial_j f\|_2 \leqslant 1$ for $j = 1, \ldots, d$ we obtain

$$\sum_{n \in \mathbb{Z}^d} \left(1 + \|n\|_2^2\right) |a_n|^2 \ll 1.$$

This implies a uniformity claim for the convergence of the Fourier series of all $f \in K$. Indeed, for any $N \geqslant 1$ we have

$$\sum_{\|n\|_2 > N} |a_n|^2 = N^{-2} \sum_{\|n\|_2 > N} N^2 |a_n|^2 \leqslant N^{-2} \sum_{\|n\|_2 > N} (1 + \|n\|_2^2)|a_n|^2 \ll N^{-2},$$

and so we see that the above tail sum goes to zero uniformly for all $f \in K$ as $N \to \infty$.

To see that K is totally bounded we fix some $\varepsilon > 0$ and choose N such that the above statement becomes

$$\sum_{\|n\|_2 > N} |a_n|^2 < \varepsilon^2/4$$

for all $f = \sum_n a_n \chi_n \in K$. Next take a finite $\varepsilon/2$-dense subset of the finite-dimensional compact set

$$\left\{ f = \sum_{\|n\|_2 \leqslant N} a_n \chi_n \mid \|f\|_2 \leqslant 1 \text{ and } \|\partial_j f\|_2 \leqslant 1 \text{ for } j = 1, \ldots d \right\}.$$

Combining these statement shows that we have found an ε-dense subset of K. As $\varepsilon > 0$ was arbitary, K is totally bounded, which implies that the closure of $\imath_{1,0}\left(B_1^{H^1(\mathbb{T}^d)}\right)$ is compact. $\square$

Exercise 6.59. Consider the map $\imath_{k,\ell} : H^k(\mathbb{T}^d) \to H^\ell(\mathbb{T}^d)$ for $k \geqslant \ell \geqslant 0$.
(a) Characterize those k and ℓ for which the map $\imath_{k,\ell}$ is compact.
(b) Characterize those k for which the map $\imath_{k,0}\imath_{k,0}^*$ is Hilbert–Schmidt class.
(c) Characterize those k for which the map $\imath_{k,0}\imath_{k,0}^*$ is trace-class.

6.4.2 A Self-Adjoint Compact Right Inverse on Open Subsets

The following provides the link between the Laplace operator and our discussion of compact self-adjoint operators in Theorem 6.27. The compactness claim is a special case of Rellich's Theorem.

Proposition 6.60 (Self-adjoint compact right inverse). *Let $U \subseteq \mathbb{R}^d$ be a bounded and open subset. Using Lemma 5.41 we equip $H_0^1(U)$ with the inner product $\langle \cdot, \cdot \rangle_1$. Then the map $\imath = \imath_{1,0} : H_0^1(U) \longrightarrow H^0(U) = L^2(U)$ has the property that $\Delta(\imath^* f) \in L^2(U)$ exists for all $f \in L^2(U)$ and equals $-f$. In other words, $\Delta \circ (-\imath^*) = I$ is the identity on $L^2(U)$. Finally, $S = -\imath^*$ is a compact self-adjoint operator $L^2(U) \longrightarrow L^2(U)$.*

PROOF. Recall the map $\imath : H_0^1(U) \longrightarrow H^0(U)$ sending $f \mapsto f$ from Proposition 5.13. The adjoint is a map $\imath^* : H^0(U) \longrightarrow H_0^1(U)$, and so the composition $\imath\imath^*$ is indeed a map from $L^2(U)$ to $L^2(U)$. By Exercise 6.17 $\imath\imath^*$ (and hence S) is self-adjoint. Now let $\phi \in C_c^\infty(U)$ and $f \in L^2(U)$. Then

$$\langle -\imath^* f, \Delta\phi \rangle_{L^2(U)} = -\sum_j \langle \imath^* f, \partial_j^2 \phi \rangle_{L^2(U)} = \sum_j \langle \partial_j \imath^* f, \partial_j \phi \rangle_{L^2(U)}$$

$$= \langle \imath^* f, \phi \rangle_1 = \langle f, \imath\phi \rangle_{L^2(U)} = \langle f, \phi \rangle_{L^2(U)}$$

shows that $\Delta \circ (-\imath^*) = I$, as claimed.

It remains to show that S is a compact operator. For this notice that $\imath$ is the composition of the operators

$$H_0^1(U) \xrightarrow{P} H^1(\mathbb{T}_R^d) \xrightarrow{\imath_{1,0}} H^0(\mathbb{T}_R^d) = L^2(\mathbb{T}_R^d) \xrightarrow{\cdot|_U} L^2(U),$$

where we choose R such that $U \subseteq B_R$, P is the periodizing operator from Lemma 5.36, $\imath_{1,0}$ is the operator from Proposition 5.3 that simply forgets regularity and is compact by Lemma 6.58, and finally $\cdot|_U$ is the restriction operator to U. We also note that we equip $H_0^1(U)$ with the norm derived from the inner product $\langle \cdot, \cdot \rangle_1$ in Lemma 5.40 but $H^1(\mathbb{T}^d)$ with the standard Sobolev norm. By Lemma 5.41 this is not an issue since the norm derived from $\langle \cdot, \cdot \rangle_1$ is equivalent to the standard Sobolev norm on $H_0^1(U)$. As $\imath$ is the composition of bounded operators and a compact operator, Lemma 6.3 applies, and it follows that $\imath$ and also S are compact operators. $\quad\square$

PROOF OF THEOREM 6.56. Let $S = -\imath\imath^* : L^2(U) \to L^2(U)$ be the self-adjoint compact operator from Proposition 6.60. Theorem 6.27 gives an orthonormal basis (f_n) of eigenvectors with $Sf_n = \mu_n f_n$ and $\mu_n \to 0$ as $n \to \infty$. By Proposition 6.60 we have $\Delta(Sf_n) = f_n$, so that $\Delta(\mu_n f_n) = f_n$. Note that the eigenvectors f_n all lie in the image of $\imath$, and so belong to H_0^1. It follows that $\mu_n \neq 0$ and $\Delta f_n = \lambda_n f_n$ with $\lambda_n = \frac{1}{\mu_n}$, and so we have $|\lambda_n| \to \infty$ as $n \to \infty$. Since $S = -\imath\imath^*$ we have $\mu_n = \langle Sf_n, f_n \rangle_{L^2(U)} = -\langle \imath^* f_n, \imath^* f_n \rangle_1 \leqslant 0$ so that $\lambda_n \to -\infty$ as $n \to \infty$.

It remains to show that f_n is smooth in U and $\Delta f_n = \lambda_n f_n$ for all $n \geqslant 1$, and this is precisely the statement of Corollary 5.47. $\quad\square$

There are very few examples of domains U for which one can write down the eigenfunctions of the Laplace operator explicitly. Important exceptions are rectangles $U = (0, a_1) \times (0, a_2) \times \cdots \times (0, a_d)$ and balls $U = B_M^{\mathbb{R}^d}$.

Exercise 6.61. (a) Let $U = (0, a_1) \times \cdots \times (0, a_d)$. Show that the functions f_n arising as eigenfunctions of the Laplace operator as in Theorem 6.56 can be chosen to take the form

$$f_n(x) = \sin(\lambda_n^{(1)} x_1) \cdots \sin(\lambda_n^{(d)} x_d).$$

(b) Let $U = \{(x_1, x_2) \in (0,1) \times (0,1) \mid x_1 + x_2 < 1\}$. Find an orthonormal basis of $L^2(U)$ consisting of eigenfunctions of the Laplace operator and satisfying the Dirichlet boundary value conditions.

Exercise 6.62. Assume that $d \geqslant 2$ (or that $d = 2$ for simplicity). Let $U \subseteq \mathbb{R}^d$ be open and $K \subseteq U$ a compact subset. Let $f \in H_0^1(U)$ be an eigenfunction of Δ (and of S as in the proof of Theorem 6.56) such that $\Delta f = \lambda f$ for some $\lambda < -1$. Show that

$$\|f\|_{K,\infty} \ll_{K,U} |\lambda|^{\frac{d}{4} + \frac{1}{2}} \|f\|_2.$$

6.4.3 Eigenfunctions on a Drum

[19] We now describe a concrete case of Theorem 6.56. As mentioned earlier, a concrete description of the Laplace eigenfunction is generally impossible unless the domain has special features. Thus a natural case beyond the open rectangle considered in Exercise 6.61 is to set U to be the open unit disc $B_1^{\mathbb{R}^2}$. For a given eigenfunction $f \in H_0^1(U)$ of Δ and some rotation matrix

$$k(\phi) = \begin{pmatrix} \cos\phi & -\sin\phi \\ \sin\phi & \cos\phi \end{pmatrix}$$

for $\phi \in [0, 2\pi)$ as in Section 1.1 we may consider the function $f^k(x) = f(kx)$. A simple calculation (which may be carried out using Proposition 1.5) shows that f^k is also an eigenfunction of Δ on U with the same eigenvalue as f. Since the eigenspace of H_0^1 functions of Δ for a given eigenvalue is finite-dimensional, it follows that we can find for any given eigenvalue a basis of the eigenspace with the property that every basis vector also has some weight $n \in \mathbb{Z}$ for the action of K on U (cf. Corollary 3.89).

Fixing the weight $n \in \mathbb{Z}$ and the eigenvalue $\lambda < 0$, the partial differential equation $\Delta f = \lambda f$ has a convenient reformulation. In fact, a calculation reveals that the Laplace operator $\Delta = \frac{\partial^2}{\partial x^2} + \frac{\partial^2}{\partial y^2}$ has the representation

$$\Delta = \frac{\partial^2}{\partial r^2} + \frac{1}{r}\frac{\partial}{\partial r} + \frac{1}{r^2}\frac{\partial^2}{\partial \theta^2}$$

in polar coordinates (we will also write f for the eigenfunction in polar coordinates), and if f has weight n then $f(r, \theta) = F(r)e^{in\theta}$ for a function F on $[0, 1]$. Since f is smooth on U, F is smooth on $(0, 1)$. Since f vanishes on ∂U we have $F(1) = 0$. Moreover, if $n \neq 0$ we must also have $F(0) = 0$ (check this). Finally, the partial differential equation $\Delta f = \lambda f$ now becomes the ordinary differential equation $\frac{d^2 F}{dr^2} + \frac{1}{r}\frac{dF}{dr} - \frac{n^2}{r^2}F = \lambda F$, or, equivalently,

$$r^2 \frac{d^2 F}{dr^2} + r\frac{dF}{dr} + \left(|\lambda|r^2 - n^2\right)F = 0, \tag{6.30}$$

with the conditions on $F(0)$ and $F(1)$ as explained above. The differential equation

$$x^2 J_n'' + x J_n' + (x^2 - n^2)J_n = 0 \tag{6.31}$$

on $(0, \infty)$ is known as Bessel's equation and the solutions are called the Bessel functions, one of a class of special functions introduced by the astronomer Bessel in 1917 in connection with the problem of three bodies moving under mutual gravitational attraction. The two equations (6.30) and (6.31) are essentially equivalent by setting $x = |\lambda|^{1/2}r$ and $J_n(x) = F(|\lambda|^{1/2}r)$.

Since (6.31) is a linear second-order differential equation there are two linearly independent real solutions for each λ and n. The function J_n is characterized up to a scalar multiple by the condition that $\lim_{x \to 0} J_n(x)$ exists (see Exercise 6.63(b)). Bessel found the integral representation

$$J_n(x) = \frac{1}{\pi}\int_0^\pi \cos(x\sin t - nt) \, dt \tag{6.32}$$

of the function J_n (we refer to Whittaker and Watson [113] for a general treatment of special functions). We will not develop this theory further, but refer to Figure 6.1–6.2 for a visualization of the resulting functions; in mod-

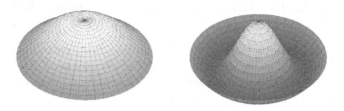

Fig. 6.1: Two eigenfunctions with weight $n = 0$

Fig. 6.2: Two eigenfunctions with weight $n = 1$

elling the behaviour of a drum the time variable is also needed, and so these illustrations may be thought of as snapshots of an oscillating drum skin (as alluded to in Section 1.2.2).

Exercise 6.63. Make the discussion of this section complete by the following steps.
(a) Prove that J_n as defined in (6.32) satisfies the differential equation (6.31).
(b) Show that the equation 6.31 has a solution Y_n with $Y_n(x) \to -\infty$ as $x \to 0$ given by

$$Y_n(x) = \frac{1}{\pi} \int_0^\pi \sin\left(x \sin t - nt\right) \, \mathrm{d}t - \frac{1}{\pi} \int_0^\infty \left(e^{nt} + (-1)^n e^{-nt}\right) e^{-x \sinh t} \, \mathrm{d}t.$$

(The solutions J_n and Y_n are referred to as Bessel functions of the first and second kind, respectively.)
(c) Show that for every $n \in \mathbb{Z}$ there is an eigenfunction of weight n.
(d) Show that for every $n \in \mathbb{Z}$ the eigenvalues (and eigenfunctions) of weight n correspond to the zeros of J_n.

6.4.4 Weyl's Law

The eigenfunctions and eigenvalues arising in Theorem 6.56 are mysterious. While their existence and some of their properties are readily proved, it is not usually possible to describe them analytically in closed form unless the open set is very special, so numerical approximations are often all that is available. It is, however, possible to count them asymptotically as expressed in this result of 1911 by Weyl [111].[20] Recall that $U \subseteq \mathbb{R}^d$ is said to be

Jordan measurable if it is bounded and its characteristic function is Riemann integrable.

Theorem 6.64 (Weyl's law). *Let $U \subseteq \mathbb{R}^d$ be open and Jordan measurable. Let $N(T) = |\{n \mid |\lambda_n| \leqslant T\}|$ denote the number of eigenvalues of Δ on U with eigenfunctions in $H_0^1(U)$ and absolute value bounded by T (with repetitions allowed, just as in Theorem 6.56). Then*

$$\lim_{T \to \infty} \frac{N(T)}{T^{d/2}} = (2\pi)^{-d} \omega_d m(U), \qquad (6.33)$$

where m is the Lebesgue measure on $\mathbb{R}^d$ and $\omega_d = m(B_1^{\mathbb{R}^d})$ is the volume of the unit ball in $\mathbb{R}^d$.

In 1966 M. Kac [50] asked 'Can one hear the shape of a drum?' As we explained in Section 1.2.2, the eigenvalues of the Laplacian on an open set U relate directly to the frequencies at which a membrane with the shape U would vibrate. Thus the notes one hears from a drum with shape U are precisely related to the eigenvalues of the Laplacian and the question raised by Kac asks whether the list of eigenvalues determines U (up to isometric motions of $\mathbb{R}^d$). One of the consequences of Theorem 6.64 is that the *size* of the drum certainly can be heard in this sense. Kac's question was answered in the negative.[21]

Our (by now well-established) approach is to first show the result for the torus, and we will then apply a technique known as Dirichlet–Neumann bracketing to extend the proof to the general case.

Proposition 6.65. *Let $R > 0$ and $U = \mathbb{T}_R^d = \mathbb{R}^d/(2R\mathbb{Z}^d)$ or $U = (0, R)^d$. Then Weyl's law holds for the eigenvalues of the Laplacian on U.*

PROOF. In both cases, write (λ_n) for the eigenvalues and (f_n) for the associated eigenfunctions in $H^1(\mathbb{T}_R^d)$ resp. $H_0^1(U)$. In the case of $\mathbb{T}_R^d$ we know that the basis of eigenfunctions is given by (χ_n), where

$$\chi_n(x) = e^{2\pi i(2R)^{-1}(n_1 x_1 + \cdots + n_d x_d)}$$

for $x \in \mathbb{T}_R^d$ and $n \in \mathbb{Z}^d$. The Laplace eigenvalue of χ_n is given by

$$\lambda_n = -(2\pi)^2 (2R)^{-2} \|n\|_2^2$$

for all $n \in \mathbb{Z}^d$. Hence

$$N(T) = \left|\{n \in \mathbb{Z}^d \mid (2\pi)^2 (2R)^{-2} \|n\|_2^2 \leqslant T\}\right|$$

$$= \left|\{n \in \mathbb{Z}^d \mid \|n\|_2^2 \leqslant T(2R)^2 (2\pi)^{-2}\}\right| = \left|\mathbb{Z}^d \cap \overline{B_{T^{1/2} 2R/(2\pi)}^{\mathbb{R}^d}}\right|$$

is determined by a lattice point counting problem.

Using the fundamental domain $F = [-\frac{1}{2}, \frac{1}{2}]^d$ for $\mathbb{Z}^d$ in $\mathbb{R}^d$ we have

$$B^{\mathbb{R}^d}_{S-\sqrt{d}} \subseteq (\mathbb{Z}^d \cap \overline{B^{\mathbb{R}^d}_S}) + F \subseteq B^{\mathbb{R}^d}_{S+\sqrt{d}}$$

for any $S > \sqrt{d}$ (see Figure 6.3). Taking the Lebesgue measure (satisfying $m(F) = 1$) we obtain $\omega_d(S - \sqrt{d})^d \leqslant \left|\mathbb{Z}^d \cap \overline{B^{\mathbb{R}^d}_S}\right| \leqslant \omega_d(S + \sqrt{d})^d$ and

$$\lim_{S \to \infty} \frac{\left|\mathbb{Z}^d \cap \overline{B^{\mathbb{R}^d}_S}\right|}{S^d} = \omega_d.$$

Combining the two discussions and setting S to be $T^{1/2}2R/2\pi$ gives

$$\lim_{T \to \infty} \frac{N(T)}{(T^{1/2}2R/2\pi)^d} = \omega_d,$$

or equivalently (6.33) for $U = \mathbb{T}^d$.

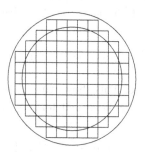

Fig. 6.3: Counting lattice points in a ball for $d = 2$.

We now extend the result to $U = (0, R)^d$. For this let $n \in \mathbb{N}_0^d$ and note that the characters (χ_m) on $\mathbb{T}_R^d$ for $m = (\pm n_1, \ldots, \pm n_d)$ all have the same eigenvalue $-4\pi^2(2R)^{-2}\|n\|_2^2$ for the Laplacian. Taking linear combinations of the characters we obtain the eigenfunctions

$$f_1(\pi R^{-1} n_1 x_1) \cdots f_d(\pi R^{-1} n_d x_d), \tag{6.34}$$

where each f_j for $j = 1, \ldots, d$ is either the sine or cosine function. If $n \in \mathbb{N}^d$ this gives 2^d linearly independent eigenfunctions, while for $n \in \mathbb{N}_0^d \backslash \mathbb{N}^d$ this gives 2^e linearly independent eigenfunctions, where e is the number of non-zero components of n. Notice that the hull of these eigenfunctions coincides with the linear hull of all characters, and that these are mutually orthogonal. We claim that the functions that only involve sine functions form the orthogonal basis of $L^2(U)$ consisting of eigenfunctions of Δ in $H_0^1(U)$. Assuming the claim, we obtain precisely one eigenfunction for every $n \in \mathbb{N}^d$, so

$$N_U(T) = \left|\{n \in \mathbb{N}^d \mid 4\pi^2 (2R)^{-2} \|n\|_2^2 \leqslant T\}\right| = \left|\mathbb{N}^d \cap \overline{B_{T^{1/2}2R/2\pi}^{\mathbb{R}^d}}\right|,$$

and so

$$\lim_{T \to \infty} \frac{N_U(T)}{T^{d/2}} = \lim_{T \to \infty} \frac{N_{\mathbb{T}_R^d}(T)}{2^d T^{d/2}} = (2\pi)^{-d} \omega_d m(U).$$

For the proof of the claim we apply the discussion of even and odd functions from Section 1.1 in d dimensions. For this we identify $L^2((0,R)^d)$ with the subspace of functions in $L^2((-R,R)^d) = L^2(\mathbb{T}_R^d)$ that are odd with respect to all coordinates. More precisely, given $f \in L^2((0,R)^d)$ we define $\tilde{f}|_U = f$ and

$$\tilde{f}(\varepsilon_1 t_1, \varepsilon_2 t_2, \ldots, \varepsilon_d t_d) = \varepsilon_1 \cdots \varepsilon_d \tilde{f}(t_1, \ldots, t_d)$$

for $\varepsilon_1, \ldots, \varepsilon_d \in \{\pm 1\}$ and $(t_1, \ldots, t_d) \in U$ (and the same formula then holds for all $t \in (-R, R)^d$). Expand $\tilde{f}$ into eigenfunctions of the form (6.34) for all $n \in \mathbb{N}^d$. If g is one of these, then either g is the product only of sine functions or it is even with respect to one or more of the variables; assume that it is even with respect to x_k. Using the substitution $x_k \to -x_k$ in the inner product we obtain

$$\langle \tilde{f}, g \rangle_{L^2((-R,R)^d)} = \langle -\tilde{f}, g \rangle_{L^2((-R,R)^d)}$$

and so $\langle \tilde{f}, g \rangle_{L^2((-R,R)^d)} = 0$. This shows that $\tilde{f}$ is expressed using products of sine functions only. We also note that for any $f, g \in L^2((0,R)^d)$ we have

$$\left\langle \tilde{f}, \tilde{g} \right\rangle_{L^2((-R,R)^d)} = 2^d \langle f, g \rangle_{L^2((0,R)^d)}$$

which may be seen by splitting $(-R,R)^d$ into 2^d smaller cubes and substituting $y_j = \pm x_j$ for $j = 1, \ldots, d$ and $x \in (0,R)^d$ on each one of them. It follows that the functions of the form $x \mapsto \sin(\pi R^{-1} n_1 x_1) \cdots \sin(\pi R^{-1} n_d x_d)$ for $n \in \mathbb{N}^d$ are an orthogonal basis of $L^2(U)$. As these functions also vanish on ∂U it follows that they belong to $H_0^1(U)$ (see Exercise 6.66), proving the proposition. The cautious reader may notice that we have only found *an* orthonormal basis of $L^2(U)$ in $H_0^1(U)$ consisting of eigenfunctions of Δ as in Theorem 6.56. However, as Δ is not a well-defined operator it is not clear whether this basis is the same as the one in Theorem 6.56. This is resolved in Lemma 6.67(a). $\square$

Essential Exercise 6.66. (a) Show that the function $x \mapsto \sin(\pi R^{-1} n x)$ lies in $H_0^1((0,R))$ for all $n \geqslant 1$.
(b) Formulate and show the analogous result for $U = (0,R)^d$.

Lemma 6.67. *Let $U \subseteq \mathbb{R}^d$ be open and bounded.*
(a) *If a function $f \in H_0^1(U) \cap C^\infty(U)$ satisfies $\Delta f = \lambda f$, then*

$$\langle f, g \rangle_1 = -\lambda \langle f, g \rangle_{L^2(U)} \tag{6.35}$$

for all $g \in H_0^1(U)$. If this holds for a non-trivial f then $\lambda < 0$ and we have $S(f) = \lambda^{-1} f$ (where $S = -\imath^*$ is as in Proposition 6.60). In particular, the eigenspaces inside $H_0^1(U)$ of Δ coincide with those of S.
(b) If $f_1, f_2, \ldots \in H_0^1(U) \cap C^\infty(U)$ are eigenfunctions of Δ with eigenvalues $0 > \lambda_1 \geqslant \lambda_2 \geqslant \cdots$ that form an orthonormal basis of $L^2(U)$, then

$$|\lambda_1|^{-\frac{1}{2}} f_1, |\lambda_2|^{-\frac{1}{2}} f_2, \ldots \tag{6.36}$$

form an orthonormal basis of $H_0^1(U)$ with respect to $\langle \cdot, \cdot \rangle_1$.
(c) Moreover, if $g \in H_0^1(U)$ and $a_n = \langle g, f_n \rangle_{L^2(U)}$ then $g = \sum_{n=1}^\infty a_n f_n$ converges in $L^2(U)$ and in $H_0^1(U)$.

PROOF. For the proof of (a), suppose that $f \in H_0^1(U) \cap C^\infty(U)$ satisfies the equation $\Delta f = \lambda f$. Let $\phi \in C_c^\infty(U)$. Then

$$\langle f, \phi \rangle_1 = \sum_j \langle \partial_j f, \partial_j \phi \rangle_{L^2(U)} = - \langle \Delta f, \phi \rangle_{L^2(U)} = -\lambda \langle f, \phi \rangle_{L^2(U)}.$$

Since $C_c^\infty(U)$ is dense in $H_0^1(U)$, we obtain (6.35) for all $g \in H_0^1(U)$. In particular, we may set $g = f$ to obtain $\langle f, f \rangle_1 = -\lambda \langle f, f \rangle_{L^2(U)}$, which implies $\lambda < 0$ if we assume that f is non-trivial. Recall that $\imath : H_0^1(U) \to L^2(U)$ is defined by $\imath(f) = f$. With this, (6.35) gives

$$\langle \imath^*(-\lambda f), g \rangle_1 = \langle -\lambda f, \imath(g) \rangle_{L^2(U)} = \langle -\lambda f, g \rangle_{L^2(U)} = \langle f, g \rangle_1$$

for all $g \in H_0^1(U)$. This implies $\imath^*(-\lambda f) = f$ and $S(f) = \lambda^{-1} f$.

For the proof of (b), let $f_1, f_2, \ldots$ be an orthonormal basis of $L^2(U)$ in $H_0^1(U) \cap C^\infty(U)$ as in the lemma. Then $\langle f_k, f_\ell \rangle_1 = -\lambda_k \langle f_k, f_\ell \rangle_{L^2(U)} = 0$ for $k, \ell \geqslant 1$ with $k \neq \ell$, and $\langle f_k, f_k \rangle_1 = |\lambda_k| \langle f_k, f_k \rangle_{L^2(U)} = |\lambda_k|$ by part (a). This shows that (6.36) forms an orthonormal list of vectors in $H_0^1(U)$ with respect to $\langle \cdot, \cdot \rangle_1$. To see that these form an orthonormal basis of $H_0^1(U)$, suppose that $f \in H_0^1(U)$ is orthogonal to all of these functions with respect to $\langle \cdot, \cdot \rangle_1$. Then
$$0 = \langle f, f_n \rangle_1 = -\lambda_n \langle f, f_n \rangle_{L^2(U)}$$
by assumption and (a). Hence $\langle f, f_n \rangle_{L^2(U)} = 0$ for all $n \geqslant 1$, and so we must have $f = 0$ since we assumed that the sequence $f_1, f_2, \ldots$ is an orthonormal basis of $L^2(U)$, and the map $\imath : H_0^1(U) \to L^2(U)$ is injective.

For the final claim, let $g = \sum_{n=1}^\infty b_n f_n$ be the expansion in $H_0^1(U)$, and notice that $\imath : H_0^1(U) \to L^2(U)$ is a bounded operator, and in particular must send a convergent series to a convergent series, so that $b_n = a_n$ for all $n \geqslant 1$. $\qquad \square$

Our proof of Weyl's law (Theorem 6.64) relies crucially on the following reformulation of the counting function.

Lemma 6.68 (Variational characterization). *Let* $U \subseteq \mathbb{R}^d$ *be open and bounded. Let* $f_1, f_2, \ldots, \lambda_1, \lambda_2, \ldots,$ *and* $N(T)$ *for* $T > 0$ *be as in Theorem 6.56. Then*

$$N(T) = \max\{\dim V \mid V \subseteq H_0^1(U), \|f\|_1 \leqslant T^{1/2}\|f\|_{L^2} \text{ for all } f \in V\} \quad (6.37)$$

for all $T > 0$.

PROOF. Fix some $T > 0$. Suppose that $|\lambda_1|, \ldots, |\lambda_n| \leqslant T$ and $|\lambda_k| > T$ for all $k > n$ (so that $n = N(T)$). Define $V_0 = \langle f_1, \ldots, f_n \rangle$. Applying Lemma 6.67(a) to each f_k for $k = 1, \ldots, n$ and $g = \sum_{\ell=1}^{n} a_\ell f_\ell$ we get

$$\|g\|_1^2 = \left\langle \sum_{k=1}^{n} a_k f_k, \sum_{\ell=1}^{n} a_\ell f_\ell \right\rangle_1 = \sum_{k=1}^{n} a_k |\lambda_k| \left\langle f_k, \sum_{\ell=1}^{n} a_\ell f_\ell \right\rangle_{L^2(U)}$$

$$= \sum_{k=1}^{n} |\lambda_k| |a_k|^2 \leqslant T \left\| \sum_{k=1}^{n} a_k f_k \right\|_{L^2(U)}^2$$

for all $(a_1, \ldots, a_n) \in \mathbb{C}^n$. This already gives $N(T) = \dim V_0 \leqslant \max \dim V$ with V as in (6.37).

For the reverse inequality, assume that $V \subseteq H_0^1(U)$ has $\dim V > N(T)$. Then there exists a non-trivial function $f \in V \cap V_0^\perp$ (because, for example, any $f \in V$ induces a linear functional on V_0 by taking the inner product, and for dimension reasons the resulting linear map $V \to V_0^*$ cannot be injective). Since $V_0^\perp = \overline{\langle f_{n+1}, f_{n+2}, \ldots \rangle}$ we may write $f = \sum_{k>n} a_k f_k$ with the sum converging in $L^2(U)$ and in $H_0^1(U)$ by Lemma 6.67(c). Therefore using Lemma 6.67(a) as above we obtain

$$\|f\|_1^2 = \left\langle \sum_{k>n} a_k f_k, \sum_{\ell>n} a_\ell f_\ell \right\rangle_1 = \sum_{k>n} |\lambda_k| |a_k|^2 > T \sum_{k>n} |a_k|^2 = T\|f\|_{L^2}^2,$$

and we see that V does not satisfy the requirement in (6.37). Hence any subspace V as in (6.37) would satisfy $\dim V \leqslant \dim V_0 = N(T)$ and the lemma follows. $\qquad \square$

PROOF OF THEOREM 6.64. Notice first that Lemma 6.68 implies for disjoint open subsets U_1 and U_2 of a bounded open set U the sub-additivity

$$N_{U_1}(T) + N_{U_2}(T) \leqslant N_U(T), \quad (6.38)$$

where we write $N_{U'}(T)$ for the counting function for an open and bounded domain $U' \subseteq \mathbb{R}^d$. Indeed, on extending functions to be zero outside U_j we may write $H_0^1(U_j) \subseteq H_0^1(U)$ for $j = 1, 2$ as in Exercise 5.27, and, once embedded, we have $H_0^1(U_1) \perp H_0^1(U_2)$, with respect to both $\langle \cdot, \cdot \rangle_1$ and $\langle \cdot, \cdot \rangle_{L^2(U)}$, since U_1 and U_2 are disjoint, so that we can take the direct sum of the subspaces realising the maximum U_1 and U_2 appearing in Lemma 6.68. We note that — although it is tempting to try — it is not possible to derive the estimate 6.38

directly by expanding eigenfunctions on U_1 and U_2 to eigenfunctions on U as there would be no reason to expect these functions to be smooth on $\partial U_1 \cap U$. The variational characterization in Lemma 6.68 avoids this issue.

Now let $U \subseteq (-R, R)^d$ be an open Jordan measurable subset. By Jordan measurability, for any $\varepsilon > 0$ we can divide $(-R, R)^d$ into finitely many cubes so that we can approximate U from the inside and from the outside by finite disjoint unions of small cubes, as illustrated in Figure 6.4.

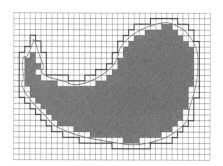

Fig. 6.4: Approximating U by two pixelated versions of U, one from inside and one from outside.

Let $I_1 \sqcup \cdots \sqcup I_k \subseteq U$ and $U \subseteq \overline{O_1 \sqcup \cdots \sqcup O_\ell}$ be the approximation from inside and outside respectively, where $I_1, \ldots, I_k$ and $O_1, \ldots, O_\ell$ are translates of the cube $(0, \frac{R}{n})^d$, and choose n so large and the approximations so well that

$$m\left((O_1 \sqcup \cdots \sqcup O_\ell) \smallsetminus (I_1 \sqcup \cdots \sqcup I_k)\right) < \varepsilon.$$

By the sub-additivity mentioned above for the various counting functions $N_U(T)$ we have

$$\liminf_{T \to \infty} \frac{N_U(T)}{T^{d/2}} \geqslant \lim_{T \to \infty} \frac{N_{I_1}(T) + \cdots + N_{I_k}(T)}{T^{d/2}}$$
$$= (2\pi)^{-d} \omega_d m(I_1 \sqcup \cdots \sqcup I_k) \geqslant (2\pi)^{-d} \omega_d \left(m(U) - \varepsilon\right)$$

by Proposition 6.65.

On the other hand, we may add extra cubes to $O_1 \sqcup \cdots \sqcup O_\ell$ to obtain

$$\overline{O_1 \sqcup \cdots \sqcup O_\ell \sqcup E_1 \sqcup \cdots \sqcup E_n} = [-R, R]^d$$

for some cubes $E_1, \ldots, E_n \subseteq [-R, R]^d$. Hence, again by sub-additivity,

$$\limsup_{T\to\infty}\left(\frac{N_U(T)}{T^{d/2}}+\sum_{j=1}^{n}\frac{N_{E_j}(T)}{T^{d/2}}\right)\leqslant \limsup_{T\to\infty}\frac{N_{\overline{O_1\sqcup\cdots\sqcup O_\ell}^o}(T)+\sum_{j=1}^{n}N_{E_j}(T)}{T^{d/2}}$$

$$\leqslant \lim_{T\to\infty}\frac{N_{(-R,R)^d}(T)}{T^{d/2}}=(2\pi)^{-d}\omega_d m((-R,R)^d).$$

Since for each j we have

$$\lim_{T\to\infty}\frac{N_{E_j}(T)}{T^{d/2}}=(2\pi)^{-d}\omega_d m(E_j)$$

it follows that

$$\limsup_{T\to\infty}\frac{N_U(T)}{T^{d/2}}\leqslant (2\pi)^{-d}\omega_d m\left(O_1\sqcup\cdots\sqcup O_\ell\right)\leqslant (2\pi)^{-d}\omega_d\left(m(U)+\varepsilon\right).$$

Since $\varepsilon>0$ was arbitrary, this and the reverse bound for the limit infimum above prove the theorem. $\qquad\square$

Exercise 6.69. Let $U\subseteq\mathbb{R}^d$ be open, bounded, and Jordan measurable. Show that

$$\lim_{n\to\infty}\frac{|\lambda_n|}{n^{2/d}}=(2\pi)^2\left(\omega_d m(U)\right)^{-2/d},$$

where $\lambda_1\geqslant\lambda_2\geqslant\cdots$ is the ordered list of eigenvalues of Δ on U.

Exercise 6.70. Assume that $d\geqslant 2$ (or, for simplicity, that $d=2$), let $U\subseteq\mathbb{R}^d$ be an open, bounded, Jordan measurable set, and let $f\in C_c^\infty(U)$. Show that the series $\sum_{n=1}^{\infty}\langle f,f_n\rangle f_n$, with (f_n) as in Theorem 6.56 ordered as in Exercise 6.69, converges pointwise on U and uniformly on any compact subset of U.

6.5 Further Topics

- In Section 8.2.2 we return one more time to the topic of Sobolev spaces and study elliptic regularity up to and including the boundary of U. For further reading in that direction, we refer to Evans [30].
- The spectral theory of compact self-adjoint operators proven here is only the starting point. We discuss spectral theory again in Chapter 9 for unitary operators, in Chapter 11 from a general perspective as a preparation for Chapter 12 for bounded normal operators, and in Chapter 13 for unbounded self-adjoint operators.

The reader should continue with Chapter 7 and Chapter 8 as these give important results for the chapters that follow.

Chapter 7
Dual Spaces

Let X be a real (or complex) normed vector space. A bounded linear operator from X into the normed space $\mathbb{R}$ (or $\mathbb{C}$) is a *(continuous) linear functional* on X. Recall that the space of all continuous linear functionals is denoted X^* or $\mathrm{B}(X, \mathbb{R})$ and it is called the *dual* or *conjugate* space of X. Lemma 2.54 shows that X^* is a Banach space with respect to the operator norm.

In Section 7.1 we prove the Hahn–Banach theorem, a fundamental tool for constructing linear functionals with prescribed properties. We also discuss several further consequences of the Hahn–Banach theorem concerned with the relationship between X and X^*. In Section 7.2 we discuss applications of these results. Finally, in Sections 7.3 and 7.4 we will identify the duals of many important Banach spaces, leading to examples and counter-examples to the property of reflexivity.

7.1 The Hahn–Banach Theorem and its Consequences

One of the most important questions one may ask of X^* is the following: are there 'enough' elements in X^*? For example, are there enough elements to separate points? This is answered in great generality using the Hahn–Banach theorem (Theorem 7.3 below); see Corollary 7.4.

7.1.1 The Hahn–Banach Lemma and Theorem

Even though in the main applications of the Hahn–Banach lemma the function p below is simply a norm, we will also see applications of this stronger form of the lemma (with the stated weaker assumptions on the function p) in Section 7.4 and Section 8.6.1.

Lemma 7.1 (Hahn–Banach lemma). *Let X be a real vector space, and assume that $p : X \to \mathbb{R}$ is a norm-like function with the properties*

© Springer International Publishing AG 2017
M. Einsiedler, T. Ward, *Functional Analysis, Spectral Theory, and Applications*,
Graduate Texts in Mathematics 276, DOI 10.1007/978-3-319-58540-6_7

$$p(x_1 + x_2) \leqslant p(x_1) + p(x_2)$$

and

$$p(\lambda x_1) = \lambda p(x_1)$$

for all $\lambda \geqslant 0$ and $x_1, x_2 \in X$. Let Y be a subspace of X, and $f : Y \to \mathbb{R}$ a linear function with $f(y) \leqslant p(y)$ for all $y \in Y$. Then there exists a linear functional $F : X \to \mathbb{R}$ such that $F(y) = f(y)$ for $y \in Y$, and $F(x) \leqslant p(x)$ for all $x \in X$.

To stress the similarities of the assumptions on p to the definition of a norm from Definition 2.1 (or semi-norm from Definition 2.11) we refer to the function p as a *norm-like* function.

PROOF OF LEMMA 7.1. Let $\mathcal{K}$ be the set of all pairs (Y_α, g_α) in which Y_α is a linear subspace of X containing Y, and g_α is a real linear functional on Y_α with $g_\alpha(y) = f(y)$ for all $y \in Y$, and $g_\alpha(x) \leqslant p(x)$ for all $x \in Y_\alpha$. We make $\mathcal{K}$ into a partially ordered set by defining $(Y_\alpha, g_\alpha) \preccurlyeq (Y_\beta, g_\beta)$ if $Y_\alpha \subseteq Y_\beta$ and $g_\alpha = g_\beta|_{Y_\alpha}$. It is clear that any totally ordered subset $\{(Y_\lambda, g_\lambda) \mid \lambda \in I\}$ has an upper bound given by the subspace $Y' = \bigcup_\lambda Y_\lambda$ and the functional defined by $g'(y) = g_\lambda(y)$ for $y \in Y_\lambda$ and $\lambda \in I$. That Y' is a subspace and that g' is well-defined both follow since $\{(Y_\lambda, g_\lambda) \mid \lambda \in I\}$ is linearly ordered.

APPLYING ZORN'S LEMMA: All of this is to prepare the ground for an application of Zorn's lemma, which roughly speaking allows us to make a transfinite induction with choices (the heart of the argument follows in the next paragraph). Indeed, by Zorn's lemma (see Section A.1), there is a maximal element (Y_0, g_0) in $\mathcal{K}$. All that remains is to check that Y_0 is all of X (so we may take F to be g_0).

EXTENDING BY ONE DIMENSION: So assume for the purposes of a contradiction that $x \in X \backslash Y_0$, and let Y_1 be the vector space spanned by Y_0 and x. Each element $z \in Y_1$ may be expressed uniquely in the form $z = y + \lambda x$ with $y \in Y_0$ and $\lambda \in \mathbb{R}$, because x is assumed not to be in the subspace Y_0. Define a linear function g_1 on Y_1 by setting $g_1(y + \lambda x) = g_0(y) + \lambda c$, where the constant c will be chosen to ensure that g_1 is bounded by p. Note that if $y_1, y_2 \in Y_0$, then

$$g_0(y_1) - g_0(y_2) = g_0(y_1 - y_2) \leqslant p(y_1 - y_2) \leqslant p(y_1 + x) + p(-x - y_2),$$

so $-p(-x - y_2) - g_0(y_2) \leqslant p(y_1 + x) - g_0(y_1)$. It follows that

$$A = \sup_{y \in Y_0} \{-p(-x - y) - g_0(y)\} \leqslant \inf_{y \in Y_0} \{p(y + x) - g_0(y)\} = B.$$

Choose c to be any number in the interval $[A, B]$. Then, by construction of A and B,

$$c \leqslant p(y + x) - g_0(y), \tag{7.1}$$

and

$$-p(-x-y) - g_0(y) \leqslant c \tag{7.2}$$

for all $y \in Y_0$. In order to show the required bound on g_1, we consider scalars of different sign separately. For $\lambda > 0$, multiply (7.1) by λ and substitute $\frac{1}{\lambda}y$ for y to obtain

$$\lambda c \leqslant p(y + \lambda x) - g_0(y) \tag{7.3}$$

from the assumed (positive) homogeneity. Similarly, for $\lambda < 0$, multiply (7.2) by λ, and substitute $\frac{1}{\lambda}y$ for y to obtain $\lambda c \leqslant |\lambda| p(-x - \frac{1}{\lambda}y) - \lambda g_0(\frac{1}{\lambda}y)$. Using the homogeneity assumption on p for $|\lambda| = -\lambda$ we obtain (7.3) again. Since the assumptions on g also give (7.3) for $\lambda = 0$, we obtain

$$g_1(y + \lambda x) = g_0(y) + \lambda c \leqslant p(y + \lambda x)$$

for all $\lambda \in \mathbb{R}$ and $y \in Y_0$.

A CONTRADICTION: Thus we have found $(Y_1, g_1) \in \mathcal{K}$ with $(Y_0, g_0) \preccurlyeq (Y_1, g_1)$ and $Y_0 \neq Y_1$. This contradicts the maximality of (Y_0, g_0) and hence $F = g_0$ is defined on all of X and satisfies the conclusion of the lemma. $\qquad\square$

Exercise 7.2. Let X be a real vector space and let $K \subseteq X$ be a convex subset. Suppose that $0 \in K$ and that for every $x \in X$ there is some $t > 0$ with $tx \in K$. Define the *gauge function* $p_K(x) = \inf\{t > 0 \mid \frac{1}{t}x \in K\}$. Show that p_K is norm-like in the sense that it is non-negative, homogeneous for positive scalars, and satisfies the triangle inequality (the latter two being assumptions in Lemma 7.1).

For real vector spaces, the Hahn–Banach theorem follows at once (for complex spaces a little more work is needed).

Theorem 7.3 (Hahn–Banach theorem). *Let X be a real or complex normed space, and Y a linear subspace. Then for any $y^* \in Y^*$ there exists an $x^* \in X^*$ such that $\|x^*\| = \|y^*\|$ and $x^*(y) = y^*(y)$ for all $y \in Y$.*

That is, any linear functional defined on a subspace may be extended to a linear functional on the whole space, without increasing the norm.

PROOF OF THEOREM 7.3. Assume first that X is a real normed space. Let $p(x) = \|y^*\|\|x\|$ and $f(x) = y^*(x)$. Apply the Hahn–Banach lemma (Lemma 7.1) to find an extension $x^* = F$ to the whole space. To check that $\|x^*\| \leqslant \|y^*\|$, write $x^*(x) = \theta|x^*(x)|$ with $\theta \in \{\pm 1\}$. Then

$$|x^*(x)| = \theta x^*(x) = x^*(\theta x) \leqslant p(\theta x) = \|y^*\|\|\theta x\| = \|y^*\|\|x\|.$$

The reverse inequality is clear, so $\|x^*\| = \|y^*\|$.

COMPLEX CASE: Now let X be a complex normed vector space, let $Y \subseteq X$ be a complex linear subspace, let $y^* \in Y^*$, and define a real linear functional $y_{\mathbb{R}}^*$ by $y_{\mathbb{R}}^*(y) = \Re(y^*(y))$ for $y \in Y$. Let $x_{\mathbb{R}}^* : X \to \mathbb{R}$ be an extension of $y_{\mathbb{R}}^*$ with $\|x_{\mathbb{R}}^*\| = \|y_{\mathbb{R}}^*\| \leqslant \|y^*\|$ (by the real case above). Now define

$$x^*(x) = x^*_{\mathbb{R}}(x) - \mathrm{i}x^*_{\mathbb{R}}(\mathrm{i}x),$$

which is once again an $\mathbb{R}$-linear map from X to $\mathbb{C}$. It is also $\mathbb{C}$-linear since

$$x^*(\mathrm{i}x) = x^*_{\mathbb{R}}(\mathrm{i}x) - \mathrm{i}x^*_{\mathbb{R}}(\mathrm{i}^2 x) = \mathrm{i}x^*_{\mathbb{R}}(x) - \mathrm{i}^2 x^*_{\mathbb{R}}(\mathrm{i}x) = \mathrm{i}x^*(x).$$

Moreover, for $y \in Y$ we have, by $\mathbb{C}$-linearity of y^*,

$$\begin{aligned}
x^*(y) &= x^*_{\mathbb{R}}(y) - \mathrm{i}x^*_{\mathbb{R}}(\mathrm{i}y) = \Re(y^*(y)) - \mathrm{i}\Re(y^*(\mathrm{i}y)) \\
&= \Re(y^*(y)) + \mathrm{i}\Im(y^*(y)) = y^*(y).
\end{aligned}$$

Finally, $|x^*(x)| = \theta x^*(x)$ for some $\theta \in \mathbb{C}$ with $|\theta| = 1$, and so

$$|x^*(x)| = \theta x^*(x) = x^*(\theta x) = x^*_{\mathbb{R}}(\theta x) \leqslant \|y^*\| \|\theta x\| = \|y^*\| \|x\|,$$

which shows that $\|x^*\| = \|y^*\|$ and hence the complex case of the theorem. $\qquad\square$

7.1.2 Consequences of the Hahn–Banach Theorem

Many useful results follow from the Hahn–Banach theorem.

Corollary 7.4 (Separation). *Let X be a non-trivial normed vector space. Then for any $x \in X$ there is a functional $x^* \in X^*$ with $\|x^*\| = 1$ and with $x^*(x) = \|x\|$. Hence, if $z \neq y \in X$ then there exists an $x^* \in X^*$ such that $x^*(y) \neq x^*(z)$.*

PROOF. Note that we may assume without loss of generality that $x \neq 0$. Apply Theorem 7.3 with Y being the linear hull of x to find an extension of the linear map $y^*(ax) = a\|x\|$ on Y. Since $|y^*(ax)| = |a|\|x\| = \|ax\|$ we have $\|y^*\| = 1$, and so we find an $x^* \in X^*$ with $\|x^*\| = 1$ and $x^*(x) = \|x\|$. For the last part, take $x = y - z$. $\qquad\square$

Notice finally that linear functionals allow us to decompose a vector space (see Exercises 3.27 and 3.28): let X be a normed vector space, and $x^* \in X^*$. The *null space* or *kernel* of x^* is the closed linear subspace

$$\ker(x^*) = \{x \in X \mid x^*(x) = 0\}.$$

If $x^* \neq 0$, then there is a point $x_0 \neq 0$ such that $x^*(x_0) = 1$. Any $x \in X$ can then be written as $x = z + \lambda x_0$, with $\lambda = x^*(x)$ and $z = x - \lambda x_0 \in \ker(x^*)$. Thus, $X = \ker(x^*) \oplus Y$, where Y is the one-dimensional space spanned by x_0.

Exercise 7.5. Show that every finite-dimensional subspace of a normed vector space has a closed linear complement.

The reader should compare the following result for a general normed vector space to the characterization of the closed linear hull in Hilbert spaces (see Corollary 3.26).

Corollary 7.6 (Closed linear hull). *Let $S \subseteq X$ be a subset of a normed vector space. Then the closed linear hull of S is precisely the set of all $x \in X$ that satisfy $x^*(x) = 0$ for all $x^* \in X^*$ with $x^*(S) = \{0\}$. Equivalently,*

$$\overline{\langle S \rangle} = \bigcap_{\substack{x^* \in X^* \\ x^*(S) = \{0\}}} \ker(x^*).$$

PROOF. The inclusion of the left-hand side in the right-hand side is clear since $\ker(x^*)$ is a closed subspace for any $x^* \in X^*$. Suppose that $x_0 \notin \overline{\langle S \rangle}$, and let $Y = \langle x_0 \rangle + \overline{\langle S \rangle}$. Then the functional y^* defined by $y^*(\alpha x_0 + z) = \alpha$ for $z \in \overline{\langle S \rangle}$ is bounded. For otherwise there would exist, for every $n \geqslant 1$, some scalar $\alpha_n \neq 0$ and some $z_n \in \overline{\langle S \rangle}$ with $|\alpha_n| \geqslant n\|\alpha_n x_0 + z_n\|$, which implies that

$$\left\| x_0 + \tfrac{1}{\alpha_n} z_n \right\| \leqslant \tfrac{1}{n},$$

forcing $x_0 \in \overline{\langle S \rangle}$. Therefore, y^* can be extended to a continuous linear functional x^* on X which satisfies $x^*(S) = \{0\}$ but $x^*(x_0) = 1$. This shows that x_0 also does not lie in the intersection of the kernels on the right-hand side, and hence proves the other inclusion. $\qquad\square$

Exercise 7.7. Let $Y \subseteq X$ be a subspace of a normed linear space X. Show that

$$\max_{\substack{\|x^*\| \leqslant 1, \\ x^*(Y) = \{0\}}} |x^*(x)| = \inf_{y \in Y} \|x - y\|$$

for all $x \in X$.

Exercise 7.8. (a) Prove that if the dual space X^* of a real normed vector space X is strictly convex (see Definition 2.17), then the Hahn–Banach extension of a continuous functional on a subspace to all of X is unique.
(b) Give an explicit example of a situation in which the extension defined by the Hahn–Banach theorem is not unique.

7.1.3 The Bidual

Corollary 7.9 (Isometric embedding into the bidual). *Let X be a normed vector space. Then*

$$\|x\| = \max_{\substack{x^* \in X^*, \\ \|x^*\| \leqslant 1}} |x^*(x)| \qquad (7.4)$$

for any $x \in X$. In particular, the natural linear map

$$\imath : X \longrightarrow X^{**} = (X^*)^*$$
$$x \longmapsto \imath(x)$$

from X into the bidual of X that sends $x \in X$ to the linear functional $\imath(x)$ defined by $\imath(x)(x^*) = x^*(x)$ for $x^* \in X^*$, is an isometric embedding.

Definition 7.10. A Banach space is called *reflexive* if the isometry $\imath$ in Corollary 7.9 is a bijection (and hence an isometric isomorphism) from X to X^{**}.

As we will see in the next section, some Banach spaces which we have already encountered are reflexive, but some are not.

PROOF OF COROLLARY 7.9. By definition, $|x^*(x_0)| \leqslant \|x^*\|\|x_0\| \leqslant \|x_0\|$ for all $x^* \in X^*$ with $\|x^*\| \leqslant 1$ and $x_0 \in X$. Moreover, we may apply Corollary 7.4 to obtain some functional $x^* \in X^*$ of norm one with $x^*(x_0) = \|x_0\|$, which proves (7.4). Now notice that

$$\sup_{\substack{x^* \in X^* \\ \|x^*\| \leqslant 1}} |x^*(x_0)|$$

is, by definition, precisely the operator norm of $\imath(x_0) \in X^{**}$. Hence we have shown that $\imath$ is an isometry, and linearity of $\imath$ is easy to check. □

Exercise 7.11. Let $Y \subseteq X$ be a closed subspace of a normed vector space.
(a) Show that $Y^\perp = \{x^* \in X^* \mid x^*(Y) = \{0\}\}$ is a closed subspace.
(b) Show that $(X/Y)^* = Y^\perp$ (that is, that there is a natural isometric isomorphism between the two).
(c) Show that $Y^* = X^*/Y^\perp$.
(d) Conclude that Y is reflexive if X is reflexive.

Exercise 7.12. Let X be a normed vector space and suppose that the dual X^* is separable. Show that X is also separable. In particular, if X is separable but X^* is not, then X cannot be reflexive. Find an example of a Banach space that is not reflexive for that reason.

The results developed above give another approach to the existence of completions.

SHORTER PROOF OF THEOREM 2.32. Let X be a normed vector space. By Corollary 7.9, X is isometric to $\imath(X) \subseteq X^{**}$. Set $B = \overline{\imath(X)}$, which is a Banach space by Lemma 2.54 and Exercise 2.26(b). □

7.1.4 An Application of the Spanning Criterion

†The description of the closed linear hull in Corollary 7.6 can be used as a spanning criterion: a subset S of a Banach space X spans X (that is, has X as its closed linear hull) if and only if there is no non-zero $x^* \in X^*$ with the property that $S \subseteq \ker(x^*)$.

This is a powerful tool, surprisingly often even without a complete description of the dual space. The following result generalizes the Stone–Weierstrass theorem on the unit interval. The full result also shows the converse, so the divergence characterises the density.

---

† The result of this subsection will not be needed in the remainder of the book.

Theorem 7.13 (Müntz [76]). *Suppose that (n_k) is a sequence in $\mathbb{N}$ with*

$$n_1 < n_2 < n_3 < \cdots$$

and with $\sum_{k=1}^{\infty} \frac{1}{n_k} = \infty$, and let $p_n(x) = x^n$ for $n \in \mathbb{N}$. Then the linear hull of $\{1, p_{n_1}, p_{n_2}, \dots\}$ is dense in $C([0,1])$.

PROOF. Let Y be the closed linear hull of the set $\{1, p_{n_1}, p_{n_2}, \dots\}$ in $C([0,1])$. By Corollary 7.6 we have to show that if $\ell \in C([0,1])^*$ has

$$\ell(1) = \ell(p_{n_k}) = 0 \tag{7.5}$$

for all $k \geqslant 1$, then $\ell = 0$. In fact, it is enough to show that if $\ell \in C([0,1])^*$ has (7.5) for all $k \geqslant 1$, then $\ell(p_n) = 0$ for all integers $n \geqslant 1$. This is because Corollary 7.6 then shows that $\mathbb{C}[x] \subseteq Y$, after which the Stone–Weierstrass theorem (Theorem 2.40) may be applied to give $Y = C([0,1])$. So assume that $\ell \in C([0,1])^*$ satisfies (7.5) for all $k \geqslant 1$, and assume also that there is some $n \in \mathbb{N}$ with $\ell(p_n) \neq 0$. We will show that this implies $\sum_{k=1}^{\infty} \frac{1}{n_k} < \infty$.

For $\zeta \in \mathbb{C}$ with $\Re(\zeta) > 0$, we define $p_\zeta(t) = t^\zeta$ for $t \in [0,1]$ (with the convention that $0^\zeta = 0$). This defines the function $p_\zeta \in C([0,1])$ satisfying $\|p_\zeta\| \leqslant 1$. Moreover, we have

$$\lim_{\mathbb{C} \ni \delta \to 0} \frac{t^{\zeta+\delta} - t^\zeta}{\delta} = \lim_{\mathbb{C} \ni \delta \to 0} t^\zeta \frac{t^\delta - 1}{\delta} = t^\zeta \log t,$$

for all $t \in [0,1]$ and in fact the convergence is with respect to the $\|\cdot\|_\infty$ norm (use the complex version of the mean value theorem to check this claimed uniformity).

Now define $f(\zeta) = \ell(p_\zeta)$ for $\zeta \in \mathbb{C}$ with $\Re(\zeta) > 0$, so $|f(\zeta)| \leqslant \|\ell\|$. Furthermore, f is analytic for $\Re(\zeta) > 0$ since

$$\lim_{\mathbb{C} \ni \delta \to 0} \frac{f(\zeta + \delta) - f(\zeta)}{\delta} = \ell(t^\zeta \log t)$$

exists by the above observation regarding uniform convergence. Finally, we have $f(n_k) = 0$ for $k \geqslant 1$ by assumption.

Now define the *Blaschke product*[22]

$$B_K(\zeta) = \prod_{k=1}^{K} \frac{\zeta - n_k}{\zeta + n_k},$$

with simple zeros $B_K(n_k) = 0$ for $k = 1, \dots, K$, $B_K(\zeta) \neq 0$ for $\Re(\zeta) > 0$ and $\zeta \notin \{n_1, \dots, n_K\}$, and the asymptotic formula

$$|B_K(\zeta)| \longrightarrow 1$$

as $\Re(\zeta) \to 0$ or as $|\zeta| \to \infty$. Together with $f(n_k) = 0$ these properties show that

$$g_K(\zeta) = \frac{f(\zeta)}{B_K(\zeta)}$$

is analytic for $\Re(\zeta) > 0$, and satisfies the estimate

$$|g_K(\zeta)| \leqslant (1 + \delta)\|\ell\| \tag{7.6}$$

for $\|\zeta\| \geqslant R(\delta)$ and $\Re(\zeta) \leqslant \varepsilon(\delta)$, the positive quantities $R(\delta)$ and $\varepsilon(\delta)$ depend on δ, and $\delta > 0$ is arbitrary.

Applying the maximum principle for g_K on the half-disk

$$\{\zeta \in \mathbb{C} \mid |\zeta - \varepsilon(\delta)| \leqslant R(\delta), \Re(\zeta) \geqslant \varepsilon(\delta)\}$$

in Figure 7.1, the function $|g_K|$ must attain its maximum on the boundary of the half-circle. As we have (7.6) on that boundary, we obtain

$$\|g_K\| \leqslant \|\ell\|(1 + \delta)$$

first on the half-disk and, by decreasing $\varepsilon(\delta)$ and increasing $R(\delta)$, on all of the right half-space. As $\delta > 0$ was arbitrary, we obtain $\|g_K\|_\infty \leqslant \|\ell\|$.

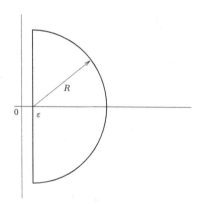

Fig. 7.1: Applying the maximum principle.

Recall that n was chosen so that $f(n) = \ell(p_n) \neq 0$. For $\zeta = n$ this shows that

$$\prod_{k=1}^{K} \left| 1 + \frac{2n}{n_k - n} \right| = \prod_{k=1}^{K} \left| \frac{n + n_k}{n - n_k} \right| = |B_K(n)|^{-1} \leqslant \frac{\|\ell\|}{|f(n)|} < \infty,$$

meaning that we have found an upper bound for the product on the left-hand side independent of K. Notice that $n_k > n$ for all but finitely many $k \in \mathbb{N}$.

Taking the logarithm and using the fact that $x \ll \log(1+x)$ for all $x \in [0,1]$, it follows that the sum $\sum_{k=1}^{K} \frac{1}{n_k - n}$ has an upper bound independent of K. Multiplying the series term-by-term with $\frac{n_k - n}{n_k}$ (and noticing that $\frac{n_k - n}{n_k} \to 1$ as $k \to \infty$), it follows that $\sum_{k=1}^{\infty} \frac{1}{n_k} < \infty$, as claimed. This contradicts our assumption, and the theorem follows. $\square$

7.2 Banach Limits, Amenable Groups, and the Banach–Tarski Paradox

In this section we start the discussion of amenability and related topics. Amenability will be one of two (quite different) functional-analytic properties of a group that we will discuss in Chapter 10.

7.2.1 Banach Limits

On the space $c(\mathbb{N}) = \{(x_n)_{n \in \mathbb{N}} \in \ell^{\infty}(\mathbb{N}) \mid \lim_{n \to \infty} x_n \text{ exists}\}$ we have the natural linear functional lim defined by

$$c(\mathbb{N}) \ni (x_n)_{n \in \mathbb{N}} \longmapsto \lim((x_n)_{n \in \mathbb{N}}) = \lim_{n \to \infty} x_n.$$

A natural question is to ask if this rather obvious functional — taking the limit of sequences that do have a limit — might have an extension to all of the much larger space $\ell^{\infty}(\mathbb{N})$. The Hahn–Banach theorem is built for just such situations, and using it we readily find such a generalized limit in the form of a linear functional.

Corollary 7.14 (Banach limit). *There exists a linear functional, which we will denote* $\mathrm{LIM} \in (\ell^{\infty}(\mathbb{N}))^*$, *with norm one, which may be thought of as a generalized limit since it satisfies the following properties:*

- $\mathrm{LIM}((a_n)) = \lim_{n \to \infty} a_n$ *if the latter limit exists;*
- $\mathrm{LIM}((a_n)) \in [\liminf_{n \to \infty} a_n, \limsup_{n \to \infty} a_n]$ *if* $a_n \in \mathbb{R}$ *for all* $n \geqslant 1$;
- $\mathrm{LIM}((a_n)) = \mathrm{LIM}((a_{n+1}))$.

The functional LIM *is called a* Banach limit.

PROOF. We work initially over $\mathbb{R}$. Let $c(\mathbb{N}) \subseteq \ell^{\infty}(\mathbb{N})$ and $\lim \in c(\mathbb{N})^*$ be as given before the statement of the corollary. Notice that $\| \lim \| = 1$ since

$$\left| \lim_{n \to \infty} a_n \right| \leqslant \sup_{n \geqslant 1} |a_n|.$$

Let $L \in (\ell^{\infty}(\mathbb{N}))^*$ be an extension as in Theorem 7.3, with $\|L\| = \| \lim \|$. We now define

$$\text{LIM}((a_n)) = L\left(a_1, \tfrac{a_1+a_2}{2}, \tfrac{a_1+a_2+a_3}{3}, \dots\right).$$

Clearly LIM is linear and extends lim on the subspace $c(\mathbb{N})$, since the Césaro averages of a convergent sequence converge to the same limit. This functional also has norm one, since

$$\left|\tfrac{a_1+\cdots+a_n}{n}\right| \leqslant \|(a_n)\|_\infty$$

for all $n \geqslant 1$, which implies that

$$\left|L\left(a_1, \tfrac{a_1+a_2}{2}, \dots\right)\right| \leqslant \|(a_n)\|_\infty.$$

Moreover,

$$\text{LIM}((a_n)_n - (a_{n+1})_n) = L\left(a_1 - a_2, \tfrac{a_1-a_3}{2}, \tfrac{a_1-a_4}{3}, \dots\right) = 0,$$

which implies the last claim in the corollary.

Let $I = \inf_{n\geqslant 1} a_n$ and $S = \sup_{n\geqslant 1} a_n$, so that $\left|a_n - \tfrac{I+S}{2}\right| \leqslant \tfrac{S-I}{2}$ for all $n \geqslant 1$, and hence

$$\left|\text{LIM}((a_n)) - \tfrac{I+S}{2}\right| \leqslant \tfrac{S-I}{2},$$

which implies that $I \leqslant \text{LIM}((a_n)) \leqslant S$. Together with the already established translation-invariance, we obtain $\inf_{n\geqslant k} a_n \leqslant \text{LIM}((a_n)) \leqslant \sup_{n\geqslant k} a_n$, and so also

$$\liminf_{n\to\infty} a_n \leqslant \text{LIM}((a_n)) \leqslant \limsup_{n\to\infty} a_n.$$

We may extend LIM from $\ell^\infty_{\mathbb{R}}(\mathbb{N})$ to $\ell^\infty_{\mathbb{C}}(\mathbb{N})$ by setting

$$\text{LIM}((a_n)) = \text{LIM}((\Re a_n)) + \mathrm{i}\,\text{LIM}((\Im a_n)) \tag{7.7}$$

for all $(a_n) \in \ell^\infty_{\mathbb{C}}(\mathbb{N})$ (see Exercise 7.15). $\square$

Exercise 7.15. Show that the extension $\text{LIM} \in \left(\ell^\infty_{\mathbb{C}}(\mathbb{N})\right)^*$ from (7.7) is $\mathbb{C}$-linear and has norm one.

By pre-composing with the projection operator from $\ell^\infty(\mathbb{Z})$ to $\ell^\infty(\mathbb{N})$ defined by $(a_n)_{n\in\mathbb{Z}} \mapsto (a_n)_{n\in\mathbb{N}}$, we can also view LIM as a translation-invariant linear functional on $\ell^\infty(\mathbb{Z})$.

7.2.2 Amenable Groups

A natural and important question to ask is which other groups G have a similar invariant functional defined on all of $\ell^\infty(G)$. We assume here that G is endowed with the discrete topology but note that the notions discussed have natural analogues for locally compact groups (see Section 10.2).

The following concept was introduced by von Neumann in 1929, and called *messbar* (measurable); the modern terminology was introduced by Day in 1949, perhaps as a pun, as these groups are 'easy to work with' and hence a-men-able (US) / a-mean-able (UK), and are groups that 'admit a mean', hence a-mean-able.

Definition 7.16. A discrete group G is called *amenable* if there exists a finitely additive (left-)invariant mean on G. That is, a map $m : \mathbb{P}(G) \to [0,1]$ defined on all subsets of G, with the following properties:

- $m(A) \geqslant 0$ for all $A \subseteq G$ and $m(G) = 1$;
- $m(A_1 \cup A_2) = m(A_1) + m(A_2)$ for disjoint sets $A_1, A_2 \subseteq G$; and
- $m(gA) = m(A)$ for all $g \in G$ and $A \subseteq G$.

One may think of a mean (which is only required to be finitely additive) as a poor substitute for a measure (which is countably additive) when a measure with the desired properties does not exist. This is the case if G is a countable infinite group, as the only translation-invariant measure in that case is the counting measure (or a scalar multiple of the counting measure), which is infinite. However, the invariant mean discussed here takes values in $[0,1]$.

Example 7.17. Corollary 7.14 together with the next lemma shows $G = \mathbb{Z}$ is amenable. Moreover, any invariant mean on $\mathbb{Z}$ (there will turn out to be many, see Exercise 7.25) will have some reassuringly natural properties. For example, if $E = \{n \in \mathbb{Z} \mid n \text{ is even}\}$, then for any invariant mean m we must have $m(E) = \frac{1}{2}$ since $2m(E) = m(E) + m(E+1) = m(\mathbb{Z}) = 1$.

Lemma 7.18. *A discrete group G is amenable if and only if there exists a positive left-invariant functional* $\mathrm{LIM} \in (\ell^\infty(G))^*$ *of norm one. Here positivity is the requirement that*

- $a \geqslant 0$ *(on all of G) implies that* $\mathrm{LIM}(a) \geqslant 0$,

and left-invariance is

- $\mathrm{LIM}(a^h) = \mathrm{LIM}(a)$ *for all* $h \in G$ *and* $a \in \ell^\infty(G)$, *where a^h is the shifted map $g \mapsto a(h^{-1}g)$.*

SKETCH OF PROOF. If LIM is given on $\ell^\infty(G)$ then we can define $m(A)$ to be $\mathrm{LIM}(\mathbb{1}_A)$ for all $A \subseteq G$, and then it is easy to check that m is a left-invariant mean on G. On the other hand, if a left-invariant mean m is given, then we may obtain every $a \in \ell^\infty(G)$ as the limit of a sequence of finite sums of the form

$$\sum_{i=1}^{\ell_n} r_i^{(n)} \mathbb{1}_{A_i^{(n)}}$$

as $n \to \infty$, where $r_i^{(n)} \in \mathbb{C}$ and $A_i^{(n)} \subseteq G$ for all $n \geqslant 1$ and i. For example, we may partition the set $\overline{B^{\mathbb{C}}_{\|a\|_\infty}}$ into finitely many sets $B_1, \ldots, B_\ell$ each of diameter less than $\frac{1}{n}$, choose $r_i^{(n)}$ in B_i and then define $A_i = a^{-1}(B_i)$ so that

$$\left\| a - \sum_{i=1}^{\ell_n} r_i^{(n)} \mathbb{1}_{A_i^{(n)}} \right\| \leqslant \frac{1}{n}.$$

Now we can define LIM by

$$\mathrm{LIM}(a) = \lim_{n \to \infty} \sum_{i=1}^{\ell_n} r_i^{(n)} m(A_i^{(n)}),$$

and check that LIM is well-defined, linear, bounded of norm one, positive, and left-invariant. ∎

Exercise 7.19. Provide a detailed proof of Lemma 7.18.

Next we show that the class of amenable groups is closed under many natural operations that allow us to give more examples.

Proposition 7.20 (Subgroups and quotients). *Let G be a discrete group.*
(a) *If G is amenable, then every subgroup $H < G$ is also amenable.*
(b) *Let $H \lhd G$ be a normal subgroup. Then G is amenable if and only if both H and G/H are amenable.*

PROOF. For (a), let LIM_G be a left-invariant positive functional on $\ell^\infty(G)$ of norm one, as in Lemma 7.18. Let $H < G$ be a subgroup and $S \subseteq G$ a set of right coset representatives, so that

$$G = \bigsqcup_{s \in S} Hs.$$

For any $a \in \ell^\infty(H)$ we define $a_G(hs) = a(h)$ for $h \in H$ and $s \in S$ and $\mathrm{LIM}_H(a) = \mathrm{LIM}_G(a_G)$, which is clearly linear. Moreover, $a \geqslant 0$ implies that $a_G \geqslant 0$, and therefore $\mathrm{LIM}_H(a) \geqslant 0$. Moreover,

$$|\mathrm{LIM}_H(a)| = |\mathrm{LIM}_G(a_G)| \leqslant \|a_G\|_\infty = \|a\|_\infty.$$

Finally, for $h_0 \in H$ and $s \in S$ we have

$$(a^{h_0})_G(hs) = a^{h_0}(h) = a(h_0^{-1}h) = a_G(h_0^{-1}hs) = (a_G)^{h_0}(hs),$$

so $\mathrm{LIM}_H(a^{h_0}) = \mathrm{LIM}_G((a_G)^{h_0}) = \mathrm{LIM}_G(a_G) = \mathrm{LIM}_H(a)$, and Lemma 7.18 shows that H is amenable.

For (b) we again use Lemma 7.18, allowing us to work with functionals on the space of bounded functions on the groups involved. Assume first that G is amenable, so that H is amenable by (a). For $a \in \ell^\infty(G/H)$, define an element $a_G \in \ell^\infty(G)$ by setting $a_G(g) = a(gH)$. Just as in (a) we define $\mathrm{LIM}_{G/H}(a) = \mathrm{LIM}_G(a_G)$ to obtain a left-invariant positive functional of norm one on $\ell^\infty(G/H)$, showing that G/H is amenable.

For the converse, assume that H and G/H are both amenable, and write LIM_H and $\text{LIM}_{G/H}$ for the associated functionals. For $a \in \ell^\infty(G)$ define the bounded function $\bar{a}$ on G by

$$\bar{a}(g) = \text{LIM}_H(h \mapsto a(gh))$$

for $g \in G$. Notice that for $h_0 \in H$

$$\bar{a}(gh_0) = \text{LIM}_H(h \mapsto a(gh_0h)) = \text{LIM}_H(h \mapsto a(gh)) = \bar{a}(g),$$

since multiplication by h_0 induces a left shift on the function $h \mapsto a(gh)$. Therefore we can think of $\bar{a}$ as a function on G/H by setting $\bar{a}(gH) = \bar{a}(g)$. We can now define

$$\text{LIM}_G(a) = \text{LIM}_{G/H}(\bar{a}) = \text{LIM}_{G/H}(g \mapsto \text{LIM}_H(h \mapsto a(gh))),$$

and it is once again straightforward to check the required properties of LIM_G to see that G is amenable. □

The following exercise gives the first indication of the existence of a maximal amenable radical subgroup: the *amenable radical* (also see Exercise 8.25).

Exercise 7.21. (a) Let G be a discrete group and let $H_1 \lhd G$ and $H_2 < G$ be two amenable subgroups with H_1 normal. Show that $H_1 H_2 = \{h_1 h_2 \mid h_1 \in H_1, h_2 \in H_2\}$ is also an amenable subgroup.
(b) Let $H_1, \ldots, H_n \lhd G$ be amenable normal subgroups for some $n \geqslant 2$. Show that the product $H_1 H_2 \cdots H_n \lhd G$ is an amenable normal subgroup.

Even though the above shows that many groups are amenable, it is also easy to give an example of a group that is not amenable.

Example 7.22. The free group $G = F_2$ generated by two elements α, β is not amenable. To see this, suppose that m is an invariant mean on G. Clearly the singletons $\{e\}, \{\alpha\}, \{\alpha\alpha\}, \ldots$ are all disjoint and are left-translates of each other, so we must have $m(\{e\}) = 0$. Now define

$$S_\alpha = \{g \in G \mid \text{the reduced form of } g \text{ starts on the left with } \alpha\},$$

and similarly define sets $S_{\alpha^{-1}}$, S_β, and $S_{\beta^{-1}}$. Since

$$G = S_\alpha \sqcup S_{\alpha^{-1}} \sqcup S_\beta \sqcup S_{\beta^{-1}} \sqcup \{e\}$$

and $m(\{e\}) = 0$, we must have

$$1 = m(S_\alpha) + m(S_{\alpha^{-1}}) + m(S_\beta) + m(S_{\beta^{-1}}). \tag{7.8}$$

However,

$$\alpha^{-1} S_\alpha = S_\alpha \sqcup S_\beta \sqcup S_{\beta^{-1}} \sqcup \{e\},$$

so

$$m(S_\alpha) = m(S_\alpha) + m(S_\beta) + m(S_{\beta^{-1}}),$$

and hence $m(S_\beta) = m(S_{\beta^{-1}}) = 0$ by positivity. Exchanging the roles of α and β also shows that $m(S_\alpha) = m(S_{\alpha^{-1}}) = 0$, which together contradict (7.8). It follows that F_2 cannot be amenable.

We note that Proposition 7.20 shows inductively that $G = \mathbb{Z}^d$ is amenable. This may also be seen by noting that a sequence (F_n) of large boxes, for example the sequence defined by $F_n = [-n, n]^d \cap \mathbb{Z}^d$ for $n \geqslant 1$, is a Følner sequence as discussed now.

Definition 7.23. A sequence $(F_n)_n$ of finite subsets of a countable group G is called a *Følner sequence* if the elements of the sequence are asymptotically translation invariant in the sense that

$$\lim_{n\to\infty} \frac{|F_n \triangle h F_n|}{|F_n|} = 0 \tag{7.9}$$

for all $h \in G$.

Lemma 7.24. *If a countable group has a Følner sequence, then it is amenable.*

PROOF. Let $\mathrm{LIM}_\mathbb{N}$ be the Banach limit from Corollary 7.14 and let (F_n) be a Følner sequence. Then for any $a \in \ell^\infty(G)$ we can define

$$\mathrm{LIM}_G(a) = \mathrm{LIM}_\mathbb{N}\left(\frac{1}{|F_n|}\sum_{g\in F_n} a(g)\right),$$

which is linear, positive, and of norm one. Then

$$\mathrm{LIM}_G(a^h) = \mathrm{LIM}_\mathbb{N}\left(\frac{1}{|F_n|}\sum_{g\in F_n} a(h^{-1}g)\right) = \mathrm{LIM}_\mathbb{N}\left(\frac{1}{|F_n|}\sum_{g\in h^{-1}F_n} a(g)\right)$$

$$= \mathrm{LIM}_\mathbb{N}\left(\frac{1}{|F_n|}\sum_{g\in F_n} a(g)\right) = \mathrm{LIM}_G(a) \qquad \text{(by (7.9))}$$

showing left-invariance.   ☐

Exercise 7.25. Show that the invariant means on $\mathbb{Z}$ constructed by using the Følner sequences defined by $F_n^{(1)} = [0, n]$, $F_n^{(2)} = [-n, 0]$, and $F_n^{(3)} = [n^2, n^2+n]$ are all different. Can you construct infinitely many different invariant means on $\mathbb{Z}$?

Exercise 7.26. Show that any countable abelian group is amenable.

Exercise 7.27. Prove that the discrete Heisenberg group

$$H = \left\{ \begin{pmatrix} 1 & k & \ell \\ 0 & 1 & m \\ 0 & 0 & 1 \end{pmatrix} \mid k, \ell, m \in \mathbb{Z} \right\}$$

with the usual matrix multiplication is amenable.

Exercise 7.28. Show that $SL_2(\mathbb{Z})$ is not amenable. You may use the fact that the group $PSL_2(\mathbb{Z}) = SL_2(\mathbb{Z})/\{\pm I\}$ is isomorphic to the free product of $\mathbb{Z}/2\mathbb{Z}$ and $\mathbb{Z}/3\mathbb{Z}$.

7.2.3 The Banach–Tarski Paradox

The following surprising consequence of the axiom of choice was one of the original motivations for the study of amenable groups and of measurable sets.

Theorem 7.29 (Banach–Tarski paradox[23]). *The closed unit ball B in $\mathbb{R}^3$ can be decomposed into finitely many disjoint sets $B = P_1 \sqcup \cdots \sqcup P_m$ with the property that there are isometric motions (that is, compositions of rotations and translations) of $\mathbb{R}^3$ sending P_i to P_i' for $1 \leqslant i \leqslant m$ such that*

$$B \sqcup \left(B + \begin{pmatrix} 3 \\ 0 \\ 0 \end{pmatrix} \right) = P_1' \sqcup \cdots \sqcup P_m'.$$

Clearly some of the sets appearing in the decomposition are of necessity non-measurable. The same result holds in $\mathbb{R}^d$ for any dimension $d \geqslant 3$, but we restrict attention to the case $d = 3$ for simplicity. However, in dimension $d = 2$ no such finite paradoxical decomposition can be found.

Notice that while the dimension d plays an essential role, other aspects of the topology of $\mathbb{R}^d$ play no real role here since the sets in the decomposition are not even measurable, and in particular cannot be described using countable unions and intersections of open and closed sets. The real drama takes place in the group of isometries of $\mathbb{R}^d$, and it is differences between the structure of this group when $d = 2$ and when $d \geqslant 3$ that lie behind the existence of the paradoxical decomposition. For this reason we will endow the group of isometries $\mathrm{Isom}(\mathbb{R}^d) = O_d(\mathbb{R}) \ltimes \mathbb{R}^d$ with the discrete topology. Doing so helps to illuminate the difference between the cases $d = 2$ and $d \geqslant 3$.

- For $d = 2$ the group $SO_2(\mathbb{R})$ is abelian, and hence amenable (we will only be able to prove this a bit later in Exercise 8.23). The translation group $\mathbb{R}^2$ and the factor $O_2(\mathbb{R})/SO_2(\mathbb{R})$ are also abelian, so by Proposition 7.20 we see that $\mathrm{Isom}(\mathbb{R}^2)$ is amenable. This helps to prevent paradoxical decompositions as in the Banach–Tarski paradox for the unit ball in 2 dimensions and using isometries (see Exercise 7.30).
- For $d \geqslant 3$ the group $SO_d(\mathbb{R})$ is far from abelian, and in fact contains a free subgroup (see Proposition 7.31) and so cannot be amenable by Proposition 7.20 and Example 7.22 (where we use the discrete topology on $SO_d(\mathbb{R})$).

Exercise 7.30. (a) Let G be a discrete amenable group, and suppose that G acts on a set X. Show that there exists a (non-canonical) finitely additive G-invariant mean m_X (that is, a function $m_X : \mathbb{P}(X) \to [0,1]$ satisfying the first two requirements of Definition 7.16 and with $m_X(g{\cdot}B) = m_X(B)$ for all $B \subseteq X$ and $g \in G$) on the set of subsets $\mathbb{P}(X)$ of X.

(b) Show that amenability of $\mathrm{Isom}(\mathbb{R}^2)$ implies that a paradoxical decomposition as in Theorem 7.29 cannot exist in 2 dimensions.

Proposition 7.31. *The rotations*

$$a = \begin{pmatrix} \frac{3}{5} & -\frac{4}{5} & 0 \\ \frac{4}{5} & \frac{3}{5} & 0 \\ 0 & 0 & 1 \end{pmatrix}, \quad b = \begin{pmatrix} 1 & 0 & 0 \\ 0 & \frac{3}{5} & -\frac{4}{5} \\ 0 & \frac{4}{5} & \frac{3}{5} \end{pmatrix}$$

in $\mathrm{SO}_3(\mathbb{R})$ *generate a free subgroup* H *of* $\mathrm{SO}_3(\mathbb{R})$.

The statement is essentially geometric: a and b are rotations by an irrational multiple of π fixing two orthogonal axes, as illustrated in Figure 7.2. The proposition means that any rotation obtained by composing a finite sequence of rotations $a^{\pm 1}$ and $b^{\pm 1}$ (in which a rotation is never followed by its inverse) cannot be obtained by any other such sequence of rotations.

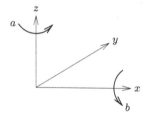

Fig. 7.2: Two rotations generating a free subgroup.

PROOF OF PROPOSITION 7.31. Let

$$\widetilde{a}_+ = 5a = \begin{pmatrix} 3 & -4 & 0 \\ 4 & 3 & 0 \\ 0 & 0 & 5 \end{pmatrix}, \widetilde{a}_- = 5a^{-1} = \begin{pmatrix} 3 & 4 & 0 \\ -4 & 3 & 0 \\ 0 & 0 & 5 \end{pmatrix},$$

and similarly define $\widetilde{b}_+ = 5b$ and $\widetilde{b}_- = 5b^{-1}$. For some part of the proof we will be working over the field $\mathbb{F}_5$ (that is, working modulo 5). The matrices arising can all be viewed as linear transformations of the vector space $\mathbb{Z}^3/5\mathbb{Z}^3 \cong \mathbb{F}_5^3$. We want to study how they act on the vectors

$$v = \begin{pmatrix} 1 \\ 1 \\ 1 \end{pmatrix}, \; w_\alpha = \begin{pmatrix} 2 \\ 1 \\ 0 \end{pmatrix}, \; w_{\alpha^{-1}} = \begin{pmatrix} 1 \\ 2 \\ 0 \end{pmatrix}, \; w_\beta = \begin{pmatrix} 0 \\ 2 \\ 1 \end{pmatrix}, \; w_{\beta^{-1}} = \begin{pmatrix} 0 \\ 1 \\ 2 \end{pmatrix}.$$

Notice that it is enough to carry out the calculations for $\widetilde{a}_+$, since the situation for the other matrices is the same up to a permutation of the basis vectors. Writing $\sim$ for proportionality, we obtain

$$\widetilde{a}_+ v = \begin{pmatrix} -1 \\ 7 \\ 5 \end{pmatrix} \equiv \begin{pmatrix} 4 \\ 2 \\ 0 \end{pmatrix} \sim \begin{pmatrix} 2 \\ 1 \\ 0 \end{pmatrix} = w_\alpha,$$

$$\widetilde{a}_+ w_\alpha = \begin{pmatrix} 3 & -4 & 0 \\ 4 & 3 & 0 \\ 0 & 0 & 0 \end{pmatrix} \begin{pmatrix} 2 \\ 1 \\ 0 \end{pmatrix} \equiv \begin{pmatrix} 2 \\ 1 \\ 0 \end{pmatrix} = w_\alpha,$$

$$\widetilde{a}_+ w_\beta = \begin{pmatrix} 3 & -4 & 0 \\ 4 & 3 & 0 \\ 0 & 0 & 0 \end{pmatrix} \begin{pmatrix} 0 \\ 2 \\ 1 \end{pmatrix} \equiv \begin{pmatrix} 2 \\ 1 \\ 0 \end{pmatrix} = w_\alpha,$$

$$\widetilde{a}_+ w_{\beta^{-1}} = \begin{pmatrix} 3 & -4 & 0 \\ 4 & 3 & 0 \\ 0 & 0 & 0 \end{pmatrix} \begin{pmatrix} 0 \\ 1 \\ 2 \end{pmatrix} \equiv \begin{pmatrix} 1 \\ 3 \\ 0 \end{pmatrix} \sim \begin{pmatrix} 2 \\ 1 \\ 0 \end{pmatrix} = w_\alpha,$$

but

$$\widetilde{a}_+ w_{\alpha^{-1}} = \begin{pmatrix} 3 & -4 & 0 \\ 4 & 3 & 0 \\ 0 & 0 & 0 \end{pmatrix} \begin{pmatrix} 1 \\ 2 \\ 0 \end{pmatrix} \equiv \begin{pmatrix} 0 \\ 0 \\ 0 \end{pmatrix}.$$

The same applies to the other matrices, which in summary means that each of the matrices $\widetilde{a}_+, \widetilde{a}_-, \widetilde{b}_+, \widetilde{b}_-$ has its own non-zero eigenvector in $\mathbb{F}_5^3$, maps the eigenvector of the matrix with the same symbol but opposite sign to the zero vector, but maps v and the other three to a multiple of its eigenvector.

Suppose now that γ is a reduced word of length $n \geqslant 1$ in F_2, the free group with generators α and β (that is, a finite string of symbols chosen from $\alpha, \alpha^{-1}, \beta, \beta^{-1}$ with the property that no symbol is immediately followed by its inverse). Define a homomorphism $\phi : F_2 \to SO_3(\mathbb{R})$ by defining it on the generators by $\phi(\alpha) = a$, $\phi(\beta) = b$ and then extending to F_2 using the homomorphism property, and use this to define $\widetilde{\phi}(\gamma) = 5^n \phi(\gamma)$. Equivalently, $\widetilde{\phi}(\gamma) \in \mathrm{Mat}_{33}(\mathbb{Z})$ may be obtained by multiplying $\widetilde{a}_+, \widetilde{a}_-, \widetilde{b}_+, \widetilde{b}_-$ in the order and multiplicities corresponding to the appearance of $\alpha, \alpha^{-1}, \beta, \beta^{-1}$ in the word γ. As the word γ is reduced, we see by induction on n and the calculations above that

$$\widetilde{\phi}(\gamma) v \in \mathbb{Z}^3$$

modulo 5 is a non-zero multiple of w_η where $\eta \in \{\alpha, \alpha^{-1}, \beta, \beta^{-1}\}$ is the left-most symbol of γ. In particular, $\widetilde{\phi}(\gamma)$ is not divisible by 5 and $\phi(\gamma) \neq I$ because $\widetilde{\phi}(\gamma) \neq 5^n I$. Thus ϕ is injective and so $\mathrm{im}\,\phi = \langle a, b \rangle \cong F_2$. $\square$

In the proof of Theorem 7.29 the free subgroup $\langle a, b \rangle < SO_3(\mathbb{R})$ from Proposition 7.31 will play a critical role. It will be convenient to define two subsets B_1, B_2 of $\mathbb{R}^3$ to be equivalent, written $B_1 \sim B_2$, if they can be decomposed as $B_1 = P_1 \sqcup \cdots \sqcup P_n$ and $B_2 = Q_1 \sqcup \cdots \sqcup Q_n$ into finitely many disjoint subsets such that Q_k is the image of P_k under some isometric motion for $k = 1, \ldots, n$.

Essential Exercise 7.32. Prove that this defines an equivalence relation.

PROOF OF THEOREM 7.29. Let $B = \overline{B_1^{\mathbb{R}^3}}$ be the closed unit ball in $\mathbb{R}^3$.

STEP 1. We claim that $B \sim B \smallsetminus \{0\}$ by using a 'Hilbert's Hotel' argument.[24] To see this, let $x_0 = (\frac{1}{2}, 0, 0)^t$ and let $\gamma : \mathbb{R}^3 \to \mathbb{R}^3$ be an irrational rotation (meaning that $\gamma^n = I$ for some $n \in \mathbb{Z}$ implies that $n = 0$) about the point x_0 in the x-y plane extended trivially to a rotation about the line parallel to the z-axis through x_0, so that the orbit $D = \{\gamma^n(0) \mid n \in \mathbb{N}_0\} \subseteq B$ is infinite. Therefore

$$B = B \smallsetminus D \sqcup D \sim B \smallsetminus D \sqcup \gamma D = B \smallsetminus \{0\},$$

proving the claim.

Now let $H < \mathrm{SO}_3(\mathbb{R})$ be the free subgroup constructed in Proposition 7.31. Since H is countable and every non-trivial rotation in $\mathrm{SO}_3(\mathbb{R})$ has a single one-dimensional eigenspace with eigenvalue 1, we can find a countable union E of lines through the origin such that $HE = E$ and with the property that no vector in $B \smallsetminus E$ is fixed by a non-trivial element of H.

STEP 2. By using countably many Hilbert Hotel arguments at once we claim that $B \smallsetminus \{0\} \sim B \smallsetminus E$. To see this, notice that the set $\mathbb{S}^2 \cap E$ is countable and so the set P of pairs of vectors $v_1, v_2 \in \mathbb{S}^2 \cap E$ with $v_1 \neq v_2$ is also countable. Therefore

$$W = \big\{ w \in \mathbb{R}^3 \mid w \perp v_1 - v_2 \text{ for some } (v_1, v_2) \in P \big\}$$

is a countable union of hyperplanes, and so a null set. Fix $x_1 \in \mathbb{R}^3 \smallsetminus (W \cup E)$ and some irrational rotation γ about the line $\mathbb{R}x_1$. If now $\gamma^m v_1 = \gamma^n v_2$ for some $m, n \in \mathbb{N}_0$ and $v_1, v_2 \in \mathbb{S}^2 \cap E$, then, since γ is an isometry fixing x_1,

$$\langle v_1, x_1 \rangle = \langle \gamma^m v_1, x_1 \rangle = \langle \gamma^n v_2, x_1 \rangle = \langle v_2, x_1 \rangle$$

gives $v_1 = v_2$ by our choice of x_1. Since $v_1 \notin \mathbb{R}x_1$ and γ is an irrational rotation about the line, we also see that $m = n$. Therefore the various sets $\gamma^n \mathbb{R}v_1 \smallsetminus \{0\}$, where we vary $n \geq 0$ and the subspace $\mathbb{R}v_1 \subseteq E$, are all disjoint. Thus

$$B \smallsetminus \{0\} = B \smallsetminus \bigsqcup_{n \geq 0} \gamma^n E \sqcup B \cap \bigsqcup_{n \geq 0} \gamma^n E \smallsetminus \{0\}$$

$$\sim B \smallsetminus \bigsqcup_{n \geq 0} \gamma^n E \sqcup B \cap \bigsqcup_{n \geq 1} \gamma^n E \smallsetminus \{0\} = B \smallsetminus E,$$

as claimed.

STEP 3. We claim that $B \smallsetminus E \sim B \smallsetminus E \sqcup (B \smallsetminus E + (3,0,0)^t)$. This is clearly the main step in the argument, and it is here that we will use the fact that H is a free group, and in particular the resulting decomposition

$$H = \{e\} \sqcup S_a \sqcup S_{a^{-1}} \sqcup S_b \sqcup S_{b^{-1}}$$

obtained by taking the image of the decomposition of F_2 in Example 7.22. In order for the decomposition of H to be useful, we have to find a *cross-section* C of $B \smallsetminus E$ satisfying

$$B \smallsetminus E = \bigsqcup_{c \in C} Hc,$$

which can be found by a direct application of the axiom of choice (and will not be measurable). We now decompose $B \smallsetminus E$ into the four disjoint sets

$$B_1 = S_a C \sqcup \bigsqcup_{n \geqslant 0} a^{-n} C, \quad B_2 = S_{a^{-1}} C \smallsetminus \bigsqcup_{n \geqslant 1} a^{-n} C, \quad B_3 = S_b C$$

and $B_4 = S_{b^{-1}} C$.

Applying a to B_2 we obtain

$$aB_2 = (S_{a^{-1}} C \sqcup C \sqcup S_b C \sqcup S_{b^{-1}} C) \smallsetminus \bigsqcup_{n \geqslant 0} a^{-n} C$$
$$= B_2 \sqcup B_3 \sqcup B_4 \sim B_2 \sqcup \left((B_3 \sqcup B_4) + (3, 0, 0)^{\mathrm{t}} \right),$$

and applying b to B_4 we obtain

$$bB_4 = (S_{b^{-1}} C \sqcup C \sqcup S_a C \sqcup S_{a^{-1}} C)$$
$$= B_4 \sqcup B_1 \sqcup B_2 \sim B_4 \sqcup \left((B_1 \sqcup B_2) + (3, 0, 0)^{\mathrm{t}} \right).$$

Leaving B_1 and B_3 untouched and taking the union this proves the claim in Step 3.

Applying Step 2 and Step 1 (twice each) backwards, the theorem follows. □

7.3 The Duals of $L^p_\mu(X)$

In this section we will present descriptions of dual spaces using a *bilinear pairing*. If X and Y are vector spaces and each $y \in Y$ induces a linear functional on X, then we often write $\langle x, y \rangle$ for the value of the functional associated to $y \in Y$ evaluated at $x \in X$. We always assume that the linear functional depends linearly on $y \in Y$, and so $\langle \cdot, \cdot \rangle$ is a bilinear functional on $X \times Y$. We use the word pairing here to signify that, for example, the map in Exercise 7.33 may be thought of in two ways. It defines on the one hand a family of functionals parameterized by elements of $\ell^1(\mathbb{N})$ defined on sequences in $c_0(\mathbb{N})$ and on the other a family of functionals parameterized by elements of $c_0(\mathbb{N})$ defined on sequences in $\ell^1(\mathbb{N})$. Even though we will prove this in many cases (indeed, it will often be the key step in an argument), when we use this notation and terminology we do not assume that Y is indeed the whole dual to X or *vice versa*. The reader may start with the following as a warm-up exercise on how dual spaces may be found.

Exercise 7.33. (a) Recall that $c_0(\mathbb{N}) = \{(a_n) \mid \lim_{n\to\infty} a_n = 0\} \subseteq \ell^\infty(\mathbb{N})$ is a Banach space with respect to the supremum norm $\|\cdot\|_\infty$. Show that there is an isometric isomorphism $(c_0(\mathbb{N}))^* \cong \ell^1(\mathbb{N})$, where the dual pairing is given by

$$\langle (a_n), (b_n) \rangle = \sum_{n=1}^\infty a_n b_n$$

for $(a_n) \in c_0(\mathbb{N})$ and $(b_n) \in \ell^1(\mathbb{N})$.
(b) Show that $\left(\ell^1(\mathbb{N})\right)^* \cong \ell^\infty(\mathbb{N})$, with the same formula for the pairing.
(c) Show that the Banach limit LIM $\in (\ell^\infty(\mathbb{N}))^*$ as in Corollary 7.14 is not in the canonical image of $\ell^1(\mathbb{N})$ in $(\ell^\infty(\mathbb{N}))^*$.
(d) Conclude that neither $c_0(\mathbb{N})$ nor $\ell^1(\mathbb{N})$ is reflexive.
(e) Now let X be any infinite discrete set, and define $c_0(X)$ (and $\ell^1(X)$) to be the space of all ($\mathbb{R}$-valued or $\mathbb{C}$-valued) functions a on X for which there exists a sequence (x_n) in X with Supp $a \subseteq \{x_1, x_2, \ldots\}$ and such that $(a(x_n))_n$ belongs to $c_0(\mathbb{N})$ (resp. $\ell_1(\mathbb{N})$). Generalize (a) and (b) to this context.

7.3.1 The Dual of $L^1_\mu(X)$

We start by generalizing the second part of Exercise 7.33.

Proposition 7.34. *Let $(X, \mathcal{B}, \mu)$ be a σ-finite measure space. Then*

$$\left(L^1_\mu(X)\right)^* \cong L^\infty_\mu(X)$$

under the pairing $\langle f, g \rangle = \int_X fg \, d\mu$ for $f \in L^1_\mu(X)$ and $g \in L^\infty_\mu(X)$. The operator norm of the functional defined by $g \in L^\infty_\mu(X)$ is the essential supremum norm $\|g\|_\infty$ (defined on p. 29).

PROOF. As indicated, we associate to every $g \in L^\infty_\mu(X)$ the functional

$$\phi(g) : f \longmapsto \int_X fg \, d\mu$$

which satisfies

$$\|\phi(g)\|_{\mathrm{op}} = \sup_{\|f\|_1 \leqslant 1} \left| \int_X fg \, d\mu \right| \leqslant \sup_{\|f\|_1 \leqslant 1} \int_X |f| |g| \, d\mu \leqslant \|g\|_\infty.$$

For the converse we assume that $g \neq 0$, let $\varepsilon \in (0, \|g\|_\infty)$ and choose a measurable set $A \subseteq \{x \in X \mid |g(x)| > \|g\|_\infty - \varepsilon\}$ with $\mu(A) > 0$ (which is possible by definition of the essential supremum) and with $\mu(A) < \infty$ (which is possible since μ is σ-finite). Now define

$$f = \frac{1}{\mu(A)} \mathbb{1}_A \frac{|g(x)|}{g(x)},$$

and notice that $\|f\|_1 = 1$ and

$$\phi(g)(f) = \int_X fg \, d\mu = \frac{1}{\mu(A)} \int_A |g| \, d\mu \geq \|g\|_\infty - \varepsilon.$$

This shows that $\|\phi(g)\|_{\mathrm{op}} = \|g\|_\infty$, so $\phi : L_\mu^\infty(X) \to (L_\mu^1(X))^*$ is an isometric embedding. It remains to show that ϕ is onto (this is often the most interesting part of the identification of a dual space).

For this, assume first that μ is finite. Then $L_\mu^2(X) \subseteq L_\mu^1(X)$ and

$$\|f\|_1 \leq \mu(X)^{1/2} \|f\|_2$$

for all $f \in L_\mu^2(X)$ by the Cauchy–Schwarz inequality. Therefore, a functional $\ell \in (L_\mu^1(X))^*$ also induces a functional $\ell' \in (L_\mu^2(X))^*$. Since $L_\mu^2(X)$ is a Hilbert space, the Fréchet–Riesz representation theorem (Corollary 3.19) now shows that there exists some $g \in L_\mu^2(X)$ with

$$\ell'(f) = \int_X fg \, d\mu$$

for all $f \in L_\mu^2(X)$. We now show that $g \in L_\mu^\infty(X)$. Let

$$A = \{x \in X \mid |g(x)| > \|\ell\|_{\mathrm{op}}\},$$

so that $f = \mathbb{1}_A \frac{|g|}{g} \in L_\mu^2(X) \subseteq L_\mu^1(X)$ and $\|f\|_1 = \mu(A)$. If $\mu(A) > 0$ then

$$\|\ell\|_{\mathrm{op}}\mu(A) < \int_A |g| \, d\mu = \left| \int_X fg \, d\mu \right| = |\ell(f)| \leq \|\ell\|_{\mathrm{op}}\mu(A)$$

gives a contradiction. Thus $\mu(A) = 0$ and so $\|g\|_\infty \leq \|\ell\|_{\mathrm{op}}$. Since ℓ and $\phi(g)$ agree on the dense subset $L_\mu^2(X) \subseteq L_\mu^1(X)$, we have $\ell = \phi(g)$, as required.

If μ is σ-finite with $X = \bigsqcup_{n=1}^\infty Y_n$ and $\mu(Y_n) < \infty$, then we may apply the argument above to $\ell|_{L_\mu^1(Y_n)}$ to find some $g_n \in L_\mu^\infty(Y_n)$ with

$$\|g_n\|_\infty = \|\ell|_{L_\mu^1(Y_n)}\|_{\mathrm{op}} \leq \|\ell\|_{\mathrm{op}}.$$

We define $g(x) = g_n(x)$ for $x \in Y_n$, and obtain a function $g \in L_\mu^\infty(X)$ with $\|g\|_\infty \leq \|\ell\|_{\mathrm{op}}$. We now extend each function in $L_\mu^1(Y_n)$ to an element of $L_\mu^1(X)$ by setting it to be zero outside Y_n. With this the linear span

$$V = \sum_{n=1}^\infty L_\mu^1(Y_n) \subseteq L_\mu^1(X)$$

contains all simple functions in its closure, so that we have $\overline{V} = L^1_\mu(X)$. By construction ℓ and $\phi(g)$ coincide on V, so once again $\ell = \phi(g)$, as required.

$\square$

7.3.2 The Dual of $L^p_\mu(X)$ for $p > 1$

Exercise 7.35. Let $p, q \in (1, \infty)$ with $\frac{1}{p} + \frac{1}{q} = 1$. Show that $(\ell^p(\mathbb{N}))^* \cong \ell^q(\mathbb{N})$.

The following provides us with many examples of reflexive spaces.

Proposition 7.36. *Let $(X, \mathcal{B}, \mu)$ be a σ-finite measure space, and assume that $p \in (1, \infty)$ has Hölder conjugate q. Then $(L^p_\mu(X))^* \cong L^q_\mu(X)$ via the pairing*

$$\langle f, g \rangle = \int_X fg \, \mathrm{d}\mu$$

for $f \in L^p_\mu(X)$ and $g \in L^q_\mu(X)$. The operator norm of the functional determined by g is precisely $\|g\|_q$.

PROOF. For $f \in L^p_\mu(X)$ and $g \in L^q_\mu(X)$ with $\frac{1}{p} + \frac{1}{q} = 1$ we have

$$|\langle f, g \rangle| \leqslant \|f\|_p \|g\|_q$$

by the Hölder inequality (Theorem B.15). It follows that the linear functional defined by g on $L^p_\mu(X)$ is bounded, with norm less than or equal to $\|g\|_q$. If we set

$$f = \begin{cases} \frac{|g|}{g} |g|^{q/p} & \text{if } g \neq 0, \\ 0 & \text{if } g = 0 \end{cases}$$

then

$$\|f\|_p = \left(\int |g|^q \, \mathrm{d}\mu \right)^{1/p} = \|g\|_q^{q/p} < \infty$$

and

$$\langle f, g \rangle = \int_X |g|^{1+\frac{q}{p}} \, \mathrm{d}\mu = \int_X |g|^q \, \mathrm{d}\mu = \|f\|_p \|g\|_q$$

shows that the norm of the functional $\phi(g) \in (L^p_\mu(X))^*$ determined by g must be equal to $\|g\|_q$. It remains to show that every bounded linear functional ℓ in $(L^p_\mu(X))^*$ is determined as above by some $g \in L^q_\mu(X)$.

Let $\ell \in (L^p_\mu(X))^*$. Replacing ℓ by $\Re(\ell)$, respectively by $\Im(\ell)$ if necessary, we may restrict to real-valued functions on X and to $\mathbb{R}$-linear functionals, as the complex case then follows by putting together the functions associated to $\Re(\ell)$ and $\Im(\ell)$.

So we work over the reals and define

$$\nu^+(B) = \sup\{\ell(\mathbb{1}_A) \mid A \subseteq B \text{ measurable}, \mu(A) < \infty\}$$

for any measurable set $B \subseteq X$. Notice that if ℓ were defined by g, then $\nu^+(B)$ would be given by $\nu^+(B) = \int_A g \, d\mu = \int_B g^+ \, d\mu$ for $A = \{x \in B \mid g(x) > 0\}$. Thus for a general ℓ we would like to show that $\nu^+ \ll \mu$ is an absolutely continuous measure on X (which then will give us g^+ as a Radon–Nikodym derivative). Clearly $\nu^+(B_2) \geqslant \nu^+(B_1) \geqslant 0$ for measurable $B_1 \subseteq B_2 \subseteq X$. For measurable disjoint $B_1, B_2 \subseteq X$ and $A_1 \subseteq B_1$, $A_2 \subseteq B_2$ as in the definition of ν^+, we have

$$\ell(\mathbb{1}_{A_1}) + \ell(\mathbb{1}_{A_2}) = \ell(\mathbb{1}_{A_1 \cup A_2}) \leqslant \nu^+(B_1 \cup B_2),$$

and taking the supremum over A_1 and A_2 gives

$$\nu^+(B_1) + \nu^+(B_2) \leqslant \nu^+(B_1 \cup B_2).$$

If, on the other hand, $A \subseteq B_1 \cup B_2$ has finite measure we define $A_i = A \cap B_i$ for $i = 1, 2$ and see that

$$\ell(\mathbb{1}_A) = \ell(\mathbb{1}_{A_1}) + \ell(\mathbb{1}_{A_2}) \leqslant \nu^+(B_1) + \nu^+(B_2).$$

Hence on taking the supremum over A we get

$$\nu^+(B_1) + \nu^+(B_2) = \nu^+(B_1 \cup B_2).$$

Suppose now that

$$B = \bigsqcup_{n=1}^{\infty} B_n.$$

Then $\sum_{n=1}^{N} \nu^+(B_n) = \nu^+(B_1 \cup \cdots \cup B_N) \leqslant \nu^+(B)$ for all $N \geqslant 1$ by the above finite additivity and monotonicity, and so

$$\sum_{n=1}^{\infty} \nu^+(B_n) \leqslant \nu^+(B).$$

To see the converse, let $A \subseteq B$ be measurable with finite measure, and define $A_n = A \cap B_n$. Clearly $\mathbb{1}_A = \sum_{n=1}^{\infty} \mathbb{1}_{A_n}$ pointwise, but by the dominated convergence theorem this convergence also holds with respect to $\|\cdot\|_p$ (since $p < \infty$). Since ℓ is continuous, it follows that

$$\ell(\mathbb{1}_A) = \sum_{n=1}^{\infty} \ell(\mathbb{1}_{A_n}) \leqslant \sum_{n=1}^{\infty} \nu^+(B_n).$$

As this holds for all $A \subseteq B$ with finite measure we get

$$\nu^+ \left(\bigsqcup_{n=1}^{\infty} B_n \right) = \sum_{n=1}^{\infty} \nu^+(B_n),$$

and so ν^+ is a measure on X.

Finally, if $B \subseteq X$ has finite μ-measure, then

$$\ell(\mathbb{1}_A) \leqslant \|\ell\|_{\mathrm{op}}\|\mathbb{1}_A\|_p = \|\ell\|_{\mathrm{op}}\mu(A)^{1/p} \leqslant \|\ell\|_{\mathrm{op}}\mu(B)^{1/p}$$

for all $A \subseteq B$, which shows that $\nu^+(B) \leqslant \|\ell\|\mu(B)^{1/p}$. It follows that $\nu^+ \ll \mu$ is absolutely continuous and is σ-finite since μ is assumed to be σ-finite. By the Radon–Nikodym theorem (see Proposition 3.29) we have $\mathrm{d}\nu^+ = g^+ \, \mathrm{d}\mu$ for some measurable function $g^+ \geqslant 0$.

We claim that $g^+ \in L^q_\mu(X)$ is the positive part of the element $g \in L^q_\mu(X)$ we are looking for. For this we first have to check that $g^+ \in L^q_\mu(X)$, which we do by estimating the L^q_μ norms of

$$g_n^+ = \min\{n, g^+\mathbb{1}_{X_n}\},$$

where (X_n), with $X_n \subseteq X_{n+1}$ for all $n \geqslant 1$, is a sequence of sets with finite measure and $X = \bigcup_{n=1}^\infty X_n$. Notice that $g_n^+ \nearrow g^+$ as $n \to \infty$. Now let $h \geqslant 0$ be a simple function of the form $h = \sum_{j=1}^m \beta_j \mathbb{1}_{B_j}$, where $\beta_j \geqslant 0$ for all j and with the sets B_j measurable and pairwise disjoint. Then

$$\int hg_n^+ \, \mathrm{d}\mu \leqslant \int hg^+ \, \mathrm{d}\mu = \sum_{j=1}^m \beta_j \sup\left\{\ell(\mathbb{1}_{A_j}) \mid A_j \subseteq B_j\right\}$$

$$= \sup\left\{\ell\left(\sum_{j=1}^m \beta_j \mathbb{1}_{A_j}\right) \mid A_j \subseteq B_j\right\},$$

but the expressions inside the last supremum we may estimate by

$$\left|\ell\left(\sum_{j=1}^m \beta_j \mathbb{1}_{A_j}\right)\right| \leqslant \|\ell\|_{\mathrm{op}}\left\|\sum_{j=1}^m \beta_j \mathbb{1}_{A_j}\right\|_p \leqslant \|\ell\|_{\mathrm{op}}\|h\|_p$$

and so

$$\int hg_n^+ \, \mathrm{d}\mu \leqslant \|\ell\|_{\mathrm{op}}\|h\|_p.$$

Using monotone convergence this estimate extends to all positive $h \in L^p_\mu(X)$. Applying the argument (for $g_n^+ \in L^q_\mu(X)$) from the beginning of the proof this shows that $\|g_n^+\|_q \leqslant \|\ell\|_{\mathrm{op}}$ and letting $n \to \infty$ also shows that $\|g^+\|_q \leqslant \|\ell\|_{\mathrm{op}}$ by monotone convergence.

Now define

$$\ell^- = \phi(g^+) - \ell \in \left(L^p_\mu(X)\right)^*$$

where $\phi(g^+)$ is the functional determined by $g^+ \in L^q_\mu(X)$. Notice that for all $B \subseteq X$ measurable with $\mu(B) < \infty$ we have

$$\ell^-(\mathbb{1}_B) = \int \mathbb{1}_B g^+ \, \mathrm{d}\mu - \ell(\mathbb{1}_B)$$

$$= \sup\{\ell(\mathbb{1}_A - \mathbb{1}_B) \mid A \text{ measurable}, A \subseteq B\}$$

$$= \sup\{-\ell(\mathbb{1}_C) \mid C \text{ measurable}, C \subseteq B\}, \tag{7.10}$$

where we used the identity $\mathbb{1}_A - \mathbb{1}_B = -\mathbb{1}_C$ for $C = B \smallsetminus A$ and $A \subseteq B$. Using $-\ell$ in the construction above we also obtain a measure $\mathrm{d}\nu^- = g^- \, \mathrm{d}\mu$ for some $g^- \in L^q_\mu(X)$. For a measurable set $B \subseteq X$ with $\mu(B) < \infty$ we now obtain from (7.10) that

$$\int \mathbb{1}_B g^+ \, \mathrm{d}\mu - \ell(\mathbb{1}_B) = \nu^-(B) = \int \mathbb{1}_B g^- \, \mathrm{d}\mu$$

or equivalently

$$\ell(\mathbb{1}_B) = \int \mathbb{1}_B(g^+ - g^-) \, \mathrm{d}\mu = \int \mathbb{1}_B g \, \mathrm{d}\mu$$

for $g = g^+ - g^- \in L^q_\mu(X)$. By the density of simple functions in $L^p_\mu(X)$ we conclude that $\ell = \phi(g)$, as required. $\qquad\square$

7.3.3 Riesz–Thorin Interpolation

[†]The Riesz–Thorin interpolation theorem (also called the Riesz–Thorin convexity theorem) bounds the norms of linear maps between L^p spaces. This can be useful because certain L^p spaces have special properties making it easier to understand properties of operators on them — this particularly applies to the cases $p = 1, 2$, and ∞.

Proposition 7.37. *Let $(X, \mathcal{B}, \mu)$ be a measure space, and assume that*

$$1 \leqslant q_0 < q < q_1 \leqslant \infty.$$

Then

$$L^{q_0}_\mu(X) \cap L^{q_1}_\mu(X) \subseteq L^q_\mu(X)$$

and $\|f\|_q \leqslant \|f\|_{q_0}^{1-t}\|f\|_{q_1}^t$ for all $f \in L^{q_0}_\mu(X) \cap L^{q_1}_\mu(X)$, where $t \in (0, 1)$ is determined by the relation $\frac{1}{q} = \frac{1-t}{q_0} + \frac{t}{q_1}$.

PROOF. We first assume that $q_1 = \infty$, and note that $|f|^q \leqslant |f|^{q_0}\|f\|_\infty^{q-q_0}$ almost everywhere for $f \in L^{q_0}_\mu(X) \cap L^\infty_\mu(X)$. Integrating over X and taking the qth root gives $\|f\|_q \leqslant \|f\|_{q_0}^{q_0/q}\|f\|_\infty^{(q-q_0)/q}$. Moreover, since $q_1 = \infty$ we have $\frac{1}{q} = \frac{1-t}{q_0}$, giving the proposition in this case.

[†] The results of this subsection conclude our discussion of L^p-spaces but will not be needed in the remainder of the book.

Now suppose that $q_1 < \infty$. In this case the numbers $\frac{q_0}{(1-t)q}$ and $\frac{q_1}{tq}$ are Hölder conjugate by definition of t. Let $f \in L_\mu^{q_0}(X) \cap L_\mu^{q_1}(X)$. Applying Hölder's inequality (Theorem B.15) gives

$$\int |f|^q \, d\mu = \int |f|^{(1-t)q} |f|^{tq} \, d\mu$$

$$\leqslant \left\| |f|^{(1-t)q} \right\|_{q_0/(1-t)q} \left\| |f|^{tq} \right\|_{q_1/tq} = \|f\|_{q_0}^{(1-t)q} \|f\|_{q_1}^{tq}.$$

Taking the qth root gives the proposition.   □

We now consider a linear map that is defined not just on one L^p space but on several, taking values in possibly different L^q spaces.

Theorem 7.38 (Riesz–Thorin interpolation). *Let $(X, \mathcal{B}, \mu)$ and $(Y, \mathcal{C}, \nu)$ be two σ-finite measure spaces and let $p_0, p_1, q_0, q_1 \in [1, \infty]$. Let*

$$T : L_\mu^{p_0}(X) \cap L_\mu^{p_1}(X) \longrightarrow L_\nu^{q_0}(Y) \cap L_\nu^{q_1}(Y)$$

be a linear map such that $\|Tf\|_{q_0} \leqslant M_0 \|f\|_{p_0}$ and $\|Tf\|_{q_1} \leqslant M_1 \|f\|_{p_1}$ for all $f \in L_\mu^{p_0}(X) \cap L_\mu^{p_1}(X)$ and some constants $M_0, M_1 \geqslant 0$. Then T has a linear extension to a linear space D of (equivalence classes of) functions on X into the space $L_\nu^{q_0}(Y) + L_\nu^{q_1}(Y)$ with the following properties. If we define p_t and q_t for any $t \in (0, 1)$ by $\frac{1}{p_t} = \frac{1-t}{p_0} + \frac{t}{p_1}$ and $\frac{1}{q_t} = \frac{1-t}{q_0} + \frac{t}{q_1}$ then we have $L_\mu^{p_t}(X) \subseteq D$ and $\|Tf\|_{q_t} \leqslant M_0^{1-t} M_1^t \|f\|_{p_t}$ for all $f \in L_\mu^{p_t}(X)$. The conclusion also holds for $t = 0$ if $p_0 < \infty$ and for $t = 1$ if $p_1 < \infty$.

Example 7.39. An interesting example to keep in mind is an application of the theorem known as the Hausdorff–Young inequality. For this, consider the map from Theorem 3.47 sending a function $f \in L^2(G)$ for a compact abelian group G to the map $a^{(f)}$ on the set $\widehat{G}$ of characters defined by

$$a^{(f)}(\chi) = a_\chi^{(f)} = \langle f, \chi \rangle$$

for every $\chi \in \widehat{G}$. For $f \in L^2(G)$ we have $a^{(f)} \in \ell^2(\widehat{G})$ and $\|a^{(f)}\|_2 = \|f\|_2$; for $f \in L^1(G)$ we have $a^{(f)} \in \ell^\infty(\widehat{G})$ with $\|a^{(f)}\|_\infty \leqslant \|f\|_1$ — or formally we have $p_0 = 2 = q_0$, $p_1 = 1$, $q_1 = \infty$, and $M_0 = M_1 = 1$. The above interpolation theorem now implies that the map is also defined for functions $f \in L^p(G)$ with $p \in [1, 2]$, taking values in a certain $\ell^q(\widehat{G})$. A short calculation reveals that in this case $q \in [2, \infty]$ is the Hölder conjugate of $p \in [1, 2]$.

PROOF OF THEOREM 7.38 IN THE CASE $p_0 = p_1$. Set $D = L_\mu^{p_0}(X)$, and notice that for $f \in L_\mu^{p_0}(X)$ we have $\|Tf\|_{q_0} \leqslant M_0 \|f\|_{p_0}$ and $\|Tf\|_{q_1} \leqslant M_1 \|f\|_{p_0}$. Applying Proposition 7.37 gives the theorem in this case.   □

For the general case we will need the following result from complex analyis.

Lemma 7.40 (Hadamard's three-lines theorem). *Let ϕ be a continuous bounded function on the strip $S = \{z \in \mathbb{C} \mid 0 \leqslant \Re(z) \leqslant 1\}$ that is holomorphic on S^o. If $|\phi(z)| \leqslant M_0$ when $\Re(z) = 0$ and $|\phi(z)| \leqslant M_1$ when $\Re(z) = 1$, then $|\phi(t + iu)| \leqslant M_0^{1-t} M_1^t$ for all $u \in \mathbb{R}$ and $t \in [0,1]$.*

PROOF. Assume first that $M_0, M_1 > 0$. For $\varepsilon > 0$, define

$$\phi_\varepsilon(z) = \phi(z) M_0^{z-1} M_1^{-z} e^{\varepsilon z^2 - \varepsilon}$$

and note that

$$\lim_{\varepsilon \to 0} \phi_\varepsilon(z) = \phi(z) M_0^{z-1} M_1^{-z} \tag{7.11}$$

for all $z \in S$. Moreover, ϕ_ε is continuous on S and holomorphic on S^o. Setting $z = t + iu \in S$ we also have $z^2 = t^2 - u^2 + 2itu$ and so

$$|\phi_\varepsilon(z)| = |\phi(z)| M_0^{t-1} M_1^{-t} e^{\varepsilon t^2 - \varepsilon u^2 - \varepsilon} \leqslant |\phi(z)| M_0^{t-1} M_1^{-t} e^{-\varepsilon u^2},$$

since $t \in [0,1]$. For $t = \Re(z) = 0$ or $t = \Re(z) = 1$ this gives $|\phi_\varepsilon(z)| \leqslant 1$ by assumption on ϕ. Moreover,

$$\lim_{|u| \to \infty} |\phi_\varepsilon(z)| = 0,$$

where $u = \Im(z)$. Applying the maximum modulus theorem on

$$\{z \in \mathbb{C} \mid 0 \leqslant \Re(z) \leqslant 1, -N_\varepsilon \leqslant \Im(z) \leqslant N_\varepsilon\}$$

for sufficiently large N_ε we see that $|\phi_\varepsilon(z)| \leqslant 1$ for all $z \in S$. By (7.11) it follows that

$$|\phi(z)| \leqslant |M_0^{1-z} M_1^z| = M_0^{1-t} M_1^t$$

for $z \in S$ with $t = \Re(z)$. If $M_0 = 0$ (or $M_1 = 0$) we may apply the argument above with M_0 (or M_1) replaced by any $\delta > 0$ and obtain the lemma by letting $\delta \to 0$. $\qquad\square$

PROOF OF THEOREM 7.38 IN THE CASE $p_0 \neq p_1$. Our first goal is the inequality

$$\|Tf\|_{q_t} \leqslant M_0^{1-t} M_1^t \|f\|_{p_t} \tag{7.12}$$

for a fixed $t \in (0,1)$ and all $f \in \Sigma_X$. Here Σ_X denotes the space of simple integrable functions on X (and Σ_Y is defined similarly). Then $\Sigma_X \subseteq L^p_\mu(X)$ for all $p \in [1,\infty]$ and in particular T is defined on Σ_X and satisfies

$$T(\Sigma_X) \subseteq L^{q_0}_\nu(Y) \cap L^{q_1}_\nu(Y) \subseteq L^{q_t}_\nu(Y)$$

by the assumption in the theorem and Proposition 7.36. Assume for the moment that $q_t \in (1, \infty]$. Then the Hölder conjugate q'_t of q_t belongs to $[1, \infty)$ and Σ_Y is dense in $L^{q'_t}_\nu(Y)$. Fix some $f \in \Sigma_X$ and assume that

$$\left| \int (Tf)g \, d\nu \right| \leqslant M_0^{1-t} M_1^t \|f\|_{p_t} \|g\|_{q_t'} \tag{7.13}$$

for all $g \in \Sigma_Y$. Then the above and Propositions 7.34 and 7.36 imply (7.12). The case $q_t = 1$ with $q_t' = \infty$ is only slightly different. Assume again (7.13) and fix some measurable set $B \subseteq Y$ with $\nu(B) < \infty$. Then

$$\{g \in \Sigma_Y \mid g(y) = 0 \text{ for } y \in Y \smallsetminus B\}$$

is dense in $L_\nu^\infty(B)$ and as before (see also Corollary 7.9) we obtain

$$\|(Tf)|_B\|_1 \leqslant M_0^{1-t} M_1^t$$

independent of B. Using the fact that ν is σ-finite, this again implies (7.12). For the proof of (7.13) it suffices to fix some $t \in (0,1)$, some $f \in \Sigma_X$ with $\|f\|_{p_t} = 1$, and some $g \in \Sigma_Y$ with $\|g\|_{q_t'} = 1$. By definition, we may express f and g as finite sums

$$f = \sum_{j=1}^m c_j \mathbb{1}_{E_j},$$

with $c_j \in \mathbb{C}$, $E_j \in \mathcal{B}$, $\mu(E_j) < \infty$ for $1 \leqslant j \leqslant m$, and

$$g = \sum_{k=1}^n d_k \mathbb{1}_{F_k},$$

with $d_k \in \mathbb{C}$, $F_k \in \mathcal{C}$, $\nu(F_k) < \infty$ for $1 \leqslant k \leqslant n$. We may also assume that the sets E_j are pairwise disjoint, and similarly for the sets F_k.

A HOLOMORPHIC FUNCTION ON THE STRIP. Next define

$$\alpha(z) = (1-z)p_0^{-1} + zp_1^{-1},$$
$$\beta(z) = (1-z)q_0^{-1} + zq_1^{-1}$$

for $z \in \mathbb{C}$ and observe that $\alpha(t) = p_t^{-1}$ and $\beta(t) = q_t^{-1}$ for the fixed $t \in (0,1)$. Notice that $\alpha(t) > 0$ (since otherwise $p_0 = p_1 = \infty$, a case already considered) and that $\beta(t) = 1$ implies that $q_0 = q_1 = q_t = 1$, $q_0' = q_1' = q_t' = \infty$, and also $\beta(z) = 1$ for all $z \in \mathbb{C}$. For any $z \in \mathbb{C}$ we now define

$$f_z = \sum_{j=1}^m |c_j|^{\alpha(z)/\alpha(t)} \arg(c_j) \mathbb{1}_{E_j},$$

$$g_z = \begin{cases} \displaystyle\sum_{k=1}^n |d_k|^{(1-\beta(z))/(1-\beta(t))} \arg(d_k) \mathbb{1}_{F_k} & \text{if } \beta(t) < 1; \\ g & \text{if } \beta(t) = 1. \end{cases}$$

Notice that $f_z \in \Sigma_X$, $g_z \in \Sigma_Y$ for all $z \in \mathbb{C}$, $f_t = f$, and $g_t = g$. Also define

$$\phi(z) = \int (Tf_z)g_z \, d\nu = \sum_{j,k} A_{jk} |c_j|^{\alpha(z)/\alpha(t)} |d_k|^{(1-\beta(z))/(1-\beta(t))},$$

where the coefficient

$$A_{jk} = \arg(c_j d_k) \int (T\mathbb{1}_{E_j}) \mathbb{1}_{F_k} \, d\nu \in \mathbb{C}$$

is independent of z for all j and k. It follows that ϕ is just a finite linear combination of exponential functions of the form $z \mapsto a^z$ for some $a > 0$, so that ϕ is entire and bounded on the strip

$$S = \{z \in \mathbb{C} \mid 0 \leqslant \Re(z) \leqslant 1\}.$$

Since $f_t = f$ and $g_t = g$ we also have

$$\phi(t) = \int (Tf)g \, d\nu,$$

which is the quantity that we wish to estimate. The desired estimate will follow from Lemma 7.40 once we establish its remaining assumptions.

BOUNDARY ESTIMATE: Consider therefore $z = iu$ with $\Re(z) = 0$ and notice that $\Re(\alpha(iu)) = p_0^{-1}$ and $\Re(1 - \beta(iu)) = 1 - q_0^{-1} = (q_0')^{-1}$. Since the sets $E_1, \ldots, E_m$, respectively $F_1, \ldots, F_n$, are disjoint, this gives

$$|f_{iu}| = |f|^{\Re(\alpha(iu)/\alpha(t))} = |f|^{p_t/p_0}$$

and

$$|g_{iu}| = \begin{cases} |g|^{\Re((1-\beta(iu))/(1-\beta(t)))} = |g|^{q_t'/q_0'} & \text{if } \beta(t) < 1; \\ |g| & \text{if } \beta(t) = 1. \end{cases}$$

Using the assumption on T this gives

$$|\phi(iu)| \leqslant \|Tf_{iu}\|_{q_0} \|g_{iu}\|_{q_0'}$$

$$\leqslant M_0 \|f_{iu}\|_{p_0} \|g_{iu}\|_{q_0'} = \begin{cases} M_0 \|f\|_{p_t}^{p_t/p_0} \|g\|_{q_t'}^{q_t'/q_0'} = M_0 & \text{if } \beta(t) < 1; \\ M_0 \|f\|_{p_t}^{p_t/p_0} \|g\|_\infty = M_0 & \text{if } \beta(t) = 1 \end{cases}$$

since $\|f\|_{p_t} = \|g\|_{q_t'} = 1$. The case of $z = 1 + iu$ with $\Re(z) = 1$ works similarly after noticing that $\Re(\alpha(1 + iu)) = p_1^{-1}$ and $\Re(\beta(1 + iu)) = q_1^{-1}$, which leads to the bound $|\phi(1 + iu)| \leqslant M_1$.

THE ESTIMATE: By Lemma 7.40 we obtain

$$|\phi(t)| = \left| \int (Tf)g \, d\nu \right| \leqslant M_0^{1-t} M_1^t.$$

Since $g \in \Sigma_Y$ was arbitrary subject to $\|g\|_{q_t'} = 1$, this implies (as explained after (7.13)) that

$$\|Tf\|_{q_t} \leqslant M_0^{1-t} M_1^t$$

for any $f \in \Sigma_X$ with $\|f\|_{p_t} = 1$. By the density of Σ_X in $L_\mu^{p_t}(X)$ and Proposition 2.59 it follows that T has a unique extension T_{p_t} to all of $L_\mu^{p_t}(X)$ with values in $L_\nu^{q_t}(Y)$ such that $\|T_{p_t} f\|_{q_t} \leqslant M_0^{1-t} M_1^t \|f\|_{p_t}$ for $f \in L_\mu^{p_t}(X)$.

ONE LINEAR MAP: It remains to find one common domain D that contains $L_\mu^{p_t}(X)$ for all t in $(0,1)$ and a linear map on D that extends the extension above. For this we let $D_0 = L_\mu^{p_0}(X) \cap L_\mu^{p_1}(X)$ and

$$D = \overline{D_0}^{p_0} + \overline{D_0}^{p_1} \subseteq L_\mu^{p_0}(X) + L_\mu^{p_1}(X),$$

where $\overline{\cdot}^{p_t}$ denotes the closure with respect to $\|\cdot\|_{p_t}$ for $t \in [0,1]$, and we have

$$D = L_\mu^{p_0}(X) + L_\mu^{p_1}(X)$$

unless $\infty \in \{p_0, p_1\}$. Applying the assumption of the theorem and Proposition 2.59 we find extensions of T (initially only defined on D_0); T_{p_0} extending T to $\overline{D_0}^{p_0}$ and T_{p_1} extending T to $\overline{D_0}^{p_1}$. We now define

$$T(f_0 + f_1) = T_{p_0}(f_0) + T_{p_1}(f_1)$$

for $f_0 \in \overline{D_0}^{p_0}$ and $f_1 \in \overline{D_0}^{p_1}$. We claim that this gives a well-defined extension of T to D with values in $L_\nu^{q_0}(Y) + L_\nu^{q_1}(Y)$. Indeed, if $f_0 + f_1 = f_0' + f_1' \in D$ with $f_0, f_0' \in \overline{D_0}^{p_0}$ and $f_1, f_1' \in \overline{D_0}^{p_1}$ then $f_0 - f_0' = f_1' - f_1 \in D_0$ and

$$T_{p_0}(f_0) - T_{p_0}(f_0') = T(f_0 - f_0') = T(f_1' - f_1) = T_{p_1}(f_1') - T_{p_1}(f_1).$$

Rearranging the terms, the claim follows.

THE MAP ON $L_\mu^p(X)$: Now let $f \in L_\mu^{p_t}(X)$ and assume without loss of generality that $p_0 < p_1$. We define the set $B = \{x \in X \mid |f(x)| \leqslant 1\}$ and use it to split $f = f^s + f^l$ into a component f^s taking on small values and a component f^l taking on large values, where

$$f^s = f \mathbb{1}_B \in L_\mu^{p_t}(X) \cap L_\mu^{p_1}(X)$$

and

$$f^l = f \mathbb{1}_{X \setminus B} \in L_\mu^{p_t}(X) \cap L_\mu^{p_0}(X).$$

If we now choose a sequence (f_n) in Σ_X with $|f_n| \leqslant |f|$ for all $n \geqslant 1$ and with $f_n \to f$ pointwise as $n \to \infty$ then $f_n^s = f_n \mathbb{1}_B \to f^s$ in $L_\mu^{p_t}(X)$ and in $L_\mu^{p_1}(X)$ by dominated convergence if $p_1 < \infty$. If however $p_1 = \infty$ then we can choose the sequence (f_n) of simple functions to also have $f_n^s \to f^s$ with respect to $\|\cdot\|_\infty$ as $n \to \infty$. Similarly, $f_n^l = f_n \mathbb{1}_{X \setminus B} \to f^l$ in $L_\mu^{p_t}(X)$ and in $L_\mu^{p_0}(X)$. Therefore, $T(f_n^s) \to T_{p_t}(f^s)$ in $L_\nu^{q_t}(Y)$ and $T(f_n^s) \to T_{p_1}(f^s)$ in $L_\nu^{q_1}(Y)$ as $n \to \infty$. Choosing a subsequence if necessary, the convergence also holds pointwise almost everywhere, which gives $T_{p_t}(f^s) = T_{p_1}(f^s)$. The same argument gives $T_{p_t}(f^l) = T_{p_0}(f^l)$, and $T_{p_t}(f) = T(f)$ follows. □

Exercise 7.41. Show that $t \mapsto \log \|T_{p_t}\|$ is convex for $t \in (0, 1)$, where

$$T_{p_t} : L_\mu^{p_t}(X) \to L_\nu^{q_t}(Y)$$

is as in the proof above.

Exercise 7.42. Let G be a locally compact, σ-compact, metrizable, abelian group. Fix some $p \in [1, \infty)$ with Hölder conjugate q and some $F \in L^p(G)$. Show (or recall) that

$$f * F(x) = \int f(t)F(x - t) \, \mathrm{d}m_G(t)$$

is well-defined almost everywhere for $f \in L^1(G)$ with $\|f * F\|_p \leqslant \|f\|_1 \|F\|_p$ and also for $f \in L^q(G)$ with $\|f * F\|_\infty \leqslant \|f\|_q \|F\|_p$. Apply the Riesz–Thorin interpolation theorem to obtain a conclusion for all $f \in L^r(G)$ with $r \in [1, q]$.

7.4 Riesz Representation: The Dual of $C(X)$

The next result is useful in many ways. It will allow us to completely describe $C(X)^*$ in Section 7.4.5, but it is more often used directly in the form presented here.

Definition 7.43. Let $\mathcal{F}(X)$ be a space of real- or complex-valued functions on some space X. Then a *positive linear functional* on $\mathcal{F}(X)$ is a linear functional Λ defined on $\mathcal{F}(X)$ with the property that any real-valued function $f \in \mathcal{F}(X)$ with $f \geqslant 0$ is mapped to $\Lambda(f) \geqslant 0$.

Theorem 7.44 (Riesz representation). *Let X be a locally compact, σ-compact metric space, and suppose that Λ is a positive linear functional defined on $C_c(X)$. Then there exists a uniquely determined locally finite (positive) Borel measure μ such that*

$$\Lambda(f) = \int_X f \, \mathrm{d}\mu$$

for all $f \in C_c(X)$.

Recall that a measure μ is *locally finite* if every point has a neighbourhood of finite measure, or equivalently if every compact subset of X has finite measure. A locally finite Borel measure is also often called a *Radon measure*. As a real-valued function in $C_c(X)$ is the difference of two non-negative functions in $C_c(X)$, a positive linear functional maps a real-valued function to a real number. Hence we may and will restrict in the proof below to the real case as this case implies the complex case as well.

Exercise 7.45. Let X be a σ-compact, locally compact metric space. Let μ be a locally finite measure on X. Show that μ is *regular*, meaning that

$$\mu(B) = \sup\{\mu(K) \mid K \subseteq B \text{ is compact}\} = \inf\{\mu(O) \mid O \supseteq B \text{ is open}\}$$

for any Borel set $B \subseteq X$.

We will prove Theorem 7.44 in several steps, first showing the claimed uniqueness of the measure, then showing existence in the totally disconnected compact case, then the compact case and finally the general case.

7.4.1 Uniqueness

We will prove uniqueness without assuming the measure to be locally finite, but this is automatic for a measure representing Λ.

PROOF OF UNIQUENESS IN THEOREM 7.44. Let Λ be a positive linear functional on $C_c(X)$ and suppose μ and ν are two positive measures with

$$\int_X f \, d\mu = \Lambda(f) = \int_X f \, d\nu$$

for all $f \in C_c(X)$. This implies that μ and ν are locally finite, since for every compact set $K \subseteq X$ there exists some function $f \in C_c(X)$ with $f \geqslant \mathbb{1}_K$ by Urysohn's lemma (Lemma A.27), which shows that $\mu(K), \nu(K) \leqslant \Lambda(f) < \infty$.

Define $m = \mu + \nu$, so that $\mu \ll m$ and $\nu \ll m$. By Proposition 3.29 there exist Radon–Nikodym derivatives $f_\mu, f_\nu \geqslant 0$ with

$$d\mu = f_\mu \, dm, \quad d\nu = f_\nu \, dm,$$

and $f_\mu + f_\nu = 1$ m-almost everywhere. Therefore

$$\int_X g f_\mu \, dm = \int_X g \, d\mu = \Lambda(g) = \int_X g \, d\nu = \int_X g f_\nu \, dm$$

for all $g \in C_c(X)$. Since $C_c(X) \subseteq L_m^1(X)$ is dense by Proposition 2.51, the functions f_μ and f_ν in $L_m^\infty(X)$ determine the same functional on $L_m^1(X)$. By Proposition 7.34 this implies $f_\mu = f_\nu$ almost everywhere with respect to m. However, this shows that $\mu = \nu$. $\square$

7.4.2 Totally Disconnected Compact Spaces

As our first step towards the existence of the measure representing a positive linear functional we consider the following kind of spaces, where the proof is quite simple.

Definition 7.46. Let X be a topological space. A set $C \subseteq X$ is called *clopen* if it is both open and closed in X. The space X is called *totally disconnected* if every open set in X is a union of clopen sets, so the topology has a basis consisting of clopen sets.

Example 7.47. Before we give the proof, let us give examples of compact metric totally disconnected spaces.

(1) $X = \{1, \ldots, a\}^{\mathbb{N}}$ is a compact metrizable space with respect to the product topology using the discrete topology on $\{1, \ldots, a\}$. It is also totally disconnected, since for any finite collection $F_1, \ldots, F_n \subseteq \{1, \ldots, a\}$ the set $\pi_1^{-1}(F_1) \cap \cdots \cap \pi_n^{-1}(F_n)$ is both open and closed (here π_j is the projection $X \to \{1, \ldots, a\}$ onto the jth coordinate).

(2) More generally, we can also take the product $X = \prod_{n=1}^{\infty} A_n$, where each A_n is a finite set equipped with the discrete topology. Note that any closed subset $Y \subseteq X$ is again totally disconnected and compact.

One way to define a metric on X as in (1) or (2) and hence also on Y as in (2) is to set

$$\mathsf{d}(x, y) = \begin{cases} 0 & \text{if } x = y, \text{ and} \\ \frac{1}{n} & \text{if } x_1 = y_1, \ldots, x_{n-1} = y_{n-1}, \text{ but } x_n \neq y_n \end{cases}$$

for all points x, y (see also Lemma A.17). In this metric the open ball of radius $\frac{1}{n}$ and centre y is given by

$$B_{\frac{1}{n}}(y) = \{x \mid x_1 = y_1, \ldots, x_n = y_n\} = \pi_1^{-1}(\{y_1\}) \cap \cdots \cap \pi_n^{-1}(\{y_n\}).$$

Also note $B_r(y) = B_{\frac{1}{n}}(y)$ if $\frac{1}{n+1} < r \leqslant \frac{1}{n}$. It follows that there are only countably many balls and that these are all clopen. As every open set $O \subseteq X$ is a union of balls it follows that every open set is actually a countable union of clopen sets. In particular, the clopen sets generate the Borel σ-algebra.

Lemma 7.48. *Let X be a totally disconnected compact metric space. Then the Borel σ-algebra is generated by the clopen sets.*

As we have already obtained a proof of the lemma in the setting of Example 7.47 and since these cases will be sufficient for the proof of Theorem 7.44 we leave the proof as an exercise.

Exercise 7.49. (a) Prove Lemma 7.48 in general by showing that in a compact totally disconnected metric space, there are only countably many clopen sets.

(b) Show that every compact totally disconnected metric space is homeomorphic to a metric space Y as in Example 7.47(2).

PROOF OF THEOREM 7.44 FOR TOTALLY DISCONNECTED COMPACT METRIC SPACES AS IN EXAMPLE 7.47. Let X be a totally disconnected compact metric space so that the algebra $\mathcal{C} = \{C \subseteq X \mid C \text{ is open and closed}\}$ generates the Borel σ-algebra of X. Let $\Lambda : C(X) \to \mathbb{R}$ be a positive linear functional. Using Λ we can already define a content $\mu_{\mathcal{C}}$ on the algebra $\mathcal{C}$. In fact, for $C \in \mathcal{C}$ we define

$$\mu_{\mathcal{C}}(C) = \Lambda(\mathbb{1}_C).$$

This is possible since $\mathbb{1}_C \in C(X)$ as C is both open and closed. It follows that

- $\mu_{\mathcal{C}}(C) \geqslant 0$ for $C \in \mathcal{C}$ (Positivity);
- $\mu_{\mathcal{C}}(C_1 \sqcup C_2) = \mu_{\mathcal{C}}(C_1) + \mu_{\mathcal{C}}(C_2)$ for disjoint $C_1, C_2 \in \mathcal{C}$ (Finite additivity).

By Caratheodory's extension theorem (see Theorem B.4) we can extend $\mu_{\mathcal{C}}$ to a measure on the Borel σ-algebra $\mathcal{B}$ of X if

$$\mu_{\mathcal{C}} \left(\bigsqcup_{n=1}^{\infty} C_n \right) = \sum_{n=1}^{\infty} \mu_{\mathcal{C}}(C_n)$$

for any disjoint sets $C_1, C_2, \dots$ in $\mathcal{C}$ with $\bigsqcup_{n=1}^{\infty} C_n \in \mathcal{C}$. In the totally disconnected compact setting this is quite easy to check. Suppose that $C_n \in \mathcal{C}$ are disjoint for $n \geqslant 1$ and $C = \bigsqcup_{n=1}^{\infty} C_n \in \mathcal{C}$. Then C is compact since $C \in \mathcal{C}$ gives that it is a closed subset of X and X is compact. On the other hand the sets $C_n \in \mathcal{C}$ are open, so $C = \bigsqcup_{n=1}^{\infty} C_n \in \mathcal{C}$ is an open cover of a compact set. It follows that $C = \bigsqcup_{n=1}^{N} C_n$ for some $N \geqslant 1$, and hence $C_n = \varnothing$ for $n > N$. Hence finite additivity gives

$$\mu_{\mathcal{C}}(C) = \sum_{n=1}^{N} \mu_{\mathcal{C}}(C_n) = \sum_{n=1}^{\infty} \mu_{\mathcal{C}}(C_n),$$

as required.

Therefore, $\mu_{\mathcal{C}}$ can be extended to a measure μ, defined on the Borel σ-algebra $\mathcal{B}$ of X. By construction

$$\int_X \mathbb{1}_C \, d\mu = \Lambda(\mathbb{1}_C)$$

for $C \in \mathcal{C}$. We wish to extend this formula to all continuous functions. For this we note that this formula extends trivially to all linear combinations of characteristic functions of clopen sets. Now note that the linear hull $\mathcal{A}$ of the characteristic functions of clopen sets is an algebra, contains the constant function, and separates points. Hence it is dense in $C(X)$ by the Stone–Weierstrass theorem (Theorem 2.40; this is also easy to see directly but the given argument is shorter). It follows that for every $f \in C(X)$ and $\varepsilon > 0$ there exists some $g \in \mathcal{A}$ (already satisfying $\Lambda(g) = \int_X g \, d\mu$) such that

$$g - \varepsilon < f < g + \varepsilon.$$

Hence we may apply Λ and $\int \cdot \, d\mu$ and obtain from the positivity of both these functionals the bounds

$$\Lambda(f), \int f \, d\mu \in [\Lambda(g) - \varepsilon\Lambda(\mathbb{1}), \Lambda(g) + \varepsilon\Lambda(\mathbb{1})]$$

and so

$$\left| \int f \, d\mu - \Lambda(f) \right| \leqslant 2\varepsilon \Lambda(\mathbb{1}).$$

As this holds for all $\varepsilon > 0$ and all $f \in C(X)$, the theorem follows. □

7.4.3 Compact Spaces

We now upgrade the result from Section 7.4.2 to the case of a general compact metric space. For this we are going to use the Hahn–Banach lemma (Lemma 7.1) and the following lemma.

Lemma 7.50 (Symbolic cover). *Let X be a compact metric space. Then there exists a totally disconnected compact metric space Y and a continuous surjective map $\phi : Y \to X$. In fact, we can choose Y as in Example 7.47(2).*

Example 7.51. A few cases of this lemma do not need a proof, and should help explain why one can think of Y as a symbolic cover.

- If $X = [0,1]$ then we may take $Y = \{0,1\}^{\mathbb{N}}$ to be the space of all binary sequences with the map $\phi((a_n)) = \sum_{n=1}^{\infty} a_n 2^{-n}$ sending the binary sequence to the real number with that binary expansion.
- Let $X \subseteq [-M, M]^d$ be a compact subset of $\mathbb{R}^d$. By composing with an affine map, we can assume without loss that $X \subseteq [0,1]^d = X'$. Define

$$Y' = \left(\{0,1\}^{\mathbb{N}} \right)^d$$

and a continuous surjective map just as above

$$\phi' : Y' \longrightarrow X' = [0,1]^d$$

$$\left((a_n^{(1)}), \dots, (a_n^{(d)}) \right) \longmapsto \left(\sum_{n=1}^{\infty} a_n^{(1)} 2^{-n}, \dots, \sum_{n=1}^{\infty} a_n^{(d)} 2^{-n} \right)$$

and finally $Y = \{y' \in Y' \mid \phi'(y') \in X\}$ with $\phi = \phi'|_Y$. Then $Y \subseteq Y'$ is closed and so again is a totally disconnected compact metric space, and $\phi : Y \to X$ is continuous and surjective.

Exercise 7.52. Suppose that X is a compact d-dimensional manifold. Construct Y and ϕ as in Lemma 7.50.

We postpone the proof of the lemma until after we have seen why it is useful for the problem at hand.

PROOF OF THEOREM 7.44 FOR COMPACT METRIC SPACES. Let X be a compact metric space, and let Y and $\phi : Y \to X$ be as in Lemma 7.50. Let $\Lambda : C(X) \to \mathbb{R}$ be a positive linear functional. For $f \in C(X)$ we have

$$\left(\sup_{x\in X} f(x)\right)\mathbb{1}_X - f \geqslant 0,$$

so

$$\Lambda\!\left(\left(\sup_{x\in X} f(x)\right)\mathbb{1}_X - f\right) \geqslant 0$$

by positivity, or equivalently

$$\Lambda(f) \leqslant \Lambda(\mathbb{1}_X)\sup_{x\in X} f(x).$$

Now let $V = \{f\circ\phi \mid f\in C(X)\} \subseteq C(Y)$, where we used the continuity of ϕ, and notice that if $f_1\circ\phi = f_2\circ\phi$ for $f_1, f_2 \in C(X)$ then $f_1 = f_2$ since ϕ is surjective. Thus we may define $\Lambda_V(f\circ\phi) = \Lambda(f)$, which is linear and satisfies

$$\Lambda_V(f\circ\phi) = \Lambda(f) \leqslant \Lambda(\mathbb{1}_X)\sup_{x\in X} f(x) = p(f\circ\phi)$$

for $p : C(Y)\to\mathbb{R}$ defined by

$$p(F) = \Lambda(\mathbb{1}_X)\sup_{y\in Y} F(y).$$

Note that p satisfies $p(F_1 + F_2) \leqslant p(F_1) + p(F_2)$ and $p(\alpha F_1) = \alpha p(F_1)$ for $F_1, F_2 \in C(Y)$ and $\alpha \geqslant 0$. These are precisely the hypotheses for the Hahn–Banach lemma (Lemma 7.1), so we conclude that Λ_V can be extended to a functional $\Lambda_Y : C(Y)\to\mathbb{R}$ which still satisfies

$$\Lambda_Y(F) \leqslant \Lambda(\mathbb{1}_X)\sup_{y\in Y} F(y).$$

If $F \geqslant 0$ then $-F \leqslant 0$ and so

$$\Lambda_Y(-F) \leqslant \Lambda(\mathbb{1}_X)\sup_{y\in Y}(-F(y)) \leqslant 0,$$

or $\Lambda_Y(F) \geqslant 0$. Hence Λ_Y is a positive linear functional on Y. By the totally disconnected compact case in Section 7.4.2 we conclude that there is a measure μ_Y on Y with

$$\Lambda_Y(F) = \int_Y F\,\mathrm{d}\mu_Y$$

for all $F \in C(Y)$. Applying this to $F = f\circ\phi$, we see that

$$\Lambda(f) = \Lambda_Y(f\circ\phi) = \int_Y f\circ\phi\,\mathrm{d}\mu_Y.$$

We now define $\mu = \phi_*\mu_Y$ by the formula $\mu(B) = \mu_Y(\phi^{-1}(B))$ for a Borel set $B \subseteq X$. Note that by this definition we have

$$\int_X \mathbb{1}_B \, d\mu = \mu(B) = \mu_Y(\phi^{-1}(B)) = \int_Y \mathbb{1}_B \circ \phi \, d\mu_Y,$$

which extends by linearity to all simple functions, then by monotone convergence to all positive measurable functions, and then to all integrable functions. In particular,

$$\Lambda(f) = \int_Y f \circ \phi \, d\mu_Y = \int_X f \, d\mu$$

for all $f \in C(X)$, proving the theorem for a compact metric space X. $\square$

We note that the argument above actually proves the following abstract principle. If $\phi : Y \to X$ is a continuous surjective map between two compact spaces, and the Riesz representation theorem holds for Y, then it also holds for X.

It remains to construct the totally disconnected symbolic cover.

PROOF OF LEMMA 7.50. Recall that since X is a compact metric space, it is also totally bounded, so for every $m \geqslant 1$ there exist finitely many points $x_1^{(m)}, \ldots, x_{n(m)}^{(m)} \in X$ with

$$X = \bigcup_{i=1}^{n(m)} B_{1/m}\big(x_i^{(m)}\big). \tag{7.14}$$

We define

$$Z = \prod_{m=1}^{\infty} \{1, \ldots, n(m)\}$$

with the product topology from the discrete topologies on each of the spaces $\{1, \ldots, n(m)\}$. Then Z is a compact metric space (see Sections A.3 and A.4). We will define Y as a closed subset of Z, and will define $\phi : Y \to X$ by

$$\phi(y) = \lim_{m \to \infty} x_{y(m)}^{(m)},$$

where $y(m) \in \{1, \ldots, n(m)\}$ is the mth coordinate of y and $x_{y(m)}^{(m)}$ is the corresponding centre of the $y(m)$-th ball in the cover (7.14). Our definition of Y will ensure that ϕ is well-defined (that is, the limit defining ϕ exists), continuous, and surjective.

THE CLOSED SET Y. Define

$$Y = \Big\{ y \in Z \mid B_{1/1}\big(x_{y(1)}^{(1)}\big) \cap \cdots \cap B_{1/m}\big(x_{y(m)}^{(m)}\big) \neq \varnothing \text{ for all } m \geqslant 1 \Big\}.$$

We will show that Y is closed by proving that its complement $Z \smallsetminus Y$ is open. So suppose that $z \in Z \smallsetminus Y$, so that

$$B_{1/1}\big(x_{z(1)}^{(1)}\big) \cap \cdots \cap B_{1/m}\big(x_{z(m)}^{(m)}\big) = \varnothing$$

for some $m \geqslant 1$. However, this means that all other sequences with the same first m coordinates also lie in $Z \setminus Y$. That is,

$$\pi_1^{-1}(\{z(1)\}) \cap \cdots \cap \pi_m^{-1}(\{z(m)\}) \subseteq Z \setminus Y,$$

and the set on the left is an open neighbourhood of z by definition, so $Z \setminus Y$ is open.

THE LIMIT DEFINING ϕ EXISTS. Let $y \in Y$ and $m > \ell$, then there exists a point

$$x \in B_{1/\ell}\big(x_{y(\ell)}^{(\ell)}\big) \cap B_{1/m}\big(x_{y(m)}^{(m)}\big)$$

and so

$$\mathsf{d}\big(x_{y(\ell)}^{(\ell)}, x_{y(m)}^{(m)}\big) \leqslant \mathsf{d}\big(x_{y(\ell)}^{(\ell)}, x\big) + \mathsf{d}\big(x, x_{y(m)}^{(m)}\big) < \tfrac{1}{\ell} + \tfrac{1}{m} < \tfrac{2}{\ell}. \qquad (7.15)$$

This shows that $\big(x_{y(m)}^{(m)}\big)$ is a Cauchy sequence in X and so has a limit in X.

CONTINUITY OF ϕ. Let $y \in Y$ and fix $\varepsilon > 0$. Choose ℓ with $\tfrac{4}{\ell} < \varepsilon$. Suppose that $z \in Y$ belongs to the neighbourhood $\pi_\ell^{-1}(\{y(\ell)\})$ defined by the ℓth coordinate of y. Letting $m \to \infty$ in (7.15) we see that

$$\mathsf{d}\big(x_{y(\ell)}^{(\ell)}, \phi(y)\big) \leqslant \tfrac{2}{\ell}$$

and similarly

$$\mathsf{d}\big(x_{z(\ell)}^{(\ell)}, \phi(z)\big) \leqslant \tfrac{2}{\ell}.$$

However, by the choice of z we have $y(\ell) = z(\ell)$ and so

$$\mathsf{d}\big(\phi(z), \phi(y)\big) \leqslant \tfrac{4}{\ell} < \varepsilon.$$

This shows the continuity of ϕ.

SURJECTIVITY. Let $x \in X$, and choose for every $m \geqslant 1$ an index $y(m)$ in $\{1, \ldots, n(m)\}$ with

$$x \in B_{1/m}\big(x_{y(m)}^{(m)}\big),$$

which is possible by (7.14). It follows directly from the definitions that $y \in Y$ and that $\phi(y) = x$. $\qquad \square$

7.4.4 Locally Compact σ-Compact Metric Spaces

Knowing Theorem 7.44 for compact metric spaces, we now extend it to σ-compact locally compact metric spaces using suitable patchworking.

PROOF OF THEOREM 7.44. Let X be a locally compact σ-compact metric space, and let $\Lambda : C_c(X) \to \mathbb{R}$ be a positive linear functional. By Lemma A.22 there exists a sequence of compact sets (K_n) with $K_n \subseteq K_{n+1}^o$ for all $n \geqslant 1$ and with $X = \bigcup_{n=1}^{\infty} K_n$.

By Urysohn's lemma (Lemma A.27) there exists a function $f_n \in C_c(X)$ with $\mathbb{1}_{K_n} \leqslant f_n \leqslant 1$ for each $n \geqslant 1$. If $f \in C_c(K_n^o)$ then

$$\left(\sup_{x \in K_n^o} f(x) \right) f_n - f \geqslant 0$$

and hence

$$\Lambda(f) \leqslant \Lambda(f_n) \sup_{x \in K_n^o} f(x).$$

We now consider $C_c(K_n^o)$ as a subspace of the space of continuous functions $C(K_n)$ on K_n. The norm-like function

$$p_n(f) = \Lambda(f_n) \sup_{x \in K_n^o} f(x)$$

for $f \in C(K_n)$ has all the properties needed to apply Lemma 7.1, so $\Lambda|_{C_c(K_n^o)}$ extends to some Λ_n defined on $C(K_n)$ and is again positive (use the argument from Section 7.4.3 to check this), and can be represented by a finite measure $\overline{\mu_n}$ defined on the Borel sets in K_n. Restricting this measure $\overline{\mu_n}$ to K_n^o, we obtain a measure $\mu_n = \overline{\mu_n}|_{K_n^o}$ on K_n^o with

$$\Lambda(f) = \int_{K_n^o} f \, \mathrm{d}\mu_n$$

for all $f \in C_c(K_n^o)$. We claim that these measures can be patched together to define a locally finite measure μ on X with the desired properties. For this, notice that μ_{n+1} is a measure on K_{n+1}^o which satisfies

$$\Lambda(f) = \int_{K_{n+1}^o} f \, \mathrm{d}\mu_{n+1} = \int_{K_n^o} f \, \mathrm{d}\mu_{n+1}$$

for all $f \in C_c(K_n^o) \subseteq C_c(K_{n+1}^o)$. By the uniqueness of the measure in Theorem 7.44 (see Section 7.4.1) this shows that $\mu_{n+1}|_{K_n^o} = \mu_n$ for all $n \geqslant 1$. Using this compatibility property we may define

$$\mu(B) = \lim_{n \to \infty} \mu_n \left(B \cap K_n^o \right) = \mu_1(B \cap K_1^o) + \sum_{n=2}^{\infty} \mu_n(B \cap K_n^o \setminus K_{n-1}^o)$$

for any measurable $B \subseteq X$. Alternatively we may also write $\mu = \sum_{n=1}^{\infty} \mu_n'$, where $\mu_1' = \mu_1$, $X_1 = K_1^o$, $\mu_n' = \mu_n|_{X_n = K_n^o \setminus K_{n-1}^o}$ for $n \geqslant 2$. By Exercise 3.30 this shows that μ is indeed a measure. Note that $\mu|_{K_n^o} = \mu_n$ for $n \geqslant 1$.

By construction $\{K_n^o \mid n \in \mathbb{N}\}$ is an open cover of X. Hence for a given compact subset $K \subseteq X$ there is a finite subcover of K so K is contained in some K_n^o, and we have $\mu(K) \leqslant \mu_n(K_n^o) < \infty$, so μ is locally finite. By the same argument any $f \in C_c(X)$ belongs to some $C_c(K_n^o)$ and hence

$$\Lambda(f) = \int_{K_n^o} f \, \mathrm{d}\mu_n = \int_{K_n^o} f \, \mathrm{d}\mu = \int_X f \, \mathrm{d}\mu$$

as required. $\qquad\qquad\qquad\qquad\qquad\qquad\qquad\qquad\qquad\qquad\qquad$ $\square$

Exercise 7.53. Let X be a σ-compact locally compact metric space, and let Λ be a positive linear functional $C_0(X) \to \mathbb{R}$ (where we do not assume that Λ is bounded). Show that $\Lambda(f) = \int f \, \mathrm{d}\mu$ for all $f \in C_0(X)$ for a finite measure μ on X.

7.4.5 Continuous Linear Functionals on $C_0(X)$

In the remainder of this section we again treat the real and the complex case simultaneously. The following result describes the dual of $C_0(X)$.

Theorem 7.54 (Riesz representation on $C_0(X)$). *Let X be a locally compact σ-compact metric space, and let $\Lambda \in (C_0(X))^*$ be a continuous linear functional on the space $C_0(X)$ of continuous functions on X that vanish at infinity. Then there exists a uniquely determined signed measure μ representing Λ. That is, there exists a positive finite measure $|\mu|$ and some measurable g with $\|g\|_\infty = 1$ such that $\mathrm{d}\mu = g \, \mathrm{d}|\mu|$ defines a signed measure with*

$$\Lambda(f) = \int_X f \, \mathrm{d}\mu = \int_X f g \, \mathrm{d}|\mu|$$

for all $f \in C_0(X)$. The operator norm of Λ is equal to $\|g\|_{L^1_{|\mu|}(X)}$, which shows that $C_0(X)^* \cong \mathcal{M}(X)$ under the pairing

$$\langle f, \mu \rangle = \int f \, \mathrm{d}\mu$$

for $f \in C_0(X)$ and $\mu \in \mathcal{M}(X)$, where $\mathcal{M}(X)$ is equipped with the norm in Exercise 3.33.

We note that in a sense Theorem 7.54 also gives a polar decomposition for complex signed measures (see Exercises 7.56 and 7.55).

In the proof below we first construct from the linear functional Λ a positive linear functional $|\Lambda|$ (which may be called the positive version of Λ) which will give rise to the positive finite measure $|\mu|$. The existence of g will then follow from Proposition 7.34. At first sight the construction of $|\Lambda|$ is surprising — we will force positivity, and then linearity is a minor miracle. Comparing this construction to our discussion of the operator norm of integration in Lemma 2.63 and its proof should make this less surprising.

PROOF OF THEOREM 7.54. Let Λ be a continuous linear functional on $C_0(X)$.

UNIQUENESS: To see the uniqueness claim in the theorem, suppose that Λ is represented by $\mathrm{d}\mu_1 = g_1 \, \mathrm{d}|\mu_1|$ and also by $\mathrm{d}\mu_2 = g_2 \, \mathrm{d}|\mu_2|$. Define

$$\mu = |\mu_1| + |\mu_2|$$

and notice that $|\mu_1|, |\mu_2| \ll \mu$. By Proposition 3.29 this implies that there is a measurable function $h_j \geqslant 0$ with $\mathrm{d}|\mu_j| = h_j \, \mathrm{d}\mu$ for $j = 1, 2$. This shows that Λ is representated by $\mathrm{d}\mu_j = g_j h_j \, \mathrm{d}\mu$ for $j = 1, 2$, so $(g_1 h_1 - g_2 h_2) \, \mathrm{d}\mu$ represents the zero functional on $C_0(X)$. By Lemma 2.63 this implies that

$$\|g_1 h_1 - g_2 h_2\|_{L^1_\mu} = 0,$$

which in turn implies that $\mu_1 = \mu_2$.

DEFINING THE POSITIVE VERSION $|\Lambda|$: To prove the existence, define the positive version of Λ by

$$|\Lambda|(f) = \sup \left\{ \Re(\Lambda(g)) \mid g \in C_0(X), |g| \leqslant f \right\}$$

for any non-negative and continuous $f \in C_{0,\mathbb{R}}(X)$. Clearly $|\Lambda(g)| \leqslant \|\Lambda\|_{\mathrm{op}} \|f\|_\infty$ for all g as in the definition of $|\Lambda|(f)$ and so

$$0 \leqslant |\Lambda|(f) \leqslant \|\Lambda\|_{\mathrm{op}} \|f\|_\infty. \tag{7.16}$$

Moreover, the definition readily implies $|\Lambda|(\alpha f) = \alpha |\Lambda|(f)$ for $\alpha \geqslant 0$. In order to extend $|\Lambda|$ to an $\mathbb{R}$-linear functional on $C_{0,\mathbb{R}}(X)$ we first consider functions $f_1, f_2 \in C_{0,\mathbb{R}}(X)$ with $f_1 \geqslant 0$ and $f_2 \geqslant 0$, and claim that

$$|\Lambda|(f_1 + f_2) = |\Lambda|(f_1) + |\Lambda|(f_2). \tag{7.17}$$

One inequality is quite easy. If $g_i \in C_0(X)$ satisfies $|g_i| \leqslant f_i$ for $i = 1, 2$, then

$$|g_1 + g_2| \leqslant |g_1| + |g_2| \leqslant f_1 + f_2$$

and so $\Re(\Lambda(g_1)) + \Re(\Lambda(g_2)) = \Re(\Lambda(g_1 + g_2)) \leqslant |\Lambda|(f_1 + f_2)$, which shows that

$$|\Lambda|(f_1) + |\Lambda|(f_2) \leqslant |\Lambda|(f_1 + f_2).$$

To show the reverse inequality, we need to take some function $g \in C_0(X)$ with $|g| \leqslant f_1 + f_2$ and split it into continuous functions $g = g_1 + g_2$ with the property that $|g_1| \leqslant f_1$ and $|g_2| \leqslant f_2$. We define

$$g_1(x) = \begin{cases} g(x) & \text{if } |g(x)| \leqslant f_1(x), \\ \frac{g(x)}{|g(x)|} f_1(x) & \text{if } |g(x)| \geqslant f_1(x) \text{ and } g(x) \neq 0, \end{cases}$$

which we claim is a continuous function satisfying $|g_1| \leqslant f_1$.

First consider the restriction h of g_1 to the set

$$D = \{ x \in X \mid |g(x)| \geqslant f_1(x) \}.$$

Clearly h is continuous wherever $g(x) \neq 0$. For the points

$$x \in D_0 = \{x \in D \mid g(x) = 0\}$$

we have $h(x) = 0$ by definition of g_1, and $f_1(x) = 0$ by definition of D. It follows that $|h| = |f_1|$ on D and, since f_1 is continuous, we see that $x_0 \in D_0$ and $x \to x_0$ inside D implies that $h(x) \to 0$ as $x \to x_0$ inside D, which proves that h is also continuous at points in D_0. Now notice that g_1 is continuous on the two closed sets D and $\{x \in X \mid |g(x)| \leqslant f_1(x)\}$, the union of which is X. It follows that g_1 is continuous and satisfies $|g_1| \leqslant f_1$ on X. Since $f_1 \in C_0(X)$ we also have $g_1 \in C_0(X)$.

We also define $g_2 = g - g_1 \in C_0(X)$ and notice that

$$|g_2(x)| = \begin{cases} 0 & \text{if } |g(x)| \leqslant f_1(x), \\ |g(x)| - f_1(x) & \text{if } |g(x)| \geqslant f_1(x) \end{cases}$$

so that $|g_2| \leqslant f_2$ by the assumption on g. Hence

$$\Re(\Lambda(g)) = \Re(\Lambda(g_1)) + \Re(\Lambda(g_2)) \leqslant |\Lambda|(f_1) + |\Lambda|(f_2),$$

which proves that

$$|\Lambda|(f_1 + f_2) \leqslant |\Lambda|(f_1) + |\Lambda|(f_2)$$

since $g \in C_0(X)$ was arbitrary satisfying $|g| \leqslant f_1 + f_2$. In particular, we now obtain (7.17).

LINEARITY OF $|\Lambda|$: Now let f be any real-valued function in $C_{0,\mathbb{R}}(X)$. We extend the definition of $|\Lambda|$ by the formula

$$|\Lambda|(f) = |\Lambda|(f^+) - |\Lambda|(f^-), \tag{7.18}$$

where $f^+ = \max\{f, 0\}$ and $f^- = \max\{-f, 0\}$ are non-negative continuous functions. We now have

$$|\Lambda|(\alpha f) = \alpha |\Lambda|(f)$$

for all $\alpha \in \mathbb{R}$ and $f \in C_{0,\mathbb{R}}(X)$. We note that (7.16) extends to all $f \in C_{0,\mathbb{R}}(X)$ since $|\Lambda|(f^+), |\Lambda|(f^-) \in [0, \|\Lambda\|_{\mathrm{op}}\|f\|_\infty]$. For linearity it remains to show that

$$|\Lambda|(f_1 + f_2) = |\Lambda|(f_1) + |\Lambda|(f_2) \tag{7.19}$$

for $f_1, f_2 \in C_{0,\mathbb{R}}(X)$. To see this, notice first that

$$(f_1 + f_2)^+ - (f_1 + f_2)^- = f_1 + f_2 = f_1^+ - f_1^- + f_2^+ - f_2^-$$

and so $(f_1 + f_2)^+ + f_1^- + f_2^- = (f_1 + f_2)^- + f_1^+ + f_2^+$. We may apply $|\Lambda|$ to the latter equation and use the non-negative linearity in (7.17) to get

$$|\Lambda|\big((f_2 + f_2)^+\big) + |\Lambda|(f_1^-) + |\Lambda|(f_2^-) = |\Lambda|\big((f_1 + f_2)^-\big) + |\Lambda|(f_1^+) + |\Lambda|(f_2^+).$$

Rearranging the terms again we get

$$|\Lambda|\big((f_2+f_2)^+\big) - |\Lambda|\big((f_1+f_2)^-\big) = |\Lambda|(f_1^+) - |\Lambda|(f_1^-) + |\Lambda|(f_2^+) - |\Lambda|(f_2^-),$$

which is precisely (7.19) by definition of $|\Lambda|$ in (7.18).

APPLYING RIESZ REPRESENTATION: We have thus shown that $|\Lambda|$ is a bounded positive linear functional on $C_{0,\mathbb{R}}(X)$. Restricting it to $C_{c,\mathbb{R}}(X)$ we may apply Theorem 7.44 and find a positive measure $|\mu|$ with

$$|\Lambda|(f) = \int_X f\,\mathrm{d}|\mu| \tag{7.20}$$

for $f \in C_{c,\mathbb{R}}(X)$. Now use local compactness, σ-compactness, and Urysohn's lemma (see Lemma A.22 and Lemma A.27) to find some non-negative function $f_n \in C_{c,\mathbb{R}}(X)$ with $f_n \nearrow 1$ as $n \to \infty$ and apply monotone convergence to obtain

$$|\mu|(X) = \lim_{n\to\infty} \int f_n\,\mathrm{d}|\mu| = \lim_{n\to\infty} |\Lambda|(f_n) \leqslant \|\Lambda\|_{\mathrm{op}} \tag{7.21}$$

by (7.16). Note that (7.20) extends now to all $f \in C_{0,\mathbb{R}}(X)$ by applying (7.20) to the sequence $(f_n f)$ in $C_{c,\mathbb{R}}(X)$ together with continuity of $|\Lambda|$ and dominated convergence.

DESCRIPTION OF Λ: We now return to the study of the original functional Λ. For any $f \in C_0(X)$ we may apply the definition of $|\Lambda|$ and $|\mu|$ to obtain

$$|\Lambda(f)| = \alpha\Lambda(f) = \Re(\Lambda(\alpha f)) \leqslant |\Lambda|(|\alpha f|) = \int_X |f|\,\mathrm{d}|\mu| \tag{7.22}$$

for some $\alpha \in \mathbb{C}$ with $|\alpha| = 1$. However, (7.22) shows that Λ is continuous with respect to $\|\cdot\|_{L^1_{|\mu|}(X)}$, hence by density of $C_c(X)$ extends to $L^1_{|\mu|}(X)$, and so by Proposition 7.34 must be of the form

$$\Lambda(f) = \int_X fg\,\mathrm{d}|\mu|$$

for some $g \in L^\infty_{|\mu|}(X)$. Moreover, $\|\Lambda\|_{\mathrm{op}} = \|g\|_{L^1(\mu)}$ by Lemma 2.63, and together with (7.21) we obtain $\|\Lambda\|_{\mathrm{op}} = \|g\|_{L^1(\mu)} \leqslant \|g\|_\infty |\mu|(X) \leqslant \|g\|_\infty \|\Lambda\|_{\mathrm{op}}$, and so $\|g\|_\infty = 1$ follows unless $\|\Lambda\|_{\mathrm{op}} = 0$. In the trivial case $\Lambda = 0$ we have $|\mu| = 0$ and may also set $g \equiv 1$. $\qquad\square$

Exercise 7.55. In the notation of Theorem 7.54 (and of its proof) show that $|g| = 1$ for $|\mu|$-almost every $x \in X$.

Exercise 7.56. (a) Recall that $\mu = \nu_1 - \nu_2$ defines a real signed measure μ on X if ν_1, ν_2 are two finite measures on a measurable space X. Show that for every real signed measure μ there exist uniquely determined positive measures $\mu_+ \perp \mu_-$ with $\mu = \mu_+ - \mu_-$. Also show that $\mu_+(B) = \sup\{\mu(A) \mid A \subseteq B \text{ measurable}\}$ and similarly for μ_-. Show the existence of μ_+, μ_- as a corollary of Proposition 3.29 and also of Theorem 7.54 if X is a locally compact σ-compact metric space.
(b) Suppose that μ is a complex signed measure defined by $\mathrm{d}\mu = h\,\mathrm{d}\nu$ for some finite positive measure ν on X and some $h \in L^1_\nu(X)$. Show that μ also has a representation in

the form $d\mu = g \, d|\mu|$ with $|g| = 1$ everywhere and $|\mu|$ being a positive finite measure. Show that $|\mu|$ is uniquely determined, as is g, $|\mu|$-almost everywhere.

Exercise 7.57. Let X be a locally compact σ-compact metric space, and let Λ be a linear functional on $C_c(X)$ with the property that for any compact $K \subseteq X$ there is a constant $C_K > 0$ such that $|\Lambda(f)| \leqslant C_K \|f\|_\infty$ for any $f \in C_c(X)$ with $\operatorname{Supp} f \subseteq K$. Show that Λ can be represented by a signed Radon measure on X, meaning that there exists a Radon measure μ on X and a locally integrable (that is, integrable on any compact subset) function g on X such that $\Lambda(f) = \int fg \, d\mu$ for all $f \in C_c(X)$.

Exercise 7.58. Let $X = [0, 1] \subseteq \mathbb{R}$ (though the reader will notice that the same conclusions holds on most compact metric spaces).
(a) Notice that every finite signed measure μ on X defines a linear functional on the space $\mathscr{L}^\infty(X) = \{f : X \to \mathbb{R} \mid \|f\|_\infty < \infty, f \text{ measurable}\}$ but that $\mathscr{L}^\infty(X)^*$ contains other functionals as well.
(b) Notice that every function $f \in \mathscr{L}^\infty(X)$ defines a linear functional on the space of finite signed measures $\mathcal{M}(X) \cong C(X)^*$. Deduce that $C(X)$ is not reflexive. Show that $\mathcal{M}(X)^*$ contains more functionals than those arising from $\mathscr{L}^\infty(X)$.

Exercise 7.59. Find a description of the dual of $C^n([0, 1])$ for all $n \in \mathbb{N}$.

7.5 Further Topics

- As we will see in Section 8.6, the Hahn–Banach lemma (Lemma 7.1) is very useful in the study of closed convex sets (even in the more general setting of locally convex vector spaces introduced in Section 8.4).
- The explicit description of the dual spaces in this chapter will give us concrete cases of the weak and the weak* topology in Chapter 8.
- The more general Marcinkiewicz interpolation theorem gives a result similar to the Riesz–Thorin interpolation theorem for certain non-linear operators, see Folland [33, Sec. 6.5].
- The Riesz representation theorem (Theorem 7.44) has numerous applications. It plays a crucial role in obtaining a point in a convex set as a generalized convex combination of extreme points of the convex set (see Section 8.6.1), in the spectral theory of unitary and self-adjoint operators on Hilbert spaces (see Chapters 9 and 12–13), in the spectral theory of unitary representations of locally compact abelian groups (see Herglotz's Theorem 9.6), and also in the construction of the Haar measure of a locally compact group (see Section 10.1).

Chapter 8
Locally Convex Vector Spaces

In this chapter we introduce the important weak and weak* topologies on Banach spaces and their duals, prove an important compactness result, introduce two more topologies on $B(V, W)$, and put these into the general context of locally convex vector spaces. Finally, we also discuss convex sets of locally convex vector spaces.

8.1 Weak Topologies and the Banach–Alaoglu Theorem

As we have seen in Proposition 2.35, the unit ball in an infinite-dimensional normed vector space is not compact in the topology induced by the norm (which is often called the norm or strong topology). Given the central importance of compactness in much of analysis, this is a significant problem. In general this is simply something that must be lived with as a price to pay for the additional power of doing analysis in infinite-dimensional spaces, but we can also improve the chance of finding compactness by studying weaker topologies than the norm topology. We also refer to Appendix A.3 since many definitions of topologies in this chapter are special cases of more general constructions discussed there. As usual, for a given normed vector space X the space X^* consists of the linear functionals that are continuous with respect to the norm topology on X.

Definition 8.1. Let X be a normed vector space with dual space X^*. The *weak topology* on X is the weakest (coarsest) topology on X for which all the elements of X^* (which are functions on X) are continuous.

Exercise 8.2. Show that the weak and norm topologies coincide for a finite-dimensional normed vector space.

By the properties of the initial topology (Definition A.15) a neighbourhood in the weak topology of $x_0 \in X$ is a set containing a set of the form

© Springer International Publishing AG 2017
M. Einsiedler, T. Ward, *Functional Analysis, Spectral Theory, and Applications*,
Graduate Texts in Mathematics 276, DOI 10.1007/978-3-319-58540-6_8

$$N_{\ell_1,\ldots,\ell_n;\varepsilon}(x_0) = \bigcap_{i=1}^{n}\{x \in X \mid |\ell_i(x) - \ell_i(x_0)| < \varepsilon\}$$

for some $\varepsilon > 0$ and functionals $\ell_1,\ldots,\ell_n \in X^*$. Note that a sequence (x_n) in X converges in the weak topology to $x \in X$ if and only if $\ell(x_n) \to \ell(x)$ for every $\ell \in X^*$. However, sequences alone are in general not sufficient to describe a topology (see Exercise 8.15); one needs to consider filters (or nets) instead. We therefore generalize this comment to that setting in the next exercise.

Exercise 8.3. Given a normed vector space X, show that a filter $\mathcal{F} \subseteq \mathbb{P}(X)$ converges in the weak topology to $x \in X$ if and only if $\lim_{\mathcal{F}} \ell = \ell(x)$ for all $\ell \in X^*$ (see Appendix A.2).

If X is infinite-dimensional, then in contrast to Exercise 8.2 the weak topology and the norm topology are different. To see this notice that

$$\bigcap_{i=1}^{n} \ker \ell_i \subseteq N_{\ell_1,\ldots,\ell_n;\varepsilon}(0),$$

which implies that no neighbourhood of 0 in the weak topology can be bounded with respect to the norm of X.

Definition 8.4. Let X be a normed vector space with dual space X^*. The weak* topology (read as 'weak star' topology) is the weakest (or coarsest) topology on X^* for which all the evaluation maps $x^* \mapsto x^*(x)$ corresponding to $x \in X$ are continuous.

Once again we can describe the weak* topology by saying that a neighbourhood of $x_0^* \in X^*$ is a set containing a set of the form

$$N_{x_1,\ldots,x_n;\varepsilon}(x_0^*) = \bigcap_{i=1}^{n}\{x^* \in X^* \mid |x^*(x_i) - x_0^*(x_i)| < \varepsilon\}$$

for some $\varepsilon > 0$ and $x_1,\ldots,x_n \in X$. As before, we can show that the weak* topology and the norm topology on X^* are different if X (and hence if X^*) is infinite-dimensional.

Example 8.5. (a) For a Hilbert space H, the weak and weak* topologies are identical. The same holds for any reflexive Banach space. However, in general there is no definition of a weak* topology on a given Banach space as there may not exist a pre-dual of X, meaning a Banach space Y with $X = Y^*$ (see Example 8.81).
(b) Let $X = [0,1]$ and consider the sequence of measures (μ_n) where

$$\mu_n = \frac{1}{n}\left(\delta_{1/n} + \delta_{2/n} + \cdots + \delta_1\right),$$

viewed (via integration) as functionals on $C(X)$ (see Theorem 7.54; here δ_t denotes the point measure defined by $\delta_t(A) = 1$ if $t \in A$ and 0 if not). Then the sequence of measures (μ_n) converges in the weak* topology to the Lebesgue measure λ, which we also identify with the functional it induces. Notice that this statement is equivalent to the beginning of the theory of the Riemann integral for continuous functions, which should help to explain why the weak* topology is a quite natural notion.

Notice however that (μ_n) does not converge in the weak topology, nor *a fortiori* in the norm topology. To see the former, notice that every function in $\mathscr{L}^\infty([0,1])$ induces a linear functional on the space $\mathcal{M}([0,1])$ of finite signed measures on $[0,1]$, and that for $f = \mathbb{1}_{\mathbb{Q} \cap [0,1]}$ we have $\int f \, \mathrm{d}\mu_n = 1$ for all n, while $\int f \mathrm{d}\lambda = 0$. Thus the weak and weak* topologies on $\mathcal{M}([0,1])$ are different.

We note that we have already seen other interesting examples of weak* convergence. In fact, Proposition 3.65 can be interpreted as saying that for every $x \in \mathbb{T}$ the measures defined by $t \mapsto F_M(x - t) \, \mathrm{d}t$ on $\mathbb{T}$ converge in the weak* topology to the Dirac measure δ_x corresponding to the unit point mass at x. The reader can analyze the proof of Proposition 3.65 and the material of this section to prove the following theorem due to Toeplitz.

Exercise 8.6 (Toeplitz). Suppose that (k_n) is a sequence of integrable complex-valued functions defined on $[0,1]$, and let x be a point in $[0,1]$. Then the measures defined by $k_n(t) \, \mathrm{d}t$ converge in the weak* topology to δ_x as $n \to \infty$ if and only if all of the following conditions hold:

(1) $\|k_n\|_1 \leqslant C$ for some constant C independent of n;

(2) $\displaystyle\int_0^1 k_n(t) \, \mathrm{d}t \longrightarrow 1$ as $n \to \infty$; and

(3) $\displaystyle\int_0^1 k_n(t) g(t) \, \mathrm{d}t \longrightarrow 0$ as $n \to \infty$ for all $g \in C^\infty([0,1])$ with $x \notin \mathrm{Supp}(g)$.

Lemma 8.7. *For a Banach space X the weak topology on X and the weak* topology on X^* are Hausdorff.*

PROOF. For the weak topology this follows from Corollary 7.4: if $y \neq z$ in X there exists some $\ell \in X^*$ with $\ell(y) \neq \ell(z)$, so that $N_{\ell;\varepsilon}(y) \cap N_{\ell;\varepsilon}(z) = \varnothing$ for $\varepsilon = \frac{|\ell(z) - \ell(y)|}{2}$. The proof for the weak* topology is similar, using the fact that for $x_1^* \neq x_2^*$ there exists some $x \in X$ with $x_1^*(x) \neq x_2^*(x)$. $\square$

Exercise 8.8. Let X be a Banach space and let (x_n) be a sequence converging to $x \in X$ in the weak topology. Show that $\sup_{n \geqslant 1} \|x_n\| < \infty$. In other words, show that weakly convergent sequences in Banach spaces are bounded.

The following exercise shows that the weak and the weak* topologies have natural compatibility properties with respect to bounded operators.

Exercise 8.9. Let $A : X \to Y$ be a bounded operator between two Banach spaces X and Y.

(a) Show that A is also a continuous operator if we equip both X and Y with the respective weak topologies.

(b) Consider the dual operator $A^* : Y^* \to X^*$ defined by $A^*(y^*) = y^* \circ A \in X^*$ for all $y^* \in Y^*$. Show that A^* is continuous if we endow both dual spaces with the weak* topologies.

(c) Suppose now that A is a compact operator. Show that $Ax_n \to Ax$ as $n \to \infty$ in the norm topology on Y whenever $x_n \to x$ as $n \to \infty$ in the weak topology on X.

8.1.1 Weak* Compactness of the Unit Ball

The importance of the weak* topology comes from the following theorem, which was alluded to in the introduction to the chapter.

Theorem 8.10 (Banach–Alaoglu). *The closed unit ball*

$$\overline{B_1^{X^*}} = \{\ell \in X^* \mid \|\ell\|_{\mathrm{op}} \leqslant 1\}$$

in the dual X^* of a normed vector space X is compact in the weak topology.*

PROOF. Let $\overline{B(r)}$ be the closed (and hence compact) ball of radius $r > 0$ in $\mathbb{R}$ or $\mathbb{C}$ depending on the field of scalars. By Tychonoff's theorem (see Theorem A.20) the space

$$Y = \prod_{x \in X} \overline{B(\|x\|)}$$

is compact with respect to the product topology (see Definition A.16). Now define the embedding

$$\phi : \overline{B_1^{X^*}} \longrightarrow Y$$
$$\ell \longmapsto (\ell(x))_{x \in X} \in Y.$$

Let $\pi_x : Y \longrightarrow \overline{B(\|x\|)}$ be the projection operator (or evaluation map) corresponding to $x \in X$ defined by $Y \ni y \mapsto \pi_x(y) = y(x)$. Then the neighbourhoods of some $y = \phi(\ell_0)$ with $\ell_0 \in \overline{B_1^{X^*}}$ in the product topology are sets containing sets of the form

$$N = \bigcap_{i=1}^{n} \pi_{x_i}^{-1} \left(B_\varepsilon(\ell_0(x_i)) \right).$$

Now notice that the pre-image of N under ϕ takes the form

$$\phi^{-1}(N) = \bigcap_{i=1}^{n} \{\ell \in \overline{B_1^{X^*}} \mid |\ell(x_i) - \ell_0(x_i)| < \varepsilon\} = N_{x_1, \dots, x_n; \varepsilon}(\ell_0),$$

which is precisely one of the neighbourhoods of $\ell_0 \in \overline{B_1^{X^*}}$ defining the weak* topology on $\overline{B_1^{X^*}}$. Therefore, ϕ is a homeomorphism from $\overline{B_1^{X^*}}$ (with the restriction of the weak* topology) to a subset of Y (with the product topology). We claim that

$$\phi\big(\overline{B_1^{X^*}}\big) \subseteq Y$$

is closed, which then implies the theorem, since any closed subset of Y is compact since Y is itself compact.

To see the claim, notice first that $\phi(\overline{B_1^{X^*}})$ consists of all linear maps in Y. This is because any element $y \in Y$ is a scalar-valued function on X with

$$y(x) \in \overline{B(\|x\|)}$$

for all $x \in X$, and so if y is linear then $\|y\| \leqslant 1$. The claim now follows easily since linearity is defined by equations and so is a closed condition, as we will now show. In fact for any scalars α_1, α_2 the set

$$D_{\alpha_1,\alpha_2} = \{(\lambda_1, \lambda_2, \lambda_3) \mid \lambda_3 = \alpha_1\lambda_1 + \alpha_2\lambda_2\}$$

is closed, and the joint evaluation map

$$\pi_{x_1,x_2,\alpha_1 x_1 + \alpha_2 x_2}(y) = (y(x_1), y(x_2), y(\alpha_1 x_1 + \alpha_2 x_2))$$

is continuous by definition of the product topology on Y. Hence the set of all linear maps in Y is given by

$$\bigcap_{x_1,x_2,\alpha_1,\alpha_2} \pi_{x_1,x_2,\alpha_1 x_1 + \alpha_2 x_2}^{-1}(D_{\alpha_1,\alpha_2})$$

and so is closed, and the theorem follows. $\square$

8.1.2 More Properties of the Weak and Weak* Topologies

The weak and weak* topologies are never metrizable for infinite-dimensional Banach spaces (see Exercise 8.12), but when restricted to the unit ball the situation is better.

Proposition 8.11. *Let $D \subseteq X$ be a dense subset of a normed vector space. Then the weak* topology restricted to $\overline{B_1^{X^*}}$ is the weakest topology on $\overline{B_1^{X^*}}$ for which the evaluation maps $\ell \mapsto \ell(x)$ are continuous for all $x \in D$. In particular, if X is separable, then the weak* topology restricted to $\overline{B_1^{X^*}}$ is metrizable.*

PROOF. Suppose that $D \subseteq X$ is dense, and suppose that

$$N_{x;\varepsilon} = \{\ell \in X^* \mid |\ell(x) - \ell_0(x)| < \varepsilon\}$$

is a neighbourhood of $\ell_0 \in \overline{B_1^{X^*}}$ defined by $\varepsilon > 0$ and some arbitrary $x \in X$. Choose some $x' \in D$ with $\|x - x'\| < \frac{\varepsilon}{3}$, and notice that for all $\ell \in \overline{B_1^{X^*}}$ we have $|\ell(x) - \ell(x')| < \frac{\varepsilon}{3}$ and so

$$N_{x';\varepsilon/3}(\ell_0) \cap \overline{B_1^{X^*}} \subseteq N_{x;\varepsilon}(\ell_0) \cap \overline{B_1^{X^*}}$$

by a simple application of the triangle inequality (check this). Thus the topologies defined on $\overline{B_1^{X^*}}$ using the evaluation maps for $x \in D$ or for $x \in X$ (the latter being the weak* topology by definition) agree.

For the last claim of the proposition, notice that if X is separable, then by definition there exists a countable dense set $D = \{x_1, x_2, \dots\} \subseteq X$. For every $x_n \in D$ the weakest topology for which $\ell \mapsto \ell(x_n)$ is continuous is the topology induced by the semi-norm $\|\ell\|_{x_n} = |\ell(x_n)|$, and so the weak* topology is the weakest topology on $\overline{B_1^{X^*}}$ that is stronger than all the topologies induced by the semi-norms $\|\cdot\|_{x_n}$ for $n \in \mathbb{N}$. By Lemma A.17 and the Hausdorff property of the weak* topology from Lemma 8.7, this topology is metrizable. $\qquad\square$

Exercise 8.12. Let X be an infinite-dimensional Banach space.
(a) Show that X is not the span of countably many elements of X. That is, show that for any $x_1, x_2, \dots \in X$ we have $X \neq \langle x_n \mid n \in \mathbb{N}\rangle$. Of course we may have $X = \overline{\langle x_n \mid n \in \mathbb{N}\rangle}$.
(b) Use part (a) to show that the weak* topology does not have a countable basis of neighbourhoods of 0. Conclude that the weak* topology on X^* cannot be metrizable.
(c) Generalize (b) to the weak topology on X.

Let us finish with the following lemma, which answers both of the following questions for a Banach space affirmatively:

- Does X^* as a vector space with the weak* topology characterize X?
- If the weak and weak* topologies on X^* agree, does it follow that X is reflexive?

Lemma 8.13. *Let X be a Banach space. A functional on X^* is continuous with respect to the weak* topology if and only if it is an evaluation map, that is a map of the form $f : x^* \mapsto x^*(x)$ for some $x \in X$.*

PROOF. Suppose f is a functional on X^* continuous with respect to the weak* topology. Then $f^{-1}(B_1^{\mathbb{C}})$ is a neighbourhood of 0 in X^*, and so there exist $x_1, \dots, x_n \in X$ and $\varepsilon > 0$ with

$$N_{x_1,\dots,x_n;\varepsilon}(0) \subseteq f^{-1}(B_1^{\mathbb{C}}).$$

If now $x^* \in X^*$ satisfies $x^*(x_1) = \cdots = x^*(x_n) = 0$ then any multiple of x^* belongs to $N_{x_1,\dots,x_n;\varepsilon}(0)$ and therefore $|f(Mx^*)| < 1$ for all scalars M. This implies that $f(x^*) = 0$, or in other words that f induces a functional on

$$Y = X^* / \bigcap_{i=1}^{n} \ker \phi_{x_i} \cong \operatorname{im} \begin{pmatrix} \phi_{x_1} \\ \phi_{x_2} \\ \vdots \\ \phi_{x_n} \end{pmatrix} = V,$$

where $\phi_x(x^*) = x^*(x)$ for $x^* \in X^*$ and $x \in X$. However, the dual of V is generated by the restrictions of the coordinate functions to V, and these correspond to the functionals $\phi_{x_1}, \ldots, \phi_{x_n}$ on Y, so that f must be a linear combination of the form

$$f = \sum_{i=1}^{n} \alpha_i \phi_{x_i} = \phi_{\sum_{i=1}^{n} \alpha_i x_i}$$

for some scalars $\alpha_1, \ldots, \alpha_n$, as claimed. □

Lemma 8.13 in particular determines the vector space X (and its weak topology) as the space of continuous functionals on X^* with respect to the weak* topology, which answers the first question above affirmatively. The second question can also be answered using the lemma. Suppose the weak and weak* topologies on X^* agree and $\ell \in X^{**}$ is a linear functional on X^*. By definition ℓ is also continuous with respect to the weak topology, and so also with respect to the weak* topology by assumption. Lemma 8.13 shows that $\ell(x^*) = x^*(x)$ for all $x^* \in X^*$ for some $x \in X$, which shows that X is reflexive since $\ell \in X^{**}$ was arbitrary.

Exercise 8.14. Let X be a reflexive Banach space. Let (x_n) in X be a bounded sequence. Show that (x_n) has a weakly convergent subsequence. Notice that this follows immediately from Theorem 8.10 and Proposition 8.11 if X^* is separable; show it in general.

Exercise 8.15. We know that the weak topology and the norm topology on infinite-dimensional Banach spaces are different. In contrast to this, show that a sequence in $\ell^1(\mathbb{N})$ converges in the weak topology if and only if it converges in the norm topology.

Exercise 8.16. Fix $p \in (1, \infty)$.
(a) Prove that a sequence (f_n) in $\ell^p(\mathbb{N})$ converges weakly to $f \in \ell^p(\mathbb{N})$ if and only if there is some M with $\|f_n\|_p \leqslant M$ for all $n \geqslant 1$ and $f_n(k) \to f(k)$ as $n \to \infty$ for each $k \in \mathbb{N}$.
(b) Find a sequence in $\ell^p(\mathbb{N})$ that converges weakly but not in norm.

Exercise 8.17. Let X, Y be normed vector spaces, and let $T : X \to Y$ be linear. Show that T is a bounded operator if and only if T is sequentially continuous with respect to the weak topology, that is, $x_n \to x$ weakly in X as $n \to \infty$ implies that $Tx_n \to Tx$ weakly in Y as $n \to \infty$.

Exercise 8.18. Let X be an infinite-dimensional normed vector space. Show that the weak closure of the unit sphere $\mathsf{S} = \{x \in X \mid \|x\| = 1\}$ is the closed unit ball

$$\overline{B_1^X} = \{x \in X \mid \|x\| \leqslant 1\}.$$

8.1.3 Analytic Functions and the Weak Topology

[†]As we have seen, weak convergence and norm convergence are in general quite different. There are, however, situations in which weak convergence can be upgraded to norm convergence. Analytic functions taking values in a Banach space provide one setting where this phenomenon is seen.

Definition 8.19. Let $G \subseteq \mathbb{C}$ be an open set, and let X be a complex Banach space. A function $f : G \to X$ is called *(strongly) analytic* if for every $\zeta \in G$ the limit

$$f'(\zeta) = \lim_{h \to 0} \frac{f(\zeta + h) - f(\zeta)}{h}$$

exists in the norm topology. Also f is called *weakly analytic* if for every $\ell \in X^*$ and $\zeta \in G$ the limit

$$(\ell \circ f)'(\zeta) = \lim_{h \to 0} \frac{\ell(f(\zeta + h)) - \ell(f(\zeta))}{h}$$

exists.

Notice that in the definition of weak analyticity we do not see immediately whether we can associate to f and ζ a weak limit of the difference quotient defining $f'(\zeta)$. What we can associate to f and $\zeta \in G$ in terms of a derivative is a weak* limit in X^{**},

$$X^* \ni \ell \longmapsto \lim_{h \to 0} \frac{\ell(f(\zeta + h)) - \ell(f(\zeta))}{h} \in \mathbb{C},$$

which is bounded by a corollary of the Banach–Steinhaus theorem (Corollary 4.3). However, much more is true.

Theorem 8.20 (Dunford). *Let $G \subseteq \mathbb{C}$ be an open set and let X be a Banach space. Then every weakly analytic function $f : G \to X$ is analytic.*

PROOF. Let $\ell \in X^*$, so that by assumption $\ell \circ f : G \to \mathbb{C}$ is analytic. Choose $\zeta \in G$, $\varepsilon > 0$ sufficiently small, and $h \in B_\varepsilon^\mathbb{C}$. Then

$$\ell \circ f(\zeta + h) = \frac{1}{2\pi \mathrm{i}} \oint_{|z - \zeta| = \varepsilon} \frac{\ell \circ f(z)}{z - (\zeta + h)} \, \mathrm{d}z$$

by the Cauchy integral formula, where the integral is a contour integral over a circular path with positive orientation winding once around ζ with radius of ε. Therefore we have

---

[†] This subsection will not be needed in the remainder of the book.

$$\ell \circ f(\zeta + h) - \ell \circ f(\zeta) = \frac{1}{2\pi i} \oint_{|z - \zeta| = \varepsilon} \ell \circ f(z) \left(\frac{1}{z - (\zeta + h)} - \frac{1}{z - \zeta} \right) dz$$

$$= \frac{h}{2\pi i} \oint_{|z - \zeta| = \varepsilon} \ell \circ f(z) \frac{1}{(z - (\zeta + h))(z - \zeta)} dz. \quad (8.1)$$

For $h \neq h'$ in $B_\varepsilon^{\mathbb{C}} \smallsetminus \{0\}$ we write

$$x(h, h') = \frac{1}{h - h'} \left(\frac{f(\zeta + h) - f(\zeta)}{h} - \frac{f(\zeta + h') - f(\zeta)}{h'} \right)$$

for the second-order difference quotient. We claim that $x(h, h')$ is uniformly bounded for $h \neq h'$ in $B_{\varepsilon/2}^{\mathbb{C}} \smallsetminus \{0\}$. This will give the theorem, since it implies that

$$h \longmapsto \frac{f(\zeta + h) - f(\zeta)}{h}$$

is a Lipschitz function on $B_{\varepsilon/2}^{\mathbb{C}} \smallsetminus \{0\}$ and so has a limit as $h \to 0$.

To prove the claim let $\ell \in X^*$ and use (8.1) to calculate that

$$\ell(x(h, h')) = \frac{1}{2\pi i (h - h')} \oint_{|z - \zeta| = \varepsilon} \left[\frac{\ell \circ f(z)}{(z - (\zeta + h))(z - \zeta)} - \frac{\ell \circ f(z)}{(z - (\zeta + h'))(z - \zeta)} \right] dz$$

$$= \frac{1}{2\pi i (h - h')} \oint_{|z - \zeta| = \varepsilon} \frac{\ell \circ f(z)(h - h')}{(z - \zeta)(z - (\zeta + h))(z - (\zeta + h'))} dz. \quad (8.2)$$

Notice that the denominator in the integral on the right-hand side of (8.2) is uniformly bounded away from zero, and the numerator is bounded above by $M \|\ell\|$ for some constant M depending only on f, ζ, and ε. It follows that

$$|\ell(x(h, h'))| \leqslant M' \|\ell\|$$

for some constant M' not depending on $h \neq h' \in B_{\varepsilon/2}^{\mathbb{C}} \smallsetminus \{0\}$. By Corollary 7.9 we see that $\|x(h, h')\| \leqslant M'$, which proves the claim and hence the theorem. $\square$

Another instance where weak convergence can be upgraded to strong convergence arises in the proof of a version of the mean ergodic theorem for a measure-preserving group action. We refer to [27, Sec. 8.7], where a simple version of an argument due to Greschonig and Schmidt [42] is presented.

8.2 Applications of Weak* Compactness

The Banach–Alaoglu theorem (Theorem 8.10) is quite helpful for the construction of the Haar measure on compact abelian groups and invariant means (see Section 3.3, Section 7.2 and Section 10.2).

Exercise 8.21. Let G be a compact metric abelian group. Show that there exists a G-invariant positive functional $\Lambda : C_{\mathbb{R}}(G) \to \mathbb{R}$ with $\Lambda(\mathbb{1}) = 1$, and deduce the existence of a Haar measure on G.

Exercise 8.22. Let $\mathcal{H}$ be a separable Hilbert space, and suppose that $A \in \mathrm{B}(\mathcal{H})$ is a compact operator on $\mathcal{H}$. Show that $A\big(\overline{B_1^{\mathcal{H}}}\big)$ is compact and that A^* is also a compact operator.

Exercise 8.23 (Discrete abelian groups are amenable). Let G be any abelian discrete group. Define for any finitely generated subgroup $H < G$ the set S_H to be the set of all positive functionals $L \in (\ell^\infty(G))^*$ which have norm one and are left-invariant under elements of H. Show that the intersection $\bigcap_{H \text{ f.g.}} S_H$ taken over all finitely generated subgroups H in G is non-empty, and deduce that G is amenable by applying Lemma 7.18.

Exercise 8.24. Use Exercise 8.23 and the Riesz representation theorem to give a different proof of the existence of Haar measure on a compact abelian group.

The next exercise generalizes Exercise 7.21(b) and shows the existence of a maximal amenable normal subgroup called the *amenable radical* of G.

Exercise 8.25 (Amenable radical). Let G be a discrete group. Let A be a set and suppose that $H_\alpha \lhd G$ is an amenable normal subgroup for any $\alpha \in A$. Show that the subgroup $\langle H_\alpha \mid \alpha \in A \rangle$ generated by these subgroups is an amenable normal subgroup.

The following exercise gives an analogue to the existence claim in Theorem 3.13.

Exercise 8.26. Let X be a Banach space and $K \subseteq X^*$ a non-empty weak* closed subset. Show that for any $x_0^* \in X^*$ we have $\|x_0^* - k_0\| = \min_{k \in K} \|x_0^* - k\|$ for some $k_0 \in K$.

8.2.1 Equidistribution

The combination of the Riesz representation theorem for functionals on $C(X)$ (Theorem 7.54) and the compactness of the unit ball in the weak* topology in the Banach–Alaoglu theorem (Theorem 8.10) provide the basic tools for studying sequences of probability measures.[(25)]

Proposition 8.27. *Let X be a compact metric space. Then the space $\mathcal{P}(X)$ of probability measures defined on the Borel σ-algebra of X forms a compact metric space in the weak* topology. The same applies to*

$$\mathcal{M}_{\leqslant T}(X) = \big\{ \mu \text{ is a positive measure on } X \text{ with } \mu(X) \leqslant T \big\}$$

for all $T \geqslant 0$.

PROOF. By the Riesz representation theorem (Theorem 7.54) we have

$$C(X)^* \cong \mathcal{M}(X) \supseteq \mathcal{P}(X),$$

where $\mathcal{M}(X)$ is the space of finite signed measures defined on the Borel σ-algebra of X. By Theorem 7.44 the set of probability measures is given by

$$\mathcal{P}(X) = \left\{\mu \in \mathcal{M}(X) \mid \int \mathbb{1}_X \, d\mu = 1\right\} \cap \bigcap_{f \geqslant 0} \left\{\mu \in \mathcal{M}(X) \mid \int f \, d\mu \geqslant 0\right\}$$

where the intersection is taken over all $f \in C(X)$ with $f \geqslant 0$. Since each of the sets in the intersection is closed in the weak* topology, we see that $\mathcal{P}(X)$ is closed as well.

By the Banach–Alaoglu theorem (Theorem 8.10), and since

$$\mathcal{P}(X) \subseteq \overline{B_1^{C(X)^*}},$$

this implies that $\mathcal{P}(X)$ is compact in the weak* topology. By Lemma 2.46 we know that $C(X)$ is separable, so by Proposition 8.11 the weak* topology on $\mathcal{P}(X)$ is metrizable. The same argument applies to $\mathcal{M}_{\leqslant T}(X)$. $\square$

Exercise 8.28. Let X be a locally compact σ-compact metric space. Show for any $T \geqslant 0$ that $\mathcal{M}_{\leqslant T}(X)$ (defined as in Proposition 8.27) is compact with respect to the weak* topology defined by $C_0(X)$ and the identification between $C_0(X)^*$ and $\mathcal{M}(X)$ in the Riesz representation (Theorem 7.54). Also show that the space of probability measures $\mathcal{P}(X)$ is necessarily not compact if X is not compact.

Definition 8.29. Let X be a compact metric space, and let (μ_n) be a sequence of probability measures in $\mathcal{P}(X)$. We say that (μ_n) *equidistributes* with respect to a probability measure $m \in \mathcal{P}(X)$ if $\mu_n \to m$ as $n \to \infty$ in the weak* topology; that is, if

$$\int_X f \, d\mu_n \longrightarrow \int_X f \, dm$$

as $n \to \infty$ for all $f \in C(X)$. A sequence (x_n) *equidistributes* with respect to $m \in \mathcal{P}(X)$ if the averages $\mu_n = \frac{1}{n}(\delta_{x_1} + \cdots + \delta_{x_n})$ of the Dirac measures at $x_1, \ldots, x_n$ equidistribute with respect to m.

One is often interested in equidistribution with respect to a natural given measure like the Lebesgue measure on $\mathbb{T}$. In that case the natural measure is often not mentioned, and we simply talk about a sequence of measures being equidistributed. For the case of the Lebesgue measure on $\mathbb{T}^d$ the following provides a characterization of equidistribution.

Lemma 8.30. *A sequence (μ_n) of probability measures on $\mathbb{T}^d$ equidistributes if and only if*

$$\int_{\mathbb{T}^d} \chi_k \, d\mu_n \longrightarrow \int_{\mathbb{T}^d} \chi_k \, dx = 0$$

for all $k \in \mathbb{Z}^d \setminus \{0\}$.

Essential Exercise 8.31. Prove Lemma 8.30 using the density of the trigonometric polynomials in $C(\mathbb{T}^d)$.

Exercise 8.32. Assume that $1, \alpha_1, \dots, \alpha_d \in \mathbb{R}$ are linearly independent over $\mathbb{Q}$. Show that

$$\frac{1}{N} \sum_{n=0}^{N-1} f\big(n(\alpha_1, \dots, \alpha_d) \pmod{\mathbb{Z}^d}\big) \longrightarrow \int_{\mathbb{T}^d} f(x)\,\mathrm{d}x$$

for any $f \in C(\mathbb{T}^d)$. Use this to generalize Exercise 2.50 to a statement about powers of 2 and 3 with the same exponent.

Exercise 8.33. Assume that $\alpha_1, \dots, \alpha_d \in \mathbb{R}$ are linearly independent over $\mathbb{Q}$. Show that

$$\frac{1}{T} \int_0^T f\big(t(\alpha_1, \dots, \alpha_d) \pmod{\mathbb{Z}^d}\big)\,\mathrm{d}t \longrightarrow \int_{\mathbb{T}^d} f(x)\,\mathrm{d}x$$

as $T \to \infty$, for any $f \in C(\mathbb{T}^d)$.

Lemma 8.30 already gives some examples of equidistributed sequences, generalizing Example 2.49 (see also Exercises 8.33 and 8.32).

Equidistribution results like this are a starting point for more general results obtained by Weyl [112]. We will only discuss a special case, and outline a proof along the lines of a slightly more recent approach due to Furstenberg [36].

Proposition 8.34. *If $\alpha \in \mathbb{R}\setminus\mathbb{Q}$, then the sequence (x_n) defined by $x_n = n^2\alpha$ modulo $\mathbb{Z}$ for all $n \geqslant 1$ is equidistributed in $\mathbb{T}$.*

The approach of Furstenberg is to study not just $n^2\alpha$ modulo $\mathbb{Z}$ but in fact orbits of points $(x, y) \in \mathbb{T}^2$ under the map $T : \mathbb{T}^2 \to \mathbb{T}^2$ defined by

$$T(x, y) = (x + \alpha, y + 2x + \alpha). \tag{8.3}$$

Notice that

$$T(0, 0) = (\alpha, \alpha),$$
$$T^2(0, 0) = (2\alpha, 4\alpha),$$
$$\vdots$$
$$T^n(0, 0) = (n\alpha, n^2\alpha),$$

so that Proposition 8.34 will certainly follow from the stronger result that the orbit $\{T^n(0, 0) \mid n \geqslant 0\}$ is equidistributed in $\mathbb{T}^2$. Dynamical questions of this sort — concerning equidistribution of an orbit under iteration of a map — are part of ergodic theory. We will briefly outline how one can use the Banach–Alaoglu theorem (Theorem 8.10) to prove Proposition 8.34 using ideas from ergodic theory without developing this theory further, and refer to [27] for a more thorough treatment.

Definition 8.35. Let X be a compact metric space, and let $T : X \to X$ be a continuous transformation. A Borel probability measure μ on X is said to be *T-invariant* (T is called *measure-preserving* with respect to μ) if $\mu(T^{-1}B) = \mu(B)$ for Borel measurable sets $B \subseteq X$. The triple (X, T, μ) is called a *measure-preserving system*. A T-invariant probability measure μ is said to be *ergodic* if any Borel measurable set $B \subseteq X$ with $\mu(T^{-1}B \triangle B) = 0$ has $\mu(B) \in \{0, 1\}$.

Ergodicity is the natural notion of indecomposability in ergodic theory (which includes the study of measure-preserving systems). To see this, notice that if $B \subseteq X$ is measurable with $\mu\left(T^{-1}B \triangle B\right) = 0$ and $\mu(B) \in (0, 1)$, then we can decompose the measure into a convex combination

$$\mu = \mu(B) \left(\frac{1}{\mu(B)} \mu|_B \right) + \mu(X \smallsetminus B) \left(\frac{1}{\mu(X \smallsetminus B)} \mu|_{X \smallsetminus B} \right),$$

where one can quickly check that $\left(\frac{1}{\mu(B)} \mu|_B \right)$ and $\left(\frac{1}{\mu(X \smallsetminus B)} \mu|_{X \smallsetminus B} \right)$ are two different T-invariant probability measures.

Thus a non-ergodic measure-preserving system can be decomposed into two disjoint measure-preserving systems (where in both systems we still consider the map $T : X \to X$). Pursuing the idea that a non-ergodic measure is one that can be decomposed in this way leads to the following alternate characterization of ergodicity.

Proposition 8.36. *Let X be a compact metric space, and let $T : X \to X$ be a continuous transformation. Then the space*

$$\mathcal{P}^T(X) = \{\mu \in \mathcal{M}(X) \mid \mu \text{ is a } T\text{-invariant probability measure}\}$$

is a weak* compact convex subset of $\mathcal{M}(X)$. The extreme points *of $\mathcal{P}^T(X)$ are precisely the ergodic measures in $\mathcal{P}^T(X)$.*

We recall that $\mu \in \mathcal{P}^T(X)$ is extremal if it cannot be expressed as a proper convex combination $\mu = s\nu_1 + (1 - s)\nu_2$ with $s \in (0, 1)$ and ν_1, ν_2 distinct elements of $\mathcal{P}^T(X)$. We will discuss extreme points from an abstract point of view in Section 8.6.1.

The characterization in Proposition 8.36 is interesting because it relates an intrinsic property of a T-invariant probability measure (ergodicity, as in Definition 8.35) with a property regarding the relative position of this measure in the space of all T-invariant probability measures.

For the proof we will make use of the following construction. Given a measurable map $\theta : (X, \mathcal{B}) \to (Y, \mathcal{C})$ between two measurable spaces the *push-forward* of a measure μ on $(X, \mathcal{B})$ is the measure $\theta_*\mu$ on $(Y, \mathcal{C})$ defined by $\theta_*\mu(A) = \mu(\theta^{-1}A)$ for all $A \in \mathcal{C}$. Notice that $\int_Y f \, d\theta_*\mu = \int_X f \circ \theta \, d\mu$ for all integrable functions f on $(Y, \mathcal{C})$, which follows for simple functions directly from the definition of $\theta_*\mu$ and then for positive measurable functions by monotone convergence.

If X and $T : X \to X$ are as in the proposition, then the uniqueness part of Riesz representation (Theorem 7.44) implies that μ is T-invariant if and only if

$$\int_X f \, \mathrm{d}T_* \mu = \int_X f \circ T \, \mathrm{d}\mu = \int_X f \, \mathrm{d}\mu \qquad (8.4)$$

for all $f \in C(X)$.

SKETCH OF PROOF OF PROPOSITION 8.36. By Proposition 8.27, $\mathcal{P}(X)$ is compact in the weak* topology. By the characterization in (8.4), the subset $\mathcal{P}^T(X) \subseteq \mathcal{P}(X)$ is therefore weak* closed and so also compact. It is easy to see that $\mathcal{P}^T(X)$ is convex, and the discussion before the proposition shows that a non-ergodic invariant measure is not extreme.

Suppose now that μ is not extreme, and write $\mu = s\nu_1 + (1-s)\nu_2$ with some $s \in (0,1)$ and ν_1, ν_2 distinct measures in $\mathcal{P}^T(X)$. Clearly $\nu_1 \ll \mu$ since s lies in $(0,1)$, so there is a measurable function $f_1 \geq 0$ with $\mathrm{d}\nu_1 = f_1 \, \mathrm{d}\mu$. We claim that f_1 is T-invariant in the sense that $f_1 \circ T = f_1$ almost everywhere with respect to μ. To see this let $B \subseteq X$ be a measurable set and note that by T-invariance of μ we have

$$\nu_1(B) = \int_B \mathbb{1}_B f_1 \, \mathrm{d}\mu = \int_X (\mathbb{1}_B f_1) \circ T \, \mathrm{d}\mu = \int_{T^{-1}B} f_1 \circ T \, \mathrm{d}\mu,$$

and, by T-invariance of ν_1,

$$\nu_1(B) = \nu_1(T^{-1}B) = \int_{T^{-1}B} f_1 \, \mathrm{d}\mu.$$

Let us assume[†] now that T has a continuous inverse (which is the case for the map on $\mathbb{T}^2$ considered above to which this result will be applied). Then the above implies that $f_1 = f_1 \circ T$ almost everywhere with respect to μ, since $T^{-1}(TB) = B$ shows that all measurable sets are pre-images.

Since $\nu_1 \neq \mu$ the function f_1 is not equal to $\mathbb{1}$ almost everywhere with respect to μ, and has

$$\int_X f_1 \, \mathrm{d}\mu = \nu_1(X) = 1.$$

Therefore, $B = f_1^{-1}([0,1))$ satisfies $\mu(B \triangle T^{-1}B) = 0$ and has $\mu(B) \in (0,1)$, so μ is not ergodic. $\qquad\square$

The compactness of $\mathcal{P}(X)$ can be used to obtain elements of $\mathcal{P}^T(X)$ from sequences of approximately invariant measures.

Essential Exercise 8.37. Let X be a non-trivial compact metric space and $T : X \to X$ be continuous. For any sequence (ν_n) in $\mathcal{P}(X)$, define a sequence (μ_n) by

---

[†] This is not necessary, we refer to [27] for the general case.

$$\mu_n = \frac{1}{n} \sum_{j=0}^{n-1} T_*^j \nu_n$$

for all $n \geqslant 1$. Show that any weak* limit μ of a subsequence of (μ_n) is T-invariant, and deduce that $\mathcal{P}^T(X)$ is non-empty.[26]

With these general facts about continuous transformations on compact metric spaces at our disposal, we will return to a consideration of the transformation (8.3). We start by explaining Example 2.49 in this language. Clearly the map $R_\alpha : \mathbb{T} \to \mathbb{T}$ defined by $R_\alpha(x) = x + \alpha$ preserves the Lebesgue measure $\lambda_{\mathbb{T}}$, so $(\mathbb{T}, R_\alpha, \lambda_{\mathbb{T}})$ is a measure-preserving system.

Lemma 8.38. *If $\alpha \in \mathbb{R} \backslash \mathbb{Q}$ then $\lambda_{\mathbb{T}}$ is ergodic for R_α.*

PROOF. Suppose that $B \subseteq \mathbb{T}$ is a measurable set with $\lambda_{\mathbb{T}}(B \triangle R_\alpha^{-1} B) = 0$. Then the characteristic function $\mathbb{1}_B$ satisfies $\mathbb{1}_B \circ R_\alpha = \mathbb{1}_{R_\alpha^{-1} B} = \mathbb{1}_B$ as elements of $L^2_{\lambda_{\mathbb{T}}}(\mathbb{T})$. Thus for the Fourier series expansion

$$\mathbb{1}_B = \sum_{m \in \mathbb{Z}} c_m \chi_m,$$

which converges in $L^2_{\lambda_{\mathbb{T}}}(\mathbb{T})$, we have

$$\mathbb{1}_B \circ R_\alpha = \sum_{m \in \mathbb{Z}} c_m \chi_m \circ R_\alpha = \sum_{m \in \mathbb{Z}} c_m \chi_m,$$

where we have used the fact that $U_{R_\alpha} : f \mapsto f \circ R_\alpha$ is an isometry of $L^2_{\lambda_{\mathbb{T}}}(\mathbb{T})$ and hence maps a convergent series to a convergent series. Notice that

$$\chi_m \circ R_\alpha(x) = e^{2\pi i m(x+\alpha)} = e^{2\pi i m \alpha} \chi_m(x),$$

so that (by uniqueness of Fourier coefficients) we must have $c_m e^{2\pi i m \alpha} = c_m$ for all $m \in \mathbb{Z}$. Since $\alpha \in \mathbb{R} \backslash \mathbb{Q}$ this implies that $c_m = 0$ for $m \in \mathbb{Z} \backslash \{0\}$, so $\mathbb{1}_B = c_0$ in $L^2_{\lambda_{\mathbb{T}}}(\mathbb{T})$, which implies that $\lambda_{\mathbb{T}}(B) \in \{0, 1\}$ as required. $\qquad\square$

In fact, R_α has a stronger property, called *unique ergodicity*: $\lambda_{\mathbb{T}}$ is the *only* measure invariant under R_α if $\alpha \in \mathbb{R} \backslash \mathbb{Q}$. This also implies Lemma 8.38 by Proposition 8.36. To see this stronger result, let μ be any R_α-invariant probability measure, and calculate

$$\int_{\mathbb{T}} \chi_m \, d\mu = \int_{\mathbb{T}} \chi_m \circ R_\alpha \, d\mu = e^{2\pi i m \alpha} \int_{\mathbb{T}} \chi_m \, d\mu,$$

which implies that

$$\int_{\mathbb{T}} \chi_m \, d\mu = \begin{cases} 1 & \text{for } m = 0; \\ 0 & \text{for } m \neq 0. \end{cases}$$

Since this is a property shared by $\lambda_{\mathbb{T}}$, and the trigonometric polynomials are dense in $C(\mathbb{T})$ by Proposition 3.65, we deduce that $\mu = \lambda_{\mathbb{T}}$. Using Exercise 8.37 together with the exercise below gives an alternative approach to Example 2.49. Of course this approach is more complicated, but it can also be used in situations where a direct calculation of the sort used in Example 2.49 is not feasible.

Essential Exercise 8.39. (a) Let Z be a topological space, let (z_n) be a sequence in Z, and let $z \in Z$. Show that the following are equivalent:

- $\lim_{n \to \infty} z_n = z$.
- For every subsequence (z_{n_k}) there is a subsequence $(z_{n_{k_\ell}})$ such that

$$\lim_{\ell \to \infty} z_{n_{k_\ell}} = z.$$

(b) Assume in addition that Z is a compact metric space, and show that the following gives another equivalent condition:

- For every convergent subsequence (z_{n_k}) we have

$$\lim_{k \to \infty} z_{n_k} = z.$$

(c) Assume now that $\alpha \in \mathbb{R} \setminus \mathbb{Q}$, and use this, together with the fact that $\mathcal{P}^{R_\alpha}(\mathbb{T})$ only contains the measure $\lambda_{\mathbb{T}}$ and Exercise 8.37, to show the equidistribution of $(n\alpha)$ in $\mathbb{T}$.

We now describe the procedure for obtaining equidistribution of the orbits of the map T defined by $T(x, y) = (x + \alpha, y + 2x + \alpha)$ on $\mathbb{T}^2$ discussed earlier, leaving some of the steps as exercises.

Essential Exercise 8.40. Show that the Lebesgue measure $\lambda_{\mathbb{T}^2}$ is T-invariant and ergodic.

SKETCH OF PROOF OF PROPOSITION 8.34. As discussed just after the statement of the proposition, it is enough to show that every T-orbit

$$(x, y), T(x, y), T^2(x, y), \ldots$$

is equidistributed with respect to $\lambda_{\mathbb{T}^2}$. Notice that the first coordinate of points in the orbit are precisely the points in the orbit

$$x, R_\alpha(x), R_\alpha^2(x), \ldots$$

in $\mathbb{T}$ for the transformation R_α. We already know that this sequence is equidistributed with respect to $\lambda_{\mathbb{T}}$. Write δ_x for the Dirac measure at $x \in \mathbb{T}$, so the equidistribution of the R_α-orbit of x is equivalent to the statement

$$\frac{1}{n}\sum_{j=0}^{n-1}T_*^j\left(\delta_x \times \lambda_{\mathbb{T}}\right) \longrightarrow \lambda_{\mathbb{T}^2} \tag{8.5}$$

as $n \to \infty$ (check this).

Fix some $\rho \in (0, \frac{1}{2})$ and write $\lambda_{y,\rho} = \lambda_{B_\rho(y)}$ for the Lebesgue measure restricted to the ρ-ball $B_\rho(y) = (y - \rho, y + \rho) \subseteq \mathbb{T}$ around $y \in \mathbb{T}$, and consider the average

$$\frac{1}{2\rho n}\sum_{j=0}^{n-1}T_*^j\left(\delta_x \times \lambda_{y,\rho}\right). \tag{8.6}$$

We want to show that these averages converge to $\lambda_{\mathbb{T}^2}$ in the weak* topology. Proposition 8.27 and Exercise 8.39 imply that for this it is enough to show that any convergent subsequence has $\lambda_{\mathbb{T}^2}$ as its limit. So assume (n_k) is the index sequence of a convergent subsequence, and denote the limit by μ_1. Using the convergence in (8.5), we see that

$$\frac{1}{(1 - 2\rho)n_k}\sum_{j=0}^{n_k-1}T_*^j(\delta_x \times (\lambda_{\mathbb{T}} - \lambda_{y,\rho})) \longrightarrow \mu_2$$

converges as $k \to \infty$. By Exercise 8.37 we have $\mu_1, \mu_2 \in \mathcal{P}^T(\mathbb{T}^2)$. We also have

$$\lambda_{\mathbb{T}^2} = 2\rho\mu_1 + (1 - 2\rho)\mu_2$$

by (8.5). Together with Exercise 8.40 and Proposition 8.36 this implies that $\lambda_{\mathbb{T}^2} = \mu_1 = \mu_2$. Using Exercise 8.39(b) this shows that the average in (8.6) converges to $\lambda_{\mathbb{T}^2}$ as $n \to \infty$.

Using the structure of the map T it is now not too difficult to upgrade the above to the statement in the proposition. Fix some function $f \in C(\mathbb{T}^2)$ and some $\varepsilon > 0$. By uniform continuity of f there is some $\rho \in (0, \frac{1}{2})$ such that

$$\mathsf{d}\big((x_1, y_1), (x_2, y_2)\big) < \rho \implies |f(x_1, y_1) - f(x_2, y_2)| < \varepsilon$$

(where d denotes the usual metric on $\mathbb{T}^2$). With this choice of ρ we have

$$\left|\frac{1}{n}\sum_{j=0}^{n-1}f\big(T^j(x, y)\big) - \frac{1}{2\rho n}\sum_{j=0}^{n-1}\int_{-\rho}^{\rho}f\big(T^j(x, y + z)\big)\,\mathrm{d}z\right| < \varepsilon$$

since $T^j(x, y + z) = T^j(x, y) + (0, z)$ has distance less than ρ from $T^j(x, y)$ for all $z \in (-\rho, \rho)$. Using the convergence of (8.6) to $\lambda_{\mathbb{T}^2}$, it follows that

$$\limsup_{n \to \infty}\left|\int_{\mathbb{T}^2}f\,\mathrm{d}\lambda_{\mathbb{T}^2} - \frac{1}{n}\sum_{j=0}^{n-1}f\big(T^j(x, y)\big)\right|$$

$$\leqslant \limsup_{n \to \infty} \left| \int_{\mathbb{T}^2} f \, d\lambda_{\mathbb{T}^2} - \frac{1}{2\rho n} \sum_{j=0}^{n-1} \int f \, dT_*^j (\delta_x \times \lambda_{y,\rho}) \right| + \varepsilon = \varepsilon.$$

Since $f \in C(\mathbb{T}^2)$ and $\varepsilon > 0$ were arbitary, the proposition follows. $\qquad \square$

8.2.2 Elliptic Regularity for the Laplace Operator

[†]We show in this and the next subsection how the Banach–Alaoglu theorem can help to prove elliptic regularity for weak solutions to equations of the form $\Delta g = u$ with g in $H_0^1(U)$, u in $L^2(U)$, and $U \subseteq \mathbb{R}^d$ open and bounded. In this section we essentially reprove Theorem 5.45 using different methods. In the next subsection we will assume that U has smooth boundary and will show the regularity (unlike in Section 5.3.2) up to and including the boundary. For convenience we will consider only $\mathbb{R}$-valued functions.

Definition 8.41 (Difference quotients). Let $U \subseteq \mathbb{R}^d$ and $V \subseteq U$ be open subsets. For any $f \in L^2(U)$, $j = 1, \ldots, d$ and $h \in \mathbb{R}$ such that $V + he_j$ is contained in U we define the *difference quotient* $D_j^h f \in L^2(V)$ by

$$D_j^h f(x) = \frac{f(x + he_j) - f(x)}{h}$$

for almost every $x \in V$.

As one might expect the difference quotient and the weak partial derivative are related. The first connection below is a direct application of our definition of the Sobolev spaces.

Lemma 8.42 (Bounding the difference quotient). *Let $V \subseteq U \subseteq \mathbb{R}^d$ be open subsets and $s > 0$ such that $V + [-s, s]e_j \subseteq U$ for some $j \in \{1, \ldots, d\}$. Then, for any function $f \in H^1(U)$,*

$$\|D_j^h f\|_{L^2(V)} \leqslant \|\partial_j f\|_{L^2(U)}$$

for $0 < |h| \leqslant s$.

PROOF. If $f \in C^\infty(U) \cap H^1(U)$ then

$$D_j^h f(x) = \frac{f(x + he_j) - f(x)}{h} = \int_0^1 \partial_j f(x + the_j) \, dt$$

for all $x \in V$ and $0 < |h| \leqslant s$. By integrating the square of this equation, applying Cauchy–Schwarz, translation invariance of the Lebesgue measure, and Fubini's theorem we obtain

---

[†] This and the next subsection finish our discussion of Sobolev spaces and the Laplace operator. In particular, this material will not be needed in the remainder of the book.

$$\int_V |D_j^h f(x)|^2 \, \mathrm{d}x = \int_V \left| \int_0^1 \partial_j f(x + th e_j) \, \mathrm{d}t \right|^2 \mathrm{d}x$$

$$\leqslant \int_V \int_0^1 |\partial_j f(x + th e_j)|^2 \mathrm{d}t \, \mathrm{d}x \leqslant \int_U |\partial_j f(x)|^2 \, \mathrm{d}x$$

whenever $0 < |h| \leqslant s$. Approximating $f \in H^1(U)$ by elements of the intersection $H^1(U) \cap C^\infty(U)$ then gives the lemma. $\qquad\square$

The above lemma will be useful but it will be more powerful when combined with its partial converse, which is a corollary of the Banach–Alaoglu theorem.

Corollary 8.43 (Existence of weak derivative). *Let $V \subseteq U \subseteq \mathbb{R}^d$ be open subsets and $s > 0$ satisfying $V + [-s, s]e_j \subseteq U$ for some $j \in \{1, \ldots, d\}$. Suppose that $f \in L^2(U)$ and $C > 0$ satisfy*

$$\|D_j^h f\|_{L^2(V)} \leqslant C \tag{8.7}$$

for all $0 < |h| \leqslant s$. Then $f|_V$ has a weak partial derivative $\partial_j f$ on V satisfying

$$\|\partial_j f\|_{L^2(V)} \leqslant C.$$

PROOF. Let $\phi \in C_c^\infty(V)$ and note that this implies that $D_j^h \phi$ converges uniformly to $\partial_j \phi$ as $h \to 0$. Indeed, we have

$$D_j^h \phi(x) = \frac{\phi(x + h e_j) - \phi(x)}{h} = \partial_j \phi(x + \xi_h h e_j) \longrightarrow \partial_j \phi(x)$$

as $h \to 0$ for some $\xi_h \in (0, 1)$ by the mean value theorem and continuity of $\partial_j \phi$. Indeed, the convergence is uniform and so for $f \in L^2(U)$ we have

$$\langle f, D_j^h \phi \rangle_{L^2(V)} \longrightarrow \langle f, \partial_j \phi \rangle_{L^2(V)} \tag{8.8}$$

as $h \to 0$. On the other hand, since ϕ has compact support within V we may also shift integration to obtain a discrete analogue of the formula defining the weak derivative in Definition 5.8,

$$\langle f, D_j^h \phi \rangle_{L^2(V)} = \frac{1}{h} \int_V f(x) \big(\phi(x + h e_j) - \phi(x) \big) \, \mathrm{d}x$$

$$= \frac{1}{h} \int_V f(y - h e_j) \phi(y) \, \mathrm{d}y - \frac{1}{h} \int_V f(x) \phi(x) \, \mathrm{d}x$$

$$= - \langle D_j^{-h} f, \phi \rangle_{L^2(V)} \tag{8.9}$$

for all sufficiently small h.

By the assumption (8.7) the functions $D_j^{-1/n} f|_V$ for $n \in \mathbb{N}$ with $\frac{1}{n} \leqslant s$ form a bounded sequence of elements of the Hilbert space $L^2(V)$ satisfying $\|D_j^{-1/n} f\|_{L^2(V)} \leqslant C$. By the Banach–Alaoglu theorem (Theorem 8.10 and Proposition 8.11) there exists a subsequence (n_k) with the property that $D_j^{-1/n_k} f|_V$ converges in the weak* topology to some function $v \in L^2(V)$ as $k \to \infty$ with $\|v\|_{L^2(V)} \leqslant C$. We claim that v is the weak partial derivative sought in the corollary. In fact, by (8.9) we now have for any $\phi \in C_c^\infty(V)$

$$\langle f, D_j^{1/n_k} \phi \rangle_{L^2(V)} = -\langle D_j^{-1/n_k} f, \phi \rangle_{L^2(V)} \longrightarrow -\langle v, \phi \rangle_{L^2(V)}$$

as $k \to \infty$. Together with (8.8) this gives $\langle f, \partial_j \phi \rangle_{L^2(V)} = -\langle v, \phi \rangle_{L^2(V)}$ for any function $\phi \in C_c^\infty(V)$, which proves the corollary. $\qquad\square$

Exercise 8.44. Using the same assumptions as in Corollary 8.43 show that the difference quotients $D_j^h f$ converge weakly in $L^2(V)$ to $\partial_j f$ as $h \to 0$. Do they also converge strongly?

In order for Corollary 8.43 to be useful we wish to upgrade the conclusion and obtain functions in H^1. For this we need some more preparations, which partially already featured in some exercises. For any open subset $U \subseteq \mathbb{R}^d$ we will identify elements $f \in L^2(U)$ with their trivial extension to all of $\mathbb{R}^d$ (by setting the extension to be equal to zero outside of U). By a slight abuse of terminology we will also say that $f \in L^2(U)$ has compact support in $\mathbb{R}^d$ if (for some representative of f) the set $\{x \in U \mid f(x) \neq 0\}$ has compact closure in $\mathbb{R}^d$.

Proposition 8.45 (Convolution and derivatives). *Let $U \subseteq \mathbb{R}^d$ be open. Let $f \in L^2(U)$ have compact support in $\mathbb{R}^d$ and let $\chi \in C_c^\infty(\mathbb{R}^d)$. Then the convolution product $f * \chi$ defined by*

$$f * \chi(x) = \int f(y) \chi(x-y) \, \mathrm{d}y = \int f(x-z) \chi(z) \, \mathrm{d}z$$

for $x \in \mathbb{R}^d$ is smooth with compact support, and its derivatives are given by $\partial_\alpha(f * \chi) = f * \partial_\alpha \chi$ for all $\alpha \in \mathbb{N}_0^d$. If $f \in L^2(U)$ has a weak α-partial derivative $f_\alpha \in L^2(U)$ for some $\alpha \in \mathbb{N}_0^d$ and the open subset $V \subseteq U$ has the property that $V - \operatorname{Supp} \chi \subseteq U$, then we also have $\partial_\alpha(f * \chi)|_V = (f_\alpha * \chi)|_V$. Similarly, if $g \in L^2(U)$ has compact support and satisfies the equation $\Delta g = u$ then $\Delta(g * \chi)|_V = (u * \chi)|_V$.

PROOF. Let f be as in the proposition and note that $f \in L^1(\mathbb{R}^d)$. Since we know $\chi \in C_c^\infty(U)$ has compact support, the set $\overline{\{x \in U \mid f(x) \neq 0\}} + \operatorname{Supp} \chi$ is compact and it is easy to see that $f * \chi$ vanishes outside of this set. Note that dominated convergence implies that $f * \chi$ is continuous. Moreover, for any $j \in \{1, \ldots, d\}$ we have

$$\partial_j(f * \chi)(x) = \lim_{h \to 0} \frac{1}{h} \big(f * \chi(x + he_j) - f * \chi(x) \big)$$

$$= \lim_{h \to 0} \int f(y) \frac{\chi(x + he_j - y) - \chi(x-y)}{h} \, \mathrm{d}y = f * \partial_j \chi(x)$$

for all $x \in \mathbb{R}^d$ by the mean value theorem and dominated convergence. Induction now shows that $f * \chi \in C_c^\infty(\mathbb{R}^d)$ and $\partial_\alpha(f * \chi) = f * \partial_\alpha \chi$ for all $\alpha \in \mathbb{N}_0^d$, as claimed.

Assume next that $f \in L^2(U)$ has the weak α-partial derivative $f_\alpha \in L^2(U)$ for some $\alpha \in \mathbb{N}_0^d$. Note that f_α also has compact support, as it vanishes by Lemma 5.10 almost everywhere on every open subset on which f vanishes almost everywhere. Also suppose for $\chi \in C_c^\infty(\mathbb{R}^d)$ that the open subset $V \subseteq U$ satisfies $V - \operatorname{Supp} \chi \subseteq U$. Now let $\phi \in C_c^\infty(V)$ and consider

$$\langle f * \chi, \partial_\alpha \phi \rangle_{L^2(V)} = \int_V \int_{\operatorname{Supp}\chi} f(x-y)\chi(y)\,\mathrm{d}y\,\partial_\alpha\phi(x)\,\mathrm{d}x$$

$$= \int_{\operatorname{Supp}\chi} \chi(y)\,\langle f, \partial_\alpha(\lambda_{-y}\phi)\rangle_{L^2(U)}\,\mathrm{d}y,$$

where $\lambda_{-y}\phi(x) = \phi(x+y)$ has support $\operatorname{Supp}\phi - y$ and defines for all y in $\operatorname{Supp}\chi$ an element of $C_c^\infty(U)$ since $V - \operatorname{Supp}\chi \subseteq U$. Using the fact that f_α is the weak partial derivative of f we obtain

$$\langle f * \chi, \partial_\alpha \phi \rangle_{L^2(V)} = (-1)^{\|\alpha\|_1} \int_{\operatorname{Supp}\chi} \chi(y)\,\langle f_\alpha, \lambda_{-y}\phi \rangle_{L^2(U)}\,\mathrm{d}y$$

$$= (-1)^{\|\alpha\|_1} \int_{\operatorname{Supp}\chi} \chi(y) \int_V f_\alpha(x-y)\phi(x)\,\mathrm{d}x\,\mathrm{d}y$$

$$= (-1)^{\|\alpha\|_1} \langle f_\alpha * \chi, \phi \rangle_{L^2(V)}$$

for any $\phi \in C_c^\infty(V)$. By uniqueness of the weak derivative (Lemma 5.10) and continuity we now obtain $\partial_\alpha(f * \chi)|_V = (f_\alpha * \chi)|_V$ (pointwise) as required.

This argument also gives the claim in the proposition for $g \in L^2(U)$ with $\Delta g = u \in L^2(U)$. Indeed, with the same arguments concerning the support of $\lambda_{-y}\phi$ with $y \in \operatorname{Supp}\chi$ we obtain

$$\langle g * \chi, \Delta\phi \rangle_{L^2(V)} = \int_V \int_{\operatorname{Supp}\chi} g(x-y)\chi(y)\,\mathrm{d}y\,\Delta\phi(x)\,\mathrm{d}x$$

$$= \int_{\operatorname{Supp}\chi} \chi(y)\,\langle g, \Delta(\lambda_{-y}\phi)\rangle_{L^2(U)}\,\mathrm{d}y$$

$$= \int_{\operatorname{Supp}\chi} \chi(y)\,\langle u, \lambda_{-y}\phi \rangle_{L^2(U)}\,\mathrm{d}y$$

$$= \int_{\operatorname{Supp}\chi} \chi(y) \int_V u(x-y)\phi(x)\,\mathrm{d}x\,\mathrm{d}y = \langle u * \chi, \phi \rangle_{L^2(V)},$$

as required. □

We will now use a non-negative function $\jmath \in C_c^\infty(\mathbb{R}^d)$ as in Exercise 5.17, so that $\operatorname{Supp}\jmath$ is the closed unit ball and $\int \jmath\,\mathrm{d}x = 1$, and define the scaled function $\jmath_\varepsilon(x) = \varepsilon^{-d}\jmath(\frac{x}{\varepsilon})$ for all $x \in \mathbb{R}^d$ and $\varepsilon > 0$.

Lemma 8.46 (Approximate identity in $L^2(\mathbb{R}^d)$). *Let $U \subseteq \mathbb{R}^d$ be open and let $f \in L^2(U)$. Then $f * \jmath_\varepsilon(x) = \int f(x - y)\jmath_\varepsilon(y)\,dy$ converges to f in $L^2(\mathbb{R}^d)$ as $\varepsilon \to 0$.*

PROOF. By Lemma 3.75 $f_\varepsilon = f * \jmath_\varepsilon$ again belongs to $L^2(\mathbb{R}^d)$ for any $\varepsilon > 0$. Next, notice that

$$f_\varepsilon(x) = \int \varepsilon^{-d}\jmath(\tfrac{y}{\varepsilon})f(x - y)\,dy = \int \jmath(z)f(x - \varepsilon z)\,dz,$$

for $x \in \mathbb{R}^d$. Therefore

$$
\begin{aligned}
\|f_\varepsilon - f\|_{L^2(\mathbb{R}^d)}^2 &= \int \left|\int \big(f(x - \varepsilon z) - f(x)\big)\jmath(z)\,dz\right|^2 dx \\
&\leqslant \iint \big|f(x - \varepsilon z) - f(x)\big|^2 \jmath(z)\,dz\,dx \\
&= \int \big\|\lambda_{\varepsilon z}f - f\big\|_{L^2(\mathbb{R}^d)}^2 \jmath(z)\,dz
\end{aligned}
$$

by Jensen's inequality (see the first paragraph of the proof of Lemma 3.75) and Fubini's theorem. By Lemma 3.74 and dominated convergence the latter converges to zero as $\varepsilon \to 0$. $\square$

We note that the following corollary to Proposition 8.45 will be combined with Corollary 8.43.

Corollary 8.47 (Weak derivatives and H^k). *Let $U \subseteq \mathbb{R}^d$ be open and let $k \geqslant 1$ be an integer. Suppose that $f \in L^2_{\mathrm{loc}}(U)$ has, for all $\alpha \in \mathbb{N}_0^d$ with $\|\alpha\|_1 \leqslant k$, a weak partial derivative $\boldsymbol{\partial}_\alpha f \in L^2_{\mathrm{loc}}(U)$. Then f lies in $H^k_{\mathrm{loc}}(U)$. In fact, we have $\chi f \in H^k(\mathbb{R}^d)$ for all $\chi \in C_c^\infty(U)$.*

PROOF. Let $\chi \in C_c^\infty(U)$. Extending the product rule for differentiation we have that $\boldsymbol{\partial}_j(\chi f) = (\partial_j \chi)f + \chi(\boldsymbol{\partial}_j f) \in L^2(U)$ for $j = 1, \ldots, d$ (check this), which generalizes inductively to the Leibniz rule for weak differentiation $\boldsymbol{\partial}_\alpha$ with $\|\alpha\|_1 \leqslant k$. In particular, χf has a weak α-partial derivative on U. We now show that χf has weak partial derivatives $\boldsymbol{\partial}_\alpha(\chi f)$ on all of $\mathbb{R}^d$, where we extend these functions trivially from U to all of $\mathbb{R}^d$. Since $\mathrm{Supp}\,\chi \subseteq U$ is compact, there exists a function $\psi \in C_c^\infty(U)$ with $\psi \equiv 1$ on $\mathrm{Supp}\,\chi$ (see Exercise 5.37). If now $\phi \in C_c^\infty(\mathbb{R}^d)$, then $\psi\phi \in C_c^\infty(U)$. Using in addition that $\mathrm{Supp}(\chi f)$ and $\mathrm{Supp}\,\boldsymbol{\partial}_\alpha(\chi f)$ are contained in $\mathrm{Supp}\,\chi \subseteq U$, we obtain

$$\langle \chi f, \partial_\alpha \phi\rangle_{L^2(\mathbb{R}^d)} = \langle \chi f, \partial_\alpha(\psi\phi)\rangle_{L^2(U)} = \langle \boldsymbol{\partial}_\alpha(\chi f), \psi\phi\rangle_{L^2(U)} = \langle \boldsymbol{\partial}_\alpha(\chi f), \phi\rangle_{L^2(\mathbb{R}^d)}$$

for all $\alpha \in \mathbb{N}_0^d$ with $\|\alpha\|_1 \leqslant k$. Applying Proposition 8.45 and Lemma 8.46 we see that $(\chi f) * \jmath_\varepsilon$ is in $C_c^\infty(\mathbb{R}^d)$ and that

$$\partial_\alpha\big((\chi f) * \jmath_\varepsilon\big) = \big(\boldsymbol{\partial}_\alpha(\chi f)\big) * \jmath_\varepsilon$$

approximates $\partial_\alpha(\chi f)$ as $\varepsilon \to 0$ for any $\alpha \in \mathbb{N}_0$ with $\|\alpha\|_1 \leqslant k$. From this it follows that $\chi f \in H^k(\mathbb{R}^d)$ and with the identification of functions we also see that $\chi f \in H^k(U)$ (see Definition 5.7). Since $\chi \in C_c^\infty(U)$ was arbitrary, this proves $f \in H_{\mathrm{loc}}^k(U)$, and hence the corollary. $\qquad\square$

The argument that we will present here gives an alternate proof of Theorem 5.45 (avoiding Fourier series).

Theorem 8.48 (Elliptic regularity). *Let $U \subseteq \mathbb{R}^d$ be open, $g \in H_{\mathrm{loc}}^1(U)$, let $k \geqslant 0$ and suppose that $\Delta g = u \in H_{\mathrm{loc}}^k(U)$. Then $g \in H_{\mathrm{loc}}^{k+2}(U)$.*

SECOND PROOF OF ELLIPTIC REGULARITY ON OPEN SETS. By definition (Definition 5.43) we have to show that $\chi g \in H^{k+2}(U)$ for all $\chi \in C_c^\infty(U)$. We initially assume that $k = 0$, fix some $\chi \in C_c^\infty(U)$, and consider $f = \chi g$, which is an element of $H^1(U)$. By Lemma 5.50 we have

$$\Delta f = v \in H^0(U) = L^2(U),$$

and we also know that v vanishes on $U \smallsetminus \mathrm{Supp}\,\chi$.

We have to show that $f \in H^2(U)$, for which, using Corollary 8.47, we need to show that $\partial_i \partial_j f$ exists in $L^2(U)$ for all $i, j = 1, \ldots, d$. It will, however, be more convenient to work on $\mathbb{R}^d$.

EXTENDING TO ALL OF $\mathbb{R}^d$. We claim that, after extending f and v trivially from U to all of $\mathbb{R}^d$, we have $f \in H^1(\mathbb{R}^d)$ and the relation $\Delta f = v$ actually holds on $\mathbb{R}^d$. In fact, the first claim follows from the assumption on g and Corollary 8.47. For the second we again apply the same argument using a function $\psi \in C_c^\infty(U)$ satisfying $\psi \equiv 1$ on $\mathrm{Supp}\,\chi$. Indeed, for $\phi \in C_c^\infty(\mathbb{R}^d)$ we have $\psi\phi \in C_c^\infty(U)$ and

$$\langle f, \Delta\phi \rangle_{L^2(\mathbb{R}^d)} = \langle f, \Delta(\psi\phi) \rangle_{L^2(U)} = \langle v, \psi\phi \rangle_{L^2(U)} = \langle v, \phi \rangle_{L^2(\mathbb{R}^d)}.$$

Thus f satisfies $\Delta f = v$ on $\mathbb{R}^d$ as $\phi \in C_c^\infty(\mathbb{R}^d)$ was arbitrary.

CONVOLUTION. Next we let $\varepsilon > 0$ and define $f_\varepsilon = f * \jmath_\varepsilon$ and $v_\varepsilon = v * \jmath_\varepsilon$. By Proposition 8.45 we have that $f_\varepsilon, v_\varepsilon \in C_c^\infty(\mathbb{R}^d)$ satisfy $\Delta f_\varepsilon = v_\varepsilon$. We now use integration by parts for smooth functions of compact support and obtain

$$\|\partial_i \partial_j f_\varepsilon\|_2^2 = \int (\partial_i \partial_j f_\varepsilon)(\partial_i \partial_j f_\varepsilon)\,\mathrm{d}x = -\int (\partial_i^2 \partial_j f_\varepsilon)(\partial_j f_\varepsilon)\,\mathrm{d}x = \langle \partial_i^2 f_\varepsilon, \partial_j^2 f_\varepsilon \rangle$$

for all $i, j \in \{1, \ldots, d\}$, which gives

$$\sum_{i,j=1}^d \|\partial_i \partial_j f_\varepsilon\|_2^2 = \sum_{i,j=1}^d \langle \partial_i^2 f_\varepsilon, \partial_j^2 f_\varepsilon \rangle = \int \left(\Delta f_\varepsilon\right)^2 \mathrm{d}x = \|v_\varepsilon\|_2^2. \qquad (8.10)$$

A UNIFORM ESTIMATE IMPLIES REGULARITY. Using Lemma 8.42 we see that

$$\|D_i^h \partial_j f_\varepsilon\|_2 \leqslant \|\partial_i \partial_j f_\varepsilon\|_2 \leqslant \|v_\varepsilon\|_2$$

for all $i, j \in \{1, \ldots, d\}$ and real numbers h with $0 < |h| \leqslant 1$. We now let $\varepsilon \to 0$ and obtain from (8.10), Proposition 8.45, and Lemma 8.46 that

$$\|D_i^h \partial_j f\|_2 \leqslant \|v\|_2$$

for all h with $0 < |h| \leqslant 1$. Applying Corollary 8.43, this implies that $\partial_i \partial_j f$ exists in $L^2(\mathbb{R}^d)$ and by Corollary 8.47 it follows that $f \in H^2_{\mathrm{loc}}(\mathbb{R}^d)$. Using the same function $\psi \in C_c^\infty(U)$ as above this implies $f = \psi f \in H^2(\mathbb{R}^d) \cap H^2(U)$, so $g \in H^2_{\mathrm{loc}}(U)$ since $f = \chi g$ and $\chi \in C_c^\infty(U)$ was arbitrary.

INDUCTION ON k. The theorem now follows by induction on $k \geqslant 0$. The case $k = 0$ is proven above. So assume now that $\Delta g = u \in H^k_{\mathrm{loc}}(U)$ for some $k \geqslant 1$. Since we then also have $u \in H^{k-1}_{\mathrm{loc}}(U)$ we obtain from the inductive hypothesis that in fact $g \in H^{k+1}_{\mathrm{loc}}(U)$. Let $\chi \in C_c^\infty(U)$. Lemma 5.50 then gives for $f = \chi g$ that $\Delta f = v \in H^k(U)$. If $\alpha \in \mathbb{N}_0^d$ satisfies $\|\alpha\|_1 \leqslant k$, then $\partial_\alpha f \in L^2(U)$ satisfies $\Delta \partial_\alpha f = \partial_\alpha v$ since

$$\langle \partial_\alpha f, \Delta \phi \rangle = (-1)^{\|\alpha\|_1} \langle f, \partial_\alpha \Delta \phi \rangle = (-1)^{\|\alpha\|_1} \langle f, \Delta \partial_\alpha \phi \rangle$$
$$= (-1)^{\|\alpha\|_1} \langle v, \partial_\alpha \phi \rangle = \langle \partial_\alpha v, \phi \rangle$$

for all $\phi \in C_c^\infty(U)$. Hence the argument above for $k = 0$ applies to $\partial_\alpha f$ and shows that $\partial_\alpha f \in H^2_{\mathrm{loc}}(U)$. As $\alpha \in \mathbb{N}_0^d$ is arbitrary with $\|\alpha\|_1 \leqslant k$, it follows that f satisfies the assumption of Corollary 8.47 for the integer $k + 2$ and hence $f = \psi f \in H^{k+2}(U)$, or equivalently that $g \in H^{k+2}_{\mathrm{loc}}(U)$ since $f = \chi g$ and $\chi \in C_c^\infty(U)$ was arbitrary. This concludes the induction and the proof of the theorem.   □

The above proof of elliptic regularity, and in particular the step in (8.10), was tailored very closely to the Laplace operator on open subsets in $\mathbb{R}^d$. In order to also obtain the regularity at the boundary we start by giving a different argument which will be more amenable for generalizations (even though it will be a bit more involved). For this we will use the following inequality, which is also known as the *Cauchy inequality with an ε*. For any measure space $(X, \mathcal{B}, \mu)$, functions $u, v \in L^2_\mu(X)$, and $\varepsilon > 0$ we have

$$\left| \langle u, v \rangle_{L^2(X,\mu)} \right| \leqslant \int_X |uv| \, d\mu \leqslant \varepsilon \|u\|_2^2 + \frac{1}{4\varepsilon} \|v\|_2^2. \tag{8.11}$$

The first inequality is the triangle inequality and the second follows from integrating the inequality $0 \leqslant \left(\sqrt{\varepsilon}|u| - \frac{|v|}{2\sqrt{\varepsilon}} \right)^2 = \varepsilon |u|^2 + \frac{|v|^2}{4\varepsilon} - |u||v|$ over X.

THIRD PROOF OF ELLIPTIC REGULARITY ON OPEN SETS. As in the second proof of elliptic regularity above, we multiply g by some $\chi \in C_c^\infty(U)$ and apply Lemma 5.50. This shows that it suffices to consider the case $U = \mathbb{R}^d$ and a function $g \in H_0^1(\mathbb{R}^d)$ with compact support satisfying $\Delta g = u \in H^k(\mathbb{R}^d)$. We also initially set $k = 0$ and will use Corollary 8.43 after bounding

$$\left\|D_\ell^h g_j\right\|_2$$

for $\ell, j = 1, \ldots, d$, where $g_j = \partial_j g$ and $h \in \mathbb{R}$ with $0 < |h| \leqslant 1$.

Recall that for any $\phi \in C_c^\infty(\mathbb{R}^d)$ we have

$$\langle g, \phi \rangle_1 = \sum_{j=1}^d \langle \partial_j g, \partial_j \phi \rangle = - \langle g, \Delta \phi \rangle = - \langle u, \phi \rangle,$$

see also Lemma 5.41. Approximating any $v \in H_0^1(\mathbb{R}^d)$ by smooth functions with compact support this formula extends to $\phi = v$. We set

$$v = -D_\ell^{-h}(D_\ell^h g) \in H_0^1(\mathbb{R}^d) \tag{8.12}$$

for some fixed $\ell \in \{1, \ldots, d\}$ and $h \in \mathbb{R}$ satisfying $0 < |h| \leqslant 1$. Therefore

$$\underbrace{\langle g, -D_\ell^{-h}(D_\ell^h g) \rangle_1}_{\mathcal{L}} = \underbrace{\langle u, D_\ell^{-h}(D_\ell^h g) \rangle}_{\mathcal{R}}, \tag{8.13}$$

where $\mathcal{L}$ denotes the left-hand side and $\mathcal{R}$ denotes the right-hand side.

STUDYING THE LEFT-HAND SIDE. By definition of $\langle \cdot, \cdot \rangle_1$, we have

$$\mathcal{L} = - \sum_{j=1}^d \langle g_j, \partial_j D_\ell^{-h}(D_\ell^h g) \rangle = - \sum_{j=1}^d \langle g_j, D_\ell^{-h}(D_\ell^h g_j) \rangle,$$

where we used the fact that ∂_j and D_ℓ^h commute (check this). Finally, we apply the same argument as in (8.9), which gives our main term

$$\mathcal{M} = \mathcal{L} = \sum_{j=1}^d \langle D_\ell^h g_j, D_\ell^h g_j \rangle = \sum_{j=1}^d \left\|D_\ell^h g_j\right\|_2^2.$$

This is precisely what we wish to estimate, and it is the only term that is quadratic in the difference quotient of the weak partial derivatives of g.

BOUNDING THE RIGHT-HAND SIDE. We are aiming to convert (8.13) into an estimate on $\mathcal{M}$ that is uniform with respect to h. For this, we need to bound the right-hand side $\mathcal{R}$ of (8.13). In fact, we have for any $\varepsilon > 0$ that

$$|\mathcal{R}| = \left| \int u D_\ell^{-h}(D_\ell^h g) \, \mathrm{d}x \right| \leqslant \int |D_\ell^{-h}(D_\ell^h g)| |u| \, \mathrm{d}x \leqslant \varepsilon \left\|D_\ell^{-h}(D_\ell^h g)\right\|_2^2 + \tfrac{1}{4\varepsilon} \|u\|_2^2$$

by (8.11) (Cauchy's inequality with an ε). In the first expression of the bound on the right we use the fact that $D_\ell^h g \in H_0^1(\mathbb{R}^d)$ and the bound on the difference quotient by the weak partial derivative in Lemma 8.42 to obtain

$$\left\|D_\ell^{-h}(D_\ell^h g)\right\|_2^2 \leqslant \left\|\partial_\ell(D_\ell^h g)\right\|_2^2 = \left\|D_\ell^h g_\ell\right\|_2^2.$$

This gives

$$|\mathcal{R}| \leqslant \varepsilon \left\| D_\ell^h g_\ell \right\|_2^2 + \tfrac{1}{4\varepsilon} \|u\|_2^2 \leqslant \varepsilon \mathcal{M} + \tfrac{1}{4\varepsilon} \|u\|_2^2, \qquad (8.14)$$

which on setting $\varepsilon = \tfrac{1}{2}$ gives $|\mathcal{R}| \leqslant \tfrac{1}{2}\mathcal{M} + \tfrac{1}{2}\|u\|_2^2$.

PUTTING THE ESTIMATES TOGETHER. Using (8.13) and the estimate for $\mathcal{R}$ in (8.14) we finally see that $\mathcal{M} \leqslant \tfrac{1}{2}\mathcal{M} + \tfrac{1}{2}\|u\|_2^2$ and so

$$\sum_{j=1}^d \left\| D_\ell^h g_j \right\|_2^2 = \mathcal{M} \leqslant \|u\|_2^2.$$

Note that this upper bound is independent of h and holds for all ℓ in $\{1,\dots,d\}$. Applying Corollary 8.43 we see that $\partial_{\ell} g_j$ exists for all ℓ, j in $\{1,\dots,d\}$. In other words, all degree two weak partial derivatives of g exist, and so g lies in $H^2(\mathbb{R}^d)$ by Corollary 8.47.

INDUCTION ON k. The theorem again follows by induction on k as in the second proof of elliptic regularity above. $\qquad\square$

8.2.3 Elliptic Regularity at the Boundary

In this subsection we will use the Banach–Alaoglu theorem (much as in the third proof of elliptic regularity starting on p. 276) to prove elliptic regularity up to the boundary. For this we need to assume that U has smooth boundary as in Definition 5.31.

Theorem 8.49 (Elliptic regularity up to the boundary). *Let $U \subseteq \mathbb{R}^d$ be bounded and open with smooth boundary. Let $g \in H_0^1(U)$, $k \geqslant 0$, and suppose that $\Delta g = u \in H^k(U)$. Then $g \in H^{k+2}(U)$.*

We define

$$C^\infty(\overline{U}) = \bigcap_{k \geqslant 0} C^k(\overline{U})$$

to consist of all smooth functions on U with the property that the function and all partial derivatives can be extended continuously to $\overline{U}$ (see also the first paragraph of Section 5.3.3). We consider $C^\infty(\overline{U})$ as a subspace of $C(\overline{U})$.

Proposition 8.50 (Sobolev embedding up to the boundary). *Let U be a bounded and open subset of $\mathbb{R}^d$ with smooth boundary. Then*

$$\bigcap_{k \geqslant 0} H^k(U) = C^\infty(\overline{U}). \qquad (8.15)$$

The above theorem and proposition together allow us to complete our discussion of the Dirichlet boundary value problem and the eigenfunctions of the Laplace operator, which previously had only weaker than desired conclusions regarding the behaviour of the functions near the boundary.

Corollary 8.51 (Smooth solutions). *Let $U \subseteq \mathbb{R}^d$ be bounded and open with smooth boundary. Then*

- *for any $f \in C^\infty(\partial U)$ the weak solution to the Dirichlet boundary value problem $\Delta g = 0$, $g|_{\partial U} = f$ from Theorem 5.51 belongs to $C^\infty(\overline{U})$ and satisfies the boundary value condition pointwise, and*
- *the Laplace eigenfunctions $f \in H_0^1(U)$ with $\Delta f = \lambda f$ from Theorem 6.56 also belong to $C^\infty(\overline{U})$ and vanish at the boundary.*

We first assume Theorem 8.49 and Proposition 8.50 and show how these imply the corollary.

PROOF OF COROLLARY 8.51. For the Dirichlet boundary value problem we recall from the proof of Theorem 5.51 that we first extended $f \in C^\infty(\partial U)$ to all of $\overline{U}$, which under our assumptions leads to a function $f \in C^\infty(\overline{U})$. Proposition 5.42 then gives a function $v \in H_0^1(U)$ with $g = f - v$ satisfying $\Delta g = 0$. In other words, $\Delta v = \Delta f \in H^k(U)$ for all $k \geqslant 0$, which by Theorem 8.49 and Proposition 8.50 gives $v, g \in C^\infty(\overline{U})$. Proposition 5.33 now implies that v vanishes at ∂U pointwise.

Similarly, suppose $f \in H_0^1(U)$ is an eigenfunction of Δ from Theorem 6.56. In this case, $\Delta f = \lambda f$ implies $f \in H^3(U)$ by Theorem 8.49, then $f \in H^5(U)$ and so on. Together with Proposition 8.50 this gives $f \in C^\infty(\overline{U})$. □

PROOF OF PROPOSITION 8.50. The claim in (8.15) is (unlike the corollary) a purely local statement. In fact, we claim that it is enough to show that every point $z^{(0)} \in \overline{U}$ has an open neighbourhood $V \subseteq \mathbb{R}^d$ with the following properties. Every function

$$f \in \bigcap_{k \geqslant 0} H^k(U \cap V)$$

with support in $\overline{U} \cap V$ is continuous and can be continuously extended to $\overline{U}$ so that the extension vanishes outside of V. Assuming this property for now we find by compactness a finite cover $V_1, \ldots, V_n$ of $\overline{U}$ comprising such neighbourhoods. By Exercise 5.52 we can find a corresponding smooth partition of unity $\psi_1, \ldots, \psi_n$, which we use to localize f to the functions

$$f_j = f\psi_j \in \bigcap_{k \geqslant 0} H^k(U \cap V_j)$$

with $\operatorname{Supp} f_j \subseteq \overline{U} \cap \operatorname{Supp} \psi_j \subseteq \overline{U} \cap V_j$ for $j = 1, \ldots, n$. The local statement then implies that f_j can be continuously extended to all of $\overline{U}$ for $j = 1, \ldots, n$ so that the extension vanishes outside of V_j. This implies that $f = f_1 + \cdots + f_n$ has an extension to $\overline{U}$ too. Applying this argument to all partial derivatives then gives (8.15).

INTERIOR POINTS. We now prove the local statement starting with the case $z^{(0)} \in U$. Here we may take $V = U$ and apply Sobolev embedding on open sets (Theorem 5.34) to conclude that $f \in C(U)$, which together with Supp $f \subseteq U$ proves that f can be extended continuously to $\overline{U}$ by setting it equal to 0 on ∂U.

BOUNDARY POINTS, FLATTENING THE BOUNDARY. So consider now some element $z^{(0)} \in \partial U$. We may translate and rotate the coordinate system so that $z^{(0)} = 0$ and use Definition 5.31 to find some $\varepsilon > 0$ so that

$$U \cap B_\varepsilon(0) = \{(x, y) \in B_\varepsilon(0) \mid y < \phi(x)\}$$

for some $\phi \in C^\infty(B_\varepsilon^{\mathbb{R}^{d-1}}(0))$. To simplify the discussion we define the new open sets $U' = (-\delta, \delta)^{d-1} \times (-\delta, 0)$, $V' = (-\delta, \delta)^d$, and the map Φ by

$$\Phi(x_1, \ldots, x_d) = (x_1, \ldots, x_{d-1}, x_d + \phi(x_1, \ldots, x_{d-1})),$$

where $\delta \in (0, \varepsilon/d)$ is chosen so that $\Phi(\overline{V'}) \subseteq B_\varepsilon(0)$. In particular, Φ and all the partial derivatives of Φ will be bounded on V' and $\Phi(U') \subseteq U \cap B_\varepsilon(0)$. We note that Φ maps the Lebesgue measure on $V' \subseteq \mathbb{R}^d$ to the Lebesgue measure on the open set $V = \Phi(V') \subseteq B_\varepsilon(0)$. This implies that every $f \in C^\infty(U \cap V)$ is of the form $f = g \circ \Phi^{-1}$ for $g = f \circ \Phi \in C^\infty(U' \cap V')$. Moreover, by the multi-dimensional chain rule and induction we have

$$\|f\|_{H^k(U \cap V)} \ll_k \|f \circ \Phi\|_{H^k(U' \cap V')} \ll_k \|f\|_{H^k(U \cap V)}$$

for all $k \geqslant 0$ and $f \in C^\infty(U \cap V)$. Therefore, the completions $H^k(U \cap V)$ and $H^k(U' \cap V')$ are also isomorphic under the map

$$H^k(U \cap V) \ni f \longmapsto f \circ \Phi \in H^k(U' \cap V')$$

for all $k \geqslant 0$. In particular, it is enough to prove the desired local statement for U' and V' instead of U and V.

Simplifying the notation further we apply a linear map, set $\delta = 1$, and may suppose in the following that $V = (-1, 1)^d$ and $U = (-1, 1)^{d-1} \times (0, 1)$.

BOUNDARY POINTS, TRACE OPERATORS ON A BOX. We define $S = (-1, 1)^{d-1}$ and will need the trace operator on the hyperplanes $S_y = S \times \{y\}$ inside U for all possible values of the height parameter $y \in (0, 1)$. The trace operators for $y \searrow 0$ will allow us to extend functions from U to $\overline{U} \cap V = U \cup S_0$ with $S_0 = S \times \{0\}$. We note that these trace operators already featured in Section 5.2.2 but (except for Exercise 5.29 and Exercise 5.35) not in the generality needed here. For completeness we quickly go through the construction of these operators once more.

To define the trace operators we note that for any $y_1, y_2 \in (0, 1)$, any function $f \in C^\infty(U)$, and $x \in S$ we have

$$f(x, y_2) = f(x, y_1) + \int_{y_1}^{y_2} \partial_d f(x, t) \, dt. \tag{8.16}$$

If $f \in C^\infty(U) \cap H^1(U)$ we may integrate over $y_1 \in (0,1)$ to obtain

$$f(x, y_2) = \int_0^1 f(x, y) \, dy + \int_0^1 \partial_d f(x, t) k(y_2, t) \, dt \qquad (8.17)$$

for some bounded measurable function k (see (5.6)), almost every $x \in S$, and $y_2 \in (0,1)$. Clearly for a fixed height $y \in (0,1)$ the trace map

$$\cdot|_{S_y} : C^\infty(U) \cap H^1(U) \longrightarrow C^\infty(S)$$
$$f \longmapsto (S \ni x \longmapsto f(x, y))$$

is linear. Using (8.17) and Cauchy–Schwarz for the integration over $t \in (0,1)$ it is also easy to see that the trace map is bounded with respect to $\|\cdot\|_{H^1(U)}$ and $\|\cdot\|_{L^2(S)}$. Therefore the trace map is defined on $H^1(U)$ and takes values in $L^2(S)$. For $y_1, y_2 \in (0,1)$ we may also use (8.16) to obtain

$$\int_S |f(x, y_2) - f(x, y_1)|^2 \, dx = \int_S \left| \int_{y_1}^{y_2} \partial_d f(x, t) \, dt \right|^2 dx$$
$$\leqslant \int_S \int_0^1 |\partial_d f(x, t)|^2 \, dt \, dx \, |y_2 - y_1|$$

by Cauchy–Schwarz for $f \in C^\infty(U) \cap H^1(U)$, which gives

$$\|f|_{S_{y_1}} - f|_{S_{y_2}}\|_{L^2(S)} \leqslant \|\partial_d f\|_{L^2(U)} \sqrt{|y_2 - y_1|} \qquad (8.18)$$

first for $f \in C^\infty(U) \cap H^1(U)$ and then for all $f \in H^1(U)$. In particular, the map $(0,1) \ni y \mapsto f|_{S_y} \in L^2(S)$ is uniformly continuous for any $f \in H^1(U)$.

We wish to combine the above with the Sobolev embedding theorem on S to obtain similar conclusions with respect to the supremum norm. Clearly if we have $f \in C^\infty(U)$ then $f|_{S_y} \in C^\infty(S)$ for all $y \in (0,1)$. Applying the trace operator to the partial derivatives of a function in $H^k(U)$ along the various directions in S now shows that $\cdot|_{S_y} : H^k(U) \to H^{k-1}(S)$ is a bounded operator for all $k \geqslant 1$ and that (8.18) implies

$$\|f|_{S_{y_1}} - f|_{S_{y_2}}\|_{H^{k-1}(S)} \leqslant \|f\|_{H^k(U)} \sqrt{|y_2 - y_1|} \qquad (8.19)$$

for $f \in H^k(U)$, $k \geqslant 1$, and $y_1, y_2 \in (0,1)$. Let now $k > 1 + \frac{d-1}{2}$. Using the Sobolev embedding theorem (Theorem 5.34) on S we obtain that $H^k(U)|_{S_y}$ belongs to $C(S)$. For $\kappa \in (0,1)$ the proof of Theorem 5.34 (see Exercise 5.39) also shows that on the compact subset $K = [-1 + \kappa, 1 - \kappa]^{d-1} \subseteq S$ we have

$$\|g\|_{K,\infty} \ll \|g\|_{H^{k-1}(S)}$$

for all $g \in H^{k-1}(S)$. Together with (8.19) we obtain

$$\sup_{x \in K} |f(x, y_1) - f(x, y_2)| \ll_\kappa \|f\|_{H^k(U)} \sqrt{|y_2 - y_1|}$$

for $f \in H^k(U)$ and $y_1, y_2 \in (0,1)$. This also shows that for any sequence (y_n) with $y_n \to 0$ as $n \to \infty$ the sequence of functions defined by $K \ni x \mapsto f(x, y_n)$ for $n \geqslant 1$ is a Cauchy sequence with respect to $\|\cdot\|_{K,\infty}$.

BOUNDARY POINTS, CONCLUSION. Thus if $k > 1 + \frac{d-1}{2}$ and $f \in H^k(U \cap V)$ has $\operatorname{Supp} f \subseteq \overline{U} \cap V$, then we can find $\kappa > 0$ so that

$$\operatorname{Supp} f \subseteq [-1 + \kappa, 1 - \kappa]^{d-1} \times [0, 1 - \kappa]$$

and apply the above to see that f can be continuously extended to $\overline{U}$. This proves the remaining local statement. As mentioned before, this discussion also applies to all partial derivatives of f and hence completes the proof of the proposition. $\square$

Much as in the proof of Proposition 8.50, the statement in Theorem 8.49 can be reduced to a purely local statement. Indeed, suppose the following holds.

(LOCAL STATEMENT) For any $z^{(0)} \in \overline{U}$ there exists a neighbourhood V of $z^{(0)}$ in $\mathbb{R}^d$ so that if $g \in H_0^1(U)$ with $\Delta g = u \in H^k(U)$ for some $k \geqslant 0$ and $\operatorname{Supp} g \subseteq \overline{U} \cap V$, then $g \in H^{k+2}(U)$.

Then we may find a finite cover of $\overline{U}$ consisting of such neighbourhoods and an associated smooth partition of unity. Together with Lemma 5.50 this reduces the proof to the local statements (check this).

Moreover, for any interior point $z^{(0)} \in U$ we may take $V = U$, apply elliptic regularity on open subsets (Theorem 5.45 or Theorem 8.48), and multiply g by a smooth $\psi \in C_c^\infty(U)$ with $\psi \equiv 1$ on $\operatorname{Supp} g$ to obtain the local statement.

We have therefore reduced the proof of Theorem 8.49 to the following:

(LOCAL STATEMENT AT BOUNDARY POINTS) For any $z^{(0)} \in \partial U$ there exists a neighbourhood $V \subseteq \mathbb{R}^d$ so that $g \in H_0^1(U)$ with $\Delta g = u \in H^k(U)$ for some $k \geqslant 0$ and $\operatorname{Supp} g \subseteq \overline{U} \cap V$ implies that $g \in H^{k+2}(U)$.

PROOF OF THEOREM 8.49, FLATTENING OF THE BOUNDARY. As in the proof of Proposition 8.50 we wish to flatten out the boundary by a diffeomorphism. Indeed, the assumption that U has smooth boundary implies for any boundary point $z^{(0)} \in \partial U$ that there exists some $\delta > 0$ and a diffeomorphism Φ defined on $V' = (-\delta, \delta)^d$ such that $\Phi(0) = z^{(0)}$ and $\Phi(U') = U \cap V$, where $U' = (-\delta, \delta)^{d-1} \times (0, \delta)$ and $V = \Phi(V')$. The proof of Proposition 8.50 also shows that we may assume that Φ extends to a diffeomorphism on a neighbourhood of $\overline{V'}$, that Φ maps the Lebesgue measure on V' to the Lebesgue measure on V, and induces an isomorphism between $H^k(U')$ and $H^k(U \cap V)$ for all $k \geqslant 0$.

Assume now $g \in H_0^1(U)$ with $\Delta g = u \in H^k(U)$ for some $k \geqslant 0$ and $\operatorname{Supp} g \subseteq \overline{U} \cap V$. Then $\langle g, \Delta\phi \rangle_{L^2(U)} = \langle u, \phi \rangle_{L^2(U)}$ for any $\phi \in C_c^\infty(U \cap V)$.

We define $g' = g \circ \Phi \in H_0^1(U')$, $u' = u \circ \Phi \in H^k(U')$, and obtain

$$\langle g', \Delta(\phi) \circ \Phi \rangle_{L^2(U')} = \langle u', \phi \circ \Phi \rangle_{L^2(U')}. \tag{8.20}$$

Here $\phi \circ \Phi \in C_c^\infty(U')$ is again an arbitrary smooth function with compact support in U', which we will denote by $\varphi \in C_c^\infty(U')$. We will use the shorthand $\Psi = \Phi^{-1}$ for the inverse diffeomorphism and wish to express $\Delta(\phi) \circ \Phi = (\Delta(\varphi \circ \Psi)) \circ \Phi$ in terms of partial derivatives of φ. By the chain rule

$$\partial_\ell(\varphi \circ \Psi) = \sum_{i=1}^d ((\partial_i \varphi) \circ \Psi)(\partial_\ell \Psi_i)$$

and the product rule

$$\partial_\ell^2(\varphi \circ \Psi) = \sum_{i,j=1}^d ((\partial_i \partial_j \varphi) \circ \Psi)(\partial_\ell \Psi_i)(\partial_\ell \Psi_j) + \sum_{i=1}^d ((\partial_i \varphi) \circ \Psi)(\partial_\ell^2 \Psi_i)$$

for all $\ell \in \{1, \ldots, d\}$, and hence

$$\Delta(\phi) \circ \Phi = \Delta(\varphi \circ \Psi) \circ \Phi = \sum_{i,j=1}^d a_{i,j} \partial_i \partial_j \varphi + \sum_{i=1}^d b_i \partial_i \varphi,$$

where $a_{i,j} = \left(\sum_{\ell=1}^d (\partial_\ell \Psi_i)(\partial_\ell \Psi_j)\right) \circ \Phi$ and $b_i = (\Delta \Psi_i) \circ \Phi$ belong to $C^\infty(\overline{U'})$ for all $i, j \in \{1, \ldots, d\}$. Hence the relationship in (8.20) between g' and u' can also be expressed as

$$\langle g', P\varphi \rangle_{L^2(U')} = \langle u', \varphi \rangle_{L^2(U')}$$

for all $\varphi \in C_c^\infty(U')$, where P is the degree two partial differential operator

$$P(\varphi) = \sum_{i,j=1}^d a_{i,j} \partial_i \partial_j \varphi + \sum_{i=1}^d b_i \partial_i \varphi$$

with smooth coefficients $a_{i,j} \in C^\infty(\overline{U'})$ and $b_i \in C^\infty(\overline{U'})$ for $i, j \in \{1, \ldots, d\}$.

We remark that the coefficients $a_{i,j}$ form a symmetric positive-definite matrix. More precisely, for any $x' \in \overline{U'}$ we have $a_{i,j}(x') = a_{j,i}(x')$ for all i, j and on setting $x = \Phi(x')$ we also have that

$$\sum_{i,j=1}^d a_{i,j}(x') v_i v_j = \sum_{\ell=1}^d \left(\sum_{i=1}^d (\partial_\ell \Psi_i)(x) v_i\right)\left(\sum_{j=1}^d (\partial_\ell \Psi_i)(x) v_j\right)$$

$$= \|D\Psi|_x v\|^2 \geq \theta \|v\|^2$$

for all $v \in \mathbb{R}^d$ and some constant θ. Here $D\Psi|_x$ is the total derivative of Ψ at $x = \Phi(x')$ and by invertibility of $D\psi|_x$ and compactness of $\overline{U'}$ the constant $\theta > 0$ can be chosen uniformly for all $x \in U'$. This makes P into a uniformly elliptic operator of degree two (as defined below), and we can apply the argument presented below. $\qquad\square$

Let $U \subseteq \mathbb{R}^d$ be open and bounded. We use functions $a_{i,j}, b_i, c \in C^\infty(\overline{U})$ for $i, j = 1, \ldots, d$ to define a partial differential operator P by

$$P(\phi) = \sum_{i,j=1}^d a_{i,j} \partial_i \partial_j \phi + \sum_{i=1}^d b_i \partial_i \phi + c\phi$$

for $\phi \in C_c^\infty(U)$. The operator P is called a *uniformly elliptic operator of degree two* if $a_{i,j} = a_{j,i}$ for all $i, j \in \{1, \ldots, d\}$ and there exists some uniform constant $\theta > 0$ with

$$\sum_{i,j=1}^d a_{i,j}(x) v_i v_j \geqslant \theta \|v\|^2 = \theta \sum_{i=1}^d v_i^2$$

for all $v \in \mathbb{R}^d$ and $x \in U$. These uniformly elliptic operators satisfy the following (by now familiar) result.

Theorem 8.52 (Elliptic regularity for P). *Let $U \subseteq \mathbb{R}^d$ be bounded and open with smooth boundary, and let P be a uniformly elliptic operator of degree two. Let $g \in H_0^1(U)$, $k \geqslant 0$, and $u \in H^k(U)$. Suppose that*

$$\langle g, P\phi \rangle_{L^2(U)} = \langle u, \phi \rangle_{L^2(U)}$$

for all $\phi \in C_c^\infty(U)$. Then $g \in H^{k+2}(U)$.

We will not prove this theorem in detail (see Exercise 8.53), but instead consider only the 'local version' needed to complete the proof of elliptic regularity of the Laplace operator in Theorem 8.49.

(LOCAL STATEMENT AT BOUNDARY OF A BOX) Let $V = (-1, 1)^d$ and $U = (-1, 1)^{d-1} \times (0, 1)$, and let $k \geqslant 0$. Assume that $g \in H_0^1(U)$ with

$$\mathrm{Supp}\, g \subseteq \overline{U} \cap V = (-1, 1)^{d-1} \times [0, 1),$$

and $u \in H^k(U)$ with

$$\langle g, P\phi \rangle_{L^2(U)} = \langle u, \phi \rangle_{L^2(U)} \tag{8.21}$$

for all $\phi \in C_c^\infty(U)$. Then

$$g \in H^{k+2}(U). \tag{8.22}$$

PROOF OF THE LOCAL STATEMENT FOR THEOREM 8.52 AT A FLAT BOUND-
ARY. We first assume that $k = 0$ and let ϕ be a function in $C_c^\infty(U)$. By
assumption,

$$\int g\Big(\sum_{i,j=1}^{d} a_{i,j}(\partial_i\partial_j\phi) + \sum_{i=1}^{d} b_i(\partial_i\phi) + c\phi \Big)\, \mathrm{d}x = \int u\phi\, \mathrm{d}x. \qquad (8.23)$$

By integration by parts and the product rule we also have

$$\int g a_{i,j}(\partial_i\partial_j\phi)\, \mathrm{d}x = - \int \partial_i(g a_{i,j})(\partial_j\phi)\, \mathrm{d}x$$

$$= - \int a_{i,j}(\partial_i g)(\partial_j\phi)\, \mathrm{d}x - \int g(\partial_i a_{i,j})(\partial_j\phi)\, \mathrm{d}x$$

$$= - \int a_{i,j}(\partial_i g)(\partial_j\phi)\, \mathrm{d}x + \int \partial_j(g\partial_i a_{i,j})\phi\, \mathrm{d}x$$

and

$$\int g b_i \partial_i\phi\, \mathrm{d}x = - \int \partial_i(g b_i)\phi\, \mathrm{d}x$$

for all $i, j \in \{1,\dots,d\}$. Using this we can rewrite (8.23) in the form

$$-\sum_{i,j=1}^{d} \int a_{i,j}(\partial_i g)(\partial_j\phi)\, \mathrm{d}x = \int \widetilde{u}\phi\, \mathrm{d}x \qquad (8.24)$$

for all $\phi \in C_c^\infty(U)$, where

$$\widetilde{u} = u - \sum_{i,j=1}^{d} \partial_j(g\partial_i a_{i,j}) + \sum_{i=1}^{d} \partial_i(g b_i) - cg \in L^2(U). \qquad (8.25)$$

This allows us to ignore the first and zero order terms in the definition of P.
Since (8.24) only involves the first derivatives of $\phi \in C_c^\infty(U)$, it also holds by
continuity for any $\phi = v \in H_0^1(U)$.

THE CHOICE OF v. We fix some $\ell \in \{1,\dots,d-1\}$. Then

$$v = D_\ell^{-h} D_\ell^h g \in H_0^1(U)$$

for small enough $h \in \mathbb{R}$ by the assumptions $g \in H_0^1(U)$, $\mathrm{Supp}\, g \subseteq \overline{U} \cap V$,
and $\ell \neq d$. With this choice, (8.24) becomes

$$-\underbrace{\sum_{i,j=1}^{d} \int a_{i,j}(\partial_i g)(\partial_j D_\ell^{-h} D_\ell^h g)\, \mathrm{d}x}_{\mathcal{L}} = \underbrace{\int \widetilde{u} D_\ell^{-h} D_\ell^h g\, \mathrm{d}x}_{\mathcal{R}}.$$

STUDYING THE LEFT-HAND SIDE. Using the assumption $\operatorname{Supp} g \subseteq \overline{U} \cap V$ and the abbreviations $g_i = \partial_i g$, and $a_{i,j}^h(x) = a_{i,j}(x + he_\ell)$, we see that

$$
\mathcal{L} = \sum_{i,j=1}^{d} \int D_\ell^h(a_{i,j} g_i) D_\ell^h g_j \, \mathrm{d}x
$$

$$
= \underbrace{\sum_{i,j=1}^{d} \int a_{i,j}^h (D_\ell^h g_i)(D_\ell^h g_j) \, \mathrm{d}x}_{\mathcal{M}} + \underbrace{\sum_{i,j=1}^{d} \int (D_\ell^h a_{i,j}) g_i (D_\ell^h g_j)}_{\mathcal{E}}
$$

since the product rule for D_ℓ^h has the form

$$
D_\ell^h(a_{i,j} g_i)(x) = \frac{1}{h} \big(a_{i,j}(x + he_\ell) g_i(x + he_\ell) - a_{i,j}(x) g_i(x) \big)
$$

$$
= \frac{1}{h} \Big(a_{i,j}(x + he_\ell) \big(g_i(x + he_\ell) - g_i(x) \big)
$$

$$
+ \big(a_{i,j}(x + he_\ell) - a_{i,j}(x) \big) g_i(x) \Big)
$$

$$
= a_{i,j}^h(x) D_\ell^h g_i(x) + (D_\ell^h a_{i,j})(x) g_i(x)
$$

for $i,j \in \{1,\ldots,d\}$ and $x \in U$. For $x \in U$ close to $\partial U \smallsetminus V$ we have $g_i(x) = 0$. Extending g_i and $a_{i,j}$ trivially, this also holds for all $x \in \mathbb{R}^d$.

The term $\mathcal{M}$ is our main term since the uniform ellipticity assumption on P implies that

$$
\mathcal{M} \geqslant \theta \sum_{i=1}^{d} \|D_\ell^h g_i\|_2^2. \tag{8.26}
$$

The extra term $\mathcal{E}$ may be bounded using the Cauchy inequality with an ε as in (8.11) to obtain

$$
|\mathcal{E}| \leqslant \sum_{i,j=1}^{d} \Big(\varepsilon \|D_\ell^h g_j\|_2^2 + \frac{1}{4\varepsilon} \|(D_\ell^h a_{i,j}) g_i\|_2^2 \Big)
$$

$$
\leqslant \varepsilon \theta^{-1} d \mathcal{M} + \frac{\kappa d}{4\varepsilon} \sum_{i=1}^{d} \|g_i\|_2^2, \tag{8.27}
$$

where we also used (8.26) and bounded the supremum norm of $D_\ell^h a_{i,j}$ on $\operatorname{Supp} g_j$ by some constant $\kappa > 0$.

BOUNDING THE RIGHT-HAND SIDE. Since our right-hand side has the same shape as the right-hand side of (8.13) the argument on p. 277 applies to give

$$
|\mathcal{R}| \leqslant \varepsilon \|D_\ell^h g_\ell\|_2^2 + \frac{1}{4\varepsilon} \|\widetilde{u}\|_2^2 \leqslant \varepsilon \theta^{-1} \mathcal{M} + \frac{1}{4\varepsilon} \|\widetilde{u}\|_2^2. \tag{8.28}
$$

PUTTING THE ESTIMATES TOGETHER. Using $\mathcal{L} = \mathcal{M} + \mathcal{E} = \mathcal{R}$ together with the estimates in (8.26)–(8.28), we obtain

$$|\mathcal{M}| = |\mathcal{R} - \mathcal{E}| \leqslant |\mathcal{R}| + |\mathcal{E}| \leqslant \varepsilon\theta^{-1}(d+1)\mathcal{M} + \tfrac{1}{\varepsilon}C,$$

where $C > 0$ depends only on $(a_{i,j})_{i,j}$, $(g_i)_i$ and $\widetilde{u}$ but not on h. We choose ε to be $\frac{\theta}{2(d+1)}$ which allows us to obtain the estimate $|\mathcal{M}| \leqslant \frac{2}{\varepsilon}C$ and hence

$$\sum_{i=1}^{d} \|D_\ell^h g_i\|_2^2 \leqslant \theta^{-1}|\mathcal{M}| \leqslant \tfrac{2}{\theta\varepsilon}C \tag{8.29}$$

for all sufficiently small $h \in \mathbb{R}\backslash\{0\}$ and $\ell \in \{0,\ldots,d-1\}$.

EXISTENCE OF SECOND WEAK PARTIAL DERIVATIVES. The uniform estimate in (8.29) and Corollary 8.43 applied to the subset

$$W_s = (-1+s, 1-s)^{d-1} \times (0,1) \subseteq U$$

for some $s > 0$ implies that $\boldsymbol{\partial}_\ell(g_i|_{W_s})$ exists for any $i \in \{1,\ldots,d\}$ and for any $\ell \in \{1,\ldots,d-1\}$. We choose $s > 0$ such that

$$\operatorname{Supp} g \subseteq (-1+2s, 1-2s)^{d-1} \times [0,1).$$

Extending $\boldsymbol{\partial}(g_i|_{W_s})$ trivially we obtain the existence of $\boldsymbol{\partial}_\ell g_i = \boldsymbol{\partial}_\ell\boldsymbol{\partial}_i g$ on U for $i \in \{1,\ldots,d\}$ and $\ell \in \{1,\ldots,d-1\}$ (check this).

To show the existence of the partial derivative $\boldsymbol{\partial}_d\boldsymbol{\partial}_d g$ we use the argument above and the operator P more directly. In fact, we note that $a_{d,d} \geqslant \theta$ on $\overline{U}$ by the uniform ellipticity assumption on P. Let $\phi \in C_c^\infty(U)$. Using $\frac{1}{a_{d,d}}\phi$ in (8.24) we obtain

$$\int a_{d,d}(\boldsymbol{\partial}_d g)\partial_d\big(\tfrac{1}{a_{d,d}}\phi\big)\,\mathrm{d}x = -\sum_{\substack{1\leqslant i,\ell\leqslant d,\\(i,\ell)\neq(d,d)}}\int a_{i,\ell}(\boldsymbol{\partial}_i g)\partial_\ell\big(\tfrac{1}{a_{d,d}}\phi\big)\,\mathrm{d}x - \int \widetilde{u}\tfrac{1}{a_{d,d}}\phi\,\mathrm{d}x$$

and hence also

$$\int \boldsymbol{\partial}_d g\partial_d\phi\,\mathrm{d}x = -\int a_{d,d}\boldsymbol{\partial}_d g\partial_d\big(\tfrac{1}{a_{d,d}}\big)\phi\,\mathrm{d}x$$
$$+\sum_{\substack{1\leqslant i,\ell\leqslant d,\\(i,\ell)\neq(d,d)}}\int \boldsymbol{\partial}_\ell\big(a_{i,\ell}\boldsymbol{\partial}_i g\big)\tfrac{1}{a_{d,d}}\phi\,\mathrm{d}x - \int \widetilde{u}\tfrac{1}{a_{d,d}}\phi\,\mathrm{d}x,$$

which shows the existence of

$$\boldsymbol{\partial}_d\boldsymbol{\partial}_d g = a_{d,d}\boldsymbol{\partial}_d g\partial_d\big(\tfrac{1}{a_{d,d}}\big) - \sum_{\substack{1\leqslant i,\ell\leqslant d,\\(i,\ell)\neq(d,d)}}\boldsymbol{\partial}_\ell\big(a_{i,\ell}\boldsymbol{\partial}_i g\big)\tfrac{1}{a_{d,d}} + \widetilde{u}\tfrac{1}{a_{d,d}} \in L^2(U). \tag{8.30}$$

Since $\operatorname{Supp} g \subseteq \overline{U} \cap V$ by assumption, we may extend g and its partial derivatives trivially from $(-1,1)^{d-1} \times (0,1)$ to the half-space $\mathbb{R}^{d-1} \times (0,\infty)$ and again obtain a function g together with all its degree one and two partial derivatives.

EXTENDING COROLLARY 8.47. From the existence of all second-order partial derivatives of g on U we would like to conclude that $g \in H^2(U)$. For this we again combine Proposition 8.45 and Lemma 8.46, much as in the argument in the proof of Corollary 8.47.

By continuity of the regular representation in Lemma 3.74 we may find, for every $\varepsilon > 0$, some $s > 0$ so that the function g_s defined by

$$g_s(x) = g(x + se_d)$$

for $x \in \mathbb{R}^{d-1} \times (-s, \infty)$ and extended trivially outside that set satisfies

$$\|\partial_\alpha g - \partial_\alpha g_s\|_2 < \varepsilon$$

for all $\alpha \in \mathbb{N}_0^d$ with $\|\alpha\| \leqslant 2$. By Proposition 8.45 and Lemma 8.46 the function $\overline{g}_\delta = g_s * \jmath_\delta$ is smooth, and satisfies $\|\overline{g}_\delta - g\|_2 < 2\varepsilon$ for $\delta \in (0,s)$ sufficiently small. Also by Proposition 8.45 and our shift of the functions derivatives $\partial_\alpha \overline{g}_\delta$ of $\overline{g}_\delta$ for all $\alpha \in \mathbb{N}_0^d$ with $\|\alpha\| \leqslant 2$ can be expressed on U by convolution of $\partial_\alpha g_s$ with $\jmath_\delta$ and so $\|\partial_\alpha \overline{g}_\delta - \partial_\alpha g\|_{L^2(U)} < 2\varepsilon$ for sufficiently small $\delta \in (0,s)$ and $\alpha \in \mathbb{N}_0^d$ with $\|\alpha\| \leqslant 2$. Therefore $g \in H^2(U)$, which concludes the case $k = 0$.

INDUCTION ON $k \geqslant 0$. The argument above gives the base of the induction on k. Suppose we already know (8.22) for $k - 1 \geqslant 0$ and assume again that $u \in H^k(U)$. By the inductive hypothesis we already know $g \in H^{k+1}(U)$. We again fix some $\ell \in \{1, \ldots, d-1\}$ and claim that $\partial_\ell g \in H_0^1(U)$ and that there exists some $u_\ell \in H^{k-1}(U)$ with

$$\langle \partial_\ell g, P\phi \rangle = \langle u_\ell, \phi \rangle \tag{8.31}$$

for all $\phi \in C_c^\infty(U)$. The inductive hypothesis then implies $\partial_\ell g \in H^{k+1}(U)$. By varying $\ell \in \{1, \ldots, d-1\}$ this shows the existence of all partial derivatives $\partial_\alpha g$ for $\alpha \in \mathbb{N}_0^d$ with $\|\alpha\|_1 \leqslant k+2$ except for $\alpha = (k+2)e_d$. To obtain this partial derivative we take the kth partial derivative of $\partial_d^2 g$ as in (8.30) with respect to the dth coordinate and obtain the existence of $\partial_{(k+2)e_d} g$ by using the other partial derivatives of degree at most $(k+2)$. In fact, if we have $u \in H^k(U)$ and $g \in H^{k+1}(U)$ then $\widetilde{u} \in H^k(U)$ by (8.25). Using $\partial_\ell g \in H^{k+1}(U)$ for each $\ell \in \{1, \ldots, d-1\}$ and (8.30) we also have $\partial_d^2 g \in H^k(U)$. As in the base case, this implies that $g \in H^{k+2}(U)$.

For the proof of the first part of the claim we note that $f \in H^1(U)$ and $\operatorname{Supp}(f) \subseteq \overline{U} \cap V$ implies

$$\left(D_\ell^h f\right)(x) = \frac{1}{h} \int_0^h \partial_\ell f(x + se_\ell)\, \mathrm{d}s \tag{8.32}$$

for sufficiently small $h \in \mathbb{R} \setminus \{0\}$ and almost every $x \in U$. Indeed, this holds for any $f \in H^1(U) \cap C^\infty(U)$ and $x, x + h e_\ell \in U$, extends by continuity to any $f \in H^1(U)$ and almost every $x \in U$ with $x + h e_\ell \in U$, and then holds for f with $\mathrm{Supp}\, f \subseteq \overline{U} \cap V$ for almost every $x \in U$ and sufficiently small $h \neq 0$. By the continuity of the regular representation in Lemma 3.74 the identity (8.32) implies that

$$\partial_\ell f = \lim_{h \to 0} D_\ell^h f.$$

Using this for $g \in H^2(U)$ and its partial derivatives $\partial_i g$ for $i = 1, \ldots, d$ we find for a given $\varepsilon > 0$ some $h > 0$ such that

$$\|\partial_\ell g - D_\ell^h g\|_2 < \varepsilon$$

and

$$\|\partial_\ell \partial_i g - D_\ell^h \partial_i g\|_2 < \varepsilon$$

for $i = 1, \ldots, d$. Since $g \in H_0^1(U)$, $\mathrm{Supp}(g) \subseteq \overline{U} \cap V$ and $\ell \in \{1, \ldots, d-1\}$ we have $D_\ell^h g \in H_0^1(U)$ for all sufficiently small h, and since $\varepsilon > 0$ was arbitrary this implies $\partial_\ell g \in H_0^1(U)$, as claimed.

For the second part (8.31) of the claim we let $\phi \in C_c^\infty(U)$ and calculate

$$
\begin{aligned}
\langle \partial_\ell g, P\phi \rangle &= -\langle g, \partial_\ell(P\phi) \rangle \\
&= -\left\langle g, \partial_\ell \Big(\sum_{i,j=1}^d a_{i,j} \partial_i \partial_j \phi + \sum_{i=1}^d b_i \partial_i \phi + c\phi \Big) \right\rangle \\
&= -\left\langle g, P\partial_\ell \phi + \sum_{i,j=1}^d (\partial_\ell a_{i,j})(\partial_i \partial_j \phi) + \sum_{i=1}^d (\partial_\ell b_i)(\partial_i \phi) + (\partial_\ell c)\phi \right\rangle \\
&= -\langle u, \partial_\ell \phi \rangle - \sum_{i,j=1}^d \langle g\partial_\ell a_{i,j}, \partial_i \partial_j \phi \rangle - \sum_{i=1}^d \langle g\partial_\ell b_i, \partial_i \phi \rangle - \langle g\partial_\ell c, \phi \rangle \\
&= \left\langle \partial_\ell u - \sum_{i,j=1}^d \partial_i \partial_j (g\partial_\ell a_{i,j}) + \sum_{i=1}^d \partial_i (g\partial_\ell b_i) - g\partial_\ell c, \phi \right\rangle \\
&= \langle u_\ell, \phi \rangle,
\end{aligned}
$$

where

$$
u_\ell = \underbrace{\partial_\ell u}_{\in H^{k-1}(U)} - \underbrace{\sum_{i,j=1}^d \partial_i \partial_j (g\partial_\ell a_{i,j})}_{\in H^{k-1}(U)} + \underbrace{\sum_{i=1}^d \partial_i (g\partial_\ell b_i)}_{\in H^k(U)} - \underbrace{g\partial_\ell c}_{\in H^{k+1}(U)}
$$

belongs to $H^{k-1}(U)$, as claimed. This concludes the induction and hence the proof of Theorem 8.49. $\qquad\square$

Exercise 8.53. Complete the proof of Theorem 8.52 following the steps below.

(a) State and give a detailed proof of the extension of Corollary 8.47 that was used in the above proof.

(b) Generalize Lemma 5.50 to allow the uniformly elliptic operator P instead of just Δ.

(c) Use the assumption that U has smooth boundary and a smooth partition of unity to localize the situation. Apply the above proof on each of the local statements.

8.3 Topologies on the space of bounded operators

Let X and Y be Banach spaces. Then we have seen that the space $\mathrm{B}(X, Y)$ of bounded linear operators from X to Y together with the operator norm is again a Banach space.

Definition 8.54. Let X and Y be Banach spaces. The topology on $\mathrm{B}(X, Y)$ induced by the operator norm is called the *uniform operator topology*.

Since any Banach space has a weak topology, there is of course also a weak topology on $\mathrm{B}(X, Y)$. There are, however, further topologies that make special use of the fact that $\mathrm{B}(X, Y)$ is a space of maps.

Definition 8.55. Let X and Y be Banach spaces. The *strong operator topology* on $\mathrm{B}(X, Y)$ is the weakest topology for which the evaluation maps

$$\mathrm{B}(X, Y) \ni L \longmapsto Lx \in Y$$

are continuous for every $x \in X$, where we use the norm topology on Y.

In other words, a neighbourhood of $L_0 \in \mathrm{B}(X, Y)$ in the strong operator topology is a set containing a set of the form

$$N_{x_1, \ldots, x_n; \varepsilon}(L_0) = \bigcap_{i=1}^{n} \left\{ L \in \mathrm{B}(X, Y) \mid \|Lx_i - L_0 x_i\| < \varepsilon \right\}$$

for some $x_1, \ldots, x_n \in X$ and $\varepsilon > 0$. Equivalently, we could define the strong operator topology by using all neighbourhoods defined by the semi-norms

$$\|L\|_{x_1, \ldots, x_n} = \max\{\|Lx_1\|, \ldots, \|Lx_n\|\}.$$

The strong operator topology is in many situations more natural than the uniform topology, and the study of unitary representations (see Definition 3.73 for the general definition) is an example.

Example 8.56. Let $\mathcal{H} = L^2(\mathbb{R})$ and define for $x \in \mathbb{R}$ the unitary map

$$\rho_x : \mathcal{H} \to \mathcal{H}$$

by $\rho_x f(t) = f(t - x)$ for $t \in \mathbb{R}$ and $f \in \mathcal{H}$. We claim that

$$\|\rho_x - \rho_y\| = 2\delta_{xy} = \begin{cases} 2 & \text{if } x \neq y; \\ 0 & \text{if } x = y. \end{cases}$$

In fact, if $x < y$ and $M > 0$ then we can define a function $f \in \mathcal{H}$ (illustrated in Figure 8.1) by

$$f(t) = \begin{cases} 0 & \text{for } t < 0; \\ (-1)^m & \text{for } t \in [m(y - x), (m + 1)(y - x)) \text{ with } 0 \leqslant m < M; \\ 0 & \text{for } t \geqslant M(y - x). \end{cases}$$

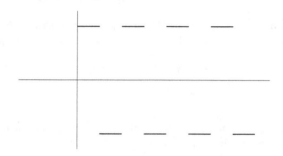

Fig. 8.1: The function f in Example 8.56.

Then $\rho_{y-x} f$ satisfies

$$|(f - \rho_{y-x} f)(t)| = |2f(t)| = 2$$

for almost every $t \in (y - x, M(y - x))$, so that

$$\|f - \rho_{y-x} f\|_2 \geqslant 2\sqrt{(M - 1)(y - x)}$$

while $\|f\|_2 = \sqrt{M(y - x)}$. As $M \geqslant 1$ was arbitrary, this shows that

$$\|\rho_x - \rho_y\| = \|I - \rho_{y-x}\| = 2$$

since $\|\rho_x\| = \|\rho_{-x}\| = 1$.

The claim implies that the map $\mathbb{R} \ni x \mapsto \rho_x \in \mathrm{B}(\mathcal{H}, \mathcal{H})$ is not continuous with respect to the uniform operator topology. However, it is continuous with respect to the strong operator topology, for if $f_1, \ldots, f_n \in \mathcal{H}$ and $\varepsilon > 0$ are given, then for y sufficiently close to x we have

$$\|\rho_y f_i - \rho_x f_i\|_2 < \varepsilon$$

for $i = 1, \ldots, n$ by the continuity property from Lemma 3.74. Thus, for y sufficiently close to x we have $\rho_y \in N_{f_1, \ldots, f_n; \varepsilon}(\rho_x)$, as required.

Exercise 8.57. Let X and Y be Banach spaces. Show that the strong operator topology on $B(X, Y)$ has the following properties:

(1) it is Hausdorff;
(2) it is weaker than the uniform operator topology (defined by the operator norm);
(3) a sequence (T_n) in $B(X, Y)$ converges to $T_0 \in B(X, Y)$ as $n \to \infty$ in the strong operator topology if and only if $T_n(v) \to T(v)$ as $n \to \infty$ for all $v \in X$; and
(4) a filter $\mathcal{F}$ on $B(X, Y)$ converges to $T_0 \in B(X, Y)$ if the filter generated by

$$\big\{ \{Tv \mid T \subset \mathcal{F}\} \mid F \in \mathcal{F} \big\}$$

converges to $T_0 v$ for all $v \in X$.

Another topology on $B(X, Y)$ is built up using functionals on Y.

Definition 8.58. Let X and Y be Banach spaces. The *weak operator topology* on $B(X, Y)$ is the weakest topology with respect to which the maps

$$B(X, Y) \ni L \longmapsto y^*(Lx)$$

are continuous for all $x \in X$ and $y^* \in Y^*$.

Equivalently, the weak operator topology can be defined using the neighbourhoods defined by the semi-norms

$$\|L\|_{x_1, y_1^*; x_2, y_2^*; \ldots; x_n, y_n^*} = \max\{|y_1^*(Lx_1)|, \ldots, |y_n^*(Lx_n)|\}.$$

Exercise 8.59. Assume that X and Y are infinite-dimensional Banach spaces. Show that the uniform topology, the weak topology, the strong operator topology, and the weak operator topology are all different Hausdorff topologies on $B(X, Y)$.

8.4 Locally Convex Vector Spaces

Even if we were initially only interested in Banach spaces, the last few sections should have left no doubt that the next definition is natural and unavoidable. It gives a class of topological vector spaces generalizing normed vector spaces.

Definition 8.60. Let X be a vector space (over $\mathbb{R}$ or $\mathbb{C}$) and suppose that

$$\{\|\cdot\|_\alpha \mid \alpha \in A\}$$

is a family of semi-norms on X with the property that for every $x \in X \smallsetminus \{0\}$ there is some $\alpha \in A$ with $\|x\|_\alpha > 0$. Then the *locally convex topology* on X induced by the semi-norms is the topology for which a neighbourhood of the point $x_0 \in X$ is a set containing a set of the form

$$N_{\alpha_1,\ldots,\alpha_n;\varepsilon}(x_0) = \bigcap_{i=1}^{n} B_\varepsilon^{\|\cdot\|_{\alpha_i}}(x_0) = \left\{ x \in X \mid \max_{i=1,\ldots,n} \|x - x_0\|_{\alpha_i} < \varepsilon \right\}.$$

The vector space X together with this topology is called a *locally convex vector space.*

Equivalently, a locally convex topology is the weakest topology that is stronger than those defined by a collection of semi-norms. Enlarging the collection of semi-norms if necessary, we may assume that for $\alpha_1, \ldots, \alpha_n \in A$ the semi-norm

$$\|x\|' = \max_{i=1,\ldots,n} \|x\|_{\alpha_i}$$

also belongs to the collection (that is, coincides with $\| \cdot \|_\alpha$ for some $\alpha \in A$). If this is the case, then the neighbourhoods of $x \in X$ are sets containing a ball of the form

$$B_\varepsilon^{\|\cdot\|_\alpha}(x_0) = \left\{ x \in X \mid \|x - x_0\|_\alpha < \varepsilon \right\}$$

for some $\alpha \in A$ and $\varepsilon > 0$.

An equivalent definition of locally convex vector spaces is obtained by requiring that the topology on the vector space X is Hausdorff, makes addition and scalar multiplication continuous, and has a basis of neighbourhoods of the point $0 \in X$ consisting of absorbent balanced convex sets. Here a convex set $C \subseteq X$ is *balanced* if for any $x \in C$ and scalar ρ with $|\rho| \leqslant 1$ we also have $\rho x \in C$, and is *absorbent* if for any $x \in X$ there exists some $\alpha > 0$ with $\alpha x \in C$. We refer to Exercise 8.72 and Conway [19, Sec. IV.1] for the equivalence, as we will not need it in this form.

Essential Exercise 8.61. Show that a locally convex vector space (as in Definition 8.60) has the property that addition and scalar multiplication are continuous, and that $0 \in X$ has a basis consisting of absorbent balanced convex sets.

As the next exercise shows, even if a locally convex vector space topology cannot be described using a norm, the locally convex structure is enough to obtain results similar to those obtained as corollaries of the Hahn–Banach theorem (Theorem 7.3).

Exercise 8.62. Let X be a locally convex vector space. Show that the space X^* of continuous linear functionals on X separates points.

We have seen many examples of locally convex vector spaces. These include normed vector spaces with their norm or weak topology, duals of Banach spaces with the weak* topology, and the space $B(X, Y)$ of operators between two Banach spaces with any of the topologies discussed in Section 8.3. However, there are further spaces that we have neglected so far because they do not fit well (or at all) into the framework of normed spaces.

Example 8.63. (1) The space $C^\infty([0,1])$ is a locally convex vector space with the semi-norms

$$\|f\|_{C^n([0,1])} = \max_{j=0,\ldots,n} \|f^{(j)}\|_\infty$$

for $n \in \mathbb{N}$. Notice that even though each of these semi-norms is already a norm, we still have to use all of them to define the locally convex topology on $C^\infty([0,1])$ we are interested in, namely the topology of uniform convergence of all derivatives. Notice that differentiation

$$C^\infty([0,1]) \ni f \xmapsto{D} f' \in C^\infty([0,1])$$

is a continuous operator on $C^\infty([0,1])$.

(2) Let $U \subseteq \mathbb{R}^d$ be an open set. Then

$$C_b^\infty(U) = \bigcap_{k \geqslant 0} C_b^k(U),$$

with $C_b^k(U)$ defined as in Example 2.24(6), is another example of a locally convex vector space if we use all of the norms $\|\cdot\|_{C_b^k(U)}$ for $k \geqslant 1$.

(3) Let $U \subseteq \mathbb{R}^d$ be an open set. Another important notion of convergence in analysis for functions on U is the notion of *uniform convergence on compact subsets*. For example, on the space $C(U)$ this notion is captured if we use the collection of semi-norms $\{\|\cdot\|_{K,\infty} \mid K \subseteq U \text{ compact}\}$, where

$$\|f\|_{K,\infty} = \sup_{x \in K} |f(x)|$$

for $f \in C(U)$ is the supremum norm of the restriction to K.

(4) Let $U \subseteq \mathbb{R}^d$ be an open set. We can also make $C_c(U)$ into a locally convex space in a natural way by endowing it with the collection of semi-norms

$$\{\|\cdot\|_F \mid F \in C(U)\},$$

where $\|f\|_F = \|fF\|_\infty$ for $f \in C_c(U)$ is the supremum norm taken after multiplication by $F \in C(U)$. The corresponding notion of convergence is less familiar but is natural for elements of $C_c(U)$ (see Exercise 8.64 below). The convergence is uniform across U, and this remains true after multiplication with any continuous function on U, however rapidly it might increase towards ∂U.

Exercise 8.64. We use the notation from Example 8.63(4) in this exercise.

(a) Show that $f \in C(U)$ belongs to $C_c(U)$ if and only if $\|f\|_F < \infty$ for all $F \in C(U)$.

(b) Suppose that (f_n) is a sequence of functions in $C_c(U)$ that converges in $C_c(U)$ to some $f \in C_c(U)$. Show that there exists a compact set $K \subseteq U$ such that

$$\mathrm{Supp}(f_n), \mathrm{Supp}(f) \subseteq K$$

for all $n \geqslant 1$.

In general, the topology of a locally convex vector space is not metrizable. One important situation in which it is metrizable is when it is sufficient to use countably many semi-norms. This is the case in Example 8.63(1), (2) and (3) (for the latter recall that an open subset of $\mathbb{R}^d$ is σ-compact), but is not for Example 8.63(4). If the locally convex topology on X is given by the semi-norms $\|\cdot\|_n$ for $n \in \mathbb{N}$, then we can define a metric on X as in Lemma A.17, leading to the following definition.

Definition 8.65. A *Fréchet space* is a locally convex vector space X whose topology is defined by countably many semi-norms $\|\cdot\|_n$ for $n \in \mathbb{N}$, such that X is complete with respect to the metric

$$d(x,y) = \sum_{n=1}^{\infty} \frac{1}{2^n} \frac{\|x-y\|_n}{1+\|x-y\|_n}. \tag{8.33}$$

Exercise 8.66. Suppose that the topology of a locally convex vector space X is induced by countably many semi-norms $\|\cdot\|_n$ for $n \in \mathbb{N}$.
(a) Show that a sequence (x_n) in X is a Cauchy sequence with respect to the metric in (8.33) if and only if (x_n) is a Cauchy sequence with respect to all of the semi-norms $\|\cdot\|_n$ for $n \in \mathbb{N}$.
(b) Show that if two families of semi-norms $\{\|\cdot\|_n\}$ and $\{\|\cdot\|'_n\}$ make X into a locally convex vector space with the same topology, then X is complete with respect to d if and only if X is complete with respect to d', where d' is defined using $\{\|\cdot\|'_n\}$ (just as in (8.33)).
(c) Show that the spaces from Example 8.63(1), (2) and (3) are Fréchet spaces.

The following exercise indicates why we restricted attention to the study of locally convex vector spaces instead of considering the larger class of topological vector spaces.

Exercise 8.67 (A topological vector space with trivial dual). Let $\mathrm{MF}([0,1])$ denote the space of all (equivalence classes of) complex-valued measurable functions on $[0,1]$, where functions are equivalent if they agree almost everywhere with respect to Lebesgue measure m on $[0,1]$. Given $f_0 \in \mathrm{MF}([0,1])$ and $\varepsilon > 0$ we define the ε-neighbourhood of f_0 by

$$U_\varepsilon(f_0) = \big\{f \in \mathrm{MF}([0,1]) \mid m\big(\{x \in [0,1] \mid |f(x) - f_0(x)| > \varepsilon\}\big) < \varepsilon\big\}.$$

(a) Show that the above system of neighbourhoods defines a basis of neighbourhoods with respect to a topology on $\mathrm{MF}([0,1])$. We note that the corresponding notion of convergence is called *convergence in measure*.
(b) Show that the vector space operations in $\mathrm{MF}([0,1])$ are continuous, making $\mathrm{MF}([0,1])$ into a so-called *topological vector space*.
(c) Show that the dual space

$$\mathrm{MF}([0,1])^* = \{\ell : \mathrm{MF}([0,1]) \to \mathbb{C} \mid \ell \text{ is linear and continuous}\}$$

is given by $\mathrm{MF}([0,1])^* = \{0\}$.
(d) Show that if a sequence (f_n) in $\mathrm{MF}([0,1])$ converges almost everywhere to f in $\mathrm{MF}([0,1])$, then it also converges in measure to f. Give an example of a sequence that converges in measure but not almost everywhere.
(e) Show that a sequence (f_n) in $\mathrm{MF}([0,1])$ converges in measure to f in $\mathrm{MF}([0,1])$ if every subsequence of (f_n) has a subsequence that converges almost everywhere to f.

8.5 Distributions as Generalized Functions

Both in applications and within mathematics it is often useful to have a generalized notion of function to allow, for example, a function F on $\mathbb{R}$ with the property that

$$\int_{\mathbb{R}} \phi(x) F(x) \, \mathrm{d}x = \phi(0) \qquad (8.34)$$

for any 'nice' function $\phi : \mathbb{R} \to \mathbb{R}$. Such an F might represent a point mass (a dimensionless object of mass 1 located at 0), or be a mathematical representation of an impulse in physics. Since F is certainly not a function, one needs to develop a new theory that includes such objects.[27] The theory of *distributions* allows for such generalized functions, and permits them to be differentiated, multiplied by smooth functions, and so on. Of course if we were only interested in expressions of the form in (8.34) then we could simply study measures, since (8.34) is simply the integral against the Dirac measure δ_0 at the origin. However, within the space of measures it does not normally make sense to take derivatives (and this is the case for δ_0), while it is possible to define a derivative map in the space of distributions.

The most direct approach to distributions superficially seems to be a cheat: We declare a distribution to be a linear continuous *functional* (that is, a linear continuous map to the base field $\mathbb{R}$ or $\mathbb{C}$) on a space of nice test functions $\{\phi\}$. Here the definition of 'nice' may vary, to give different classes of distributions. For example, we could fix an open subset $U \subseteq \mathbb{R}^d$ and all $\phi \in C_c^\infty(U)$ as test functions.

Requiring continuity of the linear functional is natural but needs a topology on $C_c^\infty(U)$. We declare the topology on $C_c^\infty(U)$ by introducing the following systems of semi-norms, which make $C_c^\infty(U)$ into a locally convex vector space. In fact, for every $\alpha \in \mathbb{N}_0^d$ and $F \in C(U)$ we define the semi-norm

$$\|f\|_{\alpha, F} = \|(\partial_\alpha f) F\|_\infty$$

for $f \in C_c^\infty(U)$. Using $F = \mathbb{1}$ and $\alpha = 0$ shows that these include $\|\cdot\|_\infty$, so that the topology is indeed Hausdorff. We define the space $\mathscr{D}(U)$ of distributions on U to be the space of continuous linear functionals on the locally convex vector space $C_c^\infty(U)$.

This definition of a distribution is a cheat because we have finessed the problem that no function F satisfies (8.34) by simply declaring F to be the distribution (that is, continuous linear functional) which sends the test function ϕ to $\phi(0)$ without giving a more direct generalization of functions on $\mathbb{R}$. We may write this formally as

$$\langle F, \phi \rangle = \phi(0),$$

where we write $\langle F, \phi \rangle$ for the action of the functional F on the test function ϕ. One sometimes also writes $\int_{\mathbb{R}} F\phi$ for $\langle F, \phi \rangle$, especially if we continue to think of F as a generalized function, but whenever one wants to prove something about F one has to go back to the formal definition of F as a functional on $C_c^\infty(U)$. Even though this may look dubious at first sight, the intuition provided by the viewpoint that F is a generalized function is often useful, and will stay consistent with the formal treatment of F as a linear functional. Our discussion of the Dirichlet boundary value problem (in Section 5.2–5.3) and the eigenfunctions of the Laplace operator (in Section 6.4) have already made use of the viewpoint provided by distribution theory. However, we will not develop the theory here, referring to the monograph of Schwartz [94, 95] for a thorough treatment.

Exercise 8.68. Show that any integrable function on an open subset $U \subseteq \mathbb{R}^d$ gives rise to a distribution. That is, any f in $L^1(U)$ defines a linear functional F_f on the space $C_c^\infty(U)$ of smooth compactly supported functions via

$$\langle F_f, \phi \rangle = \int f(x)\phi(x)\,\mathrm{d}x.$$

Prove that the resulting map $f \longmapsto F_f$ is linear and injective. Actually it is sufficient to assume that $f \in L^1_{\mathrm{loc}}(U)$, the space of *locally integrable* functions, measurable functions that are integrable on any compact set.

Exercise 8.69. Show that no measurable and locally integrable function $f : \mathbb{R} \to \mathbb{R}$ has the property (8.34) for all $\phi \in C_c^\infty(\mathbb{R})$.

Exercise 8.70. Let $U \subseteq \mathbb{R}^d$ be open and $\alpha \in \mathbb{N}_0^d$.
(a) Show that the linear map

$$\partial_\alpha : C_c^\infty(U) \longrightarrow C_c^\infty(U)$$
$$f \longmapsto \partial_\alpha f$$

is continuous with respect to the locally convex topology on $C_c^\infty(U)$.
(b) Define $\boldsymbol{\partial}_\alpha = -\partial_\alpha^* : \mathscr{D}(U) \to \mathscr{D}(U)$ by $\boldsymbol{\partial}_\alpha F = -F \circ \partial_\alpha$ for all $F \in \mathscr{D}(U)$. Show that we have $\boldsymbol{\partial}_\alpha \psi = \partial_\alpha \psi$ for $\psi \in C^\infty(U)$ if we identify ψ with the distribution

$$F_\psi : C_c^\infty(U) \ni f \longmapsto \int f\psi\,\mathrm{d}m$$

as in Exercise 8.68. In other words, the operator $\boldsymbol{\partial}_\alpha$ extends differentiation on $C^\infty(U)$ to $\mathscr{D}(U)$.
(c) Show that for $\psi \in C^\infty(U)$ and $F \in \mathscr{D}(U)$ we can define a new distribution

$$\psi \cdot F : C_c^\infty(U) \ni \phi \longmapsto \langle \psi \cdot F, \phi \rangle = \langle F, \psi\phi \rangle$$

which depends linearly on $\psi \in C^\infty(U)$ and linearly on $F \in \mathscr{D}(U)$. Prove also the product rule

$$\boldsymbol{\partial}_j(\psi \cdot F) = (\partial_j \psi) \cdot F + \psi \cdot (\boldsymbol{\partial}_j F).$$

8.6 Convex Sets

A set $K \subseteq X$ in a vector space is called *absorbent* if for any $x \in X$ there exists some $\alpha > 0$ with $\alpha x \in K$. We note that for a convex set $K \subseteq X$ with $0 \in K$ and some given $x \in X$ the set $\{t > 0 \mid \frac{1}{t}x \in K\}$ is either empty, an open interval $(p_K(x), \infty)$, or a closed interval $[p_K(x), \infty)$ for some $p_K(x) \geqslant 0$. Moreover, if $K = B_1^{\|\cdot\|}$ for some semi-norm $\|\cdot\|$ on X, then K is an absorbent convex set and $p_K(x) = \|x\|$ for any $x \in X$. A partial converse is given by the following result, which gives a solution to Exercise 7.2.

Lemma 8.71. *Let* $K \subseteq X$ *be an absorbent convex set in a vector space. Define the gauge function* $p_K : X \to \mathbb{R}_{\geqslant 0}$ *by*

$$p_K(x) = \inf\{t > 0 \mid \tfrac{1}{t}x \in K\}.$$

Then $p_K(\alpha x) = \alpha p_K(x)$ *and* $p_K(x + y) \leqslant p_K(x) + p_K(y)$ *for all* $\alpha \geqslant 0$ *and* $x, y \in X$.

PROOF. The positive homogeneity follows directly from the definition. Suppose now that $x, y \in X$ and $t_x, t_y > 0$ have

$$\frac{1}{t_x}x, \frac{1}{t_y}y \in K. \tag{8.35}$$

Then

$$\frac{1}{t_x + t_y}(x + y) = \frac{t_x}{t_x + t_y}\left(\frac{1}{t_x}x\right) + \frac{t_y}{t_x + t_y}\left(\frac{1}{t_y}y\right)$$

also lies in K, since K is convex. Thus $p_K(x + y) \leqslant t_x + t_y$, and since this holds for all t_x, t_y with (8.35), the triangle inequality follows. $\square$

Exercise 8.72. Use Lemma 8.71 to prove the converse to Exercise 8.61. More precisely, let X be a vector space endowed with a Hausdorff topology. Assume that addition and scalar multiplication are continuous and that $0 \in X$ has a basis of neighbourhoods consisting of absorbent balanced convex sets. Show that X is a locally convex space in the sense of Definition 8.60.

In the following discussion concerning convex sets we will frequently restrict to real locally convex vector spaces. This is not a severe restriction as every complex locally convex vector space X can also be considered a real vector space, and every continuous linear functional $\ell_{\mathbb{R}}$ on X has the form $\ell_{\mathbb{R}} = \Re\ell_{\mathbb{C}}$ for a continuous linear functional $\ell_{\mathbb{C}}$ on X (see the proof of the complex case in Theorem 7.3).

The next result strengthens Corollary 7.4 and Exercise 8.62, and is readily explained by Figure 8.2.

Theorem 8.73 (Separation from convex sets). *Let* X *be a locally convex vector space. Let* $K \subseteq X$ *be a closed convex set, and suppose that* $z \in X \setminus K$.

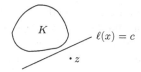

Fig. 8.2: Separation of $z \notin K$ from the convex set K by a closed hyperplane.

Then there exists a continuous linear functional $\ell \in X^*$ and a constant $c \in \mathbb{R}$ such that $\ell(y) \leqslant c < \ell(z)$ for all $y \in K$.

PROOF. Since $z \notin K$ and K is closed, $X \smallsetminus K$ is a neighbourhood of z, and in particular $N_{\alpha_1,\dots,\alpha_n;\varepsilon}(z) \subseteq X \smallsetminus K$ for some $\alpha_1,\dots,\alpha_n \in A$ and $\varepsilon > 0$ (see Definition 8.60 for the notation). We define $U = N_{\alpha_1,\dots,\alpha_n;\varepsilon/2}(0)$, so that $z + 2U \subseteq X \smallsetminus K$.

Without loss of generality we may assume that $0 \in K$ (for otherwise we can just translate both K and z by the negative of an element of K). Define

$$M = K + U = \{y + u \mid y \in K, u \in U\}$$

and notice that M is convex because both K and U are (check this) and that M is absorbent as it contains U.

We now apply Lemma 8.71 to obtain the norm-like function p_M. By definition, we have

$$p_M(\cdot) \leqslant \tfrac{2}{\varepsilon} \max\{\|\cdot\|_{\alpha_1}, \dots, \|\cdot\|_{\alpha_n}\} \tag{8.36}$$

since $U \subseteq M$.

We claim that $p_M(z) > 1$. For otherwise there exists a sequence (λ_n) with $\lambda_n \to 1$ as $n \to \infty$ and with

$$\tfrac{1}{\lambda_n} z = k_n + u_n \in M = K + U$$

for all $n \geqslant 1$. Clearly $\tfrac{1}{\lambda_n} z$ and u_n are bounded in the semi-norms

$$\|\cdot\|_{\alpha_1}, \dots, \|\cdot\|_{\alpha_n},$$

so the same holds also for k_n. Now rewrite the above equation as

$$z = k_n + (\lambda_n - 1)k_n + \lambda_n u_n$$

and notice that for large enough n we have

$$(\lambda_n - 1)k_n + \lambda_n u_n \in \tfrac{1}{2}U + \tfrac{3}{2}U = 2U,$$

since $\lambda_n \to 1$ as $n \to \infty$. However, this contradicts $z + 2U \subseteq X \smallsetminus K$. Therefore we must have $p_M(z) > 1$.

Now define $Z = \mathbb{R}z$ and a functional $\ell \in Z^*$ by $\ell(\alpha z) = \alpha p_M(z)$. Observe that $\ell(\alpha z) \leqslant p_M(\alpha z)$ for all $\alpha \in \mathbb{R}$ (for $\alpha \geqslant 0$ we have equality and for $\alpha < 0$ it follows from $\ell(\alpha z) < 0 \leqslant p_M(\alpha z)$) and $\ell(z) > 1$. By the Hahn–Banach lemma (Lemma 7.1) ℓ extends to all of X with $\ell(x) \leqslant p_M(x)$ for all $x \in X$. This implies that $\ell(y) \leqslant 1$ for $y \in K \subseteq M$. Moreover, this estimate also gives continuity of ℓ since (8.36) implies that

$$\ell(x) \leqslant \tfrac{2}{\varepsilon} \max\{\|x\|_{\alpha_1}, \ldots, \|x\|_{\alpha_n}\},$$

which upgrades to

$$|\ell(x)| \leqslant \tfrac{2}{\varepsilon} \max\{\|x\|_{\alpha_1}, \ldots, \|x\|_{\alpha_n}\}$$

by linearity of ℓ and since the right-hand side is a semi-norm. This gives the theorem (for $c = 1$). $\qquad\square$

Since the weak topology is, for infinite-dimensional vector spaces, strictly coarser than the norm topology, there is no reason why a set that is closed in the norm topology should be closed in the weak topology. However, for convex sets the situation is better.

Corollary 8.74. *Suppose that K is a convex set in a real Banach space X. Then the norm and weak closure of K agree. That is, $\overline{K}^{\text{norm}} = \overline{K}^{\text{weak}}$.*

PROOF. Suppose that $z \notin \overline{K}^{\text{norm}}$ and apply Theorem 8.73 to find a continuous linear functional $\ell \in X^*$ with $\ell(y) \leqslant c < \ell(z)$ for all $y \in \overline{K}^{\text{norm}}$ and some $c \in \mathbb{R}$. Therefore, $\ell^{-1}((c, \infty)) \subseteq X \setminus K$ is a neighbourhood of z in the weak topology and $z \notin \overline{K}^{\text{weak}}$ follows. Thus $\overline{K}^{\text{weak}} \subseteq \overline{K}^{\text{norm}}$. The reverse inclusion $\overline{K}^{\text{norm}} \subseteq \overline{K}^{\text{weak}}$ is clear as the norm topology is stronger than the weak topology. $\qquad\square$

Exercise 8.75. Generalize Corollary 8.74 to complex Banach spaces.

Exercise 8.76. Suppose that $K, L \subseteq X$ are disjoint convex sets in a locally convex vector space X over $\mathbb{R}$. Suppose one of them has non-empty interior. Show that there exists a non-trivial continuous linear functional ℓ and a constant $c \in \mathbb{R}$ such that $\ell(x) \leqslant c \leqslant \ell(y)$ for all $x \in K$ and $y \in L$.

Exercise 8.77. Let X be a normed vector space over $\mathbb{R}$, and let $K \subseteq X$ be a non-empty closed and convex subset. Show that

$$\inf_{x \in K} \|z - x\| = \sup_{\substack{\ell \in X^* \\ \|\ell\|=1}} \left(\ell(z) - \sup_{x \in K} \ell(x)\right)$$

for any $z \in X \setminus K$.

Exercise 8.78. Let X be a real Banach space and let $\imath : X \to X^{**}$ be the embedding of X into its bidual as in Corollary 7.9. Show that $\imath\left(B_1^X\right)$ is dense in $\overline{B_1^{X^{**}}}$ when X^{**} is equipped with the weak* topology.

8.6.1 Extreme Points and the Krein–Milman Theorem

An important concept for convex sets, both abstractly and for many concrete applications (see, for example, Proposition 8.36), is the notion of extreme points.

Definition 8.79. Let X be a locally convex space and let $K \subseteq X$ be a convex subset. An element $x \in K$ is an *extreme point* of K if x cannot be expressed as a proper convex combination of points of K (that is, if $x = sy + (1 - s)z$ with $y, z \in K$ and $s \in (0,1)$ then we must have $x = y = z$).

As illustrated in Figure 8.3, the set of extreme points of a convex set will not be closed in general, even in a finite-dimensional setting. In infinite-dimensional spaces, the situation is more complex still and the extreme points may even be dense (see Exercise 8.84). The smallest closed convex subset of a locally convex space X that contains $A \subseteq X$ is called the *closed convex hull* of A and is the intersection of all closed convex sets containing A, or equivalently the closure of the convex hull of A.

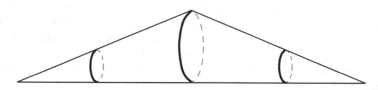

Fig. 8.3: The set of extreme points need not be closed: Here two cone-like objects are glued together at their base so that a single straight line connects the two cone points in the resulting convex set. The extreme points are the two ends together with all but one point of the central circle.

Theorem 8.80 (Krein–Milman). *Let X be a locally convex space over $\mathbb{R}$, and let $K \subseteq X$ be a compact convex subset. Then K is the closed convex hull of its extreme points. In particular, if K is non-empty then K has some extreme points.*

For the proof the following extension of the definition of extreme points will be useful. A subset $E \subseteq K$ of a convex set is called an *extremal subset* of K if E is convex, non-empty, and if $x = sy + (1 - s)z$ for $x \in E$, $y, z \in K$ and $s \in (0,1)$ forces $y, z \in E$. To better understand this notion, it may be helpful to the reader to find all extremal subsets of a polygon in $\mathbb{R}^2$ or of a polytope in $\mathbb{R}^3$.

PROOF OF THEOREM 8.80. We assume without loss of generality that $K \neq \varnothing$. The proof uses Zorn's lemma, applied to the set

$$\mathscr{F} = \{E \subseteq K \mid E \text{ is an extremal closed subset of } K\}$$

with the partial order defined by $E_1 \succcurlyeq E_2$ if $E_1 \subseteq E_2$. Note in particular that $K \in \mathscr{F}$, so that $\mathscr{F}$ is non-empty. We need to show that for any linearly ordered subset $\{E_\alpha \mid \alpha \in I\}$ there exists an element $E \in \mathscr{F}$ with $E \succcurlyeq E_\alpha$ for every $\alpha \in I$. We claim that

$$E = \bigcap_{\alpha \in I} E_\alpha$$

is such an element. For this we only need to show that $E \in \mathscr{F}$, as the fact that $E \succcurlyeq E_\alpha$ for every $\alpha \in I$ then follows directly from the definition of $\succcurlyeq$.

Since each E_α is closed and convex, the same holds for the intersection E. Since each E_α is non-empty and $\{E_\alpha \mid \alpha \in I\}$ is linearly ordered, we see that every finite intersection $E_{\alpha_1} \cap \cdots \cap E_{\alpha_n}$ is non-empty, because it must coincide with one of the sets $E_{\alpha_1}, \ldots, E_{\alpha_n}$. Since K is compact, we see that the intersection E is non-empty (see Appendix A.4). It remains to show that E is an extremal subset. Suppose therefore that $x = sy + (1-s)z \in E$ with $y, z \in K$ and $s \in (0,1)$. Then $x \in E_\alpha$ for all $\alpha \in I$ as $E \subseteq E_\alpha$. By extremality of E_α this forces $y, z \in E_\alpha$ for all $\alpha \in I$, and so $y, z \in E$ as required.

In summary, we have shown that we are in a position to use Zorn's lemma, so that there must be a maximal element E of $\mathscr{F}$. In our setting, this is a minimal closed extremal subset of K. We claim that $E = \{x\}$ is a singleton, which then implies that x must be an extreme point of K. Indeed, if E contains two points x_0, y_0, then by Theorem 8.73 there exists a continuous linear functional ℓ on X with $\ell(x_0) < \ell(y_0)$. However, by compactness this implies that

$$E' = \left\{ z \in E \mid \ell(z) = \sup \ell|_E \right\}$$

is a non-empty proper closed convex subset of E. It is also an extremal subset of K, since if $x = sy + (1-s)z \in E'$ with $y, z \in K$ and $s \in (0,1)$, then we must have $y, z \in E$ as E is extremal and so $\ell(x) = s\ell(y) + (1-s)\ell(z)$ and $\ell(y), \ell(z) \leqslant \ell(x) = \max \ell|_E$, which implies that $y, z \in E'$, as required, since $s \in (0,1)$. However, this is a contradiction since $E \subseteq K$ was supposed to be a minimal closed extremal subset of K. Therefore, $E = \{x_0\}$ is a singleton and we have shown that the set of extreme points of K is non-empty.

Now let M denote the closed convex hull of the set of all extreme points of K. Clearly $M \subseteq K$ and we need to show that $M = K$.

Suppose that x_0 lies in $K \setminus M$. By Theorem 8.73 there exists a continuous linear functional ℓ with $\ell(y) \leqslant c < \ell(x_0)$ for all $y \in M$. Now let

$$F = \{x \in K \mid \ell(x) = \max \ell|_K\},$$

and notice that $F \subseteq K \setminus M$ is a closed convex subset of K. Therefore, F is compact and by the above argument there exists an extreme point $x \in F$. We claim that x is also an extreme point of K, which gives a contradiction since then $x \in M \subseteq K \setminus F$, by definition of M.

So suppose that $x = sy + (1-s)z$ with $y, z \in K$ and $s \in (0,1)$. Then

$$\ell(x) = s\ell(y) + (1 - s)\ell(z)$$

and $\ell(y), \ell(z) \leqslant \max \ell|_K = \ell(x)$, which implies that $y, z \in F$ since $s \in (0, 1)$ and hence $x = y = z$ by extremality of x in F.

This contradiction shows that $K = M$ is the closed convex hull of the set of extreme points. $\qquad\square$

The Krein–Milman theorem, together with the Banach–Alaoglu theorem (Theorem 8.10), can produce some striking consequences.

Example 8.81. Let us show that $c_0(\mathbb{N})$ has no pre-dual. In other words, there is no Banach space X with the property that X^* is isometrically isomorphic to $c_0(\mathbb{N})$. Indeed, suppose that there is such a Banach space. Then, by the Banach–Alaoglu theorem, the unit ball of $c_0(\mathbb{N})$ would be weak* compact. Thus, by the Krein–Milman theorem,[†] the unit ball would have to contain some extreme point $(a_n)_{n \geqslant 1}$. We complete the argument by showing that there cannot be such an extreme point of the unit ball.

By definition, $|a_n| \leqslant 1$ for all $n \geqslant 1$ and $\lim_{n \to \infty} a_n = 0$. Therefore, there exists some n_0 with $|a_{n_0}| < \frac{1}{2}$ and then the sequences (b_n) and (c_n) defined by

$$b_n = \begin{cases} a_n & \text{for } n \neq n_0, \\ a_n + \frac{1}{2} & \text{for } n = n_0 \end{cases}$$

and

$$c_n = \begin{cases} a_n & \text{for } n \neq n_0, \\ a_n - \frac{1}{2} & \text{for } n = n_0 \end{cases}$$

are different, both belong to the unit ball by construction, and we have

$$(a_n) = \tfrac{1}{2}(b_n) + \tfrac{1}{2}(c_n),$$

which shows that (a_n) is not an extreme point.

Exercise 8.82. In Example 8.81 we showed that V^* is never isometrically isomorphic to $c_0(\mathbb{N})$. Generalize the result in two ways as follows. Show that there is no Banach space V with the property that V^* is isomorphic to $C_0(X)$, where X is a σ-compact, locally compact, non-compact space and the isomorphism is only assumed to be a bounded operator with a bounded inverse.

Exercise 8.83. Let X be a σ-compact, locally compact metric space.
(a) Find the extreme points of the closed unit ball $\overline{B_1^{\mathcal{M}(X)}}$ in the space of signed measures on X, and the extreme points of the convex set $\mathcal{P}(X)$ of all probability measures on X.
(b) Assume in addition that X is compact and infinite. Show that the assumptions of the Krein–Milman theorem (Theorem 8.80) hold, but that $\mathcal{P}(X)$ is not the convex hull of its extreme points. In other words, taking the closure of the convex hull is important in infinite dimensions.

[†] This assumes we are working over $\mathbb{R}$, but as mentioned before, the arguments extend easily to $\mathbb{C}$.

(c) Assume now instead that X is non-compact. Show that the conclusion of the Krein–Milman theorem (Theorem 8.80) holds for $\mathcal{P}(X)$ (despite the fact that the assumptions do not).

In many applications where convex subsets of Banach spaces or locally convex spaces appear the extreme points play a special role. One instance of this arose in our brief excursion into ergodic theory (see Section 8.2.1), where the ergodic measures (which may now be seen to exist in great generality due to the Krein–Milman theorem) are precisely the extreme points of the convex set of invariant probability measures.

The following example (or rather part (c) of it) shows how badly intuition can fail for convex sets in infinite dimensions.

Exercise 8.84. Let $K = \{f \in C_{\mathbb{R}}([0,1]) \mid f(0) = 0,\ f$ is 1-Lipschitz$\}$.
(a) Show that K is convex and compact in the norm topology.
(b) Show that any function $f \in K$ which is piecewise linear and has slope ± 1 wherever f is differentiable is an extreme point.
(c) Show that the extreme points in K are dense in K.
(d) Describe all the extreme points of K.

8.6.2 Choquet's Theorem

We now further refine the Krein–Milman theorem by showing that every point of a compact convex set can be obtained as a 'generalized convex combination' of extreme points of K. However — even after taking account of convergence questions — convex combinations alone will not be sufficiently general, as the next example shows.

Exercise 8.85. Let X and $\mathcal{P}(X) \subseteq C(X)^*$ be as in Exercise 8.83. Describe the elements of $\mathcal{P}(X)$ that can be written as a convergent (in norm, or equivalently in the weak* topology) convex combination $\sum_{n=1}^{\infty} c_n \nu_n$ of extreme points $\nu_n \in \mathcal{P}(X)$ with $c_n \geq 0$ for all $n \geq 1$ and $\sum_{n=1}^{\infty} c_n = 1$. Now let $X = [0,1]$ and give examples of Borel probability measures that cannot be obtained as such limits.

Definition 8.86. Let $K \subseteq X$ be a compact convex subset of a locally convex vector space, and suppose that the induced topology on K is metrizable. Let μ be a Borel probability measure on K and $x \in K$. We say that μ *represents* x (or that x is the *barycentre* of μ) if

$$\ell(x) = \int_K \ell \, \mathrm{d}\mu$$

for every $\ell \in X^*$.

Notice that each $\ell \in X^*$ is continuous on K and hence is integrable with respect to any μ on K as in the definition above.

Essential Exercise 8.87. Show that the barycentre of a Borel probability measure μ on a metrizable compact convex subset is uniquely determined by μ.

Throughout the discussion of this subsection we will assume that the induced topology on $K \subseteq X$ is metrizable, writing simply (as above) that K *is a metrizable subset of* X. With Proposition 8.11 and Exercise 8.12 in mind, it should be clear why we do not wish to assume that X itself is metrizable.

Lemma 8.88. *Let K be a metrizable compact convex subset of a locally convex vector space X over $\mathbb{R}$. Let μ be a probability measure on K. Then μ has a barycentre in K.*

PROOF. For every $\ell \in X^*$ we define the closed hyperplane

$$H_\ell = \left\{ x \in X \mid \ell(x) = \int_K \ell \, d\mu \right\}.$$

Notice that for a fixed $\ell \in X^*$, this hyperplane is not empty since ℓ is linear. The lemma is equivalent to the statement that

$$K \cap \bigcap_{\ell \in X^*} H_\ell \neq \varnothing.$$

However, since K is compact and the sets H_ℓ are closed, it is sufficient to show that

$$K \cap H_{\ell_1} \cap \cdots \cap H_{\ell_n} \neq \varnothing \qquad (8.37)$$

for any $\ell_1, \ldots, \ell_n \in X^*$ and $n \geqslant 1$.

For this, consider the continuous linear map

$$L : X \longrightarrow \mathbb{R}^n$$
$$x \longmapsto (\ell_1(x), \ldots, \ell_n(x))$$

and the compact convex set $L(K) \subseteq \mathbb{R}^n$. We claim that

$$b = \left(\int_K \ell_1 \, d\mu, \ldots, \int_K \ell_n \, d\mu \right) \in L(K), \qquad (8.38)$$

and note that this is equivalent to (8.37). Hence the claim implies the lemma.

Suppose therefore that (8.38) does not hold. Then by Theorem 8.73 (applied to $L(K) \subseteq \mathbb{R}^n$) there exists a functional ϕ defined by $\phi(t) = \sum_{j=1}^{n} a_j t_j$ for $t \in \mathbb{R}^n$ and some row vector $a \in \mathbb{R}^n$ such that

$$\phi(b) > \sup_{x \in K} \phi(L(x)).$$

Defining $\ell^* = \sum_{j=1}^{n} a_j \ell_j = \phi \circ L \in X^*$ we now obtain

$$\int_K \ell^* \, d\mu = \sum_{j=1}^{n} a_j \int_K \ell_j \, d\mu = \phi(b) > \sup_{x \in K} \phi(L(x)) = \sup_{x \in K} \ell^*(x),$$

which gives a contradiction since μ is assumed to be a probability measure.

$\qquad\qquad\qquad\qquad\qquad\qquad\qquad\qquad\qquad\qquad\qquad\qquad\qquad\qquad\qquad$ $\square$

Exercise 8.89. Let $K \subseteq X$ be a metrizable compact convex subset of a locally convex vector space over $\mathbb{R}$, and let $M \subseteq K$ be a closed subset. Prove that the closure of the convex hull of M is precisely the set of all barycentres of all Borel probability measures μ with $\operatorname{Supp}\mu \subseteq M$.

Theorem 8.90 (Choquet's theorem). *Let $K \subseteq X$ be a metrizable compact convex subset of a locally convex space X over $\mathbb{R}$. Then the set of extreme points $\operatorname{ext}(K)$ of K is Borel measurable. Moreover, for any $x_0 \in K$ there exists a Borel probability measure μ on K with $\mu\big(\operatorname{ext}(K)\big) = 1$ that represents x_0.*

While the issues arising in the proof are functional-analytic, the intuition behind the statement is essentially geometric, which is more visible in a finite-dimensional version illustrated in the next exercise.

Exercise 8.91 (Carathéodory's form of Minkowski's theorem). Let $K \subseteq \mathbb{R}^n$ be a compact convex subset. Show that any point $x_0 \in K$ is a convex combination of $(n+1)$ extreme points of K.

For the proof of Theorem 8.90 we will need the following lemma and some notation. We write A for the space of *affine functions* on X, that is, functions of the form $a(x) = \ell(x) + c$ for some $\ell \in X^*$ and $c \in \mathbb{R}$. Moreover, recall that $\|f\|_{K,\infty}$ denotes the supremum norm of a function f restricted to some subset $K \subseteq X$.

Lemma 8.92 (Upper envelope). *Let $K \subseteq X$ be as in Theorem 8.90. Given a bounded function $f : K \to \mathbb{R}$ we define the* upper envelope *of f by*

$$\overline{f}(x) = \inf\{a(x) \mid a \in A \text{ and } a \geqslant f \text{ on } K\}$$

for all $x \in K$. Then

(1) $f \leqslant \overline{f} \leqslant \|f\|_{K,\infty}$.
(2) $\overline{f}$ *is concave (that is, $\overline{f}\left(\lambda x + (1-\lambda)y\right) \geqslant \lambda\overline{f}(x) + (1-\lambda)\overline{f}(y)$ for all x, y in K and λ with $0 \leqslant \lambda \leqslant 1$).*
(3) $\overline{f}$ *is upper-semicontinuous (that is, the pre-image $\overline{f}^{-1}\left((-\infty, c)\right)$ is open for every $c \in \mathbb{R}$), and so in particular Borel measurable.*
(4) *If f is concave and upper-semicontinuous then $\overline{f} = f$.*
(5) *Given $r \geqslant 0$, $a \in A$, and another bounded function g on K, we have:*

$\qquad$ (a) $\overline{rf} = r\overline{f}$,
$\qquad$ (b) $\overline{f+a} = \overline{f} + a$,
$\qquad$ (c) $\overline{f+g} \leqslant \overline{f} + \overline{g}$, *and*
$\qquad$ (d) $|\overline{f} - \overline{g}| \leqslant \|f - g\|_{K,\infty}$.

PROOF. Since the constant function $\|f\|_{K,\infty}$ belongs to A, (1) follows at once from the definition of $\overline{f}$.

Given $x, y \in K$, $\lambda \in [0,1]$ and $a \in A$ with $a \geqslant f$ on K we see that

$$a\big(\lambda x + (1 - \lambda)y\big) = \lambda a(x) + (1 - \lambda)a(y) \geqslant \lambda \overline{f}(x) + (1 - \lambda)\overline{f}(y)$$

by definition of $\overline{f}(x)$ and $\overline{f}(y)$. The claim in (2) follows by taking the infimum over a.

For (3), let $c \in \mathbb{R}$ and $x_0 \in K$ such that $\overline{f}(x_0) < c$. Then there exists some $a \in A$ with $a \geqslant f$ and $a(x_0) < c$. Clearly

$$\{x \in K \mid a(x) < c\} \subseteq \{x \in K \mid \overline{f}(x) < c\}$$

and, since a is continuous, the former is an open neighbourhood of x_0. Since x_0 was an arbitrary element of $\{x \in K \mid \overline{f}(x) < c\}$ we obtain (3).

For (4), assume that f is concave and upper-semicontinuous. We claim that this implies that the set

$$M = \{(x,c) \in K \times \mathbb{R} \mid c \leqslant f(x)\}$$

of points on or underneath the graph of f is convex and closed. Here we consider M as a subset of $X' = X \times \mathbb{R}$, which is the locally convex vector space with the collection of semi-norms defined by $\|(x,c)\|'_\alpha = \max\{\|x\|_\alpha, |c|\}$ for $(x,c) \in X'$ and every semi-norm $\|\cdot\|_\alpha$ of the locally convex vector space X.

To see that M is convex, fix $(c_1, x_1), (c_2, x_2) \in M$ and $\lambda \in [0,1]$. Then we have $c_1 \leqslant f(x_1)$, $c_2 \leqslant f(x_2)$, and

$$\lambda c_1 + (1 - \lambda)c_2 \leqslant \lambda f(x_1) + (1 - \lambda)f(x_2) \leqslant f(\lambda x_1 + (1 - \lambda)x_2),$$

which shows that $\lambda(x_1, c_1) + (1 - \lambda)(x_2, c_2) \in M$.

Notice that in order to show that M is closed in X' it is enough to show that M is closed in $K \times \mathbb{R}$, since the latter is closed in X'. So suppose that $(x_0, c_0) \in (K \times \mathbb{R}) \backslash M$. By definition, this means that $f(x_0) < c_0$, so we can choose $c \in (f(x_0), c_0)$, and apply the definition of upper-semicontinuity to see that $f^{-1}((-\infty, c)) \times (c, \infty)$ is open in $K \times \mathbb{R}$, contains (x_0, c_0), and does not intersect M. This means that M is closed in $K \times \mathbb{R}$ and hence in X'.

Now let $x_0 \in K$ and $f(x_0) < c_0$, so that $z = (x_0, c_0) \notin M$. Applying Theorem 8.73 we find some functional $\ell' \in (X')^*$ with $\ell'((x,c)) < \ell'((x_0, c_0))$ for all $(x,c) \in M$. Clearly $\ell'((x,c)) = \ell(x) + ac$ for some $\ell \in X^*$ and $\alpha \in \mathbb{R}$. Since for x_0 we have $(x_0, f(x_0)) \in M$, $f(x_0) < c_0$ and

$$\ell(x_0) + \alpha f(x_0) < \ell(x_0) + \alpha c_0,$$

we see that $\alpha > 0$. Dividing ℓ' by α we may thus assume that $\alpha = 1$. Using the points $(x, f(x)) \in M$ for $x \in K$ we obtain $\ell(x) + f(x) < \ell(x_0) + c_0$ for all $x \in K$. Then $a(x) = -\ell(x) + c_0 + \ell(x_0)$ defines a function $a \in A$

with $f(x) < a(x)$ for all $x \in K$, and so $\overline{f}(x_0) \leqslant a(x_0) = c_0$. Since $x_0 \in K$ and $c_0 > f(x_0)$ were arbitrary we deduce that $\overline{f} = f$ as required.

Now let $r \geqslant 0$, $a \in A$ and g be as in (5). For $r = 0$ the statement is clear. For $r > 0$ and a function $a \in A$ on X we have $a \geqslant f$ if and only if $ra \geqslant rf$. Therefore (a) follows from standard properties of the infimum.

Next notice that $a_f \geqslant f$, $a_g \geqslant g$ and $a_f, a_g \in A$ implies that

$$A \ni a_f + a_g \geqslant f + g$$

as in the definition of $\overline{f + g}$. Hence $a_f + a_g \geqslant \overline{f + g}$, and taking the infimum over a_f and a_g separately gives $\overline{f} + \overline{g} \geqslant \overline{f + g}$, which is (c).

Applying this for $g = a \in A$ and using $\overline{a} = a$ and $\overline{-a} = -a$, we see that

$$\overline{f + a} \leqslant \overline{f} + \overline{a} = \overline{f} + a = \overline{f + a - a} + a \leqslant \overline{f + a} + \overline{-a} + a = \overline{f + a},$$

and so $\overline{f + a} = \overline{f} + a$, giving (b).

The proof of (d) is similar: In fact

$$\overline{f} = \overline{f - g + g} \leqslant \overline{f - g} + \overline{g}$$

and thus $\overline{f} - \overline{g} \leqslant \overline{f - g} \leqslant \|f - g\|_{K,\infty}$ by (1). Reversing the roles of f and g gives $|\overline{f} - \overline{g}| \leqslant \|f - g\|_{K,\infty}$ and hence the lemma. $\qquad \square$

PROOF OF THEOREM 8.90. By Lemma 2.46, $C(K)$ is separable and so the same applies to the subspace $A \subseteq C(K)$ of affine functions (when restricted to K and using the supremum norm $\| \cdot \|_{K,\infty}$ on K). Hence we may choose a dense countable subset $D = \{a_n \mid n \in \mathbb{N}\} \subseteq A$ with $\overline{D} = \overline{A} \subseteq C(K)$. Notice that D is still dense if we remove from it any (potentially contained) a_n with $a_n|_K = 0$, so we may assume that $\|a_n\|_{K,\infty} \neq 0$ for all $n \in \mathbb{N}$. Moreover, since X^* separates points by Theorem 8.73, we know that D separates points in K. Now define the function

$$F = \sum_{n=1}^{\infty} \frac{1}{n^2 \|a_n\|_{K,\infty}^2} a_n^2,$$

whose properties are crucial for the proof of the theorem. Since the series converges uniformly on K, we see that $F \in C(K)$. We claim that F is strictly convex, meaning that

$$F(\lambda x + (1 - \lambda)y) < \lambda F(x) + (1 - \lambda)yF(y) \tag{8.39}$$

for all $x \neq y \in K$ and $\lambda \in (0, 1)$. To see this, note that

$$a_n(\lambda x + (1 - \lambda)y) = \lambda a_n(x) + (1 - \lambda)a_n(y)$$

for any $n \in \mathbb{N}$, $x \neq y \in K$ and $\lambda \in (0, 1)$ so that

$$a_n(\lambda x + (1-\lambda)y)^2 \leqslant \lambda a_n(x)^2 + (1-\lambda)a_n(y)^2 \qquad (8.40)$$

by convexity of the map $t \mapsto t^2$. Also, since D separates points we have

$$a_{n_0}(x) \neq a_{n_0}(y)$$

for some $n_0 \in \mathbb{N}$ and hence a strict inequality in (8.40) for this choice of $n = n_0$ (by strict convexity of $t \mapsto t^2$). Summing over n gives (8.39).

We now fix $x_0 \in K$, which we wish to represent. Using x_0 we define the subspace

$$V = A + \mathbb{R}F \subseteq C(K),$$

the linear functional

$$\Lambda(a + cF) = a(x_0) + c\overline{F}(x_0)$$

for any $a + cF \in V$, and the function p defined by $p(f) = \overline{f}(x_0)$ for all f in $C(K)$.

By Lemma 8.92(5), the function p is norm-like, as required in the Hahn–Banach lemma (Lemma 7.1). Also, by Lemma 8.92(5) we have

$$\Lambda(a + cF) = a(x_0) + c\overline{F}(x_0) = \overline{a + cF}(x_0) = p(a + cF)$$

whenever $c \geqslant 0$. For $c < 0$ the function cF is concave and continuous, so that $\overline{cF} = cF$ by Lemma 8.92(4). Using also the inequality in Lemma 8.92(1) (multiplied by $c < 0$) and Lemma 8.92(5b) we now see that

$$\Lambda(a + cF) = a(x_0) + c\overline{F}(x_0) \leqslant a(x_0) + cF(x_0)$$
$$= a(x_0) + \overline{cF}(x_0) = \overline{a + cF}(x_0) = p(a + cF)$$

also for $c < 0$. Hence all the assumptions of the Hahn–Banach lemma are satisfied and so there is an extension of Λ (which we again denote by Λ) to all of $C(K)$ satisfying

$$\Lambda(f) \leqslant p(f) = \overline{f}(x_0) \leqslant \|f\|_{K,\infty}$$

for all $f \in C(K)$ (where the last inquality is given by Lemma 8.92(1)). For a non-negative function $f \in C(K)$ we also have $-f \leqslant 0$ and so $\Lambda(f) \geqslant 0$. Since $\mathbb{1} \in A \subseteq V$ we have $\Lambda(\mathbb{1}) = \mathbb{1}(x_0) = 1$ by definition of Λ. Therefore we may apply the Riesz representation theorem (Theorem 7.44) to obtain a Borel probability measure μ on K with

$$\Lambda(f) = \int_K f \, \mathrm{d}\mu \qquad (8.41)$$

for all $f \in C(K)$. Applying Λ to linear functionals $\ell \in X^* \subseteq A$ we see that

$$\int_K \ell \, d\mu = \Lambda(\ell) = \ell(x_0),$$

so x_0 is the barycentre of the probability measure μ. It remains to see that $\mathrm{ext}(K)$ is measurable and that $\mu\big(\mathrm{ext}(K)\big) = 1$.

For this, we claim that

$$\int_K F \, d\mu = \overline{F}(x_0) = \int_K \overline{F} \, d\mu. \tag{8.42}$$

The first of these equations follows directly by applying the functional to F since μ satisfies (8.41) and $\Lambda(F)$ was defined as $\overline{F}(x_0)$. The proof of the second equality is less direct: By Lemma 8.92(1) we have $F \leqslant \overline{F}$ and hence

$$\int_K F \, d\mu \leqslant \int_K \overline{F} \, d\mu.$$

On the other hand $a \in A$ and $a \geqslant F$ implies that $a \geqslant \overline{F}$ and therefore

$$\int_K \overline{F} \, d\mu \leqslant \int_K a \, d\mu = \Lambda(a) = a(x_0).$$

Taking the infimum over these $a \in A$ we get

$$\int_K \overline{F} \, d\mu \leqslant \overline{F}(x_0)$$

by definition of $\overline{F}$. However, since the first equality in (8.42) is already proven these inequalities together prove (8.42).

In particular, we see from (8.42) and the inequality $F \leqslant \overline{F}$ that

$$\mu\big(\{x \in K \mid F < \overline{F}\}\big) = 0. \tag{8.43}$$

We claim that

$$\mathrm{ext}(K) \supseteq \{x \in K \mid F = \overline{F}\}. \tag{8.44}$$

To see this, let $z = \lambda x + (1 - \lambda)y \in K$ be a non-extreme point (that is, with $x \neq y \in K$ and $\lambda \in (0,1)$). Since F is strictly convex, this gives

$$F(z) < \lambda F(x) + (1 - \lambda)F(y)$$
$$\leqslant \lambda \overline{F}(x) + (1 - \lambda)\overline{F}(y) \leqslant \overline{F}(z)$$

by Lemma 8.92(1) and the concavity of $\overline{F}$ from Lemma 8.92(2). Hence

$$F(z) < \overline{F}(z),$$

and the claim follows. Note that once we have shown the measurability of $\mathrm{ext}(K)$, this claim implies that $\mu\big(X \setminus \mathrm{ext}(K)\big) = 0$ by (8.43).

To see that $\mathrm{ext}(K)$ is measurable, let d denote a metric on K that induces the topology on K and notice that

$$K \smallsetminus \mathrm{ext}(K) = \{\lambda x + (1-\lambda)y \mid \lambda \in (0,1),\ x \neq y \in K\}$$
$$= \underbrace{\bigcup_{n \in \mathbb{N}} \{\lambda x + (1-\lambda)y \mid \lambda \in [\tfrac{1}{n}, 1 - \tfrac{1}{n}];\ x,y \in K \text{ with } \mathsf{d}(x,y) \geqslant \tfrac{1}{n}\}}_{=F_n}$$

is a countable union of sets F_n each of which is compact since F_n is a continuous image of the compact set

$$[\tfrac{1}{n}, 1 - \tfrac{1}{n}] \times \{(x,y) \in K^2 \mid \mathsf{d}(x,y) \geqslant \tfrac{1}{n}\}$$

for each $n \geqslant 1$. Therefore, $\mu(K \smallsetminus \mathrm{ext}(K)) = 0$ by (8.43) and (8.44) and μ represents x_0 by the argument after (8.43). This proves the theorem. $\qquad\square$

Exercise 8.93. Let K be a metrizable compact convex subset of a locally convex vector space over $\mathbb{R}$, and let $x_0 \in K$. Show that x_0 is an extreme point if and only if $\mu = \delta_{x_0}$ is the only Borel probability measure on K that represents x_0.

The material above follows the monograph of Phelps [86] loosely. We refer to those notes for many interesting applications of Choquet's theorem as well as the generalization of this result to the case of general compact convex sets in locally convex vector spaces (without the metrizability assumption) in the form of the Choquet–Bishop–de Leeuw theorem.

8.7 Further Topics

Many proofs and theories depend on weak* compactness, the notion of locally convex vector spaces, or the study of extreme points of convex subsets. We only mention a few samples and give further references.

- *Decay of Matrix Coefficients for Simple Lie Groups (the Howe–Moore Theorem):* If a simple non-compact Lie group G acts unitarily on a Hilbert space $\mathcal{H}$ without non-zero G-fixed vectors, then the matrix coefficients $\langle \pi_g v, w \rangle$ decay to zero as $g \to \infty$ in G, for any $v, w \in \mathcal{H}$. In the language of ergodic theory, this means that every measure-preserving ergodic G-action is mixing. This may sound complicated but the proof for $\mathrm{SL}_d(\mathbb{R})$ only needs as inputs the equality case of the Cauchy–Schwarz inequality, the Banach–Alaoglu theorem, and matrix multiplication. We refer to [27, Sec. 11.4] for a discussion of the easier case $G = \mathrm{SL}_2(\mathbb{R})$, and to [25] for the general case. The weak* compactness is here used on the Hilbert space $\mathcal{H}$.
- In the study of von Neumann algebras two more topologies on $B(X,Y)$ are used (particularly in the case where $X = Y$ is a Hilbert space):

the *ultra-strong operator topology* and the *ultra-weak operator topology*. We refer to von Neumann [79] for the original formulation of the ultra-strong topology, and to the monograph of Takesaki [101, Ch. II] for a full treatment.

- As discussed in Section 8.5 the locally convex vector space $C_c^\infty(U)$, also written $\mathscr{D}(U)$, is the space of *test functions for distributions* on $U \subseteq \mathbb{R}^d$ in the sense that one can define distributions as continuous linear functions on $\mathscr{D}(U)$. We refer to Folland [33, Ch. 9] for a treatment of distributions.

- The Fréchet space $\mathscr{S}(\mathbb{R}^d)$ of *Schwartz functions* on $\mathbb{R}^d$ has important connections to Fourier transforms, and is the space of test functions for tempered distributions on $\mathbb{R}^d$ (see Section 9.2.3 and Folland [33, Ch. 9]).

- There are further general classes of locally convex vector spaces. Among them are the *nuclear spaces* ($C^\infty([0,1])$, $C_b^\infty(U)$ and $\mathscr{S}(\mathbb{R}^d)$ are examples) and the *LF-spaces* (these are strict inductive limits of Fréchet spaces; examples include $C_c(U)$ and $C_c^\infty(U)$). We refer to Bourbaki [13] or Trèves [106] for more details.

- For a topological group G we will define in Section 9.3.1 the notion of a positive-definite function. If G is locally compact and abelian, the extreme points of the set of positive-definite functions (properly normalized) will be the set of characters of G. For more general groups the extreme points give rise to the irreducible unitary representations of the given group G (see Exercises 9.55 and 12.59).

Chapter 9
Unitary Operators and Flows, Fourier Transform

In this chapter we return to the topic of spectral theory by considering unitary operators. Moreover, we generalize our discussion of Fourier series by considering the Fourier transform for functions on $\mathbb{R}^d$ and use it to obtain the spectral theory of unitary flows (unitary representations of $\mathbb{R}$ or $\mathbb{R}^d$). As is natural in discussions concerning spectral theory (which will generalize eigenvalues and eigenvectors) we will only consider separable complex Hilbert spaces in this chapter.

9.1 Spectral Theory of Unitary Operators

Let $\mathcal{H}_1$, $\mathcal{H}_2$ be Hilbert spaces. Recall that a linear operator $U : \mathcal{H}_1 \to \mathcal{H}_2$ is said to be *unitary* if U is surjective and

$$\|Uv\|_{\mathcal{H}_2} = \|v\|_{\mathcal{H}_1}$$

for all $v \in \mathcal{H}_1$ (or, equivalently, if $U^* = U^{-1}$). Operators $V_1 : \mathcal{H}_1 \to \mathcal{H}_1$ and $V_2 : \mathcal{H}_2 \to \mathcal{H}_2$ on Hilbert spaces are said to be *unitarily isomorphic* if there is a unitary operator $U : \mathcal{H}_1 \to \mathcal{H}_2$ with $UV_1 = V_2U$. In contrast to the spectral theory of compact self-adjoint operators, it is not in general true that unitary operators on a Hilbert space are diagonalizable (that is, unitarily isomorphic to a diagonal operator). This may be seen in the next model example.

Example 9.1. Let μ be a finite measure on $\mathbb{T}$, let $\mathcal{H} = L^2_\mu(\mathbb{T}) = L^2(\mathbb{T}, \mu)$ and write again $\chi_1(x) = \mathrm{e}^{2\pi \mathrm{i} x}$ for $x \in \mathbb{T}$. Define the *unitary multiplication operator* $U = M_{\chi_1} : \mathcal{H} \ni f \longmapsto M_{\chi_1}(f) = \chi_1 f \in \mathcal{H}$ which gives a special case of Exercise 6.25(b). Note that any eigenvalue of a unitary operator like U must have absolute value one. Moreover, this unitary operator U has $\lambda = \mathrm{e}^{2\pi \mathrm{i} x_0}$ as an eigenvalue if and only if $f = \mathbb{1}_{\{x_0\}}$ is non-zero as an element of $\mathcal{H} = L^2_\mu(\mathbb{T})$. That is, $\lambda = \mathrm{e}^{2\pi \mathrm{i} x_0}$ is an eigenvalue if and only if x_0 is an *atom* of μ, meaning

© Springer International Publishing AG 2017
M. Einsiedler, T. Ward, *Functional Analysis, Spectral Theory, and Applications*,
Graduate Texts in Mathematics 276, DOI 10.1007/978-3-319-58540-6_9

that $\mu(\{x_0\}) > 0$. Moreover, U is diagonalizable if and only if μ is *atomic*, meaning that $\mu = \sum_{k=1}^{\infty} c_k \delta_{x_k}$ for some $c_k \geqslant 0$ and $x_k \in \mathbb{T}$.

The type of operators seen in Example 9.1 are not difficult to deal with even though they are usually not diagonalizable. Having abandoned the false hope that all unitary operators will be diagonalizable (that is, describable ultimately in terms of only countably many scalar multiplications on the ground field), the next best hope one might have is that they can be fully described in terms of multiplication by characters as in Example 9.1 (at the expense of allowing the underlying measure μ to vary). That this is in fact true is the content of the *spectral theory of unitary operators*.

Theorem 9.2 (Spectral theory of unitary operators). *Let $\mathcal{H}$ be a separable complex Hilbert space and let $U : \mathcal{H} \to \mathcal{H}$ be a unitary operator. Then $\mathcal{H}$ can be split into a countable direct sum*

$$\mathcal{H} = \bigoplus_{n \geqslant 1} \mathcal{H}_n$$

of closed mutually orthogonal subspaces $\mathcal{H}_n$, invariant under U and U^*, such that for each $n \geqslant 1$ the unitary operator $U_n = U|_{\mathcal{H}_n} : \mathcal{H}_n \to \mathcal{H}_n$ is unitarily isomorphic to the multiplication operator $M_{\chi_1} : L^2(\mathbb{T}, \mu_n) \to L^2(\mathbb{T}, \mu_n)$ for some finite measure μ_n on $\mathbb{T}$.

We will add some more information on the emerging sequence of measures in Section 9.1.3.

9.1.1 Herglotz's Theorem for Positive-Definite Sequences

Although this is not immediately apparent, a useful concept for the proof of Theorem 9.2 is the notion of a positive-definite sequence.

Definition 9.3. A sequence $(p_n)_{n \in \mathbb{Z}}$ of complex numbers is called *positive-definite* if for any finite sequence $(c_n)_{n \in \mathbb{Z}} \in c_c(\mathbb{Z})$ of complex numbers we have

$$\sum_{m,n \in \mathbb{Z}} c_m \overline{c_n} p_{m-n} \geqslant 0,$$

meaning that the sum is real and non-negative.

It is not obvious that non-trivial positive-definite sequences exist at all. There are two ways to construct examples.

Example 9.4 (First basic construction). Let μ be a finite measure on $\mathbb{T}$. Then the *Fourier coefficients $p_n(\mu)$* of μ defined by

$$p_n(\mu) = \int_{\mathbb{T}} \chi_n \, d\mu$$

for $n \in \mathbb{Z}$ form a positive-definite sequence.

Example 9.5 (Second basic construction). Let $U : \mathcal{H} \to \mathcal{H}$ be a unitary operator on a Hilbert space, and fix some $v \in \mathcal{H}$. Then the inner products

$$p_n(v) = \langle U^n v, v \rangle_{\mathcal{H}}$$

for $n \in \mathbb{Z}$ form a positive-definite sequence.

Both these claims require justification.

PROOF OF STATEMENT IN EXAMPLES 9.4 AND 9.5. Notice first that Example 9.4 is a special case of Example 9.5. Indeed, if $\mathcal{H} = L^2(\mathbb{T}, \mu)$, $U = M_{\chi_1}$, and $v = \mathbb{1}_{\mathbb{T}} = \mathbb{1}$ is the constant function, then $U^n v = \chi_n$ and so

$$p_n(\mu) = \int_{\mathbb{T}} \chi_n \, \mathrm{d}\mu = \langle U^n v, v \rangle_{L^2(\mathbb{T}, \mu)} = p_n(v)$$

for all $n \in \mathbb{Z}$. Thus it is enough to consider the sequence $(p_n(v))$ from Example 9.5. Let (c_n) be a finite complex sequence as in Definition 9.3. Then

$$\sum_{m,n \in \mathbb{Z}} c_m \overline{c_n} p_{m-n}(v) = \sum_{m,n \in \mathbb{Z}} c_m \overline{c_n} \langle U^m v, U^n v \rangle_{\mathcal{H}}$$

$$= \left\langle \sum_{m \in \mathbb{Z}} c_m U^m v, \sum_{n \in \mathbb{Z}} c_n U^n v \right\rangle_{\mathcal{H}} \geqslant 0,$$

since the inner product is positive-definite. $\square$

The main step towards the proof of Theorem 9.2 is the following description of all positive-definite sequences.

Theorem 9.6 (Herglotz's theorem). *Let $(p_n)_{n \in \mathbb{Z}}$ be a positive-definite sequence. Then there exists a uniquely determined finite measure μ on $\mathbb{T}$ for which*

$$p_n = p_n(\mu) = \int_{\mathbb{T}} \chi_n \, \mathrm{d}\mu$$

for all $n \in \mathbb{Z}$.

Note that Herglotz's theorem shows in particular that the Examples 9.4 and 9.5 both give rise to all positive-definite sequences.

PROOF OF THEOREM 9.6. Let (p_n) be a positive-definite sequence. Fix some $N \geqslant 1$. For any $\theta \in \mathbb{T}$ we may use the defining property for a positive-definite sequence with the finite sequence

$$c_n = \begin{cases} \chi_{-n}(\theta) & \text{if } 1 \leqslant n \leqslant N, \\ 0 & \text{otherwise,} \end{cases}$$

of coefficients to obtain that the function

$$F_N(\theta) = \frac{1}{N} \sum_{m,n=1}^{N} \chi_{-m+n}(\theta) p_{m-n}$$

is non-negative. Therefore, we may define a positive measure μ_N by the formula $\mathrm{d}\mu_N(\theta) = F_N(\theta)\,\mathrm{d}\theta$, that is, $\mu_N(B) = \int_B F_N(\theta)\,\mathrm{d}\theta$ for any measurable $B \subseteq \mathbb{T}$. Notice that

$$\mu_N(\mathbb{T}) = \int_{\mathbb{T}} F_N(\theta)\,\mathrm{d}\theta = \frac{1}{N} \sum_{m=1}^{N} p_0 = p_0.$$

By the Riesz representation theorem (Theorem 7.44) and the Banach–Alaoglu theorem (Theorem 8.10) in the combined form of Proposition 8.27 this sequence has a subsequence (μ_{N_k}) converging in the weak* topology to a positive measure μ. Since $\mu_{N_k}(\mathbb{T}) = p_0$ for all k we also have $\mu(\mathbb{T}) = p_0$.

Now let $\ell \in \mathbb{Z}$ and calculate

$$\int_{\mathbb{T}} \chi_\ell \,\mathrm{d}\mu = \lim_{k \to \infty} \int_{\mathbb{T}} \chi_\ell F_{N_k}(\theta)\,\mathrm{d}\theta$$

$$= \lim_{k \to \infty} \frac{1}{N_k} \sum_{m,n=1}^{N_k} p_{m-n} \int_{\mathbb{T}} \chi_{\ell-m+n}(\theta)\,\mathrm{d}\theta = \lim_{k \to \infty} \frac{N_k - |\ell|}{N_k} p_\ell = p_\ell,$$

where we used the fact that $\int_{\mathbb{T}} \chi_{\ell-m+n}\,\mathrm{d}\theta = \delta_{m,n+\ell}$ (which is 1 if $m = n + \ell$ and 0 otherwise) and $|[1, N] \cap ([1, N] + \ell) \cap \mathbb{Z}| = N - |\ell|$. This proves the existence of the finite measure μ as in the theorem. For uniqueness we note that the algebra of characters is dense in $C(\mathbb{T})$ and refer to the uniqueness statement in Theorem 7.44.   □

9.1.2 Cyclic Representations and the Spectral Theorem

Recall the notion of a unitary representation from Definition 3.73 and notice that a unitary operator U defines (and is defined by) an associated unitary representation π of the group $\mathbb{Z}$ given by $\pi_n = U^n$ for $n \in \mathbb{Z}$.

Definition 9.7. A unitary representation π of a group G on a Hilbert space $\mathcal{H}$ is called *cyclic* if

$$\mathcal{H} = \mathcal{H}_v = \overline{\langle \pi_g v \mid g \in G \rangle}$$

for some $v \in \mathcal{H}$. We will call v a *generator* of the cyclic representation. A closed subspace $\mathcal{H}' \subseteq \mathcal{H}$ is *π-invariant* if $\pi_g \mathcal{H}' \subseteq \mathcal{H}'$ for all $g \in G$. The *cyclic subspace* $\mathcal{H}_v$ generated by $v \in \mathcal{H} = \overline{\langle \pi_g v \mid g \in G \rangle}$ is the minimal π-invariant closed subspace containing v.

If $G = \mathbb{Z}$ then we also refer to a cyclic representation of $\mathbb{Z}$ as a *cyclic Hilbert space* with respect to the unitary operator $\pi_1 : \mathcal{H} \to \mathcal{H}$.

Cyclic representations can be thought of as the building blocks of more general unitary representations, as the following construction shows. If π is a unitary representation of G on a separable Hilbert space $\mathcal{H}$ (or $\pi : G \curvearrowright \mathcal{H}$ is a unitary representation), then we can write $\mathcal{H}$ as a direct sum

$$\mathcal{H} = \bigoplus_{n \geq 1} \mathcal{H}_n,$$

of pairwise orthogonal closed π-invariant subspaces $\mathcal{H}_n$, each of which is a cyclic representation. Indeed, if $w_1, w_2, \ldots$ is an orthonormal basis of $\mathcal{H}$ as in Theorem 3.39 and

$$\mathcal{H}_1 = \mathcal{H}_{w_1} = \overline{\langle \pi_g w_1 \mid g \in G \rangle},$$

then $\mathcal{H}_1$ is π-invariant. Together with the fact that $\pi_g^* = \pi_{g^{-1}}$ for all $g \in G$ and Lemma 6.30, this implies that $\mathcal{H}_1^\perp$ is also π-invariant. Define $\mathcal{H}_2 = \mathcal{H}_{w_2^\perp}$, where $w_2^\perp \in \mathcal{H}_1^\perp$ is the orthogonal projection of w_2 onto $\mathcal{H}_1^\perp$. Again the spaces $\mathcal{H}_1 \oplus \mathcal{H}_2$ and $(\mathcal{H}_1 \oplus \mathcal{H}_2)^\perp$ are π-invariant, and we can continue the process by defining $\mathcal{H}_3 = \mathcal{H}_{w_3^\perp}$, where $w_3^\perp \in (\mathcal{H}_1 \oplus \mathcal{H}_2)^\perp$ is the orthogonal projection of w_3 onto $(\mathcal{H}_1 \oplus \mathcal{H}_2)^\perp$. Clearly $w_1 \in \mathcal{H}_1$, $w_2 \in \mathcal{H}_1 \oplus \mathcal{H}_2$, and

$$w_3 \in \mathcal{H}_1 \oplus \mathcal{H}_2 \oplus \mathcal{H}_3.$$

Repeating this construction inductively gives a sequence of pairwise orthogonal, closed, π-invariant, cyclic subspaces $(\mathcal{H}_n)$ with $\mathcal{H} = \bigoplus_{n \geq 1} \mathcal{H}_n$ (see Exercise 3.37).

Therefore the following corollary to Theorem 9.6 will also be the main step towards the proof of Theorem 9.2.

Corollary 9.8 (Cyclic spaces). *Let $U : \mathcal{H} \to \mathcal{H}$ be a unitary operator on a complex Hilbert space such that $\mathcal{H}$ is cyclic with respect to U, with*

$$\mathcal{H} = \mathcal{H}_v = \overline{\langle U^n v \mid n \in \mathbb{Z} \rangle}$$

for some $v \in \mathcal{H}$. Then there is a uniquely determined finite measure μ_v on $\mathbb{T}$, called the spectral measure of v with respect to U, such that U is unitarily isomorphic to the multiplication operator M_{χ_1} on $L^2(\mathbb{T}, \mu_v)$ with the vector $v \in \mathcal{H}$ corresponding to $\mathbb{1} \in L^2(\mathbb{T}, \mu_v)$.

If $\phi : \mathcal{H} \to L^2(\mathbb{T}, \mu_v)$ is the unitary isomorphism as in the diagram then 'corresponding' here means that $\phi(v) = \mathbb{1}$. In other words, for cyclic Hilbert spaces we have a commutative diagram

$$\mathcal{H} \xrightarrow{U} \mathcal{H}$$

$$\phi \downarrow \qquad\qquad\qquad \downarrow \phi$$

$$L^2(\mathbb{T}, \mu_v) \xrightarrow[M_{\chi_1}]{} L^2(\mathbb{T}, \mu_v)$$

of unitary maps.

Together with the discussion before Theorem 9.2, we see that spectral measures should be thought of as a replacement for, or a generalization of, eigenvalues. As we will see during the proof, the spectral measure μ_v stores precisely the values of the inner products $\langle U^n v, v \rangle$ for a given unitary operator U on a Hilbert space $\mathcal{H}$ and vector $v \in \mathcal{H}$. In the case of an eigenvector with eigenvalue $\lambda_0 = e^{2\pi i x_0}$, we obtain the Dirac measure $\|v\|^2 \delta_{x_0}$. At the opposite extreme, it could be that the vectors $\ldots, U^{-1}v, v, Uv, \ldots$ are all mutually orthogonal, in which case μ_v is the multiple $\|v\|^2 m_{\mathbb{T}}$ of the Lebesgue measure.

PROOF OF COROLLARY 9.8. Let $v \in \mathcal{H}$ be a generator of

$$\mathcal{H} = \mathcal{H}_v = \overline{\langle U^n v \mid n \in \mathbb{Z} \rangle}.$$

By Example 9.5, we know that if $p_n(v) = \langle U^n v, v \rangle_{\mathcal{H}}$ for $n \in \mathbb{Z}$ then (p_n) is a positive-definite sequence. By Theorem 9.6 there exists a uniquely determined finite measure μ_v on $\mathbb{T}$ with

$$p_n(v) = p_n(\mu_v) = \int \chi_n \, \mathrm{d}\mu_v$$

for all $n \in \mathbb{Z}$. We wish to define a unitary map

$$\phi : \mathcal{H}_v = \overline{\langle U^n v \mid n \in \mathbb{Z} \rangle} \longrightarrow L^2(\mathbb{T}, \mu_v)$$

to be the unique extension of the map that sends any finite linear combinations of the vectors $U^n v$ to the corresponding trigonometric polynomial,

$$\phi : \mathcal{H}_v \longrightarrow L^2(\mathbb{T}, \mu_v)$$
$$\sum_{|n| \leqslant N} c_n U^n v \longmapsto \sum_{|n| \leqslant N} c_n \chi_n.$$

While this is a natural attempt at defining the map ϕ, it is not clear whether it produces a well-defined map. Curiously (at first encounter), this will follow from the map being an isometry: for any finite complex sequence (c_n), we have

$$\left\|\sum_{n\in\mathbb{Z}}c_nU^nv\right\|_{\mathcal{H}}^2 = \sum_{m,n\in\mathbb{Z}}\langle c_mU^mv, c_nU^nv\rangle_{\mathcal{H}} = \sum_{m,n\in\mathbb{Z}}c_m\overline{c_n}\underbrace{\langle U^{m-n}v, v\rangle_{\mathcal{H}}}_{p_{m-n}(v)=p_{m-n}(\mu_v)}$$

$$= \sum_{m,n\in\mathbb{Z}}c_m\overline{c_n}\int\chi_{m-n}\,\mathrm{d}\mu_v$$

$$= \int\left(\sum_{m\in\mathbb{Z}}c_m\chi_m\right)\left(\overline{\sum_{n\in\mathbb{Z}}c_n\chi_n}\right)\mathrm{d}\mu_v = \left\|\sum_{n\in\mathbb{Z}}c_n\chi_n\right\|_{L^2(\mathbb{T},\mu_v)}^2.$$

We now show that this implies that ϕ is well-defined on the set of finite linear combinations by the following argument. If $\sum_{n\in\mathbb{Z}}c_nU^nv = \sum_{n\in\mathbb{Z}}c_n'U^nv$ for finite complex sequences (c_n) and (c_n'), then

$$0 = \left\|\sum_{n\in\mathbb{Z}}(c_n - c_n')U^nv\right\|_{\mathcal{H}} = \left\|\sum_{n\in\mathbb{Z}}(c_n - c_n')\chi_n\right\|_{L^2(\mathbb{T},\mu_v)}$$

and so $\sum_{n\in\mathbb{Z}}c_n\chi_n = \sum_{n\in\mathbb{Z}}c_n'\chi_n$ in $L^2(\mathbb{T},\mu_v)$. Clearly the map ϕ so defined is now an isometry on a dense subspace of $\mathcal{H}_v$, and so extends by the automatic extension to the closure (Proposition 2.59) to an isometry from $\mathcal{H}_v$ into $L^2(\mathbb{T},\mu_v)$. Furthermore, the image of ϕ contains all trigonometric polynomials on $\mathbb{T}$, and which form a dense subset of $C(\mathbb{T})$ by Proposition 3.65 (and thus also of $L^2(\mathbb{T},\mu_v)$). Since $\mathcal{H}_v = \mathcal{H}$ is a Hilbert space and ϕ is an isometry, we see that $\phi(\mathcal{H}_v) \subseteq L^2(\mathbb{T},\mu_v)$ is complete and dense, and hence equal to $L^2(\mathbb{T},\mu_v)$.

It remains to check that $\phi \circ U = M_{\chi_1} \circ \phi$. Let (c_n) be a finite complex sequence. Then

$$U\left(\sum_{n\in\mathbb{Z}}c_nU^nv\right) = \sum_{n\in\mathbb{Z}}c_nU^{n+1}v,$$

and so

$$\phi\left(U\left(\sum_{n\in\mathbb{Z}}c_nU^nv\right)\right) = \sum_{n\in\mathbb{Z}}c_n\chi_{n+1} = \chi_1\sum_{n\in\mathbb{Z}}c_n\chi_n = M_{\chi_1}\phi\left(\sum_{n\in\mathbb{Z}}c_nU^nv\right).$$

That is, the desired formula holds on a dense subset of $\mathcal{H}_v$ and so by continuity on all of $\mathcal{H}_v$. This proves the corollary. $\qquad\square$

PROOF OF THEOREM 9.2. Let U be a unitary operator on a complex separable Hilbert space $\mathcal{H}$. By the discussion after Definition 9.7 $\mathcal{H}$ can be written as a direct sum of closed mutually orthogonal cyclic subspaces. Applying Corollary 9.8 to each of them proves the theorem. $\qquad\square$

Exercise 9.9. (a) Let $\mathcal{H}_v \cong L^2(\mathbb{T},\mu_v)$ be as in Corollary 9.8. Let $w \in \mathcal{H}_v$ and suppose that $f \in L^2(\mathbb{T},\mu_v)$ corresponds to w. Characterize the property $\mathcal{H}_v = \mathcal{H}_w$ in terms of f. (b) Apply (a) to the unitary operator defined by $U((a_n)) = (a_{n-1})$ on $\mathcal{H} = \ell^2(\mathbb{Z})$ and the vector (v_n) with $v_n = 0$ for $n \neq 0$ and $v_0 = 1$.

The spectral theory of unitary operators has the spectral theory of self-adjoint operators as a consequence, as the next exercise shows. However, we will also give an independent and much more detailed treatment of this theory in Chapter 12.

Exercise 9.10. (a) For any bounded operator $A : V \to V$ on a Banach space V and any power series $f(z) = \sum_{k=0}^{\infty} c_k z^k$ whose radius of convergence is bigger than $\|A\|$, show that the natural definition of $f(A)$ as the limit of the sequence of operators obtained as partial sums makes sense. Show that if $g(z) = \sum_{n=0}^{\infty} d_n(z - c_0)^n$ is the inverse function to f defined in a neighbourhood of $f(0) = c_0$ (represented by another power series) and $\sum_{n=0}^{\infty} |d_n| \left(\sum_{k=0}^{\infty} |c_k| \|A\|^k \right)^n < \infty$, then we have $g(f(A)) = A$.

(b) Let $A : \mathcal{H} \to \mathcal{H}$ be a non-zero self-adjoint operator. Replacing A by $\frac{1}{2\|A\|} A$ we may assume that $\|A\| = \frac{1}{2}$. Apply part (a) to A and the power series corresponding to e^{iz} to obtain a unitary operator $U : \mathcal{H} \to \mathcal{H}$. Show that $\|U - I\| \leqslant e^{1/2} - 1 < 1$, and that A can be recovered from U via the power series representing $\frac{1}{i} \log(z)$ in a neighbourhood of 1.

(c) Apply Theorem 9.2 to U and show that one can describe A on $\mathcal{H}$ by a direct sum of multiplication operators as in Exercise 6.25(b). In fact, for each of the direct summands the measure space can be chosen to be a copy of $\mathbb{R}$ together with a measure supported in $[-\|A\|, \|A\|]$ and the multiplication operator can be chosen to be $M_I(f)(x) = xf(x)$.

For simplicity we have been working in this section with a single unitary operator (or the group $\mathbb{Z}$) but the approach can be generalized to several commuting unitary operators (the group $\mathbb{Z}^d$) as outlined in the following exercise.

Exercise 9.11. (a) Define positive-definite functions on $\mathbb{Z}^d$ (so that the sequence case corresponds to $d = 1$), and generalize Herglotz's theorem to this context.

(b) State and prove a corollary to part (a) regarding the spectral theory of d commuting unitary operators, so that Theorem 9.2 corresponds to the case $d = 1$.

9.1.3 Spectral Measures

In this subsection we will strengthen the spectral theorem for unitary operators by studying the sequence of spectral measures appearing in it more carefully. We start with a few immediate consequences of the definition of the spectral measures.

Lemma 9.12 (Behaviour of spectral measures). *Let $U : \mathcal{H} \to \mathcal{H}$ be a unitary operator on a separable complex Hilbert space $\mathcal{H}$.*

(a) *For any $v \in \mathcal{H}$ we have $\mu_v(\mathbb{T}) = \|v\|^2$.*

(b) *If $v, w \in \mathcal{H}$ satisfy $\mathcal{H}_v \perp \mathcal{H}_w$, then $\mu_{v+w} = \mu_v + \mu_w$.*

(c) *If $w \in \mathcal{H}_v$ then $\mu_w \ll \mu_v$.*

(d) *If $v_1, v_2, \ldots \in \mathcal{H}$ are such that the cyclic spaces that they generate are orthogonal, $w = \sum_k w_k \in \bigoplus_{k=1}^{\infty} \mathcal{H}_{v_k}$, and w_k corresponds to the function $f_k \in L^2(\mathbb{T}, \mu_{v_k})$ for all $k \geqslant 1$, then $\mathrm{d}\mu_w = \sum_{k=1}^{\infty} |f_k|^2 \, \mathrm{d}\mu_{v_k}$.*

(e) *If $\mu_v \perp \mu_w$ then $\mathcal{H}_v \perp \mathcal{H}_w$.*

PROOF. (a) By definition we have $\mu_v(\mathbb{T}) = \int \chi_0 \, d\mu_v = \langle U^0 v, v \rangle = \|v\|^2$.

(b) Assume that $\mathcal{H}_v \perp \mathcal{H}_w$. Then

$$\int \chi_n \, d\mu_{v+w} = \langle U^n(v+w), v+w \rangle_{\mathcal{H}}$$

$$= \langle U^n v, v \rangle_{\mathcal{H}} + \langle U^n w, w \rangle_{\mathcal{H}} = \int \chi_n \, d\mu_v + \int \chi_n \, d\mu_w$$

for all $n \in \mathbb{Z}$, which implies that $\mu_{v+w} = \mu_v + \mu_w$ by uniqueness of the spectral measures (Corollary 9.8).

(c) Suppose that $w \in \mathcal{H}_v$ and recall that $\mathcal{H}_v \cong L^2(\mathbb{T}, \mu_v)$ with U corresponding to M_{χ_1} on $L^2(\mathbb{T}, \mu_v)$. If now $f \in L^2(\mathbb{T}, \mu_v)$ is the image of w under the unitary isomorphism, then

$$\langle U^n w, w \rangle_{\mathcal{H}} = \langle M_{\chi_1}^n f, f \rangle_{L^2(\mathbb{T}, \mu_v)} = \int \chi_n |f|^2 \, d\mu_v$$

for all $n \in \mathbb{Z}$ implies that $d\mu_w = |f|^2 \, d\mu_v$.

(d) Let $w = \sum_{k=1}^{\infty} w_k$ with $w_k \in \mathcal{H}_{v_k}$ for all $k \geqslant 1$, so that $\sum_{k=1}^{\infty} \|w_k\|^2 < \infty$. Then $\mu = \sum_{k=1}^{\infty} \mu_{w_k}$ is finite by (a). For any $k \geqslant 1$, let $f_k \in L^2(\mathbb{T}, \mu_{v_k})$ be the function corresponding to $w_k \in \mathcal{H}_{v_k} \cong L^2(X, \mu_{v_k})$, so $d\mu_{w_k} = |f_k|^2 \, d\mu_{v_k}$ by the argument in (c). We now have

$$\langle U^n w, w \rangle_{\mathcal{H}} = \sum_{k=1}^{\infty} \langle U^n w_k, w_k \rangle_{\mathcal{H}} = \sum_{k=1}^{\infty} \int \chi_n |f_k|^2 \, d\mu_{v_k} = \int \chi_n \, d\mu$$

for all $n \in \mathbb{Z}$, which implies (d).

(e) Assume that $w = x + y$ with $x \in \mathcal{H}_v$ and $y \in \mathcal{H}_v^{\perp}$. Then we have $\mathcal{H}_x \subseteq \mathcal{H}_v$ and $\mathcal{H}_y \subseteq \mathcal{H}_v^{\perp}$ so that $\mu_w = \mu_x + \mu_y$ by (b). However, (c) states that $\mu_x \ll \mu_v$ which implies that $\mu_x = 0$ since $\mu_w \perp \mu_v$. Thus $x = 0$ by (a) and $w = y \in \mathcal{H}_v^{\perp}$. $\square$

It is easy to check that the relation $\nu \ll \mu \ll \nu$ of mutual absolute continuity for measures μ and ν is an equivalence relation on the space of measures of a given measurable space, giving rise to the *measure equivalence class* of a given measure.

Corollary 9.13 (Maximal spectral type). *Let $\mathcal{H}$ and U be as in the spectral theorem (Theorem 9.2). The sequence of spectral measures in the spectral theorem can be chosen so that $\mu_1 \gg \mu_2 \gg \cdots$. In this case μ_1 is the maximal spectral measure in the sense that $\mu_v \ll \mu_1$ for any $v \in \mathcal{H}$, and this property uniquely characterizes the measure equivalence class of μ_1, which is called the* maximal spectral type *of U.*

Proof[†]. In the proof of Theorem 9.2 (more precisely, in the argument after Definition 9.7), we found a sequence of vectors $w_1, w_2, \ldots$ in $\mathcal{H}$ such that

$$\mathcal{H} = \bigoplus_{n \geqslant 1} \mathcal{H}_{w_n}$$

(where the sum is possibly finite). Each of the vectors has a spectral measure $\nu_n = \mu_{w_n}$ for $n \geqslant 1$. Applying the Lebesgue decomposition theorem (from Proposition 3.29) to ν_1 and ν_n for $n \geqslant 2$ we define

$$\nu_n = \nu_n^{\mathrm{ac}} + \nu_n^{\perp} \tag{9.1}$$

with $\nu_n^{\mathrm{ac}} \ll \nu_1$ and $\nu_n^{\perp} \perp \nu_1$. From $\nu_n^{\perp} \perp \nu_1$ it follows that there exists some measurable $B_n \subseteq \mathbb{T}$ such that $\nu_n^{\perp}(B_n) = 0$ and $\nu_1(\mathbb{T} \backslash B_n) = 0$, which implies with (9.1) that $\nu_n^{\mathrm{ac}} = \nu_n|_{B_n}$ and $\nu_n^{\perp} = \nu_n|_{\mathbb{T} \backslash B_n}$. We will use the set B_n to decompose w_n into two components. In fact, under the unitary isomorphism between $\mathcal{H}_{w_n}$ and $L^2(\mathbb{T}, \nu_n)$, let $w_n^{\perp} \in \mathcal{H}_{w_n}$ be the vector corresponding to $c_n \mathbb{1}_{\mathbb{T} \backslash B_n}$ where $c_n > 0$ is chosen so that $\|w_n^{\perp}\| \leqslant \frac{1}{2^n}$.

Let $w = w_1 + \sum_{n \geqslant 2} w_n^{\perp}$, which converges absolutely. Using Lemma 9.12(d) we now find that

$$\mathrm{d}\mu_w = \mathrm{d}\nu_1 + \sum_{n \geqslant 2} c_n^2 \mathbb{1}_{\mathbb{T} \backslash B_n} \, \mathrm{d}\nu_n,$$

or equivalently

$$\mu_w = \nu_1 + \sum_{n \geqslant 2} c_n^2 \nu_n^{\perp}.$$

From this we see that $\nu_n = \nu_n^{\mathrm{ac}} + \nu_n^{\perp} \ll \mu_w$ for all $n \geqslant 2$, and for $n = 1$ this holds trivially. In particular, Lemma 9.12(d) now shows that $\mu_v \ll \mu_w$ for any $v \in \mathcal{H} = \bigoplus_{n \geqslant 1} \mathcal{H}_{w_n}$, as claimed in the corollary. It is easy to see that this property uniquely characterizes the measure class of μ_w.

We claim now that $\mathcal{H}_{w_1} \subseteq \mathcal{H}_w$. To see this, notice that by construction $\nu_1 \perp \sum_{n \geqslant 2} c_n^2 \nu_n^{\perp}$ and let $B \subseteq \mathbb{T}$ have $\nu_1(\mathbb{T} \backslash B) = 0 = \nu_n^{\perp}(B)$ for all $n \geqslant 2$. Then by Corollary 9.8 and Lemma 9.12(d), $\mathcal{H}_w$ contains an element x corresponding to $\mathbb{1}_B \in L^2(\mathbb{T}, \mu_w)$ with spectral measure $\mu_x = \mu_w|_B = \nu_1$ and an element $y = w - x$ corresponding to $\mathbb{1}_{\mathbb{T} \backslash B}$ with spectral measure $\mu_y = \sum_{n \geqslant 2} c_n^2 \nu_n^{\perp}$. Since $\mu_y \perp \nu_1 = \mu_{w_1}$ we obtain from Lemma 9.12(e) that $y \in \mathcal{H}_{w_1}^{\perp}$. The same argument shows that $x \in \mathcal{H}_{w_1}$ even though it requires some additional thought: Recall from our construction above that $w_n^{\perp} \in \mathcal{H}_{w_n}$ satisfies $\mu_{w_n^{\perp}} \perp \mu_{w_1} = \nu_1$. Applying Lemma 9.12(c) and (d), this gives $\mu_z \perp \nu_1$ for all $z \in \bigoplus_{n \geqslant 2} \mathcal{H}_{w_n^{\perp}}$. Since $\mu_x = \nu_1$ we may apply Lemma 9.12(e) again to see that $x \perp \bigoplus_{n \geqslant 2} \mathcal{H}_{w_n^{\perp}}$. Moreover, we have $x \in \mathcal{H}_w \subseteq \mathcal{H}_{w_1} \oplus \bigoplus_{n \geqslant 2} \mathcal{H}_{w_n^{\perp}}$, which implies that $x \in \mathcal{H}_{w_1}$. Therefore, we have

---

[†] The corollary will not be needed in the remainder of the book, and the reader may skip its proof.

$$w = w_1 + \sum_{n \geqslant 2} w_n^{\perp} \qquad \text{(by construction of } w\text{)}$$

$$= x + y \qquad \text{(by construction of } x, y\text{)}$$

with $x \in \mathcal{H}_{w_1}$ and $y \in \mathcal{H}_{w_1}^{\perp}$, so that $w_1 = x \in \mathcal{H}_w$ and hence also $\mathcal{H}_{w_1} \subseteq \mathcal{H}_w$, as claimed.

The corollary follows by repeating this argument inductively. For this, we note that we can describe the space $\mathcal{H}_w^{\perp}$ in the next step as a sum of cyclic spaces such that the first space is generated by the orthogonal projection w_2' of w_2 to $\mathcal{H}_w^{\perp}$. Repeating now the argument above constructs a new vector w' such that $\mu_v \ll \mu_{w'}$ for all $v \in \mathcal{H}_w^{\perp}$ and its cyclic subspace contains the vector w_2'. It follows that w_2 belongs to the sum $\mathcal{H}_w \oplus \mathcal{H}_{w'}$. Continuing in this way, we can make sure that the direct sum of $\mathcal{H}_w, \mathcal{H}_{w'}, \ldots$ still contains $w_1, w_2, \ldots$ and so coincides with $\mathcal{H}$. By construction, the spectral measures of $w, w', w'', \ldots$ have the claimed absolute continuity property. $\square$

The next two exercises extend the discussion above, and we invite the reader to consider in particular the case in which the measures arising are atomic, so that we are simply discussing eigenvalues and their multiplicities.

Exercise 9.14. Given a unitary operator U on a separable complex Hilbert space $\mathcal{H}$ and a finite Borel measure ρ on $\mathbb{T}$, write $\mathcal{H}^{\rho} = \{v \in \mathcal{H} \mid \mu_v \ll \rho\}$. Show that $\mathcal{H}^{\rho}$ is a closed subspace of $\mathcal{H}$ that is invariant under U and U^*, and that we have $\mu_w \perp \rho$ for any vector $w \in (\mathcal{H}^{\rho})^{\perp}$.

Exercise 9.15. (a) Given finite measures μ and ν on $\mathbb{T}$ with $\mu \ll \nu \ll \mu$, show that the corresponding multiplication operators from Example 9.1 are unitarily isomorphic.
(b) Reformulate Corollary 9.13 to show the existence of a sequence of finite measures (ν_n) and a measure ν_{∞} with $\nu_m \perp \nu_n$ for all $m \neq n$ with $m, n \in \mathbb{N} \cup \{\infty\}$ with the property that $\mathcal{H} \cong \bigoplus_{n \geqslant 1} L^2(\mathbb{T}, \nu_n)^n \oplus L^2(\mathbb{T}, \nu_{\infty})^{\mathbb{N}}$ and the isomorphism carries U to the sum of the associated multiplication operators.
(c) Show that the unitary isomorphism from (b) takes $\mathcal{H}^{(1)} = \{v \in \mathcal{H} \mid \mu_v \perp \mu_w \; \forall w \in \mathcal{H}_v^{\perp}\}$ to $L^2(\mathbb{T}, \nu_1)$, and hence in particular deduce that $\mathcal{H}^{(1)}$ is a closed subspace.
(d) Show that the unitary isomorphism in (b) takes the closed subspace

$$\mathcal{H}^{(2)} = \left\{ v \in \mathcal{H} \mid \exists \, v_2 \text{ with } \mathcal{H}_v \perp \mathcal{H}_{v_2}, \mu_v = \mu_{v_2} \text{ and } \mu_w \perp \mu_v \; \forall w \in (\mathcal{H}_v \oplus \mathcal{H}_{v_2})^{\perp} \right\}$$

to $L^2(\mathbb{T}, \nu_2)^2$.
(e) Generalize (d) to higher multiplicities and conclude that the sequence of the measure classes of $\nu_1, \nu_2, \ldots, \nu_{\infty}$ and subspaces in $\mathcal{H}$ corresponding to $L^2(\mathbb{T}, \nu_n)^n$ resp. $L^2(\mathbb{T}, \nu_n)^{\mathbb{N}}$ are uniquely determined by U.

9.1.4 Functional Calculus for Unitary Operators

As in Exercise 9.10 it is relatively straightforward to obtain a definition of $h(A)$ for an analytic function h (defined by a power series) and bounded operator A (whose norm is less than the radius of convergence of the power series). For a multiplication operator M_g as in Exercise 6.25 one can go much further. For example, one could define $h(M_g)$ by setting it equal to $M_{h \circ g}$ for any bounded measurable function h. The reader should verify at this point that this definition does generalize the prior definition for analytic functions to measurable functions. Since Theorem 9.2 and Exercise 9.10 describe arbitrary unitary or self-adjoint operators in terms of multiplication operators, this allows one to also define the operators obtained by applying h to these. However, from this definition it is not clear whether the result is independent of the choices made to describe the operator on $\mathcal{H}$ as a sum of multiplication operators. As it turns out, this is the case, and we will discuss this 'functional calculus' in greater detail and in a more general setting in Chapter 12. Here we aim to give a first taste of this theory, by discussing simpler instances of the results for a single unitary operator.

For the discussion in this subsection it is convenient to use χ_1 as an isomorphism from $\mathbb{T}$ to $\mathbb{S}^1 = \{z \in \mathbb{C} \mid |z| = 1\}$, and transport the spectral measures μ_v on $\mathbb{T}$ provided by Corollary 9.8 to $\mathbb{S}^1$. We will still use the same symbol for these measures so that their characterizing property becomes

$$\langle U^n v, v \rangle_{\mathcal{H}} = \int_{\mathbb{S}^1} z^n \, \mathrm{d}\mu_v(z)$$

for $v \in \mathcal{H}$ and $n \in \mathbb{Z}$. Note that the multiplication operator in Corollary 9.8 then has the form $M_I(f)(z) = z f(z)$ for all $z \in \mathbb{S}^1$ and $f \in L^2(\mathbb{S}^1, \mu_v)$, where $I(z) = z$ denotes the identity map on $\mathbb{C}$.

To obtain a good definition of $h(U)$ the trick is to define more general spectral measures.

Definition 9.16. Let $U : \mathcal{H} \to \mathcal{H}$ be a unitary operator. A complex-valued measure $\mu_{v,w}$ on $\mathbb{S}^1$ is the *spectral measure* of $v, w \in \mathcal{H}$ with respect to U if

$$\int_{\mathbb{S}^1} z^n \, \mathrm{d}\mu_{v,w}(z) = \langle U^n v, w \rangle_{\mathcal{H}} \tag{9.2}$$

for all $n \in \mathbb{Z}$.

Proposition 9.17 (Non-diagonal spectral measures). *Let $U : \mathcal{H} \to \mathcal{H}$ be a unitary operator on a separable complex Hilbert space $\mathcal{H}$. For every v, w in $\mathcal{H}$ the spectral measure $\mu_{v,w}$ exists and is uniquely determined by v and w. Moreover, the spectral measure depends linearly on v and semi-linearly on w. For every $h \in \mathscr{L}^\infty(\mathbb{S}^1)$ there exists a bounded operator $h(U) : \mathcal{H} \to \mathcal{H}$ that is characterized by the property*

$$\langle h(U)v, w \rangle_{\mathcal{H}} = \int_{\mathbb{S}^1} h \, \mathrm{d}\mu_{v,w}$$

for all $v, w \in \mathcal{H}$. If $\mathcal{H} \cong \bigoplus_{n \geq 1} L^2(\mathbb{S}^1, \mu_n)$ as in Theorem 9.2, then $h(U)$ corresponds to the direct sum of the multiplication operators M_h on $L^2(\mathbb{S}^1, \mu_n)$ for $n \geq 1$.

The idea behind the proof of this corollary to Bochner's theorem is simple, and relies on the polarization identity

$$\langle U^n v, w \rangle_{\mathcal{H}} = \tfrac{1}{4} \sum_{\ell=0}^{3} i^\ell \left\langle U^n(v + i^\ell w), v + i^\ell w \right\rangle_{\mathcal{H}} \tag{9.3}$$

which is easily checked and gives the existence, by setting

$$\mu_{v,w} = \tfrac{1}{4} \sum_{\ell=0}^{3} i^\ell \mu_{v+i^\ell w}. \tag{9.4}$$

An important case is to consider the characteristic functions $h = \mathbb{1}_B$ for a measurable set $B \subseteq \mathbb{T}$ (see the proof of Corollary 9.13 and Exercise 9.21). In particular, for every Borel subset $B \subseteq \mathbb{S}^1$ there exists an orthogonal projection operator $E(B) = \mathbb{1}_B(U)$.

Definition 9.18. The function $E : \mathcal{B}(\mathbb{S}^1) \to \mathrm{B}(\mathcal{H})$ is called a *projection-valued measure*.

For a given measurable $B \subseteq \mathbb{S}^1$ one should think of $E(B)$ as the projection operator that projects onto the closed subspace on which all 'generalized eigenvalues belong to B'.

PROOF OF COROLLARY 9.17. Since the spectral measures $\mu_{v+i^\ell w}$ for the vectors $v + i^\ell w \in \mathcal{H}$ exist by Corollary 9.8, equation (9.3) shows that $\mu_{v,w}$ as defined in (9.4) satisfies the desired relationship. Since trigonometric polynomials (which on $\mathbb{S}^1$ are linear combinations of z^n for $n \in \mathbb{Z}$) are dense in $C(\mathbb{S}^1)$ by Proposition 3.65, and complex-valued measures are naturally identified with linear functionals on $C(\mathbb{S}^1)$ (by Theorem 7.54), it follows that the spectral measure $\mu_{v,w}$ is uniquely determined by (9.2). Since the inner product is sesqui-linear, this also implies the claimed sesqui-linearity.

We claim that

$$\|\mu_{v,w}\| \leqslant 4\|v\|\|w\| \tag{9.5}$$

(see Exercise 3.33 and Theorem 7.54 for the norm of $\mu_{v,w}$, and also Exercise 9.19 for the correct upper bound). First note that by sesqui-linearity $\mu_{tv,sw}$ is equal to $t\bar{s}\mu_{v,w}$ for all $t, s \in \mathbb{C}$, and in particular $\mu_{v,w} = 0$ if one of the vectors is zero. This allows us to assume that $v, w \in \mathcal{H}$ are non-zero and that $\|v\| = \|w\|$ after multiplying v by $\sqrt{\|w\|/\|v\|}$, the vector w by its inverse and therefore assume without loss of generality that $\|v\| = \|w\|$. Using now the definition of $\mu_{v,w}$ in (9.4) with these new vectors and the formula $\mu_{v+i^\ell w}(\mathbb{S}^1) = \|v + i^\ell w\|^2 \leqslant 4\|v\|^2$ for $0 \leqslant \ell \leqslant 3$, we obtain (9.5).

Now let $h \in \mathscr{L}^\infty(\mathbb{S}^1)$ so that

$$\left| \int_{\mathbb{S}^1} h \, d\mu_{v,w} \right| \leqslant 4\|h\|_\infty \|v\| \|w\|.$$

Fix $v \in \mathcal{H}$ and consider the function $\ell_{h,v}$ defined by

$$\ell_{h,v}(w) = \int_{\mathbb{S}^1} h(z)\,\mathrm{d}\mu_{v,w}(z)$$

for $w \in \mathcal{H}$. Then $w \mapsto \overline{\ell_{h,v}(w)}$ is a bounded linear functional on $\mathcal{H}$, and so by the Fréchet–Riesz representation theorem (Corollary 3.19) there exists some $v_h \in \mathcal{H}$ with $\ell_{h,v}(w) = \langle v_h, w\rangle$ for all $w \in \mathcal{H}$. Moreover,

$$\|v_h\| = \|\ell_{h,v}\| \leqslant 4\|h\|_\infty \|v\|$$

and v_h depends linearly on v. Hence $h(U)v = v_h$ defines a bounded linear operator with $\|h(U)\|_{\mathrm{op}} \leqslant 4\|h\|_\infty$.

We note that $v' \in \mathcal{H}_v$ and $w \in \mathcal{H}_v^\perp$ implies $\langle U^n v, w\rangle_{\mathcal{H}} = 0$ for all $n \in \mathbb{Z}$, hence $\mu_{v',w} = 0$ and so $\langle h(U)v', w\rangle_{\mathcal{H}} = 0$ for all $h \in \mathscr{L}^\infty(\mathbb{S}^1)$. As this holds for all $v' \in \mathcal{H}_v$ and $w \in \mathcal{H}_v^\perp$, we see that $h(U)\mathcal{H}_v \subseteq \mathcal{H}_v$. Thus for the proof of the description of $h(U)$ via the spectral theorem it suffices to describe the restriction of $h(U)$ to a cyclic representation, or equivalently describe $h(M_I)$ on $L^2(\mathbb{S}^1, \mu)$ for some finite measure μ on $\mathbb{S}^1$. If $f, g \in L^2(\mathbb{S}^1, \mu)$, then

$$\langle M_I^n f, g\rangle_{L^2(\mathbb{S}^1,\mu)} = \int_{\mathbb{S}^1} z^n f\overline{g}\,\mathrm{d}\mu$$

for all $n \in \mathbb{Z}$. Hence $\mathrm{d}\mu_{f,g} = f\overline{g}\,\mathrm{d}\mu$ is the spectral measure of f, g with respect to M_I, giving

$$\langle h(M_I)f, g\rangle_{L^2(\mathbb{S}^1,\mu)} = \int_{\mathbb{S}^1} hf\overline{g}\,\mathrm{d}\mu = \langle M_h f, g\rangle_{L^2(\mathbb{S}^1,\mu)}.$$

This implies the corollary. $\qquad\square$

Exercise 9.19. Improve the estimate in (9.5) to the inequality $\|\mu_{v,w}\| \leqslant \|v\|\|w\|$ for all $v, w \in \mathcal{H}$.

Exercise 9.20. Let U be a unitary operator on a separable complex Hilbert space and h a function in $\mathscr{L}^\infty(\mathbb{S}^1)$. When is $h(U)$ unitary or self-adjoint? What is the norm $\|h(U)\|_{\mathrm{op}}$?

Exercise 9.21. Let U be a unitary operator on a separable complex Hilbert space $\mathcal{H}$. Suppose that we are given a decomposition $\mathbb{S}^1 = \bigsqcup_k B_k$ for some (finite or countable) list of measurable sets B_k. Let $v \in \mathcal{H}$. Show that $v = \sum_k E(B_k)v$. Describe the spectral measure $\mu_{E(B_k)v}$ for all k in terms of the spectral measure μ_v and explain the meaning of $E(B_k)v$ in the case when μ_v is atomic.

9.1.5 An Application of Spectral Theory to Dynamics

In this subsection we will use the spectral theory of unitary operators in the context of measure-preserving systems (see Definition 8.35). In fact, we will assume here that X is a compact metric space, μ is a Borel probability measure on X, and that $T : X \to X$ is measure-preserving, continuous, and invertible, so that $T^{-1} : X \to X$ is also continuous. This guarantees that the operator $U_T : L^2(X, \mu) \to L^2(X, \mu)$ defined by $U_T f = f \circ T$ is unitary, since

$$\langle U_T f, g \rangle = \int_X (f \circ T\bar{g}) \circ T^{-1} \, d\mu = \int_X f(\bar{g} \circ T^{-1}) \, d\mu = \langle f, U_{T^{-1}} g \rangle$$

for all $f, g \in L^2(X, \mu)$.

Definition 9.22. Let (X, μ, T) be an invertible measure-preserving system as above. Then we say that T has *purely discrete spectrum* if U_T is diagonalizable (equivalently, if for any $f \in L^2(X, \mu)$ the spectral measure μ_f with respect to U_T is atomic). We say that T has *Lebesgue spectrum* if for any

$$f \in L_0^2(X, \mu) = \left\{ f \in L^2(X, \mu) \mid \int f \, d\mu = 0 \right\}$$

the spectral measure μ_f with respect to U_T is absolutely continuous with respect to the Lebesgue measure on $\mathbb{T}$.

Essential Exercise 9.23. (a) Prove that the rotation $R_\alpha : x \mapsto x + \alpha$ for $x, \alpha \in \mathbb{T}$ has purely discrete spectrum with respect to the Lebesgue measure on $\mathbb{T}$.
(b) Prove that the map

$$A : \begin{pmatrix} x \\ y \end{pmatrix} \longmapsto \begin{pmatrix} 0 & 1 \\ 1 & 1 \end{pmatrix} \begin{pmatrix} x \\ y \end{pmatrix} = \begin{pmatrix} y \\ x + y \end{pmatrix}$$

preserves Lebesgue measure on $\mathbb{T}^2$ and has Lebesgue spectrum.

We recall that if $(Z, \mathcal{B})$ and $(Z', \mathcal{B}')$ are measurable spaces, μ is a measure on Z and $\phi : Z \to Z'$ is a measurable map, then we can define the push-forward measure $\phi_* \mu$ on Z' by the formula $\phi_* \mu(B') = \mu \phi^{-1}(B')$ for all sets $B' \in \mathcal{B}'$. Using this notion, invariance of a measure μ under a map $T : X \to X$ is precisely the condition that the push-forward $T_* \mu$ coincides with the original measure μ.

Definition 9.24 (Furstenberg [37]). Let (X, ν_X, T) and (Y, ν_Y, S) be two measure-preserving systems. A *joining* of these systems is a measure ρ on the product $X \times Y$ with $(T \times S)_* \rho = \rho$, $(\pi_X)_* \rho = \nu_X$, and $(\pi_Y)_* \rho = \nu_Y$ where π_X, π_Y denote the natural projections from $X \times Y$ onto X, Y respectively. The two systems are called *disjoint*, written $T \perp S$, if $\rho = \nu_X \times \nu_Y$ is the only possible joining.

Furstenberg introduced the notion of joinings and also gave the first classes of disjoint systems.

Theorem 9.25 (Disjointness for spectral reasons). *Suppose (X, ν_X, T) and (Y, ν_Y, S) are two measure-preserving systems. Suppose moreover that for* any f in $L_0^2(X,\nu_X)$ and any g in $L_0^2(Y,\nu_Y)$ the spectral measures μ_f (with respect to U_T on $L^2(X,\nu_X)$) and μ_g (with respect to U_S on $L^2(Y,\nu_Y)$) are mutually singular. Then the two systems are disjoint.

PROOF. Let ρ be a joining of the two systems. For every $f \in L^2(X,\nu_X)$ we have $f \circ \pi_X \in L^2(X \times Y, \rho)$ with $\|f \circ \pi_X\|_{L^2(X \times Y,\rho)} = \|f\|_{L^2(X,\nu_X)}$ and

$$U_{T \times S}(f \circ \pi_X) = U_T(f) \circ \pi_X.$$

Moreover,

$$\langle U_{T \times S}^n(f \circ \pi_X), f \circ \pi_X \rangle_{L^2(X \times Y,\rho)} = \langle U_T^n f, f \rangle_{L^2(X,\nu_X)}$$

for all $n \in \mathbb{Z}$, which implies that the spectral measures $\mu_{f \circ \pi_X}$ defined using the unitary operator $U_{T \times S} : L^2(X \times Y, \rho) \to L^2(X \times Y, \rho)$ agrees with μ_f. A similar statement holds for $g \in L^2(Y,\nu_Y)$.

Applying this to $f \in L_0^2(X,\nu_X)$ and $g \in L_0^2(Y,\nu_Y)$, using the assumption in the theorem and Lemma 9.12(e) we see that $f \circ \pi_X \perp g \circ \pi_Y$. Now let $A \subseteq X$ and $B \subseteq Y$ be measurable sets and define $f = \mathbb{1}_A - \nu_X(A) \in L_0^2(X,\nu_X)$ and $g = \mathbb{1}_B - \nu_Y(B) \in L_0^2(Y,\nu_Y)$ to obtain

$$
\begin{aligned}
0 = \langle f \circ \pi_X, g \circ \pi_Y \rangle &= \langle \mathbb{1}_{A \times Y} - \nu_X(A), \mathbb{1}_{X \times B} - \nu_Y(B) \rangle \\
&= \langle \mathbb{1}_{A \times Y}, \mathbb{1}_{X \times B} \rangle - \nu_X(A) \langle \mathbb{1}, \mathbb{1}_{X \times B} \rangle - \nu_Y(B) \langle \mathbb{1}_{A \times Y}, \mathbb{1} \rangle + \nu_X(A)\nu_Y(B) \\
&= \rho(A \times B) - \nu_X(A)\nu_Y(B),
\end{aligned}
$$

where all the inner products are taken in $L^2(X \times Y, \rho)$. As this holds for all measurable $A \subseteq X$ and $B \subseteq Y$, we deduce that $\rho = \nu_X \times \nu_Y$. $\quad\square$

We wrap up the discussion of disjointness, and our excursion into ergodic theory, by discussing a consequence of disjointness for the dynamics of individual points.

Let X be a compact metric space, $T : X \to X$ a continuous map, and μ a T-invariant and ergodic probability measure on X. A consequence of the pointwise ergodic theorem (one of the fundamental results in ergodic theory, see [27, Ch. 2, Sec. 4.4.2]) is that μ-almost every point $x \in X$ satisfies

$$\frac{1}{N} \sum_{n=0}^{N-1} f(T^n x) \longrightarrow \int_X f \, \mathrm{d}\mu$$

as $N \to \infty$ for all $f \in C(X)$, in which case x is called μ-generic. Exercise 8.39 and the map in the proof of Proposition 8.34 give examples of systems in which every point is generic for the (Lebesgue) measure on the space. However, for the map $A : \mathbb{T}^2 \to \mathbb{T}^2$ in Exercise 9.23(b) it is easy to see that rational points in $\mathbb{Q}^2/\mathbb{Z}^2$ are not generic. There are also irrational non-generic points, but as the Lebesgue measure m on $\mathbb{T}^2$ is invariant and ergodic for A, one obtains from the ergodic theorem that m-almost every point in $\mathbb{T}^2$ is m-generic. With these examples in mind, we can now give a pointwise corollary of the discussion of spectral measures in the following exercise.

Essential Exercise 9.26. Let X, Y be compact metric spaces, $T : X \to X$ and $S : Y \to Y$ continuous maps, and $\nu_X \in \mathcal{P}^T(X)$, $\nu_Y \in \mathcal{P}^S(Y)$ be invariant and ergodic probability measures. Suppose that the two systems are disjoint.
(a) Let (ρ_n) be a sequence of probability measures on $X \times Y$ with the property that $(\pi_X)_* \rho_n \to \nu_X$ and $(\pi_Y)_* \rho_n \to \nu_Y$ as $n \to \infty$ with respect to the weak* topology. Suppose that any limit $\lim_{k \to \infty} \rho_{n_k}$ of a weak* convergent subsequence is $T \times S$-invariant. Show that $\lim_{n \to \infty} \rho_n = \nu_X \times \nu_Y$.
(b) Let $x \in X$ be ν_X-generic for T and let $y \in Y$ be ν_Y-generic for S. Use (a) to show that (x, y) is $\nu_X \times \nu_Y$-generic for $T \times S$.
(c) Show that the result of (b) applies in particular to the case of the rotation $T = R_\alpha : \mathbb{T} \to \mathbb{T}$ and the automorphism $S = A : \mathbb{T}^2 \to \mathbb{T}^2$ from Exercise 9.23.

9.2 The Fourier Transform

The Fourier transform generalizes the (important and satisfying) theory of Fourier series on $\mathbb{T}^d$ to an (equally important and satisfying) theory for functions on $\mathbb{R}^d$, which will in particular lead to a generalization of the spectral theory of unitary operators (Theorem 9.2) to unitary flows (Theorem 9.58). The analogue of the Fourier coefficients of a function on $\mathbb{T}^d$ will be the Fourier transform $\widehat{f}$ of a function f on $\mathbb{R}^d$, defined by

$$\widehat{f}(t) = \int_{\mathbb{R}^d} f(x) e^{-2\pi i x \cdot t} \, dx, \qquad (9.6)$$

where $x, t \in \mathbb{R}^d$ and $x \cdot t = x_1 t_1 + \cdots + x_d t_d$ is the usual inner product. The analogue of the Fourier series will be the Fourier back transform (or reverse transform) $\check{h}$ of a function h on $\mathbb{R}^d$, defined by

$$\check{h}(x) = \int_{\mathbb{R}^d} h(t) e^{2\pi i x \cdot t} \, dt. \qquad (9.7)$$

The analogue of the fact that the Fourier series represents the original function (where this is true) is a Fourier inversion formula $f = (\widehat{f})^{\vee}$. However, the way in which the optimistic identity $f = (\widehat{f})^{\vee}$ needs to be interpreted as a mathematical theorem is more involved. For example, if $f \in L^2(\mathbb{R}^d)$ then there is no reason to expect the integral defining the Fourier transform in (9.6) to exist. However, we will still be able to obtain a sensible definition of the Fourier transform as an extension of a densely defined bounded operator.

We also note that we will think of $x \in \mathbb{R}^d$ as the space variable and of $t \in \mathbb{R}^d$ as the frequency variable. In fact, any $t \in \mathbb{R}^d$ defines the wave function $x \mapsto e^{2\pi i x \cdot t}$ for which t gives the frequency and the direction of the wave. In that sense, $\widehat{f}(t)$ should be interpreted as the correlation of the function f and the wave with frequency t. The formula $f = (\widehat{f})^{\vee}$ then means that one can reconstruct the original function f as a suitable superposition of the waves with frequency t and amplitude $\widehat{f}(t)$.

We start with a concrete example.

Example 9.27 (Gaussian distribution). Let $f(x) = e^{-\pi \|x\|^2}$ for $x \in \mathbb{R}^d$. Then $\widehat{f}(t) = e^{-\pi \|t\|^2}$.

PROOF. Suppose first that $d = 1$, and start by calculating $\widehat{f}(0)$. By definition,

$$\widehat{f}(0) = \int_{\mathbb{R}} e^{-\pi x^2} \, dx.$$

Thus

$$\widehat{f}(0)^2 = \int_{\mathbb{R}^2} e^{-\pi x^2} e^{-\pi y^2} \, dx \, dy \qquad \text{(by Fubini)}$$

$$= \int_0^{\infty} \int_0^{2\pi} e^{-\pi r^2} r \, d\theta \, dr \qquad \text{(in polar coordinates)}$$

$$= 2\pi \int_0^{\infty} e^{-\pi r^2} r \, dr = \int_0^{\infty} e^{-s} \, ds = 1 \qquad \text{(where } \pi r^2 = s\text{)}$$

and as $\widehat{f}(0) > 0$ we get $\widehat{f}(0) = 1$. To verify the claimed formula for a general t in $\mathbb{R}$ we will use the Cauchy integral formula for complex path integrals applied to the holomorphic function $\mathbb{C} \ni z \mapsto e^{-\pi z^2}$. We integrate over a rectangular path γ with corners at $\pm M$ and $\pm M + it$ as illustrated in Figure 9.1.

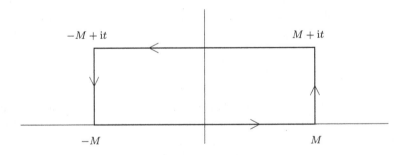

Fig. 9.1: The contour γ.

Then by Cauchy's formula $\oint_\gamma \mathrm{e}^{-\pi z^2}\mathrm{d}z = 0$. Bringing the integrals over the third and the fourth piece of the path to the other side we obtain

$$\int_{-M}^{M} \mathrm{e}^{-\pi x^2}\mathrm{d}x + \mathrm{i}\int_{0}^{t} \mathrm{e}^{-\pi(M+\mathrm{i}s)^2}\mathrm{d}s = \mathrm{i}\int_{0}^{t} \mathrm{e}^{-\pi(-M+\mathrm{i}s)^2}\mathrm{d}s + \int_{-M}^{M} \mathrm{e}^{-\pi(x+\mathrm{i}t)^2}\mathrm{d}x. \quad (9.8)$$

Now notice that

$$\mathrm{e}^{-\pi(x+\mathrm{i}s)^2} = \mathrm{e}^{-\pi(x^2 - s^2 + 2\mathrm{i}sx)}$$

which implies that for fixed $t \in \mathbb{R}$ and $|s| \leqslant |t|$ we have

$$\left| \mathrm{e}^{-\pi(\pm M + \mathrm{i}s)^2} \right| \leqslant \mathrm{e}^{-\pi(M^2 - t^2)}$$

which implies that $\mathrm{e}^{-\pi(\pm M + \mathrm{i}s)^2} \to 0$ uniformly on $[-t,t]$ as $M \to \infty$. Thus letting $M \to \infty$ in (9.8) we see that

$$1 = \int_{-\infty}^{\infty} \mathrm{e}^{-\pi x^2}\,\mathrm{d}x = \underbrace{\int_{-\infty}^{\infty} \mathrm{e}^{-\pi x^2}\mathrm{e}^{-2\pi\mathrm{i}tx}\,\mathrm{d}x}_{\widehat{f}(t)}\, \mathrm{e}^{\pi t^2},$$

which gives $\widehat{f}(t) = \mathrm{e}^{-\pi t^2}$ for all $t \in \mathbb{R}$.

For $d \geqslant 1$ notice that

$$f(x) = \mathrm{e}^{-\pi\|x\|^2} = \mathrm{e}^{-\pi x_1^2}\cdots\mathrm{e}^{-\pi x_d^2} = f_1(x_1)\cdots f_1(x_d)$$

is a product of d copies of the function $f_1(x) = \mathrm{e}^{-\pi x^2}$ discussed above, so

$$\widehat{f}(t) = \int_{\mathbb{R}^d} f_1(x_1)\cdots f_1(x_d)\mathrm{e}^{-2\pi\mathrm{i}(x_1 t_1 + \cdots + x_d t_d)}\,\mathrm{d}x$$

$$= \underbrace{\int_{\mathbb{R}} f_1(x_1)\mathrm{e}^{-2\pi\mathrm{i}x_1 t_1}\,\mathrm{d}x_1}_{=\widehat{f_1}(t_1)}\cdots\underbrace{\int_{\mathbb{R}} f_1(x_d)\mathrm{e}^{-2\pi\mathrm{i}x_d t_d}\,\mathrm{d}x_d}_{=\widehat{f_1}(t_d)}$$

$$= \mathrm{e}^{-\pi t_1^2}\cdots\mathrm{e}^{-\pi t_d^2} = \mathrm{e}^{-\pi\|t\|^2}$$

by Fubini's theorem and the case $d = 1$. $\qquad\qquad\qquad\qquad\qquad\qquad\qquad\square$

9.2.1 The Fourier Transform on $L^1(\mathbb{R}^d)$

Lemma 9.28 (Basic inequality). *The Fourier transform in (9.6) is defined for every $f \in L^1(\mathbb{R}^d)$ and $t \in \mathbb{R}^d$ and satisfies $\|\widehat{f}\|_\infty \leqslant \|f\|_1$.*

PROOF. For $f \in L^1(\mathbb{R}^d)$ and $t \in \mathbb{R}^d$ we have

$$|\widehat{f}(t)| = \left| \int_{\mathbb{R}^d} f(x) \mathrm{e}^{-2\pi \mathrm{i} x \cdot t} \, \mathrm{d}x \right| \leqslant \int_{\mathbb{R}^d} |f(x)| \, \mathrm{d}x = \|f\|_1,$$

which proves the lemma.   □

The next result is the first of many duality principles involving Fourier transforms.

Proposition 9.29 (Duality between shift and phase shift). *For x_0 and $t_0 \in \mathbb{R}^d$ we define the* shift operator *λ_{x_0} and the* multiplication operator *$M_{\chi(t_0)}$ on $L^1(\mathbb{R}^d)$ by*

$$\lambda_{x_0}(f) : x \longmapsto f(x - x_0)$$

and

$$M_{\chi(t_0)}(f) : x \longmapsto \mathrm{e}^{2\pi \mathrm{i} x \cdot t_0} f(x).$$

Then $\widehat{\lambda_{x_0}(f)} = M_{\chi(-x_0)}(\widehat{f})$ and $\widehat{M_{\chi(t_0)}(f)} = \lambda_{t_0}(\widehat{f})$.

We note that by a *phase shift* we mean multiplication by a character, the reason being this proposition.

PROOF OF PROPOSITION 9.29. By definition,

$$\widehat{\lambda_{x_0}(f)}(t) = \int_{\mathbb{R}^d} f(x - x_0) \mathrm{e}^{-2\pi \mathrm{i} x \cdot t} \, \mathrm{d}x$$

$$= \int_{\mathbb{R}^d} f(y) \mathrm{e}^{-2\pi \mathrm{i}(y + x_0) \cdot t} \, \mathrm{d}y = \mathrm{e}^{-2\pi \mathrm{i} x_0 \cdot t} \widehat{f}(t) = M_{\chi(-x_0)}(\widehat{f})(t)$$

for all $x_0, t \in \mathbb{R}^d$ and

$$\widehat{M_{\chi(t_0)}(f)}(t) = \int_{\mathbb{R}^d} \mathrm{e}^{2\pi \mathrm{i} x \cdot t_0} f(x) \mathrm{e}^{-2\pi \mathrm{i} x \cdot t} \, \mathrm{d}x$$

$$= \int_{\mathbb{R}^d} f(x) \mathrm{e}^{-2\pi \mathrm{i} x \cdot (t - t_0)} \, \mathrm{d}x = \widehat{f}(t - t_0) = \lambda_{t_0}(\widehat{f})(t)$$

for all $t_0, t \in \mathbb{R}^d$.   □

Proposition 9.30 (Duality for linear transformations). *Let $f \in L^1(\mathbb{R}^d)$ and let $A \in \mathrm{GL}_d(\mathbb{R})$ be an invertible matrix. Then $f \circ A \in L^1(\mathbb{R}^d)$, and*

$$\widehat{f \circ A} = \frac{1}{|\det A|} \widehat{f} \circ \left(A^{\mathrm{t}}\right)^{-1}.$$

PROOF. We use the definition and a substitution to get

$$\widehat{f \circ A}(t) = \int_{\mathbb{R}^d} f(Ax) e^{-2\pi i x \cdot t} \, dx$$

$$= \frac{1}{|\det A|} \int_{\mathbb{R}^d} f(Ax) |\det A| e^{-2\pi i (Ax) \cdot \left((A^t)^{-1} t \right)} \, dx$$

$$= \frac{1}{|\det A|} \widehat{f}\left((A^t)^{-1} t \right)$$

for all $t \in \mathbb{R}^d$. $\qquad\qquad\qquad\qquad\qquad\qquad\qquad\qquad\qquad\qquad\qquad\qquad$ $\square$

Proposition 9.31 (Duality of convolution and multiplication (I)). *For $f_1, f_2 \in L^1(\mathbb{R}^d)$ recall that the convolution $f_1 * f_2 \in L^1(\mathbb{R}^d)$ defined by*

$$f_1 * f_2(x) = \int_{\mathbb{R}^d} f_1(y) f_2(x - y) \, dy$$

satisfies $\|f_1 * f_2\|_1 \leqslant \|f_1\|_1 \|f_2\|_1$ and $f_1 * f_2 = f_2 * f_1$ (so that $L^1(\mathbb{R}^d)$ is a commutative Banach algebra). The Fourier transform of $f_1 * f_2$ is given by

$$\widehat{f_1 * f_2} = \widehat{f_1} \widehat{f_2}.$$

PROOF. Applying Fubini's theorem and a substitution we see that

$$\int_{\mathbb{R}^d} \int_{\mathbb{R}^d} |f_1(y) f_2(x - y)| \, dy \, dx = \int_{\mathbb{R}^d} \int_{\mathbb{R}^d} |f_1(y) f_2(x - y)| \, dx \, dy$$

$$= \int_{\mathbb{R}^d} \int_{\mathbb{R}^d} |f_1(y)| |f_2(z)| \, dz \, dy = \|f_1\|_1 \|f_2\|_1.$$

Thus the integral defining $f_1 * f_2(x)$ exists for almost every $x \in \mathbb{R}^d$, and

$$\|f_1 * f_2\|_1 \leqslant \int_{\mathbb{R}^d} \int_{\mathbb{R}^d} |f_1(y) f_2(x - y)| \, dy \, dx = \|f_1\|_1 \|f_2\|_1.$$

For commutativity we see that

$$f_1 * f_2(x) = \int f_1(y) f_2(x - y) \, dy = \int f_1(x - z) f_2(z) \, dz = f_2 * f_1(x)$$

by using the substitution $z = x - y$ for any fixed $x \in \mathbb{R}^d$.

Now let $t \in \mathbb{R}^d$ and apply Fubini's theorem to the definition of $\widehat{f_1 * f_2}(t)$ to see that

$$\widehat{f_1 * f_2}(t) = \int_{\mathbb{R}^d} \int_{\mathbb{R}^d} f_1(y) f_2(x - y) \, dy \, e^{-2\pi i x \cdot t} \, dx$$

$$= \int_{\mathbb{R}^d} f_1(y) \underbrace{\int_{\mathbb{R}^d} f_2(x - y) e^{-2\pi i (x - y) \cdot t} \, dx}_{\widehat{f_2}(t)} e^{-2\pi i y \cdot t} \, dy = \widehat{f_1}(t) \widehat{f_2}(t).$$

$\square$

Exercise 9.32. Show that the convolution in Proposition 9.31 is associative.

The impatient reader may use the propositions above together with Example 9.27 to show that the Fourier transform extends to an isometry from $L^2(\mathbb{R}^d)$ to $L^2(\mathbb{R}^d)$ via the steps of the following exercise.

Exercise 9.33. (a) Show that

$$\mathcal{A} = \left\{ x \longmapsto \sum_{\text{finite}} c_i e^{-\pi a_i \|x - x_i\|^2 + 2\pi i x \cdot t_i} \mid c_i \in \mathbb{C}, a_i > 0, x_i, t_i \in \mathbb{R}^d \right\}$$

is a sub-algebra of $C_0(\mathbb{R}^d)$ that separates points and is closed under conjugation.
(b) Show that $\widehat{\mathcal{A}} = \mathcal{A}$ and that $f = (\widehat{f})^\vee = (\check{f})^\wedge$ for all $f \in \mathcal{A}$.
(c) Show that $\mathcal{A} \subseteq C_0(\mathbb{R}^d)$ is dense with respect to $\|\cdot\|_\infty$.
(d) Show that $\mathcal{A} \subseteq L^1(\mathbb{R}^d) \cap L^2(\mathbb{R}^d)$ is dense in both $L^1(\mathbb{R}^d)$ and in $L^2(\mathbb{R}^d)$ with respect to the norms $\|\cdot\|_1$ and $\|\cdot\|_2$ respectively (which is not an immediate consequence of (c) since $\mathbb{R}^d$ has infinite Lebesgue measure). Show that if $F \in L^1(\mathbb{R}^d) \cap L^2(\mathbb{R}^d)$ and $\varepsilon > 0$ then there exists a single function $f \in \mathcal{A}$ with $\|f - F\|_1 < \varepsilon$ and $\|f - F\|_2 < \varepsilon$.
(e) Show that $\|\widehat{f}\|_2 = \|f\|_2$ for all $f \in \mathcal{A}$ so that the Fourier transform extends to a unitary map on $L^2(\mathbb{R})$ with inverse given by the Fourier back transform. Moreover, the extension agrees with the Fourier transform defined by the Lebesgue integral on $L^1(\mathbb{R}^d) \cap L^2(\mathbb{R}^d)$.

Proposition 9.34 (Riemann–Lebesgue lemma). *The Fourier transform maps $L^1(\mathbb{R}^d)$ into $C_0(\mathbb{R}^d)$.*

PROOF. If $t_n \to t$ in $\mathbb{R}^d$ as $n \to \infty$, then also $f(x)e^{-2\pi i x \cdot t_n} \to f(x)e^{-2\pi i x \cdot t}$ for almost every $x \in \mathbb{R}^d$ and so

$$\widehat{f}(t_n) = \int_{\mathbb{R}^d} f(x)e^{-2\pi i x \cdot t_n} \, dx \longrightarrow \int_{\mathbb{R}^d} f(x)e^{-2\pi i x \cdot t} \, dx = \widehat{f}(t)$$

as $n \to \infty$ by the dominated convergence theorem. Therefore $\widehat{f}$ is a bounded continuous function on $\mathbb{R}^d$ by Lemma 9.28. It remains to show that $\widehat{f}$ lies in $C_0(\mathbb{R}^d)$, which we will do by an approximation argument.

Suppose first that $f = \mathbb{1}_{[a_1, b_1] \times \cdots \times [a_d, b_d]}$ is the characteristic function of a rectangle. Then, by Fubini's theorem,

$$\widehat{f}(t) = \int_{\mathbb{R}^d} \mathbb{1}_{[a_1, b_1]}(x_1) \cdots \mathbb{1}_{[a_d, b_d]}(x_d)e^{-2\pi i x \cdot t} \, dx$$

$$= \int_{a_1}^{b_1} e^{-2\pi i x_1 t_1} \, dx_1 \cdots \int_{a_d}^{b_d} e^{-2\pi i x_d t_d} \, dx_d.$$

Each factor can be calculated explicitly, in fact

$$\int_a^b e^{-2\pi i x t} \, dx = \begin{cases} \dfrac{e^{-2\pi i b t} - e^{-2\pi i a t}}{-2\pi i t} & \text{for } t \neq 0, \text{ and} \\ b - a & \text{for } t = 0, \end{cases} \tag{9.9}$$

so each factor lies in $C_0(\mathbb{R})$. It follows that $\widehat{f} \in C_0(\mathbb{R}^d)$ if f is the characteristic function of a rectangle.

By linearity the same holds for finite linear combinations of such functions. Since $C_0(\mathbb{R}^d)$ is complete with respect to $\|\cdot\|_\infty$ and the Fourier transform is continuous from $L^1(\mathbb{R}^d)$ to $C_b(\mathbb{R}^d)$, the same holds for any element $f \in L^1(\mathbb{R}^d)$ that can be approximated in $L^1(\mathbb{R}^d)$ by such finite linear combinations, which is all of $L^1(\mathbb{R}^d)$ (see the argument on p. 172). $\qquad\square$

Exercise 9.35. Show that the Fourier transform calculated in (9.9) does not belong to $L^1(\mathbb{R})$.

As mentioned above, we will show that the Fourier back transform is the inverse of the Fourier transform. However, as we will see, this requires additional assumptions on the function, since the hypothesis $f \in L^1(\mathbb{R}^d)$ does not imply that $\widehat{f} \in L^1(\mathbb{R}^d)$ (as seen in Exercise 9.35), so there is no reason to expect that the Fourier back transform will be defined on $\widehat{f}$.

Theorem 9.36 (Fourier inversion). *If $f \in L^1(\mathbb{R}^d)$ has $\widehat{f} \in L^1(\mathbb{R}^d)$, then f agrees almost everywhere with the continuous function $(\widehat{f})^\vee \in C_0(\mathbb{R}^d)$.*

Despite the additional assumption, this theorem already implies that any L^1 function is uniquely determined by its Fourier transform. However, if the latter is not integrable, it is unclear how to recover f from $\widehat{f}$.

Corollary 9.37 (Injectivity). *If $f_1, f_2 \in L^1(\mathbb{R}^d)$ have $\widehat{f}_1 = \widehat{f}_2$, then $f_1 = f_2$.*

PROOF. Given $f_1, f_2 \in L^1(\mathbb{R}^d)$ as in the corollary, the function $f = f_1 - f_2$ satisfies $\widehat{f} = 0 \in L^1(\mathbb{R}^d)$. Applying Theorem 9.36 this implies that $f = 0$ and proves the corollary. $\qquad\square$

In order to prove Theorem 9.36 we need a preparatory lemma.

Lemma 9.38. *If $f, g \in L^1(\mathbb{R}^d)$ then $\displaystyle\int_{\mathbb{R}^d} \widehat{f} g \,\mathrm{d}x = \int_{\mathbb{R}^d} f \widehat{g} \,\mathrm{d}y$.*

PROOF. Once again, this is a simple application of Fubini's theorem, as

$$\int_{\mathbb{R}^d} \widehat{f}(x) g(x) \,\mathrm{d}x = \int_{\mathbb{R}^d} \int_{\mathbb{R}^d} f(y) e^{-2\pi i y \cdot x} \,\mathrm{d}y \, g(x) \,\mathrm{d}x$$

$$= \int_{\mathbb{R}^d} f(y) \int_{\mathbb{R}^d} g(x) e^{-2\pi i y \cdot x} \,\mathrm{d}x \,\mathrm{d}y = \int_{\mathbb{R}^d} f(y) \widehat{g}(y) \,\mathrm{d}y.$$

$\qquad\square$

PROOF OF THEOREM 9.36. Let $f \in L^1(\mathbb{R}^d)$ such that $\widehat{f} \in L^1(\mathbb{R}^d)$. By Proposition 9.34 we have $(\widehat{f})^\vee \in C_0(\mathbb{R}^d)$. We need to show that $(\widehat{f})^\vee$ agrees with f almost everywhere. To achieve this, we will use Lemma 9.38 for f and the phase-shifted stretched Gaussian distribution

$$\phi_{r,x_0}(t) = e^{2\pi i t \cdot x_0} \phi_r(t)$$

where

$$\phi_r(t) = e^{-\pi \|rt\|^2}$$

for $x_0 \in \mathbb{R}^d$ and $r > 0$. Using Lemma 9.38 we can define the function f_r in two equivalent ways by

$$f_r(x_0) = \int_{\mathbb{R}^d} \widehat{f}(t) \phi_{r,x_0}(t) \, dt = \int_{\mathbb{R}^d} f(x) \widehat{\phi_{r,x_0}}(x) \, dx \qquad (9.10)$$

for all $x_0 \in \mathbb{R}^d$. We will use the two sides of this formula to show that f_r converges as $r \to 0$ both to $(\widehat{f})^{\vee}$ and to f (in two different ways).

POINTWISE CONVERGENCE. We first show that $f_r \to (\widehat{f})^{\vee}$ pointwise as $r \to 0$ (where we will use the left-hand integral in (9.10)). Since

$$f_r(x_0) = \int_{\mathbb{R}^d} \widehat{f}(t) e^{2\pi i t \cdot x_0} e^{-\pi \|rt\|^2} \, dt$$

and $e^{-\pi \|rt\|^2} \to 1$ as $r \to 0$, we obtain

$$f_r(x_0) \longrightarrow \int_{\mathbb{R}^d} \widehat{f}(t) e^{2\pi i t \cdot x_0} \, dt = (\widehat{f})^{\vee}(x_0)$$

by the dominated convergence theorem, for any $x_0 \in \mathbb{R}^d$.

CONVERGENCE IN L^1. We next show that $f_r \to f$ in $L^1(\mathbb{R}^d)$ as $r \to 0$ (which will use the right-hand integral in (9.10)).

The proof of this step is a bit more involved and relies on the interpretation of the right-hand integral as a convolution with the approximate identity $\widehat{\phi_r}$ (it is easy to see — from the proof that follows, for example — that $\widehat{\phi_r}$ has properties similar to the Fejér kernel in Section 3.4.2 and the function used in Exercise 5.17; see also Exercise 8.6). By Example 9.27 and Proposition 9.30 we have

$$\widehat{\phi_r}(x) = r^{-d} e^{-\pi \|x/r\|^2},$$

and by Proposition 9.29 we also have

$$\widehat{\phi_{r,x_0}}(x) = r^{-d} e^{-\pi \|(x-x_0)/r\|^2}.$$

This gives

$$f_r(x_0) = \int f(x) r^{-d} e^{-\pi \|(x_0-x)/r\|^2} \, dx = f * \widehat{\phi_r}(x_0)$$

and $f * \widehat{\phi_r} \in L^1(\mathbb{R}^d)$ by Proposition 9.31. We may bring the difference of f_r and f at x_0 into the form

$$f * \widehat{\phi_r}(x_0) - f(x_0) = \int f(x_0 - x)r^{-d}e^{-\pi\|x/r\|^2} \, dx - f(x_0)$$

$$= \int (f(x_0 - rz) - f(x_0)) e^{-\pi\|z\|^2} \, dz$$

by using the substitution $z = x/r$ (and recalling that $\int_{\mathbb{R}^d} e^{-\pi\|z\|^2} \, dz = 1$). On taking the norm we obtain

$$\|f * \widehat{\phi_r} - f\|_1 = \int \Big| \int (f(x_0 - rz) - f(x_0)) e^{-\pi\|z\|^2} \, dz \Big| \, dx_0$$

$$\leqslant \iint |f(x_0 - rz) - f(x_0)| e^{-\pi\|z\|^2} \, dz \, dx_0$$

$$\leqslant \int \|\lambda_{rz}(f) - f\|_1 e^{-\pi\|z\|^2} \, dz$$

by Fubini's theorem. Now notice that $\|\lambda_{rz}(f) - f\|_1 \to 0$ as $r \to 0$ for $z \in \mathbb{R}^d$ by Lemma 3.74. Therefore

$$\|f * \widehat{\phi_r} - f\|_1 \longrightarrow 0$$

as $r \to 0$ by dominated convergence, which proves the claimed convergence in L^1.

THE LIMITS COINCIDE. Since every sequence that converges in $L^1(\mathbb{R}^d)$ has a subsequence that converges almost everywhere (see, for example, the proof of completeness of L^p spaces on p. 34), the two statements regarding the convergence properties of f_r as $r \to 0$ together prove the theorem. $\square$

9.2.2 The Fourier Transform on $L^2(\mathbb{R}^d)$

As mentioned before, the Fourier transform behaves quite well on $L^2(\mathbb{R}^d)$ (after overcoming the minor obstacle that the defining integral does not make sense).

Theorem 9.39 (Plancherel formula). *If $f \in L^1(\mathbb{R}^d) \cap L^2(\mathbb{R}^d)$, then $\widehat{f}$ lies in $L^2(\mathbb{R}^d)$ with $\|\widehat{f}\|_2 = \|f\|_2$ and the map $f \mapsto \widehat{f}$ extends continuously to a unitary operator on $L^2(\mathbb{R}^d)$ (whose inverse is the continuous extension of $f \mapsto \check{f}$).*

PROOF: A DENSE SUBSPACE. We define the space of functions

$$\mathcal{V} = \{f \in L^1(\mathbb{R}^d) \mid \widehat{f} \in L^1(\mathbb{R}^d)\}.$$

By Theorem 9.36 we have $f = (\widehat{f})^\vee \in C_0(\mathbb{R}^d)$ almost everywhere for $f \in \mathcal{V}$. Hence $\mathcal{V} \subseteq L^\infty(\mathbb{R}^d)$ consists of essentially bounded functions and so

$$|f|^2 = f\overline{f} \in L^1(\mathbb{R}^d)$$

for all $f \in \mathcal{V}$ so $\mathcal{V} \subseteq L^2(\mathbb{R}^d)$. We claim that $\mathcal{V}$ is actually dense in $L^2(\mathbb{R}^d)$. For this, notice first that $L^1(\mathbb{R}^d) \cap L^2(\mathbb{R}^d)$ is dense in $L^2(\mathbb{R}^d)$ (since it contains simple functions), so that it is enough to approximate a given function in the intersection $L^1(\mathbb{R}^d) \cap L^2(\mathbb{R}^d)$ by an element of $\mathcal{V}$ with respect to $\|\cdot\|_2$. Using the same notation as in the proof of Theorem 9.36, we already know that $f * \phi_r$ converges to f in $L^1(\mathbb{R}^d)$ as $r \to 0$. By Proposition 9.31 we see that

$$\widehat{f * \phi_r} = \widehat{f}\widehat{\phi_r}$$

belongs to $L^1(\mathbb{R}^d)$ (since it is a product of an element in $C_0(\mathbb{R}^d)$ with an element of $L^1(\mathbb{R}^d)$), which shows that $f * \phi_r \in \mathcal{V}$. Analyzing the argument, we see that only a slight modification is necessary to also obtain L^2 convergence of $f * \phi_r \to f$ as $r \to 0$. Indeed, using Jensen's inequality (see the first paragraph of the proof of Lemma 3.75) with the probability measure $\mathrm{e}^{-\pi|z|^2}\,\mathrm{d}z$ on $\mathbb{R}^d$ we get

$$\begin{aligned}
\|f * \phi_r - f\|_2^2 &= \int \left| \int (f(x_0 - rz) - f(x_0))\mathrm{e}^{-\pi|z|^2}\,\mathrm{d}z \right|^2 \mathrm{d}x_0 \\
&\leqslant \iint |\lambda_{rz}f(x_0) - f(x_0)|_2^2 \mathrm{e}^{-\pi|z|^2}\,\mathrm{d}z\,\mathrm{d}x_0 \\
&= \int \|\lambda_{rz}f - f\|_2^2 \mathrm{e}^{-\pi|z|^2}\,\mathrm{d}z \longrightarrow 0
\end{aligned}$$

as $r \to 0$, again by dominated convergence. This gives the claimed density.

UNITARITY. We will now show that the Fourier transform preserves the inner product. For this, let $f, g \in \mathcal{V}$ and define $h = \overline{\widehat{g}}$. Then

$$\widehat{h}(x) = \int_{\mathbb{R}^d} \overline{\widehat{g}}(t)\mathrm{e}^{-2\pi\mathrm{i}x\cdot t}\,\mathrm{d}t = \overline{\int_{\mathbb{R}^d} \widehat{g}(t)\mathrm{e}^{2\pi\mathrm{i}x\cdot t}\,\mathrm{d}t} = \overline{(\widehat{g})^{\vee}(x)} = \overline{g(x)}$$

almost everywhere by Theorem 9.36. Applying Lemma 9.38 we see that

$$\langle f, g \rangle_{L^2(\mathbb{R}^d)} = \int_{\mathbb{R}^d} f\overline{g}\,\mathrm{d}x = \int_{\mathbb{R}^d} f\overline{h}\,\mathrm{d}x = \int_{\mathbb{R}^d} \widehat{f}\widehat{h}\,\mathrm{d}t = \int_{\mathbb{R}^d} \widehat{f}\overline{\widehat{g}}\,\mathrm{d}t = \langle \widehat{f}, \widehat{g} \rangle_{L^2(\mathbb{R}^d)}.$$

In other words, we have shown that the Fourier transform preserves the inner product for elements in $\mathcal{V}$. It follows that the Fourier transform extends to an isometry from $L^2(\mathbb{R}^d)$ to itself, which we again denote by

$$L^2(\mathbb{R}^d) \ni f \longmapsto \widehat{f} \in L^2(\mathbb{R}^d).$$

Since $\widehat{\mathcal{V}} = \mathcal{V}$ (which follows directly from Theorem 9.36) is dense in $L^2(\mathbb{R}^d)$ (by the above), the extension is surjective. Clearly the same discussion applies to the Fourier back transform. Moreover, since $(\widehat{f})^{\vee} = f$ for $f \in V$ by Theorem 9.36, the same holds for all $f \in L^2(\mathbb{R}^d)$.

As we will use the same symbol $\widehat{f}$ for the Fourier transform of f defined for $f \in L^1(\mathbb{R}^d)$ by (9.6) and defined for $f \in L^2(\mathbb{R}^d)$ by the unique continuous extension from $V \subseteq L^2(\mathbb{R}^d)$, we still need to check that for a function $f \in L^1(\mathbb{R}^d) \cap L^2(\mathbb{R}^d)$ these definitions agree. Fortunately, most of the required work has already been done. Let $f \in L^1(\mathbb{R}^d) \cap L^2(\mathbb{R}^d)$, so that $V \ni f * \widehat{\phi_r} \to f$ as $r \to 0$ both in $L^1(\mathbb{R}^d)$ and in $L^2(\mathbb{R}^d)$. As both notions of Fourier transforms are continuous we obtain $\widehat{f * \phi_r} \to \widehat{f}^{L^1}$ in $C_0(\mathbb{R}^d)$ and $\widehat{f * \phi_r} \to \widehat{f}^{L^2}$ in $L^2(\mathbb{R}^d)$ as $r \to 0$, where we write $\widehat{f}^{L^1}$ for the Fourier transform defined using (9.6) and $\widehat{f}^{L^2}$ for the Fourier transform obtained by the above continuous extension. Taking a sequence (r_n) with $r_n \to 0$ as $n \to \infty$ such that $\widehat{f * \phi_{r_n}} \to \widehat{f}^{L^2}$ almost everywhere as $n \to \infty$ we deduce that $\widehat{f}^{L^2} = \widehat{f}^{L^1}$ almost everywhere, as desired. $\qquad\square$

Using the Plancherel formula we can give the reverse direction of the duality we first enountered in Proposition 9.31.

Corollary 9.40 (Duality of convolution and multiplication (II)). *For functions $f, g \in L^2(\mathbb{R}^d)$ the pointwise product $fg \in L^1(\mathbb{R}^d)$ has Fourier transform $\widehat{fg} = \widehat{f} * \widehat{g}$.*

PROOF. Note that since $\widehat{f}, \widehat{g} \in L^2(\mathbb{R}^d)$, the integral in

$$\widehat{f} * \widehat{g}(t) = \int_{\mathbb{R}^d} \widehat{f}(s)\widehat{g}(t - s) \, ds$$

exists for every $t \in \mathbb{R}^d$ by the Cauchy–Schwarz inequality. Also note that

$$\overline{\overline{\widehat{g}}}(t) = \overline{\int_{\mathbb{R}^d} \overline{g(x)} e^{-2\pi i x \cdot t} \, dx} = \int_{\mathbb{R}^d} g(x) e^{2\pi i x \cdot t} \, dx = \widehat{g}(-t).$$

Since the map $f \mapsto \widehat{f}$ is unitary on $L^2(\mathbb{R}^d)$, this gives

$$\widehat{f} * \widehat{g}(0) = \int_{\mathbb{R}^d} \widehat{f}(t)\widehat{g}(-t) \, dt = \langle \widehat{f}, \overline{\widehat{g}} \rangle_{L^2(\mathbb{R}^d)} = \langle f, \overline{g} \rangle_{L^2(\mathbb{R}^d)} = \int fg \, dx = \widehat{fg}(0).$$

Using Proposition 9.29 and $\lambda_{-t_0}(\widehat{g})(-t) = \widehat{g}(t_0 - t)$ for all $t \in \mathbb{R}^d$ we can extend this to

$$\widehat{f} * \widehat{g}(t_0) = \int_{\mathbb{R}^d} \widehat{f}(t)\widehat{g}(t_0 - t) \, dt = \widehat{f} * \lambda_{-t_0}(\widehat{g})(0)$$

$$= \left(\widehat{f} * \widehat{M_{\chi(-t_0)}(g)} \right)(0) = \int_{\mathbb{R}^d} f M_{\chi(-t_0)}(g) \, dx = \widehat{fg}(t_0)$$

for all $t_0 \in \mathbb{R}^d$.   □

Exercise 9.41. Show that the unitary operator $L^2(\mathbb{R}^d) \ni f \mapsto \widehat{f} \in L^2(\mathbb{R}^d)$ is completely diagonalizable and has only four eigenvalues.

Exercise 9.42. Use the Riesz–Thorin interpolation theorem to prove the Hausdorff–Young inequality. Fix $p \in (1,2)$. Show that the Fourier transform on $L^1(\mathbb{R}^d) \cap L^2(\mathbb{R}^d)$ can be extended to all $f \in L^p(\mathbb{R}^d)$ so that $\|\widehat{f}\|_q \leqslant \|f\|_p$ where $\frac{1}{p} + \frac{1}{q} = 1$.

9.2.3 The Fourier Transform, Smoothness, Schwartz Space

As with Fourier series in Section 3.4, smoothness and decay properties of the Fourier transform are closely related. For $x \in \mathbb{R}^d$ and $\alpha \in \mathbb{N}_0^d$ we write x^α for $x_1^{\alpha_1} \cdots x_d^{\alpha_d}$ and define $\big(M_{(cI)^\alpha} f\big)(x) = (cx)^\alpha f(x)$ for any function f on $\mathbb{R}^d$ and scalar c.

Proposition 9.43 (Duality between differentiation and multiplication by monomials). *If $x \mapsto x^\alpha f(x)$ lies in $L^1(\mathbb{R}^d)$ for all $\alpha \in \mathbb{N}_0^d$ with $\|\alpha\|_1 \leqslant k$, then $\widehat{f} \in C^k(\mathbb{R}^d)$, and*

$$\partial_\alpha \widehat{f} = \widehat{M_{(-2\pi i I)^\alpha}(f)}$$

for all α with $\|\alpha\|_1 \leqslant k$. If $f \in C^k(\mathbb{R}^d)$ and $\partial_\alpha f \in L^1(\mathbb{R}^d)$ for $\|\alpha\|_1 \leqslant k$ and $\partial_\alpha f \in C_0(\mathbb{R}^d)$ for $\|\alpha\|_1 \leqslant k - 1$, then

$$\widehat{\partial_\alpha f}(t) = M_{(2\pi i I)^\alpha} f$$

for all α with $\|\alpha\|_1 \leqslant k$.

PROOF. Suppose that f and $x \mapsto x_j f(x)$ lie in $L^1(\mathbb{R}^d)$. Then

$$\partial_j \widehat{f}(t) = \lim_{h \to 0} \frac{\widehat{f}(t + he_j) - \widehat{f}(t)}{h}$$

$$= \lim_{h \to 0} \int_{\mathbb{R}^d} f(x) \frac{e^{-2\pi i x \cdot (t + he_j)} - e^{-2\pi i x \cdot t}}{h}\, \mathrm{d}x$$

$$= \lim_{h \to 0} \int_{\mathbb{R}^d} f(x) \frac{e^{-2\pi i h x_j} - 1}{h} e^{-2\pi i x \cdot t}\, \mathrm{d}x$$

if the limit exists. Now notice that

$$\frac{e^{-2\pi i h x_j} - 1}{h} \longrightarrow -2\pi i x_j$$

as $h \to 0$ is bounded in absolute value by $2\pi |x_j|$ by the two-dimensional mean value theorem for differentiation. Applying the dominated convergence theorem we deduce that the above limit exists and is equal to

$$\partial_j \widehat{f}(t) = \int_{\mathbb{R}^d} f(x)(-2\pi i x_j) e^{-2\pi i x \cdot t} \, dx = \overline{M_{(-2\pi i I)^{e_j}} f}(t)$$

and so the first part of the proposition now follows by induction on k.

Now suppose $f \in C^1(\mathbb{R}^d), f \in C_0(\mathbb{R}^d)$ and $f, \partial_j f \in L^1(\mathbb{R}^d)$. Then

$$\widehat{\partial_j f}(t) = \int_{\mathbb{R}^d} \frac{\partial f}{\partial x_j}(x) e^{-2\pi i x \cdot t} \, dx.$$

By Fubini's theorem we may evaluate this integral by first integrating over x_j. Assuming as we may that this one-dimensional integral is finite, we apply integration by parts to obtain

$$\int_{\mathbb{R}} \frac{\partial f}{\partial x_j}(x) e^{-2\pi i x \cdot t} \, dx_j = \lim_{M \to \infty} f(x) e^{-2\pi i x \cdot t} \Big|_{x_j = -M}^{x_j = M}$$

$$- \lim_{M \to \infty} \int_{-M}^{M} f(x)(-2\pi i t_j) e^{-2\pi i x \cdot t} \, dx_j$$

$$= 2\pi i t_j \int_{\mathbb{R}} f(x) e^{-2\pi i x \cdot t} \, dx_j$$

by all of our assumptions on f. Integrating over the remaining variables this gives the second formula in the proposition in the case $\alpha = e_j$ and the general case now follows by induction. $\qquad\square$

Proposition 9.43 says in particular that the Fourier transform of a smooth function in $C_0(\mathbb{R}^d) \cap L^1(\mathbb{R}^d)$ whose derivatives also lie in $C_0(\mathbb{R}^d) \cap L^1(\mathbb{R}^d)$ (for example, any element of $C_c^\infty(\mathbb{R}^d)$) has a Fourier transform with *superpolynomial decay*. That is, the Fourier transform $\widehat{f}$ multiplied by any polynomial is bounded (and still decays). Similarly, a function that has superpolynomial decay has a smooth Fourier transform. Given these observations, the next definition describes a natural class of functions invariant under differentiation and under Fourier transforms.

Definition 9.44. The *Schwartz space* on $\mathbb{R}^d$ is defined by

$$\mathscr{S}(\mathbb{R}^d) = \{ f : \mathbb{R}^d \to \mathbb{C} \mid f \text{ is smooth and } \| x^\alpha \partial_\beta f \|_\infty < \infty \text{ for all } \alpha, \beta \in \mathbb{N}_0^d \}.$$

The following exercises describe the main properties of $\mathscr{S}(\mathbb{R}^d)$ and of the Fourier transform on $\mathscr{S}(\mathbb{R}^d)$.

Essential Exercise 9.45. (a) Show that $\mathscr{S}(\mathbb{R}^d)$ is a Fréchet space (see Definition 8.65) with the seminorms $\|f\|_{\alpha,\beta} = \|x^\alpha \partial_\beta f\|_\infty$ for $f \in \mathscr{S}(\mathbb{R}^d)$ and $\alpha, \beta \in \mathbb{N}_0^d$.
(b) If the seminorms $\|f\|'_{\alpha,\beta} = \|\partial_\beta(x^\alpha f(x))\|_\infty$ are used instead, do you get the same Fréchet space?
(c) What happens if we replace the supremum norms by 1-norms or 2-norms?
(d) Show that $\mathscr{S}(\mathbb{R}^d) \subseteq L^p(\mathbb{R}^d)$ for all $p \in [1, \infty]$.

Essential Exercise 9.46. Show that the Fourier transform $\widehat{}$ maps $\mathscr{S}(\mathbb{R}^d)$ to itself, is a continuous operator, and has the Fourier back transform $\widecheck{}$ as its continuous inverse.

Exercise 9.47. Prove the *Poisson summation formula*:

$$\sum_{n \in \mathbb{Z}^d} f(n) = \sum_{n \in \mathbb{Z}^d} \widehat{f}(n)$$

for $f \in \mathscr{S}(\mathbb{R}^d)$.

We will use the following in Chapters 11 and 14.

Essential Exercise 9.48. Show that $\widetilde{C_c^\infty(\mathbb{R})}$ is a dense subspace of $L^p(\mathbb{R})$ for any $p \in [1, \infty)$.

9.2.4 The Uncertainty Principle

The Fourier transform viewed as a homeomorphism $\mathscr{S}(\mathbb{R}) \to \mathscr{S}(\mathbb{R})$ (see Exercise 9.46) has multiple physical interpretations

- A function $f \in \mathscr{S}(\mathbb{R})$ with $\|f\|_2 = 1$ may describe the probability distribution of the position of a particle, so that the probability of the particle being in $B \subseteq \mathbb{R}$ is $\int_B |f(x)|^2 \, dx$. In this case $\widehat{f}$ gives a probability distribution of the momentum of the particle. Here Heisenberg's uncertainty principle from quantum mechanics states that if the position distribution f is strongly localized at one position, then the momentum distribution $\widehat{f}$ is forced to be spread out over a big range.
- A function $f \in \mathscr{S}(\mathbb{R})$ may describe a sound, in which case $\widehat{f}$ describes the frequencies present. Sampling f for a short time may be thought of as localizing f by multiplying it by a function in $\mathscr{S}(\mathbb{R})$ supported on a short interval. It is intuitively clear that if the sampling interval is very short, then we cannot get too much information about the frequencies present: f and $\widehat{f}$ cannot both be strongly localized.

It turns out that both of these observations are a consequence of the Cauchy–Schwarz inequality, and they have a convenient precise formulation as follows.[28] As before, we write $M_I(f)(x) = xf(x)$ for $x \in \mathbb{R}^d$ and $f \in L^2(\mathbb{R}^d)$.

Note that if we think of $f \in L^2(\mathbb{R})$ with $\|f\|_2 = 1$ as giving us a probability distribution for the position of a particle, then $\|M_I(f)\|^2 = \int_{\mathbb{R}} |x|^2 |f(x)|^2 \, dx$ gives the expectation of the squared distance to the origin. If that expectation were to be small, then f would certainly be quite localized near the origin. Similarly, $\|M_I(\widehat{f})\|_2$ measures how much $\widehat{f}$ is localized at zero momentum (see also Exercise 9.51 for other positions and momenta).

Theorem 9.49 (Uncertainty principle). *For $f \in \mathscr{S}(\mathbb{R})$ we have*

$$\|M_I(f)\|_2 \|M_I(\widehat{f})\|_2 \geqslant \frac{1}{4\pi} \|f\|_2^2. \tag{9.11}$$

PROOF. First notice that for $z, w \in \mathbb{C}$ we have

$$z\overline{w} + \overline{z}w = 2\Re(z\overline{w}). \tag{9.12}$$

Then

$$\begin{aligned}
\|f\|_2^2 = \int_{\mathbb{R}} |f(x)|^2 \, dx &= \int_{\mathbb{R}} f(x)\overline{f(x)} \, dx \\
&= \underbrace{x|f(x)|^2 \Big|_{-\infty}^{\infty}}_{=0 \text{ as } f \in \mathscr{S}(\mathbb{R})} - \int_{\mathbb{R}} x\big(f'(x)\overline{f(x)} + f(x)\overline{f'(x)}\big) \, dx \\
&= -2 \int_{\mathbb{R}} x\Re\big(f(x)\overline{f'(x)}\big) \, dx. \qquad \text{(by (9.12))}
\end{aligned}$$

Hence

$$\|f\|_2^2 = \left| -2 \int_{\mathbb{R}} \Re\big(xf(x)\overline{f'(x)}\big) \, dx \right| \leqslant 2\|M_I(f)\|_2 \|f'\|_2$$

by the Cauchy–Schwarz inequality. Finally we may apply (a very special case of) Proposition 9.43 to obtain

$$\begin{aligned}
\|f'\|_2 = \|\widehat{f'}\|_2 &\qquad \text{(by Theorem 9.39)} \\
&= 2\pi \|M_I(\widehat{f})\|_2, \qquad \text{(by Prop. 9.43)}
\end{aligned}$$

so that

$$\|f\|_2^2 \leqslant 4\pi \|M_I(f)\|_2 \|M_I(\widehat{f})\|_2,$$

as claimed in the theorem. $\qquad \square$

Exercise 9.50. Show that if for some $f \in \mathscr{S}(\mathbb{R})$ we have equality in (9.11) then f has the form $f(x) = Ae^{-B^2 x^2}$ for constants $A \in \mathbb{C}$ and $B > 0$.

Exercise 9.51. Extend Theorem 9.49 by showing that for any $f \in \mathscr{S}(\mathbb{R})$ and x_0, t_0 in $\mathbb{R}$ we have

$$\|M_{I-x_0}(f)\|_2 \|M_{I-t_0}(\widehat{f})\|_2 \geqslant \frac{1}{4\pi} \|f\|_2^2,$$

and that equality holds only if

$$f(x) = Ae^{2\pi i x t_0} e^{-B^2(x-x_0)^2}$$

for constants $A \in \mathbb{C}$ and $B > 0$.

We note that in Exercise 3.50 we saw another instance of an uncertainty principle for finite abelian groups.

9.3 Spectral Theory of Unitary Flows

Recall from Definition 3.73 the definition of a unitary representation of a topological group G. In the case $G = \mathbb{R}$, or more generally $G = \mathbb{R}^d$, we will also refer to it as a *unitary flow*. The important notion of a positive-definite sequence generalizes easily to this setting.

9.3.1 Positive-Definite Functions and Cyclic Representations

Definition 9.52. Let G be a topological group. A continuous function

$$p : G \to \mathbb{C}$$

is called a *continuous positive-definite function* if for any choice of constants $c_1, \ldots, c_\ell \in \mathbb{C}$ and $g_1, \ldots, g_\ell \in G$ we have

$$\sum_{m,n=1}^{\ell} c_m \overline{c_n} p(g_n^{-1} g_m) \geqslant 0.$$

Just as in Section 9.1, this notion is intimately connected to cyclic representations of the group (Definition 9.7).

Lemma 9.53 (Matrix coefficients). *Let G be a topological group and assume that $\pi : G \curvearrowright \mathcal{H}$ is a unitary representation of G on $\mathcal{H}$. For any $v \in \mathcal{H}$ the function $p_{\pi,v} : G \to \mathbb{C}$ defined by $p_{\pi,v}(g) = \langle \pi_g v, v \rangle$ for $g \in G$, also known as the* principal matrix coefficient *of v, is a continuous positive-definite function. Moreover, the function $p_{\pi,v}$ uniquely characterizes the cyclic representation $\mathcal{H}_v$ generated by the element v. More precisely, if $\pi' : G \curvearrowright \mathcal{H}'$ is another unitary representation and there is some $v' \in \mathcal{H}'$ with $p_{\pi,v} = p_{\pi',v'}$ then there is a unitary isomorphism $\Psi : \mathcal{H}_v \to \mathcal{H}'_{v'}$ with $\Psi(v) = v'$ and $\Psi \circ \pi_g = \pi'_g \circ \Psi$ for all $g \in G$.*

If there is a unitary isomorphism with the properties of Ψ in the lemma, then we say that the representations π and π' are *unitarily isomorphic*.

PROOF OF LEMMA 9.53. The argument is essentially the same as that used in the sequence case (see the justification for Example 9.5 and the proof of Corollary 9.8), so we will be brief. Let $c_1, \ldots, c_\ell \in \mathbb{C}$ and $g_1, \ldots, g_\ell \in G$. Then

$$\sum_{m,n=1}^{\ell} c_m \overline{c_n} p_{\pi,v}(g_n^{-1} g_m) = \sum_{m,n=1}^{\ell} c_m \overline{c_n} \langle \pi_{g_m} v, \pi_{g_n} v \rangle$$

$$= \left\langle \sum_{m=1}^{\ell} c_m \pi_{g_m} v, \sum_{n=1}^{\ell} c_n \pi_{g_n} v \right\rangle \geqslant 0,$$

which gives the first claim of the lemma.

Now suppose that π', $\mathcal{H}'$, and v' have the properties stated in the lemma, namely $p_{\pi,v} = p_{\pi',v'}$. Then the elements

$$\sum_{m=1}^{\ell} c_m \pi_{g_m} v \in \mathcal{H}_v$$

and

$$\sum_{m=1}^{\ell} c_m \pi'_{g_m} v' \in \mathcal{H}'_{v'}$$

have the same norms in their respective Hilbert spaces as both norms can be expressed as above in terms of the positive-definite function p. We can define Ψ on a dense subset of $\mathcal{H}_v$ by setting

$$\Psi\left(\sum_{m=1}^{\ell} c_m \pi_{g_m} v\right) = \sum_{m=1}^{\ell} c_m \pi'_{g_m} v',$$

and this is a well-defined isometry mapping from a dense subset of $\mathcal{H}_v$ onto a dense subset of $\mathcal{H}'_{v'}$. The lemma follows by taking the automatic continuous extension from Proposition 2.59.   □

Using only the definition we can prove the following elementary properties of positive-definite functions.

Lemma 9.54 (Properties of positive-definite functions). *Let G be a topological group and let $p : G \to \mathbb{C}$ be a continuous positive-definite function on G. Then $p(g^{-1}) = \overline{p(g)}$ for all $g \in G$ and $\|p\|_\infty = p(e)$.*

PROOF. Applying the defining property for a positive-definite function with the choices $x_1 = e$, $c_1 = 1$, $x_2 = g \in G$, and $c_2 = \alpha \in \mathbb{C}$, we obtain

$$p(e) + |\alpha|^2 p(e) + \alpha p(g) + \overline{\alpha} p(g^{-1}) \geqslant 0.$$

As this holds for all $\alpha \in \mathbb{C}$, we may set $\alpha = 0$ and see that $p(e) \geqslant 0$. Now use both $\alpha = 1$ and $\alpha = i$ to see that $p(g^{-1}) = \overline{p(g)}$. Finally, if $p(g) \neq 0$, we may set $\alpha = -|p(g)|/p(g)$ to see that $2p(e) - 2|p(g)| \geqslant 0$. It follows that $|p(g)| \leqslant p(e)$ (which also holds if $p(g) = 0$), with equality for $g = e$, giving the lemma.   □

Exercise 9.55. Let G be a topological group.
(a) Show that any positive-definite function p on G has the form $p = p_{\pi,v}$ for some cyclic unitary representation π on a Hilbert space $\mathcal{H}$ and generator $v \in \mathcal{H}$.
(b) Show that $\mathcal{P}_1(G) = \{p \in C_b(G) \mid p \text{ is positive-definite and } p(e) = 1\}$ is convex, and show that if $p \in \mathcal{P}_1(G)$ is extreme then the associated unitary representation in (a) is irreducible (that is, has no non-trivial proper π-invariant closed subspaces).

The converse of the statement in Exercise 9.55(b) also holds, and will be shown later (see Exercise 12.59).

9.3.2 The Case $G = \mathbb{R}^d$

The following describes all positive-definite functions for $\mathbb{R}^d$ and hence once again all cyclic representations of $\mathbb{R}^d$.

Theorem 9.56 (Bochner's theorem for $\mathbb{R}^d$). *Let $d \geqslant 1$ and suppose that $p : \mathbb{R}^d \to \mathbb{C}$ is a continuous positive-definite function. Then there exists a uniquely determined finite measure μ_p on $\mathbb{R}^d$ satisfying*

$$p(x) = \int_{\mathbb{R}^d} \mathrm{e}^{2\pi \mathrm{i} x \cdot t} \, \mathrm{d}\mu_p(t)$$

for all $x \in \mathbb{R}^d$.

We note that this means that $p(x) = \left\langle M_{\chi(x)} \mathbb{1}, \mathbb{1} \right\rangle_{L^2(\mathbb{R}^d, \mu_p)}$, where $M_{\chi(x)}$ is defined by $M_{\chi(x)}(f)(t) = \mathrm{e}^{2\pi \mathrm{i} x \cdot t} f(t)$ for all $f \in L^2(\mathbb{R}^d, \mu_p)$ and $t \in \mathbb{R}^d$.

Essential Exercise 9.57. Let μ be a finite measure on $\mathbb{R}^d$ with $d \geqslant 1$. Show that $\mathbb{R}^d \ni x \mapsto M_{\chi(x)}$ defines a unitary representation of $\mathbb{R}^d$ on $L^2(\mathbb{R}^d, \mu)$ (and, in particular, also satisfies the continuity requirement of a unitary representation in Definition 3.73).

We postpone the proof of Bochner's theorem and first discuss one of its corollaries, the spectral theorem.

Theorem 9.58 (Spectral theorem for $\mathbb{R}^d$). *Let $d \geqslant 1$ and suppose that π is a unitary representation $\mathbb{R}^d \curvearrowright \mathcal{H}$ on a separable complex Hilbert space $\mathcal{H}$. Then there is a decomposition $\mathcal{H} = \bigoplus_{n \geqslant 1} \mathcal{H}_{v_n}$ for some sequence (v_n) in $\mathcal{H}$. Moreover, for every $v \in \mathcal{H}$ the unitary representation $\pi : \mathbb{R}^d \curvearrowright \mathcal{H}_v$ is unitarily isomorphic to the unitary representation $M_{\chi(x)}$ on $L^2(\mathbb{R}^d, \mu_v)$, where μ_v is the spectral measure of $v \in \mathcal{H}$ (obtained from $p_{\pi,v}$ and Theorem 9.56) and $M_{\chi(x)}$ is the unitary multiplication operator on $L^2(\mathbb{R}^d, \mu_v)$, as above.*

PROOF. The argument after Definition 9.7 shows that $\mathcal{H}$ can be written as an orthogonal direct sum of cyclic representations $\mathcal{H}_{v_n}$ for some vectors $v_1, v_2, \ldots \in \mathcal{H}$. We apply Bochner's theorem (Theorem 9.56) to a cyclic representation, say $\mathcal{H}_v$ for $v \in \mathcal{H}$, to find the spectral measure. Lemma 9.53, the comment after Theorem 9.56, and Exercise 9.57 show that the cyclic representation is isomorphic to the cyclic representation generated by $\mathbb{1}$ inside $L^2(\mathbb{R}^d, \mu)$. It remains to show that this representation is all of the space $\mathcal{H}' = L^2(\mathbb{R}^d, \mu)$.

Suppose therefore that $f \in L^2(\mathbb{R}^d, \mu)$ belongs to the orthogonal complement of $\mathcal{H}'_{\mathbb{1}}$, so that

$$\int_{\mathbb{R}^d} f(t) \mathrm{e}^{2\pi \mathrm{i} x \cdot t} \, \mathrm{d}\mu(t) = \left\langle f, M_{\chi(-x)} \mathbb{1} \right\rangle = 0$$

for all $x \in \mathbb{R}^d$. Let $g \in \mathscr{S}(\mathbb{R}^d)$. Since $(\widehat{g})^{\vee} = g$ and $\widehat{g} \in L^1(\mathbb{R}^d)$, we obtain

$$
\begin{aligned}
\langle f, \overline{g} \rangle_{L^2(\mu)} &= \int_{\mathbb{R}^d} f g \, \mathrm{d}\mu = \int_{\mathbb{R}^d} f (\widehat{g})^{\vee} \, \mathrm{d}\mu \\
&= \int_{\mathbb{R}^d} f(t) \int_{\mathbb{R}^d} \widehat{g}(x) e^{2\pi i x \cdot t} \, \mathrm{d}x \, \mathrm{d}\mu(t) \\
&= \int_{\mathbb{R}^d} \widehat{g}(x) \int_{\mathbb{R}^d} f(t) e^{2\pi i x \cdot t} \, \mathrm{d}\mu(t) \, \mathrm{d}x = 0
\end{aligned}
$$

by Fubini's theorem. Recalling that $\mathscr{S}(\mathbb{R}^d) \supseteq C_c^\infty(\mathbb{R}^d)$ is dense in $L_\mu^2(\mathbb{R}^d)$, we see that $f = 0$.

Since this holds for all $f \in (\mathcal{H}_1')^\perp$ it follows that $\mathcal{H}_1' = \mathcal{H}' = L^2(\mathbb{R}^d, \mu)$, as required. $\qquad \square$

For the proof of Bochner's theorem it will be convenient to reformulate the defining property of positive-definite functions in terms of convolution as in the next lemma.

Lemma 9.59 (Positive-definite functions and convolutions). *Assume that $d \geqslant 1$ and suppose $p : \mathbb{R}^d \to \mathbb{C}$ is a continuous positive-definite function on $\mathbb{R}^d$. For $f \in L^1(\mathbb{R}^d)$, define $\widetilde{f} \in L^1(\mathbb{R}^d)$ by $\widetilde{f}(x) = \overline{f(-x)}$ for $x \in \mathbb{R}^d$. Then*

$$
\int_{\mathbb{R}^d} (f * \widetilde{f}) p \, \mathrm{d}x \geqslant 0.
$$

PROOF. We first suppose that $f \in C_c(\mathbb{R}^d)$ has $\mathrm{Supp}(f) \subseteq [-M, M]^d$ for some $M > 0$. In the following we write $\{P_1, \ldots, P_n\}$ for a partition of $[-M, M]^d$ into squares and assume $x_i \in P_i$ for $i = 1, \ldots, n$. Then

$$
\begin{aligned}
\int (f * \widetilde{f}) p \, \mathrm{d}x &= \int_{\mathbb{R}^d} \int_{[-M,M]^d} f(y) \overline{f(\underbrace{y - x}_{=z})} p(x) \, \mathrm{d}y \, \mathrm{d}x \\
&= \int_{[-M,M]^d} \int_{[-M,M]^d} f(y) \overline{f(z)} p(y - z) \, \mathrm{d}y \, \mathrm{d}z \\
&= \lim \sum_{i,j=1}^n f(x_i) m(P_i) \overline{f(x_j)} m(P_j) p(x_i - x_j)
\end{aligned}
$$

is a limit of Riemann sums of the form in Definition 9.52, so $\int (f * \widetilde{f}) p \, \mathrm{d}x \geqslant 0$ whenever $f \in C_c(\mathbb{R}^d)$. Approximating an arbitrary function $f \in L^1(\mathbb{R}^d)$ by such functions (using the continuity of a product in a Banach algebra from Proposition 9.31) gives the result. $\qquad \square$

PROOF OF THEOREM 9.56. Let $p : \mathbb{R}^d \to \mathbb{C}$ be a continuous positive-definite function. Let us first prove the uniqueness and assume μ_p is as in the theorem. For $f \in \mathscr{S}(\mathbb{R}^d)$ we then have

$$\int_{\mathbb{R}^d} f \, d\mu_p = \int_{\mathbb{R}^d} (\widehat{f})^{\vee} \, d\mu_p = \int_{\mathbb{R}^d} \int_{\mathbb{R}^d} \widehat{f}(x) e^{2\pi i x \cdot t} \, dx \, d\mu_p(t) = \int_{\mathbb{R}^d} \widehat{f}(x) p(x) \, dx$$

by Fubini's theorem. In particular, we see that p determines $\int_{\mathbb{R}^d} f \, d\mu_p$ uniquely for every $f \in C_c^{\infty}(\mathbb{R}^d) \subseteq \mathscr{S}(\mathbb{R}^d)$. Since $C_c^{\infty}(\mathbb{R}^d)$ is dense in $C_0(\mathbb{R}^d)$ (which contains $C_c(\mathbb{R}^d)$) it follows that p determines μ_p uniquely by the Riesz representation (Theorem 7.44).

We now turn to the existence. We will obtain the measure $\mu = \mu_p$ by defining a positive functional on $C_0(\mathbb{R}^d)$, whose properties will be proved in several steps.

FIRST STEP: DEFINITION ON THE SCHWARTZ SPACE. We initially define the functional on $\mathscr{S}(\mathbb{R}^d)$ (which only contains rapidly-decaying smooth functions). For f in $\mathscr{S}(\mathbb{R}^d)$ let

$$\Lambda(f) = \int_{\mathbb{R}^d} \widehat{f} p \, dx = \int_{\mathbb{R}^d} \int_{\mathbb{R}^d} f(t) e^{-2\pi i x \cdot t} \, dt \, p(x) \, dx,$$

which is well-defined since $\widehat{f} \in L^1(\mathbb{R}^d)$ and $\|p\|_{\infty} = p(0) < \infty$.

SECOND STEP: POSITIVITY. The assumption on p immediately gives a certain amount of positivity. Suppose $f \in \mathscr{S}(\mathbb{R}^d)$ and note that

$$\widehat{\overline{f}}(x) = \int_{\mathbb{R}^d} \overline{f(t)} e^{-2\pi i x \cdot t} dt = \overline{\int_{\mathbb{R}^d} f(t) e^{2\pi i x \cdot t} dt} = \widetilde{\overline{f}}(x)$$

(see Lemma 9.59 for the definition of $\widetilde{f}$) and that $|f|^2 \in \mathscr{S}(\mathbb{R}^d)$. By the duality of multiplication and convolution (Corollary 9.40) we obtain

$$\widehat{|f|^2} = \widehat{f} * \widehat{\overline{f}} = \widehat{f} * \widetilde{\overline{f}},$$

and so

$$\Lambda(|f|^2) = \int (\widehat{f} * \widetilde{\overline{f}}) p \, dx \geqslant 0$$

by Lemma 9.59.

We wish to upgrade the above positivity statement to say that $f \geqslant 0$ and $f \in C_c^{\infty}(\mathbb{R}^d)$ implies $\Lambda(f) \geqslant 0$. So let $f \in C_c^{\infty}(\mathbb{R}^d)$ be non-negative, and define

$$h_{\varepsilon}(t) = \sqrt{f(t) + \varepsilon e^{-\pi \|t\|^2}}.$$

Notice that $h_{\varepsilon} \in \mathscr{S}(\mathbb{R}^d)$ so that

$$\Lambda(f) + \varepsilon \Lambda(e^{-\pi \|t\|^2}) = \Lambda(|h_{\varepsilon}|^2) \geqslant 0$$

for all $\varepsilon > 0$, and hence $\Lambda(f) \geqslant 0$.

THIRD STEP: BOUNDEDNESS WITH RESPECT TO $\| \cdot \|_{\infty}$. Next we wish to obtain an estimate for $\Lambda(f)$ for any $f \in C_c^{\infty}(\mathbb{R}^d)$ that only depends on $\|f\|_{\infty}$.

We assume first that $f \in C_c^\infty(\mathbb{R}^d)$ is real-valued, and fix $\varepsilon > 0$. Then, for sufficiently large $a > 0$, we have $f(t) < (1+\varepsilon)\|f\|_\infty e^{-\pi\|t/a\|^2}$, for all $t \in \mathbb{R}$ and similarly for $-f$. Therefore, we have $(1+\varepsilon)\|f\|_\infty e^{-\pi\|t/a\|^2} - f(t) = |h|^2$ where

$$h(t) = \sqrt{(1+\varepsilon)\|f\|_\infty e^{-\pi\|t/a\|^2} - f(t)},$$

and $h \in \mathscr{S}(\mathbb{R}^d)$, which as above gives the inequality

$$(1+\varepsilon)\|f\|_\infty \Lambda\big(e^{-\pi\|t/a\|^2}\big) - \Lambda(f) \geqslant 0.$$

Note that $e^{-\pi\|t/a\|^2}$ also has a square root inside $\mathscr{S}(\mathbb{R}^d)$ which gives

$$\Lambda\big(e^{-\pi\|t/a\|^2}\big) \geqslant 0$$

and so $\Lambda(f) \in \mathbb{R}$ satisfies $\Lambda(f) \leqslant (1+\varepsilon)\|f\|_\infty \Lambda\big(e^{-\pi\|t/a\|^2}\big)$. The same estimate also holds for $\Lambda(-f)$. However,

$$\Lambda\big(e^{-\pi\|t/a\|^2}\big) = \int_{\mathbb{R}^d} \underbrace{a^d e^{-\pi\|ax\|^2}}_{\|\cdot\|_1 = 1} p(x)\,\mathrm{d}x \leqslant \|p\|_\infty = p(0)$$

can be bounded independently of $a > 0$. It follows that

$$|\Lambda(f)| \leqslant \|f\|_\infty p(0) \tag{9.13}$$

for any $\mathbb{R}$-valued $f \in C_c^\infty(\mathbb{R}^d)$.

If $f \in C_c^\infty(\mathbb{R}^d)$ is $\mathbb{C}$-valued let $\alpha \in \mathbb{C}$ have absolute value one and satisfy $|\Lambda(f)| = \alpha\Lambda(f)$. Now note $|\Lambda(f)| = \Lambda(\alpha f) = \Lambda(\Re(\alpha f))$ (since Λ maps $\mathbb{R}$-valued functions into $\mathbb{R}$) and apply (9.13) for $\Re(\alpha f)$ to obtain

$$|\Lambda(f)| = |\Lambda(\Re(\alpha f))| \leqslant \|f\|_\infty p(0),$$

showing that (9.13) holds for all $f \in C_c^\infty(\mathbb{R}^d)$.

LAST STEP: EXISTENCE AND PROPERTIES OF THE MEASURE. By the previous step the functional $\Lambda|_{C_c^\infty(\mathbb{R})}$ extends continuously to a functional Λ_0 on $C_0(\mathbb{R}^d)$. Furthermore, every non-negative function in $C_0(\mathbb{R}^d)$ can be approximated with respect to the supremum norm by non-negative functions in $C_c^\infty(\mathbb{R}^d)$. Hence the extension of Λ_0 is a positive continuous linear functional on $C_0(\mathbb{R}^d)$ and so defines, by the Riesz representation theorem (Theorem 7.44 and Theorem 7.54), a finite positive measure μ on $\mathbb{R}^d$ satisfying

$$\int_{\mathbb{R}^d} \widehat{f}p\,\mathrm{d}x = \Lambda(f) = \int_{\mathbb{R}^d} f\,\mathrm{d}\mu \tag{9.14}$$

for all $f \in C_c^\infty(\mathbb{R}^d)$.

Fix some non-negative $h \in C_c^\infty(\mathbb{R}^d)$ with $\int_{\mathbb{R}^d} h\,\mathrm{d}x = 1$ and notice that

$$h_{r,x_0}(x) = r^{-d}h\big((x-x_0)/r\big)$$

approximates the δ-measure at x_0 as $r \to 0$. Concretely, since p is continuous we may apply dominated convergence and the substitution $y = (x - x_0)/r$ to see that

$$
\begin{aligned}
p(x_0) &= \lim_{r\to 0} \int_{\mathbb{R}^d} h(y)p(x_0 + ry)\,\mathrm{d}y \\
&= \lim_{r\to 0} \int_{\mathbb{R}^d} r^{-d}h\big((x-x_0)/r\big)p(x)\,\mathrm{d}x = \lim_{r\to 0} \int_{\mathbb{R}^d} h_{r,x_0}(x)p(x)\,\mathrm{d}x. \quad (9.15)
\end{aligned}
$$

In order to be able to combine this with (9.14) we calculate the Fourier back transform of h_{r,x_0},

$$
\begin{aligned}
f_{r,x_0}(t) = \widetilde{h_{r,x_0}}(t) &= \int_{\mathbb{R}^d} r^{-d}h\big((x-x_0)/r\big)e^{2\pi i x\cdot t}\,\mathrm{d}x \\
&= \int_{\mathbb{R}^d} h(y)e^{2\pi i(x_0+ry)\cdot t}\,\mathrm{d}y = e^{2\pi i x_0\cdot t}\check{h}(rt),
\end{aligned}
$$

where we again used the substitution $y = (x - x_0)/r$. By Fourier inversion (Theorem 9.36) we also have $h_{r,x_0} = \widehat{f_{r,x_0}}$. With (9.14) and (9.15) we now obtain

$$
p(x_0) = \lim_{r\to 0} \Lambda(f_{r,x_0}) = \lim_{r\to 0} \int_{\mathbb{R}^d} e^{2\pi i x_0\cdot t}\check{h}(rt)\,\mathrm{d}\mu(t) = \int_{\mathbb{R}^d} e^{2\pi i x_0\cdot t}\,\mathrm{d}\mu(t)
$$

by dominated convergence and since $\check{h}(0) = \int_{\mathbb{R}^d} h\,\mathrm{d}x = 1$.   □

We remark that the properties of spectral measures discussed in Section 9.1.3 hold in the setting of unitary flows for essentially the same reasons, but we will not pursue this here.

9.3.3 Stone's Theorem

We already saw a connection between self-adjoint operators and unitary operators in Exercise 9.10. The spectral theorem for unitary flows allows us to expand this connection, leading to a complete description of a unitary flow in terms of a 'potentially unbounded self-adjoint' operator. For simplicity we restrict to the case of a one-parameter unitary flow (that is, a unitary representation of $G = \mathbb{R}$).

Theorem 9.60 (Stone's theorem). *Suppose that $\pi : \mathbb{R} \curvearrowright \mathcal{H}$ is a unitary representation of $\mathbb{R}$ on a separable complex Hilbert space $\mathcal{H}$. Then the subspace*

$$D = \left\{ v \in \mathcal{H} \mid \lim_{x\to 0} \frac{\pi_x v - v}{x} \text{ exists in } \mathcal{H} \right\}$$

of differentiable vectors *is dense in* $\mathcal{H}$, *and* D *is the natural domain of the closed operator*

$$D \ni v \longmapsto A(v) = \frac{1}{2\pi i} \lim_{x \to 0} \frac{\pi_x v - v}{x}.$$

Moreover, there exists an increasing sequence of closed, A-invariant, π-invariant subspaces $D_1 \subseteq D_2 \subseteq \cdots \subseteq D$ *such that* $\bigcup_{\ell \geqslant 1} D_\ell$ *is dense in* $\mathcal{H}$, $A|_{D_\ell} : D_\ell \to D_\ell$ *is self-adjoint,* $\|A|_{D_\ell}\| \leqslant \ell$, *and* $\pi_x|_{D_\ell} = \exp(2\pi i x A|_{D_\ell})$ *is defined by a convergent power series for all* $x \in \mathbb{R}$ *and* $\ell \geqslant 1$.

PROOF. We will apply the spectral theorem (Theorem 9.2) for unitary flows to describe the unitary representation in terms of multiplication operators by scalars. As a slight simplification, we assume that $\mathcal{H} = \mathcal{H}_v$ is cyclic for some $v \in \mathcal{H}$ and refer to Exercise 9.62 for the general case.

Then by the spectral theorem we have $\mathcal{H}_v \cong L^2_{\mu_v}(\mathbb{R})$, where v corresponds to $\mathbb{1}$ and π_x corresponds to $M_{\chi(x)}$ for all $x \in \mathbb{R}$. As the isomorphism between $\mathcal{H}_v$ and $L^2_{\mu_v}(\mathbb{R})$ is unitary it maps convergent sequences to convergent sequences and hence also differentiable vectors (as in the definition of D) for π precisely to the differentiable vectors for $M_{\chi(\cdot)}$. In other words, it suffices to prove the theorem in the case where $\pi = M_{\chi(\cdot)}$ and $\mathcal{H} = L^2_\mu(\mathbb{R})$ for a finite measure μ on $\mathbb{R}$. We note that the spectral theorem provides a finite measure, but the proof stays the same if μ is only locally finite. In this case, we claim that D is given by $D = \{f \in L^2_\mu(\mathbb{R}) \mid M_I(f) \in L^2_\mu(\mathbb{R})\}$. Indeed, for $f \in L^2_\mu(\mathbb{R})$ we have

$$\lim_{x \to 0} \left(\frac{\pi_x f - f}{x} \right)(t) = \lim_{x \to 0} \frac{e^{2\pi i x t} - 1}{x} f(t) = 2\pi i t f(t)$$

pointwise wherever $f(t)$ is defined. Therefore $f \in D$ forces $M_I(f) \in L^2_\mu(\mathbb{R})$, where $M_I(f)(t) = t f(t)$ for $t \in \mathbb{R}$.

For the other inclusion, note that $\left| \frac{e^{2\pi i x t} - 1}{x} \right| \leqslant 2\pi t$ by the mean value theorem for vector-valued differentiable functions. If now $f, M_I(f) \in L^2_\mu(\mathbb{R})$ then

$$\left\| \frac{\pi_x f - f}{x} - 2\pi i M_I(f) \right\|^2 = \int \underbrace{\left| \frac{e^{2\pi i x t} - 1}{x} f(t) - 2\pi i t f(t) \right|^2}_{\leqslant (4\pi t f(t))^2} d\mu(t)$$

converges to zero as $x \to 0$ by dominated convergence, which proves the claimed description of D and that $A = M_I$.

We now show that M_I is a closed operator in the sense of Definition 4.27. So suppose that $(f_n, M_I(f_n)) \in \mathrm{Graph}(M_I)$ converges to (f, g). Then (after taking a subsequence) we may assume that $f_n(t) \to f(t)$ almost everywhere and $t f_n(t) = M_I(f_n)(t) \to g(t)$ almost everywhere, and hence $g \in L^2_\mu(\mathbb{R})$ where $t f(t) = g(t)$ almost everywhere and so $f \in D$ and $M_I(f) = g$, as required.

The increasing sequence of closed subspaces can be defined by setting

$$D_\ell = L^2_{\mu|_{[-\ell,\ell]}}([-\ell,\ell]) \subseteq L^2_\mu(\mathbb{R})$$

for $\ell \geqslant 1$, where a function defined on $[-\ell, \ell]$ is extended to be defined on $\mathbb{R}$ by setting it equal to zero outside $[-\ell, \ell]$. These subspaces are clearly $M_{\chi(x)}$- and M_I-invariant for all $x \in \mathbb{R}$. Moreover, $\bigcup_{\ell \geqslant 1} D_\ell$ contains all continuous functions of compact support and therefore is dense in $L^2_\mu(\mathbb{R})$ by Proposition 2.51. Moreover, $M_I|_{D_\ell} : D_\ell \to D_\ell$ is self-adjoint, $\|M_I|_{D_\ell}\| \leqslant \ell$, and $\exp(2\pi i x M_I|_{D_\ell}) = M_{\exp(2\pi i x I)}|_{D_\ell} = M_{\chi(x)}|_{D_\ell}$ for all $x \in \mathbb{R}$ and $\ell \geqslant 1$. $\square$

Exercise 9.61. Let $\mathcal{H}$ be a separable complex Hilbert space, let $D_1 \subseteq D_2 \subseteq \cdots$ be a sequence of closed subspaces such that $\bigcup_{\ell \geqslant 1} D_\ell$ is dense in $\mathcal{H}$, and let $A : \bigcup_{\ell \geqslant 1} D_\ell \to \mathcal{H}$ be a linear map such that $A(D_\ell) \subseteq D_\ell$, $\|A|_{D_\ell}\| \leqslant \ell$, and $A|_{D_\ell} : D_\ell \to D_\ell$ is self-adjoint for all $\ell \geqslant 1$. Show that there exists a uniquely defined unitary representation π of $\mathbb{R}$ on $\mathcal{H}$ such that D_ℓ is π-invariant and $\pi_x|_{D_\ell} = \exp(2\pi i x A|_{D_\ell})$ for all $x \in \mathbb{R}$ and $\ell \geqslant 1$.

Exercise 9.62. Prove Theorem 9.60 in the general case where $\mathcal{H} = \bigoplus_{n \geqslant 1} \mathcal{H}_{v_n}$ for some sequence of vectors (v_n).

Exercise 9.63. Apply the results above to the unitary flow $(\rho_x f)(y) = f(y+x)$ for $x, y \in \mathbb{R}$ and $f \in L^2(\mathbb{R})$.
(a) Use the Fourier transform and the proof of Theorem 9.60 to show that

$$D = \{f \in L^2(\mathbb{R}) \mid t \longmapsto t\widehat{f}(t) \text{ lies in } L^2(\mathbb{R})\}.$$

(b) Show that $C_c^\infty(\mathbb{R}) \subseteq D$ and that $C_c^\infty(\mathbb{R})$ is dense in D when D is endowed with the norm in $\mathrm{Graph}(A)$ where A is defined as in Theorem 9.60.
(c) Show that $\mathrm{Graph}(A) = H_0^1(\mathbb{R})$ and that $A = \frac{1}{2\pi i}\partial_x$.
(d) Show moreover that $H^1(\mathbb{R}) = H_0^1(\mathbb{R})$.

Exercise 9.64. Generalize Theorem 9.60 to unitary representations of $\mathbb{R}^d$ by studying the space D of vectors for which all partial derivatives $\lim_{t \to 0} \frac{\pi_{t e_j} v - v}{t}$ exist.

9.4 Further Topics

- We will study unitary representations of more general topological groups in the next chapter (which in part will need the results regarding unitary flows of this chapter) and in Chapter 12.
- The theory of the Fourier transform and the Schwartz space (together with the results of Chapter 11) will be used in Chapter 14 to prove the prime number theorem.
- Our introduction to spectral theory will continue in Chapters 11, 12, and 13.

Chapter 10
Locally Compact Groups, Amenability, Property (T)

In this chapter we turn our attention to topological groups. We have seen the importance of the Haar measure for the theory of Fourier series already in Chapter 3. After proving the existence of the Haar measure[(29)] we study two different classes of groups. The notion of amenable group was already discussed in Chapter 7 but here we will drop the assumption of discreteness and discuss the property in greater detail. After this we will study groups with property (T); these are in some sense (see Exercise 10.35) opposite to amenable groups. Finally, we will link property (T) to the topic of expander graphs in Section 10.4.

10.1 Haar Measure

Using the Riesz representation theorem (Theorem 7.44) we now prove a version of the existence of Haar measures from p. 92. Throughout this section we will be working with real-valued functions.

Theorem 10.1 (Existence of Haar measure). *Let G be a locally compact, σ-compact, metrizable group. Then there exists a (left) Haar measure m_G on G: that is, there is a locally finite Borel measure m_G (that is, a Radon measure) that is positive on non-empty open sets and satisfies $m_G(gB) = m_G(B)$ for every Borel measurable set $B \subseteq G$ and all $g \in G$.*

Write λ_g for the left regular representation of G on functions (or equivalence classes of functions) on G defined by $\lambda_g(f)(h) = f(g^{-1}h)$ for $g, h \in G$. This indeed defines a representation since

$$\lambda_{g_1}\left(\lambda_{g_2}(f)\right)(h) = \lambda_{g_2}(f)(g_1^{-1}h) = f(g_2^{-1}g_1^{-1}h) = \lambda_{g_1 g_2}(f)(h)$$

for all $g_1, g_2, h \in G$ and functions f on G.

© Springer International Publishing AG 2017
M. Einsiedler, T. Ward, *Functional Analysis, Spectral Theory, and Applications*,
Graduate Texts in Mathematics 276, DOI 10.1007/978-3-319-58540-6_10

PROOF OF THEOREM 10.1. To motivate the following argument, fix some function

$$\phi \in C_c^+(G) = \{f \in C_c(G) \mid f \geqslant 0\} \backslash \{0\}$$

and think of it as a gauge[†] function. If now another non-trivial function f in $C_c^+(G)$ can be approximated (in a suitable sense) by sums of the form

$$\sum_{j=1}^{n} c_j \lambda_{g_j} \phi,$$

then we would expect that the (yet to be defined) integral

$$\int_G f \, dm_G$$

will be approximated by

$$\sum_{j=1}^{n} c_j \int_G \phi \, dm_G.$$

In particular, this would express an approximation to the integral of a general function in terms of the integral of a single chosen function.

In order to follow this through it is clear that the gauge function will need to be allowed to vary. Roughly speaking, the more localized ϕ is, the more functions can be approximated in this way. As an extreme example, if G is compact then ϕ could be a constant and only the constant functions could be approximated (also see Figure 10.1). For that reason, we will fix throughout another function $f_0 \in C_c^+(G)$ and normalize all expressions so that the (yet to be constructed) Haar measure m_G will satisfy $\int_G f_0 \, dm_G = 1$.

Hence we start the formal argument by defining for functions $\phi, f \in C_c^+(G)$ the expression

$$M(f : \phi) = \inf \left\{ \sum_{j=1}^{n} c_j \; \middle| \; f \leqslant \sum_{j=1}^{n} c_j \lambda_{g_j} \phi \text{ for some } c_1, \ldots, c_n \geqslant 0, g_1, \ldots, g_n \in G \right\}$$

and the normalized quantity

$$\Lambda_\phi(f) = \frac{M(f : \phi)}{M(f_0 : \phi)}.$$

We may think of $\sum_{j=1}^{n} c_j \lambda_{g_j} \phi$ as a ϕ-*cover* of f and of $\sum_{j=1}^{n} c_j$ as the *total weight* of the ϕ-cover. Notice that $\{g \in G \mid \phi(g) > \frac{1}{2}\|\phi\|_\infty\}$ is a non-empty open subset of G, and since $f \in C_c(G)$ has compact support it is easy to see that a cover of f as in the definition of $M(f : \phi)$ exists, and so $M(f : \phi)$ is a well-defined non-negative real number. Moreover, if $f_0 \leqslant \sum_{j=1}^{n} c_j \lambda_{g_j} \phi$

[†] The word 'gauge' means a fixed standard of measure like a ruler.

then $\|f_0\|_\infty \leqslant \sum_{j=1}^n c_j \|\phi\|_\infty$ and so $M(f_0 : \phi) \geqslant \|f_0\|_\infty \|\phi\|_\infty^{-1} > 0$, which implies that $\Lambda_\phi(f) \in \mathbb{R}_{\geqslant 0}$ is well-defined.

We collect a few immediate properties of Λ_ϕ for a scalar $\alpha > 0$ and functions $f, f_1, f_2 \in C_c^+(G)$:

- (left-invariance) $\Lambda_\phi(\lambda_g f) = \Lambda_\phi(f)$;
- (positive homogeneity) $\Lambda_\phi(\alpha f_1) = \alpha \Lambda_\phi(f_1)$;
- (monotonicity) $\Lambda_\phi(f_1) \leqslant \Lambda_\phi(f_2)$ whenever $f_1 \leqslant f_2$; and
- (sub-additivity) $\Lambda_\phi(f_1 + f_2) \leqslant \Lambda_\phi(f_1) + \Lambda_\phi(f_2)$.

These properties are immediate consequences of the definition of $M(f : \phi)$ and standard properties of the infimum. For instance, if $f_1 \leqslant \sum_{j=1}^n c_j \lambda_{g_j} \phi$ and $f_2 \leqslant \sum_{k=1}^m d_k \lambda_{h_k} \phi$ for some scalars $c_1, \ldots, c_n, d_1, \ldots, d_m \geqslant 0$ and group elements $g_1, \ldots, g_n, h_1, \ldots, h_m \in G$, then we obtain a ϕ-cover of $f_1 + f_2$ in the form $f_1 + f_2 \leqslant \sum_{j=1}^n c_j \lambda_{g_j} \phi + \sum_{k=1}^m d_k \lambda_{h_k} \phi$ and so $M(f_1 + f_2 : \phi)$ is bounded above by $\sum_{j=1}^n c_j + \sum_{k=1}^m d_k$. Since the ϕ-covers of f_1 and f_2 were arbitrary this implies that

$$M(f_1 + f_2 : \phi) \leqslant M(f_1 : \phi) + M(f_2 : \phi),$$

and the claimed sub-additivity follows after dividing by $M(f_0 : \phi)$.

The main step in the argument is to upgrade the sub-additivity of Λ_ϕ to an 'approximate additivity' property. For this we have to study not one gauge function but many (see Figure 10.1). To prepare for this, we first show that

$$\begin{cases} M(f : \phi) \leqslant M(f : f_0) M(f_0 : \phi), \\ \Lambda_\phi(f) \leqslant M(f : f_0) \end{cases} \tag{10.1}$$

whenever $\phi, f \in C_c^+(G)$. Note that the second line follows from the first on dividing by $M(f_0 : \phi)$. For the proof of the first line in (10.1), suppose that

$$f \leqslant \sum_{j=1}^n c_j \lambda_{g_j} f_0$$

and

$$f_0 \leqslant \sum_{k=1}^m d_k \lambda_{h_k} \phi$$

are an f_0-cover and a ϕ-cover of f and of f_0, respectively. We then have

$$\lambda_{g_j} f_0 \leqslant \sum_{k=1}^m d_k \lambda_{g_j h_k} \phi$$

for all j and we obtain the ϕ-cover

$$f \leqslant \sum_{j=1}^{n} \sum_{k=1}^{m} c_j d_k \lambda_{g_j h_k} \phi,$$

which gives

$$M(f\!:\!\phi) \leqslant \left(\sum_{j=1}^{n} c_j \right) \left(\sum_{k=1}^{m} d_k \right).$$

Since the f_0-cover of f and the ϕ-cover of f_0 were arbitrary, this implies (10.1).

Fig. 10.1: Using $\phi, f_1, f_2 \in C_c(\mathbb{R})$ as shown, it is clear that Λ_ϕ is not additive in general. Here this failure of additivity happens because the gauge function is not sufficiently localized to measure the functions f_1 and f_2.

We now come to the approximate additivity property mentioned above. For any two functions $f_1, f_2 \in C_c^+(G)$ and $\varepsilon > 0$ we claim that there exists a neighbourhood U of $e \in G$ with the property that

$$\Lambda_\phi(f_1 + f_2) \leqslant \Lambda_\phi(f_1) + \Lambda_\phi(f_2) \leqslant \Lambda_\phi(f_1 + f_2) + \varepsilon \qquad (10.2)$$

for any non-zero function $\phi \in C_c^+(G)$ with support contained in U.

Notice that the first inequality in (10.2) is the sub-additivity shown above, and the second inequality requires an argument that splits a ϕ-cover of $f_1 + f_2$ into two separate ϕ-covers of f_1 and f_2 without too much loss of precision. We will do this using an approximate partition of unity as follows. By Urysohn's lemma (Lemma A.27) we may find a function $F \in C_c^+(G)$ such that $F \equiv 1$ on $\mathrm{Supp}(f_1 + f_2)$. Using it we define

$$\delta = \min\left\{ 1, \tfrac{\varepsilon}{3M(f_1 + f_2 + F : f_0)} \right\} \qquad (10.3)$$

and the functions p_1 and p_2 by

$$p_k(g) = \begin{cases} \dfrac{f_k(g)}{f_1(g) + f_2(g) + \delta F(g)} & \text{for } g \in \mathrm{Supp}\, f_k; \\ 0 & \text{for } g \notin \mathrm{Supp}\, f_k \end{cases}$$

for $k = 1, 2$; notice that $p_1, p_2 \in C_c^+(G)$. By uniform continuity of p_1 and p_2 there exists some neighbourhood U of $e \in G$ such that $u \in U$ and $g \in \mathrm{Supp}\, p_k$ implies that

$$\left| p_k(gu^{-1}) - p_k(g) \right| < \delta$$

for $k = 1, 2$. Suppose now that $\phi \in C_c^+(U)$ (which we could also write as $C_c^+(G) \cap C_c(U)$ with the usual convention that functions on U may be extended to functions on G by setting them to be zero on $G \setminus U$) and

$$f_1 + f_2 + \delta F \leqslant \sum_{j=1}^{n} c_j \lambda_{g_j} \phi \tag{10.4}$$

is a ϕ-cover of $f_1 + f_2 + \delta F$. Multiplying this inequality by p_k gives

$$f_k(g) = (f_1 + f_2 + \delta F(g)) p_k(g) \leqslant \sum_{j=1}^{n} c_j p_k(g) \phi(g_j^{-1} g)$$

for all $g \in G$. Fixing $g \in G$ and one j in the sum, we see that either

$$p_k(g) \phi(g_j^{-1} g) = 0$$

or $g \in \operatorname{Supp} p_k$ and $g_j^{-1} g = u \in \operatorname{Supp} \phi \subseteq U$, which implies that $g_j = g u^{-1}$ and

$$p_k(g) \phi(g_j^{-1} g) \leqslant (p_k(g_j) + \delta) \, \phi(g_j^{-1} g)$$

in either case. Taking the sum over j gives the ϕ-cover

$$f_k \leqslant \sum_{j=1}^{n} c_j \left(p_k(g_j) + \delta \right) \lambda_{g_j} \phi$$

and so

$$M(f_k : \phi) \leqslant \sum_{j=1}^{n} c_j \left(p_k(g_j) + \delta \right)$$

for $k = 1, 2$. Taking the sum over k and using the bound $p_1 + p_2 \leqslant 1$ we obtain

$$M(f_1 : \phi) + M(f_2 : \phi) \leqslant \sum_{j=1}^{n} c_j (1 + 2\delta).$$

Taking the infimum over all ϕ-covers of $f_1 + f_2 + \delta F$ in (10.4) and dividing by $M(f_0 : \phi)$ shows that

$$\Lambda_\phi(f_1) + \Lambda_\phi(f_2) \leqslant (1 + 2\delta) \Lambda_\phi(f_1 + f_2 + \delta F)$$
$$\leqslant \Lambda_\phi(f_1 + f_2) + \delta \Lambda_\phi(F) + 2\delta \Lambda_\phi(f_1 + f_2 + \delta F).$$

Applying (10.1) for ϕ, f_0 and $f = f_1 + f_2 + F$ we arrive at

$$\Lambda_\phi(f_1) + \Lambda_\phi(f_2) \leqslant \Lambda_\phi(f_1 + f_2) + 3\delta M(f_1 + f_2 + F : f_0) \leqslant \Lambda_\phi(f_1 + f_2) + \varepsilon,$$

by our choice of δ in (10.3). This proves the second inequality in (10.2).

It remains to apply a compactness argument and the Riesz representation theorem (Theorem 7.44). For this, notice that

$$\Omega = \prod_{f \in C_c^+(G)} [0, M(f : f_0)]$$

is a compact topological space, and that by (10.1) any non-zero function ϕ in $C_c^+(G)$ defines an element $\Lambda_\phi \in \Omega$ (by thinking of the product space Ω as a space of real-valued functions on $C_c^+(G)$). For a neighbourhood U of e we can then define

$$\Omega(U) = \{\Lambda_\phi \mid \phi \in C_c^+(G) \text{ and } \operatorname{Supp} \phi \subseteq U\}.$$

It is clear that $U_1 \subseteq U_2$ implies that $\Omega(U_1) \subseteq \Omega(U_2)$, which in turn shows that $\Omega(U_1) \cap \cdots \cap \Omega(U_n)$ is non-empty for any finite collection of neighbourhoods $U_1, \ldots, U_n$ of e. It follows that the intersection

$$\bigcap_U \overline{\Omega(U)}$$

taken over all neighbourhoods U of $e \in G$ is non-empty by compactness of Ω. Let Λ be an element in this intersection. We note that left-invariance, positive homogeneity, and monotonicity of all functions Λ_ϕ implies the same properties for Λ (check this). Moreover, Λ is in addition additive. In fact, given functions $f_1, f_2 \in C_c^+(G)$ and $\varepsilon > 0$ there exists a neighbourhood U satisfying (10.2). Since $\Lambda \in \overline{\Omega(U)}$ there exists a $\phi \in C_c^+(G)$ with $\operatorname{Supp} \phi \subseteq U$ and with

$$|\Lambda_\phi(f_1) - \Lambda(f_1)| < \varepsilon,$$
$$|\Lambda_\phi(f_2) - \Lambda(f_2)| < \varepsilon,$$

and

$$|\Lambda_\phi(f_1 + f_2) - \Lambda(f_1 + f_2)| < \varepsilon.$$

Using (10.2) for Λ_ϕ we see that

$$\Lambda(f_1 + f_2) - 3\varepsilon \leqslant \Lambda(f_1) + \Lambda(f_2) \leqslant \Lambda(f_1 + f_2) + 4\varepsilon.$$

As $\varepsilon > 0$ was arbitrary, this shows that Λ is additive in the sense that

$$\Lambda(f_1 + f_2) = \Lambda(f_1) + \Lambda(f_2)$$

for all $f_1, f_2 \in C_c^+(G)$.

Now extend Λ to all of $C_c(G)$ by setting $\Lambda(0) = 0$ and

$$\Lambda(f) = \Lambda(f^+) - \Lambda(f^-), \tag{10.5}$$

where $f^+ = \max\{0, f\}$ and $f^- = \max\{0, -f\}$. Applying the argument used immediately after (7.19), we see that Λ is now a positive linear functional on $C_c(G)$, so by the Riesz representation theorem (Theorem 7.44) there exists a unique locally finite measure m with

$$\Lambda(f) = \int_G f \, dm$$

for all $f \in C_c(G)$. Since

$$\int_G \lambda_g(f) \, dm = \Lambda(\lambda_g(f)) = \Lambda(f) = \int_G f \, dm$$

for any $g \in G$ and $f \in C_c(G)$ it follows that m is left-invariant. Also note that $\Lambda(f_0) = 1 = \int f_0 \, dm$. If $O \subseteq G$ is a non-empty open subset, then every compact set can be covered by finitely many left translates of O, showing that $m(O) = 0$ would imply that $m(K) = 0$ for every compact set, and in particular that $\int_G f_0 \, dm = 0$, a contradiction. It follows that m is positive on non-empty open subsets, which completes the proof that $m = m_G$ is a left Haar measure on G. $\qquad\square$

Proposition 10.2 (Uniqueness of the Haar measure). *Let G be a locally compact, σ-compact metrizable group. Then the left Haar measure is unique up to a positive scalar multiple.*

For the proof of uniqueness, the following will be useful.

Lemma 10.3 (Positive overlaps). *Let G be as above, and suppose that m is a left Haar measure. If $B_1, B_2 \subseteq G$ are Borel measurable sets with $m(B_1)$ and $m(B_2)$ positive, then $\{g \in G \mid m(gB_1 \cap B_2) > 0\}$ is non-empty. Moreover, $m(B_1^{-1}) > 0$, where $B_1^{-1} = \{g^{-1} \mid g \in B_1\}$.*

PROOF. Let $B_1, B_2 \subseteq G$ be as in the lemma. Then

$$m(gB_1 \cap B_2) = \int \mathbb{1}_{gB_1}(h) \mathbb{1}_{B_2}(h) \, dm(h) = \int \mathbb{1}_{hB_1^{-1}}(g) \mathbb{1}_{B_2}(h) \, dm(h),$$

so by Fubini's theorem we have

$$\int m(gB_1 \cap B_2) \, dm(g) = \int \mathbb{1}_{B_2}(h) \underbrace{\int \mathbb{1}_{hB_1^{-1}}(g) \, dm(g)}_{=m(hB_1^{-1})=m(B_1^{-1})} \, dm(h)$$

$$= m(B_1^{-1}) m(B_2).$$

Setting B_2 briefly equal to G, we obtain $m(B_1)m(G) = m(B_1^{-1})m(G)$ and see that $m(B_1^{-1}) > 0$ (since in measure theory $0 \cdot \infty = 0$). With this we now obtain the lemma since we see that the set $\{g \in G \mid m(gB_1 \cap B_2) > 0\}$ must have positive measure with respect to m. □

PROOF OF PROPOSITION 10.2. Suppose that m_1, m_2 are left Haar measures on G. Define $m = m_1 + m_2$, so that m is a left Haar measure and $m_1, m_2 \ll m$. By the Radon–Nikodym theorem (Proposition 3.29) there exist measurable functions $f_1, f_2 \geqslant 0$ with $dm_i = f_i \, dm$ for $i = 1, 2$.

We claim that f_1 is constant m-almost everywhere (and so f_2 is also). This then implies that $m_1 = c_1 m$ and $m_2 = c_2 m$ for some constants $c_1, c_2 > 0$, and so the proposition follows.

Assume now that the claim does not hold. Then there exist sets $B_1, B_2 \subseteq G$ of positive m measure such that $f_1(x) < f_1(y)$ for all $x \in B_1$ and $y \in B_2$. We can find these sets, for example, as pre-images of two distinct intervals $[\frac{k}{n}, \frac{k+1}{n})$ and $[\frac{\ell}{n}, \frac{\ell+1}{n})$ for some integers $k < \ell$ and $n \geqslant 1$, for otherwise f_1 is constant m-almost everywhere. By Lemma 10.3 there exists a $g \in G$ with $m(gB_1 \cap B_2) > 0$. For any $E \subseteq G$ we also have

$$\int_E f_1(x) \, dm(x) = m_1(E) = m_1(g^{-1}E) = \int_{g^{-1}E} f_1 \, dm = \int_E f_1(g^{-1}x) \, dm(x)$$

by the left-invariance of m_1 and m. If we now take $E \subseteq gB_1 \cap B_2$ with positive and finite measure, then $y \in E$ implies $y \in B_2$ and $g^{-1}y \in B_1$, hence $f_1(y) > f_1(g^{-1}y)$ and so

$$\int_E f_1(y) \, dm(y) > \int_E f_1(g^{-1}y) \, dm(y).$$

This contradiction proves the claim that f_1 must be constant almost everywhere, and hence the proposition. □

Essential Exercise 10.4. Let G be a locally compact, σ-compact, metrizable group and let f be a measurable complex-valued function on G with the property that $m_G \left(\{g \in G \mid f(k^{-1}g) \neq f(g)\} \right) = 0$ for every $k \in G$. Show that $f = c$ almost everywhere for some constant $c \in \mathbb{C}$.

Exercise 10.5. Let G be a locally compact, σ-compact metrizable group with left Haar measure m_G. Prove the following assertions.
(a) For any continuous automorphism θ of G there exists a positive number $\mathrm{mod}_G(\theta)$ with $m_G(\theta^{-1}(B)) = \mathrm{mod}_G(\theta) m_G(B)$ for all Borel subsets $B \subseteq G$.
(b) Applying (a) to inner automorphisms θ_g defined by $\theta_g(h) = ghg^{-1}$ for all g, h in G defines a map $\Delta_G : G \to \mathbb{R}_{>0}$ by $\Delta_G(g) = \mathrm{mod}_G(\theta_g)$. Show that Δ_G, known as the *modular character*, is a continuous group homomorphism with respect to multiplication on $\mathbb{R}_{>0}$. Use this to give a formula for $\int f(gh^{-1}) \, dm(g)$.
(c) Show that

$$m_G^{(\mathrm{right})}(B) = \int_B \Delta_G(g)^{-1} \, dm_G(g)$$

defines a right-invariant Haar measure on G and show that $m_G^{(\mathrm{right})}(B) = m_G(B^{-1})$ for all Borel subsets $B \subseteq G$.

(d) Show that on a compact metrizable group a left Haar measure is also a right Haar measure.

A group G is called *unimodular* if any left Haar measure m_G is also a right Haar measure. Thus, for example, Exercise 10.5(d) says that compact groups are unimodular.

Essential Exercise 10.6 (Approximate identity). Let G be a locally compact, σ-compact, metrizable group and let (U_n) be a sequence of open neighbourhoods of the identity $e \in G$ with $\mathrm{diam}(U_n) \to 0$ as $n \to \infty$. Let (ψ_n) be a sequence of non-negative functions in $L^1(G)$ with the property that ψ_n vanishes outside of U_n and satisfies $\int_G \psi_n \, dm_G = 1$ for all $n \geqslant 1$ (for example, we may set $\psi_n = \frac{1}{m_G(U_n)} \mathbb{1}_{U_n}$). Show that

$$\lim_{n \to \infty} \psi_n * f = \lim_{n \to \infty} f * \psi_n = f$$

for all $f \in L^1(G)$.

Exercise 10.7. Show that if a locally compact σ-compact metrizable group G has the property that $m_G(G)$ is finite, then G is compact.

Exercise 10.8. A Haar measure on the additive reals $(\mathbb{R}, +)$ is (up to a scalar multiple) the Lebesgue measure dx. Show that a Haar measure on the multiplicative reals $(\mathbb{R} \smallsetminus \{0\}, \cdot)$ is given by $\frac{dx}{|x|}$.

Exercise 10.9. Let G be the group of affine transformations $x \mapsto ax + b$ with $a \neq 0$ and $a, b \in \mathbb{R}$, which may also be thought of as the matrix group

$$G = \left\{ \begin{pmatrix} a & b \\ & 1 \end{pmatrix} \mid a, b \in \mathbb{R}, a \neq 0 \right\}$$

under matrix multiplication. Show that $dm_G = \frac{da\,db}{a^2}$ defines a left Haar measure on G and $dm_G^{(\text{right})} = \frac{da\,db}{|a|}$ defines a right Haar measure on G. Compute the modular character on G (as defined in Exercise 10.5).

Exercise 10.10. Show that $dm_{\mathrm{GL}_d(\mathbb{R})}(g) = \frac{dg}{|\det g|^d}$ defines a left and right Haar measure on $\mathrm{GL}_d(\mathbb{R})$, where dg denotes Lebesgue measure on the space of real $d \times d$ matrices.

10.2 Amenable Groups

†Using the material of Chapter 8 we continue the discussion from Section 7.2.2, where the concept of amenability was introduced for discrete groups.

---

† Apart from Exercise 10.35 in Section 10.3, this section will not be used later.

10.2.1 Definitions and Main Theorem

In this section we will always assume that either

- G is a discrete (but not necessarily countable) group and $m = m_G$ denotes the counting measure defined by $m(A) = |A|$ for all $A \subseteq G$; or
- G is a locally compact σ-compact metrizable group and $m = m_G$ denotes a left Haar measure defined on the Borel σ-algebra of G.

We recall that in either case the dual space to $L^1(G)$ is precisely $L^\infty(G)$ by Proposition 7.34 resp. Exercise 7.33(c). Let us also introduce the convex set

$$\mathscr{P}(G) = \left\{ f \in L^1(G) \mid f \geqslant 0 \text{ a.e. and } \int_G f \, dm_G = 1 \right\}$$

of *probability distributions*, and the convex set $\mathscr{M}(G)$ of *means* on G defined by

$$\mathscr{M}(G) = \left\{ M \in L^\infty(G)^* \mid M \text{ is positive and } M(\mathbb{1}) = 1 \right\},$$

where $M \in L^\infty(G)^*$ is called *positive* if $\Phi \in L^\infty(G)$ with $\Phi \geqslant 0$ almost everywhere implies $M(\Phi) \geqslant 0$.

For a function f on G we write $\lambda_g f(h) = f(g^{-1}h)$ for $g, h \in G$. Notice that this definition extends to equivalence classes of functions, so λ_g is an operator on any function space $L^p(G)$ with $p \in [1, \infty]$. We can now extend the notion of amenability via a suitable form of the characterization in Lemma 7.18.

Definition 10.11. We say that G is *amenable* if there exists a left-invariant mean M on $L^\infty(G)$, meaning a mean $M \in \mathscr{M}(G)$ satisfying in addition the left-invariance property $M(\Phi) = M(\lambda_g \Phi)$ for any $\Phi \in L^\infty(G)$ and $g \in G$.

The link between amenability and geometric properties of a group seen in Lemma 7.24 also extends to this setting (see also Proposition 10.19 for a strengthening).

Definition 10.12. A group G *admits Følner sets* if for any compact subset K of G and $\varepsilon > 0$ there exists a measurable set $F \subseteq G$ of positive and finite m-measure with

$$\frac{m(kF \triangle F)}{m(F)} < \varepsilon$$

for all $k \in K$. In this case we will also call F a *Følner set* (for (K, ε)).

Exercise 10.13. Suppose that G is a locally compact σ-compact metrizable group that admits Følner sets. Show that there exists a sequence (called a *Følner sequence*) (F_n) of measurable sets with positive and finite m-measure so that for any fixed $k \in G$ we have $m(kF_n \triangle F_n)/m(F_n) \to 0$ as $n \to \infty$ and the convergence is uniform on compact subsets of G.

In the discrete case K and F are finite sets and Definition 10.12 may be thought of as follows. The Cayley graph $\Gamma(G, K)$ associated to G and the subset K (which may or may not generate G) is the graph with vertices given by elements of G, with edges joining g to kg for any $k \in K$. Then G admits Følner sets means that for any $\varepsilon > 0$ there is a finite set F such that the number of edges in $\Gamma(G, K)$ leaving F is at most $\varepsilon|F|$. This stands in stark contrast to the property of being an expander graph (see Section 10.4).

It should be clear that the two notions above — amenability and admitting Følner sets — are related. In fact, our main goal in this section is to prove Lemma 7.24 and its converse in this more general setting. For the more difficult part of the equivalence one more definition will be useful.

Definition 10.14 (Reiter's condition). A group G fulfills the *Reiter condition in L^1* if for any compact set $K \subseteq G$ and $\varepsilon > 0$ there exists some $f \in \mathscr{P}(G)$ with

$$\|\lambda_k f - f\|_1 < \varepsilon$$

for all $k \in K$. We say that $L^2(G)$ *has almost invariant vectors* (or that G *fulfills the Reiter condition in L^2*) if for any compact set $K \subseteq G$ and $\varepsilon > 0$ there exists some $f \in L^2(G)$ with $\|f\|_2 = 1$ and with

$$\|\lambda_k f - f\|_2 < \varepsilon$$

for all $k \in K$.

Theorem 10.15. *Let G be a discrete group or a locally compact σ-compact metrizable group. Then the following are equivalent:*

(1) *G is amenable;*
(2) *G admits Følner sets;*
(3) *G fulfills the Reiter condition in L^1; and*
(4) *$L^2(G)$ has almost invariant vectors.*

10.2.2 Proof of Theorem 10.15

We will restrict ourselves in the following to $\mathbb{R}$-valued function, but it is not hard to see that with a bit more work this can be avoided.

PROOF THAT $(2) \Longleftrightarrow (3) \Longleftrightarrow (4)$. Suppose (2) holds and F is a Følner set for a compact subset $K \subseteq G$ and $\varepsilon > 0$ as in Definition 10.12. Then $f = \frac{1}{m(F)} \mathbb{1}_F$ lies in $\mathscr{P}(G)$, $\lambda_k \mathbb{1}_F(g) = \mathbb{1}_F(k^{-1}g) = \mathbb{1}_{kF}(g)$, and so

$$\|\lambda_k f - f\|_1 = \frac{1}{m(F)} \int |\mathbb{1}_{kF} - \mathbb{1}_F| dm = \frac{1}{m(F)} m(kF \triangle F) < \varepsilon$$

for all $k \in K$. Similarly, if we set $f_2 = \frac{1}{\sqrt{m(F)}} \mathbb{1}_F$ we see that

$$\|\lambda_k f_2 - f_2\|_2 = \left(\int_G \frac{1}{m(F)} \left(\mathbb{1}_{kF}(g) - \mathbb{1}_F(g) \right)^2 dm(g) \right)^{1/2}$$

$$= \sqrt{\frac{m(kF \triangle F)}{m(F)}} < \sqrt{\varepsilon}.$$

Since $\varepsilon > 0$ and $K \subseteq G$ were arbitrary, we see that (2) implies (3) and (4).

ASSUMING THAT $L^2(G)$ HAS ALMOST INVARIANT VECTORS. If G satisfies (4) and $f_2 \in L^2(G)$ satisfies $\|f_2\|_2 = 1$ and $\|\lambda_k f_2 - f_2\|_2 < \varepsilon$ for all k in the compact set $K \subseteq G$, then we define $f(g) = f_2(g)^2$ for all $g \in G$ and see immediately that $f \geqslant 0$ and $\|f\|_1 = \|f_2\|_2^2 = 1$. Moreover, for $k \in K$ we also have

$$\|\lambda_k f - f\|_1 = \int_G \left| f_2(k^{-1}g)^2 - f_2(g)^2 \right| dm(g)$$

$$= \int_G \left| f_2(k^{-1}g) - f_2(g) \right| \left| f_2(k^{-1}g) + f_2(g) \right| dm(g)$$

$$= \langle |\lambda_k f_2 - f_2|, |\lambda_k f_2 + f_2| \rangle_{L^2(G)}$$

$$\leqslant \|\lambda_k f_2 - f_2\|_2 \|\lambda_k f_2 + f_2\|_2 \leqslant 2\varepsilon.$$

Therefore (4) implies (3).

ASSUMING THE REITER CONDITION IN $L^1(G)$. Assume now that (3) holds. We wish to find a Følner set as in Definition 10.12. Therefore let $K \subseteq G$ be compact and assume without loss of generality that $m_G(K) > 0$. Further fix $\varepsilon > 0$ and let $f \in \mathscr{P}(G)$ be as in Reiter's condition in Definition 10.14. For every $\alpha > 0$ we define the measurable set $F_\alpha = \{g \in G \mid f(g) \geqslant \alpha\}$, which will be a Følner set if we choose α carefully. By Fubini's theorem we have

$$\int_0^\infty m(F_\alpha) \, d\alpha = \int_0^\infty \int_G \mathbb{1}_{F_\alpha}(g) \, dm(g) \, d\alpha$$

$$= \int_G \int_0^\infty \mathbb{1}_{F_\alpha}(g) \, d\alpha \, dm(g) = \int_G f(g) \, dm(g) = \|f\|_1 = 1.$$

Moreover, for any $k \in K$ we also have

$$\int_0^\infty m \left(kF_\alpha \triangle F_\alpha \right) d\alpha = \int_0^\infty \int_G \left| \mathbb{1}_{kF_\alpha}(g) - \mathbb{1}_{F_\alpha}(g) \right| dm(g) \, d\alpha$$

$$= \int_G \int_0^\infty \left| \mathbb{1}_{F_\alpha}(k^{-1}g) - \mathbb{1}_{F_\alpha}(g) \right| d\alpha \, dm(g)$$

$$= \int_G \left| f(k^{-1}g) - f(g) \right| dm(g) = \|\lambda_k f - f\|_1 < \varepsilon.$$

Integrating this over K we obtain

$$\int_0^\infty \int_K m(kF_\alpha \triangle F_\alpha)\, dm(k)\, d\alpha < \varepsilon m(K) = \int_0^\infty \varepsilon m(K) m(F_\alpha)\, d\alpha.$$

Therefore there must exist some $\alpha \in (0, \infty)$ such that

$$\int_K m(kF_\alpha \triangle F_\alpha)\, dm(k) < \varepsilon m(K) m(F_\alpha). \tag{10.6}$$

In the case when G is discrete, K is finite, and this gives

$$|kF_\alpha \triangle F_\alpha| < \varepsilon |K||F_\alpha|$$

for all $k \in K$. Since $\varepsilon > 0$ was arbitrary this proves (2) in the discrete case.

In the non-discrete case, the statement in (10.6) is an averaged form of the inequality we are seeking, and as a result seems to be weaker than what we need. For the upgrade we use the fact that $\varepsilon > 0$ was arbitrary: we have shown that for any $\varepsilon > 0$ and $\delta > 0$ there exists a measurable set $F = F_\alpha$ such that

$$\int_K m(kF \triangle F)\, dm(k) < \varepsilon \delta m(F) < \infty.$$

In particular, we must have $m(N) < \delta$ if

$$N = \{k \in K \mid m(kF \triangle F) \geqslant \varepsilon m(F)\}.$$

Summarising, we have shown for any compact set K, any $\varepsilon > 0$, and any $\delta > 0$ that there exists a measurable set F with finite measure and a subset $N \subseteq K$ with $m(N) < \delta$ such that

$$m(kF \triangle F) < \varepsilon m(F) \tag{10.7}$$

for all $k \in K \smallsetminus N$.

We now use the group structure to upgrade this and deduce the existence of Følner sets. Define $K_1 = K \cup K^2$ and $\delta = \frac{1}{2} m(K)$. Now apply the argument above to K_1, an arbitrary $\varepsilon > 0$, and this choice of δ. This gives a measurable set $F \subseteq G$ of finite measure satisfying (10.7) for all $k_1 \in K_1 \smallsetminus N$ and some exceptional set $N \subseteq K_1$ of measure $m(N) < \frac{1}{2} m(K)$. Now fix some $k \in K$. We see that (10.7) holds for $k_1 \in K \smallsetminus N$ where $m(K \smallsetminus N) > \frac{1}{2} m(K)$, and also that (10.7) holds for all $kk_1 \in (kK) \smallsetminus N$ where $m\left((kK) \smallsetminus N\right) > \frac{1}{2} m(K)$. Since left translation by k preserves the measure m, it follows that there exists some $k_1 \in K$ such that (10.7) holds both for k_1 and for kk_1. Therefore

$$m(kF \triangle F) \leqslant m\left((kF \triangle kk_1 F) \cup (kk_1 F \triangle F)\right) \leqslant m(F \triangle k_1 F) + \varepsilon m(F) < 2\varepsilon m(F)$$

for any $k \in K$, proving (2). $\qquad\square$

To summarize, we have shown that (2), (3) and (4) are equivalent. We now turn to the equivalence between (1) and (3), which is where functional analysis will play an important role. In the following we will frequently use the left-invariance of m in the form

$$\langle \lambda_k f, \Phi \rangle = \int f(k^{-1}g)\Phi(g)\,\mathrm{d}m(g) = \int f(g')\Phi(kg')\,\mathrm{d}m(g') = \langle f, \lambda_{k^{-1}}\Phi \rangle$$

for $f \in L^1_m(G)$, $\Phi \in L^\infty(G)$ and $k \in G$.

PROOF THAT $(3) \Longrightarrow (1)$ IN THEOREM 10.15. Assume that G fulfills Reiter's condition. This shows that for a given $\varepsilon > 0$ and compact $K \subseteq G$ the function f as in Definition 10.14 satisfies

$$|\langle f, \lambda_{k^{-1}}\Phi - \Phi \rangle| = |\langle f, \lambda_{k^{-1}}\Phi \rangle - \langle f, \Phi \rangle| = |\langle \lambda_k f - f, \Phi \rangle| \leqslant \varepsilon\|\Phi\|_\infty$$

for $k \in K$ and $\Phi \in L^\infty(G)$. Taking the image of such functions under the embedding map $\imath$ into the dual of $L^\infty(G)$ we see that

$$A\left(\varepsilon, \overline{\Phi}, \overline{k}\right) = \{M \in \mathscr{M}(G) \mid |M(\Phi_i - \lambda_{k_j}\Phi_i)| \leqslant \varepsilon\|\Phi_i\|_\infty \text{ for all } i, j\}$$

is non-empty for any choice of $\varepsilon > 0$,

$$\overline{\Phi} = (\Phi_1, \ldots, \Phi_\ell) \in (L^\infty(G))^\ell,$$

$$\overline{k} = (k_1, \ldots, k_n) \in G^n,$$

and any $\ell, n \in \mathbb{N}$. By definition

$$A\left(\varepsilon, \overline{\Phi}, \overline{k}\right) \subseteq \mathscr{M}(G)$$

is weak* closed and contained in the closed unit ball of $L^\infty(G)^*$ (check this). Since any finite intersection of such sets will contain another such set we see that the collection of sets of the form $A\left(\varepsilon, \overline{\Phi}, \overline{k}\right)$ has the finite intersection property. By the Banach–Alaoglu theorem (Theorem 8.10) it follows that the intersection over these sets is non-empty. By definition, this intersection consists of all left-invariant means on $L^\infty(G)$. $\qquad\square$

For the converse, which is perhaps the most surprising part of the whole proof, we will need the following lemma.

Lemma 10.16. *Let G be as above, and let*

$$\imath : L^1(G) \longrightarrow L^1(G)^{**} = L^\infty(G)^*$$

be the natural embedding into the bidual of $L^1(G)$. Then the weak closure of the image of $\mathscr{P}(G)$ under $\imath$ in $L^\infty(G)^*$ is $\mathscr{M}(G)$.*

PROOF. Assume for the purpose of a contradiction that there is some mean $M \in \mathscr{M}(G)$ that is not in the weak* closure K of $\imath(\mathscr{P}(G))$. Applying Theorem 8.73 to $X = L^\infty(G)^*$ equipped with the weak* topology, the closed set K, and $M \notin K$ gives a continuous linear functional on X separating M from K. By Lemma 8.13 this functional is an evaluation map at some $\Phi \in L^\infty(G)$. Hence the conclusion of Theorem 8.73 is precisely that there is some $c \in \mathbb{R}$ with

$$\int_G f\Phi \, dm = \langle \Phi, \imath(f) \rangle \leqslant c < M(\Phi)$$

for all $f \in \mathscr{P}(G)$. This implies that $\Phi \leqslant c$ almost everywhere, since otherwise we could find a measurable set $B \subseteq G$ of finite positive measure with $\Phi(g) > c$ for $g \in B$, and then setting $f = \frac{1}{m(B)}\mathbb{1}_B \in \mathscr{P}(G)$ leads to a contradiction. However, $\Phi \leqslant c$ almost everywhere also implies that $M(\Phi) \leqslant c$ by the properties of $M \in \mathscr{M}(G)$. This contradiction proves the lemma. $\qquad\square$

We start with the discrete case as it is significantly easier.

PROOF OF $(1)\Longrightarrow(3)$ IN THEOREM 10.15 FOR DISCRETE G. Assume that there exists a left-invariant mean M. Using M we wish to find, for any $\varepsilon > 0$ and finite $K \subseteq G$, a function $f \in \mathscr{P}(G)$ such that

$$\|\lambda_k f - f\|_1 < \varepsilon$$

for all $k \in K$. Define the bounded linear operator

$$D : \ell^1(G) \longrightarrow \left(\ell^1(G)\right)^K$$
$$f \longmapsto (\lambda_k f - f)_{k \in K}.$$

Note that $D(\mathscr{P}(G))$ is convex, and we wish to show that $0 \in \overline{D(\mathscr{P}(G))}^{\text{norm}}$. By Corollary 8.74 we know that $\overline{D(\mathscr{P}(G))}^{\text{norm}}$ is also closed in the weak topology. Therefore it is enough to show that

$$0 \in \overline{D(\mathscr{P}(G))}^{\text{weak}} = \overline{D(\mathscr{P}(G))}^{\text{norm}}. \tag{10.8}$$

The dual of $\left(\ell^1(G)\right)^K$ is given by $(\ell^\infty(G))^K$ and it suffices to find, for every $\Phi_1, \ldots, \Phi_n \in \ell^\infty(G)$ and $\varepsilon > 0$, some $f \in \mathscr{P}(G)$ with

$$|\langle \lambda_k f - f, \Phi_j \rangle| < \varepsilon \tag{10.9}$$

for all $k \in K$ and $j = 1, \ldots, n$. The left-hand side of (10.9) may be rewritten as

$$|\langle \lambda_k f - f, \Phi_j \rangle| = |\langle f, \lambda_{k^{-1}}\Phi_j - \Phi_j \rangle| = |\langle \lambda_{k^{-1}}\Phi_j - \Phi_j, \imath(f) \rangle|. \tag{10.10}$$

Note that

$$\langle \lambda_{k^{-1}}\Phi_j - \Phi_j, M \rangle = M\left(\lambda_{k^{-1}}\Phi_j - \Phi_j\right) = 0$$

for the invariant mean M. By Lemma 10.16 we know that $\imath(\mathscr{P})$ is dense in $\mathscr{M}(G)$ with respect to the weak* topology, so there must exist an element f of $\mathscr{P}(G)$ for which (10.10) is less than ε for all $k \in K$ and $j = 1, \ldots, n$, which proves (10.9), (10.8), and hence that G fulfills the Reiter condition in L^1. $\qquad\square$

In the non-discrete case another ingredient is needed.

Lemma 10.17 (Topological left-invariant mean). *Let G be a σ-compact, locally compact, metrizable amenable group. Then there also exists a 'topologically left-invariant mean' on $L^\infty(G)$, that is, a mean M_{top} on $L^\infty(G)$ such that*

$$M_{\text{top}}(f * \Phi) = M_{\text{top}}(\Phi)$$

for any $f \in \mathscr{P}(G)$ and $\Phi \in L^\infty(G)$.

PROOF. We start by noting that the definition of $f * \Phi$ in Lemma 3.75 and Exercise 3.76 makes sense at every $g \in G$ and easily implies that

$$\|f * \Phi\|_\infty \leqslant \|f\|_1 \|\Phi\|_\infty \tag{10.11}$$

for any $f \in L^1(G)$ and $\Phi \in L^\infty(G)$.

Given $f_0, f_1 \in \mathscr{P}(G)$ and $\Phi \in L^\infty(G)$ we define

$$\Phi_0 = f_0 * \Phi$$

and claim that

$$M(f_1 * \Phi_0) = M(\Phi_0) \tag{10.12}$$

if M is a left-invariant mean on $L^\infty(G)$.

For this we first recall from Lemma 3.74 that for a given f_0 and $\varepsilon > 0$ there exists a neighbourhood U of $e \in G$ such that

$$\|\lambda_k f_0 - f_0\|_1 < \varepsilon$$

for all $k \in U$. Using the left-invariance of the Haar measure we obtain

$$\lambda_k \Phi_0(g) = (f_0 * \Phi)(k^{-1}g) = \int_G f_0(h)\Phi(h^{-1}k^{-1}g)\,dm(h)$$

$$= \int_G f_0(k^{-1}h_1)\Phi(h_1^{-1}g)\,dm(h_1) \qquad \text{(with } h_1 = kh)$$

$$= \big((\lambda_k f_0) * \Phi\big)(g). \tag{10.13}$$

Together with (10.11) we deduce that

$$\|\lambda_k \Phi_0 - \Phi_0\|_\infty = \|(\lambda_k f_0 - f_0) * \Phi\|_\infty \leqslant \|\lambda_k f_0 - f_0\|_1 \|\Phi\|_\infty \leqslant \varepsilon \|\Phi\|_\infty$$

for all $k \in U$. In other words, for Φ_0 the left regular representation satisfies the continuity claim appearing in Lemma 3.74, but with respect to the $\|\cdot\|_\infty$ norm. This property of Φ_0 is called left uniform continuity of Φ_0. As this is precisely the assumption for the strong integral discussed in Proposition 3.81, it follows that for $f_1 \in C_c(G)$ the integral $R\text{-}\int f_1(g)\lambda_g \Phi_0\,dm_G(g)$ can be obtained as a limit with respect to $\|\cdot\|_\infty$ of Riemann sums of the form

$$\sum_{P \in \xi} f_1(g_p) \lambda_{g_p} \Phi_0 m_G(P),$$

where ξ is a finite partition of $\mathrm{Supp}(f_1)$ and $g_p \in P$ for each $P \in \xi$. As convergence with respect to $\| \cdot \|_\infty$ implies pointwise convergence, we see that $_R\!\int f_1(g) \lambda_g \Phi_0 \, dm_G(g) = f_1 * \Phi_0$. Applying the continuous functional M we see that

$$M(f_1 * \Phi_0) = \lim_{\xi} \sum_{P \in \xi} f_1(g_p) m(P) M(\lambda_{g_p} \Phi_0) = M(\Phi_0) \int_G f_1 \, dm$$

since $M(\lambda_g \Phi_0) = M(\Phi_0)$ for any $g \in G$. Using the estimate (10.11) again and the density of $C_c(G)$ in $L^1(G)$ this extends to all $f_1 \in L^1(G)$. Restricting to functions in $\mathscr{P}(G)$ the claim in (10.12) follows.

We now make the definition

$$M_{\mathrm{top}}(\Phi) = M(\Phi_0) = M(f_0 * \Phi)$$

for some $f_0 \in \mathscr{P}(G)$. Note that $M_{\mathrm{top}}(\mathbb{1}) = M(\mathbb{1}) = 1$ and that $\Phi \geqslant 0$ almost surely implies $f_0 * \Phi \geqslant 0$ and $M_{\mathrm{top}}(\Phi) \geqslant 0$. We also claim that this definition is independent of f_0. Using this independence we see that

$$M_{\mathrm{top}}(f_1 * \Phi) = M(f_0 * f_1 * \Phi) = M_{\mathrm{top}}(\Phi)$$

for any $f_1 \in \mathscr{P}(G)$ by associativity of convolution (cf. the proof of Proposition 3.91), and the lemma follows.

To see the independence let $(\psi_n)_n$ be an approximate identity in $L^1(G)$ (see Exercise 10.6) so that

$$\lim_{n \to \infty} \| f_0 * \psi_n - f_0 \|_1 = 0,$$

and so by (10.11) and continuity of M also

$$\lim_{n \to \infty} M(f_0 * \psi_n * \Phi) = M(f_0 * \Phi).$$

Combining this with (10.12) and $\psi_n \in \mathscr{P}(G)$ we see that

$$\lim_{n \to \infty} M(\psi_n * \Phi) = M(f_0 * \Phi),$$

which gives the claim and the lemma. $\qquad \square$

PROOF OF (1) $\implies$ (3) IN THEOREM 10.15 WITHOUT DISCRETENESS. Fix $\psi_1, \dots, \psi_n \in \mathscr{P}(G)$. Applying the same argument as in the discrete case but to the map

$$D : L^1(G) \longrightarrow \left(L^1(G)\right)^n$$
$$f \longmapsto (\psi_j * f - f)_j$$

and using the existence of a topological left-invariant mean as in Lemma 10.17, we conclude that it is possible to find, for every $\varepsilon > 0$, some $f \in \mathscr{P}(G)$ such that

$$\|\psi_j * f - f\|_1 < \varepsilon \qquad (10.14)$$

for $j = 1, \ldots, n$.

Now fix some $\psi \in \mathscr{P}(G)$ and some dense countable subset $\{g_1, g_2, \ldots\} \subseteq G$ with $g_1 = e$. Define $\psi_j = \lambda_{g_j} \psi$ and apply the argument above to $\psi_1, \ldots, \psi_n$ and $\varepsilon = \frac{1}{n}$. This shows that there exists a sequence (f_n) in $\mathscr{P}(G)$ with

$$\|\lambda_{g_j} \psi * f_n - f_n\|_1 \longrightarrow 0$$

as $n \to \infty$ for every j.

Now let $K \subseteq G$ be a compact subset and fix $\varepsilon > 0$. By Lemma 3.74 there exists some neighbourhood U of $e \in G$ such that

$$\|\lambda_u \psi - \psi\|_1 < \varepsilon$$

for $u \in U$. By density of $\{g_1, g_2, \ldots\}$ we have $G = \bigcup_{j=1}^\infty g_j U$. By compactness of K there exists some ℓ such that

$$K \subseteq \bigcup_{j=1}^{\ell} g_j U. \qquad (10.15)$$

By construction we can choose n large enough to ensure that

$$\|\lambda_{g_j} \psi * f_n - f_n\|_1 < \varepsilon$$

for $j = 1, \ldots, \ell$. For $k \in K$ there exists by (10.15) some $j \leqslant \ell$ with $k = g_j u$ for some $u \in U$, which shows that

$$\|\lambda_k \psi - \lambda_{g_j} \psi\|_1 = \|\lambda_{g_j} (\lambda_u \psi - \psi)\|_1 < \varepsilon.$$

Thus we deduce (after recalling that $g_1 = e$ and $f_n \in \mathscr{P}(G)$) that

$$\|\lambda_k \psi * f_n - \psi * f_n\|_1 \leqslant \| (\lambda_k \psi - \lambda_{g_j} \psi) * f_n\|_1$$
$$+ \|\lambda_{g_j} \psi * f_n - f_n\|_1$$
$$+ \|f_n - \psi * f_n\| < 3\varepsilon.$$

Finally, notice that

$$\big((\lambda_k \psi) * f_n\big)(g) = \int \lambda_k \psi(h) f_n(h^{-1}g) \, dm(h)$$

$$= \int \psi(k^{-1}h) f_n(h^{-1}g) \, dm(h)$$

$$= \int \psi(h_1) f_n(h_1^{-1}k^{-1}g) \, dm(h_1) = \lambda_k \left(\psi * f_n\right)(g)$$

for all $g \in G$ and $k \in K$. Hence the function $\psi * f_n \in \mathscr{P}(G)$ satisfies Reiter's condition for $K \subseteq G$ and 3ε. $\qquad\square$

Exercise 10.18. Fill in the details of the argument leading to (10.14).

10.2.3 A More Uniform Følner Set

For subsets $A, B \subseteq G$ of a group we define $AB = \{ab \mid a \in A, b \in B\}$.

Proposition 10.19. *Let G be an amenable group as in Theorem 10.15. Then for every non-empty compact $K \subseteq G$ and every $\varepsilon > 0$ there exists a measurable Følner set $F \subseteq G$ with finite measure such that $m((KF)\triangle F) < \varepsilon m(F)$.*

PROOF. Let $U = U^{-1}$ be a compact neighbourhood of the identity $e \in G$. Given a Følner set F for (U, ε) we define a function $f : G \to \mathbb{R}$ by

$$f(g) = \frac{1}{m(U)} \int_U \mathbb{1}_F(ug) \, dm(u) = \frac{1}{m(Ug)} \int_{Ug} \mathbb{1}_F \, dm,$$

where we will think of $f(g)$ as the proportion of positive answers in the neighbourhood Ug of g to the question of whether g should belong to an improved version of F. In case G is not unimodular, we multiplied the integral and the denominator in the first expression by $\Delta_G(g)$, used $m(Ug) = \Delta_G(g)m(U)$ in the denominator and the substitution $h = ug \in Ug$ for $u \in U$ in the integral (at first reading it may be helpful to assume that G is unimodular as this simplifies some of the expressions arising). Given any majority parameter $\alpha \in (0, 1)$ we also define the set $F_\alpha = F_\alpha(F, U)$ by

$$F_\alpha = \{g \in G \mid f(g) > \alpha\} = \{g \in G \mid m(F \cap Ug) > \alpha m(Ug)\},$$

which will be a more well-rounded version of F. The defining property of F and the definition of f together with Fubini's theorem imply that

$$\|f - \mathbb{1}_F\|_1 \leqslant \frac{1}{m(U)} \int_U \|\mathbb{1}_{u^{-1}F} - \mathbb{1}_F\|_1 \, dm(u) < \varepsilon m(F).$$

This gives

$$\beta m \left(\{g \in G \mid |f - \mathbb{1}_F|(g) \geqslant \beta\}\right) < \varepsilon m(F)$$

for all $\beta \in (0,1)$. Setting $\beta = \min\{\alpha, 1-\alpha\}$ we obtain

$$m\left(F_\alpha \triangle F\right) < \frac{\varepsilon}{\min\{\alpha, 1-\alpha\}} m(F). \tag{10.16}$$

Applying this construction will give us the desired Følner set. To this end, fix some non-empty compact subset $K \subseteq G$. Since K is compact and U has non-empty interior there exist $k_1, \ldots, k_n \in K$ such that

$$K \subseteq \bigcup_{j=1}^{n} k_j U.$$

Since K is assumed to be non-empty we have $n \geqslant 1$. Suppose now that F is a Følner set for $(K \cup U^2, \frac{\varepsilon}{n})$. Set $\alpha = \frac{1}{2}$ and define the associated set

$$F' = F_{1/2}(F, U)$$

as above, so that

$$m\left(F' \triangle F\right) < \frac{2\varepsilon}{n} m(F) \tag{10.17}$$

by (10.16). Assuming $\varepsilon < \frac{1}{4}$ we have

$$m(F) \ll m(F') \ll m(F).$$

Since $n \geqslant 1$ we see, from the Følner property of F for $k_1 \in K$ and (10.17), that

$$m(F' \smallsetminus (KF')) \leqslant m(F' \smallsetminus (k_1 F')) \ll \varepsilon m(F').$$

For the second inequality we first claim that

$$UF' \subseteq F_\alpha = F_\alpha(F, U^2) \tag{10.18}$$

for the parameter

$$\alpha = \frac{m(U)}{2m(U^2)\max_{u \in U} \Delta(u)},$$

which only depends on our choice of the neighbourhood U. In fact, for $u \in U$ and $g \in F' = F_{1/2}(F, U)$ we have

$$m(F \cap U^2 ug) \geqslant m(F \cap Ug) > \tfrac{1}{2}m(Ug)$$

$$= \frac{m(U)}{2m(U^2)} m(U^2 g) \geqslant \alpha m(U^2 ug),$$

which implies (10.18).

Since F is a Følner set for $(U^2, \frac{\varepsilon}{n})$ we obtain from (10.18) and (10.16) that

$$m(UF' \smallsetminus F) \leqslant m(F_\alpha \smallsetminus F) \ll \frac{\varepsilon}{n} m(F) \ll \frac{\varepsilon}{n} m(F'),$$

where the implicit constant only depends on the choice of U. Using the fact that F is a Følner set for $(\{k_1, \ldots, k_n\}, \frac{\varepsilon}{n})$ and (10.17), this gives

$$m(k_j U F' \smallsetminus F') \leqslant m(k_j U F' \smallsetminus k_j F) + m(k_j F \smallsetminus F) + m(F \smallsetminus F') \ll \frac{\varepsilon}{n} m(F')$$

for $j = 1, \ldots, n$. Taking the union and recalling that $K \subseteq \bigcup_{j=1}^{n} k_j U$, we obtain

$$m(KF' \smallsetminus F') \leqslant m\left(\bigcup_{j=1}^{n} k_j U F' \smallsetminus F'\right) \ll \varepsilon m(F').$$

Since ε was arbitrary, this concludes the proof. $\qquad\square$

10.2.4 Further Equivalences and Properties

We conclude the discussion of amenability with a number of exercises that extend the treatment above and generalize various earlier topics to the level of generality of this section.

Exercise 10.20. Let G be a discrete group. Show that G is amenable if and only if every finitely generated subgroup of G is amenable.

Exercise 10.21. Let G be a σ-compact, locally compact, metric group.
(1) Show that Definition 10.12 and Definition 10.14 could equivalently be formulated by using only finite subsets $K \subseteq G$.
(2) Show that if G is amenable, then there exists a mean that is left-invariant and topologically left-invariant.

Unless otherwise noted G will be, as in Theorem 10.15, either a discrete group or a locally compact σ-compact metrizable group.

Exercise 10.22. Let G be an amenable group.
(1) Show that there exists a bi-invariant mean on $L^\infty(G)$, that is, one which is left-invariant and right-invariant (defined in the same way).
(2) Assume that G is in addition unimodular. Show that G admits bi-invariant Følner sets, in the sense that they are almost invariant under left and right translation by a given compact subset $K \subseteq G$.

An action of a topological group G on a locally convex vector space X is *affine* if every $g \in G$ acts via a map $V \ni v \mapsto \pi_g^{\mathrm{aff}}(v) = \pi_g^{\mathrm{lin}}(v) + w_g$, where w_g depends continuously on g and $\pi_g^{\mathrm{lin}}(v)$ depends continuously on $(g, v) \in G \times V$, and linearly on v.

Exercise 10.23. Show that the following properties are equivalent.
(1) G is amenable.
(2) If G acts continuously on a compact metric space X (see Definition 3.70), then there exists a G-invariant Borel probability measure on X.
(3) If G acts continuously by affine maps on a locally convex space V and $K \subseteq V$ is compact, convex, and G-invariant, then there exists a point $x_0 \in K$ that is fixed under all elements of G.

Exercise 10.24. Generalize Proposition 7.20 to the groups considered here:
(1) Show that if G is amenable and $H < G$ is a closed subgroup then H is amenable.
(2) Show that if $H \lhd G$ is a closed normal subgroup with the property that both H and G/H are amenable, then G is also amenable.

Exercise 10.25. Let $H < G$ be a closed subgroup with the property that $X = G/H$ supports a finite G-invariant Borel measure. Show that G is amenable if and only if H is.

In the remainder of the section we assume that G is discrete and generated by a finite symmetric set S, where a subset S of a group G is *symmetric* if $s \in S$ implies that $s^{-1} \in S$. The associated *length function* assigns to each element $g \in G$ the number $\ell_S(g) \in \mathbb{N}_0$ defined by the length of the shortest representation of g as a product of elements of S, and the associated *growth function* is defined by

$$\gamma_S(n) = |\{g \in G \mid \ell_S(g) \leqslant n\}|$$

for $n \in \mathbb{N}_0$. In order to define a growth property intrinsic to the group G rather than the pair (G, S), write $\gamma \sim \gamma'$ for functions $\gamma, \gamma' : \mathbb{N} \to \mathbb{N}$ if there exist positive constants $C_1, C_2, \kappa_1, \kappa_2$ such that

$$C_1 \gamma'(\kappa_1 n) \leqslant \gamma(n) \leqslant C_2 \gamma'(\kappa_2 n)$$

for all $n \geqslant 1$.

Exercise 10.26. Let G be generated by a symmetric set S. Show that setting

$$\mathsf{d}(g, h) = \ell_S(gh^{-1})$$

defines a metric on G.

Exercise 10.27. Show that the equivalence class $[\gamma_S]_\sim$ of the growth function of a finitely generated group is well-defined (meaning that it is independent of the choice of symmetric generating set), allowing us to write $\gamma^{(G)}$ for any representative of the equivalence class.

As a result we may make the following definition. A finitely generated infinite group G has

- *polynomial growth* if $\gamma^{(G)} \sim p_a$ for some $a > 0$, where $p_a(n) = n^a$ for all $n \geqslant 1$;
- *exponential growth* if $\gamma^{(G)} \sim \exp$, where $\exp(n) = e^n$ for $n \geqslant 1$;
- *sub-exponential growth* if $\limsup_{n \to \infty} \left(\gamma^{(G)}(n)\right)^{1/n} \leqslant 1$; and
- *intermediate growth* if it is of neither polynomial nor exponential growth.

Exercise 10.28. (1) Show that a group of sub-exponential growth is amenable.
(2) Show that the Heisenberg group in Exercise 7.27 has polynomial growth.
(3) Show that the group

$$\left\{ \begin{pmatrix} a & b \\ 0 & 1 \end{pmatrix} \mid a \in 2^{\mathbb{Z}}, b \in \mathbb{Z}[\tfrac{1}{2}] \right\}$$

is finitely generated, amenable, and has exponential growth.

10.3 Property (T)

[†]In this section we will connect the spectral theory of unitary flows in Section 9.3 to the discussion of expanders in Section 10.4. As we will see, the connection will be via another property that topological groups may have.

10.3.1 Definitions and First Properties

Let us start with some fundamental definitions where we will assume that G is a topological group and π is a unitary representation of G on a complex Hilbert space $\mathcal{H}$.

Definition 10.29 (Almost-invariant vectors). Given $\varepsilon \geqslant 0$ and a subset $Q \subseteq G$, we say that a unit vector $v \in \mathcal{H}$ is (Q, ε)-*almost invariant* if

$$\sup_{g \in Q} \|\pi_g v - v\| \leqslant \varepsilon.$$

We also say that the unitary representation π has *almost-invariant vectors* if it has (Q, ε)-almost invariant unit vectors for any $\varepsilon > 0$ and compact $Q \subseteq G$.

The case of $(G, 0)$-almost invariant vectors corresponds trivially to invariant vectors. Also note that every unit vector is trivially $(G, 2)$-invariant. Another elementary but less immediate observation is contained in the following exercise which relies on the geometry of Hilbert spaces.

Exercise 10.30. Suppose $\varepsilon \in (0, 1)$. Assume that $v \in \mathcal{H}$ is a (G, ε)-almost invariant unit vector. Show that there exists a non-zero vector that is invariant under all of G.

Definition 10.31 (Spectral gap). We say that π has *spectral gap* if π restricted to $(\mathcal{H}^G)^\perp$ does not have almost-invariant vectors, where

$$\mathcal{H}^G = \{v \in \mathcal{H} \mid \pi_g v = v \text{ for all } g \in G\}$$

is the subspace of G-invariant vectors.

Equivalently, we have spectral gap if there exists a compact subset $Q \subseteq G$ and some $\varepsilon > 0$ such that every unit vector $v \in (\mathcal{H}^G)^\perp$ is moved at least by ε by some $g \in Q$; more precisely if $\sup_{g \in Q} \|\pi_g v - v\| > \varepsilon$.

Definition 10.32 (Property (T)). Let $H < G$ be a closed subgroup of a topological group G. We say that (G, H) has *relative property* (T) if whenever a unitary representation π of G has almost-invariant vectors, then it has a non-zero vector fixed by H. We say that G has *property* (T) if (G, G) has relative property (T).

[†] This section will not be used later in the book except for Section 10.4. Amenability and property (T) will be (almost) exclusive. We note that apart from Exercise 10.35 the following will be independent of Section 10.2.

We note that the letter 'T' in property (T) stands for the trivial representation and that the parentheses indicate a neighbourhood of the trivial representation. In fact, there is a definition of a topology on the family of irreducible unitary representations of a topological group G — the Fell topology — such that property (T) is equivalent to the trivial representation being isolated in that topology.

Finding groups without property (T) is quite easy.

Example 10.33. Let $G = \mathbb{Z}$ or $G = \mathbb{R}$. Then G does not have property (T).

JUSTIFICATION OF EXAMPLE 10.33. Let $\mathcal{H} = L^2(G)$ and use the regular representation $(\lambda, \mathcal{H})$ defined by $\lambda_x f(y) = f(y - x)$ for all $x, y \in G$. Let m_G denote the Haar measure on G (that is, counting or Lebesgue measure). Let $F_n = [-n, n]$ in G for all $n \geqslant 1$. Then

$$f_n = m_G(F_n)^{-1/2} \mathbb{1}_{F_n}$$

has norm 1 and is almost-invariant in the sense that

$$\|\lambda_x f_n - f_n\| = \frac{m_G\big((F_n + x)\triangle F_n\big)^{1/2}}{m_G(F_n)^{1/2}} \longrightarrow 0$$

as $n \to \infty$, uniformly on compact sets.

If G did have property (T), then $L^2(G)$ would have to contain a G-invariant function. However, a G-invariant function on $L^2(G)$ would have to be constant (see Exercise 10.4). Since $m_G(G) = \infty$, no non-zero constant function can lie in $L^2(G)$. Therefore, G does not have property (T). □

Exercise 10.34. Show that if G is a topological group with property (T), and ϕ is a continuous homomorphism from G to G' with dense image, then G' also has property (T). Conclude that the free group F (with at least one generator) does not have property (T).

Exercise 10.35. Let G be a discrete or locally compact σ-compact metrizable group. Show that G is compact if and only if G is amenable and has property (T).

Comparing Definitions 10.31 and 10.32, we see that a unitary representation of a group with property (T) always has a spectral gap. The next lemma shows that more is true.

Lemma 10.36 (Uniform spectral gap). *Suppose G is a locally compact σ-compact metrizable group. Then it suffices to consider only separable Hilbert spaces in the definition of property (T). Moreover, assuming that G has property (T) all unitary representations of G have* uniform spectral gap *in the sense that there exists some $\varepsilon > 0$ and $Q \subseteq G$ compact such that for any unitary representation π on a Hilbert space $\mathcal{H}$ and any unit vector $v \in \big(\mathcal{H}^G\big)^{\perp}$, there is some $g \in Q$ with $\|\pi_g v - v\| > \varepsilon$.*

PROOF. Note that by Lemma A.22, G can be written as $\bigcup_{n=1}^{\infty} Q_n$ for some compact subsets $Q_n \subseteq G$ with $Q_n \subseteq Q_{n+1}^{o}$ for all $n \geqslant 1$.

Suppose that G does not satisfy the uniform spectral gap property in the lemma. Then for every $n \geqslant 1$ there is a unitary representation $(\pi_n, \mathcal{H}_n)$ without fixed vectors such that there exists a vector $v_n \in \mathcal{H}_n$ that is $(Q_n, \frac{1}{n})$-almost invariant. Since the unitary representation is continuous (and G is separable), it follows that $S_n = \{\pi_{n,g} v_n \mid g \in G\} \subseteq \mathcal{H}_n$ is separable. It follows that the closed linear hull $\mathcal{H}_n' = (\mathcal{H}_n)_{v_n}$ of S_n is a separable G-invariant subspace of $\mathcal{H}_n$. Now define $\mathcal{H} = \bigoplus_n \mathcal{H}_n'$ with the natural unitary representation $\pi = \bigoplus_n \pi_n|_{\mathcal{H}_n'}$ of G on $\mathcal{H}$ (see Exercise 3.77) and notice that $(\pi, \mathcal{H})$ has no non-zero G-invariant vectors. Moreover, it has almost-invariant vectors since for every $\varepsilon > 0$ and compact $K \subseteq G$ there exists some n such that $K \subseteq Q_n$, $\frac{1}{n} \leqslant \varepsilon$, and hence $v_n \in \mathcal{H}_n' \subseteq \mathcal{H}$ is (K, ε)-almost invariant. It follows that the failure of the uniform spectral gap property implies the existence of a unitary representation on a separable Hilbert space without spectral gap. This proves both statements of the lemma. $\qquad\square$

Exercise 10.37. Show that a discrete group with property (T) is finitely generated.

10.3.2 Main Theorems

In the following we will consider the groups $\mathrm{SL}_d(\mathbb{R})$ endowed with the topology induced by the inclusion $\mathrm{SL}_d(\mathbb{R}) \subseteq \mathrm{Mat}_{d,d}(\mathbb{R}) \cong \mathbb{R}^{d^2}$. Každan gave the definition of property (T) in 1967 and also gave the first examples of such groups.

Theorem 10.38 (Každan). $\mathrm{SL}_3(\mathbb{R})$ *has property* (T).

We note that $G = \mathrm{SL}_2(\mathbb{R})$ does not have property (T), but despite this, many of its natural (and all of its irreducible) unitary representations have spectral gap; we refer to [26] for references and a detailed discussion. The main tool for proving the above theorem is the following relative version.

Theorem 10.39 (Každan). $(\mathrm{ASL}_2(\mathbb{R}), \mathbb{R}^2)$ *has relative property* (T)*, where*

$$\mathrm{ASL}_2(\mathbb{R}) = \mathrm{SL}_2(\mathbb{R}) \ltimes \mathbb{R}^2 = \left\{ \begin{pmatrix} A & x \\ 0 & 1 \end{pmatrix} \mid A \in \mathrm{SL}_2(\mathbb{R}), x \in \mathbb{R}^2 \right\}.$$

As we will see there is a way to push property (T) from the group $\mathrm{SL}_3(\mathbb{R})$ to its discrete counterpart $\mathrm{SL}_3(\mathbb{Z})$.

Corollary 10.40 (Každan). $\mathrm{SL}_3(\mathbb{Z})$ *has property* (T).

As Margulis showed in 1988 discrete groups with property (T) quickly give rise to expander families, which we will introduce in the next section.

10.3.3 Proof of Každan's Property (T), Connected Case

For the proof of Theorem 10.38 we need the following property of unitary representations of $G = \mathrm{SL}_d(\mathbb{R})$ for $d = 2, 3$ (due to Mautner [69] and Moore [75]). For this we define the subgroup

$$U_{12} = \left\{ u_x = \begin{pmatrix} 1 & x \\ 0 & 1 \end{pmatrix} \mid x \in \mathbb{R} \right\}$$

of $\mathrm{SL}_2(\mathbb{R})$. Identifying $\mathbb{R}^2$ with the subspace $\mathbb{R}^2 \times \{0\}^{d-2}$ of $\mathbb{R}^d$, we obtain an embedding

$$g \longmapsto \begin{pmatrix} g & \\ & I_{d-2} \end{pmatrix}$$

for $g \in \mathrm{SL}_2(\mathbb{R})$ of $\mathrm{SL}_2(\mathbb{R})$ into $\mathrm{SL}_d(\mathbb{R})$ for $d \geqslant 3$ and may think of U_{12} also as a subgroup of $\mathrm{SL}_d(\mathbb{R})$. Conjugating U_{12} with permutation matrices we obtain other subgroups of $\mathrm{SL}_d(\mathbb{R})$, which we will refer to as *elementary unipotent subgroups*.

Proposition 10.41 (Mautner phenomenon). *Let $\pi : \mathrm{SL}_d(\mathbb{R}) \curvearrowright \mathcal{H}$ for some $d \geqslant 2$ be a unitary representation. Suppose that $v \in \mathcal{H}$ satisfies either*

- *$\pi_a v = v$ for a non-trivial positive diagonal matrix $a \in \mathrm{SL}_d(\mathbb{R})$, or*
- *$\pi_u v = v$ for all elements $u \in U$ of an elementary unipotent subgroup U.*

Then v is an invariant vector, meaning that $\pi_g v = v$ for all $g \in \mathrm{SL}_d(\mathbb{R})$.

For the proof we will use the following algebraic fact for $K = \mathbb{R}$. For any field K the group $\mathrm{SL}_d(K)$ is generated by the elementary unipotent subgroups (defined as above but with $x \in K$). This may be seen using a modified Gauss elimination algorithm: given any $g \in \mathrm{SL}_d(K)$ it is clear that the first column is non-zero. Multiplying g on the left by elements of U_{12} (or another elementary unipotent subgroup) corresponds to the row operation of adding a multiple of the second row to the first row (or the same with any two other rows). For example, for $d = 2$ we have

$$\begin{pmatrix} 1 & x \\ 0 & 1 \end{pmatrix} \begin{pmatrix} a & b \\ c & d \end{pmatrix} = \begin{pmatrix} a + xc & b + xd \\ c & d \end{pmatrix}$$

for all $x, a, b, c, d \in \mathbb{R}$. Using such operations we can obtain matrices g', g'' and $\widetilde{g}$ that satisfy increasingly stronger properties:

- $g'_{21} \neq 0$,
- $g''_{11} = 1$,
- $\widetilde{g}_{11} = 1$, and $\widetilde{g}_{21} = \widetilde{g}_{31} = \cdots = \widetilde{g}_{d1} = 0$.

Multiplying on the right by the same type of matrices corresponds to column operations which allows us to find now a matrix $\widehat{g} \in HgH$ satisfying

- $\hat{g}_{11} = 1$, $\hat{g}_{1k} = \hat{g}_{k1} = 0$ for $k \geqslant 2$,

where H is the subgroup of $\mathrm{SL}_d(K)$ generated by the elementary unipotent subgroups. Using induction on the number of variables we see that $\hat{g} \in H$, which implies that $g \in H$ and thus $H = \mathrm{SL}_d(K)$.

PROOF OF PROPOSITION 10.41. As we will see we only have to multiply matrices and use continuity of the unitary represenation. Let us first consider the case $d = 2$ and a diagonal positive matrix

$$ a = \begin{pmatrix} t & 0 \\ 0 & t^{-1} \end{pmatrix}, $$

where we may assume without loss of generality that $t > 1$. Let

$$ u = \begin{pmatrix} 1 & x \\ 0 & 1 \end{pmatrix} \in U_{12} $$

for some $x \in \mathbb{R}$ and notice that

$$ \lim_{n \to \infty} a^{-n} \begin{pmatrix} 1 & x \\ 0 & 1 \end{pmatrix} a^n = \lim_{n \to \infty} \begin{pmatrix} 1 & t^{-2n}x \\ 0 & 1 \end{pmatrix} = I. $$

If $\pi_a v = v$ then continuity of the unitary representation implies that

$$ \|\pi_u v - v\| = \|\pi_u \pi_{a^n} v - \pi_{a^n} v\| = \|\pi_{a^{-n}ua^n} v - v\| \longrightarrow 0 $$

as $n \to \infty$. Therefore $\pi_u v = v$ for all $u \in U_{12}$. Using the same argument with $\pi_{a^{-1}} v = v$ and the relation

$$ \lim_{n \to \infty} a^n \begin{pmatrix} 1 & 0 \\ x & 1 \end{pmatrix} a^{-n} = I, $$

we see that v is fixed by both elementary unipotent subgroups, and hence by all of $\mathrm{SL}_2(\mathbb{R})$.

Staying with the case $d = 2$, suppose now that v is fixed by the subgroup U_{12}. Define

$$ g_n = \begin{pmatrix} 1 & 0 \\ \frac{1}{n} & 1 \end{pmatrix} $$

and calculate

$$ u_n g_n u_{-n/2} = \begin{pmatrix} 1 & n \\ 0 & 1 \end{pmatrix} \begin{pmatrix} 1 & 0 \\ \frac{1}{n} & 1 \end{pmatrix} \begin{pmatrix} 1 & -\frac{n}{2} \\ 0 & 1 \end{pmatrix} = \begin{pmatrix} 2 & 0 \\ \frac{1}{n} & \frac{1}{2} \end{pmatrix}, $$

which shows that

$$ \lim_{n \to \infty} u_n g_n u_{-n/2} = \begin{pmatrix} 2 & 0 \\ 0 & \frac{1}{2} \end{pmatrix} = a_2. $$

Using continuity of the unitary representation again we see that

$$\|\pi_{a_2} v - v\| = \lim_{n \to \infty} \|\pi_{u_n} \pi_{g_n} \pi_{u_{-n/2}} v - v\|$$
$$= \lim_{n \to \infty} \|\pi_{g_n} v - \pi_{-u_n} v\| = \lim_{n \to \infty} \|\pi_{g_n} v - v\| = 0$$

since $g_n \to I$ as $n \to \infty$. Therefore, v is also invariant under a_2 and the first part of the proof shows that v is fixed by all of $\mathrm{SL}_2(\mathbb{R})$. The case of the other elementary unipotent subgroup follows by the same argument.

Let us note that the first argument above also applies for a non-trivial diagonal matrix $a \in \mathrm{SL}_d(\mathbb{R})$ with positive eigenvalues $a_1, \ldots, a_d > 0$ in the following way: If $\pi_a v = v$ and, for example, $a_1 \neq a_2$, then v is also fixed by the subgroup obtained by embedding $\mathrm{SL}_2(\mathbb{R})$ into the upper left 2-by-2 block in $\mathrm{SL}_d(\mathbb{R})$.

Suppose now that $d = 3$ and v is fixed by a non-trivial positive diagonal matrix a with eigenvalues a_1, a_2, a_3. Assume that $a_1 \neq a_2$ (the other cases are similar, or can be reduced to this one by using permutation matrices). In this case v is fixed by the subgroup H obtained by embedding $\mathrm{SL}_2(\mathbb{R})$ into the upper left 2-by-2 block in $\mathrm{SL}_3(\mathbb{R})$ and in particular by

$$a' = \begin{pmatrix} 2 & 0 & 0 \\ 0 & \frac{1}{2} & 0 \\ 0 & 0 & 1 \end{pmatrix} \in H.$$

Since the eigenvalues of a' satisfy $a_1' \neq a_3'$ and $a_2' \neq a_3'$ we may repeat the argument for $\mathrm{SL}_2(\mathbb{R})$ twice more and see that v is fixed by all elementary unipotent subgroups, which implies that v is fixed by all of $\mathrm{SL}_3(\mathbb{R})$.

Remaining with the case $d = 3$, suppose that v is fixed by an elementary unipotent subgroup U. Since U is again contained in a subgroup $H \cong \mathrm{SL}_2(\mathbb{R})$ we see that v is invariant under a non-trivial positive diagonal element to which we may apply the arguments above.

The case $d > 3$ follows similarly by induction and will not be needed later, so we leave this part of the proof to the reader (see Exercise 10.42(a)).   □

Exercise 10.42. (a) Confirm that the case $d > 3$ in Proposition 10.41 may be seen using the same argument.
(b) Suppose that $u \in \mathrm{SL}_d(\mathbb{R})$ is a non-trivial unipotent element (that is, $u \neq I$ and all eigenvalues of u are equal to 1). Show that for any unitary representation $\pi : \mathrm{SL}_d(\mathbb{R}) \curvearrowright \mathcal{H}$ any $v \in \mathcal{H}$ with $\pi_u v = v$ is invariant under all of $\mathrm{SL}_d(\mathbb{R})$.

PROOF OF THEOREM 10.38 ASSUMING THEOREM 10.39. Note that

$$\mathrm{ASL}_2(\mathbb{R}) = \mathrm{SL}_2(\mathbb{R}) \ltimes \mathbb{R}^2 = \left\{ \begin{pmatrix} A & x \\ 0 & 1 \end{pmatrix} \mid A \in \mathrm{SL}_2(\mathbb{R}), x \in \mathbb{R}^2 \right\}$$

is a closed subgroup of $\mathrm{SL}_3(\mathbb{R})$.

Suppose π is a unitary representation of $\mathrm{SL}_3(\mathbb{R})$ that has almost-invariant vectors. Restricting π to $\mathrm{ASL}_2(\mathbb{R})$ we obtain a representation of $\mathrm{ASL}_2(\mathbb{R})$ that has almost-invariant vectors. Then by Theorem 10.39, $\mathcal{H}$ contains a non-zero vector v that is fixed by

$$\begin{pmatrix} I & x \\ 0 & 1 \end{pmatrix}$$

for all $x \in \mathbb{R}^2$. By the Mautner phenomenon (Proposition 10.41) this implies that v is fixed by all of $\mathrm{SL}_3(\mathbb{R})$. It follows that $\mathrm{SL}_3(\mathbb{R})$ has property (T). $\square$

For the proof of Theorem 10.39 we will use the spectral measures from Bochner's theorem for unitary flows (Theorem 9.56).

Lemma 10.43 (Normalizer and push-forward). *Let $G = \mathrm{ASL}_2(\mathbb{R})$ and suppose π is a unitary representation of G on a complex Hilbert space $\mathcal{H}$. Let μ_v denote the spectral measure of $v \in \mathcal{H}$ with respect to the restriction of π to $\mathbb{R}^2 \lhd G$. For every $A \in \mathrm{SL}_2(\mathbb{R})$ and $v \in \mathcal{H}$ we then have $\mu_{\pi_A v} = (A^{\mathrm{t}})_*^{-1} \mu_v$ where π_A is the unitary operator obtained from the matrix A, thought of as an element of $\mathrm{ASL}_2(\mathbb{R})$.*

PROOF. Recall that for any $v \in \mathcal{H}$ the spectral measure μ_v is uniquely determined by the property

$$\langle \pi_{u(x)} v, v \rangle = \int_{\mathbb{R}^2} e^{2\pi i x \cdot t} \, d\mu_v(t)$$

for all $x \in \mathbb{R}^2$, where we use the injective homomorphism $u : \mathbb{R}^2 \to \mathrm{ASL}_2(\mathbb{R})$ defined by

$$u(x) = \begin{pmatrix} I & x \\ 0 & 1 \end{pmatrix}$$

for $x \in \mathbb{R}^2$. Applying this to $\pi_A v$, we have

$$\int_{\mathbb{R}^2} e^{2\pi i x \cdot t} \, d\mu_{\pi_A v}(t) = \langle \pi_{u(x)} \pi_A v, \pi_A v \rangle = \langle \pi_{u(A^{-1} x)} v, v \rangle$$

$$= \int_{\mathbb{R}^2} e^{2\pi i A^{-1} x \cdot t} \, d\mu_v(t) = \int_{\mathbb{R}^2} e^{2\pi i x \cdot (A^{\mathrm{t}})^{-1} t} \, d\mu_v(t)$$

$$= \int_{\mathbb{R}^2} e^{2\pi i x \cdot s} \, d(A^{\mathrm{t}})_*^{-1} \mu_v(s),$$

where we used $A^{-1} u(x) A = u(A^{-1} x)$ for all $x \in \mathbb{R}^2$. Hence $\mu_{\pi_A v} = (A^{\mathrm{t}})_*^{-1} \mu_v$ by uniqueness of the spectral measure. $\square$

Lemma 10.44 (Continuity of spectral measures). *Let π be a unitary representation of $\mathbb{R}^d$ for some $d \geqslant 1$ on a complex Hilbert space $\mathcal{H}$. If v and w in $\mathcal{H}$ have norm one, then the difference of their spectral measures satisfies $\|\mu_v - \mu_w\| \leqslant 4\|v - w\|$.*

PROOF. First decompose $w = w_1 + w_2$ with $w_1 \in \mathcal{H}_v$ and $w_2 \in \mathcal{H}_v^\perp$. By the properties of the orthogonal decomposition we have $\|v - w_1\| \leqslant \|v - w\|$ and $\|w_2\| \leqslant \|v - w\| \leqslant 2$. Just as in Lemma 9.12(b) it is easy to see that

$$\mu_w = \mu_{w_1} + \mu_{w_2}$$

which gives

$$\|\mu_v - \mu_w\| = \|\mu_v - \mu_{w_1} - \mu_{w_2}\|$$
$$\leqslant \|\mu_v - \mu_{w_1}\| + \|\mu_{w_2}\| \leqslant \|\mu_v - \mu_{w_1}\| + 2\|v - w\|$$

since $\|\mu_{w_2}\| = \|w_2\|^2$.

It remains to bound $\|\mu_v - \mu_{w_1}\|$. First recall that in the spectral theorem (Theorem 9.58) the generator v corresponds to $\mathbb{1} \in L^2_{\mu_v}(\mathbb{R}^d)$ and w_1 corresponds to some function $f \in L^2_{\mu_v}(\mathbb{R}^d)$. Also note that $\mathrm{d}\mu_{w_1} = |f|^2 \, \mathrm{d}\mu_v$ by the same argument as in the proof of Lemma 9.12(c). Therefore,

$$\|\mu_v - \mu_{w_1}\| = \int_{\mathbb{R}^d} |\mathbb{1} - |f|^2| \, \mathrm{d}\mu_v$$
$$= \int_{\mathbb{R}^d} h\left(\mathbb{1} - |f|^2\right) \, \mathrm{d}\mu_v = \langle h\mathbb{1}, \mathbb{1}\rangle_{L^2_{\mu_v}} - \langle hf, f\rangle_{L^2_{\mu_v}},$$

where $h(t) = \mathrm{sign}(1 - |f(t)|^2)$. By the Cauchy–Schwarz inequality we deduce that

$$\|\mu_v - \mu_{w_1}\| = \langle h\mathbb{1} - hf, \mathbb{1}\rangle_{L^2_{\mu_v}} + \langle hf, \mathbb{1} - f\rangle_{L^2_{\mu_v}}$$
$$\leqslant \|v - w_1\|\|v\| + \|w_1\|\|v - w_1\| \leqslant 2\|v - w_1\| \leqslant 2\|v - w\|$$

since $\mathbb{1} \in L^2_{\mu_v}(\mathbb{R}^d)$ corresponds to v and $f \in L^2_{\mu_v}(\mathbb{R}^d)$ to $w_1 \in \mathcal{H}_v$. Together with the above this gives the lemma. $\square$

The last preparatory step for the proof of Theorem 10.39 is the following negative result.

Lemma 10.45 (No invariant measures). *The natural action of* $\mathrm{SL}_2(\mathbb{R})$ *on the projective line* $\mathbb{P}^1(\mathbb{R}) = \mathbb{R}^2 \backslash \{0\} / \sim$ *has no invariant probability measures.*

The natural action here is given by

$$\mathrm{SL}_2(\mathbb{R}) \ni \begin{pmatrix} a & b \\ c & d \end{pmatrix} : \left[\begin{pmatrix} x \\ y \end{pmatrix} \right] \longrightarrow \left[\begin{pmatrix} ax + by \\ cx + dy \end{pmatrix} \right],$$

where we write $\left[\begin{pmatrix} x \\ y \end{pmatrix} \right]$ for the equivalence class (with respect to proportionality) of a vector $\begin{pmatrix} x \\ y \end{pmatrix} \in \mathbb{R}^2 \backslash \{0\}$.

PROOF OF LEMMA 10.45. Notice that $\mathrm{SO}_2(\mathbb{R}) \subseteq \mathrm{SL}_2(\mathbb{R})$ acts transitively on $\mathbb{P}^1(\mathbb{R})$, the kernel M of the action consists of $\pm I$, and that $\mathrm{SO}_2(\mathbb{R})/M$ acts simply transitively on $\mathbb{P}^1(\mathbb{R})$. Fixing an element of $\mathbb{P}^1(\mathbb{R})$, say the element corresponding to the x-axis, to correspond to the identity of $\mathrm{SO}_2(\mathbb{R})/M$, we may identify $\mathbb{P}^1(\mathbb{R})$ with $SO_2(\mathbb{R})/M$ so that the action corresponds to translation on the group. By uniqueness of the Haar measure (Proposition 10.2) there is only one $SO_2(\mathbb{R})$-invariant probability measure on $\mathbb{P}^1(\mathbb{R})$. However, other elements of $SL_2(\mathbb{R})$ do not preserve that measure. For example, the action of $\begin{pmatrix} e & \\ & e^{-1} \end{pmatrix}$ does not preserve that probability measure (check this). □

PROOF OF THEOREM 10.39. Let π be a unitary representation of $ASL_2(\mathbb{R})$ on a Hilbert space $\mathcal{H}$, and suppose it has almost invariant vectors. We note that we may assume that $\mathcal{H}$ is a complex Hilbert space, for if $\mathcal{H}$ is a real Hilbert space we may complexify it (see Exercise 6.51 and Exercise 10.46) and can extend the given representation to a unitary representation on a complex Hilbert space that will also have almost invariant vectors. Let (Q_n) be a sequence of compact subsets in $ASL_2(\mathbb{R})$ with $Q_n \subseteq Q_{n+1}$ for which

$$\bigcup_{n \geqslant 1} Q_n = ASL_2(\mathbb{R}).$$

Then for every $n \geqslant 1$ there exists some $(Q_n, \frac{1}{n})$-invariant vector $v_n \in \mathcal{H}$ with $\|v_n\| = 1$.

Let μ_{v_n} be the spectral measure of v_n with respect to $\mathbb{R}^2 \lhd ASL_2(\mathbb{R})$ for each $n \geqslant 1$. If, for some $n \geqslant 1$, we have $\mu_{v_n}(\{0\}) > 0$, then by the spectral theorem (Theorem 9.58)

$$\mathcal{H}_{v_n} \cong L^2_{\mu_{v_n}}(\mathbb{R}^2) \ni \mathbb{1}_{\{0\}}$$

contains a non-zero vector that is invariant under $\mathbb{R}^2$. This is precisely the statement that we want to prove.

So suppose that $\mu_{v_n}(\{0\}) = 0$ for all $n \geqslant 1$, and project μ_{v_n} to a measure

$$\nu_n = p_* \mu_{v_n}$$

on $\mathbb{P}^1(\mathbb{R})$, where $p : \mathbb{R}^2 \backslash \{0\} \to \mathbb{P}^1(\mathbb{R})$ denotes the natural projection map $p(v) = [v]$ for all $v \in \mathbb{R}^2 \backslash \{0\}$. Since $\mathbb{P}^1(\mathbb{R})$ is compact we may apply Proposition 8.27 and choose a subsequence (ν_{n_k}) such that $\nu_{n_k} \to \nu$ in the weak* topology as $k \to \infty$ for some probability measure ν on $\mathbb{P}^1(\mathbb{R})$. We claim that ν is invariant under the action of $SL_2(\mathbb{R})$ on $\mathbb{P}^1(\mathbb{R})$.

To show this, let $f \in C(\mathbb{P}^1(\mathbb{R}))$, consider the function

$$F = f \circ p \in \mathscr{L}^\infty(\mathbb{R}^2 \backslash \{0\}),$$

and extend it by, for example, setting $F(0) = 0$. For $A \in Q_n \cap SL_2(\mathbb{R})$ the vector v_n satisfies $\|\pi_A v_n - v_n\| \leqslant \frac{1}{n}$ and so we have

$$\int_{\mathbb{P}^1(\mathbb{R})} f \circ (A^t)^{-1} \, d\nu_n = \int_{\mathbb{R}^2} F \circ (A^t)^{-1} \, d\mu_{\nu_n} = \int_{\mathbb{R}^2} F \, d(A^t)_*^{-1} \mu_{\nu_n}$$

$$= \int_{\mathbb{R}^2} F \, d\mu_{\nu_n} + O_f\left(\tfrac{1}{n}\right) = \int_{\mathbb{P}^1(\mathbb{R})} f \, d\nu_n + O_f\left(\tfrac{1}{n}\right)$$

by Lemmas 10.43 and 10.44. Now let $n = n_k$ and take $k \to \infty$ to see that

$$\int_{\mathbb{P}^1(\mathbb{R})} f \circ (A^t)^{-1} \, d\nu = \int_{\mathbb{P}^1(\mathbb{R})} f \, d\nu$$

for all $f \in C(\mathbb{P}^1(\mathbb{R}))$ and $A \in \mathrm{SL}_2(\mathbb{R})$. However, this shows that ν is $\mathrm{SL}_2(\mathbb{R})$-invariant, which contradicts Lemma 10.45. $\qquad\square$

Exercise 10.46. Let π be a unitary representation of a topological group on a real Hilbert space $\mathcal{H}$. Let $\mathcal{H}_{\mathbb{C}}$ be the complexification of $\mathcal{H}$ as in Exercise 6.51. Show that π can be extended to a unitary representation on $\mathcal{H}_{\mathbb{C}}$, which has almost-invariant (or invariant) vectors if and only if the original representation has almost invariant (or invariant) vectors.

Exercise 10.47. Show that $\mathrm{SL}_d(\mathbb{R})$ has property (T) for all $d \geqslant 3$.

10.3.4 Proof of Každan's Property (T), Discrete Case

The connection between $\mathrm{SL}_3(\mathbb{R})$ and its discrete subgroup $\mathrm{SL}_3(\mathbb{Z})$ is largely controlled by the fact that $\mathrm{SL}_3(\mathbb{Z})$ is a *lattice* in $\mathrm{SL}_3(\mathbb{R})$. We will not discuss the important notion of lattices in detail, but instead will work with the following form of the result, which will be proved after its significance is established.

Theorem 10.48 ($\mathrm{SL}_3(\mathbb{Z})$ is a lattice). *There exists a Borel subset F of the group $G = \mathrm{SL}_3(\mathbb{R})$, called a* fundamental domain *for $\mathrm{SL}_3(\mathbb{Z})$ in $\mathrm{SL}_3(\mathbb{R})$, such that $m_G(F) < \infty$ and $G = \bigsqcup_{\gamma \in \mathrm{SL}_3(\mathbb{Z})} F\gamma$.*

Apart from this result, we will also need a simple form of induction of unitary representations which will allow us to lift a unitary representation of $\mathrm{SL}_3(\mathbb{Z})$ to a unitary representation of $\mathrm{SL}_3(\mathbb{R})$.

To explain this more generally, we let $\Gamma < G$ be a discrete subgroup of a locally compact, σ-compact, metrizable, unimodular group, and let $F \subseteq G$ be a *fundamental domain* for Γ in G, that is, a Borel subset such that

$$G = \bigsqcup_{\gamma \in \Gamma} F\gamma. \tag{10.19}$$

Simple examples include $\Gamma = \mathbb{Z}^d < G = \mathbb{R}^d$ with $F = [0,1)^d$ for any $d \geqslant 1$; we refer to [25] for more details on the properties of fundamental domains. Furthermore, let $\pi_\Gamma : \Gamma \curvearrowright \mathcal{H}_\Gamma$ be a unitary representation of Γ on a separable Hilbert space $\mathcal{H}_\Gamma$.

Using these objects, we now define a new Hilbert space $\mathcal{H}_G$ equipped with a unitary representation of G called the *induced representation*. It is possible to give this definition abstractly and in a coordinate-free way, but using a Hilbert space isomorphism between $\mathcal{H}_\Gamma$ and $\ell^2(\mathbb{N})$ we can make the definition of $\mathcal{H}_G$ more explicit. We implicitly assume here that $\mathcal{H}_\Gamma$ is infinite-dimensional. In the finite-dimensional case the construction is slightly easier and adapting the notation to this case is straightforward. In the description afforded by this isomorphism we can therefore assume that $\mathcal{H}_\Gamma = \ell^2(\mathbb{N})$ and that Γ acts unitarily on $\ell^2(\mathbb{N})$. Using the fundamental domain $F \subseteq G$ for Γ in G we give the initial definition

$$\mathcal{H}_G = \bigoplus_{n \in \mathbb{N}} L^2(F) = \left\{ (f_1, f_2, \dots) \in L^2(F)^{\mathbb{N}} \mid \sum_{n=1}^{\infty} \|f_n\|_2^2 < \infty \right\}.$$

In the following we will think of $f \in \bigoplus_{n \in \mathbb{N}} L^2(F)$ as a measurable $\ell^2(\mathbb{N})$-valued and square-integrable function $f : F \ni g \mapsto f(g) = (f_1(g), f_2(g), \dots)$. We note that $f(g)$ belongs to $\ell^2(\mathbb{N})$ for almost every $g \in F$ by Fubini's theorem. We also define the norm of $f = (f_1, f_2, \dots)$ by

$$\|f\|_{\mathcal{H}_G} = \left(\sum_{n=1}^{\infty} \|f_n\|_2^2 \right)^{1/2} = \left(\int_F \|f(g)\|_2^2 \, \mathrm{d}m_G(g) \right)^{1/2}.$$

Using the unitary representation of Γ on $\ell^2(\mathbb{N})$ and the decomposition of G in (10.19) into right translates of F we now extend the domain of definition of $f \in \bigoplus_{n \in \mathbb{N}} L^2(F)$ (resp. of the functions $f_1, f_2, \dots$ appearing in any function $f \in \mathcal{H}_G$) to all of G. Given $g \in F$ and $\gamma \in \Gamma$ we define

$$f(g\gamma) = \pi_\Gamma(\gamma)^{-1} f(g), \tag{10.20}$$

which is well-defined for every $g \in F$ with the property that

$$f(g) = (f_1(g), f_2(g), \dots) \in \ell^2(\mathbb{N}),$$

and hence for almost every $g \in F$. For those g, we have

$$f(g) = \lim_{n \to \infty} (f_1(g), \dots, f_n(g), 0, 0, \dots)$$

and

$$f(g\gamma) = \lim_{n \to \infty} \pi_\Gamma(\gamma)^{-1} (f_1(g), \dots, f_n(g), 0, 0, \dots) = \lim_{n \to \infty} \sum_{k=1}^{n} f_k(g) \pi_\Gamma(\gamma)^{-1} e_k$$

shows that the components of $f(g\gamma)$ are a convergent sum of finite linear combinations of $f_1(g), f_2(g), \dots$. In particular, $f(g\gamma)$ depends measurably on $g \in F$. We will frequently identify f on F with the extension of f to all of G

using (10.20). Moreover, we modify the definition of $\mathcal{H}_G$ so that $\mathcal{H}_G$ consists of those functions on G obtained from elements in $\bigoplus_{n\in\mathbb{N}} L^2(F)$ using (10.20), as explained above.

Even though we used the fundamental domain F in the above construction quite prominently, the 'algebraic' property of the elements of $\mathcal{H}_G$ is independent of the choice of F.

Lemma 10.49 (Equivariance property). *Let $\Gamma < G$, F, π_Γ, $\mathcal{H}_\Gamma$, and $\mathcal{H}_G$ be as above. Then for any f in $\mathcal{H}_G$ and almost every $g \in G$ we have*

$$f(g\gamma) = \pi_\Gamma(\gamma)^{-1} f(g)$$

for every $\gamma \in \Gamma$.

PROOF. Since Γ is countable it is enough to show this for a fixed $\gamma \in \Gamma$. Let $f \in \mathcal{H}_G$, $g \in G$ and $\eta \in \Gamma$ have $g\eta \in F$ and $f(g\eta) \in \ell^2(\mathbb{N})$. Then

$$f(g) = f(g\eta\eta^{-1}) = \pi_\Gamma(\eta) f(g\eta) \qquad (10.21)$$

by definition of $f \in \mathcal{H}_G$. Similarly, we also have for any $\gamma \in \Gamma$ that

$$f(g\gamma) = f(g\eta\eta^{-1}\gamma) = \pi_\Gamma(\eta^{-1}\gamma)^{-1} f(g\eta) = \pi_\Gamma(\gamma)^{-1}\pi_\Gamma(\eta) f(g\eta) = \pi_\Gamma(\gamma)^{-1} f(g)$$

by (10.21). This gives the lemma. $\qquad\qquad\square$

Just as in the algebraic property discussed above, the resulting norm on $\mathcal{H}_G$ is also independent of the choice of F.

Lemma 10.50 (Independence of norm). *Let $\Gamma < G$, F, π_Γ, $\mathcal{H}_\Gamma$, and $\mathcal{H}_G$ be as above. Let F' be another fundamental domain for Γ in G. Then*

$$\|f\|_{\mathcal{H}_G} = \left(\int_F \|f(g)\|_2^2 \, dm_G(g) \right)^{1/2} = \left(\int_{F'} \|f(g)\|_2^2 \, dm_G(g) \right)^{1/2}. \quad (10.22)$$

In the case where π_Γ is the trivial representation of Γ, the above contains the following lemma. However, as it is easier we will give an independent proof of the next lemma as a warmup for the proof of Lemma 10.50. Because of this motivation we refrain from introducing the natural viewpoint of the continuous action of G on the homogeneous space G/Γ; see Raghunathan [89] and [25] for this viewpoint.

Lemma 10.51 (Equality of measures). *Let $\Gamma < G$ be as above. Suppose $F, F' \subseteq G$ are both measurable fundamental domains as in (10.19). Then $m_G(F) = m_G(F')$ and this quantity is called the* co-volume *of Γ. Moreover, for every $g_0 \in G$ the invertible map $F \ni g \mapsto g' = g_0 g \gamma_g \in F$, with $\gamma_g \in \Gamma$ uniquely determined by g and g_0, is measure-preserving. Finally, if $B \subseteq G$ is measurable with $m_G(B) > m_G(F)$, then there exists some $g \in B$ and $\gamma \in \Gamma \setminus \{e\}$ with $g\gamma \in B$.*

PROOF. By assumption on F in (10.19) we have

$$F' = \bigsqcup_{\gamma \in \Gamma} F' \cap (F\gamma^{-1}). \qquad (10.23)$$

Multiplying $F' \cap (F\gamma^{-1})$ on the right by γ, we obtain the sets $(F'\gamma) \cap F$. These are again disjoint by assumption on F' and their union equals F. Therefore,

$$m_G(F') = \sum_{\gamma \in \Gamma} m_G(F' \cap (F\gamma^{-1})) = \sum_{\gamma \in \Gamma} m_G((F'\gamma) \cap F) = m_G(F)$$

by unimodularity of G.

Suppose now that $g_0 \in G$ and F is a measurable fundamental domain. Then $F' = g_0 F$ is another fundamental domain and (10.23) defines a map ϕ by

$$F \ni g \longmapsto g_0 g \in F' \longmapsto g_0 g \gamma_g \in F$$

(with $\gamma_g \in \Gamma$ being uniquely determined by the condition $g_0 g \gamma_g \in F$). It is clear that the inverse to this map is given by the same procedure but using g_0^{-1}. To see that these maps are measure-preserving we consider the function ϕ above. Now let $B \subseteq F$ be measurable, and note that $\phi(B)$ is defined by piecewise right translation of the set $g_0 B \subseteq F' = g_0 F$ back to F. In other words, we use (10.23) and apply the same cut-and-translate procedure to obtain the desired equality

$$m_G(B) = m_G(g_0 B) = \sum_{\gamma \in \Gamma} m_G((g_0 B) \cap (F\gamma^{-1}))$$

$$= \sum_{\gamma \in \Gamma} m_G((g_0 B\gamma) \cap F) = m_G\left(\bigsqcup_{\gamma \in \Gamma}(g_0 B\gamma) \cap F\right) = m_G(\phi(B)).$$

Strictly speaking we should prove that $m_G(\phi^{-1}(B)) = m_G(B)$ as in the definition of a measure-preserving map (Definition 8.35), but since ϕ is invertible this distinction is not important.

Suppose now that $B \subseteq G$ is measurable with $m_G(B) > m_G(F)$. Applying (10.19) we see that

$$m_G(F) < m_G(B) = \sum_{\gamma \in \Gamma} m_G(B \cap (F\gamma^{-1})) = \sum_{\gamma \in \Gamma} m_G((B\gamma) \cap F),$$

which implies the existence of $\gamma_1 \neq \gamma_2 \in \Gamma$ and $g_1, g_2 \in B$ with $g_1\gamma_1 = g_2\gamma_2$ as required. $\qquad \square$

PROOF OF LEMMA 10.50. Let F and F' again denote two fundamental domains so that $F = \bigsqcup_{\gamma \in \Gamma} F \cap (F'\gamma^{-1})$. Applying the first equality in (10.22) (which is the definition of the norm on $\mathcal{H}_G$ using F) we obtain

$$\|f\|_{\mathcal{H}_G}^2 = \sum_{\gamma \in \Gamma} \int_{F \cap (F'\gamma^{-1})} \|f(g)\|_2^2 \, dm_G(g) = \sum_{\gamma \in \Gamma} \int_{(F\gamma) \cap F'} \|f(h\gamma^{-1})\|_2^2 \, dm_G(h)$$

by using the substitution $g = h\gamma^{-1}$ for $g \in F \cap F'\gamma^{-1}$ and $h \in F\gamma \cap F'$ and unimodularity of G. Using the defining formula $f(h) = f(g\gamma) = \pi_\Gamma(\gamma)^{-1} f(g)$ we see that

$$\|f(h)\|_2 = \|f(g)\|_2 = \|f(h\gamma^{-1})\|$$

and so

$$\|f\|_{\mathcal{H}_G}^2 = \sum_{\gamma \in \Gamma} \int_{(F\gamma) \cap F'} \|f(h)\|_2^2 \, dm_G(h) = \int_{F'} \|f(h)\|_2^2 \, dm_G(h),$$

where we used the consequence $F' = \bigsqcup_{\gamma \in \Gamma} (F\gamma) \cap F'$ of (10.19). $\qquad\square$

We are now ready to prove the main properties of the unitary induction (which in a sense combines the unitary representation π_Γ and the measure-preserving maps discussed in Lemma 10.51).

Proposition 10.52. *Let G be a locally compact σ-compact metrizable unimodular group and $\Gamma < G$ a lattice (so that there exists a fundamental domain F as in (10.19) with $m_G(F) < \infty$). Given a unitary representation π_Γ of Γ on a separable Hilbert space $\mathcal{H}_\Gamma$, the Hilbert space $\mathcal{H}_G$ constructed above admits a unitary representation π_G of G defined by $\pi_{G,g_0} f(g) = f(g_0^{-1}g)$ for $g_0, g \in G$ and $f \in \mathcal{H}_G$. Moreover, $\mathcal{H}_\Gamma$ has a non-trivial Γ-fixed vector if and only if $\mathcal{H}_G$ has a non-trivial G-fixed vector, and $\mathcal{H}_G$ has almost invariant vectors if $\mathcal{H}_\Gamma$ has almost invariant vectors.*

Note that the formula defining π_{G,g_0} is the same formula as for the left regular representation on the space of functions on G, but that the space and the norm are different.

Exercise 10.53. Let $\Gamma < G$ be a discrete subgroup of a locally compact, σ-compact, metrizable, unimodular group G. Let π_Γ be the left regular representation of Γ on $\ell^2(\Gamma)$ defined by $\pi_{\Gamma,\gamma_0} f(\gamma) = f(\gamma_0^{-1}\gamma)$ for all $f \in \ell^2(\Gamma)$ and $\gamma_0, \gamma \in \Gamma$. Show that the induced representation π_G is then unitarily isomorphic to the left regular representation of G.

PROOF OF PROPOSITION 10.52. Let $g_0 \in G$ and $f \in \mathcal{H}_G$. Then

$$\|\pi_{G,g_0}(f)\|_{\mathcal{H}_G}^2 = \int_F \|f(g_0^{-1}g)\|_2^2 \, dm_G(g) = \int_{g_0^{-1}F} \|f(h)\|_2^2 \, dm_G(h)$$

by left-invariance of the Haar measure m_G. However, by Lemma 10.50 the latter is equal to $\|f\|_{\mathcal{H}_G}^2$ since $g_0^{-1}F = F'$ is also a fundamental domain. Hence π_{G,g_0} is a unitary operator on $\mathcal{H}_G$ for any $g_0 \in G$. That π_G is a homomorphism from G into the group of unitary operators of $\mathcal{H}_G$ follows by the same argument as for the regular representation on p. 353.

LIFTING INVARIANT VECTORS. Suppose now that $v \in \mathcal{H}_\Gamma$ is an invariant unit vector. Then we can define $f(g) = \frac{1}{\sqrt{m_G(F)}} v$ for all $g \in F$, and the extension of f to G will be $f(g) = \frac{1}{\sqrt{m_G(F)}} v$ for all $g \in G$. Notice that $f \in \mathcal{H}_G$ since $m_G(F) < \infty$. We therefore obtain a unit vector of $\mathcal{H}_G$ that is invariant with respect to G.

PUSHING INVARIANT VECTORS BACK TO $\mathcal{H}_\Gamma$. Suppose for the opposite direction that $\mathcal{H}_G$ has a G-invariant unit vector f. Since G acts transitively on itself, this implies that $f(g) = v$ for some non-zero v in $\mathcal{H}_\Gamma$ and almost every $g \in G$ (by using Exercise 10.4 for each component f_j of $f = (f_1, f_2, \ldots)$). Since we also have $f(g\gamma) = \pi_\Gamma(\gamma)^{-1} f(g)$ for almost every $g \in G$ and all $\gamma \in \Gamma$ we see that $v \in \mathcal{H}_\Gamma$ is a non-zero Γ-invariant vector.

LIFTING ALMOST INVARIANT VECTORS. Suppose next that $\mathcal{H}_\Gamma$ has almost invariant vectors, let $K \subseteq G$ be a compact subset and let $\varepsilon > 0$. Recall that $m_G(F) < \infty$ since $\Gamma < G$ is a lattice. By regularity of m_G there exists a compact subset $L \subseteq F$ such that $m_G(F \smallsetminus L) < \varepsilon m_G(F)$. Since $L^{-1}KL \subseteq G$ is compact and $\Gamma < G$ is discrete, the set $Q = \Gamma \cap (L^{-1}KL)$ is a finite subset of Γ. Suppose now that $v \in \mathcal{H}_\Gamma$ is a (Q, ε)-almost invariant unit vector. Much as in the discussion of invariant vectors, we define $f \in \mathcal{H}_G$ by setting

$$f(g) = \begin{cases} v & \text{for } g \in L, \\ 0 & \text{for } g \in F \smallsetminus L \end{cases}$$

and use the formula $f(g\gamma) = \pi_\Gamma(\gamma)^{-1} f(g)$ for all $g \in F$ and $\gamma \in \Gamma$ to extend f to a function $f \in \mathcal{H}_G$.

Now let $k \in K$ and $g \in F$. Then

$$\pi_{G,k} f(g) = f(k^{-1}g) = \pi_\Gamma(\gamma) f(k^{-1}g\gamma)$$

for all $\gamma \in \Gamma$. Choose $\gamma_g \in \Gamma$ such that $k^{-1}g\gamma_g = g' \in F$. Using this notation we can then write

$$\left\| \pi_{G,k} f - f \right\|_{\mathcal{H}_G}^2 = \int_F \left\| \pi_\Gamma(\gamma_g) f(\underbrace{k^{-1}g\gamma_g}_{=g' \in F}) - f(g) \right\|_2^2 dm_G(g).$$

Next we decompose the integral into the subsets:

- $g \in F \smallsetminus L$ (with $f(g) = 0$ and $\|f(g')\|_2 \leqslant \|v\|_2 = 1$);
- $g \in L$ and $g' = k^{-1}g\gamma_g \in F \smallsetminus L$ (with $f(g) = v$ and $f(g') = 0$); and
- $g, g' \in L$ (with $f(g) = f(g') = v$).

In the first case we will use $m_G(F \smallsetminus L) < \varepsilon m_G(F)$, in the second case

$$m_G \left(\{ g \in L \mid g' \in F \smallsetminus L \} \right) \leqslant m_G(F \smallsetminus L) < \varepsilon m_G(F)$$

(which follows since $F \ni g \mapsto g' \in F$ is measure-preserving by Lemma 10.51), and in the third case we see that

$$\gamma_g = g^{-1}kg' \in L^{-1}KL \cap \Gamma = Q$$

and so $\|\pi_\Gamma(\gamma_g)v - v\|_2 < \varepsilon$. Together this gives

$$\|\pi_{G,k}f - f\|_{\mathcal{H}_G}^2 < 2\varepsilon m_G(F) + \varepsilon^2 m_G(F).$$

We may assume that $\varepsilon < \frac{1}{2}$, so that

$$\|f\|_2 = \sqrt{m_G(L)} \geqslant \sqrt{\frac{m_G(F)}{2}}.$$

Since $\varepsilon > 0$ was arbitrary, this shows that $\mathcal{H}_G$ has almost invariant vectors.

CONTINUITY. The alert reader will have noticed that the arguments above do not finish the proof, because it remains to be shown that $\mathcal{H}_G$ is indeed a unitary representation, and in particular satisfies the continuity requirement. We will use Lemma 3.74 for this, but only indirectly.

Let us begin by noting that it is enough to prove continuity at the identity $e \in G$ for the following reason. Suppose that for any sequence (k_n) in G with $k_n \to e$ as $n \to \infty$ we have $\pi_{G,k_n}f \to f$ as $n \to \infty$ for any $f \in \mathcal{H}_G$. It follows that if (g_n) is a sequence in G with $g_n \to g$ as $n \to \infty$ then $\pi_{G,g_n}f - \pi_{G,g}f = \pi_{G,g}(\pi_{G,k_n}f - f) \to 0$ since $k_n = g^{-1}g_n \to e$ as $n \to \infty$ and $\pi_{G,g}$ is continuous.

In the following, we identify $\bigoplus_{n \in \mathbb{N}} L^2(F)$ with $L^2(F \times \mathbb{N}) \subseteq L^2(G \times \mathbb{N})$ and endow the latter with the unitary representation defined by

$$\lambda_g(f)(h, n) = f(g^{-1}h, n)$$

for $(h, n) \in G \times \mathbb{N}$, $g \in G$, and $f \in L^2(G \times \mathbb{N})$.

Given any $f \in \mathcal{H}_G$, we consider $f|_F \in \bigoplus_{n \in \mathbb{N}} L^2(F)$ as an element of $L^2(G \times \mathbb{N})$ satisfying $f|_F(g, n) = f_n(g)$ for $g \in F$ and $f|_F(g, n) = 0$ for $g \in G \smallsetminus F$, and $n \in \mathbb{N}$. Applying Lemma 3.74 to $f|_F \in L^2(G \times \mathbb{N})$, we see that there exists, for every $\varepsilon > 0$, a neighbourhood V of e such that

$$\left\| \lambda_g(f|_F) - f|_F \right\|_2 < \varepsilon \tag{10.24}$$

whenever $g \in V$. Given one such $g \in V$ we write

$$F = \underbrace{F \cap (g^{-1}F) \cap (gF)}_{=F_{\mathrm{in}}} \sqcup \Big(\underbrace{F \smallsetminus (gF)}_{=B_+} \cup \underbrace{F \smallsetminus (g^{-1}F)}_{=B_-} \Big) = F_{\mathrm{in}} \sqcup B,$$

that is, we decompose F into the part $F_{\mathrm{in}} \subseteq F$ that stays inside F (under the action of g and of g^{-1}) and its relative complement

$$B = F \smallsetminus (gF) \cup F \smallsetminus (g^{-1}F) = F \smallsetminus F_{\mathrm{in}}$$

(see Figure 10.2). It may help to think of B as the bad set on which λ_g and $\pi_{G,g}$ are quite different. We need to estimate its significance.

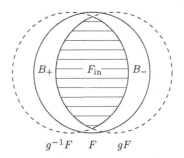

$$g^{-1}F \quad F \quad gF$$

Fig. 10.2: The circle depicts F, and the action of g translates the circle to the right, giving rise to the decomposition $F = F_{\mathrm{in}} \sqcup B$.

Using these sets, and recalling that $\mathcal{H}_G \ni f' \longmapsto f'|_F \in L^2(F \times \mathbb{N})$ is a unitary isomorphism, we also decompose $f \in \mathcal{H}_G$ into $f = f_{\mathrm{in}} + f_B$ with $f_{\mathrm{in}}, f_B \in \mathcal{H}_G$ satisfying $f_{\mathrm{in}}|_F = f|_F \mathbb{1}_{F_{\mathrm{in}}}$ and $f_B|_F = f|_F \mathbb{1}_B$. Using the identity $\lambda_g \mathbb{1}_{B_-} = \mathbb{1}_{gB_-}$ this implies that $\lambda_g(f|_F \mathbb{1}_{B_-})$ vanishes outside of gB_- and hence in particular it vanishes on F. Similarly $\lambda_g(\mathbb{1}_F) = \mathbb{1}_{gF}$ shows that $\lambda_g(f|_F)$ vanishes on B_+. Combining these with (10.24), it follows that

$$\|f|_F \mathbb{1}_{B_-}\|_2 = \|\lambda_g(f|_F \mathbb{1}_{B_-})\|_2 \leqslant \|\lambda_g(f|_F) - f|_F\|_2 < \varepsilon,$$

$$\|f|_F \mathbb{1}_{B_+}\|_2 \leqslant \|\lambda_g(f|_F) - f|_F\|_2 < \varepsilon,$$

and so

$$\|f_B\|_{\mathcal{H}_G} < 2\varepsilon. \tag{10.25}$$

For f_{in} we claim that

$$\big(\pi_{G,g} f_{\mathrm{in}}\big)\big|_F = \lambda_g\big(f_{\mathrm{in}}|_F\big). \tag{10.26}$$

Indeed, if $h \in F_{\mathrm{in}} \cup B_- \smallsetminus B_+ = F \cap gF$, then $g^{-1}h \in F$ and

$$\big(\pi_{G,g} f_{\mathrm{in}}\big)(h) = f_{\mathrm{in}}(g^{-1}h) = \lambda_g\big(f_{\mathrm{in}}|_F\big)(h)$$

as required. On the other hand, if $h \in B_+$, then $h \in F$ but $g^{-1}h \notin F$. Let $\gamma \in \Gamma \smallsetminus \{e\}$ be such that $g^{-1}h\gamma \in F$, which implies $g^{-1}h\gamma \in B_-$ and hence by Lemma 10.49

$$\big(\pi_{G,g} f_{\mathrm{in}}\big)(h) = \pi_\Gamma(\gamma)\big(f_{\mathrm{in}}(g^{-1}h\gamma)\big) = 0 = f_{\mathrm{in}}|_F(g^{-1}h) = \lambda_g\big(f_{\mathrm{in}}|_F\big)(h),$$

as claimed.

Combining (10.25)–(10.26) with (10.24) we can now obtain

$$\|\pi_{G,g}f - f\|_{\mathcal{H}_G} \leqslant \|\pi_{G,g}f_{\text{in}} - f_{\text{in}}\|_{\mathcal{H}_G} + 4\varepsilon$$
$$= \|\lambda_g(f_{\text{in}}|_F) - f_{\text{in}}|_F\|_2 + 4\varepsilon$$
$$< \|\lambda_g(f|_F) - f|_F\|_2 + 8\varepsilon < 9\varepsilon.$$

Since this holds for any $g \in V$, we obtain the continuity of the unitary representation and hence the theorem. $\qquad\square$

Proposition 10.54. *Let G be a locally compact σ-compact metrizable unimodular group and let $\Gamma < G$ be a lattice in G. If G has property (T) then Γ also has property (T).*

PROOF. Let $\pi_\Gamma : \Gamma \curvearrowright \mathcal{H}_\Gamma$ be a unitary representation on a separable Hilbert space that has almost invariant vectors. Applying Proposition 10.52 we find the unitary representation $\pi_G : G \curvearrowright \mathcal{H}_G$, also with almost invariant vectors. By assumption G has property (T) so that $\mathcal{H}_G$ has a nontrivial G-invariant vector. By Proposition 10.52, this implies that $\mathcal{H}_\Gamma$ has a non-trivial Γ-invariant vector. It follows that Γ has property (T). $\qquad\square$

It should now be clear how to combine the arguments to obtain a proof of Corollary 10.40: By Theorem 10.38, the topological group $\mathrm{SL}_3(\mathbb{R})$ has property (T). In Theorem 10.48 we claimed that the discrete subgroup $\mathrm{SL}_3(\mathbb{Z})$ is a lattice in $\mathrm{SL}_3(\mathbb{R})$. Hence Proposition 10.54 shows that $\mathrm{SL}_3(\mathbb{Z})$ also has property (T). For the proof of the lattice property in Theorem 10.48 we have to make a short excursion into the 'geometry of numbers'.

10.3.5 Iwasawa Decomposition, Geometry of Numbers, and Reduction Theory

Semi-simple groups have distinguished subgroups, some of which permit the group to be decomposed. The *Iwasawa* or *KAN decomposition* of $\mathrm{SL}_3(\mathbb{R})$ concerns the following subgroups:

- the compact *special orthogonal* group

$$K = \mathrm{SO}_3(\mathbb{R}) = \left\{ g \in \mathrm{SL}_3(\mathbb{R}) \mid gg^{\mathrm{t}} = I \right\},$$

(where we write I for the identity matrix) which is also sometimes written $\mathrm{SO}(3, \mathbb{R})$;
- the *positive diagonal* subgroup

$$A = \left\{ \begin{pmatrix} a_1 & 0 & 0 \\ 0 & a_2 & 0 \\ 0 & 0 & a_3 \end{pmatrix} \mid a_1, a_2, a_3 > 0 \text{ and } a_1 a_2 a_3 = 1 \right\};$$

- and the *unipotent* subgroup

$$N = \left\{ \begin{pmatrix} 1 & x & z \\ 0 & 1 & y \\ 0 & 0 & 1 \end{pmatrix} \mid x, y, z \in \mathbb{R} \right\}.$$

Lemma 10.55 (Iwasawa or KAN decomposition). *Any element of* $\mathrm{SL}_3(\mathbb{R})$ *can be written uniquely in the form* kan *with* $k \in K$, $a \in A$ *and* $n \in N$.

PROOF. As we will see, this is simply a reformulation of the familiar Gram–Schmidt procedure in $\mathbb{R}^3$ (see the proof of Theorem 3.39). Writing the matrix $g = (w_1, w_2, w_3) \in \mathrm{SL}_3(\mathbb{R})$ in terms of its column vectors $w_1, w_2, w_3 \in \mathbb{R}^3$ we may apply the Gram–Schmidt orthonormalization procedure to obtain

$$v_1 = a_1^{-1} w_1 \text{ with } a_1 = \|w_1\| > 0,$$

$$v_2 = a_2^{-1}(w_2 - \langle w_2, v_1 \rangle v_1) = a_2^{-1}(w_2 - n_{12}a_1 v_1) \text{ for some } a_2 > 0, n_{12} \in \mathbb{R},$$

$$v_3 = a_3^{-1}(w_3 - n_{13}a_1 v_1 - n_{23}a_2 v_2) \text{ for some } a_3 > 0 \text{ and } n_{13}, n_{23} \in \mathbb{R},$$

with the property that $\{v_1, v_2, v_3\}$ is an orthonormal basis of $\mathbb{R}^3$. This gives

$$(v_1, v_2, v_3) \begin{pmatrix} a_1 & 0 & 0 \\ 0 & a_2 & 0 \\ 0 & 0 & a_3 \end{pmatrix} \begin{pmatrix} 1 & n_{12} & n_{13} \\ 0 & 1 & n_{23} \\ 0 & 0 & 1 \end{pmatrix}$$

$$= (w_1, n_{12}a_1 v_1 + a_2 v_2, n_{13}a_1 v_1 + n_{23}a_2 v_2 + a_3 v_3) = g.$$

We define k, a, n to be the first, second, and third matrix in the equation above. Clearly $\det n = 1$ and $\det a > 0$. Since k has orthonormal column vectors we have $\det k = \pm 1$ (since $k^{\mathrm{t}}k = I$ and so $(\det k)^2 = 1$). Since we know that $\det g = 1$, this implies that $\det k = \det a = 1$, so we have established the existence of the decomposition.

If $kan = g = k'a'n'$ are both decompositions of this form, then

$$an(a'n')^{-1} = k^{-1}k' \in K \cap AN$$

since K and AN are both subgroups. Since all elements of K are diagonalizable over $\mathbb{C}$ with eigenvalues of absolute value one, we see that $K \cap AN = \{I\}$, which implies $k = k'$ and $an = a'n'$. Similarly, since $A \cap N = \{I\}$ we now see in the same way that $a = a'$ and $n = n'$. $\qquad\square$

A *lattice* in $\mathbb{R}^d$ is a subgroup of the form $\Lambda = g\mathbb{Z}^d$ for some $g \in \mathrm{GL}_d(\mathbb{R})$. Recall from Lemma 10.51 that the *co-volume* of Λ is defined by the Lebesgue measure of any fundamental domain $F \subseteq \mathbb{R}^d$ for Λ. Using $F = g[0,1)^d$ we see that the co-volume is given by $|\det g|$.

The next result is part of a theory from 1896 due to Minkowski, who also invented the descriptive name 'geometry of numbers' for it (see [74] for a reprint and the monograph of Lekkerkerker [60] for more material in this direction).

Proposition 10.56 (Choice of basis). *Let $\Lambda < \mathbb{R}^3$ be a lattice of co-volume 1. Then there exists some $g \in \mathrm{SL}_3(\mathbb{R})$ with $\Lambda = g\mathbb{Z}^3$ with the property that the matrices $a \in A$, $n \in N$, and $k \in K$ in the Iwasawa decomposition $g = kan$ satisfy $a_1 \ll a_2 \ll a_3$ and $n_{12}, n_{13}, n_{23} \in [-\frac{1}{2}, \frac{1}{2})$.*

PROOF. For the proof of the proposition we will also use a version of the conclusion in two dimensions. Let us assume first that $d \geqslant 2$ and $\Lambda \subseteq \mathbb{R}^d$ is a discrete subgroup. Let $w_1 \in \Lambda$ be a shortest non-zero vector of Λ, let V be the space $(\mathbb{R}w_1)^{\perp}$ and let $p : \mathbb{R}^d \to V$ denote the orthogonal projection. We claim that any non-zero vector $p(w) \in p(\Lambda)$ has

$$\|p(w)\| \geqslant \tfrac{\sqrt{3}}{2}\|w_1\|. \tag{10.27}$$

To prove this, suppose that $w \in \Lambda$ satisfies

$$0 < \|v\| < \tfrac{\sqrt{3}}{2}\|w_1\|$$

where $v = p(w)$. Clearly $w = v + tw_1$ for some $t \in \mathbb{R}$, and we may add an integer multiple of $w_1 \in \Lambda$ to $w \in \Lambda$ (without changing v) and suppose without loss of generality that $t \in [-\frac{1}{2}, \frac{1}{2})$. However, since $v \perp w_1$ this gives

$$0 < \|w\|^2 = \|v\|^2 + t^2\|w_1\|^2 < \tfrac{3}{4}\|w_1\|^2 + \tfrac{1}{4}\|w_1\|^2 = \|w_1\|^2,$$

which contradicts our choice of w_1.

THE TWO-DIMENSIONAL CASE. We continue with a version of the statement for $\Lambda < \mathbb{R}^2$, where we will assume that Λ is a discrete subgroup containing two linearly independendent vectors. We will also show that this implies that Λ is a lattice (not necessarily of co-volume 1).

As above we choose a non-zero $w_1 \in \Lambda$ of minimal norm, define V to be $(\mathbb{R}w_1)^{\perp}$, and $p : \mathbb{R}^2 \to V$ to be the orthogonal projection. Since Λ contains two linearly independent vectors, the kernel of p is one-dimensional, and (10.27) shows that $p(\Lambda)$ is discrete, so there exists some non-zero vector $u = p(w_2) \in p(\Lambda)$ of minimal length. As above, we may suppose that $w_2 = u + tw_1$ with $t \in [-\frac{1}{2}, \frac{1}{2})$.

Suppose now that $w \in \Lambda$ so that $p(w) \in p(\Lambda)$. Since u is of minimal length it is easy to see (by integer division with remainder) that $p(w) = n_2 u$ for some $n_2 \in \mathbb{Z}$. Now consider $w - n_2 w_2 \in \ker(p) = \mathbb{R}w_1$. Again because w_1 is of minimal length in Λ it follows that $w - n_2 w_2 = n_1 w_1$ for some $n_1 \in \mathbb{Z}$. In other words, we have shown that $\Lambda = (w_1, w_2)\mathbb{Z}^2 = \mathbb{Z}w_1 + \mathbb{Z}w_2$ is a lattice generated (as a group) by $w_1, w_2 \in \mathbb{R}^2$.

Applying the Gram–Schmidt orthonormalization procedure as in the proof of Lemma 10.55 but for $g = (w_1, w_2) \in \mathrm{GL}_2(\mathbb{R})$ we obtain $v_1 = a_1^{-1}w_1$ with $a_1 = \|w_1\|$, and $v_2 = a_2^{-1}(w_2 - \langle w_2, v_1 \rangle v_1) = a_2^{-1}(w_2 - n_{12}a_1 v_1)$ with

$$n_{12} = a_1^{-1} \langle w_2, v_1 \rangle = a_1^{-1} t \langle w_1, v_1 \rangle = t \in [-\tfrac{1}{2}, \tfrac{1}{2})$$

and $a_2 = \|w_2 - n_{12}w_1\| = \|u\|$. This gives

$$g = (w_1, w_2) = (v_1, v_2) \begin{pmatrix} a_1 & \\ & a_2 \end{pmatrix} \begin{pmatrix} 1 & n_{12} \\ & 1 \end{pmatrix}.$$

With the claim in (10.27) we now obtain $a_2 = \|u\| = \|p(w_2)\| \gg \|w_1\| = a_1$, which gives the result for dimension 2.

The three-dimensional case. Now suppose that $\Lambda < \mathbb{R}^3$ is a lattice with co-volume 1, so that Λ can be written $g_0 \mathbb{Z}^3$ for some $g_0 \in \mathrm{SL}_3(\mathbb{R})$. We again choose $w_1 \in \Lambda \setminus \{0\}$ of minimal length. To simplify the discussion we may apply some $k \in \mathrm{SO}_3(\mathbb{R})$ to rotate w_1 to the first coordinate axis. In other words, we may assume that $w_1 = \|w_1\|e_1$ is a vector of Λ with minimal length. Let $p : \mathbb{R}^3 \to \{0\} \times \mathbb{R}^2$ be the orthogonal projection with kernel $\mathbb{R}e_1$. Applying the claim in (10.27), we see that $v \in p(\Lambda)$ implies $v = 0$ or $\|v\| \gg \|w_1\|$. Since Λ contains 3 linearly independent vectors and $\ker(p)$ is one-dimensional we see that $p(\Lambda)$ contains at least two linearly independent vectors. Applying the two-dimensional case to $p(\Lambda)$ we find $w_2, w_3 \in \Lambda$ such that $v_2 = p(w_2)$ and $v_3 = p(w_3)$ satisfy $p(\Lambda) = \mathbb{Z}v_2 + \mathbb{Z}v_3$, that v_2 is the shortest vector of $p(\Lambda)$ and the size s of the orthogonal projection of v_3 onto the orthogonal complement of $\mathbb{R}v_2$ has $s \gg \|v_2\|$. For simplicity we may apply another rotation of $\{0\} \times \mathbb{R}^2$ fixing $\mathbb{R} \times \{(0,0)\}$ to Λ and suppose that $v_2 = \|v_2\|e_2$. This gives us

$$(w_1, w_2, w_3) = \begin{pmatrix} \|w_1\| & * & * \\ 0 & \|v_2\| & * \\ 0 & 0 & * \end{pmatrix}.$$

Replacing w_3 by $-w_3$ if necessary, we may assume that

$$(w_1, w_2, w_3) = \begin{pmatrix} a_1 & * & * \\ 0 & a_2 & * \\ 0 & 0 & a_3 \end{pmatrix} = \begin{pmatrix} a_1 & 0 & 0 \\ 0 & a_2 & 0 \\ 0 & 0 & a_3 \end{pmatrix} \begin{pmatrix} 1 & n_{12} & n_{13} \\ 0 & 1 & n_{23} \\ 0 & 0 & 1 \end{pmatrix} \qquad (10.28)$$

with $a_3 = s > 0$. To summarize, we have obtained $a_3 \gg a_2 \gg a_1$. Moreover, for any $w \in \Lambda$ there exist $\ell_2, \ell_3 \in \mathbb{Z}$ with $p(w) = \ell_2 v_2 + \ell_3 v_3$. Considering

$$w - \ell_2 w_2 - \ell_3 w_3 \in \mathbb{R}w_1$$

and the choice of w_1, we also find $\ell_1 \in \mathbb{Z}$ with

$$w = \ell_1 w_1 + \ell_2 w_2 + \ell_3 w_3,$$

which shows that $\Lambda = (w_1, w_2, w_3)\mathbb{Z}^3 = \mathbb{Z}w_1 + \mathbb{Z}w_2 + \mathbb{Z}w_3$. Multiplying (w_1, w_2, w_3) on the right by

$$\begin{pmatrix} 1 & \ell_1 & 0 \\ 0 & 1 & \ell_2 \\ 0 & 0 & 1 \end{pmatrix}$$

with $\ell_1, \ell_2 \in \mathbb{Z}$ allows us to modify n_{12}, n_{23} by integers while preserving the lattice. Thus we may assume that $n_{12}, n_{23} \in [-\frac{1}{2}, \frac{1}{2})$. Multiplying (w_1, w_2, w_3) after this on the right by

$$\begin{pmatrix} 1 & 0 & \ell \\ 0 & 1 & 0 \\ 0 & 0 & 1 \end{pmatrix}$$

for some $\ell \in \mathbb{Z}$ finally allows us to also obtain $n_{13} \in [-\frac{1}{2}, \frac{1}{2})$. $\qquad\square$

Lemma 10.57. *The group* $\mathrm{SL}_3(\mathbb{R})$ *is unimodular, and the Haar measure on* $\mathrm{SL}_3(\mathbb{R})$ *decomposes with respect to the Iwasawa decomposition into the product of the Haar measure* m_K *on* K *and the right Haar measure* $m_{AN}^{(r)}$ *on* AN.

PROOF. Notice first that $\mathrm{SL}_3(\mathbb{R}) \subseteq \mathrm{Mat}_{33}(\mathbb{R}) = \mathbb{R}^9$ is defined by a single equation and hence is a hypersurface. We will define the Haar measure $m_{\mathrm{SL}_3(\mathbb{R})}$ on $\mathrm{SL}_3(\mathbb{R})$ using the Lebesgue measure $m_{\mathbb{R}^9}$ by the following trick. Define a measure μ on $\mathrm{SL}_3(\mathbb{R})$ by $\mu(B) = m_{\mathbb{R}^9}(\{tg \mid t \in [0,1], g \in B\})$ for any Borel measurable set $B \subseteq \mathrm{SL}_3(\mathbb{R})$. To see that the set on the right-hand side is measurable note that $U = \{m \in \mathrm{Mat}_{33}(\mathbb{R}) \mid \det m \in (0,1)\}$ is open since the determinant map is continuous, and on U the map

$$\phi : U \longrightarrow (0,1) \times \mathrm{SL}_3(\mathbb{R})$$

$$m \longmapsto \left(\det m, \frac{1}{\sqrt[3]{\det m}} m\right)$$

is a homeomorphism. It follows that $\{tg \mid t \in [0,1], g \in B\}$ is also given by $\{0\} \cup \phi^{-1}((0,1) \times B) \cup B$ and so is measurable.

If now $B = K$ is compact, then so is $\{tg \mid t \in [0,1], g \in B\} \subseteq \mathbb{R}^9$, which gives $\mu(B) < \infty$. Also, if $B = O$ is non-empty and open, then

$$\{tg \mid t \in [0,1], g \in B\}$$

contains the non-empty open set $\{tg \mid t \in (0,1), g \in B\}$ and so in particular $\mu(B) > 0$. Now let $B \subseteq \mathrm{SL}_3(\mathbb{R})$ be measurable and $g_0 \in \mathrm{SL}_3(\mathbb{R})$. Then

$$\mu(g_0 B) = m_{\mathbb{R}^9}(\{tg_0 g \mid t \in [0,1], g \in B\})$$
$$= m_{\mathbb{R}^9}(g_0\{tg \mid t \in [0,1], g \in B\})$$
$$= m_{\mathbb{R}^9}(\{tg \mid t \in [0,1], g \in B\}) = \mu(B),$$

since left multiplication by g_0 scales Lebesgue measure on $\mathrm{Mat}_{33}(\mathbb{R})$ by the Jacobian of the map $m \mapsto g_0 m$ which is $|\det g_0|^3 = 1$, and so preserves the Lebesgue measure. This shows that μ is a left Haar measure, so we may write $\mu = m_{\mathrm{SL}_3(\mathbb{R})}$. On the other hand exactly the same argument applies to right multiplication, so μ is also a right Haar measure, and hence $\mathrm{SL}_3(\mathbb{R})$ is unimodular.

For the second claim define a map

$$\psi : K \times AN \longrightarrow \mathrm{SL}_3(\mathbb{R})$$
$$(k, an) \longmapsto k(an)^{-1},$$

and note that the Gram–Schmidt procedure in the proof of Lemma 10.55 shows that ψ is a homeomorphism. Define a measure ν on $K \times AN$ by

$$\nu(B) = m_{\mathrm{SL}_3(\mathbb{R})}(\psi(B))$$

for any measurable set $B \subseteq K \times AN$. Since ψ is a homeomorphism, μ is finite on compact sets and positive on non-empty open sets. Given some $k \in K$ and $an \in AN$ we also have

$$\nu\left((k, an)B\right) = m_{\mathrm{SL}_3(\mathbb{R})}\left(\psi\left((k, an)B\right)\right)$$
$$= m_{\mathrm{SL}_3(\mathbb{R})}\left(k\psi(B)(an)^{-1}\right) = m_{\mathrm{SL}_3(\mathbb{R})}(\psi(B)) = \nu(B).$$

Thus ν is a left Haar measure on $K \times AN$, which by uniqueness of Haar measure means that ν is a scalar multiple of $m_K \times m_{AN}$. Recalling that the inverse map $AN \to AN$ sending an to $(an)^{-1}$ maps the left Haar measure to the right Haar measure, the result follows. $\square$

For the calculation coming up we also need to know the right Haar measure on the subgroup $AN < \mathrm{SL}_3(\mathbb{R})$ explicitly in terms of coordinates.

Lemma 10.58. *Using the coordinates $(a_1, a_2, n_{12}, n_{13}, n_{23}) \in \mathbb{R}^2_{>0} \times \mathbb{R}^3$ corresponding to*

$$\begin{pmatrix} a_1 & 0 & 0 \\ 0 & a_2 & 0 \\ 0 & 0 & a_3 \end{pmatrix} \begin{pmatrix} 1 & n_{12} & n_{13} \\ 0 & 1 & n_{23} \\ 0 & 0 & 1 \end{pmatrix} \in AN \qquad (10.29)$$

with $a_3 = (a_1 a_2)^{-1}$, the right Haar measure $m_{AN}^{(r)}$ is given by

$$\frac{a_1}{a_2} \frac{a_1}{a_3} \frac{a_2}{a_3} \frac{\mathrm{d}a_1}{a_1} \frac{\mathrm{d}a_2}{a_2} \, \mathrm{d}n_{12} \, \mathrm{d}n_{13} \, \mathrm{d}n_{23}. \qquad (10.30)$$

PROOF. Multiplying the matrix in (10.29) on the right by

$$\begin{pmatrix} 1 & m_{12} & m_{13} \\ 0 & 1 & m_{23} \\ 0 & 0 & 1 \end{pmatrix}$$

gives the map (in our chosen coordinates)

$$(a_1, a_2, n_{12}, n_{13}, n_{23}) \longmapsto (a_1, a_2, n_{12}+m_{12}, m_{13}+n_{13}+n_{12}m_{23}, n_{23}+m_{23}),$$

and it is easy to see that this preserves the measure defined by (10.30). Multiplying on the right by

$$\begin{pmatrix} b_1 & 0 & 0 \\ 0 & b_2 & 0 \\ 0 & 0 & b_3 \end{pmatrix}$$

with $b_1, b_2 > 0$ and $b_3 = (b_1 b_2)^{-1}$ we obtain the map

$$(a_1, a_2, n_{12}, n_{13}, n_{23}) \longmapsto \left(a_1 b_1, a_2 b_2, \tfrac{b_2}{b_1} n_{12}, \tfrac{b_3}{b_1} n_{13}, \tfrac{b_3}{b_2} n_{23}\right).$$

Let f be a positive measurable function on $H = \mathbb{R}^2_{>0} \times \mathbb{R}^3$. Then in

$$\int_H f\left(a_1 b_1, a_2 b_2, \tfrac{b_2}{b_1} n_{12}, \tfrac{b_3}{b_1} n_{13}, \tfrac{b_3}{b_2} n_{23}\right) \tfrac{a_1}{a_2} \tfrac{a_1}{a_3} \tfrac{a_2}{a_3} \tfrac{da_1}{a_1} \tfrac{da_2}{a_2}\, dn_{12}\, dn_{13}\, dn_{23}$$

we may substitute $m_{12} = \tfrac{b_2}{b_1} n_{12}$, $m_{13} = \tfrac{b_3}{b_1} n_{13}$, and $m_{23} = \tfrac{b_3}{b_2} n_{23}$ to obtain

$$\int_H f\left(a_1 b_1, a_2 b_2, m_{12}, m_{13}, m_{23}\right) \tfrac{a_1}{a_2} \tfrac{a_1}{a_3} \tfrac{a_2}{a_3} \tfrac{b_1}{b_2} \tfrac{b_1}{b_3} \tfrac{b_2}{b_3} \tfrac{da_1}{a_1} \tfrac{da_2}{a_2}\, dm_{12}\, dm_{13}\, dm_{23}.$$

Using the substitution $c_1 = a_1 b_1$ and $c_2 = a_2 b_2$ (and setting $c_3 = a_3 b_3$) we obtain

$$\int_H f\left(c_1, c_2, m_{12}, m_{13}, m_{23}\right) \tfrac{c_1}{c_2} \tfrac{c_1}{c_3} \tfrac{c_2}{c_3} \tfrac{dc_1}{c_1} \tfrac{dc_2}{c_2}\, dm_{12}\, dm_{13}\, dm_{23}.$$

As AN is generated by these two types of elements, this proves the lemma. $\square$

The proof of Theorem 10.48 is now (essentially) reduced to a calculation.
PROOF OF THEOREM 10.48. Since $\mathrm{SL}_3(\mathbb{R})$ acts on lattices $\Lambda = g\mathbb{Z}^3$ by left multiplication, and the stabilizer of $\mathbb{Z}^3$ under this action is precisely $\mathrm{SL}_3(\mathbb{Z})$, we have the identification

$$\mathrm{SL}_3(\mathbb{R})/\mathrm{SL}_3(\mathbb{Z}) \cong \left\{ g\mathbb{Z}^3 \mid g \in \mathrm{SL}_3(\mathbb{R}) \right\}.$$

Applying Proposition 10.56 we see that there exists some $c > 0$ such that $\mathrm{SL}_3(\mathbb{R}) = B\,\mathrm{SL}_3(\mathbb{Z})$, where $B = KD$ is called a *Siegel set*, $K = \mathrm{SO}_3(\mathbb{R})$ and the Borel measurable set D consists of all matrices in AN as in (10.29) satisfying the conditions

$$0 < a_1 \leqslant ca_2 \leqslant c^2 a_3, \ a_3 = \tfrac{1}{a_1 a_2}, \ \text{and } n_{12}, n_{13}, n_{23} \in [-\tfrac{1}{2}, \tfrac{1}{2}).$$

By Lemma 10.57 we can calculate $m_{\mathrm{SL}_3(\mathbb{R})}(B)$ by calculating $m_{AN}^{(r)}(D)$. Note that the conditions on the diagonal entries a_1, a_2, a_3 imply that $a_1 \in (0, c_1]$ and $a_2 \in [c_2 a_1, c_3 a_1^{-1/2}]$ for some constants $c_1, c_2, c_3 > 0$. By Lemma 10.58

$$m_{AN}^{(r)}(D) \leqslant \int_0^{c_1} \int_{c_2 a_1}^{c_3 a_1^{-1/2}} \frac{a_1}{a_2} \frac{a_1}{(a_1 a_2)^{-1}} \frac{a_2}{(a_1 a_2)^{-1}} \frac{\mathrm{d}a_2}{a_2} \frac{\mathrm{d}a_1}{a_1}$$

$$= \int_0^{c_1} \int_{c_2 a_1}^{c_3 a_1^{-1/2}} a_1^3 a_2 \, \mathrm{d}a_2 \, \mathrm{d}a_1$$

$$= \tfrac{1}{2} \int_0^{c_1} a_1^3 a_2^2 \, \Big|_{c_2 a_1}^{c_3 a_1^{-1/2}} \, \mathrm{d}a_1 = \tfrac{1}{2} \int_0^{c_1} a_1^3 \left(c_3^2 a_1^{-1} - c_2^2 a_1^2 \right) \mathrm{d}a_1 < \infty.$$

To prove the theorem we have to show that there exists a fundamental domain contained in B, that is, a Borel measurable subset $F \subseteq B = KD$ such that $\mathrm{SL}_3(\mathbb{R}) = \bigsqcup_{\gamma \in \mathrm{SL}_3(\mathbb{Z})} F\gamma$.

For this we first prove the following injectivity claim: for any $g_0 \in \mathrm{SL}_3(\mathbb{R})$ there exists an open neighbourhood U of g_0 such that $h, h' \in U$ and $h\gamma = h'$ for some $\gamma \in \mathrm{SL}_3(\mathbb{Z})$ implies $\gamma = I$. Indeed, if this were not true, we could find two sequences $(h_n), (h'_n)$ converging to g_0 as $n \to \infty$ such that

$$h_n^{-1} h'_n = \gamma_n \in \mathrm{SL}_3(\mathbb{Z}) \backslash \{I\}.$$

However, this contradicts the fact that $\mathrm{SL}_3(\mathbb{Z})$ is a discrete subgroup of $\mathrm{SL}_3(\mathbb{R})$.

Next write $\mathrm{SL}_3(\mathbb{R}) = \bigcup_n K_n$ as a countable union of compact subsets (for example, define K_n to be the intersection of $\mathrm{SL}_3(\mathbb{R})$ with closed balls in $\mathbb{R}^9$ of radius $n \geqslant 1$ around 0). For each n choose a finite cover $U_{n,1}, \dots, U_{n,m_n}$ of K_n such that the above injectivity claim holds on each of these sets.

To simplify the notation, let us summarize the above by saying that we have found a countable list of open sets $U_1, U_2, \dots$ satisfying the injectivity claim and covering all of $\mathrm{SL}_3(\mathbb{R})$. We now define $F_1 = B \cap U_1$ and

$$F_2 = (B \cap U_2) \backslash (F_1 \, \mathrm{SL}_3(\mathbb{Z})).$$

Assuming that $F_1, \dots, F_{n-1}$ are already defined, we define

$$F_n = (B \cap U_n) \backslash \big((F_1 \cup \dots \cup F_{n-1}) \, \mathrm{SL}_3(\mathbb{Z}) \big).$$

We claim that the set $F = \bigcup_{n=1}^{\infty} F_n$ is the desired fundamental domain. First note that by construction we have $F \subseteq B$. Moreover, for a given $g \in \mathrm{SL}_3(\mathbb{R})$ we know that $(g \, \mathrm{SL}_3(\mathbb{Z})) \cap B \neq \varnothing$ and hence there exists a minimal $n \in \mathbb{N}$ such that

$$g \, \mathrm{SL}_3(\mathbb{Z}) \cap B \cap U_n \neq \varnothing.$$

By the injectivity property on U_n this intersection then consists of a single element $g\gamma$ for some $\gamma \in \mathrm{SL}_3(\mathbb{Z})$. By minimality of n we see that

$$g\gamma \notin (F_1 \cup \dots \cup F_{n-1}) \, \mathrm{SL}_3(\mathbb{Z}),$$

so that we have $\{g\gamma\} = g\,\mathrm{SL}_3(\mathbb{Z}) \cap F_n$. Finally, by construction it also follows that $g\,\mathrm{SL}_3(\mathbb{Z}) \cap F_m = \varnothing$ for all $m \neq n$, giving $\{g\gamma\} = g\,\mathrm{SL}_3(\mathbb{Z}) \cap F$. Hence F is a fundamental domain and the theorem follows.   □

10.4 Highly Connected Networks: Expanders

In designing large connected networks (for example, connecting many computers and servers) one is often confronted with two competing constraints:

- (High connectivity) Starting from any vertex, it should be easy to reach any other vertex quickly (that is, in few steps).
- (Sparsity) The network should be economical, meaning that there should not be an unnecessarily large number of edges in the network.

Clearly it is easy to achieve the first at the expense of the second by using a complete graph (in which every pair of vertices has an edge joining them), and it is easy to achieve the second at the expense of the first (by arranging the edges so that the vertices are strung along a single line, so as to achieve connectivity at the lowest possible cost).

Exercise 10.59. Analyze the number of edges as a function of the number of vertices in the two extreme constructions of connected networks from above.

Of course there is another option of creating a centre vertex with a direct connection to each of the existing vertices (something that might be called a *hub* for an airline), but the centre vertex created in this way would be very costly (or even technically impossible because of limits to the number of edges at a single vertex) and would defeat the objective of achieving sparsity.

The notion of *expander graphs* is an attempt to achieve a balance between the two constraints. In order to describe expanders, we will need some basic notation from graph theory.

A *graph* $\mathcal{G} = (\mathcal{V}, \mathcal{E})$ is a set of vertices $\mathcal{V}$ (the nodes of the network) and edges $\mathcal{E} \subseteq \mathcal{V} \times \mathcal{V}$ giving the list of direct connections between vertices. We will always assume that the graph is *undirected*, so each edge goes both ways and the set $\mathcal{E}$ is symmetric. In particular, a pair of vertices is at most connected by one edge. We will also assume that the graph is *simple*, meaning that there is never an edge from a vertex to itself. Formally, the set of edges is a subset of $(\mathcal{V} \times \mathcal{V}) \smallsetminus \{(a, a) \mid a \in \mathcal{V}\}$ with the property that $(a, b) \in \mathcal{E}$ if and only if $(b, a) \in \mathcal{E}$. We identify $(a, b) \in \mathcal{E}$ with (b, a) so that $|\mathcal{E}|$ equals the total number of edges in the graph, each of which is viewed as a two-way connection.

The requirement of sparsity is achieved by requiring that the graph $\mathcal{G}$ be k-regular for a fixed k. A graph $\mathcal{G} = (\mathcal{V}, \mathcal{E})$ is said to be k-*regular* if, for any vertex $v \in \mathcal{V}$, there are exactly k edges from v to other vertices in $\mathcal{V}$. We will fix k and look for k-regular graphs with a large number of vertices (it is easy to see that a k-regular graph on n vertices exists if and only if $n \geqslant k+1$ and nk is even). Notice that this will impose a sparsity condition on the graph, since the number of edges $|\mathcal{E}|$ will be a linear function of the number of vertices $|\mathcal{V}|$ (in contrast to the case of a complete graph, for which $|\mathcal{E}| = \frac{1}{2}|\mathcal{V}|\,(|\mathcal{V}| - 1)$).

In order to define the notion of high connectivity, we will need some preparations. A graph $\mathcal{G} = (\mathcal{V}, \mathcal{E})$ is called *connected* if for any two $v, w \in \mathcal{V}$ there exists a path from v to w in that there is a list $v = v_0, v_1, v_2, \ldots, v_n = w$ of vertices in $\mathcal{V}$ with $(v_i, v_{i+1}) \in \mathcal{E}$ for $i = 0, \ldots, n - 1$. Such a path may consist of a single vertex, so each vertex is connected to itself by a path of length zero. Notice that there is a natural metric on any connected graph: we may define $\mathsf{d}(v, w)$ to be the minimal length of a path from v to w (that is, the minimal number of edges in a path joining v to w; see Figure 10.3 and Exercise 10.60). In this metric the *diameter* of a connected graph $\mathcal{G}$ is the minimal $N \in \mathbb{N}$ with the property that for any two vertices v and w there is a path of length no more than N connecting v to w.

Exercise 10.60. Verify that the notion of distance on a graph defines a metric on the set of vertices of a connected graph.

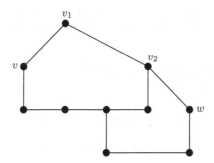

Fig. 10.3: Two points v, w at distance 3 in a connected graph.

The smaller the diameter is in comparison with $\mathcal{V}$, the better the connectivity of the graph is. The worst case with the vertices strung out on a line (or if we seek a 2-regular graph, arranged around a circle) has diameter $|\mathcal{V}| - 1$ (or $\lfloor \frac{|\mathcal{V}|}{2} \rfloor$). The other extreme case of a complete graph has diameter 1. In the case of expander graphs we will see that such families may be found with diameter $N \ll \log |\mathcal{V}|$. The implied constant will depend on k and on ξ (as in Definition 10.61) but is not allowed to depend on the particular graph $\mathcal{G}$. Considering the growth rate of the logarithm, it should be clear that this is a formulation of high connectivity.

Definition 10.61 (Expanders). A sequence of finite k-regular graphs

$$\left(\mathcal{G}_i = (\mathcal{V}_i, \mathcal{E}_i)\right)_{i \geqslant 1}$$

is an *expander family* if there exists a constant $\xi \in (0,1)$ (independent of i) with

$$|\partial S| \geqslant \xi \min\{|S|, |\mathcal{V}_i \smallsetminus S|\}$$

for any subset $S \subseteq \mathcal{V}_i$ and any $i \geqslant 1$, where

$$\partial S = \{v \in S \mid \text{there exists a } w \in \mathcal{V}_i \smallsetminus S \text{ with } (v,w) \in \mathcal{E}\}$$
$$\cup \{v \in \mathcal{V}_i \smallsetminus S \mid \text{there exists a } w \in S \text{ with } (v,w) \in \mathcal{E}\}$$

is called the *boundary* of S.

A few comments are in order. We first note that the above definition of the boundary of a subset of the vertex set of a graph does not coincide with the boundary of S considered in the metric space $\mathcal{V}_i$ (the latter is empty since $\mathcal{V}_i$ is discrete). Any finite collection of finite k-regular connected graphs (formally, a sequence as in Definition 10.61 that repeats these) is an expander family. As this is not at all interesting — and in particular does not achieve the real benefit of the slower growth rate from the logarithmic bound on the diameter — one usually requires in addition that $|\mathcal{V}_i| \to \infty$ as $i \to \infty$. Notice that we must also have $k \geqslant 3$, because $k = 2$ corresponds to a sequence of regular polygons, which we quickly see cannot be an expander family.

An expander family consists of connected graphs, but as already mentioned much more is true.

Proposition 10.62 (Small diameter). *For an expander family* $(\mathcal{G}_i)_i$, *we have* diam $\mathcal{G}_i \ll \log |\mathcal{V}_i|$.

PROOF. Given some vertex $v \in \mathcal{V}_i$ we claim that the metric ball

$$B_a(v) = \{w \in \mathcal{V}_i \mid \mathsf{d}(v,w) \leqslant a\}$$

has more than $\frac{|\mathcal{V}_i|}{2}$ elements if the integer a satisfies

$$a \geqslant D = \frac{\log(|\mathcal{V}_i|/2)}{\log\big(1 + \xi/(k+1)\big)}.$$

Assuming the claim, suppose that $v, w \in \mathcal{V}_i$ are any pair of vertices and set a equal to $\lceil D \rceil$. Then, by the claim, each of $|B_a(v)|$ and $|B_a(w)|$ is greater than $\frac{|\mathcal{V}_i|}{2}$, so that these two balls must have non-empty intersection. By the triangle inequality, it follows that

$$\mathsf{d}(v,w) \leqslant 2(D+1) \ll_{\xi,k} \log |\mathcal{V}_i|,$$

giving the proposition.

To prove the claim, let $n \geqslant 0$, set $S = B_n(v)$ and assume $|S| \leqslant \frac{|\mathcal{V}_i|}{2}$. We then have

$$B_{n+1}(v) \smallsetminus B_n(v) = \partial S \smallsetminus S.$$

Note that every element of $\partial S \cap S$ must connect to one element of $\partial S \smallsetminus S$ and at most k elements of $\partial S \cap S$ can connect to the same element of $\partial S \smallsetminus S$. We can use this to define a map from $\partial S \cap S$ to $\partial S \smallsetminus S$ that is at most k-to-1, showing that $|\partial S \cap S| \leqslant k|\partial S \smallsetminus S|$. This, together with $|\partial S| = |\partial S \cap S| + |\partial S \smallsetminus S|$, gives

$$|\partial S \smallsetminus S| \geqslant \frac{1}{k+1}|\partial S|.$$

Together with the defining property of expander graphs, and assuming as we may that $\xi \in (0,1)$, we deduce that

$$|\partial B_n(v) \smallsetminus B_n(v)| \geqslant \tfrac{\xi}{k+1}|B_n(v)|.$$

By induction we now prove $|B_0(v)| = 1$, $|B_1(v)| = k+1 > 1 + \tfrac{\xi}{k+1}$, and

$$|B_{n+1}(v)| = |B_n(v)| + |\partial B_n(v) \smallsetminus B_n(v)| \geqslant \left(1 + \tfrac{\xi}{k+1}\right)|B_n(v)| > \left(1 + \tfrac{\xi}{k+1}\right)^{n+1}$$

for all n with $|B_n(v)| \leqslant \frac{|\mathcal{V}_i|}{2}$. Since for $n = a \geqslant D$ the lower bound is greater than or equal to $\frac{|\mathcal{V}_i|}{2}$, this proves the claim. $\qquad\square$

Thus expander families achieve a balance between the two constraints of high connectivity (with logarithmic growth of the diameter) and sparsity of the graph (with only linear growth of the number of edges and a fixed number of edges at every vertex). However, several questions remain, the most pressing of which are the following.

- Do expander families exist?
- What is their connection to functional analysis?

The first examples of expander families were found by Pinsker [87] (translated in [88]) using a non-constructive probabilistic argument. The same year Margulis [67] (translation in [68]) was able to give an explicit construction[30] using Každan's Property (T) for the group $SL_3(\mathbb{Z})$.

Towards the proof of this, we now exhibit a connection between the expander property and properties of eigenvalues of linear maps associated to the graphs.

Let $\mathcal{G} = (\mathcal{V}, \mathcal{E})$ be a finite graph and identify $\mathcal{V}$ with the set $\{1, 2, \ldots, |\mathcal{V}|\}$. The *adjacency matrix* $A_\mathcal{G}$ of the graph $\mathcal{G}$ is the matrix with $|\mathcal{V}|$ rows and $|\mathcal{V}|$ columns and with entries in $\{0, 1\}$ so that $(A_\mathcal{G})_{i,j} = 1$ if and only if there is an edge from vertex i to vertex j. A simple graph $\mathcal{G}$ with adjacency matrix

$$A_\mathcal{G} = \begin{pmatrix} 0 & 1 & 0 & 1 & 1 & 0 \\ 1 & 0 & 1 & 0 & 0 & 1 \\ 0 & 1 & 0 & 1 & 0 & 1 \\ 1 & 0 & 1 & 0 & 1 & 0 \\ 1 & 0 & 0 & 1 & 0 & 1 \\ 0 & 1 & 1 & 0 & 1 & 0 \end{pmatrix}$$

is shown in Figure 10.4.

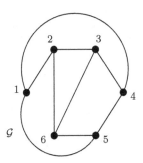

Fig. 10.4: A connected 3-regular graph on 6 vertices.

Several properties of the graph are reflected in the properties of the adjacency matrix. The matrix $A_{\mathcal{G}}$ is symmetric by our standing assumption on the graph $\mathcal{G} = (\mathcal{V}, \mathcal{E})$. We also define $M_{\mathcal{G}} = \frac{1}{k} A_{\mathcal{G}}$, which is an averaging operator in the following sense. A vector $x \in \mathbb{R}^{|\mathcal{V}|}$ may be thought of as a function on the set of vertices, and applying $M_{\mathcal{G}}$ to x gives a new function which at the vertex i is equal to the mean of the values of the function x at all the neighbours of i. By analogy with the discussion in Section 1.2, one also studies the *graph Laplace operator* $\Delta_{\mathcal{G}} = I - M_{\mathcal{G}}$. Since $M_{\mathcal{G}}$ is symmetric, it is diagonalizable and has only real eigenvalues. Moreover,

$$\sum_i |(M_{\mathcal{G}} x)_i| = \sum_i \left| \sum_j (M_{\mathcal{G}})_{i,j}\, x_j \right| \leqslant \sum_{i,j} (M_{\mathcal{G}})_{i,j}\, |x_j| = \sum_j |x_j|,$$

since

$$\sum_i (M_{\mathcal{G}})_{i,j} = 1 \tag{10.31}$$

for all j by construction. Therefore, any eigenvalue λ on $M_{\mathcal{G}}$ has $|\lambda| \leqslant 1$ and by (10.31) we see that $\lambda_1 = 1$ is an eigenvalue (with the constant vectors as eigenvectors). The relationship between the eigenvalues and connectivity is illustrated by the following elementary lemma.

Lemma 10.63 (Connectivity). *A k-regular graph is connected if and only if 1 is a simple eigenvalue of $M_{\mathcal{G}}$.*

Essential Exercise 10.64. Prove Lemma 10.63.

What we need next is a quantitative version of this relationship, which is given by the following proposition.

Proposition 10.65 (Eigenvalues and expanders). *Let $(\mathcal{G}_i = (\mathcal{V}_i, \mathcal{E}_i))_{i \geqslant 1}$ be a sequence of graphs. For each i, let $M_i = M_{\mathcal{G}_i}$ be the averaging operator* for $\mathcal{G}_i$, and order its eigenvalues $\lambda_1(M_i) = 1 \geqslant \lambda_2(M_i) \geqslant \cdots \geqslant \lambda_{|\mathcal{V}_i|}(M_i)$. Suppose that there exists some $\varepsilon > 0$ with

$$\lambda_2(M_i) \leqslant 1 - \varepsilon \tag{10.32}$$

for all $i \geqslant 1$. Then the sequence of graphs is an expander family.

The uniform estimate in (10.32) is called a *spectral gap* for the sequence of graphs. The converse of Proposition 10.65 also holds, but we will not need this direction (we refer to Lubotzky [66] for the proof).

PROOF OF PROPOSITION 10.65. Let $\varepsilon > 0$ be as in the statement of the proposition. Let $\mathcal{G} = \mathcal{G}_i$ and $M = M_i$ for some fixed i, so that $\lambda_2(M) \leqslant 1 - \varepsilon$. Also let $S \subseteq \mathcal{V}$ be any subset with $|S| \leqslant \frac{|\mathcal{V}|}{2}$. We again think of vectors in $\mathbb{R}^{|\mathcal{V}|}$ as functions on $\mathcal{V}$, and notice that $M(\mathbb{1}_S) = \mathbb{1}_S + f_{\partial S}$, where $f_{\partial S}$ is a vector that vanishes outside of ∂S and has absolute value less than or equal to 1 on the elements of ∂S. We will estimate $\|f_{\partial S}\|_2$ from above and below, and the resulting estimate will prove the claim. In fact, it follows that

$$\|M(\mathbb{1}_S) - \mathbb{1}_S\|_2 = \|f_{\partial S}\|_2 \leqslant \sqrt{|\partial S|}.$$

On the other hand, M is diagonalizable and so we may expand $\mathbb{1}_S$ into a sum of eigenvectors

$$\mathbb{1}_S = \sum_j v_j, \tag{10.33}$$

corresponding to the eigenvalues[†] $\lambda_1 = 1 > \lambda_2 \geqslant \cdots \geqslant \lambda_{|\mathcal{V}|}$. This then gives

$$M(\mathbb{1}_S) = \sum_j \lambda_j v_j,$$

and finally

$$M(\mathbb{1}_S) - \mathbb{1}_S = \sum_{j=2}^{|\mathcal{V}|} (\lambda_j - 1) v_j.$$

Furthermore, since M is symmetric we can assume that the vectors v_j are orthogonal to each other, so

$$\|M(\mathbb{1}_S) - \mathbb{1}_S\|_2 \geqslant \min_{2 \leqslant j \leqslant |\mathcal{V}|} |\lambda_j - 1| \sqrt{\sum_{j=2}^{|\mathcal{V}|} \|v_j\|_2^2} \geqslant \varepsilon \left\| \sum_{j=2}^{|\mathcal{V}|} v_j \right\|_2.$$

Thus we need to relate the last norm to the size of S. To this end, notice that a constant $\mathbb{1}$ is an eigenvector for the eigenvalue $\lambda_1 = 1$, $\|\mathbb{1}\|_2 = \sqrt{|\mathcal{V}|}$, and $\langle \mathbb{1}_S, \mathbb{1} \rangle = |S|$, so the orthogonal projection of $\mathbb{1}_S$ onto $\mathbb{1}$ is $\frac{|S|}{|\mathcal{V}|} \mathbb{1}$. Therefore, as in (10.33), we may subtract from $\mathbb{1}_S$ the vector $v_1 = \frac{|S|}{|\mathcal{V}|} \mathbb{1}$ and obtain

---

[†] If $\lambda_j = \lambda_{j+1}$ for some $j \in \{2, \ldots, |\mathcal{V}| - 1\}$ we may and will assume $v_{j+1} = 0$.

$$\left\| \sum_{j=2}^{|\mathcal{V}|} v_j \right\|_2 = \left\| \mathbb{1}_S - \tfrac{|S|}{|\mathcal{V}|} \mathbb{1} \right\|_2$$

$$\geqslant \left(1 - \tfrac{|S|}{|\mathcal{V}|} \right) \|\mathbb{1}_S\|_2 \qquad \text{(by restricting the sum to } S\text{)}$$

$$\geqslant \tfrac{1}{2} \sqrt{|S|} \qquad\qquad\qquad \text{(since } \tfrac{|S|}{|\mathcal{V}|} \leqslant \tfrac{1}{2}\text{)}$$

and putting these inequalities together gives

$$\sqrt{|\partial S|} \geqslant \|M(\mathbb{1}_S) - \mathbb{1}_S\|_2 \geqslant \varepsilon \left\| \sum_{j=2}^{|\mathcal{V}|} v_j \right\|_2 \geqslant \tfrac{\varepsilon}{2} \sqrt{|S|}.$$

As this holds for all subsets $S \subseteq \mathcal{V}$ with $|S| \leqslant \tfrac{|\mathcal{V}|}{2}$ and all graphs in the sequence $(\mathcal{G}_i)_{i \geqslant 1}$, we see that this is an expander family with $\xi = \tfrac{\varepsilon^2}{4}$. $\qquad\square$

10.4.1 Constructing an Explicit Expander Family

Corollary 10.66 (Explicit expander family). *Let $S = S^{-1}$ be a finite symmetric set of generators (not containing the identity) of $\Gamma = \mathrm{SL}_3(\mathbb{Z})$. Let $(\mathcal{V}_n)$ be a sequence of finite sets on each of which Γ acts transitively, with the property that the elements of $S^2 \backslash \{I\}$ have no fixed points for the action. We define the sequence of graphs $(\mathcal{G}_n)$ by $\mathcal{G}_n = (\mathcal{V}_n, \mathcal{E}_n)$ for all $n \geqslant 1$, where vertices v, w in $\mathcal{V}_n$ are connected in $\mathcal{G}_n$ if $s{\cdot}v = w$ for some $s \in S$. Then $(\mathcal{G}_n)$ is a sequence of $|S|$-regular graphs that form an expander family.*

We note that the generating set S, for example, could be taken to comprise the 12 matrices given by

$$\begin{pmatrix} 1 & \pm 1 & 0 \\ 0 & 1 & 0 \\ 0 & 0 & 1 \end{pmatrix}$$

together with all its conjugates by permutation matrices (a permutation matrix is one obtained by permuting the rows or the columns of the identity matrix). Furthermore, the sequence of sets with the transitive actions could, for example, be $\mathcal{V}_n = \mathrm{SL}_3(\mathbb{Z}/p_n\mathbb{Z})$, where p_n denotes the nth odd prime number.

In this section we will prove Corollary 10.66. By Corollary 10.40 we know that $\Gamma = \mathrm{SL}_3(\mathbb{Z})$ has property (T). In the following argument we let $S = S^{-1}$ be a finite symmetric set of generators of Γ (not containing the identity). Such a set exists by Exercise 10.37 (or Exercise 10.67 below).

Essential Exercise 10.67. Prove that $\mathrm{SL}_3(\mathbb{Z})$ is generated by the 12 elements given just after Corollary 10.66. (Note, however, that the argument after Proposition 10.41 does not apply directly since $\mathbb{Z}$ is not a field.)

PROOF OF COROLLARY 10.66. By Lemma 10.36 all unitary representations of Γ have uniform spectral gap, meaning that there exist a finite $Q_0 \subseteq \Gamma$ and $\varepsilon_0 > 0$ such that for any unitary representation $\pi : \Gamma \curvearrowright \mathcal{H}$ there are no (Q_0, ε_0)-almost invariant vectors in $\left(\mathcal{H}^\Gamma\right)^\perp$. Notice that $Q_0 \subseteq S^k$ for some $k \geqslant 1$ and that a $(S, \frac{\varepsilon_0}{k})$-almost invariant vector is also (Q_0, ε_0)-almost invariant. In other words, we may assume that $Q_0 = S$ and that for any unitary representation $\pi : \Gamma \curvearrowright \mathcal{H}$ there are no (S, ε_0)-almost invariant vectors in $\left(\mathcal{H}^\Gamma\right)^\perp$.

Suppose now that $(\mathcal{V}_n)$ is a sequence of finite sets with the property that for every n the group Γ acts transitively on $\mathcal{V}_n$ and there are no elements of $\mathcal{V}_n$ that are fixed by $S^2 \backslash \{I\}$.

We note that $\mathcal{V} = \mathrm{SL}_3(\mathbb{Z}/p\mathbb{Z})$ for any odd prime p satisfies these assumptions. In fact, $\mathcal{V}$ is itself a group, and we may use the reduction modulo p map $\phi : \Gamma = \mathrm{SL}_3(\mathbb{Z}) \to \mathrm{SL}_3(\mathbb{Z}/p\mathbb{Z})$ to define the action of Γ by $\gamma \cdot v = \phi(\gamma)v$ for all $v \in \mathcal{V} = \mathrm{SL}_3(\mathbb{Z}/p\mathbb{Z})$. Transitivity of this action follows since the subgroup $\phi(\mathrm{SL}_3(\mathbb{Z}))$ contains all elementary unipotent subgroups of $\mathrm{SL}_3(\mathbb{Z}/p\mathbb{Z})$ and since these generate $\mathrm{SL}_3(\mathbb{Z}/p\mathbb{Z})$ (by the argument after the statement of Proposition 10.41 applied to the field $K = \mathbb{F}_p = \mathbb{Z}/p\mathbb{Z}$).

We return to the general case and fix some $n \geqslant 1$. Since Γ acts on the finite set $\mathcal{V}_n$ we also obtain the unitary representation

$$\pi_n : \Gamma \curvearrowright \mathcal{H}_n = \ell^2(\mathcal{V}_n) = \mathbb{C}^{\mathcal{V}_n}$$

defined by $\pi_{n,\gamma} f(v) = f(\gamma^{-1} \cdot v)$ for all $\gamma \in \Gamma$ and $f \in \mathcal{H}_n$. By transitivity of Γ a Γ-invariant function in $\mathcal{H}_n$ must be constant, that is $\mathcal{H}_n^\Gamma = \mathbb{C}\mathbb{1}$. Suppose now that $f \in \left(\mathcal{H}_n^\Gamma\right)^\perp$ is a unit vector. By the uniform spectral gap property above, there exists some $\gamma \in S$ such that

$$\|\pi_{n,\gamma} f - f\| > \varepsilon_0. \tag{10.34}$$

We now show that this uniform claim implies that the sequence of graphs $(\mathcal{G}_n)$ defined by $\mathcal{G}_n = (\mathcal{V}_n, \mathcal{E}_n)$ as in Corollary 10.66 is an expander family by using Proposition 10.65. For this, let $\varepsilon > 0$ and suppose in addition that $f \in \left(\mathcal{H}_n^\Gamma\right)^\perp$ is an eigenvector for the averaging operator $M_n = M_{\mathcal{G}_n}$ associated to the graph $\mathcal{G}_n$ and eigenvalue $\lambda_2(M_n) \geqslant 1 - \varepsilon$. By definition of the graph structure and the averaging operator we have

$$M_n(f) = \frac{1}{|S|} \sum_{\gamma \in S} \pi_{n,\gamma} f,$$

and so

$$1 - \varepsilon \leqslant \lambda_2(M_n) \langle f, f \rangle = \Re \langle M_n(f), f \rangle = \frac{1}{|S|} \sum_{\gamma \in S} \Re \langle \pi_{n,\gamma} f, f \rangle.$$

Fix some $\gamma \in S$. By using in addition that $\Re \langle \pi_{n,\gamma'} f, f \rangle \leqslant \|f\|^2 = 1$ for all $\gamma' \in S \smallsetminus \{\gamma\}$, we deduce from this that $1 - \varepsilon \leqslant \frac{1}{|S|} \left(|S| - 1 + \Re \langle \pi_{n,\gamma} f, f \rangle \right)$ and therefore $1 - |S|\varepsilon \leqslant \Re \langle \pi_{n,\gamma} f, f \rangle$. However, this shows that

$$\|\pi_{n,\gamma} f - f\|^2 = \|\pi_{n,\gamma} f\|^2 - 2\Re \langle \pi_{n,\gamma}(f), f \rangle + \|f\|^2 \leqslant 2 - 2\left(1 - |S|\varepsilon\right) = 2|S|\varepsilon$$

for all $\gamma \in S$. Setting $\varepsilon = \frac{\varepsilon_0^2}{2|S|}$, this contradicts (10.34). In other words, using this ε we have $\lambda_2(M_n) < 1 - \varepsilon$ for every n, which shows that the assumptions in Proposition 10.65 are satisfied, and so Corollary 10.66 follows. □

10.5 Further Topics

- The above concludes our main discussion of non-abelian topological groups. Abelian groups will appear in Section 11.4 and Section 12.8, which together give, among other things, a complete classification of unitary representations for locally compact abelian groups. We refer to Folland [32] and [26] for more on the general theory of unitary representations of locally compact groups.

- Homogeneous spaces are quotients of the form G/Γ where $SL_3(\mathbb{R})/SL_3(\mathbb{R})$ is a very important example. These spaces have many interesting and important connections to geometry (see Helgason [44] and Ratcliffe [90]), ergodic theory and dynamical systems [27, Ch. 9–11] and [25], algebraic groups (see, for example, the monograph of Witte Morris [116]), and number theory (an introduction to this large area of interaction may be found in the monographs of Diamond and Shurman [22] and Serre [97]).

- Amenable groups play an important role in ergodic theory, see [27, Ch. 8].

- For further reading on property (T) we refer the reader to the monograph of Bekka, de la Harpe and Valette [6], to [26], and for an account of the special role played by property (T) in ergodic theory to the work of Gorodnik and Nevo [41].

Chapter 11
Banach Algebras and the Spectrum

In this chapter we will study Banach algebras as introduced in Section 2.4.2. For most of the discussion we will work over $\mathbb{C}$ and assume that the Banach algebra $\mathcal{A}$ is *unital*, meaning that there is a multiplicative unit $1_{\mathcal{A}}$. A multiplicative unit $1_{\mathcal{A}} \in \mathcal{A}$ is an element with $1_{\mathcal{A}}a = a1_{\mathcal{A}} = a$ for all $a \in \mathcal{A}$. We assume that $1_{\mathcal{A}} \neq 0$, or equivalently that $\mathcal{A} \neq \{0\}$ (in the literature one sometimes also sees the assumption $\|1_{\mathcal{A}}\| = 1$, which will hold in all the examples that we will consider). At first sight it seems as if the assumption that $\mathcal{A}$ is unital excludes the important example $(L^1(\mathbb{R}^d), +, *)$, but this may be overcome by the simple construction in the following exercise.

Essential Exercise 11.1 (Adding a unit). Let $\mathcal{A}$ be a complex Banach algebra. Define the algebra $\mathcal{A}_1 = \mathcal{A} \oplus \mathbb{C}$ with the convention that we write the elements of $\mathcal{A}_1$ in the form $a + \lambda I$ with $a \in \mathcal{A}$ and $\lambda \in \mathbb{C}$, use the norm $\|a + \lambda I\| = \|a\|_{\mathcal{A}} + |\lambda|$, the obvious linear structure as a vector space over $\mathbb{C}$, and the multiplication $(a + \lambda I)(b + \mu I) = (ab + \lambda b + \mu a) + \lambda \mu I$. Show that with these definitions $\mathcal{A}_1$ is a unital Banach algebra with $1_{\mathcal{A}_1} = I$ being its multiplicative unit.

11.1 The Spectrum and Spectral Radius

We say that an element a of a unital Banach algebra $\mathcal{A}$ is *invertible* if there exists some $b \in \mathcal{A}$ called the *inverse of* a with $ab = ba = 1_{\mathcal{A}}$.

Definition 11.2. Let $\mathcal{A}$ be a unital Banach algebra over $\mathbb{C}$. The *spectrum* of an element $a \in \mathcal{A}$ is the set $\sigma(a) = \{\lambda \in \mathbb{C} \mid a - \lambda 1_{\mathcal{A}} \text{ is not invertible}\}$. The *resolvent set* is its complement $\rho(a) = \{\lambda \in \mathbb{C} \mid a - \lambda 1_{\mathcal{A}} \text{ is invertible}\}$.

Let us note that the above generalizes the notion of an eigenvalue in the following way: If $\mathcal{A}$ is the algebra of linear maps on $\mathbb{C}^d$, then the spectrum of an element $T \in \mathcal{A}$ equals the set of eigenvalues of T.

© Springer International Publishing AG 2017

M. Einsiedler, T. Ward, *Functional Analysis, Spectral Theory, and Applications*, Graduate Texts in Mathematics 276, DOI 10.1007/978-3-319-58540-6_11

Exercise 11.3. Let X be a compact topological space, and let $\mathcal{A} = C(X)$. Find the spectrum of $f \in C(X)$ as an element of the Banach algebra $C(X)$.

Essential Exercise 11.4. In this exercise we describe the spectrum of multiplication operators $M_g : \mathcal{H} \to \mathcal{H}$ as in Exercise 6.25.
(a) Let μ be a compactly supported finite (or σ-finite) measure on $\mathbb{C}$, and write M_I for the multiplication operator corresponding to the identity map, so $(M_I(f))(z) = zf(z)$ for $f \in L^2_\mu(\mathbb{C})$. Show that the spectrum $\sigma(M_I)$ within the algebra of bounded operators $B(L^2_\mu(\mathbb{C}))$ equals the support of μ.
(b) Let $(X, \mathcal{B}, \mu)$ be a σ-finite measure space, $\mathcal{H} = L^2_\mu(X)$ and $g : X \to \mathbb{C}$ a bounded measurable function. Show that the spectrum of the multiplication operator M_g within the algebra of bounded operators coincides with the essential range of g, which is defined to consist of all $\lambda \in \mathbb{C}$ with the property that $\mu(g^{-1}(U)) > 0$ for all neighbourhoods U of λ.

The following theorem will show that the spectrum is always non-empty, and so provides us with generalized eigenvalues. Since even in finite dimensions eigenvalues may be complex, we will only consider Banach algebras over $\mathbb{C}$.

Definition 11.5. For an element a of a complex unital Banach algebra the *spectral radius* is $\max_{\lambda \in \sigma(a)} |\lambda|$.

Theorem 11.6 (Spectrum and spectral radius formula). *Let $\mathcal{A}$ be a complex unital Banach algebra. Then for every $a \in \mathcal{A}$ the spectrum $\sigma(a)$ is a non-empty compact subset of $\mathbb{C}$. Moreover, the spectral radius satisfies*

$$\max_{\lambda \in \sigma(a)} |\lambda| = \lim_{n \to \infty} \sqrt[n]{\|a^n\|}. \tag{11.1}$$

This theorem is the first of many that relate the algebraic to the topological structure in Banach algebras. The spectrum and the spectral radius of an element are defined in purely algebraic terms, whereas the limit is defined in terms of the norm. One surprising consequence is the following observation: If $\mathcal{A}$ is a unital Banach algebra contained in a larger Banach algebra $\mathcal{B}$ (with compatible structures), then it is possible for an element $a \in \mathcal{A}$ to be non-invertible in $\mathcal{A}$ but to be invertible in $\mathcal{B}$. Thus the spectrum of an element depends on the algebra it is viewed in, and $\sigma_{\mathcal{B}}(a) \subseteq \sigma_{\mathcal{A}}(a)$ with strict containment being a possibility (see Exercise 11.7). Despite this, the spectral radius of $a \in \mathcal{A}$ is not changed when it is viewed as an element of $\mathcal{B}$, since Theorem 11.6 expresses it in terms of the norms of powers of a, which are not affected by the switch from $\mathcal{A}$ to $\mathcal{B}$ (by the implicit compatibility assumption).

Exercise 11.7. Let $U : \ell^2(\mathbb{Z}) \to \ell^2(\mathbb{Z})$ be the unitary shift operator from Exercises 6.1 and 6.23(a), so $U((x_n)) = (x_{n+1})$.
(a) Show that the spectrum of U considered within the algebra $\mathcal{B}$ of all bounded operators on $\ell^2(\mathbb{Z})$ is given by $\mathbb{S}^1 = \{\lambda \in \mathbb{C} \mid |\lambda| = 1\}$.
(b) Now consider the Banach algebra $\mathcal{A}$ generated by U (obtained by taking the closed linear hull of $U^0 = I, U, U^2, \ldots$). Show that the spectrum of U within $\mathcal{A}$ is $\{\lambda \in \mathbb{C} \mid |\lambda| \leq 1\}$.

The above mentioned dependence of the spectrum of an element on the ambient algebra will not cause any confusion: In the abstract setting considered here we will only work with one algebra at a time, and in the application of these results in the context of operators on a Hilbert space $\mathcal{H}$ we will always consider the algebra $B(\mathcal{H})$ of all bounded linear operators on $\mathcal{H}$.

Exercise 11.8 (There are no interesting Banach fields). Use Theorem 11.6 to show that $\mathbb{C}$ is the only Banach algebra over $\mathbb{C}$ that is also a field.

Finally, let us comment on the precise shape of the spectral radius formula (11.1). It will be relatively straightforward to show that scalars $\lambda \in \mathbb{C}$ with $|\lambda| > \|a\|$ cannot belong to the spectrum of $a \in \mathcal{A}$. However, it is also clear that in general the norm may be much larger than the spectral radius. Self-adjoint and, more generally, normal operators on Hilbert spaces will form a nice exception to this. In fact, even in the elementary case of the algebra of two-by-two matrices (equipped with the operator norm) the norm of the matrix

$$a = \begin{pmatrix} 1 & C \\ 0 & 1 \end{pmatrix}$$

can be made arbitrarily large by increasing the value of C, but the spectrum always consists simply of $1 \in \mathbb{C}$. The right-hand side of the spectral radius formula (11.1) essentially ignores the original size of the matrix a and instead looks at the exponential growth rate of the norm of a^n. In the case at hand the norm of a^n grows linearly which makes the right-hand side equal to one (and thus equal to the left-hand side).

Exercise 11.9. Let $k \in C([0,1]^2)$ be a continuous function, so that

$$K(f)(x) = \int_0^x k(x,t)f(t)\,\mathrm{d}t$$

for $f \in C([0,1])$ defines an operator $K : C([0,1]) \to C([0,1])$. Determine $\sigma(K)$.

For the proof of Theorem 11.6 we will use Cauchy integration on the complex plane and convergent geometric series in the unital Banach algebra.

11.1.1 The Geometric Series and its Consequences

Given a unital Banach algebra $\mathcal{A}$ we will set $a^0 = 1_{\mathcal{A}}$ for any $a \in \mathcal{A}$ as is customary. Also recall that the inverse of an invertible element $a \in \mathcal{A}$ is uniquely determined by a.

Proposition 11.10. *The set $\mathcal{U}$ of invertible elements of a unital Banach algebra $\mathcal{A}$ is open. Moreover, for any $a \in \mathcal{A}$ the resolvent set $\rho(a)$ is open in $\mathbb{C}$, so the spectrum is a closed set.*

PROOF. If $a \in \mathcal{A}$ and $\|a\| < 1$, then the inverse of $1_{\mathcal{A}} - a$ is given by the geometric series

$$(1_{\mathcal{A}} - a)^{-1} = \sum_{n=0}^{\infty} a^n. \qquad (11.2)$$

Indeed, since the right-hand side converges absolutely we may take the product and obtain

$$(1_{\mathcal{A}} - a) \left(\sum_{n=0}^{\infty} a^n \right) = \left(\sum_{n=0}^{\infty} a^n \right) (1_{\mathcal{A}} - a) = \sum_{n=0}^{\infty} a^n - \sum_{n=1}^{\infty} a^n = 1_{\mathcal{A}}$$

as desired. This shows that $B_1(1_{\mathcal{A}}) \subseteq \mathcal{U}$.

Now let $a_0 \in \mathcal{U}$ be any invertible element with $\|a - a_0\| < \|a_0^{-1}\|^{-1}$. Then we claim that a is also invertible, which will then show that $\mathcal{U} \subseteq \mathcal{A}$ is open. To prove the claim, notice that

$$a = a_0 + (a - a_0) = a_0 \underbrace{\left(1_{\mathcal{A}} + a_0^{-1}(a - a_0) \right)}_{\in B_1(1_{\mathcal{A}})}$$

is a product of two elements of $\mathcal{U}$ and so lies in $\mathcal{U}$.

Finally, for any $a \in \mathcal{A}$ the resolvent set $\rho(a) = \{\lambda \in \mathbb{C} \mid a - \lambda 1_{\mathcal{A}} \in \mathcal{U}\}$ is the pre-image of an open set under a continuous mapping, and so is open. Therefore the spectrum $\sigma(a) = \mathbb{C} \setminus \rho(a)$ is closed. $\qquad \square$

Proposition 11.11. *Let $\mathcal{A}$ be a unital Banach algebra over $\mathbb{C}$, and let $a \in \mathcal{A}$. If $\lambda \in \mathbb{C}$ has $|\lambda| > \sqrt[m]{\|a^m\|}$ for some $m \geq 1$, then $\lambda \in \rho(a)$. In particular, the spectrum $\sigma(a)$ is a closed subset of $\overline{B_{\|a\|}^{\mathbb{C}}}$ and so is compact.*

PROOF. Let $a \in \mathcal{A}$ and $\lambda \in \mathbb{C}$ be as in the proposition. Then we claim that

$$(\lambda 1_{\mathcal{A}} - a)^{-1} = \lambda^{-1} \left(1_{\mathcal{A}} + \lambda^{-1} a + \cdots + \lambda^{-(m-1)} a^{m-1} \right) \sum_{n=0}^{\infty} \lambda^{-mn} a^{mn}$$

(here and below we will sometimes study $\lambda 1_{\mathcal{A}} - a$ instead of $a - \lambda 1_{\mathcal{A}}$, which clearly will not make any difference). For this notice first that by assumption

$$\sum_{n=0}^{\infty} \|\lambda^{-mn} a^{mn}\| \leq \sum_{n=0}^{\infty} \left(\underbrace{|\lambda|^{-m} \|a^m\|}_{<1} \right)^n < \infty,$$

so the series $\sum_{n=0}^{\infty} \lambda^{-mn} a^{mn}$ converges absolutely. Moreover, by combining the first three factors in the product

$$(\lambda 1_{\mathcal{A}} - a) \lambda^{-1} \left(1_{\mathcal{A}} + \lambda^{-1} a + \cdots + \lambda^{-(m-1)} a^{m-1} \right) \sum_{n=0}^{\infty} \lambda^{-mn} a^{mn}$$

we obtain that it equals

$$\left(1_{\mathcal{A}} - \lambda^{-m} a^m\right) \sum_{n=0}^{\infty} \lambda^{-mn} a^{mn} = 1_{\mathcal{A}}.$$

by (11.2). Noting that the factors commute with each other (which follows easily from continuity of multiplication in the algebra), this proves the claim, and so $\lambda \in \rho(a)$.

The case $m = 1$ gives the remaining statement about the spectrum. $\quad\square$

Exercise 11.12. Assume that (a_n) and (b_n) are sequences in a Banach algebra satisfying $a_n b_n = b_n a_n$ for all $n \geqslant 1$ and with $\lim_{n \to \infty} a_n = a$, $\lim_{n \to \infty} b_n = b$. Show that $ab = ba$.

11.1.2 Using Cauchy Integration

We have shown that $\sigma(a) \subseteq \mathbb{C}$ is compact for any element $a \in \mathcal{A}$, but have yet to show that $\sigma(a)$ is non-empty. This existence theorem uses Cauchy integration, and to prepare for this we need the following lemma concerning the resolvent.

Lemma 11.13 (Resolvent function). *Let a be an element of a unital Banach algebra $\mathcal{A}$ over $\mathbb{C}$. Then the resolvent function $R : \rho(a) \to \mathcal{A}$ defined by $R(\lambda) = (\lambda 1_{\mathcal{A}} - a)^{-1}$ is an analytic function in the sense that for any $\lambda_0 \in \rho(a)$ there is an open neighbourhood of λ_0 on which R is given by an absolutely convergent power series*

$$R(\lambda) = \sum_{n=0}^{\infty} b_n (\lambda - \lambda_0)^n$$

with coefficients $b_n \in \mathcal{A}$.

PROOF. We use essentially the same formulas as those that arise in the proof of Proposition 11.10. Let $a \in \mathcal{A}$ and $\lambda_0 \in \rho(a)$ be as in the lemma. Suppose that $\lambda \in \mathbb{C}$ satisfies $|\lambda - \lambda_0| < \|(\lambda_0 1_{\mathcal{A}} - a)^{-1}\|^{-1}$. Then

$$\begin{aligned}
\lambda 1_{\mathcal{A}} - a &= (\lambda_0 1_{\mathcal{A}} - a) - (\lambda_0 - \lambda) 1_{\mathcal{A}} \\
&= (\lambda_0 1_{\mathcal{A}} - a)\left(1_{\mathcal{A}} - (\lambda_0 - \lambda)(\lambda_0 1_{\mathcal{A}} - a)^{-1}\right),
\end{aligned}$$

which shows that

$$R(\lambda) = (\lambda_0 1_{\mathcal{A}} - a)^{-1} \sum_{n=0}^{\infty} \left((\lambda_0 1_{\mathcal{A}} - a)^{-1}\right)^n (-1)^n (\lambda - \lambda_0)^n$$

is, for $|\lambda - \lambda_0| < \|(\lambda_0 1_{\mathcal{A}} - a)^{-1}\|^{-1}$, an absolutely convergent power series, as claimed. $\quad\square$

With this analyticity we are ready to prove the first part of Theorem 11.6.

PROOF THAT $\sigma(a)$ IS NON-EMPTY. Let a be an element of a unital Banach algebra, and suppose that $\sigma(a)$ is empty. We first sketch an argument that produces a contradiction from this assumption, and then fill in the details.

AN ENTIRE FUNCTION. Since $\sigma(a)$ is empty, the resolvent function

$$R(\lambda) = (\lambda I - a)^{-1}$$

is an *entire function* (that is, is an analytic function defined on all of $\mathbb{C}$). It follows by Cauchy's integral formula that

$$\oint_\gamma R(z)\,\mathrm{d}z = 0$$

for any closed piecewise differentiable path γ in $\mathbb{C}$. The alert reader may notice that this usage of Cauchy integration is a bit unorthodox, but should read on — this will be resolved below. In particular, if γ is the closed positively oriented path with centre 0 and radius $\|a\| + 1$, then

$$R(z) = (z1_{\mathcal{A}} - a)^{-1} = \tfrac{1}{z}\left(1_{\mathcal{A}} - \tfrac{1}{z}a\right)^{-1} = \sum_{n=0}^{\infty} z^{-n-1}a^n \tag{11.3}$$

for any z on the path γ, and the sum is absolutely convergent. Therefore,

$$
\begin{aligned}
0 = \oint_\gamma R(z)\,\mathrm{d}z &= \oint_{|z|=\|a\|+1} \sum_{n=0}^{\infty} z^{-n-1}a^n\,\mathrm{d}z \\
&= \sum_{n=0}^{\infty} a^n \oint_{|z|=\|a\|+1} z^{-n-1}\,\mathrm{d}z = 2\pi\mathrm{i}1_{\mathcal{A}},
\end{aligned}
\tag{11.4}
$$

since

$$\oint_{|z|=\|a\|+1} z^{-n-1}\,\mathrm{d}z = \begin{cases} 2\pi\mathrm{i} & \text{if } n = 0, \\ 0 & \text{if } n \neq 0. \end{cases}$$

Now $1_{\mathcal{A}} \neq 0$, and so (11.4) shows that the assumption $\sigma(a) = \varnothing$ leads to a contradiction.

USING THE STANDARD CAUCHY INTEGRAL FORMULA. The difficulty with the argument sketched above is that most of the integrals are integrals of $\mathcal{A}$-valued functions. Even though it is possible to make sense of integration for $\mathcal{A}$-valued functions (see Proposition 3.81), we do not need to extend the Cauchy integral formula for $\mathcal{A}$-valued functions because of the following argument (which could be used to prove such an extension).

Let $\ell \in \mathcal{A}^*$ be a linear functional with $\ell(1_{\mathcal{A}}) \neq 0$ (such a functional is guaranteed to exist by Theorem 7.3) and consider $\ell \circ R : \rho(a) = \mathbb{C} \longrightarrow \mathbb{C}$. By Lemma 11.13, $R(z)$ can locally be represented as a power series. By continuity of ℓ, the same holds for $\ell \circ R$. It follows that $\ell \circ R : \mathbb{C} \to \mathbb{C}$ is an entire function (in the usual sense of complex analysis). Using this entire function in the calculation in (11.4) we see that

$$0 = \oint_{|z|=\|a\|+1} \ell \circ R(z) \, dz = \sum_{n=0}^{\infty} \ell(a^n) \oint_{|z|=\|a\|+1} z^{-n-1} \, dz = 2\pi i \ell(1_A) \neq 0.$$

It follows that $\ell \circ R$ cannot be defined on all of $\mathbb{C}$, and so $\sigma(a)$ is non-empty.
$\square$

For the spectral radius formula in Theorem 11.6 we also need the following elementary property of sub-additive and sub-multiplicative real-valued sequences.

Definition 11.14. A real sequence (α_n) is *sub-additive* if $\alpha_{m+n} \leqslant \alpha_m + \alpha_n$ for all $m, n \geqslant 1$, and is *sub-multiplicative* if $\alpha_n \geqslant 0$ and $\alpha_{m+n} \leqslant \alpha_m \alpha_n$ for all $m, n \geqslant 1$.

Lemma 11.15 (Fekete's lemma). *Let (α_n) be a real sequence.*

(1) *If (α_n) is sub-additive then*

$$\lim_{n \to \infty} \frac{\alpha_n}{n} = \inf_{n \geqslant 1} \frac{\alpha_n}{n}.$$

(2) *If (α_n) is non-negative and sub-multiplicative then*

$$\lim_{n \to \infty} \sqrt[n]{\alpha_n} = \inf_{n \geqslant 1} \sqrt[n]{\alpha_n}.$$

PROOF. Suppose first $\alpha_n \geqslant 0$ is sub-multiplicative. If $\alpha_{n_0} = 0$ for some n_0, then $\alpha_n = 0$ for all $n \geqslant n_0$ and in this case the claim is trivial. On the other hand, if all α_n are stricly positive the statement follows from (1) applied to the sequence $(\log \alpha_n)$.

So consider now a real-valued sub-additive sequence (α_n) and let

$$\alpha = \inf_{n \in \mathbb{N}} \frac{\alpha_n}{n},$$

so that $\frac{\alpha_n}{n} \geqslant \alpha$ for all $n \geqslant 1$. Let $\beta > \alpha$ be arbitrary and pick $k \geqslant 1$ such that $\frac{\alpha_k}{k} < \beta$. For any $n \geqslant k$ we apply division with remainder to get $n = mk + j$ for some $j \in \{0, \ldots, k-1\}$ and $m \geqslant 1$. By the sub-additivity property we then have

$$\frac{\alpha_n}{n} \leqslant \frac{\alpha_{mk}}{n} + \frac{\alpha_j}{n} \leqslant \frac{m\alpha_k}{n} + \frac{j\alpha_1}{n} = \frac{mk}{n} \frac{\alpha_k}{k} + \frac{j\alpha_1}{n}.$$

If now $n = mk + j$ is large enough we see that the right-hand side is less than β, which proves the statement.
$\square$

PROOF OF THEOREM 11.6. Notice first that the sequence (α_n) defined by

$$\alpha_n = \|a^n\|$$

for $n \geqslant 1$ is sub-multiplicative, since

$$\alpha_{m+n} = \|a^{m+n}\| \leqslant \|a^m\|\|a^n\| = \alpha_m \alpha_n$$

for all $m, n \geqslant 1$. Thus $\sqrt[m]{\|a^m\|}$ converges to $\inf_{m \geqslant 1} \sqrt[m]{\|a^m\|}$ by Lemma 11.15. Proposition 11.11 shows that, if $\lambda \in \mathbb{C}$ satisfies

$$|\lambda| > \inf_{m \geqslant 1} \sqrt[m]{\|a^m\|},$$

then $\lambda \in \rho(a)$. Thus

$$\max_{\lambda \in \sigma(a)} |\lambda| \leqslant \inf_{m \geqslant 1} \sqrt[m]{\|a^m\|} = \lim_{m \to \infty} \sqrt[m]{\|a^m\|}.$$

USING THE CAUCHY INTEGRAL FORMULA. The reverse inequality is more involved. It involves a refinement of the proof that $\sigma(a)$ is non-empty, and will use the Cauchy integral formula again. Let $s = \max_{\lambda \in \sigma(a)} |\lambda|$, so that the resolvent function $R(z) = (z1_{\mathcal{A}} - a)^{-1}$ is analytic on $\{z \in \mathbb{C} \mid |z| > s\} \subseteq \rho(a)$. Pick $\ell \in \mathcal{A}^*$ with $\|\ell\| \leqslant 1$ and fix $\varepsilon > 0$. Using the positively oriented closed path with centre 0 and radius $s + \varepsilon$, we see that

$$\left| \oint_{|z|=s+\varepsilon} \ell\big((z1_{\mathcal{A}} - a)^{-1}\big) z^m \, dz \right| \ll_{a,\varepsilon} \big((s+\varepsilon)^m\big)$$

for all $m \geqslant 0$, where $|z^m| = (s+\varepsilon)^m$ and the implicit constant only depends on the restriction of $R(z) = (z1_{\mathcal{A}}-a)^{-1}$ to $\{z \in \mathbb{C} \mid |z| = s+\varepsilon\}$, and in particular does not depend on m and ℓ. Expanding the circle to the radius $\|a\| + 1$ does not change the integral, so that we may use (11.3) again to see that

$$\oint_{|z|=s+\varepsilon} \ell\big((z1_{\mathcal{A}} - a)^{-1}\big) z^m \, dz = \oint_{|z|=\|a\|+1} \ell\big((z1_{\mathcal{A}} - a)^{-1}\big) z^m \, dz$$

$$= \sum_{n=0}^{\infty} \ell(a^n) \oint_{|z|=\|a\|+1} z^{-n+m-1} dz = 2\pi i \ell(a^m).$$

Together this gives

$$|\ell(a^m)| \ll_{a,\varepsilon} (s + \varepsilon)^m$$

for all $m \geqslant 1$ and $\ell \in \mathcal{A}^*$ with $\|\ell\| \leqslant 1$. Using Corollary 7.4 we deduce that

$$\|a^m\| \ll_{a,\varepsilon} (s + \varepsilon)^m.$$

Taking the mth root and the limit we see that the implicit constant disappears, and we get

$$\lim_{m\to\infty} \sqrt[m]{\|a^m\|} \leqslant s + \varepsilon = \max_{\lambda\in\sigma(a)} |\lambda| + \varepsilon.$$

Since this holds for any $\varepsilon > 0$ the theorem follows. □

The results of this and the following section will be used in Chapter 12 to derive the spectral theory of bounded self-adjoint operators and their functional calculus.

11.2 C^*-algebras

Definition 11.16. A Banach algebra $\mathcal{A}$ over $\mathbb{C}$ is a C^*-*algebra* if it has a *star operator* $*: \mathcal{A} \to \mathcal{A}$ with the following properties:

- $*$ is semi-linear;
- $(ab)^* = b^*a^*$ for $a, b \in \mathcal{A}$;
- $(a^*)^* = a$ for $a \in \mathcal{A}$; and
- $\|a^*a\| = \|a\|^2$ for $a \in \mathcal{A}$ (the C^*-*property* of the norm).

Example 11.17. (a) The algebra of bounded operators $\mathrm{B}(\mathcal{H})$ on a Hilbert space $\mathcal{H}$ has a star operator, namely the map that sends $A \in \mathrm{B}(\mathcal{H})$ to its adjoint $A^* \in \mathrm{B}(\mathcal{H})$ (introduced in Section 6.2.1). For this star operator we already know all the desired properties with the exception of the last (critical) property. To see this last property, let $A \in \mathrm{B}(\mathcal{H})$, and notice that A^*A is self-adjoint since $(A^*A)^* = A^*(A^*)^* = A^*A$. Hence, by Lemma 6.31,

$$\|A^*A\| = \sup_{\|x\|\leqslant 1} |\langle A^*Ax, x\rangle| = \sup_{\|x\|\leqslant 1} \langle Ax, Ax\rangle = \|A\|^2,$$

as required. Therefore $\mathrm{B}(\mathcal{H})$ is a C^*-algebra.
(b) The space of bounded functions $B(X)$, of continuous bounded functions $C_b(X)$, of measurable bounded functions $L^\infty(X)$, and of measurable essentially bounded functions $L^\infty_\mu(X)$ are all commutative unital C^*-algebras. For these multiplication is defined pointwise, and the star operator is pointwise complex conjugation.

Essential Exercise 11.18. Show that $\|a^*\| = \|a\|$ for $a \in \mathcal{A}$ if $\mathcal{A}$ is a C^*-algebra.

Definition 11.19. Let $\mathcal{A}$ be a C^*-algebra. Then an element $a \in \mathcal{A}$ is called *self-adjoint* if $a^* = a$, and is called *normal* if $a^*a = aa^*$.

Essential Exercise 11.20. Let $\mathcal{A}$ be a unital C^*-algebra. Show that the unit $1_\mathcal{A}$ is self-adjoint and has $\|1_\mathcal{A}\| = 1$.

For normal elements in a C^*-algebra the spectral radius formula simplifies.

Proposition 11.21 (Spectral radius formula for normal elements).
Let $\mathcal{A}$ be a unital C^*-algebra, and let $a \in \mathcal{A}$ be a normal element. Then the spectral radius satisfies $\max\limits_{\lambda \in \sigma(a)} |\lambda| = \|a\|$.

PROOF. We will prove by induction on n that

$$\|a^{2^n}\| = \|a\|^{2^n}. \tag{11.5}$$

The case $n = 0$ is trivial. For $n = 1$ we have

$$\|a^2\|^2 = \|(a^2)^* a^2\| = \|(a^*a)^*(a^*a)\| = \|a^*a\|^2 = \|a\|^4,$$

where we used the C^*-property of the norm for a^2, normality of a, and the C^*-property of the norm for a^*a and for a. Now suppose that (11.5) holds for a given $n \geqslant 1$ and set $b = a^{2^n}$. Then

$$\|a^{2^{n+1}}\| = \|b^2\| = \|b\|^2 = \|a^{2^n}\|^2 = \|a\|^{2^{n+1}},$$

where we used the definition of b, the case $n = 1$ for the normal element b, and the inductive hypothesis. This concludes the induction, proving (11.5) for all $n \geqslant 0$. Applying Theorem 11.6 now gives the proposition. $\qquad\square$

Starting in Section 12.5, we will use the results of this section and their refinements in Section 11.3.4 to obtain the spectral theory of commutative C^*-subalgebras of bounded operators on a Hilbert space.

11.3 Commutative Banach Algebras and their Gelfand Duals

Recall that the dual space $\mathcal{A}^*$ of a Banach algebra $\mathcal{A}$ consists of all bounded linear functionals $\mathcal{A} \to \mathbb{C}$. If $\mathcal{A}$ is in addition commutative (with $ab = ba$ for all $a, b \in \mathcal{A}$) then it is useful to study *algebra homomorphisms*. The trivial map χ defined by $\chi(a) = 0$ for all $a \in \mathcal{A}$ may also be considered an algebra homomorphism, but we will exclude this trivial map in the discussion below.

Definition 11.22. Let $\mathcal{A}$ be a commutative Banach algebra over $\mathbb{C}$. Then the *Gelfand dual* $\sigma(\mathcal{A})$ is the set of all non-trivial (equivalently, surjective) continuous algebra homomorphisms $\chi : \mathcal{A} \to \mathbb{C}$ (which are also called *characters*). That is, $\sigma(\mathcal{A}) = \left\{ \chi \in \mathcal{A}^* \smallsetminus \{0\} \mid \chi(ab) = \chi(a)\chi(b) \text{ for all } a, b \in \mathcal{A} \right\}$.

11.3.1 Commutative Unital Banach Algebras

If the Banach algebra that we consider also has a unit, then we can link the notion of algebra homomorphisms to the spectrum of the elements of the algebra. The following result establishes this link and a great deal more.

Theorem 11.23 (Properties of the Gelfand dual). *Let $\mathcal{A}$ be a commutative unital Banach algebra over $\mathbb{C}$. Then $\sigma(\mathcal{A}) \subseteq \overline{B_1^{\mathcal{A}^*}}$ is non-empty and weak* compact, and $\sigma(a) = \{\chi(a) \mid \chi \in \sigma(\mathcal{A})\}$ for every $a \in \mathcal{A}$.*

We start the proof of Theorem 11.23 by showing that any algebra homomorphism $\chi : \mathcal{A} \to \mathbb{C}$ is continuous[31] (and so strictly speaking the continuity hypothesis in Definition 11.22 could be dropped).

Lemma 11.24. *Let $\mathcal{A}$ be a commutative Banach algebra, and let $\chi : \mathcal{A} \to \mathbb{C}$ be an algebra homomorphism. Then χ is continuous and $\|\chi\| \leqslant 1$.*

PROOF. Suppose there is an element $a \in \mathcal{A}$ with $\|a\| < 1$ and with $\|\chi(a)\| > 1$. Replacing a by $a/\chi(a)$ we may assume that $\|a\| < 1$ and $\chi(a) = 1$. Then the series $b = \sum_{n=1}^{\infty} a^n$ converges and satisfies $a + ab = b$, so that

$$1 + \chi(b) = \chi(a) + \chi(a)\chi(b) = \chi(b),$$

a contradiction. This implies that χ is continuous and has $\|\chi\| \leqslant 1$. $\qquad\square$

For the next steps we will need to use some more terminology from basic algebra. Recall that an *ideal* J of a commutative algebra A is a subspace such that $AJ = \{ab \mid a \in A, b \in J\} \subseteq J$, and that for any ideal J the quotient A/J is also a commutative algebra with multiplication given by

$$(a + J)(b + J) = ab + J$$

for all $a, b \in A$. An ideal $J \subseteq A$ is *proper* if $J \neq A$. A *maximal ideal* M in a unital commutative algebra A is a proper ideal such that if J is an ideal with $M \subseteq J \subseteq A$ then $J = M$ or $J = A$. The quotient of a unital algebra by a maximal ideal M is always a field, for if $a + M \in A/M \smallsetminus \{0\}$ then $J = Aa + M$ is an ideal strictly bigger than M, and so must be A. Since A has a unit 1_A, we have $ba + m = 1_A$ for some $b \in A$ and $m \in M$, so every non-zero element of A/M has a multiplicative inverse.

The next lemma examines these general notions for Banach algebras.

Lemma 11.25. *Let $\mathcal{A}$ be a commutative unital Banach algebra. The closure of any ideal in $\mathcal{A}$ is an ideal, and any maximal ideal is closed.*

PROOF. The first claim is an easy consequence of the fact that the multiplication map $\mathcal{A} \times \mathcal{A} \to \mathcal{A}$ is continuous by the discussion in Section 2.4.2.

For the second claim, notice that a proper ideal $\mathcal{J} \subseteq \mathcal{A}$ cannot contain $1_{\mathcal{A}}$, nor indeed any invertible element. By Proposition 11.10 this implies that $1_{\mathcal{A}}$

is not an element of $\overline{\mathcal{J}}$. Since a maximal ideal $\mathcal{M}$ is proper, and its closure $\overline{\mathcal{M}}$ is also a proper ideal, we see that $\mathcal{M} = \overline{\mathcal{M}}$ is closed. $\qquad\qquad\Box$

We note that Lemma 11.25 gives a second proof that any algebra homomorphism $\chi : \mathcal{A} \to \mathbb{C}$ on a commutative unital Banach algebra is continuous: Given a non-trivial algebra homomorphism $\chi : \mathcal{A} \to C$ its kernel $\mathcal{M} = \ker \chi$ is a maximal ideal, and so is closed. Then χ equals the composition of the continuous projection $\mathcal{A} \to \mathcal{A}/\mathcal{M}$ and the isomorphism $\mathcal{A}/\mathcal{M} \to \mathbb{C}$ induced by χ (which is continuous by finite-dimensionality), hence we see that $\chi : \mathcal{A} \to \mathcal{A}/\mathcal{M} \to \mathbb{C}$ is a continuous map.

For the proof of Theorem 11.23 we need one more algebraic result.

Lemma 11.26. *Let R be a commutative ring with a unit, and let $J_0 \subseteq R$ be a proper ideal. Then there exists a maximal ideal $M \subseteq R$ containing J_0.*

PROOF. This is a direct application of Zorn's lemma. Define a set

$$\mathcal{S} = \{J \subseteq R \mid J \text{ is an ideal and } J_0 \subseteq J \subsetneq R\}$$

with the partial order defined by inclusion. If $\{J_\alpha \mid \alpha \in I\}$ is a linearly ordered subset (a chain) in $\mathcal{S}$, then

$$J = \bigcup_{\alpha \in I} J_\alpha$$

is again an ideal. Moreover, since each J_α is proper, we have $1_{\mathcal{A}} \notin J_\alpha$ for all α in I, and so $1_{\mathcal{A}} \notin J$, showing that J is also proper. By Zorn's lemma, it follows that the set $\mathcal{S}$ contains a maximal element, which by construction is a maximal ideal containing J_0. $\qquad\qquad\Box$

PROOF OF THEOREM 11.23. Note that an algebra homomorphism $\chi : \mathcal{A} \to \mathbb{C}$ is non-trivial if and only if $\chi(1_{\mathcal{A}}) = 1$. Indeed, if $\chi(a) \neq 0$ for some $a \in \mathcal{A}$ then $\chi(a) = \chi(1_{\mathcal{A}}a) = \chi(1_{\mathcal{A}})\chi(a)$ shows that $\chi(1_{\mathcal{A}}) = 1$. By the definition and Lemma 11.24 we have

$$\sigma(\mathcal{A}) = \left\{\chi \in \overline{B_1^{\mathcal{A}^*}} \mid \chi(ab) = \chi(a)\chi(b) \text{ for all } a, b \in \mathcal{A} \text{ and } \chi(1_{\mathcal{A}}) = 1\right\}$$

$$= \overline{B_1^{\mathcal{A}^*}} \cap \bigcap_{a,b \in \mathcal{A}} \left\{\chi \in \mathcal{A}^* \mid \chi(ab) = \chi(a)\chi(b)\right\} \cap \left\{\chi \in \mathcal{A}^* \mid \chi(1_{\mathcal{A}}) = 1\right\}.$$

Since the sets $\{\chi \in \mathcal{A}^* \mid \chi(1_{\mathcal{A}}) = 1\}$ and $\{\chi \in \mathcal{A}^* \mid \chi(ab) = \chi(a)\chi(b)\}$ are closed in the weak* topology for every $a, b \in \mathcal{A}$, we see that $\sigma(\mathcal{A})$ is weak* compact by the Banach–Alaoglu theorem (Theorem 8.10).

Now let $a_0 \in \mathcal{A}$ be non-invertible, so that $\mathcal{J} = \mathcal{A}a_0$ is a proper ideal. By Lemma 11.26 there is a maximal ideal $\mathcal{M} \subseteq \mathcal{A}$ containing $\mathcal{J}$. By Lemma 11.25, $\mathcal{M}$ is closed. We claim that $\mathcal{B} = \mathcal{A}/\mathcal{M}$ is also a Banach algebra. To see this, we equip $\mathcal{B}$ with the quotient norm (from Section 2.1.2) which makes $\mathcal{B}$ into a Banach space by Lemma 2.29. Since $\mathcal{M}$ is an ideal, multiplication is well-defined on $\mathcal{A}/\mathcal{M}$. Finally,

$$\|ab + \mathcal{M}\|_{\mathcal{A}/\mathcal{M}} \leqslant \|(a + m_1)(b + m_2)\|_{\mathcal{A}} \leqslant \|a + m_1\|_{\mathcal{A}}\|b + m_2\|_{\mathcal{A}}$$

for all $a, b \in \mathcal{A}$ and all $m_1, m_2 \in \mathcal{M}$, which implies that

$$\|(a + \mathcal{M})(b + \mathcal{M})\|_{\mathcal{A}/\mathcal{M}} \leqslant \|a + \mathcal{M}\|_{\mathcal{A}/\mathcal{M}}\|b + \mathcal{M}\|_{\mathcal{A}/\mathcal{M}}$$

by taking the infimum over m_1 and $m_2 \in \mathcal{M}$, as required.

Thus $\mathcal{A}/\mathcal{M}$ is a Banach algebra and a field (since $\mathcal{M}$ is maximal). We claim this implies that

$$\mathcal{A}/\mathcal{M} = \mathbb{C}(1_{\mathcal{A}} + \mathcal{M}) \cong \mathbb{C}.$$

Indeed (solving Exercise 11.8), if $a + \mathcal{M} \in \mathcal{A}/\mathcal{M}$, then $\sigma(a + \mathcal{M}) \neq \varnothing$ by Theorem 11.6, and so $a - \lambda 1_{\mathcal{A}} + \mathcal{M}$ is non-invertible for some $\lambda \in \mathbb{C}$. However, since $\mathcal{A}/\mathcal{M}$ is a field this implies that $a + \mathcal{M} = \lambda 1_{\mathcal{A}} + \mathcal{M}$ and hence the claim.

Together we have shown that if $a_0 \in \mathcal{A}$ is non-invertible, then there exists a non-trivial algebra homomorphism $\chi : \mathcal{A} \to \mathcal{A}/\mathcal{M} \cong \mathbb{C}$ with $\chi(a_0) = 0$. Applying this for any $a \in \mathcal{A}$ to $a - \lambda 1_{\mathcal{A}}$ for $\lambda \in \sigma(a)$, we see that for any such λ there is some $\chi \in \sigma(\mathcal{A})$ with $\chi(a) = \lambda$.

On the other hand, if $\chi(a) = \lambda$ for some $a \in \mathcal{A}$, $\lambda \in \mathbb{C}$ and $\chi \in \sigma(\mathcal{A})$, then

$$\chi(a - \lambda 1_{\mathcal{A}}) = 0$$

and hence $a - \lambda 1_{\mathcal{A}}$ cannot be invertible (since $\chi \neq 0$). Together we have shown the theorem. $\qquad\square$

Example 11.27 (Stone–Čech compactification). Let $\mathcal{A} = \ell^\infty(\mathbb{N})$, which is a Banach algebra with respect to the pointwise product. Clearly for any $n_0 \in \mathbb{N}$ the map defined by

$$\chi_{n_0}((a_n)) = a_{n_0}$$

is an algebra homomorphism, and $\mathbb{N} \ni n_0 \mapsto \chi_{n_0}$ defines a map from $\mathbb{N}$ to $\sigma(\mathcal{A})$. The compact (but non-metrizable) topological space $\sigma(\mathcal{A})$ is called the Stone–Čech compactification of $\mathbb{N}$ and is denoted $\beta\mathbb{N}$.

Exercise 11.28. (a) Show that the image of $\mathbb{N}$ is dense in $\beta\mathbb{N}$.
(b) Show that $\ell^\infty(\mathbb{N})$ can be canonically identified with $C(\beta\mathbb{N})$.
(c) Show that $\beta\mathbb{N}$ is non-metrizable.

11.3.2 Commutative Banach Algebras without a Unit

While the notions of invertibility and spectrum are linked to the existence of a unit, the definition of the Gelfand dual is not. However, the topological properties of $\sigma(\mathcal{A})$ are changed by the absence of a unit.

Corollary 11.29 (Properties of the Gelfand dual). *Let $\mathcal{A}$ be a commutative Banach algebra over $\mathbb{C}$. Then $\sigma(\mathcal{A}) \subseteq \overline{B_1^{\mathcal{A}^*}}$ is locally compact (and $\sigma(\mathcal{A}) \cup \{0\}$ is compact) in the weak* topology on $\mathcal{A}^*$. For any $a \in \mathcal{A}$ we also have*

$$\max_{\chi \in \sigma(\mathcal{A}) \cup \{0\}} |\chi(a)| = \lim_{n \to \infty} \sqrt[n]{\|a^n\|}.$$

Recall that if X is a compact space and $x_0 \in X$ is any point, then the space $Y = X \smallsetminus \{x_0\}$ is in general only locally compact. Moreover, in the case when Y is not compact, the one-point compactification of Y is homeomorphic to (and so can be identified with) $X = Y \cup \{x_0\}$, where the point x_0 takes the role of ∞.

Exercise 11.30. Recall the definition of the one-point compactification of a locally compact space Y and show the above claim.

PROOF OF COROLLARY 11.29. The proof that $\sigma(\mathcal{A}) \cup \{0\}$ is compact in the weak* topology is the same as in the proof of Theorem 11.23 since Lemma 11.24 implies that

$$\sigma(\mathcal{A}) \cup \{0\} = \left\{ \chi \in \overline{B_1^{\mathcal{A}^*}} \mid \chi(ab) = \chi(a)\chi(b) \text{ for all } a, b \in \mathcal{A} \right\},$$

which is easily seen to be a closed subset of $\overline{B_1^{\mathcal{A}^*}}$ in the weak* topology.

For the last claim of the corollary note that if $\mathcal{A}$ has a unit, then Theorem 11.23 applies and gives the statement. So assume that $\mathcal{A}$ does not have a unit, and consider the algebra $\mathcal{A}_1 = \mathcal{A} \oplus \mathbb{C}$ with the multiplication and norm as in Exercise 11.1. As argued in the beginning of the proof of Theorem 11.23, $\chi_1(1_{\mathcal{A}}) = 1$ for any $\chi_1 \in \sigma(\mathcal{A}_1)$ so that χ_1 is uniquely determined by $\chi = \chi_1|_{\mathcal{A}} \in \sigma(\mathcal{A}) \cup \{0\}$. Moreover, any $\chi \in \sigma(\mathcal{A}) \cup \{0\}$ can be extended to a character $\chi_1 \in \sigma(\mathcal{A}_1)$ by setting

$$\chi_1(a + \lambda 1_{\mathcal{A}}) = \chi(a) + \lambda$$

for any $a + \lambda 1_{\mathcal{A}} \in \mathcal{A}_1$, which allows us to identify $\sigma(\mathcal{A}_1)$ with $\sigma(\mathcal{A}) \cup \{0\}$. Applying Theorems 11.23 and 11.6 to $\mathcal{A}_1$ now gives

$$\max_{\chi \in \sigma(\mathcal{A}) \cup \{0\}} |\chi(a)| = \max_{\chi \in \sigma(\mathcal{A}_1)} |\chi(a)| = \lim_{n \to \infty} \sqrt[n]{\|a^n\|}.$$

$\square$

Exercise 11.31. Show that every character $\chi \in \sigma(\mathcal{A})$ can be extended to a character χ_1, as claimed in the proof of Corollary 11.29.

11.3.3 The Gelfand Transform

Definition 11.32. Let $\mathcal{A}$ be a commutative Banach algebra with Gelfand dual $\sigma(\mathcal{A})$. Then the map $(\cdot)^\circ : \mathcal{A} \to C(\sigma(\mathcal{A}))$ defined by

$$f^o(\chi) = \chi(f)$$

for $f \in \mathcal{A}$ and $\chi \in \sigma(\mathcal{A})$ is called the *Gelfand transform*.

Just as in Theorem 11.23 we will always use the weak* topology on $\sigma(\mathcal{A})$.

Proposition 11.33. *Let $\mathcal{A}$ be a commutative Banach algebra. The Gelfand transform is an algebra homomorphism from $\mathcal{A}$ into $C_0(\sigma(\mathcal{A}))$ (or $C(\sigma(\mathcal{A}))$ if $\mathcal{A}$ has a unit) so that $(f_1 f_2)^o = f_1^o f_2^o$ for all $f_1, f_2 \in \mathcal{A}$. Moreover, it satisfies $\|f^o\|_\infty \leqslant \|f\|$ for all $f \in \mathcal{A}$.*

PROOF. By definition of the weak* topology, $f^o(\chi) = \chi(f)$ depends continuously on $\chi \in \sigma(\mathcal{A})$ for each $f \in \mathcal{A}$. By Lemma 11.24,

$$|f^o(\chi)| = |\chi(f)| \leqslant \|\chi\| \|f\| \leqslant \|f\|$$

for all $\chi \in \sigma(\mathcal{A})$, and so $\|f^o\|_\infty \leqslant \|f\|$. Finally, $0 \in \mathcal{A}^*$ plays the role of infinity in the one-point compactification of $\sigma(\mathcal{A})$ in Corollary 11.29. This gives $f^o \in C_0(\sigma(\mathcal{A}))$, as required. Finally, $f_1, f_2 \in \mathcal{A}$ and $\chi \in \sigma(\mathcal{A})$ implies that $(f_1 f_2)^o(\chi) = \chi(f_1 f_2) = \chi(f_1)\chi(f_2) = f_1^o(\chi) f_2^o(\chi)$, which shows that the Gelfand transform is an algebra homomorphism from $\mathcal{A}$ into $C_0(\sigma(\mathcal{A}))$. $\square$

11.3.4 The Gelfand Transform for Commutative C^*-algebras

The Gelfand transform has good additional properties for C^*-algebras.

Corollary 11.34. *Let $\mathcal{A}$ be a commutative unital C^*-algebra. Then the Gelfand transform is an isometric algebra isomorphism from $\mathcal{A}$ onto $C(\sigma(\mathcal{A}))$ satisfying $(a^*)^o = \overline{a^o}$ for all $a \in \mathcal{A}$.*

PROOF. From Proposition 11.33 we know that the Gelfand transform is an algebra homomorphism. For $a \in \mathcal{A}$ the norm $\|a^o\|_\infty$ of the Gelfand transform equals the spectral radius of a (see Theorem 11.23). By Proposition 11.21 we get $\|a^o\|_\infty = \|a\|$, since in a commutative C^*-algebra every element $a \in \mathcal{A}$ is normal. This shows that $(\cdot)^o : \mathcal{A} \longrightarrow C(\sigma(\mathcal{A}))$ is an isometric algebra homomorphism between $\mathcal{A}$ and a complete sub-algebra of $C(\sigma(\mathcal{A}))$ which contains the unit since $1_\mathcal{A}^o = \mathbb{1}$, and separates points since $\chi_1 \neq \chi_2 \in \sigma(\mathcal{A})$ implies that there exists some $a \in \mathcal{A}$ with

$$a^o(\chi_1) = \chi_1(a) \neq \chi_2(a) = a^o(\chi_2).$$

Since $\sigma(\mathcal{A})$ is a compact space we can apply the Stone–Weierstrass theorem (Theorem 2.40) provided that we can also show that the image is closed under conjugation. Once this is done, we can conclude that the image of the Gelfand transform is both dense in $C(\sigma(\mathcal{A}))$ and complete, and therefore must be all of $C(\sigma(\mathcal{A}))$.

To show closure under conjugation, it suffices to prove that $\chi(a^*) = \overline{\chi(a)}$ for all $a \in \mathcal{A}$ and $\chi \in \sigma(\mathcal{A})$, which also implies that $(a^*)^\circ = \overline{a^\circ}$ for all $a \in \mathcal{A}$, as claimed in the corollary. This in turn follows if we know that $a = a^* \in \mathcal{A}$ implies that $\chi(a) \in \mathbb{R}$ for every $\chi \in \sigma(\mathcal{A})$. Indeed, any $a \in \mathcal{A}$ can be written as

$$a = \frac{a + a^*}{2} + i\frac{a - a^*}{2i} = a_{\Re} + i a_{\Im},$$

where both $a_{\Re} = \frac{a+a^*}{2}$ and $a_{\Im} = \frac{a-a^*}{2i}$ are self-adjoint. Assuming that $\chi(a_{\Re})$ and $\chi(a_{\Im})$ are real, we deduce that

$$\chi(a^*) = \chi(a_{\Re}) - i\chi(a_{\Im}) = \overline{\chi(a_{\Re})} + \overline{i\chi(a_{\Im})} = \overline{\chi(a_{\Re} + i a_{\Im})} = \overline{\chi(a)}.$$

The following lemma then finishes the proof of the corollary.   □

Lemma 11.35. *Let $a = a^* \in \mathcal{A}$ be a self-adjoint element of a unital C^*-algebra. Then $\sigma(a) \subseteq \mathbb{R}$.*

As we will see in the course of the proof, this can be deduced from Proposition 11.11. This might be a little confusing initially. How can a property like $\max_{\lambda \in \sigma(a)} |\lambda| \leqslant \|a\|$ imply that $\sigma(a)$ is real? One way of viewing the situation is to apply a vertical translation to the set $\sigma(a)$, as illustrated in Figure 11.1.

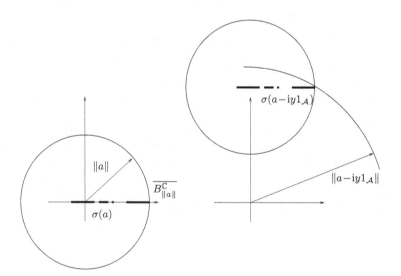

Fig. 11.1: Many possible $\lambda \in \mathbb{C}$ that satisfy the constraint $|\lambda| \leqslant \|a\|$ might not satisfy the constraint $|\lambda - iy| \leqslant \|a - iy\mathbb{1}\|$ if the norm of $a - iy\mathbb{1}_{\mathcal{A}}$ for $y \in \mathbb{R}$ is $\sqrt{\|a\|^2 + |y|^2}$ as Figure 11.1 suggests. Taking $y \to \infty$ and $y \to -\infty$ shows that the spectrum $\sigma(a)$ is a subset of $\mathbb{R}$.

PROOF OF LEMMA 11.35. By Proposition 11.11 we know that the spectral radius of $a - iy1_\mathcal{A}$ is at most $\|a - iyI\|$.

Let $\lambda \in \sigma(a)$ and $y \in \mathbb{R}$. Then $\lambda - iy \in \sigma(a - iy1_\mathcal{A})$, and

$$|\lambda - iy|^2 \leqslant \|a - iy1_\mathcal{A}\|^2 = \|(a - iy1_\mathcal{A})^*(a - iy1_\mathcal{A})\| = \|(a + iy1_\mathcal{A})(a - iy1_\mathcal{A})\|$$
$$= \|a^2 + y^2 1_\mathcal{A}\| \leqslant \|a^2\| + y^2\|1_\mathcal{A}\| = \|a\|^2 + y^2,$$

where we have used the C^*-property, the fact that a and $1_\mathcal{A}$ are self-adjoint, and the fact that $\|1_\mathcal{A}\| = 1$ (see Exercise 11.20 and its hint on p. 581). Writing $\lambda = x_0 + iy_0 \in \mathbb{C}$ with $x_0, y_0 \in \mathbb{R}$, the calculation above gives

$$x_0^2 + y^2 - 2yy_0 + y_0^2 = x_0^2 + (y - y_0)^2 \leqslant \|a\|^2 + y^2$$

for all $y \in \mathbb{R}$. However, this shows that $y_0 = 0$, and so $\sigma(a) \subseteq \mathbb{R}$, as claimed. $\qquad \square$

Exercise 11.36. Let $\mathcal{A}$ be a commutative C^*-algebra.
(a) Show that the Gelfand transform is an isometry onto $C_0(\sigma(\mathcal{A}))$.
(b) Show that $\sigma(\mathcal{A})$ is compact if and only if $\mathcal{A}$ is unital.
(c) Assume now that $\mathcal{A}$ is not unital. Show that it is possible to define a norm on $\mathcal{A}_I = \mathcal{A} \oplus \mathbb{C}$ so that $\mathcal{A}_I$ is again a C^*-algebra. (The norm from Exercise 11.1 may not do this.)

11.4 Locally Compact Abelian Groups

An important special case of the Gelfand transform is given by the following proposition, but we first need a definition. The reader may make this more familiar by assuming that $G = \mathbb{R}^d$ or $G = \mathbb{T}^d$ (cf. Exercise 1.3, which easily generalizes to $\mathbb{T}^d$ and $\mathbb{R}^d$).

Definition 11.37. Let G be a locally compact metrizable abelian group. The *dual group, character group,* or *Pontryagin dual* of G is the abelian group

$$\widehat{G} = \mathrm{Hom}(G, \mathbb{S}^1) = \{\chi : G \to \mathbb{S}^1 \mid \chi \text{ is a continuous homomorphism}\},$$

where $\mathbb{S}^1$ is the multiplicative unit circle and the group operation is pointwise multiplication.

Proposition 11.38 (Algebra homomorphisms on $L^1(G)$). *Let G be a σ-compact locally compact metrizable abelian group, which we equip with a Haar measure $m = m_G$. Then $L^1(G)$ is a separable commutative Banach algebra with respect to the convolution defined by*

$$f_1 * f_2(g) = \int_G f_1(h) f_2(g - h) \, dm(h)$$

*for $f_1, f_2 \in L^1(G)$. The Gelfand dual $\sigma(L^1(G))$ of all non-trivial algebra ho-
momorphisms from $L^1(G)$ onto $\mathbb{C}$ is a locally compact σ-compact metrizable
space which can be identified with the Pontryagin dual $\widehat{G}$. The Gelfand transform can be identified with the Fourier back transform.*

We now explain the two identifications in more detail. If $\chi_G \in \widehat{G}$ is a
continuous group homomorphism

$$\chi_G : G \longrightarrow \mathbb{S}^1 = \{z \in \mathbb{C} \mid |z| = 1\},$$

then it gives rise to an algebra homomorphism $\chi_\mathcal{A}$ on $\mathcal{A} = L^1(G)$ defined by

$$\chi_\mathcal{A}(f) = \int_G f(g)\chi_G(g)\, dm(g),$$

which is well-defined since $f \in L^1(G)$ and $\chi_G \in L^\infty(G)$. The first identification claimed is the statement that every algebra homomorphism on $L^1(G)$
has this shape. This also explains the second identification as follows. The
Fourier back transform of an element $f \in L^1(G)$ is the function $\check{f}$ on $\widehat{G}$
defined by

$$\check{f}(\chi_G) = \int_G f(g)\chi_G(g)\, dm(g)$$

for $\chi_G \in \widehat{G}$. Since we identify the Pontryagin dual $\widehat{G}$ with the Gelfand
dual $\sigma(L^1(G))$, we see that every character $\chi_G \in \widehat{G}$ corresponds precisely
to one $\chi_\mathcal{A} \in \sigma(\mathcal{A})$ and *vice versa*, and that

$$\check{f}(\chi_G) = \chi_\mathcal{A}(f) = f^\circ(\chi_\mathcal{A})$$

is the Fourier back transform and is at the same time also the Gelfand transform.

PROOF OF PROPOSITION 11.38. By Proposition 3.91, $L^1(G)$ is a separable
Banach algebra. By Exercise 3.92 (see also Lemma 3.59(1)), $L^1(G)$ is commutative. The proof that every continuous group homomorphism $\chi_G : G \to \mathbb{S}^1$
gives rise to an algebra homomorphism

$$\chi_\mathcal{A} : L^1(G) \ni f \longmapsto \int_G f\chi_G\, dm$$

is very similar to the proof for the case $G = \mathbb{R}^d$ in Proposition 9.31 and is
therefore left to the reader.

Corollary 11.29 shows that $\sigma(L^1(G))$ is locally compact in the weak* topology. The claimed metrizability follows from Proposition 8.11 since $L^1(G)$
is separable. Finally, the σ-compactness follows since $\sigma(L^1(G)) \cup \{0\}$ is also
compact and metrizable by Corollary 11.29 and Proposition 8.11.

The main claim of the proposition is therefore that every non-trivial algebra homomorphism $\chi_\mathcal{A} : L^1(G) \to \mathbb{C}$ arises from some continuous group

homomorphism $\chi_G : G \to \mathbb{S}^1$. So let $\chi_{\mathcal{A}} \in \sigma(L^1(G))$. Then by Lemma 11.24 we have $\|\chi_{\mathcal{A}}\| \leq 1$. By Proposition 7.34 there is an element $\chi \in L^\infty(G)$ with $\|\chi\|_\infty \leq 1$ such that

$$\chi_{\mathcal{A}}(f) = \int_G f\chi \, dm$$

for all $f \in L^1(G)$. We have to show that χ can be chosen in $C_b(G)$ and with the property that $\chi(gh) = \chi(g)\chi(h)$ for all $g, h \in G$ (which will also imply that $\chi(g) \in \mathbb{S}^1$ for all $g \in G$).

For this proof we apply the algebra homomorphism property of the map $\chi_{\mathcal{A}} \in \sigma(L^1(G))$ for $f, f_0 \in L^1(G)$ and obtain together with Fubini's theorem that

$$\int_G f(h)\big(\chi(h)\chi_{\mathcal{A}}(f_0)\big) \, dm(h) = \chi_{\mathcal{A}}(f)\chi_{\mathcal{A}}(f_0) = \chi_{\mathcal{A}}(f * f_0)$$
$$= \int_G \int_G f(h)f_0(g - h) \, dm(h)\chi(g) \, dm(g)$$
$$= \int_G f(h) \int_G f_0(g - h)\chi(g) \, dm(g) \, dm(h).$$

As this holds for any fixed f_0 and for all $f \in L^1(G)$, the uniqueness property in Proposition 7.34 implies that

$$\chi(h)\chi_{\mathcal{A}}(f_0) = \int_G f_0(g - h)\chi(g) \, dm(g) = \chi_{\mathcal{A}}(f_0^h) \qquad (11.6)$$

for almost every $h \in G$, where we write $f_0^h(g) = f_0(g - h)$ for $g, h \in G$ as usual. We now fix some $f_1 \in L^1(G)$ such that $\chi_{\mathcal{A}}(f_1) \neq 0$ and define

$$\chi_G(h) = \chi_{\mathcal{A}}(f_1)^{-1}\chi_{\mathcal{A}}(f_1^h)$$

for $h \in G$, so that $\chi_G = \chi$ almost everywhere and thus $\chi_{\mathcal{A}}(f) = \int_G f\chi_G \, dm$ for all $f \in L^1(G)$.

Now note that $\chi_{\mathcal{A}}(f_0^h)$ depends continuously on $h \in G$ for any $f_0 \in L^1(G)$ by the continuity claim in Lemma 3.74. Therefore χ_G is continuous and we may replace χ by its continuous representative χ_G, which implies in turn that (11.6) holds for χ_G in fact for all $h \in G$ (as both sides of the equation are now continuous with respect to $h \in G$). Applying the definition of χ_G and this version of (11.6) for $f_0 = f_1^{g_2}$ and $h = g_1$ we obtain

$$\chi_G(g_1 + g_2) = \chi_{\mathcal{A}}(f_1)^{-1}\chi_{\mathcal{A}}(f_1^{g_1+g_2}) = \chi_G(g_1)\chi_{\mathcal{A}}(f_1)^{-1}\chi_{\mathcal{A}}(f_1^{g_2})$$
$$= \chi_G(g_1)\chi_G(g_2)$$

for $g_1, g_2 \in G$. In other words, $\chi_G : G \to \mathbb{C}$ is a continuous homomorphism to the multiplicative structure of $\mathbb{C}$. Since χ_G is bounded and not identically zero, it follows that χ_G is non-zero everywhere, and that χ_G takes values in $\mathbb{S}^1$. This shows that χ_A is defined by $\chi_G \in \widehat{G}$. The identification between the Fourier back transform and the Gelfand transform follows from this, as explained before the proof.   □

The next example shows that the Fourier transform (or in general the Gelfand transform) is not an isometry.

Example 11.39. Let $G = \mathbb{R}$ and $f_1 = \mathbb{1}_{[0,1]} \in L^1(\mathbb{R})$. Then

$$\widehat{f_1}(t) = \begin{cases} \dfrac{e^{-2\pi i t} - 1}{-2\pi i t} = e^{-\pi i t}\dfrac{e^{\pi i t} - e^{-\pi i t}}{2\pi i t} = \dfrac{e^{-\pi i t}\sin(\pi t)}{\pi t} & \text{for } t \neq 0, \\ 1 & \text{for } t = 0 \end{cases}$$

so $\|\widehat{f_1}\|_\infty = 1 = \|f_1\|_1$, but the maximum value of $|\widehat{f_1}(t)|$ is attained precisely at the point $t = 0$. Now consider

$$f_2(x) = \mathbb{1}_{[0,1]}(x) - \mathbb{1}_{[-1,0]}(x) = f_1(x) - f_1(-x)$$

with

$$\widehat{f_2}(t) = \widehat{f_1}(t) - \widehat{f_1}(-t)$$

for $t \in \mathbb{R}$ and $\widehat{f}(0) = 0$. Hence $|\widehat{f_2}(t)|$ achieves its maximum for some $t_0 \neq 0$, so that

$$\|\widehat{f_2}\|_\infty = |\widehat{f_2}(t_0)| \leqslant |\widehat{f_1}(t_0)| + |\widehat{f_1}(-t_0)| < 2\|f_1\|_1 = \|f_2\|_1,$$

showing that the Fourier transform (and hence a Gelfand transform) need not be an isometry.

Exercise 11.40. Let G be as in Proposition 11.38. When does $L^1(G)$ have a unit with respect to convolution?

As the following exercise shows, the theory developed above is quite powerful. In fact the original proof of the Wiener lemma [114] was complicated and the Gelfand theory allows for a clean simple proof.

Exercise 11.41 (Wiener lemma for $C(\mathbb{T}^d)$). Let $f \in C(\mathbb{T}^d)$ be the limit of an absolutely convergent Fourier series with $f(x) \neq 0$ for all $x \in \mathbb{T}^d$. Show that $\frac{1}{f}$ is also the limit of an absolutely convergent Fourier series.

Exercise 11.42. (a) **(Wiener theorem for $L^1(\mathbb{T}^d)$)** Let $f \in L^1(\mathbb{T}^d)$. Show that the span $\langle \lambda_y f \mid y \in \mathbb{Z}^d \rangle$ is dense in $L^1(\mathbb{T}^d)$ if and only if $\widehat{f}(n) = \int f(x)\overline{\chi_n}(x)\,\mathrm{d}x \neq 0$ for all $n \in \mathbb{Z}^d$.
(b) **(Wiener theorem for $L^1(\mathbb{R}^d)$)** Let $f \in L^1(\mathbb{R}^d)$. Show that $\langle \lambda_y f \mid y \in \mathbb{R}^d \rangle$ is dense in $L^1(\mathbb{R}^d)$ if and only if $\widehat{f}(t) \neq 0$ for all $t \in \mathbb{R}^d$.

11.4.1 The Pontryagin Dual

We now show that $\widehat{G}$ is again a topological group.

Proposition 11.43 ($\widehat{G}$ is a topological group). *Let G be a σ-compact locally compact metrizable abelian group and let $\widehat{G}$ be the dual group equipped with the weak* topology from the identification with $\sigma(L^1(G))$ in Proposition 11.38. Then $\widehat{G}$ is also a locally compact σ-compact metrizable abelian group. The weak* topology on $\widehat{G}$ is the topology of uniform convergence on compact sets, that is equivalent to the topology defined by the neighbourhoods*

$$U_{K,\varepsilon}(\chi_0) = \{\chi \in \widehat{G} \mid \|\chi - \chi_0\|_{K,\infty} < \varepsilon\}$$

of $\chi_0 \in \widehat{G}$ for compact sets $K \subseteq G$ and $\varepsilon > 0$.

PROOF. By Proposition 2.51, $C_c(G)$ is dense in $L^1(G)$ which, together with Proposition 8.11, shows that the weak* topology on

$$\widehat{G} = \sigma(L^1(G)) \subseteq \overline{B_1^{L_1(G)^*}}$$

can be defined by functions in $C_c(G)$. So let $N_{f_1,\dots,f_n;\varepsilon}(\chi_0)$ be a neighbourhood of $\chi_0 \in \widehat{G}$ defined by some functions $f_1, \dots, f_n \in C_c(G) \smallsetminus \{0\}$ and $\varepsilon > 0$. Let $M = \max_{j=1,\dots,n} \|f_j\|_1$ and $K = \bigcup_{j=1}^n \operatorname{Supp}(f_j)$. If now $\chi \in U_{K,\varepsilon/M}(\chi_0)$, then

$$\left| \int_G f_j(\chi - \chi_0)\, dm \right| \leqslant \int_K |f_j| |\chi - \chi_0|\, dm < \|f_j\|_1 \tfrac{\varepsilon}{M} \leqslant \varepsilon$$

for $j = 1, \dots, n$. This shows that $U_{K,\varepsilon/M}(\chi_0) \subseteq N_{f_1,\dots,f_n;\varepsilon}(\chi_0)$, which gives one direction for the equivalence of the two topologies.

For the reverse direction, we fix some $\chi_0 \in \widehat{G}$, a compact subset $K \subseteq G$, and $\varepsilon > 0$. We need to find some $f_0, f_1, \dots, f_n \in L^1(G)$ and $\delta > 0$ so that

$$N_{f_0,f_1,\dots,f_n;\delta}(\chi_0) \subseteq U_{K,\varepsilon}(\chi_0). \tag{11.7}$$

We start by defining $f_0 = \frac{1}{m(B_0)} \mathbb{1}_{B_0}$ for some compact neighbourhood B_0 of $0 \in G$ such that

$$\left| \frac{1}{m(B_0)} \int_{B_0} \chi_0 - 1 \right| \leqslant \frac{1}{3}.$$

We will now use a similar argument as in the proof of Proposition 11.38. In fact, if $\chi \in N_{f_0;\frac{1}{3}}(\chi_0)$ then $|\check{f}_0(\chi) - 1| \leqslant \frac{2}{3}$ and so $|\check{f}_0(\chi)| \geqslant \frac{1}{3}$. Moreover, using the relation

$$\widetilde{f_0^h}(\chi) = \int_G f_0(g - h)\chi(g)\, dm(g) = \chi(h)\check{f}_0(\chi),$$

we see that

$$\chi(h) = \left(\check{f}_0(\chi)\right)^{-1} \int_G f_0^h \chi\, dm$$

for every $h \in G$. This also gives

$$|\chi(h) - \chi(0)| \leqslant 3 \|f_0^h - f_0\|_1 \tag{11.8}$$

for all $\chi \in N_{f_0;\frac{1}{3}}(\chi_0)$ and $h \in G$. In other words, the equi-continuity properties of all elements $\chi \in N_{f_0;\frac{1}{3}}(\chi_0)$ at 0 are controlled by the continuity of the map $G \ni h \mapsto f_0^h \in L^1(G)$.

We set $\delta = \min\{\frac{\varepsilon}{5}, \frac{1}{3}\}$. Using (11.8) and Lemma 3.74 we find some open neighbourhood $B \subseteq G$ of $0 \in G$ with compact closure such that $h \in B$, $g_0 \in G$, and $\chi \in N_{f_0;\frac{1}{3}}(\chi_0)$ implies

$$|\chi(g_0 + h) - \chi(g_0)| = |\chi(h) - \chi(0)| < \delta, \tag{11.9}$$

where we used the fact that $\chi \in \widehat{G}$ is a character. Since $K \subseteq G$ is compact, there exists a finite collection $g_1, \ldots, g_n \in K$ such that

$$K \subseteq \bigcup_{j=1}^{n} (g_j + B). \tag{11.10}$$

We define $f_j = \frac{1}{m(B)} \mathbb{1}_{g_j + B}$ for $j = 1, \ldots, n$ and claim that $N_{f_0, f_1, \ldots, f_n, \delta}(\chi_0)$ is the neigbourhood we were looking for. Indeed, let

$$\chi \in N_{f_0, f_1, \ldots, f_n; \delta}(\chi_0) \subseteq N_{f_0;\frac{1}{3}}(\chi_0)$$

and fix some $j \in \{1, \ldots, n\}$. Using (11.9) for $g_0 = g_j$ and $g = g_j + h \in g_j + B$ we obtain $|\chi(g) - \chi(g_j)| < \delta$ and hence also

$$\left| \check{f}_j(\chi) - \chi(g_j) \right| = \left| \int f_j \chi \, dm - \chi(g_j) \right| < \delta. \tag{11.11}$$

For any $g \in g_j + B$ we can now combine (11.9) and (11.11) for χ and χ_0, and the assumption $\chi \in N_{f_j;\delta}(\chi_0)$ to obtain

$$|\chi(g) - \chi_0(g)| \leqslant \delta + |\chi(g_j) - \chi_0(g_j)| + \delta$$
$$\leqslant \left|\chi(g_j) - \check{f}_j(\chi)\right| + \left|\check{f}_j(\chi) - \check{f}_j(\chi_0)\right| + \left|\check{f}_j(\chi_0) - \chi_0(g_j)\right| + 2\delta$$
$$\leqslant \left|\check{f}_j(\chi) - \check{f}_j(\chi_0)\right| + 4\delta < 5\delta \leqslant \varepsilon$$

for all $g \in g_j + B$. Varying j and using (11.10) this implies (11.7). Thus, the neighbourhoods of $\chi_0 \in \widehat{G}$ in the weak* topology are precisely the neighbourhoods of χ_0 with respect to the topology of uniform convergence on compact subsets of G.

With this identification of the topology on $\widehat{G}$, it is straightforward to check that the group operations are continuous. Indeed,

$$\overline{U_{K,\varepsilon}(\chi_0)} = U_{K,\varepsilon}(\overline{\chi_0})$$

for all compact $K \subseteq G$, all $\varepsilon > 0$, and $\chi_0 \in \widehat{G}$ shows that the map

$$\widehat{G} \ni \chi \mapsto \chi^{-1} = \overline{\chi}$$

is continuous. Similarly, for $\chi_0, \eta_0 \in \widehat{G}$ we therefore know that $\chi \in U_{K,\varepsilon/2}(\chi_0)$ and $\eta \in U_{K,\varepsilon/2}(\eta_0)$ imply

$$\|\chi\eta - \chi_0\eta_0\|_{K,\infty} \leqslant \|\chi\eta - \chi_0\eta\|_{K,\infty} + \|\chi_0\eta - \chi_0\eta_0\|_{K,\infty} < \tfrac{\varepsilon}{2} + \tfrac{\varepsilon}{2} = \varepsilon,$$

showing continuity of the group operation $\widehat{G} \times \widehat{G} \ni (\chi, \eta) \mapsto \chi\eta \in \widehat{G}$. □

The following exercises give further examples of the duality between a group and its dual, both viewed as topological groups.

Exercise 11.44. (a) Suppose that G is a compact metrizable abelian group. Show that $\widehat{G}$ is discrete (and countable).
(b) Suppose that G is a countable discrete abelian group. Show that $\widehat{G}$ is compact (and metrizable).

Exercise 11.45. Let G be a σ-compact locally compact metrizable abelian group.
(a) Suppose that G is connected as a topological space. Show that $\widehat{G}$ has no torsion elements, meaning that for any $\chi \in \widehat{G}$ an identity $\chi^n = \mathbb{1}$ for some $n \geqslant 1$ implies that $\chi = \mathbb{1}$ is the trivial character.
(b) Suppose that G is compact and not connected as a topological space. Show that $\widehat{G}$ has torsion elements: there is some $\chi \in \widehat{G} \setminus \{\mathbb{1}\}$ and some $n > 1$ such that $\chi^n = \mathbb{1}$.

11.5 Further Topics

- The study of $L^1(G)$ for a locally compact abelian group can lead to a vast generalization of the theory of Fourier series and the Fourier transform to all such groups. This is known as *Pontryagin duality* or *harmonic analysis on locally compact abelian groups* and will be discussed further in Section 12.8.
- Another important class of Banach algebras with additional structure are the *von Neumann algebras*. These are special C^*-sub-algebras of $B(\mathcal{H})$ for a Hilbert space $\mathcal{H}$. We refer to Blackadar [11] for an overview.

Chapter 12
Spectral Theory and Functional Calculus

In this chapter we use results from Chapter 11 to prove the spectral theorem and develop the functional calculus for single self-adjoint operators and for certain commutative C^*-algebras (arising, for example, from unitary representations of locally compact abelian groups). As an example of a self-adoint operator we discuss the Laplace operator on a regular tree.

12.1 Definitions and Basic Lemmas

In this section we will study the spectrum (as defined for abstract algebras in Section 11.1) in the context of bounded operators on a Hilbert space. More precisely, we fix a complex Hilbert space $\mathcal{H}$, let $\mathcal{A} = \mathrm{B}(\mathcal{H})$ be the Banach algebra of bounded operators, and study the spectrum of some $T \in \mathcal{A}$.

12.1.1 Decomposing the Spectrum

Since an operator with non-trivial kernel cannot be invertible, it is clear that any eigenvalue of $T \in \mathrm{B}(\mathcal{H})$ belongs to the spectrum of T. It is usual to call the set of eigenvalues the *discrete* or *point* spectrum.

Definition 12.1 (Discrete spectrum). We say that $\lambda \in \mathbb{C}$ belongs to the *discrete spectrum* of $T \in \mathrm{B}(\mathcal{H})$, and write $\lambda \in \sigma_{\mathrm{disc}}(T)$, if $\ker(T - \lambda I) \neq \{0\}$.

As we have already seen in Exercise 6.25 and Example 9.1, the discrete spectrum may well be empty for a given bounded operator. For the operators in these examples the notion of eigenvector has to be generalized to a sequence of approximate eigenvectors in the following sense.

Definition 12.2 (Approximate point spectrum). We say that $\lambda \in \mathbb{C}$ belongs to the *approximate point spectrum* of $T \in \mathrm{B}(\mathcal{H})$, and write $\lambda \in \sigma_{\mathrm{appt}}(T)$,

© Springer International Publishing AG 2017

M. Einsiedler, T. Ward, *Functional Analysis, Spectral Theory, and Applications*,
Graduate Texts in Mathematics 276, DOI 10.1007/978-3-319-58540-6_12

if there is a sequence of *approximate eigenvectors* (v_n) in $\mathcal{H}$ with $\|v_n\| = 1$ for all $n \geqslant 1$, and with $\|(T - \lambda I)v_n\| \to 0$ as $n \to \infty$.

For normal operators we will see that the approximate point spectrum coincides with the whole spectrum. We now try to describe the non-discrete part of the spectrum further.

Definition 12.3 (Approximate spectrum). We say that $\lambda \in \mathbb{C}$ belongs to the *approximate spectrum* of $T \in \mathrm{B}(\mathcal{H})$, and write $\lambda \in \sigma_{\mathrm{approx}}(T)$, if there is a sequence of *approximate eigenvectors* (v_n) with $v_n \in (\ker(T - \lambda I))^{\perp}$ and $\|v_n\| = 1$ for all $n \geqslant 1$, and with $\|(T - \lambda I)v_n\| \to 0$ as $n \to \infty$.

Definition 12.4 (Continuous spectrum). We say that $\lambda \in \mathbb{C}$ belongs to the *continuous spectrum* of $T \in \mathrm{B}(\mathcal{H})$ if $\lambda \in \sigma_{\mathrm{cont}}(T) = \sigma_{\mathrm{appt}}(T) \diagdown \sigma_{\mathrm{disc}}(T)$.

We note that the notion of continuous spectrum is quite standard, that of approximate spectrum less so. These two parts of the spectrum are similar, and should both be thought of as a 'complement' to $\sigma_{\mathrm{disc}}(T)$ inside $\sigma_{\mathrm{appt}}(T)$. The advantage of $\sigma_{\mathrm{approx}}(T)$ over $\sigma_{\mathrm{cont}}(T)$ is discussed in the exercises below. Note also that

$$\sigma_{\mathrm{appt}}(T) = \sigma_{\mathrm{disc}}(T) \sqcup \sigma_{\mathrm{cont}}(T) = \sigma_{\mathrm{disc}}(T) \cup \sigma_{\mathrm{approx}}(T).$$

In general the approximate point spectrum may not yet describe the whole spectrum, which motivates the next definition.

Definition 12.5 (Residual spectrum). We say that $\lambda \in \mathbb{C}$ belongs to the *residual spectrum* of $T \in \mathrm{B}(\mathcal{H})$, and write $\lambda \in \sigma_{\mathrm{resid}}(T)$, if $\lambda \notin \sigma_{\mathrm{disc}}(T)$ and

$$\overline{\mathrm{im}(T - \lambda I)} \neq \mathcal{H}.$$

Exercise 12.6. (a) Show that $\sigma_{\mathrm{appt}}(T)$ is a closed subset of $\mathbb{C}$ for any $T \in \mathrm{B}(\mathcal{H})$, and that $\sigma_{\mathrm{approx}}(T)$ is a closed subset of $\mathbb{C}$ for any normal operator $T \in \mathrm{B}(\mathcal{H})$.
(b) Let $\mathcal{H} = L^2([0,1])^2$ and define $T \in \mathrm{B}(\mathcal{H})$ by $T(f, g) = (M_I f, f)$ for all $(f, g) \in \mathcal{H}$ (so that $T(f,g)(x) = (xf(x), f(x))$ for $x \in (0,1)$). Show that $\sigma_{\mathrm{approx}}(T)$ is equal to $(0,1]$, and in particular is not closed.
(c) Find an example of an operator $T \in \mathrm{B}(\mathcal{H})$ for which $\sigma_{\mathrm{disc}}(T)$ and $\sigma_{\mathrm{cont}}(T)$ are not closed subsets of $\mathbb{C}$. More specifically, find an example of a self-adjoint operator for which $\sigma_{\mathrm{disc}}(T)$ is countable and dense in $\sigma_{\mathrm{approx}}(T) = \sigma_{\mathrm{appt}}(T) = [0,1]$.

Exercise 12.7. Suppose $T_j \in \mathrm{B}(\mathcal{H}_j)$ for $j = 1, 2$ are bounded operators on two Hilbert spaces $\mathcal{H}_1, \mathcal{H}_2$. Let $T = T_1 \times T_2 \in \mathrm{B}(\mathcal{H}_1 \times \mathcal{H}_2)$. Show that $\sigma_{\mathrm{disc}}(T) = \sigma_{\mathrm{disc}}(T_1) \cup \sigma_{\mathrm{disc}}(T_2)$, and similarly for σ_{appt} and σ_{approx}. Find an example of a pair of self-adjoint operators showing that the corresponding statement does not hold for the continuous spectrum.

Roughly speaking, for multiplication operators the discrete spectrum corresponds to atoms, and we would expect the continuous spectrum to correspond to the continuous part of the measure, as discussed in the following refinement of Exercise 11.4.

Exercise 12.8. (a) Let μ be a compactly supported σ-finite measure on $\mathbb{C}$, and let

$$(M_I(v))(z) = zv(z)$$

for $v \in L^2_\mu(\mathbb{C})$ be the multiplication operator corresponding to the identity map on $\mathbb{C}$. Show that

$$\sigma_{\mathrm{disc}}(M_I) = \{\lambda \in \mathbb{C} \mid \mu(\{\lambda\}) > 0\},$$
$$\sigma_{\mathrm{appt}}(M_I) = \sigma(M_I) = \mathrm{Supp}(\mu),$$
$$\sigma_{\mathrm{approx}}(M_I) = \{\lambda \in \mathbb{C} \mid \mu(U \smallsetminus \{\lambda\}) > 0 \text{ for every neighbourhood } U \text{ of } \lambda\},$$
$$\overline{\sigma_{\mathrm{cont}}(M_I)} \supseteq \mathrm{Supp}(\mu_{\mathrm{cont}}),$$

and that $\sigma_{\mathrm{resid}}(M_I)$ is empty (here μ_{cont} is the measure determined by the decomposition $\mu = \mu_{\mathrm{cont}} + \mu_{\mathrm{disc}}$, where μ_{cont} has no atoms and μ_{disc} is purely atomic).
(b) Let $(X, \mathcal{B}, \mu)$ be a σ-finite measure space, and let $g : X \to \mathbb{C}$ be a bounded measurable function. Generalize (a) to the multiplication operator M_g on $L^2_\mu(X)$.
(c) Let $X = [0,1] \subseteq \mathbb{R}$, and let λ_{count} be the counting measure on $\mathbb{Q} \cap [0,1]$ considered as a σ-finite measure on X. Let M_I be as in part (a). Describe each of the parts of the spectrum of M_I.

Example 12.9. Let $T : \ell^2(\mathbb{N}) \to \ell^2(\mathbb{N})$ be the operator from Exercise 6.23(c) defined by $T(v_n) = (0, v_1, v_2, \dots)$. Then $\|Tv\| = \|v\|$ for any $v \in \ell^2(\mathbb{N})$, and so

$$0 \notin \sigma_{\mathrm{appt}}(T) = \sigma_{\mathrm{disc}}(T) \cup \sigma_{\mathrm{approx}}(T).$$

However, the image of T is the proper closed subspace $\{v \in \ell^2(\mathbb{N}) \mid v_1 = 0\}$, so $0 \in \sigma_{\mathrm{resid}}(T)$.

Exercise 12.10. For the operator T from Example 12.9, show that

$$\sigma_{\mathrm{disc}}(T) = \varnothing,$$
$$\sigma_{\mathrm{approx}}(T) = \sigma_{\mathrm{cont}}(T) = \mathbb{S}^1 = \{\lambda \in \mathbb{C} \mid |\lambda| = 1\}, \text{ and}$$
$$\sigma_{\mathrm{resid}}(T) = B_1^{\mathbb{C}} = \{\lambda \in \mathbb{C} \mid |\lambda| < 1\}.$$

The next lemma gives the main relationship between the parts of the spectrum from this section and the spectrum in the sense of Definition 11.2 for $\mathcal{A} = \mathrm{B}(\mathcal{H})$.

Lemma 12.11 (Decomposition of spectrum). *Let $\mathcal{H}$ be a complex Hilbert space, and let $T \in \mathrm{B}(\mathcal{H})$. Then*

$$\sigma(T) = \sigma_{\mathrm{appt}}(T) \cup \sigma_{\mathrm{resid}}(T) = \sigma_{\mathrm{disc}}(T) \cup \sigma_{\mathrm{approx}}(T) \cup \sigma_{\mathrm{resid}}(T).$$

Moreover, the residual spectrum is empty if T is normal, so in this case the spectrum coincides with the approximate point spectrum.

PROOF. If $\lambda \notin \sigma(T)$ then, by definition, $(T - \lambda I)^{-1} \in \mathrm{B}(\mathcal{H})$ and so $v_n \in \mathcal{H}$ with $\|v_n\| = 1$ for all $n \geqslant 1$ implies

$$1 = \|v_n\| = \|(T - \lambda I)^{-1}(T - \lambda I)v_n\| \leqslant \|(T - \lambda I)^{-1}\| \|(T - \lambda I)v_n\|.$$

This shows that (v_n) cannot be a sequence of approximate eigenvectors, and hence $\lambda \notin \sigma_{\mathrm{appt}}(T)$. Finally, if $\lambda \notin \sigma(T)$ then $T - \lambda I$ is an onto map, and so $\lambda \notin \sigma_{\mathrm{resid}}(T)$.

The reverse inclusion can be shown almost as directly. Suppose that

$$\lambda \notin \sigma_{\mathrm{appt}}(T) \cup \sigma_{\mathrm{resid}}(T).$$

Then $T - \lambda I$ is injective, since in particular $\lambda \notin \sigma_{\mathrm{disc}}(T)$, and there exists some $\varepsilon > 0$ with

$$\varepsilon \|v\| \leqslant \|(T - \lambda I)v\| \tag{12.1}$$

for all $v \in \mathcal{H}$ since $\lambda \notin \sigma_{\mathrm{approx}}(T)$. Therefore $(T - \lambda I) : \mathcal{H} \to \mathrm{im}(T - \lambda I)$ is bijective and has an inverse $(T - \lambda I)^{-1} : \mathrm{im}(T - \lambda I) \to \mathcal{H}$ that is continuous by (12.1). This implies that $\mathrm{im}(T - \lambda I)$ is complete (check this), and so is a closed subspace of $\mathcal{H}$. Since $\lambda \notin \sigma_{\mathrm{resid}}(T)$, it follows that $\mathrm{im}(T - \lambda I) = \mathcal{H}$ and so $T - \lambda I$ is invertible, so that $\lambda \notin \sigma(T)$.

Suppose now that $T : \mathcal{H} \to \mathcal{H}$ is a normal operator, and that $\lambda \in \mathbb{C}$ has

$$V = \overline{\mathrm{im}(T - \lambda I)} \neq \mathcal{H}.$$

By normality, $T^*(T - \lambda I)v = (T - \lambda I)(T^*v)$ for $v \in \mathcal{H}$, which implies in particular that V is T^*-invariant. By Lemma 6.30 we deduce that $V^\perp$ is T-invariant. Now let $v \in V^\perp \backslash \{0\}$. Then $(T - \lambda I)v \in V^\perp$ by T-invariance, and $(T - \lambda I)v \in V$ by definition. This implies that $(T - \lambda I)v = 0$, and so $\lambda \in \sigma_{\mathrm{disc}}(T)$. It follows that $\sigma_{\mathrm{resid}}(T) = \varnothing$. $\square$

12.1.2 The Numerical Range

The following definition is useful because it gives an 'upper bound' for the spectrum.

Definition 12.12. The *numerical range* of $T \in \mathrm{B}(\mathcal{H})$ is the set

$$N(T) = \{\langle Tv, v \rangle \mid v \in \mathcal{H}, \|v\| = 1\}.$$

Lemma 12.13. *The spectrum of $T \in \mathrm{B}(\mathcal{H})$ is contained in the closure of the numerical range of T.*

PROOF. We have to show that $\lambda \in \mathbb{C} \backslash \overline{N(T)}$ implies that $\lambda \notin \sigma(T)$. By assumption, $|\langle (T - \lambda I)v, v \rangle| = |\langle Tv, v \rangle - \lambda| \geqslant \varepsilon$ for some fixed $\varepsilon > 0$ and all $v \in \mathcal{H}$ with $\|v\| = 1$. This shows that $\lambda \notin \sigma_{\mathrm{approx}}(T)$, and that any vector $v \in \mathcal{H}$ with $\|v\| = 1$ is not orthogonal to $\mathrm{im}(T - \lambda I)$, so $\lambda \notin \sigma_{\mathrm{resid}}(T)$. By Lemma 12.11, we deduce that $\lambda \notin \sigma(T)$. $\square$

Exercise 12.14. Show that $N(T)$ is really only an upper bound for the spectrum of an operator $T \in \mathrm{B}(\mathcal{H})$ by showing that $N(T)$ is the convex hull of the eigenvalues of T if T is diagonalizable, that is, if $\mathcal{H}$ admits an orthonormal basis consisting of eigenvectors of T.

The following is a direct consequence of Lemma 12.13 (giving an easy alternative to the argument used in Lemma 11.35 in the setting of bounded operators on a Hilbert space).

Lemma 12.15. *If $T \in B(\mathcal{H})$ is self-adjoint then $\sigma(T) \subseteq \mathbb{R}$.*

PROOF. For any $v \in \mathcal{H}$ we have $\overline{\langle Tv, v \rangle} = \langle v, Tv \rangle = \langle T^*v, v \rangle = \langle Tv, v \rangle$ so that $\langle Tv, v \rangle \in \mathbb{R}$ and hence $\sigma(T) \subseteq \overline{N(T)} \subseteq \mathbb{R}$ by Lemma 12.13. $\square$

12.1.3 The Essential Spectrum

In this section we describe another notion of spectrum through a series of exercises. Recall the definition of the Calkin algebra $B(\mathcal{H})/K(\mathcal{H})$ from Exercise 6.8. The spectrum of an operator $T \in B(\mathcal{H})$ when considered in the Calkin algebra is the *essential spectrum* of T, denoted $\sigma_{\mathrm{ess}}(T)$, and the spectral radius of T in this algebra is the *essential radius*.

Definition 12.16. A bounded operator $T \in B(\mathcal{H})$ on a separable Hilbert space $\mathcal{H}$ is called a *Fredholm operator* if T is almost injective in the sense that $\dim(\ker(T)) < \infty$, and almost surjective in the sense that $T(\mathcal{H})$ is closed and $\dim(\mathcal{H}/T(\mathcal{H})) < \infty$.

Clearly invertible operators are Fredholm, as is the operator in Example 12.9.

Exercise 12.17. Show that $I - A$ is Fredholm for any compact operator $A \in K(\mathcal{H})$ on a separable Hilbert space $\mathcal{H}$.

Exercise 12.18 (Atkinson's theorem). Let $T \in B(\mathcal{H})$ be a bounded operator on a separable Hilbert space $\mathcal{H}$. Prove that T is Fredholm if and only if there exists some operator $S \in B(\mathcal{H})$ such that $ST - I$ and $TS - I$ are both compact.

Exercise 12.19. Let (X, μ) be a σ-finite measure space and $g \in L^\infty_\mu(X)$. Show that the essential spectrum $\sigma_{\mathrm{ess}}(M_g)$ of the normal operator M_g is given by

$$\sigma_{\mathrm{ess}}(M_g) = \sigma_{\mathrm{approx}}(M_g) \cup \{\lambda \in \sigma_{\mathrm{disc}}(M_g) \mid \dim(\ker(M_g - \lambda I)) = \infty\}.$$

12.2 The Spectrum of a Tree

In this section we want to study the spectrum of the Laplace operator on a $(p + 1)$-regular tree (which has strong connections to the properties of the random walk on the tree).

Let us recall that a graph is a set of vertices $\mathcal{V}$ together with a set of edges $\mathcal{E} \subseteq \mathcal{V} \times \mathcal{V}$. We will assume that the graph is undirected, meaning that $(v, w) \in \mathcal{E}$ if and only if $(w, v) \in \mathcal{E}$ for all $v, w \in \mathcal{V}$. We write $v \sim w$ if there is an edge $(v, w) \in \mathcal{E}$ joining v to w.

More concretely, we fix an integer $p \geqslant 2$ (the case $p = 1$ is quite different and much easier, see Exercise 12.21) and suppose that $(\mathcal{V}, \mathcal{E})$ is a $(p + 1)$-regular tree. This means that $\mathcal{V}$ is countably infinite, connected, every vertex $v \in \mathcal{V}$ is connected to exactly $(p + 1)$ further vertices by edges in $\mathcal{E}$, and there are no loops (see Figure 12.1).

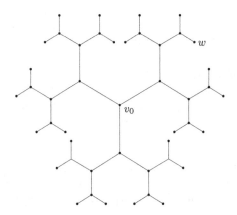

Fig. 12.1: The 3-regular tree is illustrated here by showing all vertices of distance no more than 4 from a given initial vertex v_0 (also called the *root*). Of course the pattern repeats indefinitely, from w and from all the other vertices at distance 4 from our chosen root v_0.

At first sight there are three natural operators that we can define on $\ell^2(\mathcal{V})$ using the tree structure (and our discussion will also involve a fourth). In the following we fix $p \geqslant 2$ and a $(p + 1)$-regular tree $(\mathcal{V}, \mathcal{E})$.

Definition 12.20. The *averaging operator* on $\ell^2(\mathcal{V})$ is defined by

$$T(f)(v) = \frac{1}{p+1} \sum_{w \sim v} f(w),$$

for $f \in \ell^2(\mathcal{V})$. It replaces the value of a function at a vertex v by the average $T(f)(v)$ of all values $f(w)$ at the direct neighbours $w \sim v$ in the tree. The *summing operator* is defined by $S = (p+1)T$, simply summing the values at the immediate neighbours. Finally, the *Laplace operator* $\Delta = I - T$ compares the value at each vertex with the average over all its immediate neighbours.

Clearly T, S, and Δ are essentially equivalent. If one is understood well, then the same applies to the other two.

Exercise 12.21. Set $p = 1$, so that we may think of the $(p+1)$-regular graph as having vertex set $\mathcal{V} = \mathbb{Z}$ and edge set $\mathcal{E} = \{(n, n \pm 1) \mid n \in \mathbb{Z}\}$. Show that the summing operator S is self-adjoint, and describe its spectrum.

Exercise 12.22. Show that the summing operator $S : \ell^2(\mathcal{V}) \to \ell^2(\mathcal{V})$ on a $(p+1)$-regular tree is a self-adjoint bounded operator with $\|S\| \leqslant p + 1$. Show that there is no eigenvalue $\lambda \in \sigma_{\mathrm{disc}}(S)$ of absolute value $|\lambda| = p + 1$.

12.2.1 The Correct Upper Bound for the Summing Operator

While it is not difficult to see that $\|S\| \leqslant p + 1$, one might also guess that this upper bound is not the real value of $\|S\|$. Indeed, the proof of the last statement of Exercise 12.22 already hints at this. Due to the very rapid growth in the number of vertices in balls $B_n^{\mathcal{V}}(v_0)$ (measured with respect to the natural path length on the tree), elements of $\ell^2(\mathcal{V})$ must decay rather rapidly. We start by calculating $\|S\|$ and go on to discuss the spectrum of S on $\ell^2(\mathcal{V})$.

Theorem 12.23. [32] Let $p \geqslant 2$ and let $(\mathcal{V}, \mathcal{E})$ be a $(p+1)$-regular tree. The summing operator $S : \ell^2(\mathcal{V}) \to \ell^2(\mathcal{V})$ satisfies $\|S\| = 2\sqrt{p} < p + 1$.

PROOF OF THE LOWER BOUND IN THEOREM 12.23. We first show $\|S\| \geqslant 2\sqrt{p}$ by considering the function $f = f_N$ defined by

$$f(v) = \begin{cases} p^{-\frac{1}{2}\mathrm{d}(v,v_0)} & \text{if } \mathrm{d}(v, v_0) \leqslant N; \\ 0 & \text{if } \mathrm{d}(v, v_0) > N, \end{cases}$$

where $\mathrm{d}(\cdot, \cdot)$ denotes the distance function on $\mathcal{V}$ (cf. p. 401), $v_0 \in \mathcal{V}$ is a fixed initial vertex of $\mathcal{V}$, and $N \geqslant 1$ is an arbitrary integer. First note that

$$\|f\|_2^2 = 1 + \sum_{v \sim v_0} p^{-1} + \cdots + \sum_{\mathrm{d}(v,v_0)=N} p^{-N}$$

$$= 1 + (p+1)p^{-1} + \cdots + (p+1)p^{N-1}p^{-N} = 1 + N\left(1 + \tfrac{1}{p}\right). \quad (12.2)$$

Now calculate

$$S(f)(v) = p^{-(n-1)/2} + \sum_{\substack{w \sim v \\ \mathrm{d}(w,v_0)=n+1}} p^{-(n+1)/2} = \sqrt{p}\,p^{-n/2} + pp^{-(n+1)/2} = 2\sqrt{p}f(v)$$

whenever $1 \leqslant n = \mathrm{d}(v, v_0) < N$. This gives

$$\|S(f)\|_2^2 \geqslant \sum_{n=1}^{N-1} \left(2\sqrt{p}\right)^2 \sum_{\mathrm{d}(v,v_0)=n} |f(v)|^2 > \left(2\sqrt{p}\right)^2 (N-1)\left(1 + \tfrac{1}{p}\right)$$

by using the same calculation as for $\|f\|_2^2$ again. On dividing this lower bound by (12.2) and letting $N \to \infty$ we deduce that $\|S\| \geqslant 2\sqrt{p}$. $\qquad\square$

Exercise 12.24. Show that the sequence $\left(\frac{1}{\|f_N\|} f_N\right)$ in the previous proof is a sequence of approximate eigenvectors of S in the sense of Definition 12.2.

For the proof of the upper bound we use an argument that goes back to work of Gabber and Galil [38], which can also be used for other graphs.

PROOF OF THE UPPER BOUND IN THEOREM 12.23. Let $\mathcal{G} = (\mathcal{V}, \mathcal{E})$ be an undirected graph with the property that every vertex $v \in \mathcal{V}$ has at most N neighbours for some fixed $N \in \mathbb{N}$. The summing operator S is again defined by

$$S(f)(v) = \sum_{w \sim v} f(w)$$

for $v \in \mathcal{V}$. Notice that

$$\|S(f)\|_2^2 = \sum_{v \in \mathcal{V}} \left| \sum_{w \sim v} f(w) \right|^2 \leqslant \sum_{v \in \mathcal{V}} N^2 \max_{w \sim v}(|f(w)|^2) \leqslant \sum_{w \in \mathcal{V}} N^3 |f(w)|^2$$

for all $f \in \ell^2(\mathcal{V})$, so that $\|S\| < \infty$. Given $f_1, f_2 \in \ell^2(\mathcal{V})$ we have

$$\begin{aligned}
\langle f_1, S f_2 \rangle &= \sum_{v \in \mathcal{V}} f_1(v) \sum_{w \sim v} \overline{f_2}(w) \\
&= \sum_{w \sim v} f_1(v) \overline{f_2}(w) = \sum_{w \in \mathcal{V}} \sum_{v \sim w} f_1(v) \overline{f_2}(w) = \langle S f_1, f_2 \rangle,
\end{aligned}$$

which shows that $S = S^*$ is self-adjoint.

We let $\overrightarrow{\mathcal{E}}$ denote the set of edges in the directed graph $\overrightarrow{\mathcal{G}} = (\mathcal{V}, \overrightarrow{\mathcal{E}})$, where we replace each edge in $\mathcal{G}$ by two edges going in either direction, so formally $\overrightarrow{\mathcal{E}} = \{(v, w) \in \mathcal{V} \times \mathcal{V} \mid v \sim w\}$. Suppose that $\lambda : \overrightarrow{\mathcal{E}} \to \mathbb{R}_{>0}$ is a function satisfying $\lambda\big((w, v)\big) = \lambda\big((v, w)\big)^{-1}$ for each pair of neighbours $(v, w) \in \mathcal{E}$, and suppose that

$$\rho = \sup_{v \in \mathcal{V}} \sum_{w \sim v} \lambda(v, w) < \infty.$$

We claim that this implies $\|S\| \leqslant \rho$.

To prove the claim fix some $f \in \ell^2(\mathcal{V})$. Then for any two neighbours $v \sim w$ in $\mathcal{G}$ we have

$$|f(v)|^2 \lambda(v, w) + |f(w)|^2 \lambda(w, v) \mp \left(f(v)\overline{f(w)} + f(w)\overline{f(v)} \right)$$

$$= \left| f(v)\sqrt{\lambda(v, w)} \mp f(w)\sqrt{\lambda(w, v)} \right|^2 \geqslant 0$$

or equivalently

$$\pm 2 \Re\big(f(v)\overline{f(w)} \big) \leqslant |f(v)|^2 \lambda(v, w) + |f(w)|^2 \lambda(w, v).$$

Summing over all neighbouring vertices we obtain from this

$$\pm 2\Re \langle f, Sf \rangle = \pm 2\Re \sum_{v \in \mathcal{V}} f(v) \sum_{w:w\sim v} \overline{f(w)} = \pm 2\Re \sum_{v,w:v\sim w} f(v)\overline{f(w)}$$

$$\leqslant \sum_{v} |f(v)|^2 \sum_{w:w\sim v} \lambda(v,w) + \sum_{w} |f(w)|^2 \sum_{v:v\sim w} \lambda(w,v)$$

$$\leqslant 2\rho \|f\|_2^2.$$

Since S is self-adjoint we have $\langle f, Sf \rangle = \langle Sf, f \rangle \in \mathbb{R}$ and therefore

$$|\langle Sf, f \rangle| \leqslant \rho \|f\|_2^2$$

for any $f \in \ell^2(\mathcal{V})$. Using Lemma 6.31 this implies that $\|S\| \leqslant \rho$.

It remains to define $\lambda(v,w)$ in the context of the $(p+1)$-regular tree so that $\rho = 2\sqrt{p}$. We again use a root $v_0 \in \mathcal{V}$ and define

$$\lambda(v,w) = \begin{cases} p^{-1/2} & \text{if } \mathsf{d}(w,v_0) = \mathsf{d}(v,v_0) + 1; \\ p^{1/2} & \text{if } \mathsf{d}(w,v_0) = \mathsf{d}(v,v_0) - 1. \end{cases}$$

With this we obtain

$$\sum_{w\sim v_0} \lambda(v_0, w) = (p+1)p^{-1/2} = p^{1/2} + p^{-1/2} < 2\sqrt{p}$$

in the case $v = v_0$ and

$$\sum_{w\sim v} \lambda(v, w) = p^{1/2} + pp^{-1/2} = 2\sqrt{p}$$

in the case $\mathsf{d}(v,v_0) \geqslant 1$. $\qquad\square$

12.2.2 The Spectrum of S

We outline in this section how to obtain the complete description of the spectrum of S.

Proposition 12.25. *The spectrum of the summing operator S on a $(p+1)$-regular tree is the interval $[-2\sqrt{p}, 2\sqrt{p}]$.*

We will leave the details of the proof as an exercise, explaining just the crucial ideas. Since S is self-adjoint, $\sigma(S) \subseteq \mathbb{R}$ by Lemma 12.15. By Theorem 12.23 we know that $\|S\| = 2\sqrt{p}$, so $\sigma(S) \subseteq [-2\sqrt{p}, 2\sqrt{p}]$. For the reverse inclusion we generalize Exercise 12.24 and give for each $\theta \in [0, \pi]$ a sequence of approximate eigenvectors (f_N) for $\lambda = 2\sqrt{p}\cos\theta$. So we again fix a root vertex v_0 and define f_N by

$$f_N(v) = \begin{cases} \left(e^{i\theta}p^{-1/2}\right)^{\mathsf{d}(v,v_0)} & \text{if } \mathsf{d}(v,v_0) \leqslant N; \\ 0 & \text{otherwise.} \end{cases}$$

Exercise 12.26. Calculate Sf_N, and show that $\left(\frac{1}{\|f_N\|}f_N\right)$ is a sequence of approximate eigenvectors of S for $\lambda = 2\sqrt{p}\cos\theta$.

12.2.3 No Eigenvectors on the Tree

We outline in this subsection, via a series of exercises, a proof of the fact that the summing operator S on the $(p+1)$-regular tree has no discrete spectrum. For this proof we will use yet another normalization of the averaging and summing operators. We refer to this as the *unitarily normalized* summation,

$$\mathbf{U}_1 = \tfrac{1}{\sqrt{p}}S.$$

In fact we will also need the operators $\mathbf{U}_n$ for $n \geqslant 0$ as defined in the next exercise.

Exercise 12.27. For any $n \geqslant 0$, let $\mathbf{U}_n$ be the operator that maps any function f on a $(p+1)$-regular tree to the function $\mathbf{U}_n(f)$ defined by

$$\mathbf{U}_n(f)(v) = \frac{1}{p^{n/2}} \sum_{\substack{k \leqslant n, \\ k \equiv n(\mathrm{mod}2)}} \sum_{w \sim_k v} f(w),$$

where $w \sim_k v$ means that w and v have distance k in the $(p+1)$-regular tree. Then the sequence of operators $(\mathbf{U}_n)$ satisfies $\mathbf{U}_0 = I$, $\mathbf{U}_1 = \frac{1}{\sqrt{p}}S$, and

$$\mathbf{U}_{n+1} = \mathbf{U}_1 \circ \mathbf{U}_n - \mathbf{U}_{n-1}$$

for $n \geqslant 1$.

The recurrence relation above is classical.[33]

Definition 12.28. The *Chebyshev polynomials of the second kind* are the polynomials $U_n \in \mathbb{Z}[x]$ defined recursively by $U_0(x) = 1$, $U_1(x) = 2x$, and $U_{n+1}(x) = 2xU_n(x) - U_{n-1}(x)$ for $n \geqslant 1$.

This sequence of polynomials has the following concrete connection to trigonometric functions.

Exercise 12.29. Let $x = \cos\theta$ for some $\theta \in (0, \pi)$. Show that

$$U_n(x) = \frac{\sin\big((n+1)\theta\big)}{\sin\theta} \tag{12.3}$$

for all $n \geqslant 1$.

Exercise 12.30. Suppose that $f \in \ell^2(\mathcal{V})$ is an eigenfunction for $\mathbf{U}_1$ with corresponding eigenvalue $\lambda \in [-2, 2]$, and derive a contradiction as follows.
(a) Show that

$$\sum_{w \sim_{2n} v} |f(w)|^2 \geqslant \frac{1}{2}\left(U_{2n}(\cos\theta) - \frac{1}{p}U_{2n-2}(\cos\theta)\right)^2 |f(v)|^2$$

for any $n \geqslant 1$.

(b) Show that it is enough to consider the case $\lambda \geqslant 0$ so that we may write $\lambda = 2\cos\theta$ for some $\theta \in [0, \frac{\pi}{2}]$.

(c) Assuming $f(v) \neq 0$, show that

$$\sum_{d(w,v) \leqslant n} |f(w)|^2 \longrightarrow \infty$$

as $n \to \infty$, and conclude that $\mathbf{U}_1$ (or, equivalently, S) has no eigenfunctions in $\ell^2(\mathcal{V})$.

12.3 Main Goals: The Spectral Theorem and Functional Calculus

The main goal of this chapter is to establish two related theorems about normal operators, the first of which gives a complete classification of normal operators in terms of operators as in the next example (which featured in other forms before).

Example 12.31. Let $\mathcal{H} = L^2(X, \mu)$ for a σ-finite measure space (X, μ), and let $g : X \to \mathbb{C}$ be a bounded measurable function. The multiplication operator M_g is then normal on $\mathcal{H}$. We claim that the spectrum $\sigma(M_g)$ is the *essential range* of g, consisting of all $z \in \mathbb{C}$ with the property that $\mu(g^{-1}U) > 0$ for any neighbourhood U of z. Note first that we have $M_g - \lambda I = M_{g-\lambda}$. If $X_\lambda = \{x \in X \mid g(x) = \lambda\}$ has positive measure (which clearly implies that λ belongs to the essential range), then λ lies in $\sigma_{\mathrm{disc}}(M_g)$ since, for example, $\mathbb{1}_B \in \ker(M_g - \lambda I) \smallsetminus \{0\}$ for any measurable $B \subseteq X_\lambda$ of positive finite measure. If on the other hand λ lies in $\sigma_{\mathrm{disc}}(M_g)$ and $v \in \ker(M_g - \lambda I) \smallsetminus \{0\}$, then $\mu(\{x \in X \mid v(x) \neq 0\} \smallsetminus X_\lambda) = 0$, so that $\mu(X_\lambda) > 0$.

So suppose now $g(x) \neq \lambda$ almost everywhere. Then we can solve the equation $(M_g - \lambda I)u = v$, formally, for any $v \in L^2(X, \mu)$, by putting

$$u = (g - \lambda)^{-1}v,$$

and this is in fact the only solution as a set-theoretic function on X. It follows that $\lambda \notin \sigma(M_g)$ if and only if the operator

$$v \longmapsto (g - \lambda)^{-1}v$$

is a bounded linear map on $L^2(X, \mu)$. By Corollary 4.30, we know this is equivalent to asking that $(g - \lambda)^{-1}$ be an L_μ^∞ function on X. This translates to the condition that there exists some $C > 0$ such that

$$\mu(\{x \in X \mid |(g(x) - \lambda)^{-1}| > C\}) = 0,$$

or equivalently that $\mu(\{x \in X \mid |g(x) - \lambda| < \frac{1}{C}\}) = 0$, which means that λ is not in the essential range of g. This chain of equivalences gives the claim.

It is convenient to observe that the essential range coincides with the support of the push-forward measure $\nu = g_*\mu$ on $\mathbb{C}$ so that the above can be reformulated as

$$\sigma(M_g) = \operatorname{Supp}(g_*\mu). \tag{12.4}$$

In particular, if X is a bounded subset of $\mathbb{C}$ and $g(z) = I(z) = z$, then the spectrum of M_I is the support of μ.

Exercise 12.32. Let (X, μ) be a σ-finite measure space. Show that there exists a finite measure ν on X with $\nu \ll \mu \ll \nu$ and a unitary isomorphism $\Phi : L^2(X, \mu) \to L^2(X, \nu)$ such that $\Phi \circ M_g = M_g \circ \Phi$ whenever $g : X \to \mathbb{C}$ is measurable and bounded and M_g acts on the spaces $L^2(X, \mu)$ and $L^2(X, \nu)$, respectively.

Theorem 12.33 (Spectral theorem for normal operators). *Let $\mathcal{H}$ be a separable complex Hilbert space, and let $T \in B(\mathcal{H})$ be a normal operator on $\mathcal{H}$. Then there exists a finite measure space (X, μ), a bounded measurable function $g : X \to \mathbb{C}$, and a unitary isomorphism $\phi : \mathcal{H} \to L^2_\mu(X)$ such that*

$$
\begin{array}{ccc}
\mathcal{H} & \xrightarrow{\ \ T\ \ } & \mathcal{H} \\
{\scriptstyle \phi}\downarrow & & \downarrow{\scriptstyle \phi} \\
L^2_\mu(X) & \xrightarrow[\ M_g\ ]{} & L^2_\mu(X)
\end{array}
$$

commutes.

As we will see, we can always choose $X = \sigma(T) \times \mathbb{N}$, which we will identify with the countable disjoint union $\bigsqcup_{n \in \mathbb{N}} \sigma(T)$. Moreover, the measure μ on X will be obtained from countably many *spectral measures* which we will define using the *continuous functional calculus*. Finally, g will be the bounded continuous map $g(z, n) = z$ on X.

The second goal is to establish the *measurable functional calculus*, which allows us to obtain normal operators $f(T)$ from any normal $T \in B(\mathcal{H})$ and any bounded measurable $f \in \mathscr{L}^\infty(\sigma(T))$. Notice that the function f in this formulation lies in the space $\mathscr{L}^\infty(\sigma(T))$ defined in Example 2.24(8) rather than $L^\infty_\mu(\sigma(T))$ for some measure μ, because we do not have, at this stage, any preferred measure (see Exercise 12.71 for more on this). For a given normal $T \in B(\mathcal{H})$ this assignment

$$\mathscr{L}^\infty(\sigma(T)) \ni f \longmapsto f(T) \in B(\mathcal{H}) \tag{12.5}$$

has many natural functorial properties:

(FC1) *(Polynomials)* If $f(z) = \sum_{j=0}^{n} a_j z^j$ for all $z \in \sigma(T)$, then $f(T) = \sum_{j=0}^{n} a_j T^j$.

(FC2) *(Continuity)* The map in (12.5) is continuous, with $\|f(T)\|_{\mathrm{op}} \leqslant \|f\|_\infty$ for $f \in \mathscr{L}^\infty(\sigma(T))$, and is an isometry on $C(\sigma(T))$, meaning that

$$\|f(T)\|_{\mathrm{op}} = \|f\|_\infty \tag{12.6}$$

for $f \in C(\sigma(T))$.

(FC3) *(Algebra)* The map in (12.5) is an algebra homomorphism. In particular, $f_1(T)$ commutes with $f_2(T)$ for $f_1, f_2 \in \mathscr{L}^\infty(\sigma(T))$. Moreover,

$$(f(T))^* = \overline{f}(T^*)$$

for $f \in \mathscr{L}^\infty(\sigma(T))$.

(FC4) *(Multiplication operators)* If $\mathcal{H}$ is unitarily isomorphic to $L^2_\mu(X)$ and T in $\mathrm{B}(\mathcal{H})$ corresponds (via ϕ and the commutative diagram) to M_g on $L^2_\mu(X)$ as in Theorem 12.33, then $f \circ g$ is defined almost everywhere and $f(T)$ corresponds to $M_{f \circ g}$ for any $f \in \mathscr{L}^\infty(\sigma(T))$.

(FC5) *(Commuting operators)* If $V \subseteq \mathcal{H}$ is a closed subspace that is invariant under both T and T^*, then V is also $f(T)$-invariant for all $f \in \mathscr{L}^\infty(\sigma(T))$. Moreover, if $S \in \mathrm{B}(\mathcal{H})$ commutes with the normal operator $T \in \mathrm{B}(\mathcal{H})$ and its adjoint, then S also commutes with $f(T)$ for all $f \in \mathscr{L}^\infty(\sigma(T))$.

(FC6) *(Iteration)* If $f \in \mathscr{L}^\infty(\sigma(T))$ and $h \in \mathscr{L}^\infty\big(\overline{f(\sigma(T))}\big)$, then

$$h(f(T)) = (h \circ f)(T).$$

Theorem 12.34 (Measurable functional calculus for normal operators). *Let $\mathcal{H}$ be a complex Hilbert space, and $T \in \mathrm{B}(\mathcal{H})$ a normal operator. Then there exists a functional calculus for T — that is, a uniquely determined map as in (12.5) with the properties (FC1)–(FC6).*

Example 12.35. (a) Suppose $T \in \mathrm{B}(\mathcal{H})$ is a normal operator on a complex Hilbert space $\mathcal{H}$, and suppose that $f(z) = \sum_{n \geqslant 0} a_n z^n$ is a power series with radius of convergence R such that $\sigma(T) \subseteq B^{\mathbb{C}}_R$. Then we may restrict the function f to $\sigma(T)$ so that the power series converges uniformly and absolutely. By combining (FC1) with (FC2), it follows that $f(T) = \sum_{n \geqslant 0} a_n T^n$ is also defined by the absolutely converging power series.

(b) Let (X, μ) be a finite (or σ-finite) measure space, $g : X \to \mathbb{C}$ a bounded measurable function, and let

$$M_g : L^2_\mu(X) \to L^2_\mu(X)$$

be the multiplication operator as in Example 12.31. For $f \in \mathbb{C}[z]$ it follows from (FC1) that $f(M_g) = M_{f \circ g}$. Hence it is reasonable to expect that this holds more generally for $f \in C(\sigma(M_g))$ or even $f \in \mathscr{L}^\infty(\sigma(M_g))$ as in (FC4). Note that the composition $f \circ g$ is well-defined in $L^\infty_\mu(X)$, although the image of g might not lie entirely in $\sigma(M_g)$ (and so in the domain of the continuous or measurable function). In fact, the description of the spectrum $\sigma(M_g)$ in (12.4) shows that

$$\mu(\{x \in X \mid g(x) \notin \sigma(M_g)\}) = g_*\mu(\mathbb{C} \smallsetminus \sigma(M_g)) = 0,$$

(the complement of the support being the largest open set with measure 0) so that, for almost every $x \in X$, $g(x)$ lies in $\sigma(M_g)$ and therefore $f(g(x))$ is defined for almost every x (we can set the value of the function $f(g(x))$ on the zero-measure subset where $g(x) \notin \sigma(M_g)$ to be 0). This shows that all the expressions in property (FC4) make sense.

It is tempting to say that in view of (FC4) the existence of the map in (12.5) is simply a consequence of Theorem 12.33. However, if we really used Theorem 12.33 and (FC4) as the definition of the functional calculus then we would not know whether it is *canonical* — that is, independent of the isomorphism ϕ in Theorem 12.33. The fact that we will define the functional calculus independently of the isomorphism ϕ, but nonetheless obtain (FC4) as one of its properties, demonstrates that there is only one reasonable way to define $f(T)$ for $f \in \mathscr{L}^\infty(\sigma(T))$. In particular, we obtain the uniqueness claimed in Theorem 12.34.

Exercise 12.36. Show that Theorem 12.33, (FC4), and (FC5) together imply the uniqueness claim in Theorem 12.34 (despite the difference in the assumptions on $\mathcal{H}$).

For simplicity we will start with the case of self-adjoint operators, which only needs the material from Section 11.1 and Section 11.2. In Section 12.5 we will start the discussion of commutative C^*-sub-algebras of $B(\mathcal{H})$, which includes the case of finitely many commuting normal operators and builds on Section 11.3.

Before continuing, let us note a slightly confusing point in the notation for the functional calculus. As usual I denotes the identity map $x \mapsto x$ and $\mathbb{1}$ denotes the constant function $x \mapsto 1$. Thus (FC1) states in particular that $\mathbb{1}(T) = I$ and $I(T) = T$ for any normal operator $T \in B(\mathcal{H})$. The connection to multiplication operators should help to explain why this makes sense.

12.4 Self-Adjoint Operators

The goal of this section is to show how to define an operator $f(T)$ where the operator $T \in B(\mathcal{H})$ is a self-adjoint operator and $f \in C(\sigma(T))$ and to use this functional calculus to clarify the relationship between the spectrum of an operator and its action on vectors. This will imply Theorem 12.33 for these operators.

12.4.1 Continuous Functional Calculus

For certain functions f the definition of $f(T)$ for an operator T on a Hilbert space $\mathcal{H}$ is clear. For example, if

$$p(z) = \sum_{j=0}^{d} a_j z^j$$

is a polynomial with coefficients in $\mathbb{C}$ restricted to $\sigma(T)$, then the only reasonable definition for $p(T)$ is

$$p(T) = \sum_{j=0}^{d} a_j T^j \in \mathrm{B}(\mathcal{H}),$$

where T^0 is defined to be the identity I on $\mathcal{H}$.

In fact, this polynomial definition makes sense for any $T \in \mathrm{B}(\mathcal{H})$, not only for T normal, but there is a technical point which explains why only normal operators are really suitable here. If $\sigma(T)$ is finite, then a polynomial of unknown degree is not uniquely determined by its restriction to $\sigma(T)$. Thus in this case the definition above *a priori* only gives a map $\mathbb{C}[T] \to \mathrm{B}(\mathcal{H})$, not one defined on $C(\sigma(T))$. We cannot hope to have a functional calculus only depending on the spectrum if this dependency is real, and simple examples show that it sometimes is. Consider, for example, the operator $A \in \mathrm{B}(\mathbb{C}^2)$ given by the matrix

$$A = \begin{pmatrix} 0 & 1 \\ 0 & 0 \end{pmatrix}.$$

Clearly $\sigma(A) = \{0\}$, so the polynomials $p_1(z) = z$ and $p_2(z) = z^2$ coincide when restricted to $\sigma(A)$, but $p_1(A) = A \neq 0 = p_2(A)$.

However, if we assume that T is normal, this problem does not arise, because then (as we will show) for every $p \in \mathbb{C}[z]$ we have $\|p(T)\|_{\mathrm{op}} = \|p\|_{\infty,\sigma(T)}$, as claimed in (12.6). This suggests that we should attempt to define, for any function $f \in C(\sigma(T))$, the functional calculus for T applied to f by

$$\mathsf{FC}_T(f) = \lim_{n \to \infty} p_n(T), \tag{12.7}$$

where (p_n) is a sequence of polynomials with $\|f - p_n\|_{\infty,\sigma(T)} \to 0$ as $n \to \infty$, which will then allow us to define $f(T)$ to be $\mathsf{FC}_T(f)$.

This definition is indeed sensible and possible, and the basic properties of this construction are given in the following theorem. Roughly speaking, any operation on (or property of) the function f which is reasonable corresponds to an analogous operation on (or property of) $f(T)$.

Theorem 12.37 (Continuous functional calculus). *Let $\mathcal{H}$ be a complex Hilbert space and $T \in \mathrm{B}(\mathcal{H})$ a self-adjoint bounded operator. Then there exists* *a unique linear map*

$$\mathsf{FC} = \mathsf{FC}_T : C(\sigma(T)) \longrightarrow \mathrm{B}(\mathcal{H}),$$

which we will also denote by $f \mapsto f(T) = \mathsf{FC}(f)$, with the following properties:

(1) *For any polynomial $p(z) = \sum_{j=0}^{d} a_j z^j \in \mathbb{C}[z]$ we have*

$$\mathsf{FC}(p) = p(T) = \sum_{j=0}^{d} a_j T^j$$

   (that is, $\mathsf{FC}(f)$ is just $f(T)$ when f is a polynomial, extending the definition above).

(2) *For any $f \in C(\sigma(T))$ we have*

$$\|\mathsf{FC}(f)\|_{\mathrm{op}} = \|f\|_{\infty,\sigma(T)}. \tag{12.8}$$

(3) *The map FC is a Banach algebra homomorphism, meaning that*

$$\mathsf{FC}(f_1 f_2) = \mathsf{FC}(f_1)\mathsf{FC}(f_2)$$

   for $f_1, f_2 \in C(\sigma(T))$ and $\mathsf{FC}(\mathbb{1}) = I$. For any $f \in C(\sigma(T))$, we have

$$\mathsf{FC}(f)^* = \mathsf{FC}(\bar{f})$$

   (that is, $f(T)^* = \bar{f}(T)$), and in particular $f(T)$ is normal.

(4) *If $\lambda \in \sigma_{disc}(T)$ is in the point spectrum then*

$$\ker(T - \lambda I) \subseteq \ker(f(T) - f(\lambda)I).$$

   If $T = M_g$ for some bounded measurable $g : X \to \mathbb{C}$ and a finite measure space (X, μ), then $f(M_g) = M_{f \circ g}$ for all $f \in C(\sigma(M_g))$.

As already observed, the essence of the proof of existence of FC is to show that (12.7) is a valid definition. After having established its existence we will again simply write $f(T) = \mathsf{FC}_T(f)$ for all $f \in C(\sigma(T))$.

Lemma 12.38. *Let $\mathcal{H}$ be a complex Hilbert space.*

(1) *For $T \in \mathrm{B}(\mathcal{H})$ and a polynomial $p \in \mathbb{C}[z]$, define $p(T) \in \mathrm{B}(\mathcal{H})$ as before. Then*

$$\sigma(p(T)) = p(\sigma(T)). \tag{12.9}$$

(2) *Let $T \in \mathrm{B}(\mathcal{H})$ be normal and let $p \in \mathbb{C}[z]$ be a polynomial. Then*

$$\|p(T)\|_{\mathrm{op}} = \|p\|_{\infty,\sigma(T)}. \tag{12.10}$$

PROOF. For (1), observe first that the statement is trivially true if p is a constant. If p has degree at least one, then fix $\lambda \in \mathbb{C}$ and factor the polynomial $p(z) - \lambda$ in $\mathbb{C}[z]$ to give

$$p(z) - \lambda = \alpha \prod_{1 \leqslant i \leqslant d} (z - \lambda_i),$$

for some $\alpha \in \mathbb{C} \setminus \{0\}$ and complex numbers $\lambda_1, \ldots, \lambda_d \in \mathbb{C}$ (not necessarily distinct). Since $p \mapsto p(T)$ is an algebra homomorphism, it follows that

$$p(T) - \lambda I = \alpha \prod_{1 \leqslant i \leqslant d} (T - \lambda_i I).$$

If $\lambda \notin p(\sigma(T))$, then the solutions λ_i to the equation $p(z) = \lambda$ are *not* in $\sigma(T)$, so each factor $T - \lambda_i I$ is invertible, and hence $p(T) - \lambda I$ is invertible. It follows that

$$\sigma(p(T)) \subseteq p(\sigma(T)).$$

Conversely, if $\lambda \in p(\sigma(T))$, then one of the λ_i must lie in $\sigma(T)$. Because the factors commute, we can assume without loss of generality that either $i = 1$ if $T - \lambda_i I$ is not surjective — in which case $p(T) - \lambda I$ is not surjective either, or $i = d$ if $T - \lambda_i I$ is not injective — in which case neither is $p(T) - \lambda I$. In all situations, $\lambda \in \sigma(p(T))$, proving the reverse inclusion by Proposition 4.25. The use of Proposition 4.25 here can be avoided by using the fact that

$$\sigma(T) = \sigma_{\mathrm{appt}}(T) \cup \sigma_{\mathrm{resid}}(T)$$

in Lemma 12.11 and arguing for $\lambda_i \in \sigma_{\mathrm{appt}}(T)$ as in the case where $T - \lambda_i I$ is not injective.

For (2), we note first that $\mathsf{FC}(p) = p(T)$ is normal if T is. By the improved spectral radius formula (Proposition 11.21), we have

$$\|p(T)\|_{\mathrm{op}} = \max_{\lambda \in \sigma(p(T))} |\lambda|,$$

and by (12.9), we get

$$\|p(T)\|_{\mathrm{op}} = \max_{\lambda \in p(\sigma(T))} |\lambda| = \max_{\lambda \in \sigma(T)} |p(\lambda)| = \|p\|_{\infty, \sigma(T)},$$

as desired. $\square$

PROOF OF THEOREM 12.37. Let $T \in \mathsf{B}(\mathcal{H})$ be self-adjoint so that $\sigma(T)$ is a compact subset of $\mathbb{R}$. By Lemma 12.38, we deduce that the map

$$\mathsf{FC} : (\mathbb{C}[z], \| \cdot \|_{\infty, \sigma(T)}) \longrightarrow \mathsf{B}(\mathcal{H})$$

sending p to $p(T)$ is linear and continuous (indeed, is an isometry). Hence it extends uniquely, using the Stone–Weierstrass theorem (Theorem 2.40) and the automatic extension to the closure (Proposition 2.59), to a map defined on $C(\sigma(T))$, and the extension remains isometric, as claimed in (2).

By continuity, the properties $\mathsf{FC}(f_1 f_2) = \mathsf{FC}(f_1)\mathsf{FC}(f_2)$ and $\mathsf{FC}(f)^* = \mathsf{FC}(\bar{f})$, which are valid for polynomials (using $T = T^*$ for the latter), pass to the limit and are true for all $f \in C(\sigma(T))$. It follows that $\mathsf{FC}(C(\sigma(T)))$ is commutative and closed under taking adjoints, and in particular $f(T)$ is normal for all functions $f \in C(\sigma(T))$. This proves (1), (2), and (3).

To prove (4) we choose for $f \in C(\sigma(T))$ a sequence (p_n) in $\mathbb{C}[z]$ such that $\|p_n - f\|_{\infty, \sigma(T)} \to 0$ as $n \to \infty$. If now $v \in \ker(T - \lambda I)$ then $Tv = \lambda v$, and by induction and linearity $p_n(T)v = p_n(\lambda)v$, for all $n \geqslant 1$, and we deduce that $f(T)(v) = f(\lambda)v$.

Finally, assume that $T = M_g$. Then by (1) we have $p(M_g) = M_{p \circ g}$ for all $p \in \mathbb{C}[x]$ and by (2), $M_{p_n \circ g} = p_n(M_g) \to f(M_g)$ as $n \to \infty$. However, by Corollary 4.30 and the discussion in Example 12.35(b) we also have

$$\|M_{p_n \circ g} - M_{f \circ g}\| = \|M_{p_n \circ g - f \circ g}\| = \|p_n \circ g - f \circ g\|_{\text{esssup}} = \|p_n - f\|_{\infty, \sigma(T)} \longrightarrow 0$$

as $n \to \infty$. Together these imply that $f(M_g) = M_{f \circ g}$. $\square$

Exercise 12.39. Analyze the proof of Theorem 12.37 above and find out where the argument fails for a normal operator that is not self-adjoint.

The following definition will help us introduce spectral measures in the next section.

Definition 12.40. Let $T \in B(\mathcal{H})$ be a bounded operator on a complex Hilbert space. We say that T is a *positive operator*, written $T \geqslant 0$, if it is self-adjoint and has $\langle Tv, v \rangle \geqslant 0$ for all $v \in \mathcal{H}$.

The requirement that T is self-adjoint is redundant, as the next exercise shows.

Exercise 12.41. Let $\mathcal{H}$ be a complex Hilbert space. Show that any $T \in B(\mathcal{H})$ with the property that $\langle Tv, v \rangle \in \mathbb{R}$ for all $v \in \mathcal{H}$ is self-adjoint.

Corollary 12.42. *Let $\mathcal{H}$ be a complex Hilbert space, and let $T \in B(\mathcal{H})$ be self-adjoint. If $f \in C(\sigma(T))$ is non-negative, then $f(T)$ is a positive operator.*

PROOF. If $f \in C(\sigma(T))$ satisfies $f \geqslant 0$, then we can write $f = (\sqrt{f})^2 = g^2$ where $g = \sqrt{f} \geqslant 0$ is also continuous on $\sigma(T)$. Then $g(T)$ is well-defined by the continuous functional calculus in Theorem 12.37, is self-adjoint (by Theorem 12.37(3) since g is real-valued), and

$$\langle f(T)v, v \rangle = \langle g(T)^2 v, v \rangle = \langle g(T)v, g(T)v \rangle \geqslant 0,$$

for all $v \in V$, which shows that $f(T) \geqslant 0$. $\square$

In the case of a self-adjoint compact operator the continuous functional calculus discussed above is quite straightforward.

Example 12.43. Let $\mathcal{H}$ be a separable complex Hilbert space, and let T in $K(\mathcal{H})$ be a compact self-adjoint operator. Applying Theorem 6.27 we have

$$Tv = \sum_{n \geqslant 1} \lambda_n \langle v, e_n \rangle e_n,$$

where (λ_n) is the sequence of (real) eigenvalues of T with (e_n) the sequence of corresponding eigenvectors. If $\dim \mathcal{H} < \infty$ then the spectrum simply consists of the eigenvalues, and if $\dim \mathcal{H} = \infty$ then

$$\sigma(T) = \{0\} \cup \{\lambda_1, \lambda_2, \dots\}$$

and for $f \in C(\sigma(T))$ we have by Theorem 12.37(4)

$$f(T)v = \sum_{\lambda \in \sigma(T)} f(\lambda) P_\lambda v,$$

where $P_\lambda \in B(\mathcal{H})$ is the orthogonal projection onto $\ker(T - \lambda I)$.

12.4.2 Corollaries to the Continuous Functional Calculus

†The following exercise generalizes Lemma 12.38(1) to any continuous function.

Exercise 12.44 (Spectral mapping theorem). Let $\mathcal{H}$ be a complex Hilbert space, and let $T \in B(\mathcal{H})$ be a self-adjoint operator. Show that $\sigma(f(T)) = f(\sigma(T))$ for any $f \in C(\sigma(T))$.

Corollary 12.45 (Positive roots). *Let $T \in B(\mathcal{H})$ be a positive operator. For any $n \geqslant 1$, there exists a positive operator, denoted $T^{1/n}$, with the property that $(T^{1/n})^n = T$.*

We note that such an operator is unique, but we will only prove this a little later (see Exercise 12.72).

PROOF. Since $T \geqslant 0$, we have $\sigma(T) \subseteq [0, \infty)$, so the function $f : x \mapsto x^{1/n}$ is defined and continuous on $\sigma(T)$. Since $f(x)^n = x$ for all $x \geqslant 0$, the functional calculus implies that $f(T)^n = T$. Moreover $f \geqslant 0$, and hence $f(T) \geqslant 0$ by Corollary 12.42. $\qquad\square$

The case $n = 2$ is sufficient to prove Lemma 6.38, which claimed that any bounded operator on a separable complex Hilbert space may be written as a sum of four unitary operators and was used in our discussion of trace-class operators.

PROOF OF LEMMA 6.38. Let B be a bounded operator on the complex Hilbert space $\mathcal{H}$. Then

---

† The results of this subsection help to explain the functional calculus and how it can be used further, but will not be needed later.

$$B = \tfrac{1}{2}(B + B^*) + i\tfrac{1}{2i}(B - B^*)$$

and both $\tfrac{1}{2}(B + B^*)$ and $\tfrac{1}{2i}(B - B^*)$ are self-adjoint. Thus it remains to show that every self-adjoint operator can be written as a linear combination of two unitary operators.

So let A be a self-adjoint operator on $\mathcal{H}$ and assume without loss of generality that $\|A\|_{\mathrm{op}} \leqslant 1$. Then $I - A^2$ is positive since it is clearly self-adjoint and

$$\langle (I - A^2)v, v \rangle = \|v\|^2 - \|Av\|^2 \geqslant 0$$

for all $v \in \mathcal{H}$. By Corollary 12.45 we find an operator $U = A + i(I - A^2)^{1/2}$ satisfying

$$UU^* = U^*U = A^2 + (I - A^2) = I$$

and $\tfrac{1}{2}(U + U^*) = A$, which shows that A is the linear combination of two unitary operators. $\qquad\square$

The next corollary, which will be generalized later, starts to show how the functional calculus can be used to provide detailed information about the spectrum.

Corollary 12.46 (Isolated points). *Let $\mathcal{H}$ be a complex Hilbert space and let $T \in B(\mathcal{H})$ be a bounded self-adjoint operator. Let $\lambda \in \sigma(T)$ be an* isolated point *meaning that there is some $\varepsilon > 0$ for which $\sigma(T) \cap (\lambda - \varepsilon, \lambda + \varepsilon) = \{\lambda\}$. Then $\lambda \in \sigma_{\mathrm{disc}}(T)$.*

PROOF. The fact that λ is isolated implies that the function

$$f = \mathbb{1}_{\{\lambda\}} : \sigma(T) \to \mathbb{C}$$

which maps λ to 1 and $\sigma(T) \backslash \{\lambda\}$ to 0 is a continuous function on $\sigma(T)$. Hence we can define an operator $P = f(T) \in B(\mathcal{H})$. We claim that P is non-zero, and is a projection to $\ker(T - \lambda I)$. This will show that λ is in the discrete spectrum.

Firstly, $P \neq 0$ because $\|P\|_{\mathrm{op}} = \|f\|_{\infty, \sigma(T)} = 1$ by the functional calculus. Clearly $f = f^2$ in $C(\sigma(T))$, so

$$P = f(T) = f(T)^2 = P^2,$$

and

$$P = f(T) = \overline{f}(T) = P^*$$

since f is real-valued, which shows that P is an orthogonal projection. Moreover, we have an identity of continuous functions

$$[(I - \lambda \mathbb{1})f](z) = (z - \lambda)f(z) = 0$$

for all $z \in \sigma(T)$, so by the functional calculus we get $(T - \lambda I)P = 0$, which shows that $0 \neq \mathrm{im}(P) \subseteq \ker(T - \lambda I)$. $\qquad\square$

Exercise 12.47. Extend Corollary 12.46 by showing that $\lambda \notin \sigma_{\mathrm{approx}}(T)$.

Exercise 12.48 (Polar decomposition). Let $\mathcal{H}_1, \mathcal{H}_2$ be complex Hilbert spaces and suppose that $T : \mathcal{H}_1 \to \mathcal{H}_2$ is a bounded operator. Show that there exist a positive self-adjoint operator $A \in B(\mathcal{H}_1)$ and a bounded operator $U : \mathcal{H}_1 \to \mathcal{H}_2$ with the property that $U|_{\ker(T)} = 0, A|_{\ker(T)} = 0, U|_{(\ker(T))^{\perp}} : (\ker(T))^{\perp} \to \mathcal{H}_2$ is an isometry, and $T = UA$.

12.4.3 Spectral Measures

Using the functional calculus, we can now clarify how the spectrum represents an operator T and its action on vectors $v \in \mathcal{H}$.

Proposition 12.49 (Spectral measure). *Let $T \in B(\mathcal{H})$ be a self-adjoint operator on a complex Hilbert space $\mathcal{H}$. Then for any $v \in \mathcal{H}$ there exists a uniquely determined measure μ_v on $\sigma(T)$, depending on T and on v, such that*

$$\int_{\sigma(T)} f(x) \, \mathrm{d}\mu_v(x) = \langle f(T)v, v \rangle$$

for all $f \in C(\sigma(T))$. In particular, we have

$$\mu_v(\sigma(T)) = \|v\|^2, \tag{12.11}$$

so μ_v is a finite measure. This measure is called the spectral measure *associated to v (with respect to T).*

PROOF. This is a direct application of the Riesz representation theorem (Theorem 7.44). Indeed, the linear functional

$$\ell : C(\sigma(T)) \longrightarrow \mathbb{C}$$
$$f \longmapsto \langle f(T)v, v \rangle$$

is well-defined and positive, since if $f \geqslant 0$ we have $f(T) \geqslant 0$ by Corollary 12.42, and so $\langle f(T)v, v \rangle \geqslant 0$ by definition. Hence there exists a uniquely determined positive locally finite measure μ_v on $\sigma(T)$ such that

$$\langle f(T)v, v \rangle = \ell(f) = \int_{\sigma(T)} f(x) \, \mathrm{d}\mu_v(x)$$

for all $f \in C(\sigma(T))$. Moreover, taking $f = \mathbb{1}$, we obtain (12.11) (which also implies that $\|\ell\| = \|v\|^2$). $\qquad\square$

Example 12.50. Let $T : \mathcal{H} \to \mathcal{H}$ be as in Example 12.43. Then we have

$$\int_{\sigma(T)} f(x) \, \mathrm{d}\mu_v(x) = f(0)\|P_0(v)\|^2 + \sum_{n \geqslant 1} f(\lambda_n)\|P_{\lambda_n}(v)\|^2$$

for all continuous functions f on $\sigma(T)$. Therefore, as a measure on $\sigma(T)$, μ_v is a series of Dirac measures at the eigenvalues λ_n (including 0 if 0 is an eigenvalue) with $\mu_v(\{\lambda_n\})$ equal to $\|P_{\lambda_n}(v)\|^2$.

This example indicates how, roughly speaking, one can think of μ_v in general. The spectral measure indicates how the vector v is spread out across the spectrum; in general, any individual point $\lambda \in \sigma(T)$ carries a vanishing proportion of the vector, because $\mu_v(\{\lambda\})$ is often zero. However, $\mu_v(U) > 0$ for a subset $U \subseteq \sigma(T)$ indicates that a positive proportion of the vector belongs to the 'generalized eigenspace' corresponding to that part of the spectrum. We will discuss this interpretation of the spectrum again in Section 12.7.

Essential Exercise 12.51. Let (X, μ) be a finite measure space and let T be the multiplication operator M_g for some bounded measurable $g : X \to \mathbb{R}$. Describe the spectral measure of $v \in L^2_\mu(X)$.

12.4.4 The Spectral Theorem for Self-Adjoint Operators

Using spectral measures, we can now give a complete description of a self-adjoint operator in $B(\mathcal{H})$ (essentially by adapting the arguments from Section 9.1.2).

To see how this works, consider first some $v \in \mathcal{H}$ and the associated spectral measure μ_v, so that

$$\langle f(T)v, v \rangle = \int_{\sigma(T)} f(x) \, d\mu_v(x)$$

for all continuous functions f defined on the spectrum of T. In particular, if we apply this to $|f|^2 = f\overline{f}$ and use the properties of the continuous functional calculus in Theorem 12.37, we get

$$\|f(T)v\|^2 = \langle f(T)v, f(T)v \rangle = \langle (\overline{f}f)(T)v, v \rangle$$
$$= \int_{\sigma(T)} |f(x)|^2 \, d\mu_v(x) = \|f\|^2_{L^2(\sigma(T), \mu_v)}.$$

In other words, the map ϕ defined by

$$\phi : \{f(T)v \mid f \in C(\sigma(T))\} \longrightarrow L^2(\sigma(T), \mu_v)$$
$$f(T)v \longmapsto f$$

is an isometry. We note that the above also implies that ϕ is well-defined, since $f_1(T)v = f_2(T)v$ for $f_1, f_2 \in C(\sigma(T))$ implies

$$0 = \|(f_1 - f_2)(T)v\| = \|f_1 - f_2\|_{L^2(\sigma(T), \mu_v)}$$

and so $f_1 = f_2$ in $L^2(\sigma(T), \mu_v)$. Using the automatic extension to the closure (Proposition 2.59) we can extend the above map to an isometry, again denoted by ϕ, from the closed subspace

$$\mathcal{H}_v = \overline{\{f(T)v \mid f \in C(\sigma(T))\}}$$

into $L^2(\sigma(T), \mu_v)$. Since μ_v is a finite positive measure, continuous functions are dense in the Hilbert space $L^2_{\mu_v}(\sigma(T))$ (by Proposition 2.51), which implies that ϕ is onto (since the image is complete due to the isometry property).

Next we show that the subspace $\mathcal{H}_v$ is invariant under T. Indeed, to see that $T(\mathcal{H}_v) \subseteq \mathcal{H}_v$, it is enough to show that $T(f(T)v) \in \mathcal{H}_v$ for $f \in C(\sigma(T))$. For this let us again write I for the function $\sigma(T) \ni x \mapsto x$. In this notation the functional calculus in Theorem 12.37 gives $I(T) = T$ and

$$(If)(T) = Tf(T).$$

Applying this operator to v gives

$$Tf(T)v = (If)(T)(v)$$

and hence

$$\phi \circ T(f(T)v) = If = M_I \phi(f(T)v)$$

for all $f \in C(\sigma(T))$. By the density of the vectors $f(T)v \in \mathcal{H}_v$ for functions $f \in C(\sigma(T))$ we obtain

$$\phi \circ T = M_I \circ \phi. \tag{12.12}$$

Thus the above discussion proves a special case of Theorem 12.33, namely the case where T is self-adjoint and there exists some vector v with $\mathcal{H}_v = \mathcal{H}$.

It is important in this reasoning to keep track of the measure μ_v, which depends on the vector v, and to remember that elements of L^2 are actually equivalence classes of functions. Indeed, it could well be that μ_v has support which is much smaller than the spectrum, and then the values of a continuous function f outside the support are irrelevant in viewing f as an element of $L^2_{\mu_v}$. In particular, the map $C(\sigma(T)) \to L^2_{\mu_v}(\sigma(T))$ is not necessarily injective.

Definition 12.52. Let $\mathcal{H}$ be a Hilbert space and $T \in B(\mathcal{H})$. The *cyclic subspace* generated by a vector $v \in \mathcal{H}$ (also called the *cyclic vector for $\mathcal{H}_v$*) equals the closure

$$\mathcal{H}_v = \overline{\{f(T)v \mid f \in C(\sigma(T))\}} = \overline{\langle T^n v \mid n \in \mathbb{N}_0 \rangle}.$$

For a unital sub-algebra $\mathcal{A} \subseteq B(\mathcal{H})$ the *cyclic subspace* generated by v is defined by $\mathcal{H}_v = \overline{\mathcal{A}v}$.

The equivalence of the two definitions of $\mathcal{H}_v$ follows from the density of the subspace of polynomials in $C(\sigma(T))$ and Theorem 12.37. We also note that $\mathcal{H}_v$ is separable.

It is not always the case that T admits a cyclic vector for all of $\mathcal{H}$. However, we have the following lemma which allows us to reduce many questions to the cyclic case.

Lemma 12.53. *Let $\mathcal{H}$ be a Hilbert space, and let $T \in B(\mathcal{H})$ be a self-adjoint operator. Then there exists a family $(\mathcal{H}_i)_{i \in I}$ of non-zero, pairwise orthogonal, closed subspaces of $\mathcal{H}$ such that $\mathcal{H} = \bigoplus_{i \in I} \mathcal{H}_i$ is the orthogonal direct sum of the $\mathcal{H}_i$, $T(\mathcal{H}_i) \subseteq \mathcal{H}_i$ for all i, and T restricted to $\mathcal{H}_i$ is, for all i, a self-adjoint bounded operator in $B(\mathcal{H}_i)$ with a cyclic vector.*

Essential Exercise 12.54. Prove Lemma 12.53.

Notice that if $\mathcal{H}$ is separable, the index set in the above result is either finite or countable, since each $\mathcal{H}_i$ is non-zero.

We can now prove Theorem 12.33 for a single self-adjoint operator.

Theorem 12.55 (Spectral theorem for self-adjoint operators). *Let $\mathcal{H}$ be a separable complex Hilbert space and $T \in B(\mathcal{H})$ a continuous self-adjoint operator. Then there exists a finite measure space (X, μ), a unitary isomorphism*

$$\phi : \mathcal{H} \to L^2_\mu(X)$$

and a bounded measurable function $g : X \to \mathbb{R}$, such that

$$M_g \circ \phi = \phi \circ T.$$

In fact, we can set $X = \sigma(T) \times \mathbb{N}$ and $g(z, n) = z$ for $z \in \sigma(T)$ and $n \in \mathbb{N}$.

PROOF. Consider a (possibly finite) family $(\mathcal{H}_n)_{n \geqslant 1}$ of pairwise orthogonal non-zero closed subspaces of $\mathcal{H}$, spanning $\mathcal{H}$, for which $T(\mathcal{H}_n) \subseteq \mathcal{H}_n$ and T has a cyclic vector $v_n \neq 0$ on $\mathcal{H}_n$ as in Lemma 12.53. By replacing v_n with $n^{-1} \|v_n\|^{-1} v_n$, we can assume that $\|v_n\|^2 = n^{-2}$ (without changing $\mathcal{H}_n$). Let $\mu_n = \mu_{v_n}$ be the spectral measure associated to v_n (and T), so that

$$\mu_n(\sigma(T)) = \|v_n\|^2 = n^{-2}$$

for all $n \geqslant 1$. If the list of subspaces is finite, $\mathcal{H}_1, \ldots, \mathcal{H}_{n_0}$ say, then we set $\mathcal{H}_n = \{0\}$ for $n > n_0$ and still work with the index set $\mathbb{N}$.

By the argument at the beginning of this section, we have unitary maps

$$\phi_n : \mathcal{H}_n \to L^2_{\mu_n}(\sigma(T)),$$

such that $\phi_n \circ T = M_I \circ \phi_n$ and M_I is the multiplication operator corresponding to the function I defined by $I(z) = z$ for $z \in \sigma(T)$. Now define $X = \sigma(T) \times \mathbb{N}$ with the product topology, and define the locally finite positive measure μ by $\mu(A \times \{n\}) = \mu_n(A)$ for $n \geqslant 1$ and measurable $A \subseteq \sigma(T)$. It is easily checked (see Exercise 3.30) that this indeed defines a measure on X. Moreover, in this context measurable functions on X correspond one-to-one with sequences of measurable functions (f_n) on $\sigma(T)$ by mapping f to (f_n) with $f_n(z) = f(z, n)$ for all $(z, n) \in \sigma(T) \times \mathbb{N}$, and

$$\int_X f(x) \, d\mu(x) = \sum_{n \geqslant 1} \int_{\sigma(T)} f_n(z) \, d\mu_n(z)$$

whenever this makes sense (for example, if $f \geqslant 0$, equivalently $f_n \geqslant 0$ for all n, or if f is integrable, which is equivalent to f_n being μ_n-integrable for all n and the sum of the integrals of $|f_n|$ being convergent). In particular,

$$\mu(X) = \sum_{n \geqslant 1} \mu_n(\sigma(T)) = \sum_{n \geqslant 1} n^{-2} < \infty,$$

so that (X, μ) is a finite measure space. Moreover, the map

$$\bigoplus_n L^2(\sigma(T), \mu_n) \longrightarrow L^2(X, \mu)$$

$$(f_n) \longmapsto f$$

is a unitary isomorphism (cf. Exercise 3.37) which we will use implicitly in the following. Now recall that $\mathcal{H} = \bigoplus_n \mathcal{H}_n$ so that we can construct ϕ by defining

$$\phi\Big(\sum_n w_n\Big) = \big(\phi_n(w_n)\big)_n$$

for all $\sum_n w_n \in \bigoplus_n \mathcal{H}_n = \mathcal{H}$. Since $\|\phi_n(w_n)\|_{L^2(\sigma(T), \mu_n)} = \|w_n\|_{\mathcal{H}}$, this defines a unitary map with inverse given by

$$\phi^{-1}(f) = \sum_{n \geqslant 1} \phi_n^{-1}(f_n) \in \bigoplus_n \mathcal{H}_n = \mathcal{H}$$

for $f = (f_n)_n \in L^2(X, \mu)$.

Now consider the map $g : X \to \mathbb{C}$ sending (z, n) to z, which is bounded and measurable. Then by (12.12) we have

$$\phi(T(w_n)) = \phi_n(T(w_n)) = M_I(\phi_n(w_n)) = M_g(\phi(w_n))$$

for all $w_n \in \mathcal{H}_n$. As this holds for all $n \geqslant 1$ we see that $\phi \circ T = M_g \circ \phi$. $\quad\square$

This spectral theorem is extremely useful. It immediately implies a number of results which could also be proved directly from the continuous functional calculus, but less transparently so.

Note that the method of proof (treating first the case of cyclic operators, and then extending the result to direct sums) may also be a shorter approach to some of the other corollaries, since in the cyclic case one knows that the multiplication operator M_I can be taken to correspond to the identity function $I : \sigma(T) \ni x \mapsto x$ on the spectrum.

Corollary 12.56 (Positivity). *Let $\mathcal{H}$ be a separable complex Hilbert space and let $T \in B(\mathcal{H})$ be a self-adjoint operator. Then for any $f \in C(\sigma(T))$ we have $f(T) \geqslant 0$ if and only if $f \geqslant 0$.*

PROOF. Because of Corollary 12.42, we only need to check that $f(T) \geqslant 0$ implies that $f \geqslant 0$. Now two unitarily equivalent operators are simultaneously either positive or not, so it suffices to consider an operator of the form $T = M_g$ acting on $L^2_\mu(X)$ for a finite measure space (X, μ). Without loss of generality we may assume that $X = \sigma(T) \times \mathbb{N}$ and $g(z, n) = z$ as in Theorem 12.55. Recall that $f(M_g) = M_{f \circ g}$ for any function $f \in C(\sigma(T))$ by Theorem 12.37(4). Now set $v = \mathbb{1}_{\{(z,n)|f(z)<0\}}$ to obtain

$$\int_{\{(z,n)|f(z)<0\}} f(z) \, d\mu(z, n) = \langle f(M_g)v, v \rangle \geqslant 0,$$

which implies that $g_*\mu(\{z \mid f(z) < 0\}) = \mu(\{(z,n) \mid f(z) < 0\}) = 0$ and so $f \geqslant 0$, since $\sigma(T) = \sigma(M_g) = \mathrm{Supp}(g_*\mu)$ by Example 12.31. $\qquad\square$

12.4.5 Consequences for Unitary Representations

As the following exercises show, the material above is also useful for the study of unitary representations.

Exercise 12.57. Let G be a topological group, $\mathcal{H}_1$ and $\mathcal{H}_2$ complex Hilbert spaces, and let $\pi_1 : G \curvearrowright \mathcal{H}_1$ and $\pi_2 : G \curvearrowright \mathcal{H}_2$ be unitary representations of G. Suppose that π_1 and π_2 are isomorphic in the sense that there exists a bijective bounded operator T from $\mathcal{H}_1$ to $\mathcal{H}_2$ with $T\pi_1(g) = \pi_2(g)T$ for all $g \in G$. Show that this implies that π_1 and π_2 are also unitarily isomorphic, meaning that T can be chosen to be in addition a unitary isomorphism $T : \mathcal{H}_1 \to \mathcal{H}_2$.

Exercise 12.58 (Schur's lemma). (a) Let $\pi_1 : G \curvearrowright \mathcal{H}_1$ and $\pi_2 : G \curvearrowright \mathcal{H}_2$ be unitary representations of a topological group G, and let $B : \mathcal{H}_1 \to \mathcal{H}_2$ be a bounded operator with $B\pi_1(g) = \pi_2(g)B$ for all $g \in G$. Show that if π_1 is irreducible (that is, there are no closed π_1-invariant subspaces in $\mathcal{H}_1$ other than $\{0\}$ and $\mathcal{H}_1$) then $B^*B = \lambda I_{\mathcal{H}_1}$ for some $\lambda \geqslant 0$ and if π_2 is also irreducible then $BB^* = \lambda I_{\mathcal{H}_2}$.
(b) Suppose now that $\pi_1 = \pi_2$ is irreducible and deduce that $B = \lambda I_{\mathcal{H}_1}$ for some $\lambda \in \mathbb{C}$.

Exercise 12.59. Let G be a topological group, and recall the set $\mathcal{P}_1(G)$ of normalized positive-definite functions in $C_b(G)$ from Exercise 9.55. Show that $p \in \mathcal{P}_1(G)$ is extreme in $\mathcal{P}_1(G)$ if and only if the associated unitary representation from Exercise 9.55(a) is irreducible.

12.5 Commuting Normal Operators

The following is a natural generalization of the spectral theorem for normal operators (Theorem 12.33), showing that any commutative C^*-sub-algebra of $B(\mathcal{H})$ is unitarily equivalent to a C^*-sub-algebra of multiplication operators.

Theorem 12.60 (Spectral theorem for commuting normal operators). *Let $\mathcal{H}$ be a separable complex Hilbert space, and let $\mathcal{A} \subseteq B(\mathcal{H})$ be a separable commutative unital C^*-sub-algebra of $B(\mathcal{H})$. Then there exists a finite measure space (X, μ), a unitary isomorphism $\phi : \mathcal{H} \to L^2_\mu(X)$ and for every $a \in \mathcal{A}$ a bounded function $g_a \in L^\infty_\mu(X)$ such that*

$$
\begin{array}{ccc}
\mathcal{H} & \xrightarrow{\ \ a\ \ } & \mathcal{H} \\
\phi \downarrow & & \downarrow \phi \\
L^2_\mu(X) & \xrightarrow[\ M_{g_a}\ ]{} & L^2_\mu(X)
\end{array}
$$

commutes. In fact, we can choose $X = \sigma(\mathcal{A}) \times \mathbb{N}$ and $g_a = a^o$, where we identify the function a^o with the function defined by $a^o(x, n) = a^o(x)$ for $(x, n) \in \sigma(\mathcal{A}) \times \mathbb{N}$ and all $a \in \mathcal{A}$ and the map that sends $a \in \mathcal{A}$ to $g_a \in L^\infty_\mu(X)$ is a C^*-isomorphism preserving products, the adjoint operation, and norms.

Clearly Theorem 12.33 is a special case of Theorem 12.60 (cf. Exercise 12.61). Using Section 12.4 the following proof will be much shorter than the proof of the case of a single self-adjoint operator above. By Corollary 11.34 the Gelfand transform

$$
\mathcal{A} \ni a \longmapsto a^o \in C(\sigma(\mathcal{A}))
$$

is an isometry and an algebra isomorphism satisfying

$$
(a^*)^o = \overline{a^o} \tag{12.13}
$$

for all $a \in \mathcal{A}$. (We note that Lemma 11.35 in the proof of Corollary 11.34 can here be replaced by Lemma 12.15.) We recall that $\sigma(\mathcal{A})$ is the generalization of the spectrum of a single operator and note that in the following the inverse map

$$
C(\sigma(\mathcal{A})) \ni f = a^o \longmapsto a \in \mathcal{A}
$$

should be thought of as a generalized continuous functional calculus.

PROOF OF THEOREM 12.60. Fix $v \in \mathcal{H}$ and define a linear functional

$$
\Lambda : C(\sigma(\mathcal{A})) \to \mathbb{C}
$$

by

$$\Lambda(a^o) = \langle av, v \rangle$$

for every $a \in \mathcal{A}$. We claim that Λ is a positive functional on $C(\sigma(\mathcal{A}))$. Suppose that $a \in \mathcal{A}$ with $a^o \geqslant 0$. Then there exists some $b = b^* \in \mathcal{A}$ (defined using the Gelfand transform by $b^o = \sqrt{a^o}$) with $b^2 = a$. The claimed positivity now follows, since

$$\Lambda(a^o) = \langle av, v \rangle = \langle bv, bv \rangle \geqslant 0.$$

By the Riesz representation theorem (Theorem 7.44) there exists a positive finite measure μ_v on $\sigma(\mathcal{A})$ such that

$$\langle av, v \rangle = \int_{\sigma(\mathcal{A})} a^o \, d\mu_v$$

and it follows that

$$\|av\|^2 = \langle av, av \rangle = \langle a^*av, v \rangle = \int (a^*a)^o \, d\mu_v = \int |a^o|^2 \, d\mu_v$$

for all $a \in \mathcal{A}$ by (12.13). Just as in Sections 9.1.2 and 12.4.4, this induces a unitary isomorphism between the cyclic subspace $\mathcal{H}_v = \overline{\mathcal{A}v}$ and $L^2_{\mu_v}(\sigma(\mathcal{A}))$ which sends $av \in \mathcal{A}v$ to $a^o \in C(\sigma(\mathcal{A}))$. In particular, for $a, b \in \mathcal{A}$ we have

$$\phi(abv) = (ab)^o = a^o b^o = a^o \phi(bv).$$

Fixing $a \in \mathcal{A}$, this extends by continuity to the statement

$$\phi(aw) = M_{a^o} \phi(w)$$

for all $w \in \mathcal{H}_v$. As in Sections 9.1.2 and 12.4.4 this extends to a proof of Theorem 12.60 as follows. If $w_1, w_2, \ldots$ is an orthonormal basis of $\mathcal{H}$ then we define $\mathcal{H}_1 = \mathcal{H}_{w_1}$, $\mathcal{H}_2 = \mathcal{H}_{w_2^\perp}$ where $w_2 \in \mathcal{H}_1^\perp$ is the orthogonal projection to $\mathcal{H}_1^\perp$, similarly $\mathcal{H}_3 = \mathcal{H}_{w_3^\perp}$ with $w_3 \in (\mathcal{H}_1 \oplus \mathcal{H}_2)^\perp$, and so on. Replacing w_n by a scalar multiple for each $n \geqslant 1$ we may assume that $\sum_{n \geqslant 1} \|w_n^\perp\|^2 < \infty$. Define

$$X = \sigma(\mathcal{A}) \times \mathbb{N} \cong \bigsqcup_{n \in \mathbb{N}} \sigma(\mathcal{A})$$

with measure

$$\mu = \bigsqcup_{n \in \mathbb{N}} \mu_{w_n^\perp},$$

where the disjoint union notation indicates that we consider $\mu_{w_n^\perp}$ as a measure on $\sigma(\mathcal{A}) \times \{n\}$ and then take the sum to obtain the measure μ on X.

With this

$$L^2_\mu(X) \cong \bigoplus_{n \in \mathbb{N}} L^2(X, \mu_{w_n^\perp}) \cong \bigoplus_{n \in \mathbb{N}} \mathcal{H}_{w_n^\perp} = \mathcal{H}, \qquad (12.14)$$

and application of $a \in \mathcal{A}$ leaves each subspace $\mathcal{H}_{w_n^\perp}$ invariant and corresponds to multiplication by $a^\circ \in C(\sigma(\mathcal{A}))$ on $L^2(\sigma(\mathcal{A}), \mu_{w_n^\perp}) \subseteq L^2(X, \mu)$. As this holds for all $n \in \mathbb{N}$ the map in (12.14) gives the unitary isomorphism

$$\phi : \mathcal{H} \to L^2_\mu(X),$$

with the required properties. $\square$

Exercise 12.61. (a) Suppose that $\mathcal{A}$ is the unital commutative C^*-algebra generated by T, a normal operator on a complex Hilbert space $\mathcal{H}$, in the sense that

$$\mathcal{A} = \overline{\langle T^m (T^*)^n \mid m, n \geqslant 0 \rangle}$$

where $T^0 = (T^*)^0 = I$. Show that $\imath : \sigma(\mathcal{A}) \ni \phi \mapsto \phi(T) \in \sigma(T)$ defines a homeomorphism between compact metric spaces $\mathcal{A} = \langle T^m (T^*)^n \mid m, n \geqslant 0 \rangle$ and use this to deduce Theorem 12.33 from Theorem 12.60.
(b) Now consider the algebra $\mathcal{A} = \langle I, T_1, T_1^*, T_2, T_2^* \rangle$ generated by two commuting normal operators $T_1, T_2 \in B(\mathcal{H})$ on a complex Hilbert space. Show that

$$\imath : \sigma(\mathcal{A}) \ni \phi \longmapsto (\phi(T_1), \phi(T_2)) \in \sigma(T_1) \times \sigma(T_2)$$

is continuous and injective. Give a concrete example to show that the image of $\imath$ may not be all of $\sigma(T_1) \times \sigma(T_2)$.

Exercise 12.62. State and prove a spectral theorem for normal compact operators as a corollary of Theorem 12.60.

12.6 Spectral Measures and the Measurable Functional Calculus

For the proof of Theorem 12.34 we now discuss some more general spectral measures. As it makes little difference whether we consider a single (self-adjoint or normal) operator or a commutative Banach algebra (as in the previous section), we will do the latter. The reader only interested in the case of a single self-adjoint operator T may replace the use of Theorem 12.60 below by Theorem 12.55, set $\sigma(\mathcal{A}) = \sigma(T)$ as in Exercise 12.61(a) and replace the operation $a^\circ \mapsto a \in \mathcal{A}$ by the continuous functional calculus

$$C(\sigma(T)) \ni f \longmapsto f(T).$$

12.6.1 Non-Diagonal Spectral Measures

Definition 12.63. Let $\mathcal{H}$ be a complex Hilbert space and let $\mathcal{A} \subseteq B(\mathcal{H})$ be a separable commutative unital C^*-sub-algebra. For $v, w \in \mathcal{H}$ a *non-diagonal spectral measure* is a finite complex-valued measure $\mu_{v,w}$ on $\sigma(\mathcal{A})$ with

$$\int_{\sigma(\mathcal{A})} a^o \, d\mu_{v,w} = \langle av, w \rangle \tag{12.15}$$

for all $a^o \in C(\sigma(\mathcal{A}))$.

Proposition 12.64. *Let $\mathcal{H}$ be a complex Hilbert space, and assume that $\mathcal{A}$ is a separable commutative unital C^*-sub-algebra of $B(\mathcal{H})$. Then for every pair $v, w \in \mathcal{H}$ there exists a uniquely determined finite complex-valued spectral measure $\mu_{v,w}$ on $\sigma(\mathcal{A})$ satisfying (12.15). Moreover, the measure $\mu_{v,w}$ depends sesqui-linearly on $v, w \in \mathcal{H}$ and satisfies $\|\mu_{v,w}\| \leqslant \|v\| \|w\|$.*

PROOF. We may assume without loss of generality that $\mathcal{H}$ is separable, for otherwise we may replace $\mathcal{H}$ by the closure of $\mathcal{A}v + \mathcal{A}w$, which is separable by the assumption on $\mathcal{A}$. Recall from Corollary 11.34 that

$$C(\sigma(\mathcal{A})) = \{a^o \mid a \in \mathcal{A}\}.$$

Recall that linear functionals on $C(\sigma(\mathcal{A}))$ can be uniquely identified with complex-valued measures by Theorem 7.54. We now apply this to the linear functional

$$\Lambda_{v,w} : C(\sigma(\mathcal{A})) \cong \mathcal{A} \ni a^o \longmapsto \langle av, w \rangle$$

which satisfies

$$|\langle av, w \rangle| \leqslant \|av\| \|w\| \leqslant \|a\| \|v\| \|w\| = \|a^o\|_\infty \|v\| \|w\|$$

by Cauchy–Schwarz, the definition of the operator norm, and the isometry claim in Corollary 11.34. This shows that $\Lambda_{v,w} \in C(\sigma(\mathcal{A}))^*$ is a bounded functional and we obtain the uniquely defined complex-valued measure $\mu_{v,w}$ on $\sigma(\mathcal{A})$ satisfying (12.15) and $\|\mu_{v,w}\| \leqslant \|v\| \|w\|$.

Since uniqueness and existence are now shown, the sesqui-linearity follows easily from the sesqui-linearity of the inner product on $\mathcal{H}$. $\qquad\square$

The following exercise clarifies how the more general non-diagonal spectral measures can be constructed from the diagonal spectral measures $\mu_v = \mu_{v,v}$ appearing in the spectral theorem.

Exercise 12.65. Let $\mathcal{H}$ and $\mathcal{A}$ be as in Proposition 12.64. Let $v, w \in \mathcal{H}$ and decompose w into $w = w_0 + w^\perp$ with $w_0 \in \mathcal{H}_v$ and $w^\perp \in \mathcal{H}_v^\perp$. Use the spectral theorem to express $\mu_{v,w}$ in terms of μ_v and the vector $f_0 \in L^2_{\mu_v}(\sigma(\mathcal{A}))$ corresponding to w_0.

12.6.2 The Measurable Functional Calculus

Using the spectral measures from above, we can now define for every measurable function $f \in \mathscr{L}^\infty(\sigma(\mathcal{A}))$ a corresponding operator $f_\mathcal{H} \in B(\mathcal{H})$. In the case of $\mathcal{A}$ being generated by I and a normal operator T this gives a definition of $f(T)$ for a function $f \in \mathscr{L}^\infty(\sigma(T))$.

Proposition 12.66. *Let $\mathcal{H}$ be a complex Hilbert space and let $\mathcal{A} \subseteq B(\mathcal{H})$ be a separable commutative unital C^*-sub-algebra. For any $f \in \mathscr{L}^\infty(\sigma(\mathcal{A}))$ there exists a bounded operator $f_{\mathcal{H}}$ which is uniquely characterized by the property*

$$\langle f_{\mathcal{H}} v, w \rangle = \int_{\sigma(\mathcal{A})} f \, d\mu_{v,w} \qquad (12.16)$$

for all $v, w \in \mathcal{H}$. Moreover, the operator norm of $f_{\mathcal{H}}$ satisfies $\|f_{\mathcal{H}}\| \leqslant \|f\|_\infty$.

PROOF. Since $\mu_{v,w}$ is a finite complex-valued measure, and $f \in \mathscr{L}^\infty(\sigma(\mathcal{A}))$ is bounded, the integral $\int_{\sigma(\mathcal{A})} f \, d\mu_{v,w}$ exists. Moreover,

$$\left| \int_{\sigma(\mathcal{A})} f \, d\mu_{v,w} \right| \leqslant \|f\|_\infty \|\mu_{v,w}\| \leqslant \|f\|_\infty \|v\| \|w\|$$

by Proposition 12.64. Thus for a fixed $v \in \mathcal{H}$ the map

$$w \longmapsto \overline{\int_{\sigma(\mathcal{A})} f \, d\mu_{v,w}}$$

is linear and bounded with operator norm bounded by $\|f\|_\infty \|v\|$. Therefore, by Fréchet–Riesz representation (Corollary 3.19) there exists some uniquely determined v_f with

$$\|v_f\| \leqslant \|f\|_\infty \|v\| \qquad (12.17)$$

for which

$$\overline{\langle v_f, w \rangle} = \langle w, v_f \rangle = \overline{\int_{\sigma(\mathcal{A})} f \, d\mu_{v,w}}$$

for all $w \in \mathcal{H}$. By linearity of $v \mapsto \mu_{v,w}$ and the bound (12.17), we see that

$$v \longmapsto v_f = f_{\mathcal{H}} v$$

defines a bounded operator $f_{\mathcal{H}}$ with (12.16), and $\|f_{\mathcal{H}}\| \leqslant \|f\|_\infty$. $\qquad \square$

Proposition 12.66 defines the measurable functional calculus. We now discuss its main properties, which will also give the proof of the existence claim in Theorem 12.34 (recall that uniqueness was the content of Exercise 12.36).

Proposition 12.67. *Let $\mathcal{H}$ be a complex Hilbert space, and let $\mathcal{A} \subseteq B(\mathcal{H})$ be a separable commutative unital C^*-sub-algebra. The measurable functional calculus $\mathscr{L}^\infty(\sigma(\mathcal{A})) \ni f \mapsto f_{\mathcal{H}} \in B(\mathcal{H})$ has the following properties:*
(FC1) If $f = a^\circ \in C(\sigma(\mathcal{A}))$, then $f_{\mathcal{H}} = a \in \mathcal{A}$.
(FC2) $\|f_{\mathcal{H}}\| = \|f\|_\infty$ for any f in $C(\sigma(\mathcal{A}))$ and $\|f_{\mathcal{H}}\| \leqslant \|f\|_\infty$ for any f in $\mathscr{L}^\infty(\sigma(\mathcal{A}))$.
(FC3) $(f_{\mathcal{H}})^* = (\overline{f})_{\mathcal{H}}$ and $(f_1 f_2)_{\mathcal{H}} = (f_1)_{\mathcal{H}}(f_2)_{\mathcal{H}}$ for $f_1, f_2, f \in \mathscr{L}^\infty(\sigma(\mathcal{A}))$.
In particular, properties (FC1)–(FC3) in Theorem 12.34 hold.

PROOF. Recall that Corollary 11.34 gives the existence of a C^*-algebra isomorphism $C(\sigma(\mathcal{A})) \ni f = a^o \mapsto f_{\mathcal{H}} = a \in \mathcal{A}$ (see Theorem 12.37 in the case of a single self-adjoint operator).

Also recall that in Proposition 12.64 we derived the existence of the family of finite complex-valued measures $\{\mu_{v,w}\}$ on $\sigma(\mathcal{A})$ with

$$\langle f_{\mathcal{H}} v, w \rangle = \langle av, w \rangle = \int_{\sigma(\mathcal{A})} a^o \, d\mu_{v,w} = \int_{\sigma(\mathcal{A})} f \, d\mu_{v,w} \tag{12.18}$$

for all $f = a^o \in C(\sigma(\mathcal{A}))$, which in Proposition 12.66 we turned around to use (12.18) as the definition of $f_{\mathcal{H}}$ for $f \in \mathscr{L}^\infty(\sigma(\mathcal{A}))$. Hence this definition of the measurable functional calculus extends the definition of the continuous functional calculus (that is, of the map $C(\sigma(\mathcal{A})) \ni a^o \mapsto a \in \mathcal{A}$), and hence satisfies (FC1). By Corollary 11.34 and Proposition 12.66 above we also have (FC2).

To prove (FC3) we argue in the following way. First, by Corollary 11.34 we already know (FC3) for continuous functions. We will use this to encode the properties in (FC3) into properties of the non-diagonal spectral measures $\mu_{v,w}$, which in turn will give the same properties for measurable functions.

Let us start with $f_{\mathcal{H}}^* = (\overline{f})_{\mathcal{H}}$, which we know by Corollary 11.34 for

$$f = a^o \in C(\sigma(\mathcal{A})).$$

We claim this implies that $\mu_{v,w} = \overline{\mu_{w,v}}$. To see this, let $a \in \mathcal{A}$ and notice that

$$\int \overline{a^o} \, d\mu_{v,w} = \langle a^* v, w \rangle = \langle v, aw \rangle = \overline{\langle aw, v \rangle} = \overline{\int a^o \, d\mu_{w,v}} = \int \overline{a^o} \, d\overline{\mu_{w,v}},$$

for any $v, w \in \mathcal{H}$, and since this holds for all $f = a^o \in C(\sigma(\mathcal{A}))$ the claim follows. Now we use essentially the same identity (in a slightly different order, and with a different logic) to deduce that $f_{\mathcal{H}}^* = \overline{f}_{\mathcal{H}}$ for $f \in \mathscr{L}^\infty(\sigma(\mathcal{A}))$. So let $f \in \mathscr{L}^\infty(\sigma(\mathcal{A}))$. Then

$$\langle f_{\mathcal{H}}^* v, w \rangle = \langle v, f_{\mathcal{H}} w \rangle = \overline{\langle f_{\mathcal{H}} w, v \rangle}$$

$$= \overline{\int f \, d\mu_{w,v}} = \int \overline{f} \, d\mu_{v,w} = \langle (\overline{f})_{\mathcal{H}} v, w \rangle$$

for all $v, w \in \mathcal{H}$, as required.

We now show that

$$(f_1 f_2)_{\mathcal{H}} = (f_1)_{\mathcal{H}} (f_2)_{\mathcal{H}} \tag{12.19}$$

for $f_1, f_2 \in \mathscr{L}^\infty(\sigma(\mathcal{A}))$. Again we know this property for $f_1, f_2 \in C(\sigma(\mathcal{A}))$. We claim that this implies

$$d\mu_{(f_2)_{\mathcal{H}} v, w} = f_2 \, d\mu_{v,w} \tag{12.20}$$

for $f_2 \in \mathscr{L}^\infty(\sigma(\mathcal{A}))$ and $v, w \in \mathcal{H}$.

For $f_1, f_2 \in C(\sigma(\mathcal{A}))$ we have

$$\int_{\sigma(\mathcal{A})} f_1 \, \mathrm{d}\mu_{(f_2)_{\mathcal{H}}v,w} = \langle (f_1)_{\mathcal{H}}(f_2)_{\mathcal{H}} v, w \rangle$$

$$= \langle (f_1 f_2)_{\mathcal{H}} v, w \rangle = \int_{\sigma(T)} f_1 f_2 \, \mathrm{d}\mu_{v,w}.$$

As this holds for all $f_1 \in C(\sigma(\mathcal{A}))$ we obtain (12.20) for $f_2 \in C(\sigma(\mathcal{A}))$ and for all $v, w \in \mathcal{H}$. Using $\mu_{v,w} = \overline{\mu_{w,v}}$ this also shows that

$$\mathrm{d}\mu_{v,(\overline{f_1})_{\mathcal{H}}w} = \overline{\mathrm{d}\mu_{(\overline{f_1})_{\mathcal{H}}w,v}} = f_1 \overline{\mathrm{d}\mu_{w,v}} = f_1 \, \mathrm{d}\mu_{v,w}$$

for all $f_1 \in C(\sigma(\mathcal{A}))$. For general $f_2 \in \mathscr{L}^\infty(\sigma(\mathcal{A}))$ we now see that

$$\int_{\sigma(\mathcal{A})} f_1 \, \mathrm{d}\mu_{(f_2)_{\mathcal{H}}v,w} = \langle (f_1)_{\mathcal{H}}(f_2)_{\mathcal{H}} v, w \rangle = \langle (f_2)_{\mathcal{H}} v, (\overline{f_1})_{\mathcal{H}} w \rangle$$

$$= \int_{\sigma(\mathcal{A})} f_2 \, \mathrm{d}\mu_{v,(\overline{f_1})_{\mathcal{H}}w} = \int_{\sigma(\mathcal{A})} f_1 f_2 \, \mathrm{d}\mu_{v,w}$$

for all $f_1 \in C(\sigma(\mathcal{A}))$ and $v, w \in \mathcal{H}$, which implies the claim in (12.20).

We now derive (12.19) from (12.20) for $f_1, f_2 \in \mathscr{L}^\infty(\sigma(\mathcal{A}))$. Indeed, applying (12.20) we see that

$$\langle (f_1)_{\mathcal{H}}(f_2)_{\mathcal{H}} v, w \rangle = \int_{\sigma(\mathcal{A})} f_1 \, \mathrm{d}\mu_{(f_2)_{\mathcal{H}}v,w}$$

$$= \int_{\sigma(\mathcal{A})} f_1 f_2 \, \mathrm{d}\mu_{v,w} = \langle (f_1 f_2)_{\mathcal{H}} v, w \rangle.$$

As $v, w \in \mathcal{H}$ were arbitrary, we derive (12.19) and conclude (FC3). $\qquad\square$

Proposition 12.68. *Under the same hypotheses as in Proposition 12.64, the measurable functional calculus has the following properties:*

(FC4) *Suppose μ is a finite measure on $X = \sigma(\mathcal{A}) \times \mathbb{N}$ so that the operators a in $\mathcal{A}$ correspond to multiplication operators M_{a° via a unitary isomorphism $\phi : \mathcal{H} \to L_\mu^2(X)$ as in Theorem 12.60. Then for any f in $\mathscr{L}^\infty(\sigma(\mathcal{A}))$ the operator $f_{\mathcal{H}}$ corresponds to M_f (with $f(x,n) = f(x)$ for $(x,n) \in \sigma(\mathcal{A}) \times \mathbb{N}$).*

(FC5) *If $V \subseteq \mathcal{H}$ is a closed subspace such that $aV \subseteq V$ for all $a \in \mathcal{A}$, then V is also invariant under $f_{\mathcal{H}}$ for all $f \in \mathscr{L}^\infty(\sigma(\mathcal{A}))$. Moreover, if $S \in \mathrm{B}(\mathcal{H})$ commutes with all $a \in \mathcal{A}$, then S commutes with $f_{\mathcal{H}}$ for all $f \in \mathscr{L}^\infty(\sigma(\mathcal{A}))$.*

In the case of a normal operator $T \in \mathrm{B}(\mathcal{H})$ the measurable functional calculus satisfies (FC4)–(FC6) in Theorem 12.34.

PROOF. We first prove (FC5). Suppose that $S \in \mathrm{B}(\mathcal{H})$ commutes with all elements $a \in \mathcal{A}$. We extend this again to $f_{\mathcal{H}}$ for $f \in \mathscr{L}^\infty(\sigma(\mathcal{A}))$ using the spectral measures. For these, we have that $\mu_{Sv,w} = \mu_{v,S^*w}$ since

$$\int_{\sigma(\mathcal{A})} a^\circ \, d\mu_{Sv,w} = \langle aSv, w \rangle = \langle av, S^*w \rangle = \int_{\sigma(\mathcal{A})} a^\circ \, d\mu_{v,S^*w}$$

for all $a \in \mathcal{A}$. Now let $f \in \mathscr{L}^\infty(\sigma(\mathcal{A}))$, and notice that

$$\langle f_{\mathcal{H}} S v, w \rangle = \int_{\sigma(\mathcal{A})} f \, d\mu_{Sv,w} = \int_{\sigma(\mathcal{A})} f \, d\mu_{v,S^*w} = \langle S f_{\mathcal{H}} v, w \rangle$$

for all $v, w \in \mathcal{H}$, showing that $f_{\mathcal{H}} S = S f_{\mathcal{H}}$.

To complete the proof of (FC5), we still have to consider an invariant subspace $V \subseteq \mathcal{H}$. By Lemma 6.30 the closed subspace $V^\perp$ is a^*-invariant for all $a \in \mathcal{A}$. This implies that every $a \in \mathcal{A}$ commutes with the orthogonal projection $P_V : \mathcal{H} \to \mathcal{H}$ onto V, since for $v \in V$ we have $av \in V$ and

$$a P_V(v) = av = P_V(av),$$

and since for $w \in V^\perp$ we also have $aw \in V^\perp$ and

$$a P_V(w) = 0 = P_V(aw).$$

By what we have already proved, P_V commutes with $f_{\mathcal{H}}$ for $f \in \mathscr{L}^\infty(\sigma(\mathcal{A}))$. If now $v \in V$ then

$$f_{\mathcal{H}} v = f_{\mathcal{H}} \circ P_V(v) = P_V \circ f_{\mathcal{H}}(v) \in V$$

shows the remaining claim in (FC5).

For the proof of (FC4) it suffices, because of (FC5), to prove the cyclic case. That is, to prove (FC4) in the case where $X = \sigma(\mathcal{A})$ and μ is the spectral measure on $\sigma(\mathcal{A})$ corresponding to the generator of $\mathcal{H}$. For $v, w \in \mathcal{H}$ we then have

$$d\mu_{v,w} = \phi(v)\overline{\phi(w)} \, d\mu$$

since

$$\langle av, w \rangle = \int a^\circ \phi(v)\overline{\phi(w)} \, d\mu$$

for all $a \in \mathcal{A}$ by the spectral theorem (Theorem 12.60). Therefore

$$\langle f_{\mathcal{H}} v, w \rangle = \int_{\sigma(\mathcal{A})} f \, d\mu_{v,w} = \int_{\sigma(\mathcal{A})} f \phi(v)\overline{\phi(w)} \, d\mu = \langle M_f \phi(v), \phi(w) \rangle_{L^2(\sigma(\mathcal{A}),\mu)}$$

for any $v, w \in \mathcal{H}$ and for $f \in \mathscr{L}^\infty(\sigma(\mathcal{A}))$. This proves (FC4) as in the proposition.

It remains to prove (FC4) and (FC6) in Theorem 12.34 (since (FC4) above differs slightly from (FC4) in Theorem 12.34).

So suppose that T is unitarily isomorphic to the multiplication operator $M_g : L^2_\mu(X) \to L^2_\mu(X)$ for some bounded measurable $g : X \to \mathbb{C}$, and let $f \in \mathscr{L}^\infty(\sigma(M_g))$. Then $f \circ g$ is defined almost everywhere (specifically, on $g^{-1}(\sigma(M_g))$; see Examples 12.31 and 12.35(b)). For $v, w \in L^2_\mu(X)$ we see that $d\mu_{v,w}$ is the push-forward of $v\overline{w}\,d\mu$ under g (by solving Exercise 12.51) since

$$\langle f(M_g)v, w \rangle = \int_X f \circ g v \overline{w}\,d\mu = \int_{\sigma(M_g)} f\,d\mu_{v,w}$$

first for $f \in \mathbb{C}[z]$ and then for all $f \in C(\sigma(M_g))$ by the properties of the functional calculus in Theorem 12.37. Therefore,

$$\langle f(M_g)v, w \rangle = \int_{\sigma(M_g)} f\,d\mu_{v,w} = \int_X (f \circ g)v\overline{w}\,d\mu = \langle M_{f \circ g}v, w \rangle$$

for all $v, w \in L^2_\mu(X)$ and $f \in \mathscr{L}^\infty(\sigma(M_g))$, which proves (FC4).

For the proof of (FC6) we assume first that $\mathcal{H}$ is cyclic. By Theorem 12.55 there is a finite measure space (X, μ), some bounded measurable $g : X \to \mathbb{C}$, and a unitary isomorphism $\phi : \mathcal{H} \to L^2_\mu(X)$ such that

$$\phi \circ T = M_g \circ \phi.$$

By (FC4) we have $\phi \circ f(T) = f(M_g) \circ \phi = M_{f \circ g} \circ \phi$ for all $f \in \mathscr{L}^\infty(\sigma(T))$. Next note that by Example 12.31 (specifically, by (12.4)) we have

$$\sigma(M_{f \circ g}) \subseteq \overline{f(\sigma(T))}.$$

Let $h \in \mathscr{L}^\infty(\overline{f(\sigma(T))})$ and apply (FC4) twice more to see that

$$\phi \circ h(f(T)) = h(M_{f \circ g}) \circ \phi = M_{h \circ f \circ g} \circ \phi = h \circ f(M_g) \circ \phi = \phi \circ (h \circ f)(T),$$

which gives $h(f(T)) = (h \circ f)(T)$, as claimed in (FC6). If $\mathcal{H}$ is not cyclic, then we decompose $\mathcal{H}$ into a direct sum of cyclic subspaces and apply (FC5) and the above case. $\square$

Exercise 12.69. Suppose that $\mathcal{H}$ is a separable complex Hilbert space and that

$$\mathcal{A} \subseteq \mathcal{A}' \subseteq B(\mathcal{H})$$

are two separable commutative unital C^*-sub-algebras.
(a) Suppose that $\mathcal{H}$ and the action of $\mathcal{A}$ on $\mathcal{H}$ is described as in Theorem 12.60. Generalize (12.4) from Example 12.31 to this context by showing that

$$\sigma(\mathcal{A}) = \mathrm{Supp}\big((\pi_{\sigma(\mathcal{A})})_* \mu\big),$$

where $\pi_{\sigma(\mathcal{A})} : X = \sigma(\mathcal{A}) \times \mathbb{N} \longrightarrow \sigma(\mathcal{A})$ is the projection to the first factor.
(b) Show that the restriction map $\pi : \sigma(\mathcal{A}') \ni \phi' \mapsto \phi'|_{\mathcal{A}} \in \sigma(\mathcal{A})$ is continuous.
(c) Let $\mu_{v,w}$ be the spectral measure for $\mathcal{A}$ on $\sigma(\mathcal{A})$ and let $\mu'_{v,w}$ be the spectral measure for $\mathcal{A}'$ on $\sigma(\mathcal{A}')$ for $v, w \in \mathcal{H}$. Show that $\pi_* \mu'_{v,w} = \mu_{v,w}$.

(d) Show that the two notions of measurable calculus are compatible in the sense that any $f \in \mathscr{L}^\infty(\sigma(\mathcal{A}))$ defines some $f' \in \mathscr{L}^\infty(\sigma(\mathcal{A}'))$ by $f' = f \circ \pi$ which satisfies $f_\mathcal{H} = f'_\mathcal{H}$.
(e) Show that π is surjective.

Exercise 12.70. Generalize the results of Section 9.1.3 to the context of a single normal operator $T \in \mathrm{B}(\mathcal{H})$ or a separable commutative unital C^*-sub-algebra of $\mathrm{B}(\mathcal{H})$.

Exercise 12.71. In the notation of Theorem 12.34, fix a normal operator T, suppose it has a description as a multiplication operator M_g on some measure space $L^2_\mu(X)$, and let $\nu = g_* \mu$ be the push-forward measure on $\mathbb{C}$. Show that $f(T)$ is now well-defined with $f \in L^\infty_\nu(\sigma(T))$ by proving that if $f_1, f_2 \in \mathscr{L}^\infty(\sigma(T))$ agree ν-almost everywhere, then $f_1(T) = f_2(T)$.

Exercise 12.72. Let $T \in \mathrm{B}(\mathcal{H})$ be a positive self-adjoint bounded operator on a complex separable Hilbert space. Show that for every $n \geqslant 1$ there is only one positive operator S in $\mathrm{B}(\mathcal{H})$ with $S^n = T$.

12.7 Projection-Valued Measures

In this section, we describe another version of the spectral theorem, which is essentially equivalent but sometimes more convenient. Moreover, it allows us to examine some concepts from Section 9.1.4 in greater detail. The idea is to generalize the following interpretation of the spectral theorem (Theorem 6.27) for a compact self-adjoint operator $T \in \mathrm{K}(\mathcal{H})$. If we denote by P_λ the orthogonal projection onto $\ker(T - \lambda I)$ for $\lambda \in \mathbb{R}$ as in Example 12.43, then we have

$$v = \sum_{\lambda \in \mathbb{R}} P_\lambda(v),$$

$$T(v) = \sum_{\lambda \in \mathbb{R}} \lambda P_\lambda(v),$$

$$f(T)(v) = \sum_{\lambda \in \mathbb{R}} f(\lambda) P_\lambda(v)$$

for all $v \in \mathcal{H}$ and $f \in C(\sigma(T))$, where the series are well-defined because P_λ is 0 for $\lambda \notin \sigma(T)$.

To generalize this, it is natural to expect that one must replace the summations with appropriate integrals. Thus some form of integration for functions taking values in $\mathrm{B}(\mathcal{H})$ is needed. Moreover, $\ker(T - \lambda I)$ may be zero for all λ, and so the projections must be generalized. We start by considering these two questions abstractly.

Definition 12.73 (Projection-valued measure). Let $\mathcal{H}$ be a complex Hilbert space and let $\mathrm{P}(\mathcal{H})$ denote the set of orthogonal projections onto closed subspaces in $\mathrm{B}(\mathcal{H})$. A *(finite) projection-valued measure* Π on $\mathcal{H}$ is a map

$$\mathcal{B} \longrightarrow \mathrm{P}(\mathcal{H})$$
$$B \longmapsto \Pi_B$$

defined on the Borel σ-algebra $\mathcal{B}$ of a compact metric space X and taking values in the set of projections, such that the following hold:

(1) $\Pi_\varnothing = 0$ and $\Pi_X = I$.
(2) If (B_n) is a sequence (or finite list) of pairwise disjoint Borel subsets of X, and

$$B = \bigsqcup_{n \geqslant 1} B_n,$$

then

$$\Pi_B = \sum_{n \geqslant 1} \Pi_{B_n} \tag{12.21}$$

where the series converges in the *strong operator topology* (see Section 8.3).

We note that (12.21) simply means that $\Pi_B(v) = \sum_{n \geqslant 1} \Pi_{B_n}(v)$ for $v \in \mathcal{H}$ (see Exercise 8.57). In the study of a single normal operator T on $\mathcal{H}$ we will set $X = \sigma(T) \subseteq \mathbb{C}$, and more generally $X = \sigma(\mathcal{A})$ in the study of a commutative separable unital C^*-sub-algebra $\mathcal{A} \subseteq \mathrm{B}(\mathcal{H})$. Also notice that Definition 12.73 resembles in some ways the definition of a (finite) Borel measure on X. The discussion below will reveal further parallels to Lebesgue integration.

Lemma 12.74. *Let $\mathcal{H}$ be a complex Hilbert space and Π a projection-valued measure on $\mathcal{H}$ defined on the σ-algebra $\mathcal{B}$ of Borel subsets of a compact metric space X. If $X = \bigsqcup_{j=1}^n B_j$ is a disjoint decomposition of X into measurable subsets $B_1, \ldots, B_n \in \mathcal{B}$, then $\mathcal{H} = \bigoplus_{j=1}^n \mathrm{im}\, \Pi_{B_j}$ is an orthogonal direct sum of the closed subspaces $\mathrm{im}\, \Pi_{B_1}, \ldots, \mathrm{im}\, \Pi_{B_n} \subseteq \mathcal{H}$.*

PROOF. By the defining properties of projection-valued measures we have

$$v = \sum_{j=1}^n \Pi_{B_j} v$$

with $\Pi_{B_j} v \in \mathcal{H}_j = \mathrm{im}\, \Pi_{B_j}$ and it remains to show that

$$\mathcal{H}_j \perp \mathcal{H}_k$$

for $1 \leqslant j \neq k \leqslant n$. So suppose $w = \Pi_{B_j} v$ so that $w = \Pi_{B_j} w$. Since

$$B_j \cap B_k = \varnothing$$

we have $\Pi_{B_j \cup B_k} = \Pi_{B_j} + \Pi_{B_k}$ by the properties of Π. Applying this operator to w we obtain $\Pi_{B_j \cup B_k} w = w + \Pi_{B_k} w$, and taking the inner product with w gives

$$\left\|\Pi_{B_j \cup B_k} w\right\|^2 = \langle \Pi_{B_j \cup B_k} w, w \rangle = \|w\|^2 + \langle \Pi_{B_k} w, w \rangle = \|w\|^2 + \|\Pi_{B_k} w\|^2 \geqslant \|w\|^2.$$

However $\|\Pi_{B_j \cup B_k} w\| \leqslant \|w\|$, and it follows that $\Pi_{B_k} w = 0$ or equivalently that $w \in \ker \Pi_{B_k} = \mathcal{H}_k^\perp$. Since this holds for all $w \in \mathcal{H}_j = \operatorname{im} \Pi_{B_j}$ and all $1 \leqslant j \neq k \leqslant n$, the lemma follows. $\square$

Exercise 12.75. Let $\mathcal{H}$, X, and Π be as in Lemma 12.74. Show that

$$\Pi_{B_1} \Pi_{B_2} = \Pi_{B_1 \cap B_2} = \Pi_{B_2} \Pi_{B_1}$$

for any $B_1, B_2 \in \mathcal{B}$.

As expected, the point of projection-valued measures is that one can integrate with respect to them, and construct operators in $B(\mathcal{H})$ using this formalism.

Proposition 12.76 (Integration and uniform convergence). *Let $\mathcal{H}$ be a complex Hilbert space and let Π be a projection-valued measure on $\mathcal{H}$ defined on the Borel σ-algebra $\mathcal{B}$ of a compact metric space X. For any $f \in \mathscr{L}^\infty(X)$ there exists a bounded operator*

$$T = \int_X f(\lambda) \, d\Pi_\lambda,$$

which can be constructed as the uniform limit of the following simple approximation. For any $\varepsilon > 0$ and measurable partition $\xi = \{P_1, \ldots, P_m\}$ of $\overline{B_{\|f\|_\infty}^{\mathbb{C}}}$ with $\operatorname{diam} P_j \leqslant \varepsilon$ and a choice of sample points $\lambda_j \in P_j$ for $j = 1, \ldots, m$ we define the simple function

$$f_\xi = \sum_{j=1}^m \lambda_j \mathbb{1}_{f^{-1}(P_j)} \tag{12.22}$$

and its integral

$$\int f_\xi(\lambda) \, d\Pi_\lambda = \sum_{j=1}^m \lambda_j \Pi_{f^{-1}(P_j)},$$

which satisfies $\|T - \int f_\xi(\lambda) \, d\Pi_\lambda\| \leqslant \varepsilon$.

PROOF OF PROPOSITION 12.76. As indicated in the proposition we define the integral of a simple function

$$f = \sum_{j=1}^m \lambda_j \mathbb{1}_{B_j}$$

when $\lambda_j \neq \lambda_k$ for $1 \leqslant j \neq k \leqslant m$ and $X = \bigsqcup_{j=1}^m B_j$ with $B_1, \ldots, B_m \in \mathcal{B}$ by

$$\int_X f(\lambda) \, d\Pi_\lambda = \sum_{j=1}^m \lambda_j \Pi_{B_j} \in B(\mathcal{H}).$$

This definition makes sense since the additional assumption on $\lambda_1, \ldots, \lambda_m$ and $B_1, \ldots, B_m$ as above ensure that the presentation of f as a sum is unique.

Suppose now that $\xi = \{P_1, \ldots, P_m\}$ and $\zeta = \{Q_1, \ldots, Q_n\}$ are finite partitions of $\overline{B^{\mathbb{C}}_{\|f\|_\infty}}$, $f \in \mathscr{L}^\infty(X)$ and $\varepsilon > 0$ as in the proposition. Choose also two collections of sample points $(\lambda_j)_{j=1,\ldots,m}$ and $(\lambda'_k)_{k=1,\ldots,n}$ with $\lambda_j \in P_j$ and $\lambda'_k \in Q_k$ for all $1 \leqslant j \leqslant m$, $1 \leqslant k \leqslant n$, so that we may define f_ξ by (12.22) and f_ζ similarly. Let $\eta = \{P_j \cap Q_k \mid 1 \leqslant j \leqslant m, 1 \leqslant k \leqslant n\}$ be the common refinement of ξ and ζ. Applying the defining properties of Π we see that

$$\Pi_{f^{-1}(P_j)} = \sum_{k=1}^n \Pi_{f^{-1}(P_j \cap Q_k)}$$

for every $1 \leqslant j \leqslant m$ and similarly for $f^{-1}(Q_k)$. Write

$$A_{\xi,\zeta} = \int f_\xi(\lambda)\,d\Pi_\lambda - \int f_\zeta(\lambda)\,d\Pi_\lambda$$

$$= \sum_{j=1}^m \lambda_j \Pi_{f^{-1}(P_j)} - \sum_{k=1}^n \lambda'_k \Pi_{f^{-1}(Q_k)}$$

$$= \sum_{(j,k):P_j \cap Q_k \neq \varnothing} (\lambda_j - \lambda'_k)\,\Pi_{f^{-1}(P_j \cap Q_k)}.$$

If now $v \in \mathcal{H}$ with $\|v\| \leqslant 1$, then

$$\|A_{\xi,\zeta} v\|^2 = \sum_{(j,k):P_j \cap Q_k \neq \varnothing} |\lambda_j - \lambda'_k|^2 \left\| \Pi_{f^{-1}(P_j \cap Q_k)} v \right\|^2 \tag{12.23}$$

by Lemma 12.74. Using the assumption that $\operatorname{diam} P_j, \operatorname{diam} Q_k \leqslant \varepsilon$, we see that

$$|\lambda_j - \lambda'_k| \leqslant 2\varepsilon \tag{12.24}$$

whenever $P_j \cap Q_k \neq \varnothing$. Putting this into (12.23) gives $\|A_{\xi,\zeta} v\| \leqslant 2\varepsilon \|v\| \leqslant 2\varepsilon$, again by Lemma 12.74.

If we now choose a sequence of partitions (ξ_N) with $\max_{P \in \xi_N} \operatorname{diam} P \to 0$ as $N \to \infty$, then the argument above shows that $\left(\int f_{\xi_N}(\lambda)\,d\Pi_\lambda \right)$ forms a Cauchy sequence with respect to the operator norm (for any choice of the sample points). Just as in the last part of the proof of Proposition 3.81, this also implies that the limit is independent of the choice of sample points and the choice of the sequence of partitions.

In order to improve the estimate from 2ε (as above) to ε in the last part of the proposition, we fix a partition ξ and construct the partitions ξ_N as above so that they are finer than ξ (that is, every partition element of ξ_N is contained in one of the partition elements of ξ). This allows us to make the cosmetic improvement of the 2ε in (12.24) to ε, giving the estimate

$$\left\| \int f_\xi(\lambda)\,\mathrm{d}\Pi_\lambda - \int f_{\xi_N}(\lambda)\,\mathrm{d}\Pi_\lambda \right\| \leqslant \varepsilon,$$

and letting $N \to \infty$ gives the proposition. $\square$

Exercise 12.77. Let $\mathcal{H}$, X, and Π be as in Proposition 12.76. Show that $\int f(\lambda)\,\mathrm{d}\Pi_\lambda$ depends linearly on $f \in \mathscr{L}^\infty(X)$ and that

$$\left\| \int_X f(\lambda)\,\mathrm{d}\Pi_\lambda \right\| \leqslant \|f\|_\infty,$$

$$\left(\int_X f(\lambda)\,\mathrm{d}\Pi_\lambda \right)^* = \int_X \overline{f}(\lambda)\,\mathrm{d}\Pi_\lambda,$$

and

$$\left(\int_X f_1(\lambda)\,\mathrm{d}\Pi_\lambda \right) \left(\int_X f_2(\lambda)\,\mathrm{d}\Pi_\lambda \right) = \int_X (f_1 f_2)(\lambda)\,\mathrm{d}\Pi_\lambda$$

for any $f, f_1, f_2 \in \mathscr{L}^\infty(X)$.

Exercise 12.78 (Strong convergence). Let $\mathcal{H}, X, \Pi$ be as in Proposition 12.76, and assume that (f_n) is a sequence of functions in $\mathscr{L}^\infty(X)$ with $\sup_{n \geqslant 1} \|f_n\|_\infty < \infty$ and $f_n(x) \to f(x)$ as $n \to \infty$ for every $x \in X$. Show that

$$\int_X f_n(\lambda)\,\mathrm{d}\Pi_\lambda \longrightarrow \int_X f(\lambda)\,\mathrm{d}\Pi_\lambda$$

as $n \to \infty$ in the strong operator topology.

It remains to establish the connection between the functional calculus as in Section 12.6 and the projection-valued measures considered here.

Theorem 12.79. *Let $\mathcal{H}$ be a complex Hilbert space and let $\mathcal{A} \subseteq \mathrm{B}(\mathcal{H})$ be a separable commutative unital C^*-sub-algebra (for example, the unital algebra generated by a single normal operator T). Then there exists a projection-valued measure Π on $\mathcal{H}$ defined on the σ-algebra $\mathcal{B}$ of Borel subsets of the space $X = \sigma(\mathcal{A})$ (respectively, $\sigma(T)$ in the case of a single normal operator) such that*

$$f_\mathcal{H} = \int_X f(\lambda)\,\mathrm{d}\Pi_\lambda$$

for any $f \in \mathscr{L}^\infty(X)$.

PROOF. We define the projection-valued measure Π_B for a Borel set $B \in \mathcal{B}$ in X using the functional calculus by setting

$$\Pi_B = (\mathbb{1}_B)_\mathcal{H}$$

(resp. $\Pi_B = \mathbb{1}_B(T)$ in the case of a single normal operator). To show that Π satisfies the property in Definition 12.73, suppose that $B = \bigsqcup_{n \geqslant 1} B_n$ with $B_n \in \mathcal{B}$ for all $n \geqslant 1$ and fix some $v \in \mathcal{H}$. Then

$$\left\|\left(\Pi_B - \sum_{n=1}^{N} \Pi_{B_n}\right) v\right\|^2 = \left\|\left(\mathbb{1}_B - \sum_{n=1}^{N} \mathbb{1}_{B_n}\right)_{\mathcal{H}} v\right\|^2$$

$$= \left\|\left(\mathbb{1}_{B \smallsetminus \sqcup_{n=1}^{N} B_n}\right)_{\mathcal{H}} v\right\|^2 = \left\langle \left(\mathbb{1}_{B \smallsetminus \sqcup_{n=1}^{N} B_n}\right)_{\mathcal{H}} v, v \right\rangle$$

$$= \int_X \mathbb{1}_{B \smallsetminus \sqcup_{n=1}^{N} B_n} \, \mathrm{d}\mu_v = \mu_v \left(B \smallsetminus \bigsqcup_{n=1}^{N} B_n\right) \longrightarrow 0$$

as $N \to \infty$, which shows (12.21).

Now let $f \in \mathscr{L}^\infty(X)$ with f_ξ a simple approximation to f as in Proposition 12.76. Then

$$\int_X f_\xi(\lambda) \, \mathrm{d}\Pi_\lambda = \sum_{P_j \in \xi} \lambda_j \left(\mathbb{1}_{f^{-1}(P_j)}\right)_{\mathcal{H}} = (f_\xi)_{\mathcal{H}}$$

by linearity of the functional calculus. Using a sequence of partitions (ξ_n) with the property that $f_{\xi_n} \to f$ uniformly as $n \to \infty$ as in the proof of Proposition 12.76, we see that

$$\int_X f_{\xi_n}(\lambda) \, \mathrm{d}\Pi_\lambda \longrightarrow \int_X f(\lambda) \, \mathrm{d}\Pi_\lambda$$

as $n \to \infty$ by Proposition 12.76, and

$$(f_{\xi_n})_{\mathcal{H}} \longrightarrow f_{\mathcal{H}}$$

as $n \to \infty$ by the continuity bound of the functional calculus ((FC2) in Proposition 12.67). This gives the theorem. $\qquad\square$

12.8 Locally Compact Abelian Groups and Pontryagin Duality

In this section we study the relationship between the unitary representations of a locally compact σ-compact metric abelian group G and its dual or character group as defined in Definition 11.37. As a consequence of the results of the previous and the current chapter we will also prove the completeness of characters claimed on p. 92.

As we have seen, a surprising and satisfying fact is that the dual group $\widehat{G}$ of a locally compact abelian group G is also a locally compact abelian group. This in turn allows us to repeat the operation of forming the dual group to $\widehat{G}$, giving the *bidual* of G, which will be canonically isomorphic to G as a topological group. This *duality* or *reflexivity* of locally compact abelian groups is called *Pontryagin duality*.

12.8.1 The Spectral Theorem for Unitary Representations

Let π be a unitary representation of G on $\mathcal{H}$ as in the corollary, and recall the definition of the operator $f \underset{\pi}{*}$ from Section 3.5.4 for any $f \in L^1(G)$.

Essential Exercise 12.80. Suppose that $\pi : G \curvearrowright \mathcal{H}$ is a unitary representation of a locally compact σ-compact metric abelian group G on a Hilbert space $\mathcal{H}$. Show that $(f \underset{\pi}{*})^* = \widetilde{f} \underset{\pi}{*}$ for any $f \in L^1(G)$, where $\widetilde{f}$ is defined by $\widetilde{f}(g) = \overline{f(-g)}$ for all $g \in G$.

Corollary 12.81. *Let G be a locally compact σ-compact metric abelian group and let $\widehat{G}$ be its dual group (as in Definition 11.37, and equipped with the weak* topology as in Proposition 11.38). Let $\pi : G \to \mathrm{B}(\mathcal{H})$ be a unitary representation of G on a separable complex Hilbert space $\mathcal{H}$. Then there exists a finite measure μ on $X = \widehat{G} \times \mathbb{N}$ (resp. $X = \widehat{G}$ if $\mathcal{H}$ is cyclic) and a unitary isomorphism $\phi : \mathcal{H} \to L^2_\mu(X)$ such that*

$$\phi \circ \pi_g = M_g \circ \phi \tag{12.25}$$

for all $g \in G$, where M_g is the multiplication operator defined by the function $X \ni (\chi, n) \mapsto \chi(g)$. Moreover, a unitary isomorphism $\phi : \mathcal{H} \to L^2_\mu(X)$ satisfies (12.25) if and only if it satisfies $\phi \circ (f \underset{\pi}{*}) = M_{\widehat{f}} \circ \phi$ for any $f \in L^1(G)$.

The proof of the corollary consists largely of assembling the evidence that we have already proved it.

PROOF OF COROLLARY 12.81. By Proposition 11.38, $L^1(G)$ is a separable commutative Banach algebra. Applying Exercise 11.1 we obtain the separable commutative unital Banach algebra $L^1(G) \oplus \mathbb{C}$, whose elements we will write as $f + \lambda I$ where I denotes the multiplicative unit of the algebra.

Using Exercise 3.86 we define the bounded operator

$$\imath : L^1(G) \oplus \mathbb{C} \ni f + \lambda I \longmapsto f \underset{\pi}{*} + \lambda I \in \mathrm{B}(\mathcal{H}).$$

By Proposition 3.91 and Exercise 3.92 it follows that the closure $\mathcal{A}$ of the image $\imath(L^1(G) \oplus \mathbb{C})$ is a separable commutative unital sub-algebra of $\mathrm{B}(\mathcal{H})$. By Exercise 12.80, we see that $\mathcal{A}$ is also a C^*-sub-algebra. Applying Theorem 12.60 we find a unitary isomorphism

$$\phi : \mathcal{H} \to L^2_\mu(X)$$

for some finite measure μ on $X = \sigma(\mathcal{A}) \times \mathbb{N}$ satisfying $\phi \circ a = M_{a^\circ} \circ \phi$ for all $a \in \mathcal{A}$.

We will deduce (12.25) from this formula. However, instead of carrying the factor $\mathbb{N}$ around in the following discussions we simplify the notation and assume that the unitary representation is cyclic, and hence $X = \sigma(\mathcal{A})$ (see Exercise 12.82). The general case follows easily from this by either dropping that assumption after one has understood the argument below, or by putting the various cyclic subspaces back together as we have done many times before.

Since $\imath : L^1(G) \oplus \mathbb{C} \to \mathcal{A}$ has dense image, a linear functional on $\mathcal{A}$ is uniquely determined by its restriction to the image $\imath(L^1(G) \oplus \mathbb{C})$. Equivalently, the dual map

$$\imath^* : \mathcal{A}^* \longrightarrow \left(L^1(G) \oplus \mathbb{C}\right)^*$$

is injective. By Exercise 8.9(b) (see also the hint on p. 574), $\imath^*$ is continuous with respect to the weak* topology. By Theorem 11.23, $X = \sigma(\mathcal{A})$ is compact, and hence the restriction of $\imath^*$ to X is a homeomorphism to

$$\imath^*(X) \subseteq X' = \sigma(L^1(G) \oplus \mathbb{C}).$$

Next we define the measure $\mu' = (\imath^*)_* \mu$, which also gives us the identification $L^2_\mu(X) = L^2_{\mu'}(X')$. Fix some $f \in L^1(G)$, then we have $\phi \circ (f \underset{*}{*}) = M_{a^\circ} \circ \phi$ with $a = \imath(f) = f \underset{*}{*}$. We use $\imath^*$ to identify X with the subset $\imath^*(X) \subseteq X'$, and claim that in this sense the function f° extends the function a°. Indeed if $\chi \in X = \sigma(\mathcal{A})$, then

$$a^\circ(\chi) = \chi(a) = \chi(\imath(f)) = \chi \circ \imath(f) = f^\circ(\chi \circ \imath) = f^\circ(\imath^*(\chi)).$$

In other words, we obtain the following slightly more convenient description: the unitary isomorphism $\phi : \mathcal{H} \to L^2_{\mu'}(X')$ satisfies $\phi \circ (f \underset{*}{*}) = M_{f^\circ} \circ \phi$ for all $f \in L^1(G)$.

By Corollary 11.29 (and its proof) we have the identification

$$X' = \sigma(L^1(G) \oplus \mathbb{C}) = \sigma(L^1(G)) \cup \{0\},$$

which is also a homeomorphism. We claim that μ' actually gives full measure to $L^1(G)$ and zero measure to the extra point 0. This follows from continuity of the unitary representation. Indeed, if $\phi(v) = \mathbb{1}_{\{0\}}$ and $\mu'(\{0\})$ is positive then $v \in \mathcal{H}$ is non-zero and there exists a compact neighbourhood B of $0 \in G$ such that $\Re(\langle \pi_g v, v \rangle) > 0$ for all $g \in B$. This implies that $\mathbb{1}_B \underset{*}{*} v$ is non-zero, since

$$\Re\langle \mathbb{1}_B \underset{*}{*} v, v \rangle = \int_B \Re\langle \pi_g v, v \rangle \, dm(g) > 0.$$

On the other hand, $(\mathbb{1}_B)^\circ(0) = 0$ and hence $M_{\mathbb{1}_B^\circ} \mathbb{1}_{\{0\}} = 0$. Recalling the formula $\phi \circ (f \underset{*}{*}) = M_{f^\circ} \circ \phi$ we derive a contradiction and see that $\mu'(\{0\}) = 0$.

Finally, we recall from Proposition 11.38 that the Gelfand dual $\sigma(L^1(G))$ can be identified with the Pontryagin dual group $\widehat{G}$ and the Gelfand transform can be identified with the Fourier back transform. Simplifying the notation, we can summarize the above by saying that the spectral theorem shows the existence of a finite measure μ on $X = \widehat{G}$ and a unitary isomorphism

$$\phi : \mathcal{H} \to L^2_\mu(X)$$

such that $\phi \circ (f \underset{\pi}{*}) = M_{\tilde{f}} \circ \phi$ for all $f \in L^1(G)$.

This implies (12.25) by using an approximate identity (which also gives the first direction of the claimed equivalence in the corollary). Let (B_k) be a decreasing sequence of open neighbourhoods of $0 \in G$ that form a basis of the topology at 0, define $\psi_k = \frac{1}{m(B_k)} \mathbb{1}_{B_k}$, and fix some $g_0 \in G$. Applying the argument above to the function f defined by

$$f(g) = \psi_k^{g_0}(g) = \psi_k(g - g_0),$$

we obtain

$$\phi \circ (\psi_k^{g_0} \underset{\pi}{*}) = M_{\widetilde{\psi_k^{g_0}}} \circ \phi \tag{12.26}$$

for all k. Fix some $v \in \mathcal{H}$. We will prove that $M_{\widetilde{\psi_k^{g_0}}}(\phi(v))$ converges to $M_{g_0}(\phi(v))$ (as defined in the corollary) as $k \to \infty$ and that $\phi(\psi_k^{g_0} \underset{\pi}{*} v)$ converges to $\phi(\pi_{g_0} v)$, which then gives (12.25).

To see that $M_{\widetilde{\psi_k^{g_0}}}(\phi(v))$ converges to $M_{g_0}(\phi(v))$, we note that

$$\left| \widetilde{\psi_k^{g_0}}(\chi) \right| = \left| \int_G \psi_k^{g_0} \chi \, \mathrm{d}m \right| \leqslant \|\psi_k^{g_0}\|_1 = 1,$$

for all $\chi \in \widehat{G}$, so that $\|\widetilde{\psi_k^{g_0}}\|_\infty \leqslant 1$, that

$$\lim_{k \to \infty} \widetilde{\psi_k^{g_0}}(\chi) = \lim_{k \to \infty} \frac{1}{m(B_k)} \int_{B_k + g_0} \chi \, \mathrm{d}m = \chi(g_0),$$

and that with this dominated convergence implies that

$$\lim_{k \to \infty} \left\| M_{\widetilde{\psi_k^{g_0}}}(\phi(v)) - M_{g_0}(\phi(v)) \right\|_2^2 = \lim_{k \to \infty} \int_X \left| \widetilde{\psi_k^{g_0}}(\chi) - \chi(g_0) \right|^2 |\phi(v)|^2 \mathrm{d}\mu(\chi) = 0,$$

as claimed.

Since ϕ is a unitary isomorphism, (12.26) shows that $(\psi_k^{g_0} \underset{\pi}{*} v)$ converges in $\mathcal{H}$. In order to identify the limit $\widetilde{v} \in \mathcal{H}$, fix some $w \in \mathcal{H}$ and use the definition of $\psi_k^{g_0} \underset{\pi}{*}$ to see that

$$\langle \widetilde{v}, w \rangle = \lim_{k \to \infty} \langle \psi_k^{g_0} \underset{\pi}{*} v, w \rangle = \lim_{k \to \infty} \int_G \psi_k^{g_0}(h) \langle \pi_h v, w \rangle \, \mathrm{d}m(h)$$

$$= \lim_{k \to \infty} \frac{1}{m(B_k)} \int_{B_k + g_0} \langle \pi_h v, w \rangle \, \mathrm{d}m(h) = \langle \pi_{g_0} v, w \rangle$$

by the continuity property of unitary representations. Since $w \in \mathcal{H}$ is arbitrary, it follows that $\psi_k^{g_0} \underset{\pi}{*} v$ converges to $\pi_{g_0} v$ as $k \to \infty$. Therefore (12.25) follows by taking the limit of the equation (12.26) as $k \to \infty$.

To prove the corollary it remains to show that (12.25) for some unitary isomorphism $\phi : \mathcal{H} \to L^2_\mu(X)$ for a finite measure on $X = \widehat{G} \times \mathbb{N}$ implies that

$$\phi \circ (f \underset{*}{*}) = M_{\tilde{f}} \circ \phi$$

for any $f \in L^1(G)$. This follows from the definition of $f \underset{*}{*}$ and Fubini's theorem. Indeed, using (12.25) in the form $\phi(\pi_g u)(\chi, n) = \chi(g)\phi(u)(\chi, n)$ for all $(\chi, n) \in \widehat{G} \times \mathbb{N}$ we obtain

$$
\begin{aligned}
\langle \phi(f \underset{*}{*} u), f_1 \rangle = \langle f \underset{*}{*} u, \phi^* f_1 \rangle &= \int_G f(g) \, \langle \pi_g u, \phi^* f_1 \rangle \, dm(g) \\
&= \int_G \int_X \phi(\pi_g u)(\chi, n) \overline{f_1}(\chi, n) f(g) \, d\mu(\chi, n) \, dm(g) \\
&= \int_X \int_G f(g) \chi(g) \, dm(g) \phi(u)(\chi, n) \overline{f_1(\chi, n)} \, d\mu(\chi, n) \\
&= \left\langle M_{\tilde{f}}(\phi(u)), f_1 \right\rangle
\end{aligned}
$$

for all $u \in \mathcal{H}$ and $f_1 \in L^2_\mu(X)$. $\square$

Essential Exercise 12.82. In the third paragraph of the proof of Corollary 12.81 we assumed 'without loss of generality' that the Hilbert space $\mathcal{H}$ is cyclic. However, this contained a small cheat as we did not clarify whether we meant cyclic with respect to the unitary representation π (as in Definition 9.7) or cyclic with respect to the sub-algebra $\mathcal{A}$ obtained from $L^1(G)$ and convolution (as in Definition 12.52). Show that these two notions are equivalent.

Exercise 12.83. Given a unitary representation π of a locally compact σ-compact metric abelian group G apply Theorem 12.79 to the closure of the image of the algebra $L^1(G) \oplus \mathbb{C}$ to obtain a projection-valued measure on $\widehat{G}$ such that π_g is given by $\int_{\widehat{G}} \chi(g) \, d\Pi_\chi$ for all $g \in G$.

12.8.2 Characters Separate Points

Using the spectral theorem for unitary representations as in the last corollary, we turn to the question of whether the Pontryagin dual group is sufficiently rich to separate points, as claimed on p. 92.

Theorem 12.84 (Completeness of characters). *On every locally compact σ-compact metric abelian group G there are enough characters to separate points. That is, if $g, h \in G$ have $g \neq h$, then there exists a character $\chi \in \widehat{G}$ with $\chi(g) \neq \chi(h)$.*

PROOF. Let G be a locally compact σ-compact metric abelian group, and let $g_0 \in G \setminus \{0\}$ be non-trivial. Then it is easy to see (for example, by using characterstic functions of sufficiently small compact neighbhorhoods of 0 in G) that the unitary operator

$$\lambda_{g_0} : L^2(G) \longrightarrow L^2(G)$$
$$f \longmapsto (f^{g_0} : h \mapsto f(h - g_0))$$

is not the identity map. Therefore, if we apply Corollary 12.81 to the regular representation λ of G on $L^2(G)$ we see that there exists some $\chi \in \widehat{G}$ with $\chi(g_0) \neq 1$. Applying this to $g_0 = g - h$ for some $g, h \in G$ with $g \neq h$, we obtain completeness of characters, as claimed.   □

12.8.3 The Plancherel Formula

Recall from Proposition 11.43 that $\widehat{G}$ is a σ-compact locally compact metric abelian group if G is and let us note that in the next subsection we will use this to establish Pontryagin duality. We show in this subsection that by applying Corollary 12.81 to the regular representation of G we obtain a generalization of the Fourier transform to more general abelian groups. Because of these results it is natural to treat G and $\widehat{G}$ in the same way, and in particular to use the same additive notation in both groups. This is a familiar process in functional analysis, where notation supports abstraction of ideas: The dual group $\widehat{G}$ is initially defined as a collection of maps taking values in $\mathbb{S}^1$ with the operation of pointwise multiplication, which is therefore most naturally written multiplicatively. However, given the developed structure of $\widehat{G}$ it is now natural to think of the dual group operation additively as follows. However, this is really just a change of notation and not an isomorphism between two differently defined objects. Hence nothing needs to be proved, but the relation between the old and the new notation needs to be clarified.

Let us write t for an element of $\widehat{G}$ (as we did in Section 9.2), use additive notation for the group operation, and write $\chi_t : G \to \mathbb{S}^1$ for the character on G corresponding to $t \in \widehat{G}$. In particular, this means that $\chi_0 = \mathbb{1}$ and

$$\chi_{t_1+t_2} = \chi_{t_1} \chi_{t_2}$$

for $t_1, t_2 \in \widehat{G}$. If we want to remove the discrimination between G and $\widehat{G}$ even further (as we will do, for example, in the next subsection) we also write

$$\langle g, t \rangle = \chi_t(g) \in \mathbb{S}^1$$

for the *dual pairing* of $g \in G$ and $t \in \widehat{G}$. For $t \in \widehat{G}$ write $M_t : L^2(G) \to L^2(G)$ for the multiplication operator defined by $M_t(f)(g) = \langle g, t \rangle f(g)$ for $g \in G$. Finally, we will write $\widehat{\lambda}$ for the regular representation of $\widehat{G}$ on (equivalence classes of) functions f on $\widehat{G}$, so that

$$\widehat{\lambda}_{t_0}(f)(t) = f(t - t_0)$$

for all $t, t_0 \in \widehat{G}$.

Theorem 12.85 (Plancherel formula). *Let G be a σ-compact locally compact metrizable abelian group with Haar measure m_G. Then there exists a normalization of the Haar measure $m_{\widehat{G}}$ on $\widehat{G}$ and a unitary isomorphism $\phi : L^2(G) \to L^2(\widehat{G})$ which extends the Fourier back transform $f \mapsto \check{f}$ on $L^1(G) \cap L^2(G)$ to all of $L^2(G)$ and satisfies $\phi \circ \lambda_g = M_g \circ \phi$ as well as $\phi \circ M_t = \widehat{\lambda}_{-t} \circ \phi$ for all $g \in G$ and $t \in \widehat{G}$.*

We note that this generalizes Theorem 9.39 and Proposition 9.29, except that we work here with the Fourier back transform. We split the proof of the theorem into several steps. Our argument below may not be the most direct approach, but will also help us to prove Pontryagin duality in the next subsection. We will assume the hypotheses of Theorem 12.85 throughout.

Lemma 12.86 (A Gaussian on G). *There exists a $\psi \in \mathcal{V} = L^1(G) \cap L^2(G)$ with $\check{\psi}(t) > 0$ for all $t \in \widehat{G}$.*

PROOF. Recall that for an approximate identity $\psi_k = \frac{1}{m(B_k)} \mathbb{1}_{B_k} \in \mathcal{V}$ (as in the proof of Corollary 12.81), we have $\widetilde{\psi_k}(t) \longrightarrow 1$ as $k \to \infty$ for any element t of $\widehat{G}$. In particular, for every $t \in \widehat{G}$ there exists some $k \in \mathbb{N}$ with $\widehat{\psi_k}(g) \neq 0$.
Moreover, with $\widetilde{\psi_k} = \overline{\psi_k(-g)}$ we have

$$\widetilde{\widetilde{\psi_k}}(t) = \int_G \overline{\overline{\psi_k(-g)}\chi_t(g)}\, dm(g) = \int_G \overline{\overline{\psi_k(h)}\chi_t(h)}\, dm(h) = \overline{\widetilde{\psi_k}(t)}$$

for every $t \in \widehat{G}$ and $k \in \mathbb{N}$. Therefore, $\widetilde{\psi_k \ast \widetilde{\psi_k}} = |\check{\psi_k}|^2 \geqslant 0$ for every $k \in \mathbb{N}$. Setting $\psi = \sum_{k=1}^{\infty} c_k \psi_k \ast \widetilde{\psi_k}$ for some rapidly decaying positive sequence $(c_k)_k$ we obtain $\psi \in L^1(G) \cap L^2(G)$. By the above we have $\check{\psi} = \sum_{k=1}^{\infty} c_k |\widetilde{\psi_k}| > 0$ and the lemma follows. □

Lemma 12.87 (Correcting measure and isomorphism). *If we apply Corollary 12.81 to the regular representation λ of G we may assume without loss of generality that the resulting measure space is defined using $X = \widehat{G}$ (instead of $\widehat{G} \times \mathbb{N}$). Moreover, assuming the conclusions of Corollary 12.81 it is possible to replace the original spectral measure by an absolutely continuous σ-finite measure μ such that ϕ also satisfies*

$$\phi(f)(t) = \check{f}(t) \tag{12.27}$$

for $t \in \widehat{G}$ and all $f \in \mathcal{V} = L^1(G) \cap L^2(G)$.

PROOF. Applying Corollary 12.81 to the unitary representation λ, we obtain a finite measure μ_0 on $X = \widehat{G} \times \mathbb{N}$ and a unitary isomorphism

$$\phi_0 : L^2(G) \longrightarrow L^2_{\mu_0}(X)$$

such that $\phi_0 \circ (f *) = M_{\check{f}} \circ \phi_0$ for all $f \in L^1(G)$.

Suppose that $f_1, f_2, f \in V = L^1(G) \cap L^2(G)$. Then, by definition of $f_1 * f_2$ and Fubini's theorem,

$$
\begin{aligned}
\langle f_1 * f_2, f \rangle &= \int f_1(g)\langle \lambda_g(f_2), f \rangle \, dm(g) \\
&= \iint f_1(g) f_2(\underbrace{h - g}_{k}) f(h) \, dm(g) \, dm(h) \\
&= \iint f_1(h - k) f_2(k) f(h) \, dm(k) \, dm(h) = \langle f_2 * f_1, f \rangle,
\end{aligned}
$$

which proves $f_1 * f_2 = f_2 * f_1$ by density of $V \subseteq L^2(G)$ (see also Exercises 3.86 and 3.92). Applying the unitary isomorphism ϕ_0 this gives

$$
\check{f_1}\phi_0(f_2) = \check{f_2}\phi_0(f_1) \tag{12.28}
$$

almost everywhere with respect to μ_0.

We now set $f_1 = \psi$ with $\psi \in V$ as in Lemma 12.86 and define

$$
w(t, n) = \frac{\phi_0(\psi)(t, n)}{\check{\psi}(t)}
$$

for $(t, n) \in \widehat{G} \times \mathbb{N}$. Setting $f_2 = f \in V$ and dividing (12.28) by $\widehat{\psi}$ we obtain

$$
\phi_0(f)(t, n) = w(t, n)\check{f}(t) \tag{12.29}
$$

for all $f \in V$ and μ_0-almost every $(t, n) \in X$. This represents the main step towards the lemma, which we will obtain by modifying the unitary isomorphism and the measure as follows.

Since $\phi_0 : L^2(G) \to L^2_{\mu_0}(X)$ is an isomorphism and $V = L^1(G) \cap L^2(G)$ is dense in $L^2(G)$, we see from (12.29) that $w(t, n) \neq 0$ μ_0-almost everywhere. Using this we define the σ-finite measure μ_1 on X by

$$
\frac{d\mu_1}{d\mu_0} = |w|^2,
$$

and the map $\phi_1 = M_{w^{-1}} \circ \phi_0$ (with inverse $\phi_0^{-1} \circ M_w$) which satisfies

$$
\begin{aligned}
\|\phi_1(f)\|^2_{L^2_{\mu_1}(X)} &= \int_X |w|^{-2}|\phi_0(f)|^2 \, d\mu_1 = \int_X |\phi_0(f)|^2 |w|^{-2}\frac{d\mu_1}{d\mu_0} \, d\mu_0 \\
&= \|\phi_0(f)\|^2_{L^2_{\mu_0}(X)} = \|f\|_2
\end{aligned}
$$

for all $f \in L^2(G)$. Hence ϕ_1 is a unitary isomorphism $\phi_1 : L^2(G) \to L^2_{\mu_1}(X)$, and

$$
\phi_1(f)(t, n) = \check{f}(t) \tag{12.30}
$$

for all $f \in \mathcal{V}$ and μ_1-almost every $(t, n) \in X$. Since any two multiplication operators on X commute, the new unitary isomorphism still satisfies the conclusions of the spectral theorems.

To summarize, $\phi_1 : L^2(G) \to L^2_{\mu_1}(X)$ is a unitary isomorphism satisfying (12.30). Finally, since $\phi_1(\mathcal{V})$ is dense in $L^2_{\mu_1}(X)$, this implies that every element of $L^2_{\mu_1}(X)$ can be expressed as a pointwise limit of a sequence in $\phi_1(\mathcal{V})$ and so has a representative that only depends on $t \in \widehat{G}$. Let $p : X \to \widehat{G}$ denote the projection to the first coordinate of $X = \widehat{G} \times \mathbb{N}$, and write $X = \bigcup_{n \geqslant 1} X_n$ as a union of sets $X_n \subseteq X$ with finite measure. Then for every $n \geqslant 1$ we may use the above observation for the function $\mathbb{1}_{X_n} \in L^2_{\mu_1}(X)$ and see that there exists a measurable set $Y_n \subseteq \widehat{G}$ with $\mu_1 (X_n \triangle (Y_n \times \mathbb{N})) = 0$. It follows that

$$\mu_1 \left(X \smallsetminus \bigcup_{n \geqslant 1} Y_n \times \mathbb{N} \right) = 0$$

and so $p_* \mu_1$ is a σ-finite measure on $\widehat{G}$ with $L^2_{p_* \mu_1}(\widehat{G}) = L^2_{\mu_1}(X)$.

Simplifying the notation, we may assume that $\phi : L^2(G) \to L^2_\mu(\widehat{G})$ is a unitary isomorphism, that μ is a σ-finite measure on $\widehat{G}$, and that in addition to the claims of the spectral theorem it also satisfies (12.27). □

Essential Exercise 12.88. Show that $\widetilde{L^1(G)} \subseteq C_0(\widehat{G})$ is dense.

Lemma 12.89 (Spectral theorem produces Haar measure). *Let μ be a σ-finite measure on $\widehat{G}$ satisfying the conclusion of Lemma 12.87. Then the measure $\mu = m_{\widehat{G}}$ is a Haar measure on $\widehat{G}$.*

PROOF. Throughout the proof we will use the function ψ from Lemma 12.86. We first claim that μ is locally finite. Notice that $\check{\psi} \in C_0(\widehat{G})$ by definition of the topology on $\widehat{G}$ in Propositions 11.33 and 11.38. Hence

$$O_{t_0} = \{ t \in \widehat{G} \mid |\check{\psi}|^2(t) > \tfrac{1}{2} |\check{\psi}|^2(t_0) \}$$

is a neighbourhood of $t_0 \in \widehat{G}$. Together with $\check{\psi} = \phi(\psi) \in L^2_\mu(\widehat{G})$ (as assumed in the lemma) and $|\check{\psi}|^2 \in L^1_\mu(\widehat{G})$, it follows that $\mu(O_{t_0}) < \infty$. Since $t_0 \in \widehat{G}$ was arbitrary, it follows that μ is locally finite, as claimed.

Next we note that for $t_0 \in \widehat{G}$ and $f \in L^1(G)$ we have $\widetilde{\chi_{t_0} f} = \widehat{\lambda}_{-t_0} f$ since

$$(\widetilde{\chi_{t_0} f})(t) = \int_G (\chi_{t_0} f) \chi_t \, dm = \check{f}(t_0 + t) = \widehat{\lambda}_{-t_0}(\check{f})(t) \tag{12.31}$$

for all $t \in \widehat{G}$.

Below we combine (12.31) for $f = \psi \in L^1(G)$ with a similar claim for the spectral measure of $\psi \in L^2(G)$ and will obtain the lemma from this. By the assumptions in the lemma we can define the spectral measures μ_F on $\widehat{G}$ for the algebra $L^1(G)$ acting on functions $F \in L^2(G)$ by $d\mu_F = |\phi(F)|^2 \, d\mu$ since

$$\langle f \underset{\lambda}{*} F, F\rangle_{L^2(G)} = \int_{\widehat{G}} \check{f} |\phi(F)|^2 \, d\mu$$

for all $f \in L^1(G)$.

We will now show that the spectral measures satisfy

$$\mu_{\chi_{t_0}\psi} = (T_{t_0})_* \mu_\psi \tag{12.32}$$

for all $t_0 \in \widehat{G}$, where we use the translation defined by $T_{t_0}(t) = t - t_0$ for all $t \in \widehat{G}$ and the push-forward of the measure (as defined on p. 265). Indeed, for $f \in L^1(G)$ we have (by our definitions)

$$
\begin{aligned}
\int_{\widehat{G}} \check{f} \, d\mu_{\chi_{t_0}\psi} &= \langle f \underset{\lambda}{*} (\chi_{t_0}\psi), \chi_{t_0}\psi\rangle_{L^2(G)} \\
&= \int_G f(g)\langle \lambda_g(\chi_{t_0}\psi), \chi_{t_0}\psi\rangle_{L^2(G)} \, dm(g) \\
&= \int_G f(g)\chi_{t_0}(-g) \underbrace{\langle \chi_{t_0}\lambda_g\psi, \chi_{t_0}\psi\rangle_{L^2(G)}}_{=\langle \lambda_g\psi,\psi\rangle} \, dm(g) \\
&= \langle (\chi_{-t_0}f) \underset{\lambda}{*} \psi, \psi\rangle_{L^2(G)} = \int_{\widehat{G}} \widecheck{\chi_{-t_0}f} \, d\mu_\psi = \int_{\widehat{G}} \check{f} \, d(T_{t_0})_* \mu_\psi,
\end{aligned}
$$

since $(\widecheck{\chi_{-t_0}f})(t) = \widehat{\lambda}_{t_0}\check{f}(t) = \check{f}(t - t_0) = \check{f} \circ T_{t_0}(t)$ for all $t \in \widehat{G}$ by (12.31). This proves the claim (12.32) by the uniqueness properties of the spectral measures (which follow from Exercise 12.88) as $f \in L^1(G)$ was arbitrary.

We now combine (12.31), (12.32), and the assumption that $\phi(f) = \check{f}$ for any $f \in \mathcal{V}$. For some test function $F \in C_c(\widehat{G})$ we then have

$$
\begin{aligned}
\int_{\widehat{G}} F|\check{\psi}|^2 \, d\mu &= \int_{\widehat{G}} F \, d\mu_\psi = \int_{\widehat{G}} F \circ T_{-t_0} \, d(T_{t_0})_* \mu_\psi \\
&= \int_{\widehat{G}} \widehat{\lambda}_{-t_0} F \, d\mu_{\chi_{t_0}\psi} = \int_{\widehat{G}} \widehat{\lambda}_{-t_0} F |\widecheck{\chi_{t_0}\psi}|^2 \, d\mu \\
&= \int_{\widehat{G}} \widehat{\lambda}_{-t_0} F \lambda_{-t_0} |\check{\psi}|^2 \, d\mu = \int_{\widehat{G}} \widehat{\lambda}_{-t_0} (F|\check{\psi}|^2) \, d\mu.
\end{aligned}
$$

Replacing F by $F|\check{\psi}|^{-2} \in C_c(\widehat{G})$ we also obtain

$$\int_{\widehat{G}} F \, d\mu = \int_{\widehat{G}} \widehat{\lambda}_{-t_0} F \, d\mu$$

for any $t_0 \in \widehat{G}$ and $F \in C_c(\widehat{G})$. By the uniqueness property of the measure in the Riesz representation theorem (Theorem 7.54) we deduce that μ is invariant under translation.

Finally, note that $\mu(O) > 0$ for any non-empty open subset, since otherwise every compact subset could be covered by finitely many translates of O and hence would have measure 0. Since $\mu \neq 0$ we deduce that μ is a Haar measure. $\square$

PROOF OF THEOREM 12.85. By Lemma 12.87 we may apply Corollary 12.81 and assume that $X = \widehat{G}$ and $\phi(f) = \check{f}$ for all $f \in \mathcal{V} = L^1(G) \cap L^2(G)$. Applying Lemma 12.89 we also see that the measure is given by the Haar measure $\mu = m_{\widehat{G}}$. The formula $\phi \circ \lambda_g = M_g \circ \phi$ holds by Corollary 12.81. Finally, $\phi(M_{t_0} f) = \widehat{\lambda}_{-t_0}(\phi(f))$ holds initially for $f \in \mathcal{V}$, but knowing that μ is the Haar measure on $\widehat{G}$ extends easily to all of $L^2(G)$. This concludes the proof of the theorem. $\square$

12.8.4 Pontryagin Duality

We are now ready to establish a complete symmetry between G and its dual group $\widehat{G}$. Using Proposition 11.43 we can define the dual group $\widehat{\widehat{G}}$ of the dual group $\widehat{G}$ of G and are led to the question of reflexivity of locally compact abelian groups. Fortunately, the situation here is much better than that for Banach spaces in Chapter 7, as the next result shows. Let us prepare for it with the following exercise.

Essential Exercise 12.90. Let G be a locally compact σ-compact metric abelian group with dual group $\widehat{G}$.
(a) Show that $G \times \widehat{G} \ni (g, t) \mapsto \langle g, t \rangle \in \mathbb{S}^1$ is continuous.
(b) Show that the map $\imath : G \to \widehat{\widehat{G}}$ defined by $\imath(g)(t) = \langle g, t \rangle = \chi_t(g)$ for $g \in G$ and $t \in \widehat{G}$ is a continuous and injective homomorphism of groups.
(c) Suppose $g_n \to \infty$ as $n \to \infty$. Show that for any $f \in L^2(G)$ the sequence $\lambda_{g_n} f$ converges weakly to 0.
(d) Show that $\imath$ is a proper map, meaning that $g_n \to \infty$ as $n \to \infty$ implies that $\imath(g_n) \to \infty$ as $n \to \infty$.
(e) Show that $\imath$ is closed (that is, the image of every closed set is again closed).

Corollary 12.91 (Pontryagin duality). *Let G be a locally compact σ-compact metric abelian group. Then the canonical map $\imath : G \to \widehat{\widehat{G}}$ is an isomorphism of topological groups.*

PROOF. By Exercise 12.90 the map $\imath$ is a continuous closed injective homomorphism of topological groups. It only remains to show that it is surjective; continuity of its inverse will then follow from $\imath$ being a closed map. Let $\phi : L^2(G) \to L^2(\widehat{G})$ be as in Theorem 12.85, which satisfies

$$\phi \circ M_t = \widehat{\lambda}_{-t} \circ \phi$$

for all $t \in \widehat{G}$. We will now read this formula backwards and derive the corollary from it. For this we define a unitary isomorphism

$$U : L^2(\widehat{G}) \to L^2(\widehat{G})$$

by reflecting functions through $0 \in \widehat{G}$, that is by defining $U(f)(t) = f(-t)$ for $f \in L^2(\widehat{G})$ and $t \in \widehat{G}$. Using U and ϕ we also define

$$\psi = \phi^{-1} \circ U : L^2(\widehat{G}) \to L^2(G),$$

which is also a unitary isomorphism. Now notice that $U \circ \widehat{\lambda}_t = \widehat{\lambda}_{-t} \circ U$ since

$$\left(U(\widehat{\lambda}_t(f)) \right)(t') = \widehat{\lambda}_t(f)(-t') = f(-t' - t)$$

and

$$\left(\widehat{\lambda}_{-t}(U(f)) \right)(t') = U(f)(t' + t) = f(-t' - t)$$

for all $f \in L^2(\widehat{G})$ and $t, t' \in \widehat{G}$. Since $\phi^{-1} \circ \widehat{\lambda}_{-t} = M_t \circ \phi^{-1}$ it follows that

$$\psi \circ \widehat{\lambda}_t = \phi^{-1} \circ U \circ \widehat{\lambda}_t = \phi^{-1} \circ \widehat{\lambda}_{-t} \circ U = M_t \circ \phi^{-1} \circ U = M_t \circ \psi$$

for all $t \in \widehat{G}$, which is the conclusion of Corollary 12.81 (if we were to apply it to $\widehat{G}$ and $\widehat{\lambda}$).

We now apply Lemma 12.87, which says in our context (and using the injection $\imath : G \to \widehat{\widehat{G}}$) that the image measure $\mu = \imath_* m_G$ can be modified by a density so that $\psi(f) = \check{f}$ for f in $L^1(\widehat{G}) \cap L^2(\widehat{G})$. By Lemma 12.89, this new measure $\mathrm{d}m_{\widehat{\widehat{G}}} = |w|^2 \, \mathrm{d}\mu$ is a Haar measure on $\widehat{\widehat{G}}$. However, by Exercise 12.90 the image $\imath(G) \subseteq \widehat{\widehat{G}}$ is a closed subgroup, so $\mathrm{Supp}\, \mu \subseteq \imath(G)$. Since $m_{\widehat{\widehat{G}}}(\widehat{\widehat{G}} \smallsetminus \imath(G)) = 0$ we see that $\widehat{\widehat{G}} = \imath(G)$, as claimed in the corollary. $\square$

We close with several exercises developing certain functorial aspects of Pontryagin duality. Throughout these exercises the groups arising are assumed to be locally compact σ-compact metric abelian groups, as usual.

Exercise 12.92. Show that if $H < G$ is a closed subgroup, then G/H is also a locally compact σ-compact metric abelian group with respect to the quotient topology.

Exercise 12.93. Let $H < G$ be a closed subgroup, and define the *annihilator group*

$$H^\perp = \{t \in \widehat{G} \mid \langle h, t \rangle = 1 \text{ for all } h \in H\}.$$

(a) Show that $\widehat{G/H} \cong H^\perp$.
(b) Using the canonical isomorphism between G and $\widehat{\widehat{G}}$ we can also define the double annihilator $(H^\perp)^\perp$ as a subgroup of G. Show that $(H^\perp)^\perp = H$.
(c) Deduce from this that $\widehat{H} \cong \widehat{G}/H^\perp$.

Exercise 12.94. Let $\theta : H \to G$ be a continuous homomorphism.

(a) Show that $\widehat{\theta}(\chi_t) = \chi_t \circ \theta$ for $t \in \widehat{G}$ defines a continuous homomorphism $\widehat{\theta}$ from $\widehat{G}$ to $\widehat{H}$.
(b) Show that θ is injective (or has dense image) if and only if $\widehat{\theta}$ has dense image (respectively is injective).

Exercise 12.95. (a) Show that $\widehat{G_1 \times G_2} \cong \widehat{G_1} \times \widehat{G_2}$ for any groups G_1, G_2 as above.
(b) Let (G_n) be a sequence of compact groups. Show that the direct product $\prod_{n \geqslant 1} G_n$ is again a compact metric abelian group, and that its dual is given by the direct sum

$$\widehat{\prod_{n \geqslant 1} G_n} \cong \bigoplus_{n \geqslant 1} \widehat{G_n} = \left\{ (t_n) \in \prod_{n \geqslant 1} \widehat{G_n} \,\middle|\, t_n = 0 \text{ for all but finitely many } n \geqslant 1 \right\}$$

with the discrete topology.

Exercise 12.96. Let (G_n) be a sequence of compact groups and suppose in addition that there is a surjective continuous homomorphism $\phi_n : G_{n+1} \to G_n$ for each $n \geqslant 1$. The *projective limit* of the system (G_n, ϕ_n) is defined by

$$\varprojlim(G_n, \phi_n) = \left\{ (g_n) \in \prod_{n \geqslant 1} G_n \,\middle|\, \phi_n(g_{n+1}) = g_n \text{ for all } n \geqslant 1 \right\}.$$

Show that this is again a compact metric abelian group (with the topology inherited from the product topology), and that

$$\widehat{\varprojlim(G_n, \phi_n)} = \bigcup_{n \geqslant 1} \widehat{G_n},$$

where we use the injective continuous homomorphism $\widehat{\phi_n} : \widehat{G_n} \to \widehat{G_{n+1}}$ to identify the group $\widehat{G_n}$ with a subgroup of $\widehat{G_{n+1}}$; this *direct limit* is also written $\varinjlim(\widehat{G_n}, \widehat{\phi_n})$.

Exercise 12.97. Formulate and prove the dual statements to Exercise 12.95–12.96 (starting with direct sums, respectively direct limits).

12.9 Further Topics

- Spectral theory will be developed further in Chapter 13, where we study the spectral theory of unbounded self-adjoint operators.
- We refer to Folland [32] and [26] for more on the theory of abstract harmonic analysis and unitary representations of non-abelian groups.
- For more material on abelian harmonic analysis and the structure of abelian topological groups, we refer to Hewitt and Ross [45].

Chapter 13
Self-Adjoint and Symmetric Operators

13.1 Examples and Definitions

In this chapter we will generalize the spectral theorem from Chapter 12 to the case of unbounded self-adjoint operators (the formal definition will be given below). The model case for such an operator is again a multiplication operator.

Example 13.1. Let $(X, \mathcal{B}, \mu)$ be a σ-finite measure space, and let $g : X \to \mathbb{R}$ be measurable. The multiplication operator $M_g : f \mapsto gf$ has the natural domain

$$D_{M_g} = \left\{ f \in L^2_\mu(X) \mid gf \in L^2_\mu(X) \right\}.$$

Clearly

$$\langle M_g(f_1), f_2 \rangle = \int_X g f_1 \overline{f_2} \, \mathrm{d}\mu = \langle f_1, M_g(f_2) \rangle$$

for $f_1, f_2 \in D_{M_g}$. This suggests that M_g is a self-adjoint operator, which is unbounded if $g \notin L^\infty_\mu(X)$ (though this statement requires a proof after we have seen the formal definitions).

The example above as well as the following ones and Exercise 4.29 show that unbounded self-adjoint operators cannot reasonably be required to be defined on the whole Hilbert space. In contrast to Definition 4.27, we will in this chapter always assume that $X = \mathcal{H}$ and $Y = \mathcal{H}'$ are complex Hilbert spaces, and that the domain $D_T \subseteq \mathcal{H}$ is dense.

Definition 13.2. Let $\mathcal{H}$ and $\mathcal{H}'$ be complex Hilbert spaces, let $D_T \subseteq \mathcal{H}$ be a subspace, and let $T : D_T \to \mathcal{H}'$ be a linear operator. Then we write

$$(D_T, T) : \mathcal{H} \longrightarrow \mathcal{H}'.$$

If D_T is a dense subspace then we say that T is a *densely defined* operator from $\mathcal{H}$ to $\mathcal{H}'$. We say that T is *closable* if $\overline{\mathrm{Graph}(T)}$ is again the graph of a

© Springer International Publishing AG 2017

M. Einsiedler, T. Ward, *Functional Analysis, Spectral Theory, and Applications*, Graduate Texts in Mathematics 276, DOI 10.1007/978-3-319-58540-6_13

densely defined operator $(D_{\overline{T}}, \overline{T}) : \mathcal{H} \to \mathcal{H}'$. We say that T is a *closed operator* if $\text{Graph}(T)$ is closed. If $(D_T, T) : \mathcal{H} \to \mathcal{H}'$ and $(D_S, S) : \mathcal{H} \to \mathcal{H}'$ are linear operators, then we say that T is *equal to* S if $D_T = D_S$ and $T = S$, and say that S is an *extension* of T, written $T \subseteq S$, if $D_T \subseteq D_S$ and $S|_{D_T} = T$.

Of course bounded operators between two Hilbert spaces are special cases of this definition and in this case we will keep using the notation $B : \mathcal{H} \to \mathcal{H}'$. We note that the inverse and composition of operators will be understood here as in set theory: If

$$(D_T, T) : \mathcal{H} \to \mathcal{H}'$$

is injective, then

$$(D_{T^{-1}}, T^{-1}) : \mathcal{H}' \to \mathcal{H}$$

is simply the inverse map, and it is densely defined if $D_{T^{-1}} = T(D_T)$ is dense in $\mathcal{H}'$. If $(D_T, T) : \mathcal{H} \to \mathcal{H}'$ and $(D_S, S) : \mathcal{H}' \to \mathcal{H}''$ are densely defined operators, then

$$D_{ST} = \{v \in D_T \mid Tv \in D_S\}$$

is a subspace and $ST : D_{ST} \to \mathcal{H}''$ is linear, but in general it is not clear whether this defines a densely defined operator.

Lemma 13.3 (Adjoint operator). *Let $(D_T, T) : \mathcal{H} \to \mathcal{H}'$ be a densely defined operator between complex Hilbert spaces. Then there exists a closed operator $(D_{T^*}, T^*) : \mathcal{H}' \to \mathcal{H}$, called the* adjoint, *satisfying*

$$\langle Tv, w \rangle_{\mathcal{H}'} = \langle v, T^*w \rangle_{\mathcal{H}}$$

for all $v \in D_T$ and all vectors w belonging to the domain

$$D_{T^*} = \{w \in \mathcal{H}' \mid D_T \ni v \longmapsto \langle Tv, w \rangle_{\mathcal{H}'} \text{ is bounded}\}.$$

Moreover, T^* is densely defined if and only if T is closable. In this case the adjoint of the adjoint, T^{**}, is equal to the closure $\overline{T}$ of the operator T.

We will prove this lemma together with Lemma 13.8, but only after we have seen a few more examples. We note again that in the lemma above and in the following definition equality of operators entails equality of their domains.

Definition 13.4. Let $(D_T, T) : \mathcal{H} \to \mathcal{H}$ be a densely defined operator on a complex Hilbert space. If $T = T^*$ then T is said to be *self-adjoint*.

Essential Exercise 13.5. (a) Check that Example 13.1 indeed defines a self-adjoint operator in the sense of Definition 13.4.
(b) When does a complex-valued measurable function on a σ-finite measure space $(X, \mathcal{B}, \mu)$ define a densely defined, closable, closed, self-adjoint, or bounded multiplication operator?

Example 13.6. Let $\frac{\mathrm{d}}{\mathrm{d}x} : C_c^\infty(\mathbb{R}) \longrightarrow \mathcal{H} = L^2(\mathbb{R})$ be the differentiation operator, and define an operator T by

$$\mathrm{Graph}(T) = \overline{\mathrm{Graph}\big(\tfrac{\mathrm{d}}{\mathrm{d}x}\big)}.$$

By Definitions 5.7, 5.14, and the properties of the weak derivative (Lemma 5.10 applied with $d = k = 1$) this indeed defines a map

$$T : D_T = H_0^1(\mathbb{R}) \longrightarrow L^2(\mathbb{R}).$$

The map $\mathrm{i}T : D_{\mathrm{i}T} = D_T \to L^2(\mathbb{R})$ can be checked to be an unbounded self-adjoint operator which is conjugate to an unbounded self-adjoint multiplication operator as in Example 13.1. The unitary isomorphism is given by the Fourier transform: by Proposition 9.43 we have

$$\widehat{\tfrac{\mathrm{d}}{\mathrm{d}x}f}(t) = 2\pi \mathrm{i} t \widehat{f}(t)$$

for $f \in C_c^\infty(\mathbb{R})$. From this one can deduce that

$$\widehat{D_T} = \widehat{H_0^1(\mathbb{R})} = D_{M_g} = \{f \in L^2(\mathbb{R}) \mid \|tf(t)\|_2 < \infty\}$$

where $g(t) = -2\pi t$, and that the diagram

$$
\begin{array}{ccc}
D_{\mathrm{i}T} = D_T & \xrightarrow{\ \mathrm{i}T\ } & L^2(\mathbb{R}) \\
\Big\downarrow & & \Big\downarrow \\
D_{M_g} = \widehat{D_T} & \xrightarrow[\ M_g\ ]{} & L^2(\mathbb{R})
\end{array}
$$

commutes and completely describes $\mathrm{i}T$ (and hence T and $D_T = H_0^1(\mathbb{R})$) in terms of a multiplication operator (and its domain). We refer to Exercise 9.63 and its hints on p. 578.

The following exercise shows that we have to be more careful about the domain of unbounded operators, in contrast to the discussions in the previous chapters which mostly involved bounded operators. In fact, the principle of automatic extension (Proposition 2.59) has been used extensively throughout the text but fails in many ways for the unbounded operators we have just introduced.

Exercise 13.7. Let $X = (0,1)$ and consider again the operator

$$\frac{\mathrm{d}}{\mathrm{d}x} : C_c^\infty((0,1)) \longrightarrow L^2((0,1)).$$

(a) Recall that $T_0 : H_0^1((0,1)) \to L^2((0,1))$ sending f to $\partial_1 f$ extends the operator $\frac{\mathrm{d}}{\mathrm{d}x}$ to a closed operator $(D_{T_0}, T_0) : L^2((0,1)) \to L^2((0,1))$.
(b) Recall that $T_p : H^1(\mathbb{T}) \to L^2(\mathbb{T})$ sending f to $\partial_1 f$ also extends the operator $\frac{\mathrm{d}}{\mathrm{d}x}$ to a closed operator $(D_{T_p}, T_p) : L^2((0,1)) \to L^2((0,1))$.

(c) Show that $T_0 \subsetneq T_p = -T_p^* \subsetneq -T_0^*$, and describe $(D_{T_0^*}, T_0^*)$.

We will base our discussion of the spectral theory of self-adjoint operators on the following lemma.

Lemma 13.8 (Orthogonal decomposition into two graphs). *Let*

$$(D_T, T) : \mathcal{H} \to \mathcal{H}'$$

be a closed densely defined operator between two complex Hilbert spaces. The orthogonal complement of the closed set

$$\mathrm{Graph}(T) \subseteq \mathcal{H} \times \mathcal{H}'$$

is given by $\widetilde{\mathrm{Graph}(T^*)}$, *where*

$$\sim \, : \mathcal{H}' \times \mathcal{H} \longrightarrow \mathcal{H} \times \mathcal{H}'$$
$$(w, v) \longmapsto (v, -w).$$

PROOF OF LEMMAS 13.3 AND 13.8. Let $(D_T, T) : \mathcal{H} \to \mathcal{H}'$ be a densely defined operator. Notice that if $w \in D_{T^*}$, so that the linear map

$$D_T \ni v \mapsto \langle Tv, w \rangle_{\mathcal{H}'}$$

is bounded by definition, then this linear functional can be uniquely extended from the dense subset D_T to $\mathcal{H}$. In particular, by Fréchet–Riesz representation (Corollary 3.19) there exists a uniquely defined $T^* w \in \mathcal{H}$ with

$$\langle Tv, w \rangle_{\mathcal{H}'} = \langle v, T^* w \rangle_{\mathcal{H}} \tag{13.1}$$

for all $v \in D_T$. It is easy to check that D_{T^*} is a linear subspace and that this defines the linear operator $T^* : D_{T^*} \to \mathcal{H}$.

For the proof of Lemma 13.3 we wish to show next that T^* is closed. For this it is useful to first prove that

$$\mathrm{Graph}(T)^\perp = \widetilde{\mathrm{Graph}(T^*)}, \tag{13.2}$$

which in particular will imply Lemma 13.8. Let $w \in D_{T^*}$ so that (13.1) holds for all $v \in D_T$. By definition,

$$(T^* w, -w) \in \widetilde{\mathrm{Graph}(T^*)}$$

and

$$\langle (v, Tv), (T^* w, -w) \rangle_{\mathcal{H} \times \mathcal{H}'} = \langle v, T^* w \rangle_{\mathcal{H}} - \langle Tv, w \rangle_{\mathcal{H}'} = 0$$

for all $v \in D_T$. On the other hand, if $(v', -w) \in \mathrm{Graph}(T)^\perp$ so that

$$\langle (v, Tv), (v', -w) \rangle_{\mathcal{H} \times \mathcal{H}'} = \langle v, v' \rangle_{\mathcal{H}} - \langle Tv, w \rangle_{\mathcal{H}'} = 0$$

for all $v \in D_T$, then $D_T \ni v \mapsto \langle Tv, w \rangle_{\mathcal{H}'}$ is bounded. Thus we have $w \in D_{T^*}$ and $v' = T^*w$, so

$$(v', -w) = (T^*w, -w) \in \widetilde{\mathrm{Graph}(T^*)}.$$

Hence Lemma 13.8 follows, and in particular T^* is a closed operator.

We now show that T^* is densely defined if and only if T is closable. For this, note that $w_0 \in (D_{T^*})^{\perp}$ if and only if

$$\langle (0, w_0), (T^*w, -w) \rangle_{\mathcal{H} \times \mathcal{H}'} = 0$$

for all $w \in D_{T^*}$, or equivalently $(0, w_0) \in \widetilde{\mathrm{Graph}(T^*)}^{\perp}$. By (13.2) and the characterization of the closed linear hull in Corollary 3.26 this is in turn equivalent to

$$(0, w_0) \in \overline{\mathrm{Graph}(T)}.$$

If now T is closable, then $(0, w_0) \in \mathrm{Graph}(\overline{T})$ implies that $w_0 = 0$ and so $(D_{T^*})^{\perp} = \{0\}$ and thus $\overline{D_{T^*}} = \mathcal{H}'$. On the other hand, if T is not closed, then there exists a non-zero vector $(0, w_0) \in \overline{\mathrm{Graph}(T)}$ and so the element $w_0 \in (D_{T^*})^{\perp}$ shows that T^* is not densely defined.

For the final remark of Lemma 13.3 we apply (13.2) to T and to T^* (and also note that the operator $\sim$ is unitary and $((v, w)^\sim)^\sim = -(v, w)$ for all $(v, w) \in \mathcal{H} \times \mathcal{H}'$) to see that

$$\mathrm{Graph}(\overline{T}) = \mathrm{Graph}(T)^{\perp\perp} = \left(\widetilde{\mathrm{Graph}(T^*)} \right)^{\perp}$$

$$= \left(\mathrm{Graph}(T^*)^{\perp} \right)^\sim = \left(\widetilde{\mathrm{Graph}(T^{**})} \right)^\sim = \mathrm{Graph}(T^{**}),$$

as claimed. $\square$

13.2 Operators of the Form T^*T

As we have seen, differentiation can often be used to define a closed operator T which sends a function to its total derivative. Moreover, T^* is then often the negative of the divergence on vector fields, so that T^*T is often some kind of Laplace operator. This observation also holds true in other cases more general than those considered here, and this motivates the following discussion.

Theorem 13.9 (Spectral theory of T^*T). *Let $(D_T, T) : \mathcal{H} \to \mathcal{H}'$ be a densely defined closed linear operator. Then $(D_{T^*T}, T^*T) : \mathcal{H} \to \mathcal{H}$ is a densely defined self-adjoint operator which is unitarily isomorphic to a multiplication operator*

$$(D_{M_g}, M_g) : L^2_\mu(X) \longrightarrow L^2_\mu(X)$$

for some finite measure space (X, μ) and measurable function $g : X \to [0, \infty)$.

PROOF. The proof of the theorem essentially comprises a careful analysis of Figure 13.1.

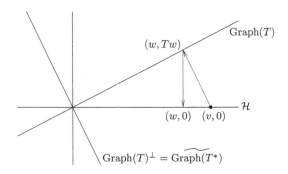

Fig. 13.1: We obtain a bounded operator $B : \mathcal{H} \to \mathcal{H}$ by sending v to $w = Bv$ using two orthogonal projections.

Let us write $P_{\mathrm{Graph}} : \mathcal{H} \times \mathcal{H}' \to \mathcal{H} \times \mathcal{H}'$ for the orthogonal projection onto the closed subspace $\mathrm{Graph}(T) \subseteq \mathcal{H} \times \mathcal{H}'$, $\imath_{\mathcal{H}} : \mathcal{H} \to \mathcal{H} \times \mathcal{H}'$ for the embedding map $v \mapsto (v, 0)$, and $P_{\mathcal{H}} : \mathcal{H} \times \mathcal{H}' \to \mathcal{H}$ for the projection map $(v, w) \mapsto v$. Note that $v, w \in \mathcal{H}$ and $w' \in \mathcal{H}'$ implies

$$\langle \imath_{\mathcal{H}}(v), (w, w') \rangle_{\mathcal{H} \times \mathcal{H}'} = \langle v, w \rangle_{\mathcal{H}} = \langle v, P_{\mathcal{H}}(w, w') \rangle_{\mathcal{H}}$$

so that $\imath_{\mathcal{H}}^* = P_{\mathcal{H}}$. Also note that $P_{\mathrm{Graph}}^* = P_{\mathrm{Graph}} = P_{\mathrm{Graph}}^2$. Now define

$$B = P_{\mathcal{H}} \circ P_{\mathrm{Graph}} \circ \imath_{\mathcal{H}},$$

so that

$$B^* = \imath_{\mathcal{H}}^* \circ P_{\mathrm{Graph}}^* \circ P_{\mathcal{H}}^* = P_{\mathcal{H}} \circ P_{\mathrm{Graph}} \circ \imath_{\mathcal{H}} = B$$

is self-adjoint. Moreover,

$$\|B\| \leqslant \|P_{\mathcal{H}}\| \|P_{\mathrm{Graph}}\| \|\imath_{\mathcal{H}}\| = 1. \tag{13.3}$$

Also, by definition,

$$\begin{aligned} \langle Bv, v \rangle_{\mathcal{H}} &= \langle P_{\mathcal{H}} P_{\mathrm{Graph}} \imath_{\mathcal{H}}(v), v \rangle_{\mathcal{H}} \\ &= \langle P_{\mathrm{Graph}} \imath_{\mathcal{H}}(v), P_{\mathrm{Graph}} \imath_{\mathcal{H}}(v) \rangle_{\mathcal{H} \times \mathcal{H}'} \geqslant 0 \end{aligned} \tag{13.4}$$

for any $v \in \mathcal{H}$. To summarize, $B : \mathcal{H} \to \mathcal{H}$ is a positive self-adjoint bounded operator with spectrum in $[0, 1]$.

We now relate B to T^*T, after which we can simply apply Theorem 12.55 to B and obtain the spectral theorem for T^*T. In fact, we claim that[†]

$$B = (I + T^*T)^{-1}, \tag{13.5}$$

or more precisely that

(a) $(I + T^*T)B = I$ and, in particular, $\text{im}(B) \subseteq D_{T^*T}$;
(b) $B(I + T^*T) = I_{D_{T^*T}}$ and, in particular, $D_{T^*T} \subseteq \text{im}(B)$.

Together this implies that $D_{T^*T} = \text{im}(B)$, that B is injective, and finally that

$$T^*T = B^{-1} - I$$

is completely determined by the operator B.

To prove (a) we chase the equations defining B (see Figure 13.1). Let $v \in \mathcal{H}$ and $w = Bv$ so that (by definition) $w \in D_T$, $(w, Tw) \in \text{Graph}(T)$, and

$$(w, Tw) - (v, 0) \in \text{Graph}(T)^\perp = \widetilde{\text{Graph}(T^*)}$$

by Lemma 13.8. This gives

$$(w - v, Tw) = (w, Tw) - (v, 0) = (-T^*Tw, Tw),$$

so $w \in D_{T^*T}$ and $w - v = -T^*Tw$, or equivalently

$$(I + T^*T)Bv = w + T^*Tw = v.$$

To prove (b), we essentially use the same formulas. Fix $w \in D_{T^*T}$ and define $v = w + T^*Tw$. Then

$$(w, Tw) \in \text{Graph}(T),$$

$$(T^*Tw, -Tw) \in \widetilde{\text{Graph}(T^*)},$$

and

$$(v, 0) = (w, Tw) + (T^*Tw, -Tw),$$

which implies that $w = Bv = B(I + T^*T)w$, as claimed.

Now apply Theorem 12.55 to B to find a finite measure space (X, μ) and some bounded measurable function $h \in L_\mu^\infty(X)$ so that B and M_h are unitarily isomorphic. Since B is injective and satisfies (13.3)–(13.4), h takes values in $(0, 1]$ μ-almost everywhere. After modifying h on a null set we may therefore assume that h takes values in $(0, 1]$ everywhere.

Using the same isomorphism ϕ we claim that T^*T is isomorphic to M_g for $g = \frac{1}{h} - 1$. Indeed,

$$\phi(D_{T^*T}) = \phi(\text{im}(B)) = \text{im}(M_h) = \left\{ f \in L_\mu^2(X) \mid \tfrac{1}{h} f \in L_\mu^2(X) \right\} = D_{M_g},$$

[†] The alert reader may at this point feel a sense of *déjà vu* (cf. Exercise 13.12).

and, since $T^*Tw = B^{-1}w - w$ for all $w \in D_{T^*T}$, we also see that

$$\phi(T^*Tw) = \phi(B^{-1}w) - \phi(w) = M_h^{-1}\phi(w) - \phi(w) = M_g\phi(w)$$

for all such w. Applying Exercise 13.5(a) or 13.10 gives the theorem. $\square$

Exercise 13.10. Let $B : \mathcal{H} \to \mathcal{H}$ be an injective self-adjoint bounded operator on a Hilbert space $\mathcal{H}$. Show directly that the inverse $(D_{B^{-1}}, B^{-1}) : \mathcal{H} \to \mathcal{H}$ is a self-adjoint operator with domain $D_{B^{-1}} = \mathrm{im}(B)$.

Exercise 13.11 (The influence of the domain). Let $\mathcal{H} = L^2((0, 1))$.
(a) Let $(D_{T_0}, T_0) = \left(H_0^1((0, 1)), \frac{d}{dx}\right)$ be the weak derivative map restricted to the space $H_0^1((0, 1))$. Show that $T_0^*T_0$ equals the negative of the second weak derivative on

$$D_{T_0^*T_0} = H_0^1((0, 1)) \cap H^2((0, 1))$$

(these are the Dirichlet boundary conditions), and that its eigenfunctions are (scalar multiples of) the functions $x \mapsto \sin(\pi n x)$ for $n \in \mathbb{N}$.
(b) Let $(D_T, T) = \left(H^1((0, 1)), \frac{d}{dx}\right)$. Show that T^*T coincides with the negative of the second weak derivative on

$$D_{T^*T} = \{f \in H^2((0, 1)) \mid f' \in H_0^1((0, 1))\},$$

(which are the Neumann boundary conditions), and that its eigenfunctions are the functions $x \mapsto \cos(\pi n x)$ for $n \in \mathbb{N}_0$.
(c) Let $(D_{T_p}, T_p) = \left(H^1(\mathbb{T}), \frac{d}{dx}\right)$. Show that $T_p^*T_p$ coincides with $-\Delta$ on $H^2(\mathbb{T})$ (which corresponds to the periodic boundary conditions).
(d) Show that $T_0^*T_0$, T^*T, and $T_p^*T_p$ are all different and no one extends any other.

Exercise 13.12. Compare the general construction of this section to the arguments of Section 6.4.2.

Exercise 13.13. Let $\mathcal{G} = (\mathcal{V}, \mathcal{E})$ be an undirected simple graph as in Section 10.4 (but possibly infinite) such that any $v \in \mathcal{V}$ has finitely many neighbours, and let $\overrightarrow{\mathcal{E}}$ be the set of oriented edges as in Section 12.2. Let $\mathcal{H} = L^2(\mathcal{V})$ and $\mathcal{H}' = L^2(\overrightarrow{\mathcal{E}})$, where we simply use the counting measure on the vertices in $\mathcal{V}$ and the edges in $\overrightarrow{\mathcal{E}}$. Now define

$$(Tf)((v_1, v_2)) = f(v_2) - f(v_1)$$

for any edge $(v_1, v_2) \in \overrightarrow{\mathcal{E}}$ (with $v_1 \neq v_2 \in \mathcal{V}$) and for any function f on $\mathcal{V}$, giving an operator $(D_T, T) : L^2(\mathcal{V}) \to L^2(\overrightarrow{\mathcal{E}})$.
(a) Show that T is a bounded operator if and only if there exists some $N \in \mathbb{N}$ such that every $v \in \mathcal{V}$ has at most N neighbours.
(b) Describe the operators T^* and T^*T where they are defined.

Exercise 13.14. Let $\mathcal{G} = (\mathcal{V}, \mathcal{E})$ be a finite undirected graph as in Section 10.4, but now glue for every edge $e \in \mathcal{E}$ connecting two vertices $v_1, v_2 \in \mathcal{V}$ a compact line segment S_e of length $\ell_e > 0$ between v_1 and v_2. We assume that the graph is undirected and we put for any two vertices at most one line segment linking them directly. This defines a topological space Q, called a *metric graph*, consisting of a network of compact line segments (one for each edge in the graph) that are glued together at the vertices of the graph. Endow Q with the measure obtained from using the Lebesgue measure on each line segment S_e, which in particular leads to

$$L^2(Q) = \sum_{e \in \mathcal{E}} L^2(S_e).$$

Define $H^1(Q)$ to be the space of all continuous functions on Q such that the restriction to the compact line segment $S_e \subseteq Q$ belongs to $H^1(S_e)$ for every edge $e \in \mathcal{E}$. Define the operator $T : H^1(Q) \to L^2(Q)$ by setting $(Tf)_{S_e} = \partial f_e$, where $f_e = f|_{S_e}$ and the weak derivative is taken in $H^1(S_e)$ with respect to the fixed orientation on S_e. Then

$$(H^1(Q), T) : L^2(Q) \to L^2(Q)$$

is a densely defined operator (check this). The study of the eigenfunctions of T^*T is called the theory of *quantum graphs*.

(a) Describe the operators T^* and T^*T and their domains, especially in relationship to the behaviour of the functions in the domain at the vertices.

(b) Show that there exists an orthonormal basis of $L^2(Q)$ consisting of eigenfunctions of T^*T.

(c) Assume now that $\mathcal{G}$ consists of four vertices with one vertex in the centre and three vertices connected to it. Prove a version of Weyl's law for the operator T^*T on the associated quantum graph.

13.3 Self-Adjoint Operators

Using the construction from the last section we can also prove the spectral theorem for general self-adjoint operators.

Theorem 13.15. *Let* $(D_T, T) : \mathcal{H} \to \mathcal{H}$ *be a densely defined self-adjoint operator. Then there exists a finite measure space* (X, μ) *and a real-valued measurable function* $g : X \to \mathbb{R}$ *such that* (D_T, T) *is unitarily isomorphic to* (D_{M_g}, M_g)*, meaning that there is a unitary isomorphism* $\phi : \mathcal{H} \to L^2_\mu(X)$ *such that* $\phi(D_T) = D_{M_g}$ *and the diagram*

$$
\begin{array}{ccc}
\mathcal{H} \supseteq D_T & \xrightarrow{\;\;T\;\;} & \mathcal{H} \\[2pt]
{\scriptstyle\phi}\downarrow & & \downarrow{\scriptstyle\phi} \\[4pt]
L^2_\mu(X) \supseteq D_{M_g} & \xrightarrow[\;M_g\;]{} & L^2_\mu(X)
\end{array}
$$

commutes.

Since a self-adjoint operator T as in Theorem 13.15 is also closed, it is clear that we could directly apply the method of the previous section to T. Note, however, that a simple application of Theorem 13.9 only gives a description of T^2, which does not allow a description of T. In fact, T^2 has a potentially smaller domain, and may have lost some information about T (namely the sign of eigenvalues or approximate eigenvalues). To compensate we will study two operators: B as in the previous section, and $A = TB$, as in Figure 13.2.

PROOF OF THEOREM 13.15. Let $B = (I + T^*T)^{-1} = (I + T^2)^{-1}$ be as in the proof of Theorem 13.9. We also define

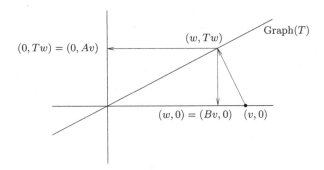

Fig. 13.2: For the proof of Theorem 13.15 we study the operators A and B.

$$A = TB = P_{\mathcal{H},2} \circ P_{\text{Graph}} \circ \imath_{\mathcal{H}},$$

where $P_{\mathcal{H},2}(v, w) = w$ is the projection to the second copy of $\mathcal{H}$ in $\mathcal{H} \times \mathcal{H}$, see Figure 13.2.

Below we will apply Theorem 12.60 to the bounded operators A and B, and to do this we first have to show that A is normal (in fact, it is self-adjoint; this is something we already know for B by the proof of Theorem 13.9), and that A and B commute. To prepare for this, we first claim that

$$BT \subseteq TB = A. \tag{13.6}$$

To prove the claim, fix $w \in D_T$ and define $v = Bw$ so that

$$(I + T^2)v = w$$

by (13.5). Since $w \in D_T$ this shows that $v \in D_{T^3}$ and

$$(I + T^2)Tv = T(I + T^2)v = Tw,$$

which by (13.5) means that $BTw = Tv$. Since $v = Bw$, we have shown that

$$BTw = TBw$$

for all $w \in D_T$ and hence the claim in (13.6).

To prove that $A^* = A$ we argue as follows. For $w \in D_T$ and $v \in \mathcal{H}$ we have

$$\langle Av, w \rangle = \langle TBv, w \rangle = \langle Bv, Tw \rangle = \langle v, BTw \rangle = \langle v, Aw \rangle$$

since T and B are self-adjoint and by (13.6). Thus $A^*w = Aw$ for all elements w of the dense subset $D_T \subseteq \mathcal{H}$. Since A is a bounded operator, it follows that $A^* = A$.

Moreover,

$$BA = BTB \subseteq TBB = AB$$

by (13.6). Since both AB and BA are defined on all of $\mathcal{H}$ this shows that A and B commute.

Next we have to show that A and B together uniquely determine T (so that when A and B are realized as multiplication operators we have some hope of deducing a similar realization for T). We claim that $T = B^{-1}A$ and, in particular,

$$D_T = D_{B^{-1}A} = \{v \in \mathcal{H} \mid Av \in \operatorname{im}(B)\}.$$

To see this, note that $B^{-1}B = I$ since B is injective and hence

$$T = B^{-1}BT \subseteq B^{-1}TB = B^{-1}A$$

by (13.6). For the converse recall the construction of B in the proof of Theorem 13.9 (see also Figures 13.2 and 13.3) and the definition of A. With these we obtain

$$(Bv, Av) = (Bv, TBv) \in \operatorname{Graph}(T), \qquad (13.7)$$
$$(Bv - v, Av) = (Bv - v, TBv) \in \operatorname{Graph}(T)^{\perp}$$

for any $v \in \mathcal{H}$. Let $v \in D_{B^{-1}A}$, and replace the latter instance of v with $B^{-1}Av$ to obtain

$$(Av - B^{-1}Av, TAv) \in \operatorname{Graph}(T)^{\perp}.$$

Since T is self-adjoint, Lemma 13.8 shows that $\operatorname{Graph}(T)^{\perp} = \widetilde{\operatorname{Graph}(T)}$ so that we have equivalently

$$(T^2Bv, -TBv + B^{-1}Av) \in \operatorname{Graph}(T). \qquad (13.8)$$

Taking the sum of (13.7) and (13.8) and using the identity $(I + T^2)B = I$ gives

$$(v, B^{-1}Av) \in \operatorname{Graph}(T).$$

Thus $v \in D_T$ and $Tv = B^{-1}Av$, as claimed (see Figure 13.3).

Now we apply Theorem 12.60 to A and B to obtain a finite measure space (X, μ) and two functions $g_A : X \to \mathbb{R}$ and $g_B : X \to (0, \infty)$ such that A and B are conjugate to M_{g_A} and M_{g_B}, respectively. Since we have shown that D_T and T are purely defined in terms of A and B, we can finally use the same unitary isomorphism ϕ to describe (D_T, T) as follows:

$$\begin{aligned}
\phi(D_T) &= \phi(\{v \in \mathcal{H} \mid Av \in \operatorname{im}(B)\}) \\
&= \{f \in L^2_\mu(X) \mid M_{g_A}(f) \in \operatorname{im}(M_{g_B})\} \\
&= \left\{f \in L^2_\mu(X) \mid \tfrac{g_A}{g_B}f \in L^2_\mu(X)\right\} = D_{M_g},
\end{aligned}$$

where we set $g = \frac{g_A}{g_B}$, and also

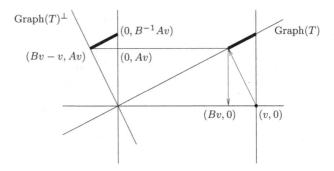

Fig. 13.3: As the proof of Theorem 13.15 shows, the two marked segments are translates of each other.

$$\phi(Tv) = \phi(B^{-1}Av) = M_{g_B}^{-1} M_{g_A} \phi(v) = M_g \phi(v)$$

for all $v \in D_T$.   □

Exercise 13.16 (Schur's lemma for densely defined closed operators). Assume that $\pi_1 : G \curvearrowright \mathcal{H}_1$ and $\pi_2 : G \curvearrowright \mathcal{H}_2$ are unitary representations of a topological group G such that π_1 is irreducible. Moreover, assume that $(D_T, T) : \mathcal{H}_1 \to \mathcal{H}_2$ is a densely defined closed operator satisfying $\pi_1(g)D_T \subseteq D_T$ and $T\pi_1(g) = \pi_2(g)T$ on D_T for all $g \in G$. Show that $D_T = \mathcal{H}_1$ and that T is bounded, and deduce that the conclusions of Schur's lemma (Exercise 12.58) holds in this setting.

13.4 Symmetric Operators

In this section we will discuss another class of unbounded operators appearing in applications, which is closely related to the class of self-adjoint operators discussed above. The requirement that the operator be densely defined in Definition 13.17 is sometimes dropped, and in the physics literature the term *Hermitian* is sometimes used for symmetric.

Definition 13.17. A densely defined operator $(D_S, S) : \mathcal{H} \to \mathcal{H}$ on a Hilbert space $\mathcal{H}$ is called *symmetric* if $\langle Su, v \rangle = \langle u, Sv \rangle$ for all $u, v \in D_S$, or equivalently if $S \subseteq S^*$.

Because of the satisfyingly complete description of self-adjoint operators in the previous section, it is often useful to extend a given symmetric operator to a self-adjoint operator. As we will see, this is sometimes but not always possible.

13.4.1 The Friedrichs Extension

Theorem 13.18. *Let $\mathcal{H}$ be a complex Hilbert space and let $(D_S, S) : \mathcal{H} \to \mathcal{H}$ be a densely defined symmetric operator that is also positive in the sense that $\langle Su, u \rangle \geqslant 0$ for all $u \in D_S$. Then there exists a positive self-adjoint extension $S_F \supseteq S$.*

Exercise 13.19 (Quantum harmonic oscillator). Define an operator H with domain $D_H = \mathscr{S}(\mathbb{R}) \subsetneq L^2(\mathbb{R})$ by $H(f)(x) = -\frac{1}{2}\frac{d^2}{dx^2}f(x) + \frac{1}{2}x^2 f(x)$. Show that H is positive and symmetric but unbounded.

PROOF OF THEOREM 13.18. Let $(D_S, S) : \mathcal{H} \to \mathcal{H}$ be as in the theorem. For any $u, v \in D_S$ we define the semi-inner product $\langle \cdot, \cdot \rangle_S$ by

$$\langle u, v \rangle_S = \langle Su, v \rangle = \langle u, Sv \rangle,$$

and let $\|u\|_S = \sqrt{\langle u, u \rangle_S}$ be the induced semi-norm. We let $V_0 \subseteq D_S$ denote the kernel of $\|\cdot\|_S$ and define $\mathcal{H}_S$ to be the completion of D_S/V_0 with respect to $\|\cdot\|_S$. We denote the extension of $\langle \cdot, \cdot \rangle_S$ (and of $\|\cdot\|_S$) to the completion again by $\langle \cdot, \cdot \rangle_S$ (and $\|\cdot\|_S$), and the canonical map from D_S to $\mathcal{H}_S$ by $\imath_0$. We claim that $\imath_0$ is closable (as in Definition 13.2) and will write $\imath$ for the closed operator with the property that

$$\mathrm{Graph}(\imath) = \overline{\mathrm{Graph}(\imath_0)} \subseteq \mathcal{H} \times \mathcal{H}_S.$$

To see that $\overline{\mathrm{Graph}(\imath_0)}$ is indeed a graph we assume that a sequence $((u_n, \imath_0 u_n))$ in $\mathrm{Graph}(\imath_0)$ converges to $(0, v)$ in $\mathcal{H} \times \mathcal{H}_S$. For any $w \in D_S$ we then have

$$\langle v, \imath_0 w \rangle_S = \lim_{n \to \infty} \langle \imath_0 u_n, \imath_0 w \rangle_S = \lim_{n \to \infty} \langle u_n, Sw \rangle = 0$$

since $\lim_{n \to \infty} u_n = 0$ with respect to $\|\cdot\|$. Since $\imath_0(D_S)$ is dense in $\mathcal{H}_S$ we see that $v = 0$, as required.

Since $(D_\imath, \imath) : \mathcal{H} \to \mathcal{H}_S$ is densely defined and closed, it follows from Theorem 13.9 that $\imath^*\imath$ is a densely defined self-adjoint operator. We claim that $S \subseteq \imath^*\imath$. Suppose therefore that $u \in D_S$. Then

$$\langle Su, v \rangle = \langle \imath_0 u, \imath_0 v \rangle_S = \langle \imath u, \imath v \rangle_S$$

for all $v \in D_S = D_{\imath_0}$. However, by the density of $\mathrm{Graph}(\imath_0)$ in $\mathrm{Graph}(\imath)$ this equality extends, giving

$$\langle Su, v \rangle = \langle \imath u, \imath v \rangle_S$$

for all $v \in D_\imath$. Hence for $u \in D_S$ the map

$$D_\imath \ni v \longmapsto \langle \imath v, \imath u \rangle_S = \langle v, Su \rangle$$

is bounded with respect to $\|\cdot\|$. This implies that $\imath u \in D_{\imath^*}$ so $u \in D_{\imath^*\imath}$ and

$$Su = \imath^*\imath u,$$

as required. Thus $S_F = \imath^*\imath$ is a self-adjoint extension of S, which is clearly positive. □

The following exercise gives some properties of the Friedrichs extension.

Exercise 13.20. Using the same notation as in the proof of Theorem 13.18, show in turn the following statements.
(a) Graph($\imath$) is isomorphic to the completion $\mathcal{H}_1$ of D_S with respect to the norm derived from the inner product $\langle u, v \rangle_1 = \langle u, v \rangle + \langle Su, v \rangle$ for $u, v \in D_S$, and $\mathcal{H}_1$ can be identified with the subspace $D_\imath$ of $\mathcal{H}$ allowing us to write $D_S \subseteq \mathcal{H}_1 \subseteq \mathcal{H}$.
(b) The domain of $S_F = \imath^*\imath$ consists of all $u \in \mathcal{H}_1$ such that $\mathcal{H}_1 \ni v \longmapsto \langle v, u \rangle_1$ is bounded with respect to $\|\cdot\|$, and in that case $\langle v, u \rangle_1 = \langle v, u + S_F u \rangle$ for all $v \in \mathcal{H}_1$.

13.4.2 Cayley Transform and Deficiency Indices

We finish this chapter (and hence our discussion of spectral theory) with a series of exercises concerning work going back to von Neumann on the existence of self-adjoint extensions of a general symmetric operator. The main tool for this discussion is the *Cayley transform*, the definition of which may at first be a little surprising. To motivate the definition we recall that the spectrum of a self-adjoint bounded operator is a compact subset of $\mathbb{R}$. Generalizing this definition we suppose now that $(D_T, T) : \mathcal{H} \to \mathcal{H}$ is a self-adjoint operator on a complex Hilbert space $\mathcal{H}$ and define its resolvent set by

$$\rho(T) = \{\lambda \in \mathbb{C} \mid T - \lambda I = B^{-1} \text{ for some } B \in \mathrm{B}(\mathcal{H})\}$$

and its spectrum by $\sigma(T) = \mathbb{C} \setminus \rho(T)$.

Exercise 13.21. Let $(D_T, T) : \mathcal{H} \to \mathcal{H}$ be a self-adjoint operator on a complex Hilbert space $\mathcal{H}$. Show that $\sigma(T) \subseteq \mathbb{R}$.

Next note that the function $\phi(z) = \frac{z-\mathrm{i}}{z+\mathrm{i}}$ maps $\mathbb{R}$ bijectively into the unit circle with the point 1 removed, suggesting a way to associate a unitary operator to a self-adjoint operator.

Exercise 13.22. Let $(D_T, T) : \mathcal{H} \to \mathcal{H}$ be a self-adjoint operator on a complex Hilbert space $\mathcal{H}$.
(a) Show that $T + \mathrm{i}I$ is injective and $\mathrm{im}(T + \mathrm{i}I) = \mathcal{H}$.
(b) Show that $U(Tv + \mathrm{i}v) = Tv - \mathrm{i}v$ for $v \in D_T$ defines a unitary operator $U : \mathcal{H} \to \mathcal{H}$.

The Cayley transform generalizes Exercise 13.22 to the setting of symmetric operators, where we will have to stop relying on the spectral theorem.

Let $(D_S, S) : \mathcal{H} \to \mathcal{H}$ be a symmetric operator on a complex Hilbert space. The *Cayley transform* of S is the operator

$$U_S(Sv + \mathrm{i}v) = Sv - \mathrm{i}v$$

for all $v \in D_S$ with natural domain $D_{U_S} = \{Sv + iv \mid v \in D_S\}$. The Cayley transform is a partial isometry with the properties that $I - U_S$ is injective and that $\mathrm{im}(I - U_S) = \{v - U_S v \mid v \in D_S\}$ is dense in $\mathcal{H}$. In summary, we may write $U_S = (S - iI)(S + iI)^{-1}$.

Essential Exercise 13.23. Let $\mathcal{H}$ and (D_S, S) be as above. Show that
(a) $D_S \ni v \mapsto Sv + iv$ is injective,
(b) U_S is an isometry on its domain,
(c) $I - U_S$ is injective, and
(d) $\mathrm{im}(I - U_S)$ is dense in $\mathcal{H}$.

The Cayley transform has an inverse operation, which allows us to associate to any given partially defined isometry $(D_U, U) : \mathcal{H} \to \mathcal{H}$ for which $I - U$ is injective and $\mathrm{im}(I - U)$ is dense in $\mathcal{H}$ a densely defined symmetric operator $(D_{S_U}, S_U) : \mathcal{H} \to \mathcal{H}$ by setting $D_{S_U} = \mathrm{im}(I - U)$ and

$$S_U(w - Uw) = \mathrm{i}(w + Uw)$$

for all $w \in D_U$. Thus we may write $S_U = \mathrm{i}(I + U)(I - U)^{-1}$.

Essential Exercise 13.24. Let $\mathcal{H}$ and (D_U, U) be as above. Show that S_U as defined above is a densely defined symmetric operator.

Essential Exercise 13.25. Show that the procedure above is indeed the inverse to the Cayley transform by the following steps.
(a) Given a densely defined symmetric operator $(D_S, S) : \mathcal{H} \to \mathcal{H}$, show that $S = S_{U_S}$.
(b) Given a partially defined isometry $(D_U, U) : \mathcal{H} \to \mathcal{H}$ for which $I - U$ is injective and $\mathrm{im}(I - U)$ is dense, show that $U = U_{S_U}$.

Essential Exercise 13.26. Suppose that $U : \mathcal{H} \to \mathcal{H}$ is unitary. Show that $I - U$ is injective if and only if $\mathrm{im}(I - U)$ is dense in $\mathcal{H}$. If U has these equivalent properties, show that $S = S_U$ is self-adjoint.

If S and S' are densely defined symmetric operators, it is clear from the definition of

$$D_{U_S} = \{Sv + iv \mid v \in D_S\}$$

and the relation $U_S = (S - iI)(S + iI)^{-1}$ that $S \subseteq S'$ implies $U_S \subseteq U_{S'}$. Similarly, if U and U' are partial isometries with the properties above, one sees that $U \subseteq U'$ implies $S_U \subseteq S_{U'}$. If U' is even unitary, then the density of $\mathrm{im}(I - U')$ follows from the corresponding property of U, so that by Exercise 13.26 we also know that $I - U'$ is injective. Hence in the case of a unitary extension U' of U, the properties above are automatically satisfied. Using this, the remaining part of Exercise 13.26, and Exercise 13.22 we deduce that the problem of finding self-adjoint extensions of symmetric operators is equivalent to finding unitary extensions of (certain) partial isometries.

Exercise 13.27. Give an example of a densely defined symmetric operator that does not have a self-adjoint extension.

The following exercise presents the main results of this section.

Essential Exercise 13.28. Let $(D_S, S) : \mathcal{H} \to \mathcal{H}$ be a densely defined symmetric operator on a separable complex Hilbert space $\mathcal{H}$. We define the *deficiency indices* n_+ and n_- by

$$n_\pm(S) = \dim\big(\{Sv \pm iv \mid v \in D_S\}^\perp\big) \in \mathbb{N}_0 \cup \{\infty\}.$$

(a) Show that S has a self-adjoint extension if and only if $n_+(S) = n_-(S)$.

(b) A symmetric operator is called *essentially self-adjoint* if it has a unique self-adjoint extension. Show that S is essentially self-adjoint if and only if $n_+(S) = n_-(S) = 0$.

Exercise 13.29. Find an example of an essentially self-adjoint operator that is not self-adjoint.

Exercise 13.30. Show that we have $n_+(S) = n_-(S) = 1$ for the operator $S = iT_0$ from Exercise 13.7. Deduce that we can parameterise the self-adjoint extensions S_α of S (in a natural manner) by elements $\alpha \in \mathbb{S}^1$.

13.5 Further Topics

- In the formulation of quantum mechanics due to Dirac and von Neumann, physical observables (momentum, position, spin, angular momentum and so on) are represented by self-adjoint operators on a Hilbert space; we refer to Reed and Simon [91] for this.
- As we have already seen in Chapters 5 and 6, many boundary value problems in the study of partial differential equations have natural descriptions in terms of self-adjoint operators.

Chapter 14
The Prime Number Theorem

It would be difficult to overstate the importance of functional analysis in number theory, as almost all the topics discussed in this volume play a foundational role in modern number theory. It would be impossible to really justify that statement here, and we instead follow Tao [103] and give a proof of the classical prime number theorem using Banach algebras (as in Chapter 11), Fourier analysis (as in Chapter 9), the weak* topology on the dual of $C_c(\mathbb{R})$ (using Chapter 8), and some elementary number theory. By the last phrase we mean a version of Mertens' theorem in Section 14.2.5 (which predates the first proof of the prime number theorem) and Selberg's symmetry formula in Section 14.2.2 (which was a key ingredient of the 'elementary' proof of the prime number theorem due to[34] Selberg [96]; see also Erdös [29]). In Section 14.5 we discuss a generalization of the prime number theorem to primes in arithmetic progressions.

14.1 Two Reformulations

Gauss observed in 1792–93 (at the age of 15 or 16), that the density of primes close to x seemed to be approximately $\frac{1}{\log x}$, leading to the suggestion that the prime counting function

$$\pi(x) = |\{p \leqslant x \mid p \text{ is a prime in } \mathbb{N}\}|$$

has the asymptotic growth rate $\frac{x}{\log x}$, and this statement is now called the prime number theorem. After many partial and weaker results, Hadamard and (independently) de la Vallée-Poussin extended work in which Riemann introduced complex-analytic methods to give the first proofs of the prime number theorem in 1896. We will not concern ourselves with the error rate in this approximation; the best conjectured error rate is a reformulation of the famous Riemann hypothesis.[35]

© Springer International Publishing AG 2017 503
M. Einsiedler, T. Ward, *Functional Analysis, Spectral Theory, and Applications*,
Graduate Texts in Mathematics 276, DOI 10.1007/978-3-319-58540-6_14

Recall that given positive functions f, g on an interval $[a, \infty)$ we say that f and g are *asymptotic*, written $f \sim g$, if $\lim_{x \to \infty} \frac{f(x)}{g(x)} = 1$ or equivalently if $f(x) = g(x) + o(g(x))$ as $x \to \infty$.

Theorem 14.1 (Prime number theorem). *For the prime counting function* $\pi(x) = \sum_{p \leqslant x} 1$ *(where p runs over the primes in* $\mathbb{N}$*) we have* $\pi(x) \sim \frac{x}{\log x}$ *as* $x \to \infty$.

As it turns out, it is more convenient to work with the *von Mangoldt function* Λ, defined by

$$\Lambda(n) = \begin{cases} \log p & \text{if } n = p^k \text{ for a prime } p \text{ and some } k \geqslant 1; \\ 0 & \text{otherwise} \end{cases}$$

for $n \geqslant 1$. In the remainder of the chapter $p, p_1, p_2, \ldots$ will always denote primes in $\mathbb{N}$ and the letters d, m, n, q will usually denote positive integers. For brevity we will refer to 'the prime number theorem' as 'PNT' throughout this chapter.

Lemma 14.2 (First reformulation of PNT). *If*

$$\sum_{1 \leqslant n \leqslant x} \Lambda(n) \sim x \tag{14.1}$$

as $x \to \infty$ then Theorem 14.1 *follows.*

PROOF. Suppose that (14.1) holds, and fix some small $\delta > 0$. Then we also have

$$\sum_{1 \leqslant n \leqslant x^{1-\delta}} \Lambda(n) = x^{1-\delta} + o(x^{1-\delta}) = o_\delta(x)$$

as $x \to \infty$ (which is also easy to see directly). Taking the difference we see that

$$\sum_{x^{1-\delta} < p \leqslant x} \log p \; + \sum_{\substack{x^{1-\delta} < p^k \leqslant x \\ k > 1}} \log p = \sum_{x^{1-\delta} < n \leqslant x} \Lambda(n) = x + o_\delta(x).$$

The first sum on the left is the one we are interested in, so we wish to estimate the second sum. For this, notice that $n = p^k \in (x^{1-\delta}, x]$ with $k > 1$ implies that $k \leqslant \frac{\log x}{\log p} \leqslant \frac{\log x}{\log 2}$ and $p \leqslant x^{1/2}$. Therefore

$$\sum_{\substack{x^{1-\delta} < p^k \leqslant x \\ k > 1}} \log p \leqslant \frac{\log x}{\log 2} \sum_{1 \leqslant n \leqslant x^{1/2}} \log(x^{1/2}) = O(x^{1/2} \log^2 x) = o(x),$$

and substituting this into the above we see that

$$\sum_{x^{1-\delta} < p \leqslant x} \log p = x + o_\delta(x).$$

Since $\log p \leqslant \log x$ for $p \in (x^{1-\delta}, x]$ we obtain

$$x + o_\delta(x) = \sum_{x^{1-\delta} < p \leqslant x} \log p \leqslant \log x \sum_{x^{1-\delta} < p \leqslant x} 1,$$

which, after dividing by $\log x$, gives the lower bound in

$$\frac{x}{\log x} + o_\delta\left(\frac{x}{\log x}\right) \leqslant \sum_{x^{1-\delta} < p \leqslant x} 1 \leqslant \frac{x}{(1-\delta)\log x} + o_\delta\left(\frac{x}{\log x}\right).$$

The upper bound follows similarly using $\log p \geqslant (1-\delta)\log x$ for $p \in (x^{1-\delta}, x]$. Since

$$\sum_{p \leqslant x^{1-\delta}} 1 \leqslant x^{1-\delta} = o_\delta\left(\frac{x}{\log x}\right)$$

we may add the small primes back in to obtain

$$\frac{x}{\log x} + o_\delta\left(\frac{x}{\log x}\right) \leqslant \pi(x) \leqslant \frac{x}{(1-\delta)\log x} + o_\delta\left(\frac{x}{\log x}\right).$$

As $\delta > 0$ is arbitrary, this gives the lemma. $\qquad\square$

Exercise 14.3. Prove that Theorem 14.1 implies (14.1) in Lemma 14.2.

In the second reformulation below due to Tao we will start to see the connection to functional analysis.

Proposition 14.4 (Second reformulation of PNT). *Define the Λ-seminorm $\| \cdot \|_\Lambda$ by*

$$\|f\|_\Lambda = \limsup_{h \to \infty} \left| \sum_{n \geqslant 1} \frac{\Lambda(n)}{n} f(\log n - h) - \int_\mathbb{R} f(t)\,dt \right| \qquad (14.2)$$

for $f \in C_c(\mathbb{R})$. If $\|f\|_\Lambda = 0$ for every $f \in C_c(\mathbb{R})$, then Theorem 14.1 holds.

It is easy to see from the properties of the limit supremum that

$$\|\lambda f\|_\Lambda = |\lambda|\|f\|_\Lambda$$

and

$$\|f_1 + f_2\|_\Lambda \leqslant \|f_1\|_\Lambda + \|f_2\|_\Lambda$$

for any $f_1, f_2 \in C_c(\mathbb{R})$ and $\lambda \in \mathbb{C}$. In other words, $\| \cdot \|_\Lambda$ defines a seminorm on $C_c(\mathbb{R})$ once we have checked that it is well-defined in the sense that $\|f\|_\Lambda < \infty$ for every $f \in C_c(\mathbb{R})$. We will prove this in the next section.

PROOF OF PROPOSITION 14.4. Supposing that the semi-norm is identically zero amounts to assuming that

$$\sum_{n \geqslant 1} \frac{\Lambda(n)}{n} f \left(\log \frac{n}{x} \right) = \int_{\mathbb{R}} f(t) \, \mathrm{d}t + \mathrm{o}(1)$$

as $x = \mathrm{e}^h \to \infty$ for any $f \in C_c(\mathbb{R})$. If now $g \in C_c((0, \infty))$ then we may set $f(t) = \mathrm{e}^t g(\mathrm{e}^t)$ to define a function $f \in C_c(\mathbb{R})$ and conclude that

$$\sum_{n \geqslant 1} \frac{\Lambda(n)}{n} \frac{n}{x} g \left(\frac{n}{x} \right) = \sum_{n \geqslant 1} \frac{\Lambda(n)}{n} f \left(\log \frac{n}{x} \right)$$

$$= \int_{\mathbb{R}} g(\mathrm{e}^t) \mathrm{e}^t \, \mathrm{d}t + \mathrm{o}(1) = \int_0^\infty g(u) \, \mathrm{d}u + \mathrm{o}(1),$$

or equivalently

$$\sum_{n \geqslant 1} \Lambda(n) g \left(\frac{n}{x} \right) = x \int_0^\infty g(u) \, \mathrm{d}u + \mathrm{o}(x).$$

Applying this to compactly supported functions $0 \leqslant g_- \leqslant \mathbb{1}_{[\frac{1}{2}, 1]} \leqslant g_+ \leqslant 1$ satisfying $\int (g_+ - g_-) \, \mathrm{d}x < \delta$ for some $\delta > 0$ (see Figure 2.3 on p. 50) we obtain

$$\sum_{\frac{1}{2}x < n \leqslant x} \Lambda(n) \leqslant \sum_{n \geqslant 1} \Lambda(n) g_+ \left(\frac{n}{x} \right) \leqslant (\tfrac{1}{2} + \delta)x + \mathrm{o}_\delta(x)$$

and

$$\sum_{\frac{1}{2}x < n \leqslant x} \Lambda(n) \geqslant \sum_{n \geqslant 1} \Lambda(n) g_- \left(\frac{n}{x} \right) \geqslant (\tfrac{1}{2} - \delta)x + \mathrm{o}_\delta(x).$$

As $\delta > 0$ is arbitrary, this also gives

$$\sum_{\frac{1}{2}x < n \leqslant x} \Lambda(n) = \tfrac{1}{2}x + \mathrm{o}(x). \tag{14.3}$$

We sum estimates of this form to get the desired claim. To handle the error term carefully we fix some $\varepsilon > 0$ and suppose $M \geqslant 1$ is such that the error term is bounded in absolute value by εx whenever $x \geqslant M$. Also note that for a fixed ε both M and $\sum_{n \leqslant M} \Lambda(n)$ are $\mathrm{o}_\varepsilon(x)$. Hence we may write

$$\sum_{n \leqslant x} \Lambda(n) = \sum_{\frac{1}{2}x < n \leqslant x} \Lambda(n) + \sum_{\frac{1}{4}x < n \leqslant \frac{1}{2}x} \Lambda(n) + \cdots + \sum_{\frac{1}{2^{\ell+1}}x < n \leqslant \frac{1}{2^\ell}x} \Lambda(n) + \mathrm{o}_\varepsilon(x),$$

where $\ell \geqslant 0$ is chosen maximally with $\frac{1}{2^\ell}x \geqslant M$. Applying the asymptotic in (14.3) to each sum we see that the main terms add to $x - \frac{1}{2^{\ell+1}}x = x + \mathrm{o}_\varepsilon(x)$

and the error terms add up to no more than $2\varepsilon x$ by choice of M. As $\varepsilon > 0$ was arbitrary, we see that (14.1) in Lemma 14.2 follows. $\square$

Exercise 14.5. Show that PNT implies that the Λ-semi-norm vanishes on $C_c(\mathbb{R})$.

14.2 The Selberg Symmetry Formula and Banach Algebra Norm

Theorem 14.6. *The function* $f \longmapsto \|f\|_\Lambda$ *on* $C_c(\mathbb{R})$ *defined in* (14.2) *is a semi-norm with*

$$\|f\|_\Lambda \leqslant \|f\|_1 \qquad (14.4)$$

and $\|f_1 * f_2\|_\Lambda \leqslant \|f_1\|_\Lambda \|f_2\|_\Lambda$ *for* $f, f_1, f_2 \in C_c(\mathbb{R})$.

This will be an important step towards the proof of PNT. Assuming that the semi-norm is not identically zero will allow us to construct a Banach algebra homomorphism from $L^1(\mathbb{R})$ to the completion $\mathcal{A}_\Lambda$ of $C_c(\mathbb{R})$ with respect[†] to the semi-norm $\| \cdot \|_\Lambda$, which will induce a dual homomorphism from the space of characters $\mathcal{A}_\Lambda^o$ of $\mathcal{A}_\Lambda$ into $L^1(\mathbb{R})^o \cong \mathbb{R}$. This will eventually lead to a contradiction.

For the proof of Theorem 14.6 we will need some elementary tools from number theory: the Selberg symmetry formula and Mertens' theorem.

14.2.1 Dirichlet Convolution and Möbius Inversion

Given two functions $f_1, f_2 : \mathbb{N} \to \mathbb{R}$ we define the *Dirichlet convolution*[‡] by

$$(f_1 \underset{D}{*} f_2)(n) = \sum_{d \mid n} f_1(d) f_2\left(\frac{n}{d}\right),$$

where the sum is taken over all divisors d of $n \in \mathbb{N}$ (including both 1 and n itself).

The special case of convolution with the constant function $\mathbb{1}$ is of particular interest as it simply corresponds to taking the sum over all divisors,

$$(f \underset{D}{*} \mathbb{1})(n) = \sum_{d \mid n} f(d).$$

[†] The reader may easily check that the formal mechanism of taking the completion with respect to a semi-norm gives the same result as first forming the quotient with respect to the kernel of the semi-norm and then taking the completion with respect to the norm.

[‡] This is often just denoted $f_1 * f_2$ as it is a multiplicative convolution on the semigroup $\mathbb{N}$. However, as we make more use of the additive convolution in this volume we reserve the unadorned $*$ for the latter.

In the case of the von Mangoldt function we claim that

$$(\Lambda \underset{D}{*} \mathbb{1})(n) = \sum_{d|n} \Lambda(d) = \log n. \tag{14.5}$$

In fact, if $n = p^k$ is a prime power, then

$$\sum_{d|n} \Lambda(d) = \sum_{\ell=0}^{k} \Lambda(p^\ell) = k \log p = \log n,$$

and if $n = p_1^{k_1} \cdots p_a^{k_a}$, then

$$\sum_{d|n} \Lambda(d) = \sum_{d|p_1^{k_1}} \Lambda(d) + \cdots + \sum_{d|p_a^{k_a}} \Lambda(d) = \log n.$$

The map $f \longmapsto f \underset{D}{*} \mathbb{1}$ has an inverse operation, known as *Möbius inversion*. We define the Möbius function $\mu : \mathbb{N} \to \mathbb{Z}$ by

$$\mu(n) = \begin{cases} 0 & \text{if } p^2 | n \text{ for some prime } p \text{ and} \\ (-1)^\ell & \text{if } n = p_1 \cdots p_\ell, \end{cases}$$

where $p_1, \ldots, p_\ell$ denote distinct primes. Notice that the second case includes the statement that $\mu(1) = (-1)^0 = 1$ as 1 is taken to be a product of no primes.

Proposition 14.7 (Möbius inversion). *Given functions $f, g : \mathbb{N} \to \mathbb{R}$ we have $g = f \underset{D}{*} \mathbb{1}$ if and only if $f = g \underset{D}{*} \mu$. Moreover $\delta_1 = \mathbb{1} \underset{D}{*} \mu$, where*

$$\delta_1(n) = \begin{cases} 1 & \text{for } n = 1, \\ 0 & \text{otherwise.} \end{cases}$$

We note that in particular this allows us to reformulate (14.5), giving an alternate definition of the von Mangoldt function as

$$\Lambda(n) = \sum_{d|n} \mu(d) \log\left(\frac{n}{d}\right).$$

PROOF OF PROPOSITION 14.7 As we will see the proposition is a consequence of unique prime factorization and is a number-theoretic version of the inclusion–exclusion principle. We start by noting that Dirichlet convolution is commutative and associative since

$$(f_1 \underset{D}{*} f_2)(n) = \sum_{n=de} f_1(d) f_2(e) = (f_2 \underset{D}{*} f_1)(n)$$

and

$$(f_1 \underset{D}{*} f_2) \underset{D}{*} f_3(n) = \sum_{n=def} f_1(d)f_2(e)f_3(f) = f_1 \underset{D}{*} (f_2 \underset{D}{*} f_3)(n)$$

for any three functions f_1, f_2, f_3 on $\mathbb{N}$ and all $n \in \mathbb{N}$. Also note that the function δ_1 is an identity for Dirichlet convolution since

$$(f \underset{D}{*} \delta_1)(n) = \sum_{n=de} f(d)\delta_1(e) = f(n).$$

Thus it is sufficient to show that $\mathbb{1} \underset{D}{*} \mu = \delta_1$, or equivalently that

$$\sum_{d|n} \mu(d) = \begin{cases} 1 & \text{for } n = 1, \\ 0 & \text{otherwise.} \end{cases}$$

For $n = 1$ this is clear. Suppose that $n = p_1^{k_1} \cdots p_\ell^{k_\ell} > 1$ with $p_1, \ldots, p_\ell$ distinct primes. Then

$$\sum_{d|n} \mu(d) = \mu(1) + \sum_{r=1}^{\ell} \sum_{j_1,\ldots,j_r} \mu(p_{j_1} \cdots p_{j_r}) = \sum_{r=0}^{\ell} \binom{\ell}{r}(-1)^r = (1-1)^\ell = 0,$$

where the inner sum over $j_1, \ldots, j_r$ runs over all different r-tuples of distinct indices within $\{1, \ldots, \ell\}$. $\qquad\square$

14.2.2 The Selberg Symmetry Formula

Summarizing the above discussions for the von Mangoldt function we have

$$\Lambda(n) = (\mu \underset{D}{*} \log)(n) = \begin{cases} \log p & \text{if } n = p^k \text{ for a prime } p \text{ and some } k \geqslant 1; \\ 0 & \text{otherwise,} \end{cases}$$

for every $n \in \mathbb{N}$. By analogy we define the *second von Mangoldt function* by

$$\Lambda_2 = \mu \underset{D}{*} \log^2.$$

We start by describing the second von Mangoldt function more carefully.

Lemma 14.8 (Second von Mangoldt function). *For every $n \in \mathbb{N}$ we have*

$$\Lambda_2(n) = \Lambda(n) \log n + (\Lambda \underset{D}{*} \Lambda)(n) = \begin{cases} (2k-1)\log^2 p & \text{if } n = p^k; \\ 2\log p_1 \log p_2 & \text{if } n = p_1^{k_1} p_2^{k_2}; \\ 0 & \text{otherwise,} \end{cases}$$

where p_1 and p_2 denote different primes and we assume $k, k_1, k_2 \geqslant 1$.

PROOF. Below we will use the fact that $\Lambda(n) = \log p$ for $n = p^k$ and $k \geqslant 1$ and $\Lambda(n) = 0$ otherwise without explicit reference. Define f to be the second expression in the lemma, that is

$$f(n) = \Lambda(n) \log n + \sum_{d|n} \Lambda(d) \Lambda\left(\frac{n}{d}\right)$$

for $n \in \mathbb{N}$. We first claim that f also equals the third expression (defined by the three cases). In fact, $f(1) = 0$ and if $n = p^k$ then

$$f(n) = \log p \log p^k + \sum_{\ell=1}^{k-1} \log^2 p = (2k-1) \log^2 p.$$

If $n = p_1^{k_1} p_2^{k_2}$ for two different primes p_1, p_2 and $k_1, k_2 \geqslant 1$, then $\Lambda(n) = 0$ and

$$f(n) = \Lambda(p_1^{k_1})\Lambda(p_2^{k_2}) + \Lambda(p_2^{k_2})\Lambda(p_1^{k_1}) = 2 \log p_1 \log p_2.$$

Finally, if n has three or more prime factors, then clearly $f(n) = 0$ since for $d|n$ either d or $\frac{n}{d}$ must have at least two prime divisors.

Now let $n = p_1^{k_1} \cdots p_\ell^{k_\ell}$ and calculate

$$(f \underset{D}{*} \mathbb{1})(n) = \sum_{d|n} f(d) = \sum_j \sum_{a=1}^{k_j} f(p_j^a) + \sum_{j_1 \neq j_2} \sum_{a_1=1}^{k_{j_1}} \sum_{a_2=1}^{k_{j_2}} f(p_{j_1}^{a_1} p_{j_2}^{a_2})$$

$$= \sum_j \sum_{a=1}^{k_j} (2a-1) \log^2 p_j + 2 \sum_{j_1 \neq j_2} k_{j_1} k_{j_2} \log p_{j_1} \log p_{j_2}$$

$$= \sum_j k_j^2 \log^2 p_j + 2 \sum_{j_1 \neq j_2} k_{j_1} k_{j_2} \log p_{j_1} \log p_{j_2}$$

$$= \left(\sum_j k_j \log p_j\right)^2 = \log^2 n,$$

where the sum over j, respectively $j_1 \neq j_2$, is always from 1 to ℓ. Using Proposition 14.7 we see that $f = \mu \underset{D}{*} \log^2 = \Lambda_2$, as claimed. □

Recalling the reformulation (14.1) of PNT in Lemma 14.2 and thinking of Λ_2 as a modified version of Λ we might be interested in $\sum_{n \leqslant x} \Lambda_2(n)$. As Selberg noticed, the latter sum is much easier to understand, but its asymptotic description is still useful for obtaining PNT. This was an important ingredient in Selberg's elementary proof of PNT, and is also crucial for the argument of Tao presented here.

While not strictly relevant, it might be helpful to provide a reason why the Selberg symmetry formula is easier to obtain than PNT. The second formula in Lemma 14.8 shows that Λ_2 might be thought of as a weighted counting function for products of primes, and since this set is larger than the set of primes it might be easier to study. Whatever the true rationale may be, it is still surprising that the following argument relies only on elementary analysis and Möbius inversion.

Proposition 14.9 (Selberg symmetry formula). *We have*

$$\sum_{n \leqslant x} \Lambda_2(n) = 2x \log x + O(x)$$

for $x \geqslant 1$.

PROOF. We fix some $x \geqslant 1$. The proof will use Möbius inversion and the following (elementary) asymptotic estimates, which we will prove below.

$$\sum_{m \leqslant y} \left(\frac{y}{m}\right)^{-1} \frac{1}{m} = \frac{\lfloor y \rfloor}{y} = 1 + O\left(\frac{1}{y}\right); \tag{14.6}$$

$$\sum_{m \leqslant y} \frac{1}{m} = \log y + c_1 + O\left(\frac{1}{y}\right); \tag{14.7}$$

$$\sum_{m \leqslant y} \frac{\log(y/m)}{m} = \frac{1}{2} \log^2 y + c_1 \log y + c_2 + O\left(\frac{\log(1+y)}{y}\right); \tag{14.8}$$

$$\frac{1}{y} \sum_{m \leqslant y} \log^2 m = \log^2 y - 2\log y + 2 + O\left(\frac{\log^2(1+y)}{y}\right), \tag{14.9}$$

for some constants $c_1, c_2 \in \mathbb{R}$ and all $y \geqslant 1$.

USING MÖBIUS INVERSION. Given a function $g : \mathbb{N} \to \mathbb{R}$ we may define f by

$$f(n) = (\mu \underset{D}{*} g)(n) = \sum_{n=dm} \mu(d)g(m)$$

for all $n \geqslant 1$, which gives

$$\sum_{n \leqslant x} f(n) = \sum_{n \leqslant x} \sum_{n=dm} \mu(d)g(m) = \sum_{d \leqslant x} \mu(d) \sum_{m \leqslant \frac{x}{d}} g(m),$$

where we changed the order of summation. Applying this with $g(n) = \log^2 n$ and $f(n) = \Lambda_2(n)$ gives

$$\frac{1}{x} \sum_{n \leqslant x} \Lambda_2(n) = \sum_{d \leqslant x} \frac{\mu(d)}{d} \frac{1}{x/d} \sum_{m \leqslant \frac{x}{d}} \log^2 m, \tag{14.10}$$

and we see that the estimate (14.9) might be useful for $y = \frac{x}{d}$. Multiplying (14.8) by 2, (14.7) by a constant c_3, (14.6) by a constant c_4, and summing we can choose the constants to match[†] the right-hand side of the asymp-

---

[†] Explicitly, $c_3 = -2 - 2c_1$ and $c_4 = 2 - c_2 - c_1 c_3$.

totic (14.9) to obtain

$$\frac{1}{y} \sum_{m \leqslant y} \log^2 m = \sum_{m \leqslant y} \frac{2\log(\frac{y}{m}) + c_3 + c_4(\frac{y}{m})^{-1}}{m} + O\left(\frac{\log^2(1+y)}{y}\right).$$

To simplify the expressions we introduce the shorthand

$$F(t) = 2\log t + c_3 + c_4 t^{-1}$$

and put the above approximate identity with $y = \frac{x}{d}$ into (14.10) to obtain

$$\frac{1}{x} \sum_{n \leqslant x} \Lambda_2(n) = \sum_{d \leqslant x} \frac{\mu(d)}{d} \sum_{m \leqslant \frac{x}{d}} \frac{F(\frac{x}{dm})}{m} + \sum_{d \leqslant x} \frac{\mu(d)}{d} O\left(\frac{\log^2(1+\frac{x}{d})}{x/d}\right)$$

$$= \sum_{d \leqslant x} \mu(d) \sum_{dm \leqslant x} \frac{F(\frac{x}{dm})}{dm} + O\left(\frac{1}{x} \sum_{d \leqslant x} \log^2\left(1 + \frac{x}{d}\right)\right)$$

$$= \sum_{n \leqslant x} \frac{F(\frac{x}{n})}{n} \sum_{d \mid n} \mu(d) + O\left(\frac{1}{x} \sum_{d \leqslant x} \log^2\left(1 + \frac{x}{d}\right)\right),$$

where we set $n = dm$ and exchanged the order of summation again. Next we recall that $\sum_{d \mid n} \mu(d) = (\mathbb{1} \underset{D}{*} \mu)(n) = \delta_1(n)$ for all $n \geqslant 1$ by Proposition 14.7, and claim that

$$\sum_{d \leqslant x} \log^2\left(1 + \frac{x}{d}\right) = O(x). \qquad (14.11)$$

Together we obtain

$$\frac{1}{x} \sum_{n \leqslant x} \Lambda_2(n) = F(x) + O(1) = 2\log x + c_3 + c_4 \frac{1}{x} + O(1) = 2\log x + O(1)$$

and the proposition follows.

RIEMANN SUMS. It remains to prove the estimates (14.7)–(14.9) and (14.11) (the bound (14.6) is clear), all of which are simple exercises in Riemann integration.

Recall that a non-decreasing function $f : [1, \infty) \to \mathbb{R}_{\geqslant 0}$ satisfies

$$\int_1^y f(t)\,dt - f(y) \leqslant \int_1^{\lfloor y \rfloor} f(t)\,dt \leqslant \sum_{m=2}^{\lfloor y \rfloor} f(m) \leqslant \int_2^{\lfloor y \rfloor + 1} f(t)\,dt \leqslant \int_2^y f(t)\,dt + f(y+1),$$

and so

$$\left| \sum_{m=1}^{\lfloor y \rfloor} f(m) - \int_1^y f(t)\,dt \right| \leqslant f(1) + f(y+1).$$

Applying this with $f(t) = \log^2 t$ gives

$$\sum_{m \leqslant y} \log^2 m = y \log^2 y - 2y \log y + 2y + O(\log^2(1+y)),$$

which then gives (14.9) after dividing by y.

A similar estimate holds for non-increasing functions, so that

$$\left| \sum_{d \leqslant x} \log^2\left(1 + \frac{x}{d}\right) - \int_1^x \log^2\left(1 + \frac{x}{t}\right) dt \right| \leqslant \log^2(1+x) + \log 2 = O(x).$$

Using the substitution $u = \frac{x}{t}$ (with $du = x(-t^{-2}) dt$) we see that

$$\int_1^x \log^2\left(1 + \frac{x}{t}\right) dt = \int_1^x \log^2(1 + u) \frac{x}{u^2} du \leqslant x \int_1^\infty \frac{\log^2(1+u)}{u^2} du = O(x),$$

which gives (14.11).

For the statements in (14.7) and (14.8) we have to be more careful as these involve finer estimates. For (14.7), let $f(t) = \frac{1}{\lfloor t \rfloor}$ and note that

$$\sum_{m \leqslant y} \frac{1}{m} = \int_1^{\lfloor y \rfloor + 1} f(t) \, dt.$$

Also note that $\left| f(t) - \frac{1}{t} \right| \ll \frac{1}{t^2}$. Hence

$$\sum_{m \leqslant y} \frac{1}{m} = \log y + \int_1^{\lfloor y \rfloor + 1} f(t) \, dt - \int_1^y \frac{1}{t} \, dt$$

$$= \log y + \int_1^\infty \left(f(t) - \frac{1}{t} \right) dt - \int_y^\infty \left(f(t) - \frac{1}{t} \right) dt + \int_y^{\lfloor y \rfloor + 1} f(t) \, dt$$

$$= \log y + c_1 + O\left(\frac{1}{y}\right)$$

proves (14.7).

In light of (14.7), it is sufficient to consider the sum $\sum_{m \leqslant y} \frac{\log m}{m}$ for the proof of (14.8). We define $f(t) = \frac{\log \lfloor t \rfloor}{\lfloor t \rfloor}$, and by differentiating $t \mapsto \frac{\log t}{t}$ the mean value theorem gives

$$\left| f(t) - \frac{\log t}{t} \right| \ll \frac{1 + \log t}{t^2}.$$

Note that

$$\int_1^y \frac{\log t}{t} \, dt = \frac{1}{2} \log^2 y$$

and that

$$\int_y^\infty \frac{\log t}{t^2}\, dt = \int_{\log y}^\infty ue^{-u}\, du = \left. \left(-ue^{-u} - e^{-u} \right) \right|_{\log y}^\infty = \frac{\log y + 1}{y}.$$

Hence, if we set

$$-c_2 = \int_1^\infty \left(f(t) - \frac{\log t}{t} \right) dt$$

then we have, similarly,

$$\sum_{m \leqslant y} \frac{\log m}{m} = \frac{1}{2}\log^2 y + \int_1^y f(t)\, dt - \int_1^y \frac{\log t}{t}\, dt + \int_y^{\lfloor y \rfloor + 1} f(t)\, dt$$

$$= \frac{1}{2}\log^2 y - c_2 - \int_y^\infty \left(f(t) - \frac{\log t}{t} \right) dt + O\left(\frac{\log(1+y)}{y} \right)$$

$$= \frac{1}{2}\log^2 y - c_2 + O\left(\frac{\log(1+y)}{y} \right).$$

Multiplying this formula by -1 and (14.7) by $\log y$ we obtain (14.8) as the sum.   □

14.2.3 Convolution of Measures and a Measure-Theoretic Reformulation of Selberg's Symmetry Formula

In order to get closer to functional analysis we associate to any given function $f : \mathbb{N} \to \mathbb{R}$ a Radon measure ν_f on $[0, \infty)$ defined by

$$\nu_f = \sum_{n=1}^\infty \frac{f(n)}{n} \delta_{\log n}.$$

Lemma 14.10 (Convolution of measures). *For two functions f_1, f_2 on $\mathbb{N}$ we have*

$$\nu_{f_1} * \nu_{f_2} = \nu_{f_1 \underset{D}{*} f_2},$$

where the convolution of two Radon measures ν_1, ν_2 on $[0, \infty)$ is defined by

$$(\nu_1 * \nu_2)(B) = \iint \mathbb{1}_B(t_1 + t_2)\, d\nu_1(t_1)\, d\nu_2(t_2)$$

for any Borel subset $B \subseteq \mathbb{R}$, and is again a Radon measure on $[0, \infty)$.

PROOF. For any $x \geqslant 0$ we have

$$\nu_1 * \nu_2([-x, x]) = \nu_1 * \nu_2([0, x]) \leqslant \nu_1([0, x])\nu_2([0, x]) < \infty,$$

which shows that $\nu_1 * \nu_2$ is again a Radon measure. Let $B \subseteq \mathbb{R}$ be measurable. Then

$$
\begin{aligned}
(\nu_{f_1} * \nu_{f_2})(B) &= \sum_{m_1=1}^{\infty} \sum_{m_2=1}^{\infty} \mathbb{1}_B(\log m_1 + \log m_2) \frac{f_1(m_1)}{m_1} \frac{f_2(m_2)}{m_2} \\
&= \sum_{n=1}^{\infty} \mathbb{1}_B(\log n) \frac{1}{n} \sum_{d|n} f_1(d) f_2\left(\frac{n}{d}\right) \\
&= \int \mathbb{1}_B \, d\nu_{f_1 \underset{D}{*} f_2} = \nu_{f_1 \underset{D}{*} f_2}(B),
\end{aligned}
$$

as claimed. $\qquad\square$

For a Radon measure ν on $\mathbb{R}$ and some $h \in \mathbb{R}$ we define the shifted measure $\lambda_h \nu$ by

$$
\int_{\mathbb{R}} f \, d\lambda_h \nu = \int_{\mathbb{R}} f(t - h) \, d\nu(t).
$$

Clearly any Radon measure defines a functional on $C_c(\mathbb{R})$. If we endow $C_c(\mathbb{R})$ with a locally convex vector space structure as in Example 8.63(4) then these functionals are in fact continuous. However, we will not need this, even though we are interested in the associated weak* topology on the space of Radon measures (which is also referred to as the *vague topology*).

Exercise 14.11. Show that a Radon measure defines a continuous functional on the locally convex vector space $C_c(\mathbb{R})$ defined by the semi-norms from Example 8.63(4).

Lemma 14.12 (Third reformulation of PNT via weak* convergence). *The semi-norm $\|\cdot\|_\Lambda$ as in Proposition 14.4 vanishes on all of $C_c(\mathbb{R})$ if and only if the measures $\lambda_h \nu_\Lambda$ weak* converge to Lebesgue measure m as $h \to \infty$.*

PROOF. Fix some $f \in C_c(\mathbb{R})$ and $h \in \mathbb{R}$. Then

$$
\int_{\mathbb{R}} f \, d\lambda_h \nu_\Lambda = \int_{\mathbb{R}} f(t - h) \, d\nu_\Lambda(t) = \sum_{n \geqslant 1} \frac{\Lambda(n)}{n} f(\log n - h).
$$

Hence we see that $\|f\|_\Lambda = 0$ is equivalent to

$$
\int_{\mathbb{R}} f \, d\lambda_h \nu_\Lambda \longrightarrow \int_{\mathbb{R}} f \, dm
$$

as $h \to \infty$. $\qquad\square$

We will not use this lemma except as motivation for the next step, which relies on the Selberg symmetry formula.

Corollary 14.13 (Measure-theoretic Selberg symmetry). *We define the measure $\nu_{\mathrm{sym}} = \nu_{\Lambda_2/\log}$. Then*

$$d\nu_{\text{sym}}(t) = d\nu_\Lambda(t) + \frac{1}{t}\, d\big(\nu_\Lambda * \nu_\Lambda\big)(t)$$

and $\lambda_h \nu_{\text{sym}}$ converges to $2m$ in the weak* topology as $h \to \infty$.

PROOF. By Lemma 14.10 we have $\nu_\Lambda * \nu_\Lambda = \nu_{\Lambda \underset{D}{*} \Lambda}$ and so

$$\nu_{\text{sym}}(B) = \nu_{\Lambda_2/\log}(B) = \sum_{n \geqslant 1} \mathbb{1}_B(\log n)\frac{\Lambda_2(n)}{n \log n}$$

$$= \sum_{n \geqslant 1} \mathbb{1}_B(\log n)\left(\frac{\Lambda(n)}{n} + \frac{(\Lambda \underset{D}{*} \Lambda)(n)}{n \log n}\right)$$

$$= \nu_\Lambda(B) + \int \mathbb{1}_B(t)\frac{1}{t}\, d\nu_{\Lambda \underset{D}{*} \Lambda}(t)$$

by the properties of Λ_2 in Lemma 14.8. This gives the identity in the corollary.

To obtain the claimed convergence we apply the Selberg symmetry formula (Proposition 14.9). By this formula,

$$\sum_{n \leqslant cx} \Lambda_2(n) = 2cx \log x + \mathrm{O}(x) = 2cx \log x + \mathrm{o}(x \log x)$$

for any constant $c > 0$. Dividing by $x \log x$ this gives the asymptotic

$$\frac{1}{x \log x} \sum_{ax < n \leqslant bx} \Lambda_2(n) = 2(b - a) + \mathrm{o}(1) \qquad (14.12)$$

as $x \to \infty$ for any $0 < a < b$. Now fix $\varepsilon > 0$ and assume that x is large enough to ensure that $ax < n \leqslant bx$ implies that

$$1 - \varepsilon \leqslant \frac{\log x}{\log n} \leqslant 1 + \varepsilon.$$

Multiplying this by $\frac{1}{\log x}\Lambda_2(n)$ and summing over $n \in (ax, bx]$ leads to

$$1 - \varepsilon \leqslant \frac{\frac{1}{x}\sum_{ax < n \leqslant bx} \frac{\Lambda_2(n)}{\log n}}{\frac{1}{x \log x}\sum_{ax < n \leqslant bx} \Lambda_2(n)} \leqslant 1 + \varepsilon.$$

By (14.12) we know an asymptotic for the denominator, so the same asymptotic holds for the numerator since $\varepsilon > 0$ was arbitrary. Therefore, we have

$$\sum_{n \geqslant 1} \mathbb{1}_{(\log a, \log b]}(\log n - \log x)\, e^{\log n - \log x}\frac{\Lambda_2(n)}{n \log n} = \sum_{ax < n \leqslant bx} \frac{n}{x}\frac{\Lambda_2(n)}{n \log n}$$

$$= 2(b - a) + \mathrm{o}(1)$$

as $x \to \infty$. Here the left-hand side equals the integral of the function defined by $f(t) = \mathbb{1}_{(\log a, \log b]}(t)e^t$ with respect to $\lambda_{\log x}\nu_{\text{sym}}$, so that

$$\int \mathbb{1}_{(\log a, \log b]}(t)e^t \, d\lambda_{\log x}\nu_{\text{sym}} = 2(b-a) + o(1) = 2\int \mathbb{1}_{(\log a, \log b]}(t)e^t \, dt + o(1)$$

as $x \to \infty$ (or, equivalently, as $h = \log x \to \infty$). In other words, the convergence claim already holds for $f_{a,b}(t) = \mathbb{1}_{(\log a, \log b]}(t)e^t$ for any $a < b$ in $(0, \infty)$. Finally, we can approximate any $f \in C_c(\mathbb{R})$ uniformly from above and below by finite linear combinations of such functions. The corollary follows. $\square$

14.2.4 A Density Function and the Continuity Bound

Notice that Corollary 14.13 in particular gives $0 \leqslant \nu_\Lambda \leqslant \nu_{\text{sym}}$ and so

$$0 \leqslant \lambda_h \nu_\Lambda \leqslant \lambda_h \nu_{\text{sym}} \tag{14.13}$$

for any $h \in \mathbb{R}$. Using the Banach–Alaoglu theorem and Radon–Nikodym derivatives we can conclude from this the following result.

Proposition 14.14. *For any sequence* (h_n) *with* $h_n \to \infty$ *as* $n \to \infty$ *there exists a subsequence* (h_{n_k}) *and a Borel measurable function* $D : \mathbb{R} \to [0,2]$ *such that for any* $f \in C_c(\mathbb{R})$ *we have*

$$\lim_{k \to \infty} \int f \, d\lambda_{h_{n_k}} \nu_\Lambda = \int f D \, dm.$$

PROOF. Fix some integer $\ell \geqslant 1$. Since by Corollary 14.13

$$\lambda_h \nu_{\text{sym}}([-\ell, \ell]) \leqslant \int f_\ell \, d\lambda_h \nu_{\text{sym}} \longrightarrow 2(2\ell + 1)$$

as $h \to \infty$ for the function defined by

$$f_\ell(x) = \begin{cases} 1 & \text{if } |x| \leqslant \ell, \\ \ell + 1 - |x| & \text{if } |x| \in [\ell, \ell + 1], \text{ and} \\ 0 & \text{if } |x| \geqslant \ell + 1, \end{cases}$$

we see that $(\lambda_{h_n}\nu_\Lambda)|_{[-\ell,\ell]}$ can be identified with a bounded sequence of functionals on $C([-\ell, \ell])$. Let $R_\ell = \sup_{n \geqslant 1} \lambda_{h_n}\nu_\Lambda([-\ell, \ell])$. By the Banach–Alaoglu theorem (Theorem 8.10 and Proposition 8.11) the closed ball

$$B_\ell = \overline{B_{R_\ell}}^{C([-\ell,\ell])^*}$$

of radius R_ℓ is compact and metrizable in the weak* topology. Therefore

$$X = \prod_{\ell \geqslant 1} B_\ell$$

is compact and metrizable in the product topology (see Appendix A.3). Unfolding the definitions, it follows that $(\lambda_{h_n} \nu_\Lambda)$ has a subsequence $(\lambda_{h_{n_k}} \nu_\Lambda)$ such that $\int f \, d\lambda_{h_{n_k}} \nu_\Lambda$ converges as $k \to \infty$ for any $f \in C_c(\mathbb{R})$. As integration and the limit are linear, by taking the limit we obtain a linear functional on $C_c(\mathbb{R})$. Moreover, this functional is non-negative for any non-negative $f \in C_c(\mathbb{R})$ and so the Riesz representation theorem (Theorem 7.44) shows that there is a Radon measure μ on $\mathbb{R}$ such that

$$\lim_{k \to \infty} \int f \, d\lambda_{h_{n_k}} \nu_\Lambda = \int f \, d\mu,$$

for all $f \in C_c(\mathbb{R})$. By Corollary 14.13, $f \geqslant 0$ also implies that

$$\int f \, d\mu = \lim_{k \to \infty} \int f \, d\lambda_{h_{n_k}} \nu_\Lambda \leqslant \lim_{k \to \infty} \int f \, d\lambda_{h_{n_k}} \nu_{\text{sym}} = 2 \int f \, dm.$$

Using the density of $C_c(\mathbb{R})$ in $L^1_{\mu+m}(\mathbb{R})$ we can approximate the characteristic function $\mathbb{1}_B$ of any bounded measurable set by a non-negative function in $C_c(\mathbb{R})$ simultaneously with respect to both μ and m, which implies that $\mu(B) \leqslant 2m(B)$. Using Proposition 3.29 it follows that μ is absolutely continuous with respect to m and that the Radon–Nikodym derivative $D = \frac{d\mu}{dm}$ takes values in $[0, 2]$ almost everywhere.   □

Recall from Lemma 14.12 that we wish to show that $D \equiv \mathbb{1}$.

PROOF OF FIRST INEQUALITY IN THEOREM 14.6. Fix some $f \in C_c(\mathbb{R})$ and choose a sequence (h_n) with $h_n \to \infty$ as $n \to \infty$ such that

$$\|f\|_\Lambda = \lim_{n \to \infty} \left| \sum_{n \geqslant 1} \frac{\Lambda(n)}{n} f(\log n - h_n) - \int_{\mathbb{R}} f \, dm \right|.$$

Applying Proposition 14.14 we can choose a subsequence such that $\lambda_{h_{n_k}} \nu_\Lambda$ converges to the measure $D \, dm$ with D taking values in $[0, 2]$. Therefore

$$\|f\|_\Lambda = \left| \int f(D - 1) \, dm \right| \leqslant \int |f| \, dm = \|f\|_1.$$

As $f \in C_c(\mathbb{R})$ was arbitrary the inequality (14.4) in Theorem 14.6 follows. □

14.2.5 Mertens' Theorem

The second number-theoretic input needed is one of several results known as Mertens' theorem, which we state in the following form.

Theorem 14.15 (Mertens' theorem). *We have*

$$\sum_{n \leqslant x} \frac{\Lambda(n)}{n} = \log x + \mathrm{O}(1) \tag{14.14}$$

for $x \geqslant 1$.

PROOF. We first claim that

$$\sum_{n \leqslant x} \Lambda(n) = \mathrm{O}(x), \tag{14.15}$$

which will allow us to control error terms in the calculation below. The bound (14.15) is a trivial consequence of PNT (in the form of the statement (14.1)), but fortunately the results developed above are sufficient to prove (14.15) quite directly. In fact, the continuity bound $\|f\|_\Lambda \leqslant \|f\|_1$ in (14.4) (proven above) implies that

$$\limsup_{x \to \infty} \left| \sum_{n \geqslant 1} \frac{\Lambda(n)}{n} f\left(\log \frac{n}{x}\right) \right| \leqslant 2\|f\|_1$$

for every $f \in C_c((0, \infty))$. Using this for $f(t) = e^t g_+(e^t)$ and a non-negative function $g_+ \in C_c(\mathbb{R})$ with $\mathbb{1}_{[\frac{1}{2}, 1]} \leqslant g_+$ we obtain (much as in the proof of Proposition 14.4) that

$$\limsup_{x \to \infty} \frac{1}{x} \sum_{\frac{1}{2}x < n \leqslant x} \Lambda(n) \leqslant 2 \int_0^\infty g_+(t)\, dt,$$

and so

$$\sum_{\frac{1}{2}x < n \leqslant x} \Lambda(n) \leqslant Cx$$

for some $C \geqslant 1$ and all $x \geqslant 1$. Therefore

$$\sum_{n \leqslant x} \Lambda(n) = \sum_{\frac{1}{2}x < n \leqslant x} \Lambda(n) + \sum_{\frac{1}{4}x < n \leqslant \frac{1}{2}x} \Lambda(n) + \cdots + \sum_{\frac{1}{2^{\ell+1}}x < n \leqslant \frac{1}{2^\ell}x} \Lambda(n)$$

$$\leqslant Cx + \tfrac{1}{2}Cx + \cdots + \tfrac{1}{2^\ell}Cx \leqslant 2Cx,$$

where $\ell = \lfloor \log_2 x \rfloor$, so that $\frac{1}{2^\ell}x \geqslant 1$ but $\frac{1}{2^{\ell+1}}x < 1$. This proves (14.15).

Working towards (14.14), we first recall from (14.5) that $\log n = \sum_{d|n} \Lambda(d)$, which implies that

$$\sum_{n \leqslant x} \log n = \sum_{n \leqslant x} \sum_{d|n} \Lambda(d) = \sum_{d \leqslant x} \Lambda(d) \left\lfloor \frac{x}{d} \right\rfloor.$$

On the right-hand side we use $\lfloor \frac{x}{d} \rfloor = \frac{x}{d} + \mathrm{O}(1)$ and (14.15) to obtain

$$\sum_{n \leqslant x} \log n = x \sum_{d \leqslant x} \frac{\Lambda(d)}{d} + \mathrm{O}(x). \qquad (14.16)$$

On the left-hand side we again use monotonicity of $t \mapsto \log t$ to replace the sum by an integral

$$\sum_{n \leqslant x} \log n = \int_1^x \log t \,\mathrm{d}t + \mathrm{O}\big(\log(1+x)\big) = x \log x + \mathrm{O}(x) \qquad (14.17)$$

as in the proof of the Selberg symmetry formula above. Combining (14.16) and (14.17) and dividing by x gives the theorem. $\qquad\qquad\qquad\qquad\Box$

14.2.6 Completing the Proof

PROOF OF BANACH ALGEBRA INEQUALITY IN THEOREM 14.6. It remains to show that for any $f_1, f_2 \in C_c(\mathbb{R})$ we have

$$\left| \int (f_1 * f_2) \,\mathrm{d}\lambda_h \nu_\Lambda - \int (f_1 * f_2) \,\mathrm{d}m \right| \leqslant \|f_1\|_\Lambda \|f_2\|_\Lambda + \mathrm{o}_{f_1, f_2}(1) \qquad (14.18)$$

as $h \to \infty$. Note that the left-hand side is the integral of $f = f_1 * f_2$ with respect to the measure[†] $\lambda_h \nu_\Lambda - m$. Next we define $m_+ = m|_{[0,\infty)}$ and notice that $\lambda_h m_+ \to m$ as $h \to \infty$. For that reason we can equivalently consider the integral of f with respect to λ_h applied to $\nu_\Lambda - m_+$. It will be helpful to write the latter measure in a different way as a sum of measures.

For this, notice first that the convolution of $m_+ = m|_{[0,\infty)}$ with itself is given by

$$(m_+ * m_+)(B) = \int_0^\infty \int_0^\infty \mathbb{1}_B(t+s) \,\mathrm{d}t \,\mathrm{d}s = \int_0^\infty \mathbb{1}_B(u) \int_0^u \mathrm{d}s \,\mathrm{d}u = \int_0^\infty \mathbb{1}_B(t) t \,\mathrm{d}t$$

for any Borel subset $B \subseteq \mathbb{R}$, where we used the substitution $u = t + s$ (for a fixed s) and Fubini's theorem. In particular, we can write the relationship above in the convenient but notationally illogical form

$$\frac{1}{t} \,\mathrm{d}(m_+ * m_+)(t) = m_+.$$

SPLITTING INTO THREE SIGNED MEASURES. Working in the dual space of $C_c(\mathbb{R})$, it follows from Corollary 14.13 that we can write

---

[†] Note that this 'signed Radon measure' may be undefined on unbounded Borel sets, but interpreting these measures as functionals on $C_c(\mathbb{R})$ gives the right viewpoint.

$$\nu_\Lambda - m_+ = \underbrace{\nu_{\text{sym}} - 2m_+}_{=\rho_1} - \tfrac{1}{t}(\, \mathrm{d}\nu_\Lambda * \nu_\Lambda)(t) + m_+$$

$$= \underbrace{\rho_1 - \tfrac{1}{t}\,\mathrm{d}\big((\nu_\Lambda - m_+) * (\nu_\Lambda - m_+)\big)(t)}_{=\rho_2} + \underbrace{2m_+ - 2\tfrac{1}{t}\,\mathrm{d}(\nu_\Lambda * m_+)(t)}_{=\rho_3}$$

as a sum of three signed measures. For the first of these we recall from Corollary 14.13 that

$$\lambda_h \rho_1 = \lambda_h(\nu_{\text{sym}} - 2m_+) \longrightarrow 2m - 2m = 0 \qquad (14.19)$$

as $h \to \infty$. For the third we calculate $\nu_\Lambda * m_+$ and obtain

$$(\nu_\Lambda * m_+)(B) = \sum_{n \geqslant 1} \int_0^\infty \mathbb{1}_B(t + \log n)\frac{\Lambda(n)}{n}\,\mathrm{d}t$$

$$= \sum_{n \geqslant 1} \int_0^\infty \mathbb{1}_B(s)\mathbb{1}_{[\log n,\infty)}(s)\frac{\Lambda(n)}{n}\,\mathrm{d}s$$

$$= \int_0^\infty \mathbb{1}_B(s)\sum_{n \leqslant e^s}\frac{\Lambda(n)}{n}\,\mathrm{d}s$$

for a Borel subset $B \subseteq \mathbb{R}$. Therefore, $\tfrac{1}{t}\mathrm{d}(\nu_\Lambda * m_+)$ is equal to the absolutely continuous measure whose Radon–Nikodym derivative with respect to m is given by

$$\frac{1}{t}\sum_{n \leqslant e^t}\frac{\Lambda(n)}{n}.$$

By Mertens' theorem (Theorem 14.15) this function is $1 + o(1)$ as $t \to \infty$. Hence

$$\lambda_h\left(\frac{1}{t}\,\mathrm{d}(m_+ * \nu_\Lambda)\right) \longrightarrow m \qquad (14.20)$$

as $h \to \infty$, and we also have

$$\lambda_h \rho_3 \longrightarrow 0 \qquad (14.21)$$

as $h \to \infty$. To summarize, we have split $\nu_\Lambda - m_+$ into the sum $\rho_1 - \rho_2 + \rho_3$, where ρ_1 and ρ_3 satisfy (14.19) resp. (14.21).

SHIFTING ρ_2. We claim that in addition

$$\left|\int f_1 * f_2 \,\mathrm{d}\lambda_h\rho_2\right| \leqslant \|f_1\|_\Lambda\|f_2\|_\Lambda + o(1) \qquad (14.22)$$

as $h \to \infty$ for $f_1, f_2 \in C_c(\mathbb{R})$. Together with what we have proved above about ρ_1 and ρ_3, this will imply (14.18).

Hence it remains to prove (14.22) for $f_1, f_2 \in C_c(\mathbb{R})$. We set $f = f_1 * f_2$ and recall that by definition of λ_h and of ρ_2 we have

$$\int f \, d\lambda_h \rho_2 = \int f(t-h) \, d\rho_2(t) = \int f(t-h) \tfrac{1}{t} \, d\big((\nu_\Lambda - m_+) * (\nu_\Lambda - m_+)\big)(t).$$

To be able to operate with this expression we note that

$$\tfrac{1}{t} \, d\big((\nu_\Lambda + m_+) * (\nu_\Lambda + m_+)\big)$$

is bounded from above by

$$\tfrac{1}{t}\big(\, d(\nu_\Lambda + m_+) * (\nu_\Lambda + m_+)\big) + \nu_\Lambda = \nu_{\mathrm{sym}} + \tfrac{2}{t} \, d(\nu_\Lambda * m_+) + m_+$$

which when shifted by λ_h converges to $5m$ as $h \to \infty$ by Corollary 14.13 and (14.20). If $g_h, g \in C_c(\mathbb{R})$ with $g \geqslant 0$ satisfy $|g_h| \leqslant o(1)g$ as $h \to \infty$, then the above gives

$$\left| \int g_h \, d\lambda_h \rho_2 \right| \leqslant o(1) \int g(t-h) \tfrac{1}{t} \, d\big((\nu_\Lambda + m_+) * (\nu_\Lambda + m_+)\big)(t) = o_g(1)$$

as $h \to \infty$. We apply this to $g = |f|$ and

$$g_h(t) = f(t)\left(\tfrac{t+h}{h} - 1\right).$$

Note that the requirement above is satisfied, since $\tfrac{t+h}{h} = 1 + o(1)$ as $h \to \infty$ uniformly for $t \in \mathrm{Supp}\, f$ by compactness. We obtain

$$\int f(t-h)\left(\tfrac{t}{h} - 1\right)\underbrace{\tfrac{1}{t} \, d\big((\nu_\Lambda - m_+) * (\nu_\Lambda - m_+)\big)(t)}_{\mathrm{d}\rho_2(t)} = \int g_h \, d\lambda_h \rho_2 = o_f(1)$$

as $h \to \infty$, and so we also have

$$\int f \, d\lambda_h \rho_2 = \frac{1}{h} \int f(t-h) \, d\big((\nu_\Lambda - m_+) * (\nu_\Lambda - m_+)\big)(t) + o_f(1). \qquad (14.23)$$

Finally, we recall that $f = f_1 * f_2$ and calculate

$$\frac{1}{h} \int (f_1 * f_2)(t-h) \, d\big((\nu_\Lambda - m_+)\big) * (\nu_\Lambda - m_+)(t)$$

$$= \frac{1}{h} \iint (f_1 * f_2)(t_1 + t_2 - h) \, d(\nu_\Lambda - m_+)(t_1) \, d(\nu_\Lambda - m_+)(t_2)$$

$$= \frac{1}{h} \iiint f_1(u) f_2(t_1 + t_2 - h - u) \, du \, d(\nu_\Lambda - m_+)(t_1) \, d(\nu_\Lambda - m_+)(t_2)$$

$$= \frac{1}{h} \iiint f_1(t_1 - r) f_2(t_2 + r - h) \, dr \, d(\nu_\Lambda - m_+)(t_1) \, d(\nu_\Lambda - m_+)(t_2)$$

$$= \frac{1}{h} \int \underbrace{\iint f_1 \, d\lambda_r(\nu_\Lambda - m_+)}_{=I_1(r)} \underbrace{\int f_2 \, d\lambda_{h-r}(\nu_\Lambda - m_+)}_{=I_2(h-r)} \, dr,$$

where we used the substitution $u = t_1 - r$ and Fubini's theorem (extended by linearity to signed Radon measures and functions with compact support). Depending on the support of f_1 and f_2 there exists some $R \in \mathbb{R}$ with the property that the first inner integral $I_1(r)$ vanishes if $r < R$ and the second inner integral $I_2(h - r)$ vanishes if $h - r < R$. Given $\varepsilon > 0$ there also exists some S such that

$$|I_1(r)| \leqslant \|f_1\|_\Lambda + \varepsilon$$

for $r > S$ and

$$|I_2(h - r)| \leqslant \|f_2\|_\Lambda + \varepsilon$$

for $h - r > S$. Together with the bound (14.23) this gives

$$\left| \int (f_1 * f_2) \, d\lambda_h \rho_2 \right| \leqslant \frac{1}{h} \left| \int_R^{h-R} I_1(r) I_2(h - r) \, dr \right| + o_{f_1,f_2}(1)$$

$$\leqslant \frac{1}{h} o_{f_1,f_2}(1) + \frac{1}{h} \left| \int_S^{h-S} I_1(r) I_2(h - r) \, dr \right| + o_{f_1,f_2}(1)$$

$$\leqslant \frac{h - 2S}{h} \left(\|f_1\|_\Lambda + \varepsilon \right) \left(\|f_2\|_\Lambda + \varepsilon \right) + o_{f_1,f_2}(1)$$

as $h \to \infty$. As $\varepsilon > 0$ was arbitrary, this proves the claim in (14.22) and hence the theorem. $\qquad\square$

14.3 Non-Trivial Spectrum of the Banach Algebra

We assume in this section that the semi-norm $\|\cdot\|_\Lambda$ defined in Proposition 14.4 is non-trivial. Let $\mathcal{A}_\Lambda$ be the completion of $C_c(\mathbb{R})$ with respect to $\|\cdot\|_\Lambda$ and note that Theorem 14.6 shows that there is a Banach algebra homomorphism

$$\Phi : L^1(\mathbb{R}) \to \mathcal{A}_\Lambda.$$

Essential Exercise 14.16. Give the details of the argument that deduces the existence of a Banach algebra homomorphism Φ as above from Theorem 14.6.

Theorem 14.17 (Spectrum of $\mathcal{A}_\Lambda$). *Suppose that $\|\cdot\|_\Lambda$ is a non-trivial semi-norm on $C_c(\mathbb{R})$, so that the associated Banach algebra $\mathcal{A}_\Lambda$ is non-trivial. Then there exists some $\xi \in \mathbb{R}$ such that*

$$C_c(\mathbb{R}) \ni f \longmapsto \widehat{f}(\xi) = \int_{\mathbb{R}} f(t) e^{-2\pi i t \xi} \, dt$$

is continuous with respect to $\|\cdot\|_\Lambda$. Given such a ξ, if $f \in C_c(\mathbb{R})$ and $f(t)e^{-2\pi i t \xi}$ is non-negative for all $t \in \mathbb{R}$, then $\|f\|_\Lambda = \|f\|_1$.

PROOF. Notice that every character χ of $\mathcal{A}_\Lambda$ gives rise to the character $\chi \circ \Phi$ on $L^1(\mathbb{R})$. If χ is a non-trivial character, then $\chi \circ \Phi$ is also non-trivial since Φ has dense image (by definition of $\mathcal{A}_\Lambda$). As every non-trivial character of $L^1(\mathbb{R})$ has the form

$$f \longmapsto \widehat{f}(\xi)$$

for some $\xi \in \mathbb{R}$ (see Proposition 11.38), it is sufficient to show that $\mathcal{A}_\Lambda$ has a non-trivial character.

By the spectral radius formula (Corollary 11.29) the existence of a non-trivial character follows if we can find some element $f \in \mathcal{A}_\Lambda$ whose spectral radius $\lim_{n \to \infty} \|f^{*n}\|_\Lambda^{1/n}$ is non-zero.

Suppose now that $g_0 \in C_c(\mathbb{R})$ has $\|g_0\|_\Lambda > 0$. By the density of $\mathscr{S}(\mathbb{R})$ in $L^1(\mathbb{R})$ we can find some $g_1 \in \mathscr{S}(\mathbb{R})$ with $\|g_1 - g_0\|_1 < \|g_0\|_\Lambda$, so that by Theorem 14.6 we also have $\|g_1\|_\Lambda > 0$. In fact, we may even assume that $\widehat{g_1}$ lies in $C_c^\infty(\mathbb{R})$ since these functions are also dense in $L^1(\mathbb{R})$ (see Exercise 9.48). Now let $f \in \mathscr{S}(\mathbb{R})$ be chosen so that $\widehat{f} \equiv 1$ on Supp $\widehat{g_1}$. By Proposition 9.31 this implies that

$$\widehat{g_1 * f^{*n}} = \widehat{g_1}\widehat{f}^n = \widehat{g_1}$$

and so $g_1 * f^{*n} = g_1$ by Fourier inversion (Theorem 9.36). That is, f behaves like an identity for the element g_1. Applying the continuous algebra homomorphism $\Phi : L^1(\mathbb{R}) \to \mathcal{A}_\Lambda$ we see that

$$0 < \|g_1\|_\Lambda = \|g_1 * f^{*n}\|_\Lambda \leqslant \|g_1\|_\Lambda \|f^{*n}\|_\Lambda$$

and so $\|f^{*n}\|_\Lambda \geqslant 1$ for all $n \geqslant 1$. As argued above, this gives the existence of $\xi \in \mathbb{R}$ as in the theorem.

Now let one such $\xi \in \mathbb{R}$ be fixed and suppose that $f \in C_c(\mathbb{R})$ has

$$f(t)e^{-2\pi it\xi} \geqslant 0$$

for all $t \in \mathbb{R}$. We then have

$$\|f\|_1 = \int |f(t)| \, \mathrm{d}t = \int f(t)e^{-2\pi it\xi} \, \mathrm{d}t = \widehat{f}(\xi) \leqslant \|f\|_\Lambda$$

since the norm of a character is at most one by Lemma 11.24. By Theorem 14.6 this shows that $\|f\|_1 = \|f\|_\Lambda$, as claimed.   □

14.4 Trivial Spectrum of the Banach Algebra

Theorem 14.18 (Trivial spectrum). *For every $\xi \in \mathbb{R}$ there exists some f in $C_c(\mathbb{R})$ such that $f(t)e^{-2\pi it\xi} \geqslant 0$ for all $t \in \mathbb{R}$ and $\|f\|_\Lambda < \|f\|_1$.*

Notice that Theorems 14.17 and 14.18 together show that $\|f\|_A = 0$ for every $f \in C_c(\mathbb{R})$. Proposition 14.4 then gives the PNT (Theorem 14.1).

PROOF OF THEOREM 14.18 FOR $\xi \neq 0$. In this case we will only use that the density function D in Proposition 14.14 takes values in $[0,2] \subseteq \mathbb{R}$ almost surely. Let

$$
f_0(t) = \begin{cases} 1 & \text{if } |t| \leqslant 1, \\ 2 - |t| & \text{if } |t| \in [1,2], \\ 0 & \text{otherwise,} \end{cases}
$$

and for a fixed $\xi \neq 0$ we define $f(t) = f_0(t)\mathrm{e}^{2\pi \mathrm{i} t \xi}$. Choose a sequence (h_n) with $h_n \to \infty$ as $n \to \infty$ for which

$$
\|f\|_A = \lim_{n \to \infty} \left| \int f \, \mathrm{d}\lambda_{h_n} \nu_A - \int f \, \mathrm{d}m \right|.
$$

By Proposition 14.14 we may find a Borel measurable function $D : \mathbb{R} \to [0,2]$ with

$$
\|f\|_A = \left| \int f(D-1) \, \mathrm{d}m \right|.
$$

Choose $\theta \in \mathbb{R}$ with

$$
\|f\|_A = \mathrm{e}^{\mathrm{i}\theta} \int f(D-1) \, \mathrm{d}m,
$$

so that with the bound $|D(t) - 1| \leqslant 1$ for all $t \in \mathbb{R}$ we obtain

$$
\|f\|_A = \int_{-2}^{2} \Re\big(\mathrm{e}^{\mathrm{i}\theta}\mathrm{e}^{2\pi \mathrm{i} t \xi}\big) f_0(t)(D(t) - 1) \, \mathrm{d}t
$$

$$
= \int_{-2}^{2} \cos(\theta + 2\pi t \xi) f_0(t)(D(t) - 1) \, \mathrm{d}t
$$

$$
< \int_{-2}^{2} f_0(t) \, \mathrm{d}t = \|f_0\|_1 = \|f\|_1,
$$

where the strict inequality follows from $\xi \neq 0$. Thus the theorem follows in this case. $\qquad \square$

PROOF OF THEOREM 14.18 FOR $\xi = 0$. In this case we will use Mertens' theorem one more time. Define a function $f_0 \in C_c(\mathbb{R})$ by

$$
f_0(t) = \begin{cases} 1 & \text{if } t \in [-N, 0], \\ 1 - t & \text{if } t \in [0, 1], \\ N + 1 + t & \text{if } t \in [-(N+1), -N], \\ 0 & \text{otherwise,} \end{cases}
$$

for some N to be determined later. By Mertens' theorem (Theorem 14.15) we have

$$\sum_{n \geqslant 1} \frac{\Lambda(n)}{n} f_0(\log n - h) \leqslant \sum_{e^{h-(N+1)} \leqslant n \leqslant e^{h+1}} \frac{\Lambda(n)}{n}$$

$$= \log e^{h+1} - \log e^{h-(N+1)} + O(1) = N + O(1)$$

for $h \geqslant N + 1$ and

$$\sum_{n \geqslant 1} \frac{\Lambda(n)}{n} f_0(\log n - h) \geqslant \sum_{e^{h-N} \leqslant n \leqslant e^h} \frac{\Lambda(n)}{n}$$

$$= \log e^h - \log e^{h-N} + O(1) = N + O(1)$$

for $h \geqslant N$. Choosing N sufficiently large and using $\int f_0(t)\,dt = N + 1$, we find that

$$\left| \sum_{n \geqslant 1} \frac{\Lambda(n)}{n} f_0\left(\log \frac{n}{h}\right) - \int f_0(t)\,dt \right| < N < \|f_0\|_1$$

for all sufficiently large h. Thus $\|f_0\|_\Lambda < \|f_0\|_1$ and the theorem follows. $\square$

14.5 Primes in Arithmetic Progressions

The prime number theorem generalizes to give an asymptotic density for primes in arithmetic progressions, strengthening Dirichlet's classical result that an arithmetic progression $\{nq + a \mid n \in \mathbb{N}\}$ contains infinitely many primes if $\gcd(q, a) = 1$. In this section we once again follow Tao's blog [103] and indicate, largely through a sequence of exercises, how the arguments above can be adapted to obtain additional asymptotic results which combine to prove the following.

Theorem 14.19 (PNT in arithmetic progressions). *Fix* $q \geqslant 1$. *Then*

$$\sum_{\substack{p \leqslant x; \\ p \equiv a \pmod q}} 1 \sim \frac{x}{\phi(q) \log x} \tag{14.24}$$

as $x \to \infty$, *where* $a \in \mathbb{Z}$ *has* $\gcd(a, q) = 1$ *and* $\phi(q) = |(\mathbb{Z}/q\mathbb{Z})^\times|$ *is the* Euler totient function *of* q.

Essential Exercise 14.20. Show that in order to prove Theorem 14.19 it is enough to show that

$$\sum_{\substack{n \leqslant x; \\ n \equiv a \pmod q}} \Lambda(n) \sim \frac{x}{\phi(q)}$$

for any $a \in \mathbb{Z}$ with $\gcd(a, q) = 1$.

The sum in Exercise 14.20 lacks a certain structure, so we decompose the characteristic function of the arithmetic progression into more convenient expressions. Characters of the multiplicative group $(\mathbb{Z}/q\mathbb{Z})^{\times}$ are usually called *multiplicative characters*, so a function $\chi : (\mathbb{Z}/q\mathbb{Z})^{\times} \to \mathbb{S}^1$ is a multiplicative character if $\chi(ab) = \chi(a)\chi(b)$ for all $a, b \in (\mathbb{Z}/q\mathbb{Z})^{\times}$. These are of course simply characters on this abelian multiplicative group in the sense of Fourier analysis. As is customary, we think of a multiplicative character χ as defining a function $\chi' : \mathbb{Z} \to \mathbb{C}$ called a *Dirichlet character of level or modulus q* by defining $\chi'(k) = 0$ if $\gcd(q, k) \neq 1$ and $\chi'(k) = \chi(k + q\mathbb{Z})$ if $\gcd(q, k) = 1$ for $k \in \mathbb{Z}$. For convenience we will again write χ for χ'.

Essential Exercise 14.21. (a) For any $a \in \mathbb{Z}$ with $\gcd(q, a) = 1$ show that the function $f_a = \mathbb{1}_{\{k \in \mathbb{Z} \mid k \equiv a \ (\mathrm{mod}\ q)\}}$ can be expressed as a linear combination of Dirichlet characters of modulus q.
(b) Show that in order to prove Theorem 14.19 it is enough to show that

$$\sum_{n \leqslant x} \chi(n)\Lambda(n) = o(x) \tag{14.25}$$

for any non-trivial Dirichlet character χ of modulus q.

Now fix a non-trivial Dirichlet character χ of modulus q.

Essential Exercise 14.22. Adapt Proposition 14.4 and deduce that it is sufficient to show that

$$\|f\|_{\chi} = \limsup_{h \to \infty} \left| \sum_{n \geqslant 1} \frac{\chi(n)\Lambda(n)}{n} f(\log n - h) \right|$$

vanishes for all $f \in C_c(\mathbb{R})$.

Essential Exercise 14.23. Show that $\| \cdot \|_{\chi}$ from Exercise 14.22 defines a semi-norm on $C_c(\mathbb{R})$ satisfying $\|f\|_{\chi} \leqslant \|f\|_1$ for every $f \in C_c(\mathbb{R})$.

We now study a 'twisted' version of ν_Λ from Section 14.2.3, namely

$$\nu_{\chi\Lambda} = \sum_{n=1}^{\infty} \frac{\chi(n)\Lambda(n)}{n} \delta_{\log n},$$

which is a complex-valued Radon measure on $[0, \infty)$.

Essential Exercise 14.24. Using the fact that $\lambda_h \nu_\Lambda \to m$ in the weak* topology as $h \to \infty$, show that for any sequence (h_n) with $h_n \to \infty$ as $n \to \infty$ there exists a subsequence (h_{n_k}) such that $\lambda_{h_{n_k}} \nu_{\chi\Lambda}$ converges in the weak* topology to $D_\chi \, dm$ for some density function D_χ taking values in the convex hull of the values of χ in $\mathbb{C}$.

Working towards the proof of the algebra inequality, we replace the second von Mangoldt function with the twisted version $\chi\Lambda_2$, which by Lemma 14.8 satisfies

$$\chi(n)\Lambda_2(n) = \sum_{d|n} \chi(d)\mu(d)\chi\left(\tfrac{n}{d}\right)\log^2\left(\tfrac{n}{d}\right) \tag{14.26}$$

$$= \chi(n)\Lambda(n)\log n + \sum_{d|n} \chi(d)\Lambda(d)\chi\left(\tfrac{n}{d}\right)\Lambda\left(\tfrac{n}{d}\right)$$

$$= \chi(n)\Lambda(n)\log n + (\chi\Lambda \underset{D}{*} \chi\Lambda)(n),$$

by Lemma 14.8. By using a complex analogue of Lemma 14.10, we can define $\nu_{\mathrm{sym}}^{\chi}$ by

$$\mathrm{d}\nu_{\mathrm{sym}}^{\chi} = \mathrm{d}\nu_{\chi\Lambda_2/\log} = \mathrm{d}\nu_{\chi\Lambda} + \frac{1}{t}\,\mathrm{d}(\nu_{\chi\Lambda} * \nu_{\chi\Lambda}),$$

where the second equality follows from the formula above, just as in the proof of Corollary 14.13.

Essential Exercise 14.25. (a) Show that

$$\frac{1}{y}\sum_{n\leqslant y} \chi(n)\log^2 n = \mathrm{O}\left(\frac{\log^2(1+y)}{y}\right)$$

for $y \geqslant 1$.
(b) Deduce the twisted version of the Selberg symmetry formula,

$$\sum_{n\leqslant x} \chi(n)\Lambda_2(n) = \mathrm{O}(x)$$

for $x \geqslant 1$.
(c) Show that $\lambda_h \nu_{\mathrm{sym}}^{\chi} \to 0$ in the weak* topology as $h \to \infty$.

Essential Exercise 14.26. Show that $\|f_1 * f_2\|_{\chi} \leqslant \|f_1\|_{\chi}\|f_2\|_{\chi}$ for all functions $f_1, f_2 \in C_c(\mathbb{R})$.

Essential Exercise 14.27. Show that Theorem 14.17 also holds in a similar way for the semi-norm $\|\cdot\|_{\chi}$.

Essential Exercise 14.28. Use Exercise 14.27 to prove Theorem 14.18 for the semi-norm $\|\cdot\|_{\chi}$ and $\xi \neq 0$.

It remains to establish the analogue of Theorem 14.18 for $\|\cdot\|_{\chi}$ and $\xi = 0$. In this case we previously used the full force of Mertens' theorem (Theorem 14.15). Here we replace this with the statement

$$\sum_{n\leqslant x} \frac{\chi(n)\Lambda(n)}{n} = \mathrm{O}(1) \tag{14.27}$$

for $x \geqslant 1$, which we prove in the following subsection (this is also due to Dirichlet).

Essential Exercise 14.29. Assuming (14.27), prove Theorem 14.18 for $\|\cdot\|_\chi$ and $\xi = 0$, and conclude the proof of Theorem 14.19.

14.5.1 Non-Vanishing of Dirichlet L-function at 1

In this section we will prove (14.27). This will require a brief excursion into the beginnings of analytic number theory; we refer to Serre [97] for more details. The tools needed are basic properties of Dirichlet series and the Abel summation formula. Following a convention going back to Riemann, we write $s = \sigma + it$ with $\sigma, t \in \mathbb{R}$ for any $s \in \mathbb{C}$. There are shorter proofs of (14.27) which do not use complex analysis; we refer, for example, to Tao [103] for the details.

Essential Exercise 14.30. Given a sequence (a_n) of complex numbers, associate to it a formal Dirichlet series $\sum_{n \geqslant 1} \frac{a_n}{n^s}$. Suppose that the set of $s \in \mathbb{C}$ for which the series converges absolutely is neither the empty set nor all of $\mathbb{C}$. Show that there exists some $\sigma_a \in \mathbb{R}$, the *abscissa of absolute convergence*, such that the series converges absolutely if $\sigma > \sigma_a$ but does not converge absolutely if $\sigma < \sigma_a$. Moreover, show that the convergence is uniform on compact subsets of $\{s \in \mathbb{C} \mid \Re(s) > \sigma_a\}$, so that the Dirichlet series defines a holomorphic function on this half-plane.

A function $\theta : \mathbb{N} \to \mathbb{C}$ is said to be *completely multiplicative* if

$$\theta(mn) = \theta(m)\theta(n)$$

for all $m, n \geqslant 1$ and is *multiplicative* if the same property holds for all $m, n \geqslant 1$ with $\gcd(m, n) = 1$. In particular, Dirichlet characters are completely multiplicative and the Möbius function is multiplicative.

Theorem 14.31 (Dirichlet L-functions). *Fix $q \geqslant 1$ and let χ be a Dirichlet character of modulus q. Define*

$$L(s, \chi) = \sum_{n \geqslant 1} \frac{\chi(n)}{n^s} = \sum_{n : \gcd(n,q)=1} \frac{\chi(n)}{n^s},$$

which converges absolutely for $\Re(s) > 1$.

(1) *If χ is non-trivial, then the series defining $L(s, \chi)$ converges uniformly on any compact subset of $H_+ = \{s \in \mathbb{C} \mid \Re(s) > 0\}$, so that the series for $L(s, \chi)$ defines a holomorphic function on the right half-plane H_+. Moreover,*

$$\left| L(1,\chi) - \sum_{n\leqslant x} \frac{\chi(n)}{n} \right| \ll x^{-1} \qquad (14.28)$$

for $x \geqslant 1$.

(2) *Writing χ_0 for the trivial character, $L(s,\chi_0)$ has a meromorphic extension to the half-plane H_+ which has a simple pole at $s = 1$, and is holomorphic on $H_+\setminus\{1\}$.*

(3) *If χ is a non-trivial character, then $L(1,\chi) \neq 0$.*

Using in particular the last statement of this theorem, we will be able to prove the remaining step for the proof of Theorem 14.19.

Corollary 14.32. *The bound (14.27) holds for any non-trivial Dirichlet character χ.*

The following is a rather simple but useful tool for our discussions.

Lemma 14.33 (Abel summation). *For any sequences (a_n) and (b_n),*

$$S_m = \sum_{n=1}^{m-1} A_n(b_n - b_{n+1}) + A_m b_m$$

where $A_m = \sum_{n=1}^{m} a_n$ and $S_m = \sum_{n=1}^{m} a_n b_n$, for all $m \geqslant 1$.

PROOF. Notice that $a_n = A_n - A_{n-1}$ for $n \geqslant 1$ and $A_0 = 0$, so that

$$S_m = \sum_{n=1}^{m}(A_n - A_{n-1})b_n = \sum_{n=1}^{m} A_n b_n - \sum_{n=1}^{m-1} A_n b_{n+1}.$$

$\square$

PROOF OF COROLLARY 14.32. We now use $L(1,\chi) \neq 0$ to deduce (14.27). We calculate

$$\sum_{n\leqslant x} \frac{\chi(n)\log n}{n} = \sum_{n\leqslant x} \frac{\chi(n)}{n} \sum_{n=dm} \Lambda(d) = \sum_{d\leqslant x} \frac{\chi(d)\Lambda(d)}{d} \underbrace{\sum_{m\leqslant \frac{x}{d}} \frac{\chi(m)}{m}}_{S_k},$$

where $S_k = \sum_{m\leqslant k} \frac{\chi(m)}{m}$ with $k = \lfloor \frac{x}{d} \rfloor$ is the partial sum appearing in Theorem 14.31(1). By (14.28) we then have $|S_k - L(1,\chi)| \ll \frac{1}{k}$. Substituting this into the expression above gives

$$\sum_{n \leqslant x} \frac{\chi(n) \log n}{n} = \sum_{d \leqslant x} \frac{\chi(d) \Lambda(d)}{d} \left(L(1, \chi) + O\left(\frac{d}{x}\right) \right)$$

$$= L(1, \chi) \sum_{d \leqslant x} \frac{\chi(d) \Lambda(d)}{d} + O\left(\frac{1}{x} \sum_{d \leqslant x} \Lambda(d) \right)$$

$$= L(1, \chi) \sum_{d \leqslant x} \frac{\chi(d) \Lambda(d)}{d} + O(1)$$

by (14.15).

We want to show that the left-hand side in the last calculation is also $O(1)$, as then (14.27) follows since $L(1, \chi) \neq 0$. For this we use Lemma 14.33 with $a_n = \chi(n)$ and $b_n = \frac{\log n}{n}$. Note that

$$A_m = \sum_{n=1}^{m} a_n$$

satisfies $|A_m| \leqslant \phi(q)$ since $\sum_{n=0}^{q-1} \chi(n) = 0$ and $\chi(n+q) = \chi(n)$ for all $n \in \mathbb{N}$. This gives

$$\left| \sum_{n \leqslant x} \frac{\chi(n) \log n}{n} \right| \leqslant \phi(q) \sum_{n=1}^{\ell-1} |b_n - b_{n+1}| + \phi(q) b_\ell,$$

where $\ell = \lfloor x \rfloor$. Since the sequence (b_n) is monotonically decreasing for sufficiently large n with limit 0, we obtain a telescoping sum and deduce that this term is indeed $O(1)$. As mentioned above, this finishes the proof since $L(1, \chi) \neq 0$. $\qquad \square$

We split the proof of Theorem 14.31 into several steps.

PROOF OF THEOREM 14.31(1) AND (2). Since $\left| \frac{\chi(n)}{n^s} \right| \leqslant \frac{1}{n^\sigma}$, the series defining $L(s, \chi)$ converges absolutely for $\sigma > 1$.

UNIFORM CONVERGENCE ON COMPACT SUBSETS OF H_+. Let χ be a nontrivial character, $K \subseteq H_+$ a compact subset, and choose σ_0 so as to ensure that $\Re(s) \geqslant \sigma_0 > 0$ for all $s \in K$. We apply Abel summation with $a_n = \chi(n)$ and $b_n = n^{-s}$ to obtain

$$\sum_{n=1}^{m} \frac{\chi(n)}{n^s} = S_m = \sum_{n=1}^{m-1} A_n \left(n^{-s} - (n+1)^{-s} \right) + A_m m^{-s}. \tag{14.29}$$

Note that $|m^{-s}| \leqslant m^{-\sigma_0} \to 0$ as $m \to \infty$. Since the derivative of $x \mapsto x^{-s}$ with respect to x is $-s x^{-s-1}$, we see that

$$\left| n^{-s} - (n+1)^{-s} \right| \leqslant R n^{-\sigma_0 - 1}$$

for all $s \in K$, where $R > 0$ satisfies $K \subseteq \overline{B_R^{\mathbb{C}}}$. Together with $|A_n| \leqslant \phi(q)$ for all $n \geqslant 1$ we see that (14.29) converges uniformly on K. This shows that

$$L(s, \chi) = \sum_{n=1}^{\infty} A_n \left(n^{-s} - (n+1)^{-s}\right)$$

is a holomorphic function on H_+, and setting $s = 1$ we also obtain

$$|L(1, \chi) - S_m| = \left| \sum_{n=m}^{\infty} A_n \left(n^{-1} - (n+1)^{-1}\right) - A_m m^{-1} \right| \ll m^{-1},$$

after again using the triangle inequality and $|A_n| \leqslant \phi(q)$, since the sum telescopes. This gives the claim in (14.28).

PROPERTIES OF $L(s, \chi_0)$. Let χ_0 be the trivial character of modulus q. It will be convenient to start by recalling some properties of the Riemann zeta function, defined for $\Re(s) > 1$ by

$$\zeta(s) = \sum_{n=1}^{\infty} \frac{1}{n^s}.$$

One easily see that this series converges absolutely for $\Re(s) > 1$, and so defines a holomorphic function there by Exercise 14.30. To obtain the extension to H_+ and the pole at $s = 1$, we write

$$\zeta(s) - \frac{1}{s-1} = \sum_{n=1}^{\infty} n^{-s} - \int_1^{\infty} x^{-s} \, dx = \sum_{n=1}^{\infty} \int_n^{n+1} \left(n^{-s} - x^{-s}\right) \, dx,$$

and as in the proof of the first part of the theorem, we see that the series on the right-hand side converges uniformly on any compact subset $K \subseteq H_+$.

Returning to the trivial Dirichlet character χ_0 of modulus q, we will see that the difference between $L(s, \chi_0)$ and ζ (or more precisely, their ratio) is relatively benign. Let $p_1, \ldots, p_\ell$ be the finite list of primes that divide q. Using unique factorization and

$$L(s, \chi_0) = \sum_{n:\gcd(n,q)=1} n^{-s}$$

we obtain

$$L(s, \chi_0)(1 - p_1^{-s})^{-1} \cdots (1 - p_\ell^{-s})^{-1} =$$

$$\left(\sum_{n:\gcd(n,q)=1} n^{-s} \right) \prod_{j=1}^{\ell} \left(1 + p_j^{-s} + p_j^{-2s} + \cdots\right) = \sum_{n \geqslant 1} n^{-s} = \zeta(s) \quad (14.30)$$

for $\Re(s) > 1$ by absolute convergence of all the series involved. Since the functions $s \mapsto (1 - p^{-s})^{\pm 1}$ are holomorphic on H_+, we may use the results for ζ above and deduce the same properties for $L(s, \chi_0)$. $\square$

The last part of the proof of Theorem 14.31 requires some facility with Dirichlet series provided by the following exercise and lemma.

Essential Exercise 14.34. Show that if $f(s) = \sum_{n \geqslant 1} \frac{a_n}{n^s}$ converges absolutely for $\Re(s) > \sigma_0$, then $-\sum_{n \geqslant 1} \frac{a_n \log n}{n^s}$ converges absolutely and uniformly on compact subsets of $\{s \in \mathbb{C} \mid \Re(s) > \sigma_0\}$, and $f'(s) = -\sum_{n \geqslant 1} \frac{a_n \log n}{n^s}$ there.

Lemma 14.35. *Let (a_n) be a real sequence with $a_n \geqslant 0$ for all $n \geqslant 1$ and suppose the Dirichlet series $f(s) = \sum_{n \geqslant 1} \frac{a_n}{n^s}$ converges for $\Re(s) > 1$. Suppose that f can be extended to a meromorphic function on H_+, also denoted f. Then either*

- $\sum_{n \geqslant 1} \frac{a_n}{n^s}$ *converges absolutely for $\Re(s) > 0$ and f is holomorphic on H_+,* or
- *there exists some $\sigma_0 > 0$ such that $\sum_{n \geqslant 1} \frac{a_n}{n^{\sigma_0}} = \infty$, $\sum_{n \geqslant 1} \frac{a_n}{n^s}$ converges absolutely for $\Re(s) > \sigma_0$, and f has a pole at σ_0.*

PROOF. Define
$$\sigma_0 = \inf \Big\{ \sigma > 0 \mid \sum_{n \geqslant 1} \frac{a_n}{n^\sigma} < \infty \Big\}.$$

By non-negativity of the coefficients a_n and monotonicity of $\sigma \mapsto n^{-\sigma}$, we see that the series $\sum_{n \geqslant 1} \frac{a_n}{n^s}$ converges absolutely for $\Re(s) > \sigma_0$, and so defines a holomorphic function there (see Exercise 14.30), which must therefore coincide with f. If $\sigma_0 = 0$ then we are in the first case of the lemma. If $\sigma_0 > 0$ and $\sum_{n \geqslant 1} \frac{a_n}{n^{\sigma_0}} = \infty$, then $f(s) = \sum_{n \geqslant 1} \frac{a_n}{n^s}$ for $\Re(s) > \sigma_0$. Moreover, it is easy to see (for example, using the monotone convergence theorem) by non-negativity of the coefficients that
$$\lim_{\sigma \searrow \sigma_0} f(\sigma) = \lim_{\sigma \searrow \sigma_0} \sum_{n \geqslant 1} \frac{a_n}{n^\sigma} = \sum_{n \geqslant 1} \frac{a_n}{n^{\sigma_0}} = \infty,$$

which shows that f must have a pole at σ_0.

It remains to rule out the possibility that $\sigma_0 > 0$ and $\sum_{n \geqslant 1} \frac{a_n}{n^{\sigma_0}} < \infty$. Assuming this is the case and using the assumption that f is meromorphic on H_+ we will deduce that $\sum_{n \geqslant 1} \frac{a_n}{n^{\sigma_0 - \varepsilon}} < \infty$ for some $\varepsilon \in (0, \sigma_0)$, which will be a contradiction to the definition of σ_0. Since the coefficients are non-negative we have
$$\sum_{n \geqslant 1} \frac{a_n}{n^{\sigma_0}} = \lim_{\sigma \searrow \sigma_0} \sum_{n \geqslant 1} \frac{a_n}{n^\sigma} = \lim_{\sigma \searrow \sigma_0} f(\sigma),$$

which shows that f must be holomorphic at σ_0 (since if σ_0 were to be a pole of f we would have $\lim_{\sigma \searrow \sigma_0} |f(\sigma)| = \infty$). By the same argument and Exercise 14.34 we then have

$$\sum_{n \geqslant 1} \frac{a_n (\log n)^k}{n^{\sigma_0}} = (-1)^k f^{(k)}(\sigma_0) < \infty$$

for all $k \geqslant 0$. By the Taylor expansion of f at σ_0, this gives for sufficiently small $\varepsilon > 0$ that

$$f(\sigma_0 - \varepsilon) = \sum_{k=0}^{\infty} \frac{1}{k!} \left(f^{(k)}(\sigma_0) \right) (-\varepsilon)^k = \sum_{k=0}^{\infty} \frac{1}{k!} \sum_{n \geqslant 1} \frac{a_n (\log n)^k}{n^{\sigma_0}} \varepsilon^k$$

$$= \sum_{n \geqslant 1} a_n \sum_{k=0}^{\infty} \frac{1}{k!} \left((-\log n)^k n^{-\sigma_0} (-\varepsilon)^k \right)$$

since all terms are again non-negative. The inner sum is precisely the Taylor expansion of $s \mapsto n^{-s}$ at σ_0 and so gives $n^{-(\sigma_0 - \varepsilon)}$ and hence

$$\sum_{n \geqslant 1} \frac{a_n}{n^{\sigma_0 - \varepsilon}} = f(\sigma_0 - \varepsilon) < \infty,$$

which contradicts the definition of σ_0. $\square$

PROOF OF THEOREM 14.31(3): NON-VANISHING OF $L(1, \chi)$. Summarizing the arguments above, we have obtained the meromorphic function

$$s \mapsto L(s, \chi_0)$$

on H_+ for the trivial Dirichlet character χ_0 with a simple pole at $s = 1$, and the holomorphic functions $s \mapsto L(s, \chi)$ on H_+ for any non-trivial Dirichlet character χ. If $L(1, \chi) = 0$ for some non-trivial Dirichlet character χ, then the function ζ_q defined by

$$\zeta_q(s) = \prod_{\chi} L(s, \chi)$$

would be holomorphic on H_+. Here the product is taken over all the characters.

We will see that ζ_q has a pole at 1 by using Euler product expansions. Unique factorization in the integers and complete multiplicativity of χ show that

$$L(s, \chi) = \sum_{n \geqslant 1} \frac{\chi(n)}{n^s} = \prod_p \left(1 - \frac{\chi(p)}{p^s} \right)^{-1} = \prod_{p : \gcd(p, q) = 1} \left(1 - \frac{\chi(p)}{p^s} \right)^{-1}$$

for $\Re(s) > 1$. This may be seen by extending the argument for (14.30) to all primes and using absolute convergence. Taking the product over all Dirichlet characters χ again gives the function

$$\zeta_q(s) = \prod_{p:\gcd(p,q)=1} \prod_\chi \left(1 - \frac{\chi(p)}{p^s}\right)^{-1}. \qquad (14.31)$$

Clearly the set of Dirichlet characters of modulus q forms a group with respect to pointwise multiplication, and for a fixed p with $\gcd(p,q) = 1$ the image of the homomorphism $\chi \mapsto \chi(p) \in \mathbb{S}^1 \subseteq \mathbb{C}$ is a subgroup consisting of all roots of unity of order $f(p)|\phi(q)$, and the kernel of this homomorphism has $g(p) = \frac{\phi(q)}{f(p)}$ elements. It follows that

$$\prod_\chi \left(1 - \chi(p)p^{-s}\right) = \prod_{k=1}^{f(p)} \left(1 - \omega_{f(p)}^k p^{-s}\right)^{g(p)} = \left(1 - p^{-f(p)s}\right)^{g(p)},$$

where $\omega_{f(p)}$ is a primitive $f(p)$th root of unity. Using this in the expression (14.31) for $\zeta_q(s)$ gives

$$
\begin{aligned}
\zeta_q(s) &= \prod_{p:\gcd(p,q)=1} \left(1 - p^{-f(p)s}\right)^{-g(p)} \\
&= \prod_{p:\gcd(p,q)=1} \left(1 + p^{-f(p)s} + p^{-2f(p)s} + \cdots\right)^{g(p)} \qquad (14.32) \\
&= \sum_{n:\gcd(n,q)=1} \frac{a_n}{n^s} = \sum_n \frac{a_n}{n^s}
\end{aligned}
$$

for $\Re(s) > 1$, where we expanded the Euler product once again into a convergent Dirichlet series with certain coefficients a_n. Notice that the precise form of (14.32) shows that $a_n \in \mathbb{N}_0$ for all $n \in \mathbb{N}$. By Lemma 14.35 the series $\sum_{n\geq 1} \frac{a_n}{n^s}$ either converges absolutely for $\Re(s) > 0$ and is holomorphic on H_+ or there exists some $\sigma_0 > 0$ such that ζ_q has a pole at σ_0. In the latter case it follows that $\sigma_0 = 1$ and $L(1, \chi) \neq 0$ for every non-trivial Dirichlet character $\chi \neq \chi_0$.

It remains to show that the former case cannot occur. To see this, notice that for any prime p with $\gcd(p,q) = 1$ and $\sigma > 0$ we have

$$\left(1 + p^{-f(p)\sigma} + p^{-2f(p)\sigma} + \cdots\right)^{g(p)} \geqslant \left(1 + p^{-\phi(q)\sigma} + p^{-2\phi(q)\sigma} + \cdots\right)$$

and hence under the assumption that $\zeta_q(\sigma) = \sum_{n\geq 1} \frac{a_n}{n^\sigma}$ converges for $\sigma > 0$,

$$\zeta_q(\sigma) = \sum_{n \geqslant 1} \frac{a_n}{n^\sigma} = \prod_{p:\gcd(p,q)=1} \left(1 + p^{-f(p)\sigma} + p^{-2f(p)\sigma} + \cdots\right)^{g(p)}$$

$$\geqslant \prod_{p:\gcd(p,q)=1} \left(1 + p^{-\phi(q)\sigma} + p^{-2\phi(q)\sigma} + \ldots\right) = L(\phi(q)\sigma, \chi_0) \quad (14.33)$$

also for all $\sigma > 0$. Using (14.33) we obtain

$$\lim_{\sigma \searrow \frac{1}{\phi(q)}} \zeta_q(\sigma) = \lim_{\sigma \searrow \frac{1}{\phi(q)}} L(\phi(q)\sigma, \chi_0) = \infty,$$

a contradiction to ζ_q being holomorphic on H_+. $\qquad\square$

In this chapter we have discussed a single result of great historical importance, which involved, in addition to functional analysis, some ideas from analytic number theory. For a thorough introduction to this important subject
we refer to the guided course in the work of Murty [77] and the monograph
of Iwaniec and Kowalski [49].

Appendix A: Set Theory and Topology

A.1 Set Theory and the Axiom of Choice

We will be using naive set theory, and in particular will use without specific reference the axioms of Zermelo–Fraenkel set theory with the axiom of choice (we refer to Kelley [51] for a good general source on all of the material in this appendix). This does require some caution. For example, it does not permit there to be a set that contains all sets, for if there were such a 'universal' set $\mathcal{V}$ then its subset $\mathcal{C} = \{A \in \mathcal{V} \mid A \notin A\}$ forces the statement $\mathcal{C} \in \mathcal{C} \iff \mathcal{C} \notin \mathcal{C}$, which is contradictory.

Here are some basic properties of sets that we will use without comment.

(1) A set will never contain itself.
(2) For every set $\mathcal{S}$ of sets there is a set $\bigcup_{A \in \mathcal{S}} A$, the union, containing all elements that are contained in some $A \in \mathcal{S}$.
(3) For every set A there is a power set $\mathbb{P}(A)$ containing all subsets of A.
(4) Any condition on the elements of a set can be used to define a new set, namely the subset of all elements that satisfy the condition.

Examples of sets include the empty set $\varnothing$, the natural numbers $\mathbb{N}$, the real numbers $\mathbb{R}$, the set of functions $\mathbb{R} \to \mathbb{C}$, which may also be written as $\mathbb{C}^{\mathbb{R}}$, and so on.

The following axiom of set theory is less intuitive than those above, but it plays a central role in analysis.

Axiom of Choice. *Suppose that A_i is a non-empty set for all $i \in I$. Then there is a function*

$$ f : I \longrightarrow \bigcup_{i \in I} A_i $$

with $f(i) \in A_i$ for all $i \in I$.

In other words, the Cartesian product $\prod_{i \in I} A_i$ (which by definition comprises all such functions) is non-empty.

© Springer International Publishing AG 2017
M. Einsiedler, T. Ward, *Functional Analysis, Spectral Theory, and Applications*,
Graduate Texts in Mathematics 276, DOI 10.1007/978-3-319-58540-6

While this axiom appears quite innocent (indeed, it appears almost obvious), it turns out to have a number of exotic consequences.[36] The axiom of choice has many equivalent formulations, one of which is Zorn's lemma, which is particularly useful in analysis. In order to state this, recall that a *partial order* on a set S is a relation $\preccurlyeq$ with the *reflexivity* property that $a \preccurlyeq a$ for all $a \in S$, the *transitivity* property that $a \preccurlyeq b, b \preccurlyeq c \implies a \preccurlyeq c$ for all $a, b, c \in S$, and the *anti-symmetry property* that $a \preccurlyeq b, b \preccurlyeq a \implies a = b$ for all $a, b \in S$. A partial order is a *linear order* if for every pair $a, b \in S$ we have either $a \preccurlyeq b$ or $b \preccurlyeq a$. A *maximal element* in a partially ordered set $(S, \preccurlyeq)$ is an element $m \in S$ for which $m \preccurlyeq a$ for some $a \in S$ implies that $a = m$.

Zorn's lemma. *Let $(S, \preccurlyeq)$ be a partially ordered set, and suppose that for every linearly ordered subset $L \subseteq S$ there exists an element $m \in S$ with $\ell \preccurlyeq m$ for all $\ell \in L$. Then there exists a maximal element $m \in S$.*

One might imagine setting out to prove Zorn's lemma inductively along the following lines. Starting with a single element (which certainly forms a linearly ordered set) one can build larger and larger linearly ordered subsets. If the current linearly ordered subset L has a maximal element, then it may also be a maximal element for S, in which case we are done. Otherwise, one can use the assumed property and add an element to L which is bigger than every element of L. Repeating this inductively (by transfinite induction, and noting that this procedure only ends once a maximal element in S is found), Zorn's lemma follows. However, in the course of the proof one has to make (potentially uncountably) many choices, and doing this carefully reveals that the argument needs the axiom of choice.

A.2 Basic Definitions in Topology

The notion of an *open set* is fundamental for defining continuity and convergence.

Definition A.1. Let X be a set. A family $\mathcal{T} \subseteq \mathbb{P}(X)$ of subsets of X is called a *topology* on X if

- $\varnothing, X \in \mathcal{T}$;
- if $O_1, O_2 \in \mathcal{T}$ then $O_1 \cap O_2 \in \mathcal{T}$;
- if $O_i \in \mathcal{T}$ for all $i \in I$, where I is an arbitrary index set, then $\bigcup_{i \in I} O_i \in \mathcal{T}$.

The pair $(X, \mathcal{T})$ is called a *topological space*. The elements of a topology are called *open sets* and a set $A \subseteq X$ with $X \smallsetminus A \in \mathcal{T}$ is called *closed*. A set that is both open and closed is called a *clopen* set.

Given a point x in a topological space, a *neighbourhood* of x is a set V containing an open set U that contains x. We will often want to assume that neighbourhoods are open sets, in which case we will speak of *open neighbour-hoods*. A topological space is called *Hausdorff* if for any points $x_1 \neq x_2$ in X there exist neighbourhoods U_1 of x_1 and U_2 of x_2 such that $U_1 \cap U_2 = \varnothing$.

Many of the topological spaces that we will study are particularly well-behaved ones arising from a metric.

Definition A.2. A function $\mathsf{d} : X \times X \to \mathbb{R}$ is called a *metric* if it satisfies the following properties:

- *(strict positivity)* $\mathsf{d}(x, y) \geqslant 0$ and $\mathsf{d}(x, y) = 0$ if and only if $x = y$, for all $x, y \in X$;
- *(symmetry)* $\mathsf{d}(x, y) = \mathsf{d}(y, x)$ for all $x, y \in X$;
- *(triangle inequality)* $\mathsf{d}(x, y) \leqslant \mathsf{d}(x, z) + \mathsf{d}(z, y)$ for all $x, y, z \in X$.

The pair (X, d) is called a *metric space*. A set $O \subseteq X$ in a metric space is called *open* if for any $x \in O$ there is some $\varepsilon > 0$ such that

$$B_\varepsilon(x) = \{y \in X \mid \mathsf{d}(x, y) < \varepsilon\} \subseteq O.$$

The set $B_\varepsilon(x)$ is called an open ε-ball around x.

It is easy to check that the collection of all open sets in a metric space defines a topology on the metric space. If instead of strict positivity we only have

- *(positivity)* $\mathsf{d}(x, y) \geqslant 0$ for all $x, y \in X$

then we say that d is a *pseudo-metric*; this also gives rise to a topology in the same way.

Definition A.3. A function $f : X \to Y$ between topological spaces $(X, \mathcal{T}_X)$ and $(Y, \mathcal{T}_Y)$ is *continuous* if $f^{-1}(O) \in \mathcal{T}_X$ for all $O \in \mathcal{T}_Y$.

Definition A.4. Let X be a set and suppose that $\mathcal{T}_1$ and $\mathcal{T}_2$ are two topologies on X. If the identity map $I : X \to X$ viewed as a map from $(X, \mathcal{T}_1)$ to $(X, \mathcal{T}_2)$ is continuous (which means that $\mathcal{T}_2 \subseteq \mathcal{T}_1$), then $\mathcal{T}_2$ is said to be *weaker* or *coarser* than $\mathcal{T}_1$, and $\mathcal{T}_1$ is called *stronger* or *finer* than $\mathcal{T}_2$.

As is well-known from analysis, we say that a sequence (x_n) in a Hausdorff topological space X *converges to* x, written $\lim_{n \to \infty} x_n = x$, if for every neighbourhood U of x there exists some N such that $x_n \in U$ for all $n \geqslant N$. While this notion is sufficient for metric spaces, for more general topological spaces we also need the notions of *filters* and *convergent filters* (see also Exercise A.12). If the topology is given by a metric d, then this is equivalent to the property that for any $\varepsilon > 0$ there is some N such that $n \geqslant N \implies \mathsf{d}(x_n, x) < \varepsilon$. Sufficiency means, for example, that we can characterize continuity for functions between metric spaces using convergence of sequences.

Definition A.5. Let X be a set. A family $\mathcal{F} \subseteq \mathbb{P}(X)$ of subsets of X is a *filter* if

- $X \in \mathcal{F}$ but $\varnothing \notin \mathcal{F}$;
- if $F_1, F_2 \in \mathcal{F}$ then $F_1 \cap F_2 \in \mathcal{F}$; and
- if $F \in \mathcal{F}$ and $F \subseteq B \subseteq X$, then $B \in \mathcal{F}$.

Example A.6. (a) Let $(X, \mathcal{T})$ be a topological space and $x \in X$. Then

$$\mathcal{U}_x = \{U \in \mathbb{P}(X) \mid \text{ there exists some } O \in \mathcal{T} \text{ with } x \in O \subseteq U\}$$

is a filter, called the *neighbourhood filter of* x.
(b) Let $X = \mathbb{N}$ and set

$$\mathcal{F}_\infty = \{B \subseteq \mathbb{N} \mid \text{there exists some } N \in \mathbb{N} \text{ with } n \in B \text{ for all } n \geqslant N\}.$$

Then $\mathcal{F}_\infty$ is a filter, called the *tail filter* of $\mathbb{N}$.
(c) While this is not needed here, we mention that a directed set (as in the definition of nets) gives rise to a generalization of tail filters.

Definition A.7. Let $\mathcal{F}_1, \mathcal{F}_2 \subseteq \mathbb{P}(X)$ be filters on a set X. Then $\mathcal{F}_1$ is *finer* than $\mathcal{F}_2$, or $\mathcal{F}_2$ is *coarser* than $\mathcal{F}_1$, if $\mathcal{F}_1 \supseteq \mathcal{F}_2$.

Definition A.8. Let X be a Hausdorff topological space, and let $\mathcal{F} \subseteq \mathbb{P}(X)$ be a filter. We say that $\mathcal{F}$ *converges* to $x \in X$, written $x = \lim \mathcal{F}$, if $\mathcal{F}$ is finer than the neighbourhood filter $\mathcal{U}_x$.

Exercise A.9. Let X be a Hausdorff topological space, and let $\mathcal{F} \subseteq \mathbb{P}(X)$ be a filter. Show that the limit $\lim \mathcal{F}$ is unique if it exists.

Definition A.10. Let M be a set, $\mathcal{F} \subseteq \mathbb{P}(M)$ a filter, X a topological space, and $f : M \to X$ a map. We say that f *converges along* $\mathcal{F}$ *to* $x \in X$, written as $\lim_{\mathcal{F}} f = x$, if the image filter

$$f(\mathcal{F}) = \{B \subseteq X \mid \text{there exists some } A \in \mathcal{F} \text{ with } f(A) \subseteq B\}$$

is finer than $\mathcal{U}_x$ (that is, the image filter converges to x).

Exercise A.11. Let $M = \mathbb{N}$, let X be a Hausdorff topological space, and let the function $f : \mathbb{N} \to X$ correspond to a sequence $(f(n))$. Show that the sequence $(f(n))$ converges in the usual sense if and only if f converges along $\mathcal{F}_\infty$, and in this case

$$\lim_{n \to \infty} f(n) = \lim_{\mathcal{F}_\infty} f,$$

where $\mathcal{F}_\infty$ is the tail filter from Example A.6(b).

Exercise A.12. Let X, Y be Hausdorff topological spaces, and let $f : X \to Y$ be a map. Show that f is continuous if and only if for all $x \in X$ we have $\lim_{\mathcal{U}_x} f = f(x)$, where $\mathcal{U}_x$ is the neighbourhood filter from Example A.6(a).

A more direct generalization of sequences and their convergence properties is given by nets. These are also known as Moore–Smith sequences, and while they are an intuitive route to generalize sequences, they are a less natural starting point for the important notion of ultrafilters (see Definition A.23).

To define nets we first need to define the directed sets that will replace the use of $\mathbb{N}$ and its order $\leqslant$ that give the domain of a sequence. A *directed set* is a set D together with a binary relation $\lesssim$ that satisfies the following properties:

- (Reflexivity) We have $n \lesssim n$ for all $n \in D$.
- (Transitivity) If $\ell \lesssim m$ and $m \lesssim n$ for some $\ell, m, n \in D$, then $\ell \lesssim n$.
- (Filter property) If $\ell_1, \ell_2 \in D$, then there exists some $n \in D$ with $\ell_1 \lesssim n$ and $\ell_2 \lesssim n$.

Exercise A.13. Show that Example A.6(b) can be generalized as follows. For any directed set $(D, \lesssim)$ the set

$$\{B \subseteq D \mid \text{there exists some } \ell \in D \text{ such that } \{n \in D \mid \ell \lesssim n\} \subseteq B\}$$

forms a filter: the *tail filter* of $(D, \lesssim)$.

A function $f : D \to X$ whose domain D is a directed set taking values in a topological space is called a *net*. A net $f : D \to X$ *converges* to $x_0 \in X$ if for every neighbourhood $U \ni x_0$ of x_0 there exists some $\ell_0 \in D$ such that $n \in D$ and $\ell_0 \lesssim n$ implies that $f(n) \in U$.

Exercise A.14. (a) Show that a net $f : D \to X$ converges to $x_0 \in X$ if and only if f converges to x_0 along the tail filter of $(D, \lesssim)$.
(b) Conversely, given a function $f : M \to X$ on some set M taking values in a topological space X and a filter $\mathcal{F} \subseteq \mathbb{P}(M)$, show how to define a directed set $(D, \lesssim)$ and a net so that the net converges to $x_0 \in X$ if and only if f converges to $x_0 \in X$ along $\mathcal{F}$.

A.3 Inducing Topologies

If $(X, \mathcal{T})$ is a topological space and $Y \subseteq X$ is any subset, then the topology on Y induced from the topology on X is the weakest topology on Y for which the identity inclusion map $Y \hookrightarrow X$ is continuous. Equivalently, the induced topology on Y is $\{Y \cap O \mid O \in \mathcal{T}\}$.

Suppose that $f : X \to Y$ is a map between two sets. If Y has a topology $\mathcal{T}_Y$, then there is a *weakest* topology on X which makes f continuous. This topology is given by $f^{-1}(\mathcal{T}_Y) = \{f^{-1}(O) \mid O \in \mathcal{T}_Y\}$. If on the other hand X has a topology $\mathcal{T}_X$, then there is a *strongest* topology on Y which makes f continuous. It is given by $\{O \subseteq Y \mid f^{-1}(O) \in \mathcal{T}_X\}$. The former case has an important generalization as follows.

Definition A.15. Let X be a set, and let $f_i : X \to Y_i$ for $i \in I$ be a family of maps from X to topological spaces $(Y_i, \mathcal{T}_{Y_i})$. Then the *initial* (or *weak*, or *limit*, or *projective*) topology induced by these maps is the weakest topology for which all of the maps are continuous.

The open sets in the initial topology are arbitrary unions of finite intersections of elements of $f_\imath^{-1}(\mathcal{T}_{Y_\imath})$ for various $\imath \in I$. The initial topology can also be characterized by the following universal property. A function $g : Z \to X$ is continuous if and only if $f_\imath \circ g : Z \to Y_\imath$ is continuous for each $\imath \in I$.

A particular case of the initial topology is the product topology.

Definition A.16. Suppose that $(Y_\imath, \mathcal{T}_\imath)$ for $\imath \in I$ is a collection of topological spaces. Define

$$X = \prod_{\imath \in I} Y_\imath.$$

The *product topology* on X is the initial topology induced by the projection maps

$$\pi_\jmath : (y_\imath)_{\imath \in I} \longmapsto y_\jmath$$

from X to $Y_\jmath$ for all $\jmath \in I$.

Another case is given by the topology generated by a family of topologies. Suppose that X is a set, and for all $\imath \in I$ we have a topology $\mathcal{T}_\imath$ on X. Then we may consider the identity map $I : X \to X$ as a map from X to the topological space $(X, \mathcal{T}_\imath)$ for each $\imath \in I$, and associate to X the weakest topology that is finer than all the topologies $\mathcal{T}_\imath$ for $\imath \in I$.

Notice that the product topology, or the weakest topology that is finer than a given family of topologies, may not be metric (that is, derived from a metric) even if the original topologies were metric. However, there is a special situation in which the metric property is preserved by taking products.

Lemma A.17. *Let X be a set and suppose that $\mathsf{d}_n : X \times X \to \mathbb{R}$ is a sequence of pseudo-metrics. Then the weakest topology that is finer than the topologies induced by d_n for $n \in \mathbb{N}$ is itself induced by a pseudo-metric. In particular, the countable product of metric spaces is a metric space in the product topology.*

PROOF. For the main part of the argument it is important to know that we may assume that d_n only takes on values in $[0, 1)$. To see this, we claim that if d_n is any pseudo-metric then $\overline{\mathsf{d}_n} = \frac{\mathsf{d}_n}{1 + \mathsf{d}_n}$ is a pseudo-metric that defines the same topology as d_n does.

Positivity and symmetry of $\overline{\mathsf{d}_n}$ are clear since they hold for d_n. Hence it is enough to check the triangle inequality for $\overline{\mathsf{d}_n}$. For this, notice first that the function $u \mapsto \frac{u}{1+u}$ maps from $[0, \infty)$ to $[0, 1)$, is monotone increasing and satisfies

$$\frac{u+v}{1+u+v} \leqslant \frac{u}{1+u} + \frac{v}{1+v} \tag{A.1}$$

for $u, v \in [0, \infty)$. The inequality (A.1) follows from the inequality

$$(u + v)(1 + u)(1 + v) = (u + v)(1 + u + v + uv)$$
$$\leqslant (1 + u + v)(u + v + 2uv)$$
$$= (1 + u + v)\left(u(1 + v) + v(1 + u)\right)$$

after dividing by $(1 + u + v)(1 + u)(1 + v)$. It follows that if $x, y, z \in X$, then

$$\overline{d_n}(x, y) = \frac{d_n(x, y)}{1 + d_n(x, y)} \leqslant \frac{d_n(x, z) + d_n(z, y)}{1 + d_n(x, z) + d_n(z, y)}$$
$$\leqslant \frac{d_n(x, z)}{1 + d_n(x, z)} + \frac{d_n(z, y)}{1 + d_n(z, y)} = \overline{d_n}(x, z) + \overline{d_n}(z, y),$$

as required. It is clear that $d_n(x, y) < \varepsilon$ for $x, y \in X$ implies that $\overline{d_n}(x, y) < \varepsilon$. For the converse, notice that $\overline{d_n}(x, y) < \frac{\varepsilon}{1+\varepsilon}$ implies that $d_n(x, y) < \varepsilon$ for all $\varepsilon > 0$ and $x, y \in X$, since $u \mapsto \frac{u}{1+u}$ is strictly monotonely increasing. This implies that d_n and $\overline{d_n}$ define the same open sets.

So suppose that $d_n : X \times X \to [0, 1)$ is a pseudo-metric for each $n \geqslant 1$. We define

$$d(x, y) = \sum_{n=1}^{\infty} \frac{1}{2^n} d_n(x, y).$$

Since this sum converges on $X \times X$, it defines another pseudo-metric on X. We claim that the topology induced by d is precisely the weakest topology that is finer than all the topologies induced by d_n for $n \geqslant 1$.

Suppose first that $O \subseteq X$ is an open set with respect to d, and let $x \in O$. By definition there exists an $\varepsilon > 0$ with

$$B_\varepsilon^d(x) = \{y \in X \mid d(x, y) < \varepsilon\} \subseteq O.$$

Now choose N with $\sum_{n=N+1}^{\infty} \frac{1}{2^n} < \frac{\varepsilon}{2}$. Then

$$\bigcap_{n=1}^{N} B_{\varepsilon/2N}^{d_n}(x) \subseteq B_\varepsilon^d(x) \subseteq O$$

since if $y \in X$ satisfies $d_n(y, x) < \frac{\varepsilon}{2N}$ for $n = 1, \ldots, N$ then

$$d(x, y) \leqslant \sum_{n=1}^{N} d_n(x, y) + \sum_{n=N+1}^{\infty} \frac{1}{2^n} < \varepsilon.$$

As this holds for all $x \in O$, we see that O is a union of finite intersections of sets that are open with respect to the topology induced by d_n.

The converse is similar. Suppose O is a union of finite intersections of sets that are open with respect to d_n. Let $x \in O$ and suppose that

$$x \in \bigcap_{n=1}^{N} O_n \subseteq O,$$

where O_n is open with respect to d_n for $n = 1, \ldots, N$. Then we may as well assume $O_n = B_\varepsilon^{\mathsf{d}_n}(x)$ for some $\varepsilon > 0$. We claim that

$$B_{\varepsilon/2^N}^{\mathsf{d}}(x) \subseteq \bigcap_{n=1}^{N} O_n \subseteq O,$$

which then implies that O is open with respect to the pseudo-metric d. So suppose $y \in X$ satisfies

$$\sum_{n=1}^{\infty} \frac{1}{2^n} \mathsf{d}_n(y, x) = \mathsf{d}(y, x) < \frac{\varepsilon}{2^N},$$

then $\frac{1}{2^n} \mathsf{d}_n(y, x) < \frac{\varepsilon}{2^N}$ and for $n \in \mathbb{N}$ with $1 \leqslant n \leqslant N$ this implies that $\mathsf{d}_n(y, x) < \varepsilon$, hence $y \in O_n$ and so the claim. The first part of the lemma follows.

Now suppose that

$$X = \prod_{n=1}^{\infty} X_n$$

where each (X_n, d_n) is a metric space, and we define

$$\mathsf{d}\big((x_n), (y_n)\big) = \sum_{n=1}^{\infty} \frac{1}{2^n} \frac{\mathsf{d}_n(x_n, y_n)}{1 + \mathsf{d}_n(x_n, y_n)}.$$

Then d is a pseudo-metric by the argument above. However,

$$\mathsf{d}\big((x_n), (y_n)\big) = 0 \implies \mathsf{d}_n(x_n, y_n) = 0$$

for all $n \geqslant 1$, which implies that $(x_n) = (y_n)$, so d is a metric on X. The topology induced by the pseudo-metric

$$((x_k)_k, (y_k)_k) \longmapsto \mathsf{d}_n(x_n, y_n)$$

is precisely the weakest topology for which the projection to X_n is continuous. By the first part of the lemma, this shows that the topology induced by d is the weakest topology for which all the projections are continuous, so d induces the product topology. $\qquad\square$

A.4 Compact Sets and Tychonoff's Theorem

Compactness is a fundamental notion for all of analysis, and in particular for functional analysis. It plays a role in analysis a little like finiteness does in combinatorics.

Definition A.18. Let $(X, \mathcal{T})$ be a Hausdorff topological space. A family of sets $\mathcal{U}$ is called an *open cover* if $\mathcal{U}$ consists of open sets and

$$X \subseteq \bigcup_{O \in \mathcal{U}} O.$$

The space $(X, \mathcal{T})$ is called *compact* if every open cover has a *finite subcover*, that is, a finite subset $\mathcal{V} \subseteq \mathcal{U}$ which is also an open cover.

An alternative and equivalent condition for compactness can be given in terms of closed sets. A collection of sets $\{A_\imath \mid \imath \in I\}$ has the *finite intersection property* if

$$\bigcap_{\ell=1}^{k} A_{\imath_\ell} \neq \varnothing$$

for any finite subset $\{\imath_1, \dots, \imath_k\} \subseteq I$, and has the *infinite intersection property* if

$$\bigcap_{\imath \in I} A_\imath \neq \varnothing.$$

Then a Hausdorff topological space $(X, \mathcal{T})$ is compact if and only if every family of closed sets with the finite intersection property also has the infinite intersection property.

Recall that a metric space (X, d) is called *complete* if every sequence (x_n) with the Cauchy property that for every $\varepsilon > 0$ there is some $N = N(\varepsilon)$ for which

$$m, n \geqslant N \implies \mathsf{d}(x_m, x_n) < \varepsilon$$

is convergent, meaning that there is some $x^* \in X$ with the property that for any $\varepsilon > 0$ there is some $N = N(\varepsilon)$ such that

$$n \geqslant N \implies \mathsf{d}(x_n, x^*) < \varepsilon.$$

For metric spaces there are further equivalent properties characterizing compactness.

- A metric space (X, d) is *sequentially compact* if any sequence (x_n) in X has a convergent subsequence.
- A metric space (X, d) is compact if and only if it is complete and *totally bounded*, meaning that for every $\varepsilon > 0$ there is a finite set of points $\{x_1, \dots, x_n\}$ in X with

$$X = \bigcup_{i=1}^{n} B_\varepsilon(x_i).$$

Exercise A.19. Recall the proofs that the different notions of compactness coincide for metric spaces.

Compactness is closed under taking products in the following sense.

Theorem A.20 (Tychonoff). *Let I be an index set, and suppose for $\imath \in I$ that $(X_\imath, \mathcal{T}_\imath)$ is a compact topological space. Then $\prod_{\imath \in I} X_\imath$ is compact with respect to the product topology.*

The notion of compactness has many useful extensions and generalizations. We will only need two of these.

Definition A.21. A Hausdorff topological space is called *locally compact* if every point has a neighbourhood which is compact in the induced topology. A topological space is called *σ-compact* if it can be written as $\bigcup_{n=1}^{\infty} K_n$ with each K_n compact in the induced topology.

Lemma A.22. *Let X be a locally compact space. Then for every compact subset $K \subseteq X$ there exists an open subset $O \subseteq X$ with compact closure that contains K. If X is in addition σ-compact, then there exists a sequence of compact sets (K_n) such that $X = \bigcup_{n=1}^{\infty} K_n$ and $K_n \subseteq K_{n+1}^{o}$ for all $n \geqslant 1$.*

PROOF. Let $K \subseteq X$ be compact. Since X is locally compact, any $x \in K$ has an open neighbourhood U_x with compact closure. Applying compactness to the open cover $\{U_x \mid x \in K\}$ we get $K \subseteq O = U_{x_1} \cup \cdots \cup U_{x_m}$ for some finite collection of points $x_1, \ldots, x_m \in K$. By construction, O is open with compact closure.

Suppose now X is σ-compact. Then there exists a sequence of compact sets (Q_n) with $X = \bigcup_{n=1}^{\infty} Q_n$. We first define $K_1 = Q_1$ and then construct K_n inductively as follows. Suppose $K_n \supseteq Q_n$ has already been constructed. Applying the above argument to K_n gives some open set O_n with compact closure that contains K_n. Now define $K_{n+1} = \overline{O_n} \cup Q_{n+1}$. The sequence constructed in this way satisfies all the desired properties. $\square$

Compactness can also be characterized in terms of filters, and for this another notion is useful.

Definition A.23. Let X be a set and $\mathcal{F} \subseteq \mathbb{P}(X)$ a filter. Then $\mathcal{F}$ is an *ultrafilter* if for every $B \subseteq X$ we have $B \in \mathcal{F}$ or $X \setminus B \in \mathcal{F}$.

Proposition A.24. *Let X be a Hausdorff topological space. Then the following are equivalent.*

(1) X is compact.

(2) *Every filter on X has a finer filter that converges to some $x \in X$.*
(3) *Every ultrafilter converges.*

The implication (3) $\implies$ (1) uses the axiom of choice in the form of Zorn's lemma (to show that any filter is contained in an ultrafilter; see Exercise A.25).

Exercise A.25. (a) Use Zorn's lemma to show that every filter has a finer filter that is an ultrafilter.
(b) Prove Proposition A.24.
(c) Use Proposition A.24 to prove Tychonoff's theorem.

A.5 Normal Spaces

A circle of useful constructions concerns ways to approximate functions with continuous functions. The appropriate level of generality is provided by *normal spaces*; as the name suggests, many of the topological spaces that arise in mathematics have this property (in particular, any metric space is normal and any compact space is normal).

Definition A.26. A topological space $(X, \mathcal{T})$ is said to be *normal* if for any closed sets A, B in X with $A \cap B = \varnothing$ there are open sets $U \supseteq A$ and $V \supseteq B$ with $U \cap V = \varnothing$.

This definition, which says that disjoint closed sets can be separated by open sets, may be thought of as requiring that there are 'enough' open sets. An important consequence is that there are 'enough' continuous functions in the following sense (this presentation is taken from Tao's blog [103]).

Lemma A.27 (Urysohn's lemma). *Let $(X, \mathcal{T})$ be a topological space. Then the following properties of X are equivalent.*

(1) *X is a normal space.*
(2) *For every closed set $K \subseteq X$ and every open set $U \supseteq K$, there is an open set V and a closed set L with $U \supseteq L \supseteq V \supseteq K$.*
(3) *For every pair of closed sets K and L in X with $K \cap L = \varnothing$, there exists a continuous function $f : X \to [0,1]$ with*

$$f(x) = \begin{cases} 1 & \text{if } x \in K, \\ 0 & \text{if } x \in L. \end{cases}$$

(4) *For every closed set $K \subseteq X$ and every open set $U \supseteq K$, there exists a continuous function $f : X \to [0,1]$ with $\mathbb{1}_K \leqslant f \leqslant \mathbb{1}_U$.*

For metric spaces the proof of the difficult step below is rather simple: Given two disjoint closed sets K and L we can define the function f as in (3) by

$$f(x) = \frac{\mathsf{d}(x, L)}{\mathsf{d}(x, K) + \mathsf{d}(x, L)}$$

for $x \in X$, where $\mathsf{d}(\cdot, K)$ and $\mathsf{d}(\cdot, L)$ are the continuous distance functions defined in (2.29).

PROOF. The implications (3) $\iff$ (4) and (1) $\iff$ (2) are clear, since a set is closed if and only if its complement is open.

Assume now that (3) holds. Given disjoint closed sets $K, L \subseteq X$, let f be the function given by (3). Then the open sets $U = \{x \in X \mid f(x) > 0.9\}$ and $V = \{x \in X \mid f(x) < 0.1\}$ show (1).

Assume next that (2) holds, let $K = K_1$ be a closed set, and let $U = U_0$ be an open set with $K_1 \subseteq U_0$. By (2), we can find a closed set $K_{1/2}$ and an open set $U_{1/2}$ with

$$U_0 \supseteq K_{1/2} \supseteq U_{1/2} \supseteq K_1.$$

Applying (2) again twice gives closed sets $K_{1/4}, K_{3/4}$ and open sets $U_{1/4}, U_{3/4}$ with

$$U_0 \supseteq K_{1/4} \supseteq U_{1/4} \supseteq K_{1/2} \supseteq U_{1/2} \supseteq K_{3/4} \supseteq U_{3/4} \supseteq K_1.$$

Continuing in exactly the same way and setting $K_0 = X$ and $U_1 = \varnothing$, we construct for every rational $q \in D = \{\frac{a}{2^n} \mid n \geqslant 0, a \in \mathbb{Z}, 0 \leqslant a \leqslant 2^n\}$ a closed set K_q and an open set U_q with $K_q \supseteq U_q$ for $q \in D$ and with $U_{q_1} \supseteq K_{q_2}$ for all $q_1, q_2 \in D$ with $q_1 < q_2$. Now define

$$f(x) = \begin{cases} 0 & \text{for } x \notin U_0, \\ \sup\{q \in D \mid x \in U_q\} & \text{otherwise.} \end{cases}$$

Notice that we also have

$$f(x) = \begin{cases} 1 & \text{for } x \in K_1, \\ \inf\{q \in D \mid x \notin K_q\} & \text{otherwise.} \end{cases}$$

It is easy to check that

$$f^{-1}((s, \infty)) = \{x \in X \mid f(x) > s\} = \bigcup_{q > s} U_q$$

for any $s \geqslant 0$ and $f^{-1}((s, \infty)) = X$ for $s < 0$. Similarly,

$$f^{-1}((-\infty, s)) = \{x \in X \mid f(x) < s\} = \bigcup_{q < s} X \setminus K_q$$

for $s \leqslant 1$ and $f^{-1}((-\infty, s)) = X$ for $s > 1$. Hence both $f^{-1}((s, \infty))$ and $f^{-1}((-\infty, s))$ are open sets for any real s, so f is continuous and (4) follows. □

For a continuous function f on a topological space X with values in $\mathbb{R}$, $\mathbb{C}$, or a vector space, we define the *support* of f to be

$$\operatorname{Supp} f = \overline{\{x \in X \mid f(x) \neq 0\}}$$

and say that f is *supported on a subset* $Y \subseteq X$ if $\operatorname{Supp} f \subseteq Y$.

Lemma A.28 (Partition of unity). *Let X be a normal topological space, and assume that $\{K_\alpha \mid \alpha \in A\}$ is a collection of closed sets covering X, and $\{U_\alpha \mid \alpha \in A\}$ is an open cover of X with the property that $U_\alpha \supseteq K_\alpha$ for each $\alpha \in A$. Suppose further that each $x \in X$ has an open neighbourhood that non-trivially intersects only finitely many U_α. Then for each $\alpha \in A$ there exists a continuous function $f_\alpha : X \to [0, 1]$ supported on U_α such that*

$$\sum_{\alpha \in A} f_\alpha(x) = 1$$

for all $x \in X$.

PROOF. By Urysohn's lemma (Lemma A.27), there exists for each $\alpha \in A$ an open set V_α containing K_α with $\overline{V_\alpha} \subseteq U_\alpha$. Furthermore, there is a continuous function $g_\alpha : X \to [0, 1]$ which is equal to 1 on K_α and equal to 0 on $X \setminus V_\alpha$. In particular, g_α is supported on U_α. Then $g = \sum_{\alpha \in A} g_\alpha$ is well-defined (by the finite intersection property) and is bounded below by 1. Setting $f_\alpha = g_\alpha / g$ for $\alpha \in A$ gives the result. □

Proposition A.29 (Tietze's extension theorem). *Let X be a normal topological space, $A \subseteq X$ a closed subset, and let $f : A \to \mathbb{R}$ (or $\mathbb{C}$) be a bounded continuous function. Then there exists a bounded continuous function $F : X \to \mathbb{R}$ (or $\mathbb{C}$) with $F|_A = f$. If in addition S is locally compact and A is compact, then we can find such an extension F in $C_c(X)$.*

We note that the assumption of boundedness is not essential, but does simplify the proof and is sufficient for our purposes.

PROOF OF PROPOSITION A.29. If f is complex-valued then we may use the following argument for $\Re(f)$ and $\Im(f)$ separately, so it is enough to consider the real-valued case. If $|f(x)| \leqslant M$ for all $x \in X$ then we may also apply the following argument to $\frac{1}{M} f$, so we may assume without loss of generality that f is a continuous function from A to $[-1, 1]$.

Define sets $B_- = f^{-1}([-1, -\frac{1}{3}])$ and $B_+ = f^{-1}([\frac{1}{3}, 1])$. By definition and by continuity of f, $B_- \subseteq A$ and $B_+ \subseteq A$ are disjoint closed sets. By Urysohn's lemma (Lemma A.27) there exists a continuous function $g : X \to [0, 1]$ with $g|_{B_-} = 0$ and $g|_{B_+} = 1$. Define $h_1 = \frac{2}{3}(g - \frac{1}{2})$. We claim that $\|f - h_1|_A\|_\infty \leqslant \frac{2}{3}$ by considering each possibility in turn:

- If $x \in B_-$ then $f(x) \in [-1, -\frac{1}{3}]$ and $h_1(x) = -\frac{1}{3}$, so $|f(x) - h_1(x)| \leqslant \frac{2}{3}$.
- If $x \in B_+$ then $f(x) \in [\frac{1}{3}, 1]$ and $h_1(x) = \frac{1}{3}$, so $|f(x) - h_1(x)| \leqslant \frac{2}{3}$.
- Finally, if $x \in A \smallsetminus (B_- \cup B_+)$, then $f(x) \in (-\frac{1}{3}, \frac{1}{3})$ and $|h_1(x)| \leqslant \frac{1}{3}$, so $|f(x) - h_1(x)| \leqslant \frac{2}{3}$ again.

We interpret the argument above as follows. Every continuous function

$$f : A \to [-1, 1]$$

has an approximation $h_1|_A$ which is the restriction of a continuous function $h_1 : X \to [-1, 1]$ to A, with $\|f - h_1|_A\|_\infty \leqslant \frac{2}{3}$. Applying this general statement to $f_2 = \frac{3}{2}(f - h_1|_A)$ we find some continuous function $h_2 : X \to [-1, 1]$ with $\|f_2 - h_2|_A\|_\infty \leqslant \frac{2}{3}$ or, equivalently, with

$$\left\| f - \left(h_1 + \tfrac{2}{3}h_2 \right) |_A \right\|_\infty = \tfrac{2}{3} \| \underbrace{\tfrac{3}{2}\left(f - h_1|_A \right)}_{=f_2} - h_2|_A \|_\infty \leqslant \left(\tfrac{2}{3} \right)^2 .$$

Continuing inductively starting with $f_3 = \left(\frac{3}{2} \right)^2 (f_2 - h_2|_A)$, we find functions $h_1, h_2, \ldots, h_n : X \to [-1, 1]$ with

$$\left\| f - \left(h_1 + \tfrac{2}{3}h_2 + \cdots + (\tfrac{2}{3})^{n-1}h_n \right) |_A \right\|_\infty \leqslant \left(\tfrac{2}{3} \right)^n . \tag{A.2}$$

We set

$$F = \sum_{n=1}^{\infty} \left(\tfrac{2}{3} \right)^{n-1} h_n,$$

and notice that

$$\left| F(x) - \sum_{n=1}^{m} \left(\tfrac{2}{3} \right)^{n-1} h_n(x) \right| \leqslant 3 \left(\tfrac{2}{3} \right)^m$$

for any $m \geqslant 1$ and $x \in X$, so F is bounded, the convergence is uniform, and $F \in C(X)$ (see the proof of Example 2.24(3) on p. 30). By (A.2) we have $f = F|_A$, as required.

If X is also assumed to be locally compact and $A \subseteq X$ is compact, then by Lemma A.22 there exists an open set $O \supseteq A$ with compact closure. Now extend f, first by using the definition

$$\widetilde{f}(x) = \begin{cases} f(x) & \text{for } x \in A, \\ 0 & \text{for } x \in X \smallsetminus O. \end{cases}$$

Then $\widetilde{A} = A \cup X \smallsetminus O$ is closed, and $\widetilde{f}$ is a continuous function on $\widetilde{A}$ which (by the argument above) can be extended to a continuous function $F \in C(X)$. By construction $\mathrm{Supp}(F) \subseteq \overline{O}$ is compact, so $F \in C_c(X)$, as required. $\qquad \square$

Appendix B: Measure Theory

Measure theory is one approach to making rigorous the idea of the size (or length, volume, and so on) of a set in an abstract setting. We refer to the notes of Tao [105] for a good general introduction to measure theory. By carefully controlling the complexity of the sets allowed in the theory, the basic intuition (for example, that the volume of the disjoint union of two sets is the sum of their volumes) can be developed into a powerful theory, indispensable in several fields including functional analysis and probability.

B.1 Basic Definitions and Measurability

The path to the definition of the Lebesgue integral starts with a discussion about which sets (and hence which functions) are allowed in the theory.

Definition B.1. Let X be a set. A family $\mathcal{A} \subseteq \mathbb{P}(X)$ of subsets of X is called an *algebra* if it satisfies the following properties:

- $\varnothing, X \in \mathcal{A}$;
- if $A \in \mathcal{A}$ then $A^c = X \smallsetminus A \in \mathcal{A}$;
- if $A_1, \ldots, A_n \in \mathcal{A}$ then $\bigcup_{i=1}^{n} A_i \in \mathcal{A}$;

and if, in addition,

- if $A_1, A_2, \cdots \in \mathcal{A}$ then $\bigcup_{n=1}^{\infty} A_n \in \mathcal{A}$

then $\mathcal{A}$ is a *σ-algebra*.

If $\mathcal{A}$ is a σ-algebra, then we call the pair $(X, \mathcal{A})$ a *measurable space* and the elements of $\mathcal{A}$ *measurable sets* or *$\mathcal{A}$-measurable sets*.

It is straightforward to check that the intersection of any collection of σ-algebras is also a σ-algebra. Hence for any family $\mathcal{C} \subseteq \mathbb{P}(X)$ of subsets there is a unique smallest σ-algebra containing $\mathcal{C}$, called the σ-algebra *generated by $\mathcal{C}$*, and denoted $\sigma(\mathcal{C})$. If X is a topological space, then the σ-algebra generated by all open subsets of X is called the *Borel σ-algebra*, and is denoted $\mathcal{B}$ or $\mathcal{B}(X)$.

© Springer International Publishing AG 2017

M. Einsiedler, T. Ward, *Functional Analysis, Spectral Theory, and Applications*,
Graduate Texts in Mathematics 276, DOI 10.1007/978-3-319-58540-6

Definition B.2. A function $\phi : X \to Y$ between measurable spaces $(X, \mathcal{A}_X)$ and $(Y, \mathcal{A}_Y)$ is called *measurable* if $\phi^{-1}(A) \in \mathcal{A}_X$ for all $A \in \mathcal{A}_Y$.

If Y is a topological space and ϕ is a map from a measurable space $(X, \mathcal{A})$ to Y then we will usually assume unless explicitly indicated otherwise that we are dealing with the Borel σ-algebra on Y to define measurability of ϕ. In particular, in such a setting ϕ is measurable if and only if $\phi^{-1}(O) \in \mathcal{A}$ for every open set $O \subseteq Y$, since ϕ^{-1} preserves all set-theoretic operations and the Borel σ-algebra is generated by the open sets in Y. This applies in particular to the cases $Y = \mathbb{R}$ and $Y = \mathbb{C}$.

Pointwise limits of sequences of measurable functions are measurable in the following sense. If (f_n) is a sequence of measurable functions $f_n : X \to Y$ for each $n \geqslant 1$, where Y is a metric locally compact σ-compact space, and for each $x \in X$ the sequence $(f_n(x))$ converges to some $f(x)$ in Y, then the pointwise limit function $f : X \to Y$ is measurable.

In order to define the integral of a measurable function one needs a precise notion of size or *measure* of a measurable set.

Definition B.3. A function $\mu : \mathcal{A} \to \mathbb{R} \cup \{\infty\}$ defined on a σ-algebra $\mathcal{A}$ of subsets of a set X is called a (positive) *measure* if it has the following properties:

- *(positivity)* $\mu(A) \geqslant 0$ for $A \in \mathcal{A}$;
- *(σ-additivity)* if $A_n \in \mathcal{A}$ for all $n \geqslant 1$ and $A_n \cap A_m = \varnothing$ for all $m \neq n$, then

$$\mu \left(\bigcup_{n=1}^{\infty} A_n \right) = \sum_{n=1}^{\infty} \mu(A_n),$$

where the sum on the right-hand side may or may not converge.

Theorem B.4 (Carathéodory extension [15]). *Let $\mathcal{A}$ be an algebra of subsets of X, and assume that $\mu : \mathcal{A} \to [0, \infty]$ is a function satisfying the following properties:*

(1) *if $A_1, A_2, \ldots$ are disjoint members of $\mathcal{A}$ with $\bigcup_{n=1}^{\infty} A_n \in \mathcal{A}$, then*

$$\mu \left(\bigcup_{n=1}^{\infty} A_n \right) = \sum_{n=1}^{\infty} \mu(A_n);$$

(2) *there is a countable collection $\{A_n \mid n \in \mathbb{N}\}$ with $A_n \in \mathcal{A}$ and $\mu(A_n) < \infty$ for all $n \geqslant 1$, and with $X = \bigcup_{n=1}^{\infty} A_n$.*

Then there is a measure $\overline{\mu}$ on the smallest σ-algebra containing $\mathcal{A}$ (equivalently, on the σ-algebra generated by $\mathcal{A}$) that extends μ in the sense that

$$\mu(A) = \overline{\mu}(A)$$

for any $A \in \mathcal{A}$.

The triple $(X, \mathcal{A}, \mu)$ consisting of a space X, a σ-algebra $\mathcal{A}$ of subsets of X, and a measure μ is called a *measure space*, and is called a *probability space* if $\mu(X) = 1$. If the measure μ of a measure space $(X, \mathcal{A}, \mu)$ satisfies condition (2) of Theorem B.4 then we say that the measure and the measure space are σ-*finite*. We will assume from now on that we are given some σ-finite measure μ on a measurable space $(X, \mathcal{B})$.

Definition B.5. A measurable function $f : X \to \mathbb{C}$ is called *simple* if

$$\mu(\{x \in X \mid f(x) \neq 0\}) < \infty$$

and we have finite range $|f(X)| < \infty$. In other words, f is simple if

$$f = \sum_{n=1}^{N} a_n \mathbb{1}_{B_n} \tag{B.1}$$

for some constants $a_n \in \mathbb{C}$ and $B_n \in \mathcal{B}$ with $\mu(B_n) < \infty$ for $1 \leqslant n \leqslant N$. The *integral* of the function f in (B.1) is defined to be

$$\int_X f \, \mathrm{d}\mu = \sum_{n=1}^{N} a_n \mu(B_n). \tag{B.2}$$

One can show rather easily that the integral defined in (B.2) is independent of the particular description of f as a finite sum as in (B.1). For the next definition, the analogous claim is an important step in the theory (this is essentially the monotone convergence theorem discussed below).

Definition B.6. Suppose that $f : X \to \mathbb{R}_{\geqslant 0} \cup \{\infty\}$ is measurable. We define the integral of f as the limit

$$\int_X f \, \mathrm{d}\mu = \lim_{n \to \infty} \int_X f_n \, \mathrm{d}\mu,$$

where (f_n) is a sequence of simple measurable functions $f_n : X \to \mathbb{R}_{\geqslant 0}$ with

$$0 \leqslant f_m \leqslant f_n \leqslant f$$

for $m \leqslant n$, and $f(x) = \lim_{n \to \infty} f_n(x)$ for all $x \in X$.

Implicit in this definition is the fact that (on σ-finite measure spaces) any non-negative measurable function is a pointwise limit of simple functions. Notice also that we permit sets to have infinite measure and functions to have infinite integral. If $f : X \to [0, \infty] = \mathbb{R}_{\geqslant 0} \cup \{\infty\}$, then we define

$$\int_X f \, \mathrm{d}\mu = \infty$$

if $\mu\left(\{x \in X \mid f(x) = \infty\}\right) > 0$, and

$$\int_X f \, \mathrm{d}\mu = \int_X f \mathbb{1}_{\{x \in X \mid f(x) < \infty\}} \, \mathrm{d}\mu$$

otherwise. Here the product $\infty \cdot 0$ is defined to be 0. The function f is called *integrable* if

$$\int_X f \, \mathrm{d}\mu < \infty.$$

If $f : X \to \mathbb{R}$ and both $f^+ = \max\{0, f\}$ and $f^- = \max\{0, -f\}$ are integrable, then we define

$$\int_X f \, \mathrm{d}\mu = \int_X f^+ \, \mathrm{d}\mu - \int_X f^- \, \mathrm{d}\mu$$

and say that f is integrable. Finally, if $f : X \to \mathbb{C}$ and $\Re(f)$, $\Im(f)$ are integrable, then we define

$$\int f \, \mathrm{d}\mu = \int \Re(f) \, \mathrm{d}\mu + \mathrm{i} \int \Im(f) \, \mathrm{d}\mu$$

and once again say that f is integrable.

B.2 Properties of the Integral

The space of integrable functions forms a vector space, and the integral is a linear function on that vector space. Moreover, the integral satisfies the following fundamental continuity properties, each of which is a consequence of the σ-additivity of the measure μ.

Theorem B.7 (Monotone convergence). *Let $(X, \mathcal{B}, \mu)$ be a measure space, and let (f_n) be a sequence of measurable functions $f_n : X \to [0, \infty]$ for $n \geqslant 1$ with $f_n \nearrow f$ as $n \to \infty$. That is, $f_m \leqslant f_n$ for $m \leqslant n$, and*

$$f(x) = \lim_{n \to \infty} f_n(x)$$

for all $x \in X$. Then f is measurable and

$$\int f \, \mathrm{d}\mu = \lim_{n \to \infty} \int f_n \, \mathrm{d}\mu.$$

Theorem B.8 (Dominated convergence). *Let $(X, \mathcal{B}, \mu)$ be a measure space, and let (f_n) be a sequence of complex-valued measurable functions with*

$$f(x) = \lim_{n \to \infty} f_n(x)$$

for all $x \in X$. Assume that there is an integrable function $g : X \to \mathbb{R}_{\geqslant 0}$ with $|f_n| \leqslant g$ for $n \in \mathbb{N}$. Then f is integrable and

$$\int f \, d\mu = \lim_{n \to \infty} \int f_n \, d\mu.$$

Definition B.9. We write

$$\mathscr{L}^1_\mu(X) = \{f : X \to \mathbb{C} \mid f \text{ is integrable}\}$$

for the space of integrable functions on a measure space $(X, \mathcal{B}, \mu)$, and define

$$\|f\|_1 = \int |f| \, d\mu$$

for any measurable function $f : X \to \mathbb{C}$.

Notice that $f \in \mathscr{L}^1_\mu \iff \|f\|_1 < \infty$, and

$$\left| \int f \, d\mu \right| \leqslant \|f\|_1$$

for all $f \in \mathscr{L}^1_\mu(X)$. It is easy to check that $\|\lambda f\|_1 = |\lambda| \|f\|_1$ and

$$\|f + g\|_1 \leqslant \|f\|_1 + \|g\|_1$$

for all $f, g \in \mathscr{L}^1_\mu$ and $\lambda \in \mathbb{C}$.

Definition B.10. A measurable set $N \subseteq X$ is called a *null set* if $\mu(N) = 0$. We say that a property holds *almost everywhere* (also written a.e., or where the measure is not obvious from the context, μ-almost everywhere) if it holds on the complement of a null set.

Thus, for example, if $f \in \mathscr{L}^1_\mu(X)$ then $\|f\|_1 = 0 \iff f = 0$ almost everywhere with respect to μ. It is often convenient to relax the requirement that a null set be measurable and call a set N a null set if there is a measurable set N' with $\mu(N') = 0$ and $N' \supseteq N$.

Definition B.11. We define

$$L^1_\mu(X) = \mathscr{L}^1_\mu(X)/\sim,$$

where the equivalence relation $\sim$ is μ-almost everywhere equality, meaning that $f \sim g$ if and only if $f = g$ μ-almost everywhere.

While the natural notation for the element of $L^1_\mu(X)$ containing a function f in $\mathscr{L}^1_\mu(X)$ is $[f]_\sim$, it is conventional to simply write $f \in L^1_\mu(X)$, with the understanding that such a function f is only defined up to equivalence under $\sim$.

Finally, we turn to integration of functions of several variables.

Theorem B.12 (Fubini's theorem). *For σ-finite measure spaces $(X, \mathcal{B}, \mu)$ and $(Y, \mathcal{C}, \nu)$ there exists a unique σ-finite measure $\mu \times \nu$ on $X \times Y$ such that*

$$(\mu \times \nu)(B \times C) = \mu(B)\nu(C)$$

for all $B \in \mathcal{B}$ and $C \in \mathcal{C}$. If now f is a measurable function on $X \times Y$, integrable in the sense that

$$\int_{X \times Y} |f(x,y)| \, \mathrm{d}(\mu \times \nu) < \infty,$$

then for almost every $y \in Y$ and $x \in X$, the integrals

$$h(x) = \int_Y f(x,y) \, \mathrm{d}\nu(y), \quad g(y) = \int_X f(x,y) \, \mathrm{d}\mu(x)$$

are integrable, and

$$\int_{X \times Y} f \, \mathrm{d}(\mu \times \nu) = \int_X h \, \mathrm{d}\mu = \int_Y g \, \mathrm{d}\nu. \tag{B.3}$$

This may also be written in a more familiar form as

$$\int_{X \times Y} f(x,y) \, \mathrm{d}(\mu \times \nu)(x,y) = \left\{ \begin{array}{l} \displaystyle\int_X \left(\int_Y f(x,y) \, \mathrm{d}\nu(y) \right) \mathrm{d}\mu(x) \\[12pt] \displaystyle\int_Y \left(\int_X f(x,y) \, \mathrm{d}\mu(x) \right) \mathrm{d}\nu(y). \end{array} \right\} \tag{B.4}$$

Theorem B.13 (Tonelli's theorem). *Let $f : X \times Y \to [0, \infty]$ be a non-negative measurable function on the product of two σ-finite measure spaces $(X, \mathcal{B}, \mu)$ and $(Y, \mathcal{C}, \nu)$. Then (B.4) holds again.*

We will simply refer to 'Fubini's theorem' whenever an interchange of the order of integration is justified by an application of either of these theorems; the reader may check in each case that the application is justified.

B.3 The p-Norm

Definition B.14. For any $p \in [1, \infty)$ we define

$$\|f\|_p = \left(\int_X |f|^p \, \mathrm{d}\mu \right)^{1/p}$$

for any measurable $f : X \to \mathbb{C}$,

$$\mathscr{L}_\mu^p(X) = \{f : X \to \mathbb{C} \mid \|f\|_p < \infty\},$$

and $L_\mu^p(X) = \mathscr{L}_\mu^p(X)/\sim$, where again $f \sim g$ if $f = g$ μ-almost everywhere.

The measure μ is often clear from the context of the space X, and in that case we will simply write $L^p(X)$. Thus, for example, $L^p(\mathbb{R})$ denotes $L_m^p(\mathbb{R})$ where m is Lebesgue measure on $\mathbb{R}$.

Once again it is clear that $\|\lambda f\|_p = |\lambda|\|f\|_p$ and $\|f + g\|_p \leqslant \|f\|_p + \|g\|_p$ holds as well for any measurable f, g and $\lambda \in \mathbb{C}$. We will review the proof of this triangle inequality, starting with the following important step.

Theorem B.15 (Hölder's inequality). *Let* $p, q \in (1, \infty)$ *satisfy* $\frac{1}{p} + \frac{1}{q} = 1$ *(in which case* q *is called the* conjugate exponent *of* p*). Then*

$$\int_X |fg| \, d\mu \leqslant \|f\|_p \|g\|_q$$

for any measurable functions $f, g : X \to \mathbb{C}$.

For $p = q = 2$ this is the Cauchy–Schwarz inequality (see also Proposition 3.2).

PROOF OF THEOREM B.15. For $\|f\|_p = 0$ or $\|g\|_q = 0$ we have $fg = 0$ μ-almost everywhere, and so $\int |fg| \, d\mu = 0$. Assume that $\|f\|_p > 0$ and $\|g\|_q > 0$. If either is infinite, then the inequality holds trivially. So it is enough to consider the case $\|f\|_p, \|g\|_q \in (0, \infty)$. Dividing by $\|f\|_p$ and by $\|g\|_q$ we may also assume that $\|f\|_p = \|g\|_q = 1$.

Suppose now that $x \in X$ satisfies $|f(x)| > 0$ and $|g(x)| > 0$. Then we may choose $s, t \in \mathbb{R}$ with $|f(x)| = e^{s/p}$ and $|g(x)| = e^{t/q}$. By convexity of the function $v \mapsto e^v$ on $\mathbb{R}$, we see that

$$|fg|(x) = e^{s/p+t/q} \leqslant \tfrac{1}{p}e^s + \tfrac{1}{q}e^t = \tfrac{1}{p}|f(x)|^p + \tfrac{1}{q}|g(x)|^q, \qquad (B.5)$$

and the inequality between the left-hand side and the right-hand side of (B.5) also holds trivially if $f(x) = 0$ or $g(x) = 0$. Integrating (B.5) over $x \in X$ gives

$$\int |fg| \, d\mu \leqslant \tfrac{1}{p}\|f\|_p^p + \tfrac{1}{q}\|g\|_q^q = 1,$$

proving the theorem. $\qquad\square$

Theorem B.16 (Triangle inequality). *For measurable functions* f *and* g *from* X *to* $\mathbb{C}$ *we have* $\|f + g\|_p \leqslant \|f\|_p + \|g\|_p$ *for any* $p \in [1, \infty)$.

PROOF. For $p = 1$ the inequality follows by integrating the triangle inequality $|f(x) + g(x)| \leqslant |f(x)| + |g(x)|$ for complex numbers over $x \in X$. Assume from now on that $p > 1$. Then we have $|f + g|^p \leqslant |f||f + g|^{p-1} + |g||f + g|^{p-1}$, and integrating over X and applying Hölder's inequality (Theorem B.15) gives

$$\|f + g\|_p^p \leqslant \|f\|_p \|(f + g)^{p-1}\|_q + \|g\|_p \|(f + g)^{p-1}\|_q, \qquad (\text{B.6})$$

where q is the conjugate exponent of p. Notice that $q(p - 1) = p$ implies

$$\|(f + g)^{p-1}\|_q = \left(\int |f + g|^p \, d\mu \right)^{1/q} = \|f + g\|_p^{p/q}.$$

If now $\|f + g\|_p \in (0, \infty)$ we can divide (B.6) by $\|f + g\|_p^{p/q}$ and the theorem follows since $p - p/q = 1$.

However, if $\|f + g\|_p = 0$ then the inequality in the theorem is trivially satisfied. Finally, note that

$$|f + g|^p \leqslant (|f| + |g|)^p \leqslant 2^p \max\{|f|^p, |g|^p\} \leqslant 2^p(|f|^p + |g|^p)$$

implies $\|f + g\|_p^p \leqslant 2^p(\|f\|_p^p + \|g\|_p^p)$. Hence if $\|f + g\|_p = \infty$, then $\|f\|_p = \infty$ or $\|g\|_p = \infty$ and the theorem also holds in this case. $\qquad \square$

B.4 Near-Continuity of Measurable Functions

Even though measurable functions are typically very far from being continuous, if we are working with a finite measure on the Borel σ-algebra of a metric space then they are nearly continuous in the following sense.

Theorem B.17 (Lusin: near-continuity of measurable functions). *Let X be a metric space, let μ be a finite measure on the Borel σ-algebra of X, let Y be a separable metric space, and let $f : X \to Y$ be (Borel) measurable. Then for every $\varepsilon > 0$ there exists a closed set $K \subseteq X$ with $\mu(X \setminus K) < \varepsilon$ such that $f|_K$ is continuous. If X is σ-compact, then K can be chosen to be compact.*

As the proof will show, we will in essence produce the continuity of $f|_K$ by removing very small open subsets around every possible discontinuity. To do this we will use the following regularity property of measures on metric spaces.

Lemma B.18 (Regularity of measures). *Let (X, d) be a metric space and let μ be a finite Borel measure on X. Then for every Borel set $B \subseteq X$ and every $\varepsilon > 0$ there exists a closed set $K \subseteq X$ and an open set $O \subseteq X$ with*

$$K \subseteq B \subseteq O$$

and $\mu(O \setminus K) < \varepsilon$.

PROOF. Consider the family $\mathcal{A} \subseteq \mathcal{B}$ of sets $B \in \mathcal{B}$ with the property that for every $\varepsilon > 0$ there exists a closed set $K \subseteq B$ and an open set $O \supseteq B$

with $\mu(O \smallsetminus K) < \varepsilon$. The statement of the lemma is then $\mathcal{A} = \mathcal{B}$, which we will prove in stages. By definition of the Borel σ-algebra $\mathcal{B}$, it is enough to show that $\mathcal{A}$ is a σ-algebra containing all the open sets.

CLOSURE UNDER COMPLEMENTS. Since taking complements switches open and closed sets, $\mathcal{A}$ is closed under taking complements. Explicitly, if B lies in $\mathcal{A}$ and for a given $\varepsilon > 0$ we have $K \subseteq B \subseteq O$ as in the definition of $\mathcal{A}$, then $X \smallsetminus O \subseteq X \smallsetminus B \subseteq X \smallsetminus K$ and $\mu((X \smallsetminus K) \smallsetminus (X \smallsetminus O)) = \mu(O \smallsetminus K) < \varepsilon$, which shows that $X \smallsetminus B \in \mathcal{A}$.

OPEN SETS. Using the continuous distance function $x \mapsto \mathsf{d}(x, A)$ for a closed subset A of X from (2.29) it follows that

$$A = \bigcap_{n \geqslant 1} \underbrace{\{x \in X \mid \mathsf{d}(x, A) < \tfrac{1}{n}\}}_{=O_n,\text{ an open set}}$$

is a decreasing countable intersection of open sets (that is, a G_δ-set). From the properties of the measure it now follows that $B = A$ satisfies the claim of the lemma with $K = A$ and $O = O_n$ with n depending on $\varepsilon > 0$. Since closed sets belong to $\mathcal{A}$, open sets also belong to $\mathcal{A}$ by the previous step.

FINITE UNIONS. Suppose $B_1, B_2 \in \mathcal{A}$ and $\varepsilon > 0$. Then there exist

$$K_1 \subseteq B_1 \subseteq O_1$$

and

$$K_2 \subseteq B_2 \subseteq O_2$$

as in the definition of $\mathcal{A}$, with $\mu(O_1 \smallsetminus K_1) < \varepsilon$ and $\mu(O_2 \smallsetminus K_2) < \varepsilon$. Now define $K = K_1 \cup K_2$, $B = B_1 \cup B_2$ and $O = O_1 \cup O_2$ so that K is closed, O is open, and $K \subseteq B \subseteq O$. Moreover, $\mu(O \smallsetminus K) \leqslant \mu(O_1 \smallsetminus K_1) + \mu(O_2 \smallsetminus K_2) < 2\varepsilon$, and since $\varepsilon > 0$ was arbitrary we deduce that $B = B_1 \cup B_2 \in \mathcal{A}$. By induction, the same holds for any finite union.

COUNTABLE UNIONS. By the steps above, $B_1 \cup \cdots \cup B_n \in \mathcal{A}$ if $B_1, \ldots, B_n \in \mathcal{A}$ for any $n \geqslant 1$. Therefore, and since we are interested in the union of these sets, we may assume that $B_1, B_2, \cdots \in \mathcal{A}$ satisfy $B_n \subseteq B_{n+1}$ for all $n \geqslant 1$. Define $B_1' = B_1$ and $B_{n+1}' = B_{n+1} \smallsetminus B_n \in \mathcal{A}$ for all $n \geqslant 1$, so that

$$\sum_{n=1}^{\infty} \mu(B_n') = \mu\left(\bigcup_{n=1}^{\infty} B_n\right) < \infty$$

by our assumption that μ is a finite measure. Therefore, for any $\varepsilon > 0$ there exists some $m \geqslant 1$ with

$$\sum_{n=m}^{\infty} \mu(B_{n+1}') < \varepsilon.$$

Since $B_m \in \mathcal{A}$, there exists some closed $K \subseteq B_m$ and open $O' \supseteq B_m$ with $\mu(O' \smallsetminus K) < \varepsilon$. Since $B_n' \in \mathcal{A}$ there must exist open sets $O_n \supseteq B_n'$

with $\mu(O_n \setminus B'_n) < \varepsilon/2^n$ for $n > m$. Now define

$$O = O' \cup \bigcup_{n=m+1}^{\infty} O_n$$

and notice that

$$K \subseteq \bigcup_{n=1}^{\infty} B_n \subseteq O$$

and

$$\mu(O \setminus K) \leqslant \mu(O' \setminus K) + \mu\left(\bigcup_{n=m+1}^{\infty} O_n\right)$$

$$< \varepsilon + \mu\left(\bigcup_{n=m+1}^{\infty} (O_n \setminus B'_n)\right) + \mu\left(\bigcup_{n=m+1}^{\infty} B'_n\right) < 3\varepsilon.$$

It follows that

$$\bigcup_{n=1}^{\infty} B_n \in \mathcal{A}.$$

CONCLUSION. By the above $\mathcal{A}$ is a σ-algebra that contains all open sets, and as mentioned above this forces $\mathcal{A} = \mathcal{B}$. $\qquad\square$

PROOF OF THEOREM B.17. Let $f : X \to Y$ be as in the statement of the proposition. By definition of measurability and of the Borel σ-algebra, the pre-image of every open set is Borel measurable in X. We wish to find, for every $\varepsilon > 0$, a closed set $K \subseteq X$ with $\mu(X \setminus K) < \varepsilon$ such that

$$(f|_K)^{-1}(U) = K \cap f^{-1}(U)$$

is open in K for every open set $U \subseteq Y$.

By our assumptions on Y, there exists a countable basis of the topology of Y (for example, using all balls of radius $\frac{1}{n}$ for $n \geqslant 1$ with centres at the points of a countable dense subset). Let $\{U_n \mid n \geqslant 1\}$ be such a basis. Now apply Lemma B.18 to each of the sets $f^{-1}(U_n) \subseteq X$ to find a closed set K_n and an open set O_n with $K_n \subseteq f^{-1}(U_n) \subseteq O_n$ and with $\mu(O_n \setminus K_n) < \varepsilon/2^n$ for $n \geqslant 1$. Now define

$$K = \bigcap_{n=1}^{\infty} (K_n \cup X \setminus O_n).$$

Notice first that $K_n \cup X \setminus O_n$ is closed for all $n \geqslant 1$, and so K is also closed. Second, we have

$$\mu(X \smallsetminus K) = \mu\left(X \smallsetminus \bigcap_{n=1}^{\infty}(K_n \cup X \smallsetminus O_n)\right) \leqslant \sum_{n=1}^{\infty} \mu\left(\underbrace{X \smallsetminus (K_n \cup X \smallsetminus O_n)}_{=O_n \smallsetminus K_n}\right) < \varepsilon$$

by construction of K_n and O_n. Finally, notice that

$$f|_K^{-1}(U_n) = K \cap f^{-1}(U_n) = K \cap O_n$$

is an open subset of K (in the induced topology). Since this holds for all the sets U_n in the basis, it follows that $f|_K$ is continuous.

If now in addition $X = \bigcup_{n=1}^{\infty} L_n$ is a countable union of compact sets, then $K' = K \cap \bigcup_{n=1}^{N} L_n$ satisfies the final claim of the proposition if N is sufficiently large. □

In the setting considered in this section there is a convenient formulation of the support of a measure as follows.

Definition B.19. The *support* Supp μ of a Borel measure μ on the Borel σ-algebra of a metric space X is the set of all points $x \in X$ with the property that every neighbourhood of x has positive measure.

Notice that with this definition $\mu(X \smallsetminus \text{Supp}\,\mu) = 0$ for the spaces considered in this section.

B.5 Signed Measures

Let $(X, \mathcal{B})$ be a measurable space. We define a (real- or complex-)valued signed measure ν by a finite measure μ and some (real- or complex-)valued function $g \in \mathscr{L}_\mu^1(X)$ and the formula $\mathrm{d}\nu = g\,\mathrm{d}\mu$. More concretely, ν is in that case the (real- or complex-)valued function on $\mathcal{B}$ defined by $\nu(B) = \int_B g\,\mathrm{d}\mu$. By dominated convergence (applied for the measure μ), ν is σ-additive on $\mathcal{B}$. The signed measure can also be used to integrate bounded measurable functions by setting

$$\int f\,\mathrm{d}\nu = \int fg\,\mathrm{d}\mu$$

for $f \in \mathscr{L}^\infty(X)$. Using dominated convergence and a sequence of simple functions to approximate $f \in \mathscr{L}^\infty(X)$, we see that $\int f\,\mathrm{d}\nu$ only depends on the function ν (that is, on the signed measure) and not on the choices involved in the representation of ν in terms of μ and $g \in \mathscr{L}_\mu^1(X)$.

Exercise B.20 (Polar decomposition). Let ν be a signed measure on a measurable space $(X, \mathcal{B})$. Show that the representation $\mathrm{d}\nu = g\,\mathrm{d}\mu$ consisting of μ and $g \in \mathscr{L}_\mu^1(X)$ can be chosen so that $|g(x)| = 1$ for all $x \in X$.

Using the Radon–Nikodym theorem (Proposition 3.29) it can be shown that the set of signed measures as defined above forms a vector space. To see this, suppose that $\nu_1 = g_1 \, d\mu_1$ and $\nu_2 = g_2 \, d\mu_2$ are signed measures as above, and λ_1, λ_2 are scalars. Then we may define the finite measure $\mu = \mu_1 + \mu_2$ which satisfies $\mu_1, \mu_2 \ll \mu$ and so $d\mu_1 = f_1 \, d\mu$ and $d\mu_2 = f_2 \, d\mu$ for some nonnegative functions $f_1, f_2 \in \mathscr{L}^1_\mu(X)$. This gives the presentation $d\nu_j = g_j f_j \, d\mu$ for $j = 1, 2$, and so $d(\lambda_1 \nu_1 + \lambda_2 \nu_2) = (\lambda_1 g_1 f_1 + \lambda_2 g_2 f_2) \, d\mu$ defines the linear combination $\lambda_1 \nu_1 + \lambda_2 \nu_2$ of the signed measures ν_1 and ν_2.

Hints for Selected Problems

Exercise 1.3 (p. 3): One way to start the proof is to lift ϕ to a function $\tilde{\phi} : \mathbb{R} \to \mathbb{C}$ and to show that $\tilde{\phi}$ must be differentiable and satisfies a differential equation by comparing $\tilde{\phi}(x)$ with

$$\psi(x) = \int_x^{x+\varepsilon} \tilde{\phi}(t)\,\mathrm{d}t = \int_0^\varepsilon \tilde{\phi}(x+t)\,\mathrm{d}t = \tilde{\phi}(x)c$$

(for ε small enough to ensure that $c = \int_0^\varepsilon \tilde{\phi}(t)\,\mathrm{d}t \neq 0$).

Exercise 1.7 (p. 11): Extend the given function first to an odd function on $(-1, 1)$ and then by periodicity to a function on $\mathbb{R}/2\mathbb{Z}$. Then use the Fourier series.

Exercise 2.7 (p. 20): For the first part consider rapidly oscillating functions, which can have $\|\cdot\|_{C([0,1])}$ small and $\|\cdot\|_{C^1([0,1])}$ large. For the second, use the fundamental theorem of calculus.

Exercise 2.9 (p. 20): Find a way to use Proposition 2.6.

Exercise 2.18 (p. 24): Suppose the unit ball is strictly convex and $v, w \in V \smallsetminus \{0\}$ satisfy $\|v + w\| = \|v\| + \|w\|$ and $\|v\| \leqslant \|w\|$. Use the estimate

$$\left\| \|v\|^{-1}v + \|v\|^{-1}w \right\| \geqslant \left\| \|v\|^{-1}v + \|v\|^{-1}w \right\| - \left\| \|v\|^{-1}w - \|w\|^{-1}w \right\|$$

$$= \|v\|^{-1}\|v + w\| - \left(\|v\|^{-1} - \|w\|^{-1} \right)\|w\|$$

$$= \|v\|^{-1}\left(\|v\| + \|w\| \right) - \left(\|v\|^{-1} - \|w\|^{-1} \right)\|w\| = 2$$

to conclude from strict convexity of the unit ball that $\|v\|^{-1}\|v\| = \|w\|^{-1}\|w\|$.

Exercise 2.26 (p. 29): For (a) assume that (y_n) is a sequence in Y converging to $x \in X$, and note that (y_n) must be a Cauchy sequence. For the reverse implication in (b) assume that (y_n) is a Cauchy sequence in Y and note that it then is also a Cauchy sequence in X.

Exercise 2.39 (p. 42): For (b) we note that in the formulation of the compactness criterion in $C_0(X)$ (which is not given in the exercise) an extra uniformity condition regarding decay at infinity is necessary.

Exercise 2.43 (p. 47): Notice first that without constants the given proof cannot be applied. Add a point x_{new}, forming $X_{\text{new}} = X \sqcup \{x_{\text{new}}\}$. Extend functions in $\mathcal{A}$ to X_{new} by setting $f(x_{\text{new}}) = 0$ for all $f \in \mathcal{A}$. Define $\mathcal{A}_{\text{new}} = \mathcal{A} + \mathbb{R}\mathbb{1}$ (or $\mathcal{A} + \mathbb{C}\mathbb{1}$) and apply

M. Einsiedler, T. Ward, *Functional Analysis, Spectral Theory, and Applications*,
Graduate Texts in Mathematics 276, DOI 10.1007/978-3-319-58540-6

Theorem 2.40. For (a) let x_{new} be an isolated point of X_{new}. For (b) define the topology on X_{new} so that this space is the one-point compactification of X.

Exercise 2.48 (p. 51): Recall that a Riemann integrable function is a function that can be approximated from above and below by step functions such that the integral of the difference is arbitrary small. With this in mind repeat the argument for (2) $\Longrightarrow$ (1) to show that (1) $\Longrightarrow$ (3).

Exercise 2.50 (p. 51): Express this in terms of the orbit of 0 under $t \mapsto t + \log_{10} 2$ modulo 1.

Exercise 2.56 (p. 56): Apply (2.31) for $\|v\| \leqslant 1$ and consider $L(\|v\|^{-1}v)$ for non-zero vectors $v \in V$.

Exercise 2.61 (p. 59): Use the Cauchy integral formula to see that E_z and

$$\imath_{\overline{O}} : V \ni f \mapsto f|_{\overline{O}} \in C(\overline{O})$$

are continuous. To prove injectivity, define $D_r = \{z \in \mathbb{C} \mid |z| < r\}$ for $r < 1$, $V_r = V(D_r)$ and its completion $H^p(D_r)$. Now consider the maps

$$H^p(D) \ni f \longmapsto \left(\imath_{D_r}(f) \mid r < 1\right) \in \prod_{r<1} C(\overline{D_r})$$

and

$$\prod_{r<1} C(\overline{D_r}) \ni (f_r \mid r < 1) \longmapsto (f_r \mid r < 1) \in \prod_{r<1} H_p(D_r),$$

and notice that for $f \in V$ the composition of the two maps is given by

$$V \ni f \mapsto (f|_{D_r} \mid r < 1)$$

which satisfies $\|f\|_{H^p(D)} = \sup_{r<1} \|f|_{D_r}\|_{H^p(D_r)}$. Extend this to all $f \in H^p(D)$ and consider now the case where $f \in H^p(D)$ satisfies $\imath_{D_r}(f) = 0$ for all $r < 1$.

Exercise 2.70 (p. 69): Using the discussion concerning (2.47) the assumption is equivalent to $f'' = \lambda f$ with the boundary conditions $f(0) = f(1) = 0$.

Exercise 3.4 (p. 73): Use the same argument as in (2.34).

Exercise 3.5 (p. 74): For (b) express $\langle x_1, y \rangle + \langle x_2, y \rangle$ in terms of the norm using the definition and apply the parallelogram identity separately to the positive and the negative parts to obtain $\frac{1}{2}\langle x_1 + x_2, 2y \rangle$. Setting $x_2 = 0$, this gives $\langle x_1, y \rangle = \frac{1}{2}\langle x_1, 2y \rangle$. Now consider rational multiples of x_1 to prove linearity of $\langle \cdot, \cdot \rangle$. For part (c) verify first the complex polarization identity

$$\langle x, y \rangle = \frac{1}{4} \sum_{k=0}^{3} \mathrm{i}^k \|x + \mathrm{i}^k y\|^2$$

for elements $x, y \in \mathcal{H}$ of a complex inner product space $\mathcal{H}$.

Exercise 3.9 (p. 75): Use the polarization identity from Exercise 3.5.

Exercise 3.10 (p. 76): For (a), analyze the proof of the triangle inequality using the equality case of the Cauchy–Schwarz inequality in Proposition 3.2. For (b) show that the closed unit ball is strictly convex and apply Exercise 2.18.

Exercise 3.15 (p. 77): For (a), use the inequality $(a^2/(a^2+b^2))^q + (b^2/(a^2+b^2))^q \leqslant 1$. For (b) use Jensen's inequality.

Exercise 3.21 (p. 81): For (a) apply Corollary 3.19 to the linear functional sending y to $\overline{B(x,y)}$ for a fixed $x \in \mathcal{H}$. For (b), notice that $\|Tx\| \geqslant c\|x\|$ and show that this implies that $T(\mathcal{H}) \subseteq \mathcal{H}$ is closed. Finally, $x \in T(\mathcal{H})^{\perp}$ implies $\langle Tx, x \rangle = 0 \geqslant c\|x\|^2$.

Exercise 3.22 (p. 81): Define $\langle \phi(x), \phi(y) \rangle_{\mathcal{H}^*} = \overline{\langle x, y \rangle_{\mathcal{H}}}$.

Exercise 3.23 (p. 81): Either use Section 2.2.2, or define an inner product on the closure of the image of $\mathcal{H}$ in the double dual.

Exercise 3.28 (p. 82): For (1) simply use the evaluation maps for every $n \in \mathbb{N}$. For (2) use the axiom of choice to fix for every $i \in I$ a sequence $(y_m^{(i)})$ of rationals approaching i and define $x_n^{(i)}$ to be equal to 1 if the nth rational number appears in the sequence $(y_m^{(i)})$ and to be equal to 0 otherwise. Now use (2) for proving (3), (3) for (4), and that I is uncountable to conclude.

Exercise 3.30 (p. 85): For (b) recall (from analysis or as a trivial case of Fubini's theorem) that if $a_{mn} \geqslant 0$ then $\sum_{m=1}^{\infty} \sum_{n=1}^{\infty} a_{mn} = \sum_{n=1}^{\infty} \sum_{m=1}^{\infty} a_{mn}$ (where the sum is also allowed to be infinity).

Exercise 3.31 (p. 85): One approach is to use the result for the case of a finite measure space and apply Exercise 3.30.

Exercise 3.33 (p. 85): First show that ν can be identified with a linear functional on $\mathscr{L}^{\infty}(X)$ and show that $\|\nu\|$ is precisely the operator norm. For (b) use Lemma 2.28, Exercise B.20, and the fact that $\mu = \sum_{n=1}^{\infty} \mu_n$ defines a finite measure if each μ_n is a finite measure for $n \geqslant 1$ and $\sum_{n=1}^{\infty} \mu_n(X) < \infty$.

Exercise 3.34 (p. 86): Working first with real-valued L^2 functions, start with the projection operator $P : L_{\mu}^2(X, \mathcal{B}) \to L_{\mu}^2(X, \mathcal{A})$ and show that $\|P(f)\|_1 \leqslant \|f\|_1$.

Exercise 3.37 (p. 87): For (a), show first that the inner product is well-defined on $\bigoplus_n \mathcal{H}_n$ and satisfies all the properties of an inner product. Then show that a Cauchy sequence in the sum gives rise to Cauchy sequences in each $\mathcal{H}_n$ for all n. For (b) use (a) and the canonical map $(v_n) \mapsto \sum_n v_n$ from the abstract Hilbert space sum $\bigoplus_n \mathcal{H}_n$ into $\mathcal{H}$.

Exercise 3.48 (p. 94): Recall that $\chi_{e_1}, \ldots, \chi_{e_d}$ are sufficient to separate points on $\mathbb{T}^d$, and that the group of characters generated by these are all the characters of the stated form.

Exercise 3.49 (p. 95): The characters already appeared implicitly in Section 1.1.

Exercise 3.50 (p. 95): For (d), write $L^p(G)$ for functions on G with respect to normalized Haar measure, and $\ell_p(\widehat{G})$ for functions on $\widehat{G}$ with respect to counting measure. Notice that $f = \mathbb{1}_{\mathrm{Supp}(f)} f$ so that

$$\|f\|_1 = \|\mathbb{1}_{\mathrm{Supp}(f)} f\|_1 \leqslant \|\mathbb{1}_{\mathrm{Supp}(f)}\|_2 \|f\|_2 = |\mathrm{Supp}(f)|^{\frac{1}{2}} |G|^{-\frac{1}{2}} \|f\|_2$$

by the Cauchy–Schwarz inequality. Similarly

$$\|\widehat{f}\|_2 = \|\mathbb{1}_{\mathrm{Supp}(\widehat{f})} \widehat{f}\|_2 \leqslant \|\mathbb{1}_{\mathrm{Supp}(\widehat{f})}\|_2 \|\widehat{f}\|_{\infty} = |\mathrm{Supp}(\widehat{f})|^{\frac{1}{2}} \|\widehat{f}\|_{\infty}.$$

Combine these inequalities with the estimate from (c).

Exercise 3.53 (p. 95): For (a) prove first that $\mathbb{T}^\Gamma$ is a metric compact abelian group and show that G is a closed subgroup. For (b) notice that G can be identified with the group of characters on G and that every $\gamma_0 \in \Gamma$ defines the continuous group homomorphism $\chi_{\gamma_0} : G \ni (z_\gamma) \mapsto e^{2\pi i z_{\gamma_0}} \in \mathbb{S}^1$, which is non-trivial by the theorem on completeness of characters (applied to Γ). If now χ is a character on G, then there exists a neighbourhood U of $0 \in G$ such that $\chi(U) \subseteq B^{\mathbb{S}^1}_{1/10}(1)$. By the definition of the product topology there exist finitely many $\gamma_1, \ldots, \gamma_d \in \Gamma$ such that $H = \{(z_\gamma) \in G \mid z_{\gamma_1} = \cdots = z_{\gamma_d} = 0\} \subseteq U$. Using that H is a subgroup and χ is a homomorphism, show that $\chi(H) = 1$, which shows that χ is well-defined on G/H and depends only on the coordinates $z_{\gamma_1}, \ldots, z_{\gamma_d}$ for any $(z_\gamma) \in G$. Combine this with Exercise 3.48 to conclude that χ can be expressed in terms of $\chi_{\gamma_1}, \ldots, \chi_{\gamma_d}$.

Exercise 3.55 (p. 96): For part (b), consider the odd extension of a given function f in $L^2((0,1))$ and apply part (a) rephrased for $L^2((-1,1))$. For part (c) consider the even extension.

Exercise 3.56 (p. 96): Use de Moivre's formula $(e^{2\pi i\phi})^n = \cos(2\pi n\phi) + i\sin(2\pi n\phi)$.

Exercise 3.69 (p. 106): Localize to a small open subset $B_\delta(x) \subseteq U$ by multiplying by a function $C_c^\infty(B_\delta(x))$ which is equal to 1 on $B_{\delta/2}(x)$. Treat the new localized function as an element on $\mathbb{T}^2$. Now generalize Theorem 3.57 to give an inequality concerning (and as a result, the existence of) $\partial_1\partial_2 f$ at x. This exercise should become easier after reading Theorem 5.6.

Exercise 3.72 (p. 107): Do this via a familiar sequence of approximations, first for indicator functions of measurable sets, then for simple functions, then for non-negative functions by monotone convergence, and finally for all integrable functions.

Exercise 3.76 (p. 110): First use the case $p = 1$ to see that

$$\left\{ x \in X \mid \int_G \phi(g)f(g^{-1}\cdot x)\, dm_G(g) \neq 0 \right\}$$

is a null set for any $f \in \mathscr{L}^\infty(X)$ with $f = 0$ μ-almost everywhere. Therefore, if two bounded measurable functions f_1 and f_2 on X are equivalent modulo μ, then

$$\int_G \phi(g)f_1(g^{-1}\cdot x)\, dm_G(g) = \int_G \phi(g)f_2(g^{-1}\cdot x)\, dm_G(g)$$

for almost every $x \in X$. Use this to see that $\phi * f$ is well-defined for an equivalence class of functions $f \in L_\mu^\infty(X)$.

Exercise 3.77 (p. 110): For the continuity requirement approximate (v_n) by a finitely supported vector $(v_1, \ldots, v_k, 0, \ldots)$ for some $k \geqslant 1$ and then use continuity of the unitary representations $\pi_1, \ldots, \pi_k$.

Exercise 3.82 (p. 113): Prove the same statements first for the Riemann sums.

Exercise 3.86 (p. 115): Use Proposition 3.83 for the first part. For part (b) go through the proof of Lemma 3.75 to see that it also works for a measure ν. Then take a second function $f' \in L_\mu^2(X)$ and apply Fubini's theorem to $\langle f', \nu * f \rangle$ (and similarly to $\langle f', \phi * f \rangle$).

Exercise 3.90 (p. 118): Use integration by parts just as in the proof of Theorem 3.57 to bound $\overline{\chi_n} * f$ uniformly on compact subsets of $\mathbb{R}^2$.

Exercise 3.92 (p. 119): Repeat the argument for Lemma 3.59(1).

Exercise 3.93 (p. 119): For $\mu, \nu \in \mathcal{M}(G)$ and any Borel measurable $B \subseteq G$ define

$$\mu * \nu(B) = \iint \mathbb{1}_B(gh) \, d\mu(g) \, d\nu(h).$$

Exercise 4.4 (p. 123): See Example 8.56 for the counter-example.

Exercise 4.16 (p. 128): Consider the closed sets

$$X_n = \{x \in X \mid \|T_\alpha x\| \leqslant n \text{ for all } \alpha \in A\}$$

for $n \geqslant 1$.

Exercise 4.18 (p. 128): Define the *oscillation* of f at $x \in X$ by

$$\text{osc}_f(x) = \inf_{\varepsilon > 0} \text{diam}(f(B_\varepsilon(x))),$$

so that f is continuous at x if and only if $\text{osc}_f(x) = 0$. Show that the set

$$\{x \in X \mid \text{osc}_f(x) < c\}$$

is open for $c > 0$, and that

$$\bigcap_{n \geqslant 1} \{x \in X \mid \text{osc}_f(x) < \tfrac{1}{n}\}$$

gives the set of points where f is continuous.

Exercise 4.19 (p. 128): Show that

$$O_{(a,b),n} = \left\{ f \in L^1((0,1)) \mid \int_a^b |f| dx > n(b-a) \right\}$$

is open in $L^1((0,1))$ for all $0 \leqslant a < b \leqslant 1$ and $n \geqslant 1$, and show that

$$D_{(a,b),n} = \bigcup_{(c,d):a<c<d<b} O_{(c,d),n}$$

is open and dense. Now take the intersection over $(a,b) \in \mathbb{Q}^2$ with $a < b$ and $n \in \mathbb{N}$.

Exercise 4.20 (p. 128): Consider for every $n \in \mathbb{N}$ the set B_n^+ of functions f in $C([0,1])$ with the property that there exists some $x \in [0, \tfrac{1}{2}]$ such that

$$\left| \frac{f(x+h)-f(x)}{h} \right| \leqslant n$$

for all $h \in (0, \tfrac{1}{2}]$. Use compactness of $[0, \tfrac{1}{2}]$ to show that each B_n^+ is closed. Show that $C([0,1]) \backslash B_n^+$ is dense, for example by using piecewise linear functions. Repeat the argument considering difference quotients for $x \in [\tfrac{1}{2}, 1]$ and $h \in [-\tfrac{1}{2}, 0)$ to define B_n^-. Conclude that $C([0,1]) \backslash \bigcup_{n \in \mathbb{N}} (B_n^+ \cup B_n^-)$ is dense and consists of functions that are nowhere differentiable.

Exercise 4.23 (p. 130): Use the argument from the proof of Lemma 4.22.

Exercise 5.11 (p. 142): Recall first that $C_c(U) \subseteq L^p(U)$ is dense by Proposition 2.51. Given some $f \in C_c(U)$ choose some open V with compact closure $\overline{V} \subseteq U$. Now apply the Stone–Weierstrass theorem (in the form of Exercise 2.43(b)) to $C_c^\infty(V) \subseteq C_0(V)$.

Exercise 5.15 (p. 143): Describe the relationship between the Fourier coefficients of f and of $\partial_\alpha f$ and use Lemma 5.4. Alternatively, convolve with a suitable version of $\jmath_\varepsilon$ from Exercise 5.17 and show that the resulting smooth (or, using Exercise 5.17, L^2) function actually approximates f with respect to the norm on $H^k(\mathbb{T}^d)$.

Exercise 5.16 (p. 144): Show by induction that the derivatives of ψ for $t < 0$ are of the form $p(\frac{1}{t})\psi(t)$ for some real-valued polynomials p and show that such functions converge to 0 as $t \nearrow 0$. Use this and the mean value theorem for differentiation to show that all derivatives of ψ at $t = 0$ vanish.

Exercise 5.17 (p. 144): For (a) use Exercise 5.16. For (b) argue as in the proof of Theorem 3.54. For (c), differentiate under the integral (which may be justified by dominated convergence). In (e) the appropriate convergence is with respect to $\| \cdot \|_p$, which can be obtained using the density of $C_c(U) \subseteq L^p(U)$, Lemma 3.75, and parts (b) and (d).

Exercise 5.18 (p. 144): Localize the function $f \in C(U)$ to a small set using a smooth function of compact support (for example, replace f with $f(x)\jmath_\varepsilon(x - x_0)$ for $\jmath_\varepsilon$ from Exercise 5.17) and consider it as a smooth function on $\mathbb{T}^d$ (see also Lemma 5.36). Then use (3.18), $\sum_{j=1}^d n_j^{2k} \asymp \|n\|_2^{2k}$ for $n \in \mathbb{Z}^d$ and $k \geqslant 1$, Lemma 5.4, and Theorem 5.6.

Exercise 5.19 (p. 144): Note that for $\lambda \in (0, 1)$ the function $f^\lambda(x) = f(\lambda x)$ is defined on a slightly larger version of the set U. Let $\varepsilon > 0$ and recall the function $\jmath_\varepsilon$ from Exercise 5.17. Show that the restriction of $f_j * \jmath_\varepsilon$ is the weak e_j-partial derivative of $f * \jmath_\varepsilon$ on the set $U_\varepsilon = \{x \in U \mid x + \overline{B_\varepsilon} \subseteq U\}$. Now choose some $\lambda < 1$ sufficiently close to 1 and $\varepsilon > 0$ sufficiently small and show that the smooth function $f^\lambda * \jmath_\varepsilon$ is defined on U, its restriction is close to f in L^2, and that the same applies to the weak partial derivatives.

Exercise 5.25 (p. 146): Either convolve with an approximate identity (that is, with $\jmath_\varepsilon$ from Exercise 5.17) or show that the sequence of functions (f_n) defined by $f_n = \min\{n, f\}$ all lie in $H^1(B_{1/2})$.

Exercise 5.27 (p. 146): For (a) (and (b)) consider first functions in $C^\infty(U) \cap H^k(U)$ (respectively $C_c^\infty(V)$). For (c) consider for instance $d = 1$, $V = (0, \frac{1}{2}) \subseteq U = (0, 1)$, find some $\chi \in C_c^\infty(U)$ with $\chi(\frac{1}{2}) \neq 0$ and show that $\chi|_V \in H^1(V) \setminus H_0^1(V)$ using the argument in Example 5.20.

Exercise 5.30 (p. 147): Use the regular map ϕ to pull back any function

$$f_0 \in C^\infty(U) \cap H^1(U)$$

(or $f_0 \in H^1(U)$) to an element

$$f \in C^\infty((0, 1)^d) \cap H^1((0, 1)^d)$$

(or $f \in H^1((0, 1)^d)$) and then apply Example 5.28.

Exercise 5.35 (p. 150): Here elements of function spaces on the closed cube are defined to have the claimed degree of smoothness in the interior of the cube, and in addition have the property that all the claimed partial derivatives extend continuously to the closure. For (a) apply the trace operator in Example 5.28 (see Exercise 5.29) for every $\alpha \in \mathbb{N}_0^{d-1}$ with $\|\alpha\|_1 \leqslant k - 1$ to see that the map

$$H^k(U) \ni f \mapsto \partial_\alpha f|_{S_y} \in L^2(S_y)$$

is a bounded operator. Together these show that $H^k(U) \ni f \mapsto f|_{S_y} \in H^{k-1}(U)$ is also bounded. For (b) first generalize (a) to prove that

$$\|f|_{S_{y_1}} - f|_{S_{y_2}}\|_{H^{k-1}(S)} \ll \|f\|_{H^k(U)} \sqrt{|y_1 - y_2|}.$$

Next set $\ell = 0$ and use induction on the dimension to prove that $\|f\|_\infty \ll \|f\|_{H^d(U)}$. Finally, take $\ell \geqslant 1$ and apply the first part to bound $\|\partial_\alpha f\|_\infty$ for all α with $\|\alpha\|_1 \leqslant \ell$. For (c) use the arguments in (b) together with $y \searrow 0$.

Exercise 5.37 (p. 151): Choose $\varepsilon > 0$ such that $\overline{K + B_{3\varepsilon}} \subseteq U$. Let $\jmath_\varepsilon$ be the function from Exercise 5.17 and consider $\jmath_\varepsilon * \mathbb{1}_{K+B_\varepsilon}$.

Exercise 5.39 (p. 152): Apply the arguments behind Lemma 5.36 and Theorem 5.34 using some fixed $\chi \in C_c^\infty(U)$ with $\chi|_K \equiv 1$.

Exercise 5.52 (p. 161): Set $K_j = V_j \setminus \left(\bigcup_{i \neq j} V_i \right)$ for $j = 0, \ldots, k$ and apply Lemma A.28 to find a continuous partition of unity. Combine this with Exercise 5.17 to obtain the smooth partition of unity.

Exercise 5.54 (p. 163): Average the function over large balls with different centres and use Proposition 5.53 and the boundedness assumption to estimate the difference between the values at the two centres.

Exercise 5.57 (p. 165): Assume first either that U is a set of the form $U \cap B_\varepsilon(z^{(0)})$ or is convex as in Definition 5.31. Show that for $\phi \in C^\infty(\overline{U})$ we have

$$\langle \partial_j g, \phi \rangle_{L^2(U)} = -\langle g, \partial_j \phi \rangle_{L^2(U)}$$

so g satisfies the usual integration by parts formula but even for $\phi \in C^\infty(\overline{U})$. Then for $\lambda > 1$ show that the function g^λ defined by

$$g^\lambda(x) = \begin{cases} g(\lambda x) & \text{for } \lambda x \in U, \\ 0 & \text{for } \lambda x \notin U \end{cases}$$

is in $H_0^1(U)$ (for example, by using similar arguments to Exercise 5.19) take $\lambda \searrow 1$, and conclude that $g \in H_0^1(U)$. Finally, use Lemmas 5.40, 5.41, and $\Delta g = 0$. For more general sets as in Definition 5.31 use a smooth partition of unity to localize g to sets of the form $U \cap B_\varepsilon(z^{(0)})$ (without destroying the feature that g vanishes in the square-mean sense at the boundary).

Exercise 6.1 (p. 167): For (a) note that an eigenvalue would have absolute value one and the eigenvector would have to be a sequence with constant absolute value. For (b) consider geometric sequences.

Exercise 6.6 (p. 169): Use Hölder's inequality and the Arzela–Ascoli theorem to prove compactness for $p > 1$. For $p = 1$ compactness fails as one sees from studying a sequence (f_n) of positive functions with integral one and support $[\frac{1}{2} - \frac{1}{n}, \frac{1}{2} + \frac{1}{n}]$ for $n \geqslant 3$.

Exercise 6.9 (p. 170): The special case of $k = 0$ in (a) is treated in detail in Lemma 6.58 and the method there also works for general $k \geqslant 0$. For (b) use the Arzela–Ascoli theorem (Theorem 2.38) in the case of compact closure and consider a fixed non-zero function and all its shifts in the case of $U = \mathbb{R}$. In (c) the answer is negative, for example because the closure of the image of the unit ball contains all characters.

Exercise 6.12 (p. 171): Use the first part of the proof of Proposition 6.11, Proposition 2.51, and Lemma 6.7.

Exercise 6.16 (p. 174): Consider the images under K of the functions $f_n = \mathbb{1}_{[3n,3n+1]}$ for $n \geqslant 1$, all of which have L^2 norm one.

Exercise 6.24 (p. 176): Set $V = U(\mathcal{H})^{\perp}$ and show that $U^n(V) \perp U^m(V)$ for all integers m, n with $0 \leqslant n < m$. Define $\mathcal{H}_{\text{shift}} = \bigoplus_{n \geqslant 0} U^n V$ and show that

$$\mathcal{H}_{\text{unitary}} = \mathcal{H}_{\text{shift}}^{\perp} = \bigcap_{n \geqslant 0} U^n \mathcal{H}$$

satisfies the claims in the exercise.

Exercise 6.29 (p. 178): Expand ϕ and f in terms of the orthonormal basis of Theorem 6.27, and compare coefficients.

Exercise 6.33 (p. 181): In both cases show that $\bigcap_{n \in J} \ker(A_n - \lambda_n I)$ is invariant under $A_1, A_2, \ldots$ for any choice of λ_n and any choice of index set $J \subseteq \mathbb{N}$. Show moreover that this intersection is finite-dimensional if $\lambda_1 \neq 0$. Now apply Theorem 6.27 to A_1 and to A_n restricted to the eigenspaces of A_1.

Exercise 6.34 (p. 182): To prove the second inequality (6.14) first take the linear hull V_0 of the eigenvectors $v_1, \ldots, v_k$ corresponding to the first k positive eigenvalues (assume first that there are at least k positive eigenvalues) and calculate the minimum. Then let W be the linear hull of $V_0^{\perp}$ and the k-th eigenvector v_k (also belonging to V_0), and note that any k-dimensional subspace V will intersect W non-trivially.

Exercise 6.45 (p. 192): Consider the self-adjoint compact operator $A^* A$ and apply Theorem 6.27. Using that basis, define P such that $P^2 = A^* A$, and define Q so that $A = QP$.

Exercise 6.47 (p. 193): Calculate the trace-class norm for k a character (it will be 1) and use absolute convergence of Fourier series (Theorem 6.47).

Exercise 6.49 (p. 195): For a fixed compact set $Y \subseteq X$ use the proof of Proposition 6.48 to show that $\int_Y |k(x,x)| \, d\mu(x) \leqslant \|K\|_{\text{tc}}$. Conclude that k is integrable along the diagonal. Fix an increasing sequence of compact sets with $X = \bigcup_{n \geqslant 1} Y_n$. For every n consider a sequence of partitions $(\xi_{n,\ell})_{\ell \geqslant 1}$ of $Y_n \setminus Y_{n-1}$ as in the proof of Proposition 6.48 (where we set $Y_0 = \varnothing$). Use this sequence (by enumerating $\mathbb{N}^2$ in some fashion) to define an orthonormal basis. To conclude, use in addition Lemma 6.41.

Exercise 6.50 (p. 195): For (a), suppose that (A_k) is a Cauchy sequence with respect to $\|\cdot\|_{\mathrm{tc}}$. Since $\|\cdot\|_{\mathrm{op}} \leqslant \|\cdot\|_{\mathrm{tc}}$ we have $\lim_{k\to\infty} A_k = A \in B(\mathcal{H})$. For given $k, \ell, N \geqslant 1$ and any list of orthonormal vectors $(v_n)_{n=1,\dots,N}$ and $(w_n)_{n=1,\dots,N}$ we have

$$\sum_{n=1}^{N} |\langle (A_k - A_\ell)v_n, w_n\rangle| \leqslant \|A_k - A_\ell\|_{\mathrm{tc}}.$$

Fixing (v_n) and (w_n) and letting $\ell \to \infty$ gives

$$\sum_{n\geqslant 1} |\langle (A_k - A)v_n, w_n\rangle| < \infty.$$

For $k = 1$ this shows $A \in \mathrm{TC}(\mathcal{H})$, and taking $k \to \infty$ then gives $\|A_k - A\|_{\mathrm{tc}} \to 0$.

Exercise 6.52 (p. 195): Let (v_n) and (w_n) be lists of orthonormal vectors, and notice that

$$\sum_{n=1}^{N} |\langle A v_n, w_n\rangle| = \sum_{n=1}^{N} \left| \int_T \langle A_t v_n, w_m\rangle \, \mathrm{d}\mu(t) \right| \leqslant \int_T \|A_t\|_{\mathrm{tc}} \, \mathrm{d}\mu(t)$$

for all $N \geqslant 0$. This gives the first claim; the argument for the trace of A is similar.

Exercise 6.53 (p. 195): Part (a) follows quickly from the identity

$$|\langle A e_j, e_k\rangle| = |\langle A^* e_k, e_j\rangle|$$

for all j, k. For (b), let (f_n) be a different orthonormal basis and note that

$$\sum_{k\geqslant 1} |\langle A e_j, e_k\rangle|^2 = \|A e_j\|^2 = \sum_{k\geqslant 1} |\langle A e_j, f_k\rangle|^2,$$

and so

$$\sum_{j,k\geqslant 1} |\langle A e_j, e_k\rangle|^2 = \sum_{j,k\geqslant 1} |\langle A e_j, f_k\rangle|^2 = \sum_{j,k\geqslant 1} |\langle e_j, A^* f_k\rangle|^2.$$

Arguing similarly one can also replace e_j by f_j. For (c), suppose first that $B = U$ is unitary and apply Lemma 6.38 to conclude the argument. For (d) define

$$\langle A_1, A_2\rangle_{\mathrm{HS}} = \sum_{j,k\geqslant 1} \langle A_1 e_j, e_k\rangle \overline{\langle A_2 e_j, e_k\rangle}$$

and show that the $(a_{mn}) \in \ell^2(\mathbb{N}^2)$ correspond precisely to operators $A \in \mathrm{HS}(\mathcal{H})$ by setting

$$A\Big(\sum_{j\geqslant 1} c_j e_j\Big) = \sum_{k\geqslant 1} \Big(\sum_{j\geqslant 1} a_{jk} c_j\Big) e_k$$

(Proposition 6.11 with $X = Y = \mathbb{N}$ shows this is a well-defined bounded operator). For (e) suppose $A, B \in \mathrm{HS}(\mathcal{H})$ and calculate

$$\|AB\|_{\mathrm{HS}}^2 = \sum_{j,k\geqslant 1} |\langle AB e_j, e_k\rangle|^2 = \sum_{j,k\geqslant 1} |\langle B e_j, A^* e_k\rangle|^2$$

$$\leqslant \sum_{j,k\geqslant 1} \|B e_j\|^2 \|A^* e_k\|^2 = \|B\|_{\mathrm{HS}}^2 \|A\|_{\mathrm{HS}}^2$$

by part (a) and (b) and its proof. For (f) assume that $\mathcal{H}$ is infinite-dimensional, define $B e_n$ to be $n^{-1/2} e_n$ and show that $B \in \overline{\mathrm{HS}(\mathcal{H})}$. For (g) and (h) apply Proposition 6.11.

Exercise 6.54 (p. 196): Let $A, B \in \mathrm{HS}(\mathcal{H})$ and let (v_n) and (w_n) be two orthonormal lists. Then

$$\sum_{n=1}^{N} |\langle ABv_n, w_n \rangle| \leqslant \sum_{n=1}^{N} \|Bv_n\| \|A^* w_n\| \leqslant \left(\sum_{n=1}^{N} \|Bv_n\|^2 \right)^{1/2} \left(\sum_{n=1}^{N} \|A^* w_n\|^2 \right)^{1/2}$$

shows that $\|AB\|_{\mathrm{tc}} \leqslant \|A\|_{\mathrm{HS}} \|B\|_{\mathrm{HS}}$ by Exercise 6.53(a) and (b) (and its proof above). For (b) suppose first that P is positive, self-adjoint and trace-class, and find $A \in \mathrm{HS}(\mathcal{H})$ with $P = A^2$. Then apply Exercise 6.45.

Exercise 6.55 (p. 196): For (a), show that for any $w \in \mathcal{H}$ the map $\mathcal{H} \ni v \mapsto \langle v, w \rangle_0$ is a bounded linear operator that depends semi-linearly on w. Conclude that it must be of the form $\langle v, w \rangle_0 = \langle v, Aw \rangle_{\mathcal{H}}$ for a bounded operator A. Use the properties of $\langle \cdot, \cdot \rangle_0$ to show that A is positive and self-adjoint. For (b) recall that $\mathcal{H} \ni f \mapsto f(x)$ is a bounded functional and show that A as in (a) is of the form $A(v) = \langle v, v_x \rangle v_x$ for some $v_x \in \mathcal{H}$. For (c) show that $\sup_{x \in K} \|v_x\|$ is finite for all compact subsets K of U (for example, by analyzing the arguments leading to Theorem 5.34).

Exercise 6.62 (p. 199): Apply the argument behind Theorem 5.45 to prove that

$$\|\chi f\|_{H^k(U)} \ll_{\chi,k} |\lambda|^{k/2} \|f\|_2$$

for some fixed $\chi \in C_c^\infty(U)$ and $k \geqslant 1$. Apply Exercise 5.39.

Exercise 6.63 (p. 201): For (a), differentiate under the integral sign to express J_n' and J_n'' as integrals. Simplify

$$x^2 J_n''(x) + (x^2 - n^2) J_n$$

using the identities $\sin^2 t + \cos^2 t = 1$ and $a^2 - b^2 = (a - b)(a + b)$ and integration by parts. Notice that the resulting expression coincides with $-x J_n'(x)$. For (b), repeat the argument for the first integral in the expression for Y_n. The boundary terms from the partial integration cancel with the corresponding expression arising from treating the second integral in the same way (differentiating under the integral needs to be justified as the domain is unbounded), via the identity $\sinh^2 t + 1 = \cosh^2 t$. For (c), notice that if $f \in L^2(U)$ has weight n then f is orthogonal to all eigenfunctions of weight $m \in \mathbb{Z} \setminus \{n\}$.

Exercise 6.66 (p. 204): Given $f(x) = \sin(\pi R^{-1} n_1 x_1) \cdots \sin(\pi R^{-1} n_d x_d)$, for $x \in (0, R)^d$ and $f(x) = 0$ for $x \in \mathbb{R}^d \setminus (0, R)^d$, define

$$f_\lambda(x) = f\left(\left(\tfrac{R}{2}, \ldots, \tfrac{R}{2} \right) + \lambda \left(x - \left(\tfrac{R}{2}, \ldots, \tfrac{R}{2} \right) \right) \right)$$

for $\lambda > 1$ and $\widetilde{f} = f_\lambda * \jmath_\varepsilon$ (cf. Exercise 5.17; also see the proof of Corollary 8.47).

Exercise 6.70 (p. 208): Use $\Delta f_n = \lambda_n f_n$ and $f \in C_c^\infty(U)$ to first show that

$$|\langle f, f_n \rangle| \ll_{f,k} |\lambda_n|^{-k}$$

for any $k \geqslant 1$. Then fix some compact $K \subseteq U$ and use Exercise 6.62 to bound $\|f_n\|_{K,\infty}$ in terms of $|\lambda_n|$, whose growth rate we know.

Exercise 7.5 (p. 212): Construct the complement as a kernel of a linear map, using the Hahn–Banach theorem.

Exercise 7.12 (p. 214): Consider a dense countable subset $\{\ell_1, \ell_2, \dots\}$ of X^* and choose for every ℓ_n some $x_n \in X$ with $\|x_n\| = 1$ and $|\ell_n(x_n)| \geqslant \|\ell_n\|/2$. Now take the $\mathbb{Q}$-linear (or $\mathbb{Q}(i)$-linear) hull of $\{x_n\}$, which is countable, and show that it is dense.

Exercise 7.15 (p. 218): After establishing linearity over $\mathbb{C}$ choose $\theta \in \mathbb{C}$ with $|\theta| = 1$ and $\theta \operatorname{LIM}((a_n)) = \operatorname{LIM}((\theta a_n)) \geqslant 0$ to prove that the complex extension has norm one.

Exercise 7.26 (p. 222): If H is abelian and finitely generated, then H is a quotient of some $\mathbb{Z}^d$ and so has Følner sequences. Use this for finitely generated subgroups of a countable abelian group G to find a Følner sequence for G.

Exercise 7.27 (p. 222): One approach is to construct a box-like (not cube-like) Følner sequence. An alternative is to write the group as a semi-direct product and use Proposition 7.20.

Exercise 7.28 (p. 223): Emulate the strategy used to show that a free group is not amenable in Example 7.22.

Exercise 7.30 (p. 223): For (a), let m_G be a finitely additive left-invariant mean on G. Let $x_0 \in X$ and $B \subseteq X$ and define $m_X(B) = m_G(\{g \in G \mid g \cdot x_0 \in B\})$. For (b), note that setting $X = \mathbb{R}^2$ does not immediately work in order to prove that (a) implies (b), as one would have $m_X(K) = 0$ for any bounded set K in $\mathbb{R}^2$. Instead, use m_X as in (a) to construct a finitely additive function m defined on all bounded sets by setting

$$m(B) = c_n m_X(2^{n+1}\mathbb{Z}^2 + B)$$

for any subset B of $[-2^n, 2^n)^2$ and define $c_n > 0$ so that $m([0,1)^2) = 1$, and show that the definition does not depend on n.

Exercise 7.45 (p. 240): Check the claim first for open subsets, and then argue along the lines used to prove Proposition 2.51.

Exercise 7.49 (p. 241): For (a) note that X has a countable base $\{U_n\}$ for the topology, and write every clopen set as a union of finitely many U_n. For (b) use this to construct an injective continuous map from X to $\{0,1\}^{\mathbb{N}}$.

Exercise 7.53 (p. 248): Apply Theorem 7.44 to obtain a locally finite measure representing the restriction of Λ to $C_0(X)$. Assuming that $\mu(X) = \infty$, find some function $f \in C_0(X)$ for which $\int_X f \, d\mu = \infty$, and then use positivity to obtain a contradiction. Finally, show that μ represents Λ on all of $C_0(X)$ by showing that Λ is necessarily bounded.

Exercise 7.57 (p. 252): Combine the argument in Section 7.4.4 with Theorem 7.54.

Exercise 7.58 (p. 252): For (a), notice that $\ell^\infty(\mathbb{N})$ can be embedded into the Banach space $\mathscr{L}^\infty(X)$ using the subset $\{\frac{1}{n} \mid n \in \mathbb{N}\}$. Now extend the Banach limit from $\ell^\infty(\mathbb{N})$ to $\mathscr{L}^\infty(X)$ and show that it does not arise from a signed measure on X. For (b), if f is a non-measurable bounded function $X \to \mathbb{R}$ then f induces a linear functional on the space

$$\{\mu \in \mathcal{M}(X) \mid |\mu|(B) = 0 \text{ for all } B \subseteq X \smallsetminus D \text{ for some countable set } D \subseteq X\},$$

since for each such measure one can define $\int f \, d\mu$ as a countable sum. Now extend this functional to all of $\mathcal{M}(X)$.

Exercise 8.6 (p. 255): To see that (1) is necessary apply Theorem 4.1.

Exercise 8.8 (p. 255): Apply Theorem 4.1.

Exercise 8.9 (p. 256): For (b) prove $(A^*)^{-1}N_{x_1,\dots,x_n;\varepsilon}(A^*y_0^*) = N_{Ax_1,\dots,Ax_n;\varepsilon}(y_0^*)$.

Exercise 8.12 (p. 258): For (a) use the Baire category theorem (Theorem 4.12). For (b) assume that the neighbourhoods of the form $N_{x_1,\dots,x_n;1/n}(0)$ form a basis of the weak* topology neighbourhoods of $0 \in X^*$ and conclude that X is the linear hull of $\{x_1, x_2, \dots\}$ by using the same argument as in the proof of Lemma 8.13.

Exercise 8.14 (p. 259): Apply Exercises 7.11–7.12 to reduce to the separable case.

Exercise 8.15 (p. 259): Suppose that there is a sequence that converges weakly but not in norm. Show that this implies that there is a sequence (f_n) in $\ell^1(\mathbb{N})$ such that $\|f_n\|_1 = 1$ for all $n \geqslant 1$ but for which f_n converges weakly to 0 as $n \to \infty$. Use this to construct a strictly increasing sequence of natural numbers (I_j) and a subsequence (f_{n_j}) such that $\sum_{k=1}^{I_j-1} |f_{n_j}(k)| \leqslant \frac{1}{5}$ and $\sum_{k=I_j+1}^{\infty} |f_{n_j}(k)| \leqslant \frac{1}{5}$ for all $j \geqslant 1$, where we set $I_0 = 0$. Using this partition, construct an element h in $\ell^\infty(\mathbb{N})$ for which $\sum_{k=1}^{\infty} f_{n_j}(k)h(k)$ does not converge to 0 as $j \to \infty$.

Exercise 8.21 (p. 261): For every $g \in G$ consider the map $L_g : C_{\mathbb{R}}(G) \to C_{\mathbb{R}}(G)$ defined by $(L_g f)(x) = f(gx)$. Show that $\{\Lambda \in C(G)^* \mid \Lambda = \Lambda \circ L_{g_1} = \dots = \Lambda \circ L_{g_n}, \Lambda \geqslant 0, \Lambda(\mathbb{1}) = 1\}$ is a closed non-empty subset of the unit ball in $C(G)^*$ for any $g_1, \dots, g_n \in G$. To see that these sets are non-empty, use induction and suppose that Λ_0 belongs to the set defined by $g_1, \dots, g_{n-1} \in G$. Then any weak* limit of $\frac{1}{K}\sum_{k=0}^{K-1} \Lambda_0 \circ L_{g_n}^k$ will belong to the set defined by $g_1, \dots, g_n \in G$. See also Exercise 8.37 and the discussion there.

Exercise 8.22 (p. 262): For both parts of the exercise, let (v_n) be any sequence in $\mathcal{H}$ with $\|v_n\| \leqslant 1$, assume without loss of generality that $v_n \to v \in \mathcal{H}$ as $n \to \infty$ in the weak* topology, and recall Exercise 8.6. Use compactness of A to prove that $\|AA^*(v_n - v)\| \to 0$ as $n \to \infty$ and consider

$$\|A^*(v_n - v)\|^2 = \langle A^*(v_n - v), A^*(v_n - v)\rangle = \langle (v_n - v), AA^*(v_n - v)\rangle.$$

Exercise 8.23 (p. 262): Show first that S_H is weak* closed and non-empty. Using Theorem 8.10 deduce that the intersection is only empty if some finite intersection $S_{H_1} \cap \dots \cap S_{H_n}$ is empty. However, $H = H_1 + \dots + H_n$ is another finitely generated subgroup and so $S_{H_1} \cap \dots \cap S_{H_n} = S_H$ is non-empty.

Exercise 8.24 (p. 262): Give G the discrete topology, so that by amenability there is a Banach limit in $(\ell^\infty(G))^*$. Restrict this to $C(G)$ and deduce the existence of a translation-invariant measure from the Riesz representation theorem.

Exercise 8.26 (p. 262): Suppose without loss of generality that $x_0^* = 0$. Now apply weak* compactness to the weak* closed subsets $\overline{B_s^{X^*}} \cap K$ for $s > \inf_{k \in K} \|k\|$.

Exercises 8.32–8.33 (p. 264): Both exercises require the generalization of Section 2.3.3 to $\mathbb{T}^d$.

Exercise 8.40 (p. 268): For ergodicity, use Fourier series as in the proof of Lemma 8.38.

Exercise 8.44 (p. 272): Apply Exercise 8.39 to obtain weak* convergence on the set

$$V + B_{s/2}.$$

To obtain strong convergence, express the difference quotient at a point $x \in V$ and for h, $0 < |h| < \frac{s}{2}$ as an integral of shifts of the weak derivative (cf. Lemma 8.42) and apply Lemma 3.74.

Exercise 8.59 (p. 292): The uniform operator topology is the only topology that has neighbourhoods that are bounded with respect to the operator norm. If x_n in X and y_n in Y^* have norm one for all $n \geqslant 1$, then

$$L \longmapsto \sum_{n=1}^{\infty} \frac{1}{2^n} y_n^*(Lx_n)$$

is a continuous functional on $B(X, Y)$ and so also continuous with respect to the weak topology. Choosing the sequence (x_n) carefully makes this functional not continuous with respect to the strong, nor the weak, operator topology. Finally, notice that for the strong operator topology and $x \in X \smallsetminus \{0\}$ there exists a neighbourhood, namely $N_{x;1}(0)$, such that $\{Lx \mid L \in N_{x;1}(0)\} \subseteq Y$ is bounded while there is no such neighbourhood in the weak operator topology.

Exercise 8.62 (p. 293): Apply the Hahn–Banach lemma (Lemma 7.1).

Exercise 8.67 (p. 295): For (c) suppose that $\ell : \mathrm{MF}([0,1])^* \to \mathbb{C}$ is continuous and linear. Suppose $\varepsilon > 0$ is chosen so that $f \in U_\varepsilon(0)$ implies that $|\ell(f)| < 1$. Given any $f \in \mathrm{MF}([0,1])$ use a partition of $[0,1]$ to split f into a finite sum $f = \sum_{k=1}^{n} f_k$ such that $\lambda f_k \in U_\varepsilon(0)$ for all $\lambda \in \mathbb{C}$.

Exercise 8.75 (p. 300): Look at the proof of Theorem 7.3 to see how to obtain a complex-linear functional from a real-linear functional.

Exercise 8.76 (p. 300): Without loss of generality we may assume that 0 is an interior point of K. Fix some $y_0 \in L$ so that 0 is an interior point of $M = K - L + y_0$. Since K and L are disjoint, $K - L$ cannot contain 0 and M does not contain y_0. Now let g be the gauge function of M so $g(y_0) \geqslant 1$. Define $f(\lambda y_0) = \lambda g(y_0)$ for all scalars λ. Extend f to the whole space with $f(x) \leqslant g(x)$ for all x using the Hahn–Banach lemma, and notice that $f(x) \leqslant 1$ for all $x \in M$ and $f(y_0) \geqslant 1$.

Exercise 8.77 (p. 300): Let $r = \inf_{x \in K} \|z - x\|$. Hence for every $\varepsilon > 0$ there is an $x_0 \in K$ such that $v = z - x_0$ satisfies $\|v\| < r + \varepsilon$ and hence for $\ell \in X^*$ with $\|\ell\| = 1$ we have

$$r + \varepsilon \geqslant \ell(v) = \ell(z) - \ell(x_0) \geqslant \ell(z) - \sup_{x \in K} \ell(x).$$

For the converse set $L = B_r(z)$ and apply Exercise 8.76 for K and L.

Exercise 8.78 (p. 300): Consider the compact convex set $K = \imath \overline{\left(B_1^X \right)}$ with the closure taken with respect to the weak* topology, assume that $\ell \in \overline{(B_1^X)^{**}} \smallsetminus K$, and apply Theorem 8.73 using the weak* topology on X^{**}.

Exercise 8.84 (p. 304): For (a) use the Arzela–Ascoli theorem. For (c) use piecewise linear functions as in (b) to approximate a given function $f \in K$. For (d) use the fact that any $f \in K$ is almost everywhere differentiable with derivative in $[-1, 1]$.

Exercise 8.87 (p. 304): Notice that two possible barycentres of μ cannot be separated by X^*.

Exercise 8.89 (p. 306): To see that the set of barycentres is closed, show that

$$\{(\mu, x) \in C(M)^* \times K \mid \mu \text{ is a probability measure on } M, x \text{ is the barycentre of } \mu\}$$

is a compact subset of $C(M)^* \times X$ and consider the projection map to K.

Exercise 8.91 (p. 306): Use induction on the dimension n. If x_0 is a boundary point, then there exists a hyperplane V that contains x_0 with the property that K lies in one of the closed half-spaces with boundary V. If x_0 is an interior point, take any extreme point y and find a boundary point z such that x_0 is in the line segment from y to z.

Exercise 8.93 (p. 311): One direction is clear using Theorem 8.90. For the other direction, suppose that μ represents x_0 and $\mu \neq \delta_{x_0}$. Then there exists some y in $\mathrm{Supp}(\mu) \smallsetminus \{x_0\}$, a linear functional $\ell \in X^*$, and an open neighbourhood U of y with $\mu(U) \in (0,1)$ and $\sup_{z \in U} \ell(z) < \ell(x_0)$. Now use $\mu = \lambda \frac{1}{\mu(U)} \mu|_U + (1-\lambda) \frac{1}{\mu(K \smallsetminus U)} \mu|_{K \smallsetminus U}$ with $\lambda = \mu(U)$ and the existence of barycentres (Lemma 8.88) to see that x_0 is not extreme.

Exercise 9.9 (p. 319): For (a) the precise condition is $f(x) \neq 0$ for μ_v-almost every x. Simplifying the notation, assume that $\mathcal{H} = L^2(\mathbb{T}, \mu)$. If f vanishes on a set B of positive measure, then clearly $\mathcal{H}_f \perp \mathbb{1}_B$. So suppose that $f \neq 0$ almost everywhere. Then clearly $gf \in \mathcal{H}_f$ for g any character, hence for g any trigonometric polynomial, hence for $g \in C(\mathbb{T})$, hence for $g = \mathbb{1}_O$ for any open set O by dominated convergence, and finally for $g = \mathbb{1}_G$ for any G_δ-set $\bigcap_{n \geq 1} O_n$. Since any measurable set coincides modulo μ with a G_δ-set, we may apply dominated convergence once again to obtain the case $g \in L^\infty_\mu(\mathbb{T})$. Apply this to the function defined by $g_n = \frac{1}{f} \mathbb{1}_{\{x \in \mathbb{T} \mid |f(x)| \geq 1/n\}}$ to obtain $\mathbb{1}_{\{x \in \mathbb{T} \mid |f(x)| \geq 1/n\}} \in \mathcal{H}_f$ for all $n \geq 1$ and conclude that $\mathbb{1} \in \mathcal{H}_f$. In (b) the spectral measure is given by the Lebesgue measure.

Exercise 9.10 (p. 320): For (a) note first that $\sum_{n=0}^\infty d_n \left(\sum_{k=1}^\infty c_k z^k\right)^n = z$ for sufficiently small z, and as an identity in the ring $\mathbb{C}[[z]]$ of formal power series. Using the assumption that $\sum_{n=0}^\infty |d_n| \left(\sum_{k=1}^\infty |c_k| \|A\|^k\right)^n < \infty$ we see first that

$$g\big(f(A)\big) = \sum_{n=0}^\infty d_n \left(\sum_{k=1}^\infty c_k A^k\right)^n \approx \sum_{n=0}^N d_n \left(\sum_{k=1}^K c_k A^k\right)^n$$

if N and K are sufficiently large, so

$$\sum_{n=0}^N d_n \left(\sum_{k=1}^K c_k A^k\right)^n = A + \sum_{\ell=\min\{N,K\}+1}^{NK} e_{N,K,\ell} A^\ell,$$

where the last sum can be made arbitrarily small if N and K are sufficiently large.

Exercise 9.14 (p. 323): Use Theorem 9.2 to obtain

$$\mathcal{H} = \bigoplus_{n \geq 1} \mathcal{H}_{w_n} \cong \bigoplus_{n \geq 1} L^2(\mathbb{T}, \mu_{w_n})$$

and prove that $\mathcal{H}^\rho$ can be expressed as the direct sum of certain subspaces of $L^2(\mathbb{T}, \mu_{w_n})$.

Exercise 9.15 (p. 323): For (a), multiply by the square root of the Radon–Nikodym derivative of one measure with respect to the other. Using (a), we may modify the measures in Corollary 9.13 to satisfy $\mu_n = \mu_1|_{B_n}$ for some nested sequence of Borel sets

$$B_1 = \mathbb{T} \supseteq B_2 \supseteq \cdots,$$

hence (b) follows by repeating that argument. For (c), show that $f \in L^2(\mathbb{T}, \nu_1)$ satisfies the property defining $\mathcal{H}^{(1)}$. For the converse note that we can define in a measurable way for any $u \in \mathbb{C}^n$ with $n \geqslant 2$ or $u \in \ell^2(\mathbb{N})$ a vector u' in the same space with $u' \perp u$ and with $\|u'\| = \|u\|$, for example by first projecting onto $\mathbb{C}^2 \subseteq \mathbb{C}^n \subseteq \ell^2(\mathbb{N})$ and there using the orthogonal direction or a suitable multiple of the first basis vector if the projection is zero. Now consider a general function $F = (f_1, f_2, \ldots, f_\infty)$ with $f_n : (\mathbb{T}, \nu_n) \to \mathbb{C}^n$ and $f_\infty : (\mathbb{T}, \nu_\infty) \to \ell^2(\mathbb{N})$. If $f_n \neq 0$ for some $n \in \{\infty, 2, 3, \ldots\}$ then, using the argument above, construct a function f'_n with $f_n(x) \perp f'_n(x)$ and with $\|f_n(x)\| = \|f'_n(x)\|$ for ν_n-almost every x. Set $f'_m = 0$ for all $m \neq n$ and conclude that F does not satisfy the property defining $\mathcal{H}^{(1)}$, showing the reverse inclusion. For (d) argue in a similar way: For F given by $(0, f_2, 0, \ldots, 0)$, define F_2 by rotating f_2 and show the defining property for $\mathcal{H}^{(2)}$. We note again that for $n \geqslant 3$ we can measurably define for every $u_1, u_2 \in \mathbb{C}^n$ or in $\ell^2(\mathbb{N})$ a vector u' orthogonal to both u_1 and u_2 with $\|u'\| = \|u_1\|$. Using this argue as before.

Exercise 9.19 (p. 326): Fix some $v \in \mathcal{H}$ and describe the unitary operator U on $\mathcal{H}_v$ by a multiplication operator on $L^2(\mathbb{S}^1, \mu_v)$. Now calculate the spectral measures $\mu_{v,w} = \mu_{v, P(w)}$ in that context, where $P : \mathcal{H} \to \mathcal{H}_v$ is the orthogonal projection.

Exercise 9.23 (p. 327): For R_α note that the characters are eigenfunctions. For A show that a character is mapped to a character but that the orbit of any non-trival character is infinite, and then apply Lemma 9.12.

Exercise 9.33 (p. 334): For (c), apply the Stone–Weierstrass theorem (see Exercise 2.43). For (d), first approximate g simultaneously in $L^1(\mathbb{R}^d)$ and $L^2(\mathbb{R}^d)$ by some function f_0 in $C_c(\mathbb{R}^d)$. Then approximate $e^{\pi \|x\|^2} f_0(x)$ by some function $f_1 \in \mathcal{A}$ with respect to $\|\cdot\|_\infty$, and notice that $f(x) = e^{-\pi \|x\|^2} f_1(x)$ will then approximate g with respect to $\|\cdot\|_1$ and $\|\cdot\|_2$. For (e) consider $f_1, f_2 \in \mathcal{A}$ and express the inner product in the form

$$\langle f_1, f_2 \rangle = \int f_1 \overline{f_2} \, dx = \widehat{(f_1 \overline{f_2})}(0)$$

and use part (b) and Proposition 9.34.

Exercise 9.41 (p. 340): Show that the four-fold Fourier transform of a function is again the original function, and apply the argument used in Section 1.1 (or Theorem 3.80). Also consider the function f in Example 9.27 together with $\lambda_{x_0} f$ and products of such functions to prove that all four possible eigenvalues appear.

Exercise 9.47 (p. 342): Consider the associated (well-defined) function g in $C^\infty(\mathbb{T}^d)$ defined by $g(x) = \sum_{n \in \mathbb{Z}^d} f(n + x)$.

Exercise 9.48 (p. 342): First show that we can approximate any function in $L^p(\mathbb{R})$ by a function of compact support, so it is enough to approximate a compactly supported function $f \in L^p(\mathbb{R})$ (so that $\widehat{f} \in C^\infty(\mathbb{R})$). Let $h_1 \in C_c^\infty(\mathbb{R})$ be a non-trivial real-valued function with $h_1(x) = h_1(-x)$, define h to be $h_1 * h_1$, multiply by a scalar so that $h(0) = 1$, and set $h_r(x) = h(rx)$ for all $r > 0$. Now prove that $\widehat{h_r} * f \to f$ in $L^p(\mathbb{R})$ as $r \to \infty$ using Jensen's inequality as in the proof of Theorem 9.39.

Exercise 9.50 (p. 343): Use the condition for equality in the Cauchy–Schwarz inequality to deduce that f must satisfy a differential equation of the form $f'(x) = \lambda x f(x)$ for some $\lambda \in \mathbb{R}$.

Exercise 9.51 (p. 343): Notice that if for f we have equality, then we have equality in Exercise 9.50 for $g(x) = e^{-2\pi i x t_0} f(x) + x_0$.

Exercise 9.55 (p. 345): For (a) let $\mathbb{C}[G]$ be the space of finitely supported complex-valued measures and use p to define a semi-inner product on $\mathbb{C}[G]$ with

$$p(g) = \langle \delta_g, \delta_e \rangle = \langle \delta_{hg}, \delta_h \rangle$$

for all $g, h \in G$. Then show that $\pi_h(\delta_g) = \delta_{hg}$ extends to a unitary representation on the completion of $\mathbb{C}[G]$ modulo the kernel of the semi-norm induced by the semi-inner product. For (b) assume that p is extreme and the unitary representation in (a) is reducible, decompose the generator into the components corresponding to an invariant subspace and its orthocomplement, and study the matrix coefficient of these three vectors.

Exercise 9.63 (p. 352): For (b) notice that this is equivalent to $\widehat{C_c^\infty}(\mathbb{R})$ being dense in $\overline{\mathrm{Graph}(A)} = \{(f, g) \in L^2(\mathbb{R}) \times L^2(\mathbb{R}) \mid g(t) = tf(t) \text{ for } t \in \mathbb{R}\}$. For this improve the argument for Exercise 9.48 (also see its hint on p. 577) by proving that

$$M_I\big(\widehat{h_r} * f\big) = \big(M_I \widehat{h_r}\big) * f + \widehat{h_r} * (M_I f)$$

and showing that $\big(M_I \widehat{h_r}\big) * f \to 0$ in $L^2(\mathbb{R})$ as $r \to \infty$. For (d) extend Proposition 9.43 to weak derivatives.

Exercise 10.4 (p. 360): Set $B_1 = f^{-1}\big(\big\{z \in \mathbb{C} \mid \Re(z) \in [\frac{k}{n}, \frac{k+1}{n}), \Im(z) \in [\frac{\ell}{n}, \frac{\ell+1}{n})\big\}\big)$ for some $k, \ell \in \mathbb{Z}$ and $n \in \mathbb{N}$, and set $B_2 = G \setminus B_1$. Combine the assumption in the exercise and Lemma 10.3 to conclude that $m_G(B_1) = 0$ or $m_G(B_2) = 0$. Vary k, ℓ, n to conclude the proof.

Exercise 10.5 (p. 360): For (a) show that $\theta_* m_G(B) = m_G(\theta^{-1} B)$ defines a left Haar measure and use Proposition 10.2. For the continuity in (b) let $B = K$ be a fixed compact set and use the regularity of the measure m_G. For (c) use the substitution formula

$$\int f \circ \theta \, dm_G = \int f \, d\theta_* m_G.$$

For (d) apply (a) with $B = G$.

Exercise 10.6 (p. 361): For $f \in C_c(G)$ we may use uniform continuity to argue that

$$\psi_n * f(g) = \int \psi_n(h) \underbrace{f(h^{-1}g)}_{\approx f(g)} \, dm_G(h)$$

since ψ_n vanishes outside U_n. For the convolution on the right a different argument is needed as follows. Write

$$f * \psi_n(g) = \int_G f(h)\psi_n(\underbrace{h^{-1}g}_{=k^{-1}}) \, dm_G(h) = \int_G f(gk)\psi_n(k^{-1}) \, dm_G(k),$$

using the subsitution $gk = h$. From here (depending on how much one wishes to assume about the sequence of functions) one could assume that each ψ_n is symmetric in the sense that $\psi_n(g) = \psi_n(g^{-1})$, or use the fact that the modular character is itself a continuous function so the difference between integrating against $\psi_n(k^{-1})$ and $\psi_n(k)$ for $k \in U_n$ is small. To deduce the result for $f \in L^1(G)$ use the usual approximation arguments.

Exercise 10.7 (p. 361): Let $U = U^{-1}$ be a compact neighbourhood of the identity e in the group G. Find a maximal collection of disjoint left translates $g_1 U, g_2 U, \ldots$ and show that this collection must be finite. Now show that $G = \bigcup g_i U^2$.

Exercise 10.20 (p. 373): Use Proposition 7.20(a) and the Følner condition. For the converse use the same argument as in Exercise 8.23.

Exercise 10.21 (p. 373): For (1) notice that the proof that (3) $\implies$ (1) in Theorem 10.15 only uses finite sets. For (2) show, for example, that a function f in $\mathscr{P}(G)$ satisfying Reiter's condition in Definition 10.14 for a compact $K \subseteq G$ and $\varepsilon > 0$ also satisfies a topological version for all $f_0 \in \mathscr{P}(G)$ that vanish outside of K. Use this to induce a left-invariant mean that is also topologically left-invariant.

Exercise 10.22 (p. 373): For (1) show that $G \times G$ is amenable. Then use the left-invariant mean M_2 on $G \times G$ to define $M(\phi) = M_2((g_1, g_2) \mapsto \phi(g_1 g_2^{-1}))$. For (2) convolve a function f as in Reiter's condition with its flipped version $\tilde{f}(g) = f(g^{-1})$ and show that $f * \tilde{f}$ satisfies Reiter's condition for left- and right-multiplication.

Exercise 10.23 (p. 373): If G is discrete and uncountable, then combine the following with the conclusion in Exercise 10.20. So assume that G is σ-compact, locally compact, and metric. For (1) $\implies$ (2), define for $f \in C(X)$ the functional $\Lambda(f) = M(g \mapsto f(gx))$ for some left-invariant mean M on G and some $x \in X$. For (2) $\implies$ (3) one would like to use the action of G on K to find an invariant measure μ and then apply Lemma 8.88 to find a G-invariant barycentre of μ in K. However, as K is not assumed to be metrizable this requires a small work-around as follows. Let $\| \cdot \|_1, \ldots, \| \cdot \|_m$ be a finite collection of semi-norms on V and write $G = \bigcup_{n=1}^{\infty} G_n$ with G_n compact and G_n contained in G_{n+1}^o for all $n \geqslant 1$. Show that the semi-norms

$$\|v\|_{k,n} = \sup_{g \in G_n} \|\pi_g^{\mathrm{lin}}(v)\|_k$$

are finite for any $1 \leqslant k \leqslant m$ and $n \geqslant 1$ and are compatible with the topology on V. Define $V_0 = \{v \in V \mid \|v\|_{k,n} = 0 \text{ for all } 1 \leqslant k \leqslant m \text{ and } n \geqslant 1\}$, set $W = V/V_0$ and define $p : V \to W$ by $p(v) = v + V_0$. Equip W with the collection of quotient semi-norms induced by $\| \cdot \|_{k,n}$ and show that p is continuous. Show that G acts continuously on W and that p is equivariant for the G-action. By applying the argument for the metrizable case outlined above, show that the set

$$\{v \in K \mid \|\pi_g^{\mathrm{aff}}(v) - v\|_1 = \cdots = \|\pi_g^{\mathrm{aff}}(v) - v\|_m = 0 \text{ for all } g \in G\}$$

is closed and non-empty for any finite collection $\| \cdot \|_1, \ldots, \| \cdot \|_m$ of semi-norms on V. Finally apply compactness. For (3) $\implies$ (1) we would like to use the compact convex set

$$\mathscr{M}(G) \subseteq (L^\infty(G))^*$$

and the linear action λ_g^* for the left regular representation $\lambda_g : L^\infty(G) \longrightarrow L^\infty(G)$. If G is discrete this is possible, but in general this does not define a continuous affine action of the sort considered in (3). For this reason, define the subspace

$$\mathrm{LUC}(G) = \{\phi \in \mathscr{L}^\infty(G) \mid \|\lambda_g \phi - \phi\|_\infty \longrightarrow 0 \text{ as } g \to e\}$$

of left uniformly continuous functions on G, and its dual space $X = (\mathrm{LUC}(G))^*$ with the topology induced by the semi-norms $\|M\|_{K,\phi} = \sup_{g \in K} |M(\lambda_g \phi)|$ for $M \in (\mathrm{LUC}(G))^*$, where $K \subseteq G$ is a non-empty compact subset and $\phi \in \mathrm{LUC}(G)$. Show that on any bounded subset B of $(\mathrm{LUC}(G))^*$ the topology induced by these semi-norms agrees with the weak* topology on B. Show that the action λ_g^* for $g \in G$ on X satisfies the assumptions of (3) and deduce that there exists a left-invariant mean on $\mathrm{LUC}(G)$. Use the argument from the proof of Lemma 10.17 and the step (1) $\implies$ (3) in Theorem 10.15 to complete the argument.

Exercise 10.24 (p. 374): For (1), show that the Reiter condition for G implies the Reiter condition for H. Let $f \in C_c(G) \cap \mathscr{P}(G)$ satisfy the Reiter condition for $\varepsilon > 0$ and a finite $K \subseteq H$. Define the space $X = H\backslash G$ with the usual map $g \mapsto Hg$ from $G \to X$, and the probability measure ν on X by

$$\nu(B) = \int_G \mathbb{1}_B(Hg) f(g) \, dm(g).$$

It suffices to find $g \in G$ such that

$$F(g) = \sum_{k \in K} \frac{\int |f(k^{-1}hg) - f(hg)| \, dm_H(h)}{\int f(hg) \, dm_H(h)}$$

is defined and bounded above by $|K|\varepsilon$. Show that $F(h_0 g) = F(g)$ for every h_0 in H (even if H is not unimodular), that $\nu\left(\{Hg \mid \int f(hg) \, dm_H(h) = 0\}\right) = 0$, and choose a compact subset $L \subseteq H$ so that $f(g) > 0$ and $f(hg) > 0$ (or $f(k^{-1}hg) > 0$) implies $h \in L$. Then

$$\int_X F(Hg) \, d\nu = \int_G F(Hg) f(g) \, dm_G(g)$$

$$= \sum_{k \in K} \frac{1}{m_H(L)} \int_G F(Hg) \int_L f(hg) \, dm_H(h) \, dm_G(g)$$

$$= \sum_{k \in K} \frac{1}{m_H(L)} \int_G \int_L |f(k^{-1}hg) - f(hg)| \, dm_H(h) \, dm_G(g)$$

$$= \sum_{k \in K} \|\lambda_k f - f\|_1 < |K|\varepsilon.$$

Finally, use Exercise 10.21. For (2) use Exercise 10.23 (either (2) or (3)).

Exercise 10.25 (p. 374): Use Exercises 10.23(2) and 10.24(1).

Exercise 10.28 (p. 374): For (1), fix a generating set and use metric open balls of increasing radius to define a Følner sequence.

Exercise 10.30 (p. 375): Take the closed convex hull K of $\{\pi_g v \mid g \in G\}$ and apply Theorem 3.13 with $v_0 = 0$.

Exercise 10.35 (p. 376): Apply the definitions and Exercises 10.7 and 10.30.

Exercise 10.37 (p. 377): For any finite subset $F \subseteq G$ define the subgroup $H_F = \langle F \rangle$ generated by F. Note that G acts on the quotient space G/H_F and also unitarily on $\ell^2(G/H_F)$. Consider the direct product representation of G on $\bigoplus_F \ell^2(G/H_F)$.

Exercise 10.47 (p. 384): Combine Theorem 10.38 and Proposition 10.41.

Exercise 10.53 (p. 388): Given a function $f \in L^2(G)$ show that for almost every $g \in G$ the function $f_g : \Gamma \ni \gamma \mapsto f(g\gamma)$ belongs to $\ell^2(\Gamma)$. Define $\phi : L^2(G) \to \mathcal{H}_G$ by $\phi(f)(g) = f_g$ for all $g \in G$.

Exercise 10.64 (p. 404): Show that the characteristic function of a 'connected component' is also an eigenfunction for eigenvalue one, and use the level sets of a non-constant eigenfunction for the converse.

Exercise 10.67 (p. 406): Combine the argument after Proposition 10.41 with division with remainder in $\mathbb{Z}$.

Exercise 11.4 (p. 410): For $\lambda = 0$. Use Corollary 4.30 (or Exercise 6.25) to see that M_g is invertible if and only if $\mu(\{0\}) = 0$ (respectively, g is non-zero μ-almost everywhere) and $z \mapsto \frac{1}{z}$ (resp. $\frac{1}{g}$) is essentially bounded with respect to μ.

Exercise 11.7 (p. 410): For (a) use the isomorphism between $\ell^2(\mathbb{Z})$ and $L^2(\mathbb{T})$ provided by Fourier series (Theorem 3.54). For (b), you may show that $\mathcal{A}$ is isomorphic (as a Banach algebra) to the algebra generated by S with S as in Exercise 6.1(b).

Exercise 11.9 (p. 411): Use Lemma 2.67 and Theorem 11.6.

Exercise 11.12 (p. 413): Recall from Section 2.4.2 that multiplication is continuous.

Exercise 11.18 (p. 417): Use the C^*-property of the norm to first show $\|a\| \leqslant \|a^*\|$ for all $a \in \mathcal{A}$.

Exercise 11.20 (p. 417): Start with the identity $1_{\mathcal{A}}^* 1_{\mathcal{A}} = 1_{\mathcal{A}} 1_{\mathcal{A}}^* = 1_{\mathcal{A}}^*$, apply the star operator and then use the C^*-property.

Exercise 11.36 (p. 425): For (a), combine Proposition 11.21, Corollary 11.29, and Exercise 2.43(b). For (c), use (a) and the fact that $C_0(\sigma(\mathcal{A})) \oplus \mathbb{C} \cong C(\sigma(\mathcal{A}) \cup \{\infty\})$.

Exercise 11.41 (p. 428): The Banach algebra of limits of absolutely convergent Fourier series with pointwise multiplication is isometrically isomorphic to $\ell^1(\mathbb{Z}^d)$ with convolution. Apply Theorem 11.23 and Proposition 11.38 to $\mathbb{Z}^d$ and $\widehat{\mathbb{Z}^d} \cong \mathbb{T}^d$.

Exercise 11.42 (p. 428): Show first that if G is a locally compact metrizable abelian group then
$$V = \overline{\langle \lambda_y f \mid y \in G \rangle}$$
cannot be $L^1(G)$ if $\widehat{f}(t) = 0$ for some $t \in \widehat{G}$. Next show that $L^1(G) * f \subseteq V$ (for example, using Propositions 2.51, 3.81, and 3.91). For (a), take $G = \mathbb{T}^d$ and show that $\chi_n \in V$ for all $n \in \mathbb{Z}^d$ by Lemma 3.59(2). For (b), set $\mathcal{A} = L^1(G) \oplus \mathbb{C}$ as in Exercise 11.1. Replace f by $\widetilde{f} * f$ if necessary to assume $\widehat{f} > 0$. Fix some $g \in \mathscr{S}(\mathbb{R}^d)$ such that $\widehat{g} \in C_c^\infty(\mathbb{R}^d)$. Let $h \in \mathscr{S}(\mathbb{R}^d)$ have $\widehat{h} \in C_c^\infty(\mathbb{R}^d)$, $\widehat{h} \in [0,1]$ and $\widehat{h} \equiv 1$ on $\mathrm{Supp}(\widehat{g})$. Show that $h * g = g$ and that $1_{\mathcal{A}} - h + f \in \mathcal{A}$ is invertible. Use this to show that $g \in V$ and apply Exercise 9.48.

Exercise 11.45 (p. 431): For (a), suppose $\chi \in \widehat{G} \smallsetminus \{\mathbb{1}\}$ has $\chi^n = 1$ for some $n > 1$ and notice that χ then takes values in a discrete subgroup of $\mathbb{S}^1$. For (b), suppose that $G = O_1 \sqcup O_2$ is a partition into two non-empty clopen sets. Show that there exists a neighbourhood U

of $e \in G$ with $U + O_j = O_j$ for $j = 1, 2$. Define $H = \langle U \rangle$ and show that H is a proper open subgroup of G, and that G/H is a finite abelian group.

Exercise 12.10 (p. 435): For the description of $\sigma_{\text{resid}}(T)$, prove that

$$(\text{im}(T - \lambda I))^{\perp} = \ker\left(T^* - \overline{\lambda}I\right)$$

and then use this together with an explicit description of T^* (see Exercises 6.23(c) and 6.1(b)).

Exercise 12.17 (p. 437): Since the kernel of $I - A$ is the eigenspace of A for eigenvalue 1, almost injectivity follows directly from compactness of A (see, for example, Exercise 3.40). The proof that $\text{im}(I - A)$ is closed is a little more involved. Assume first that

$$(I - A)v_n = v_n - Av_n \to w$$

as $n \to \infty$ with $v_n \in \ker(I - A)^{\perp}$. Show that (v_n) is bounded (for example, by assuming that $\|v_n\| \to \infty$ as $n \to \infty$ and applying compactness for $v_n' = \|v_n\|^{-1}v_n$). Finally, use compactness of A to conclude $w \in \text{im}(I - A)$. To prove that $T = I - A$ is almost surjective assume that $V = (T(\mathcal{H}))^{\perp}$ is infinite-dimensional, and let (v_n) be an orthonormal basis of V so that $\langle v_n, v_n - Av_n \rangle = 0$ for all $n \geq 1$. Now choose a subsequence (v_{n_k}) with $Av_{n_k} \to w$ as $k \to \infty$ and derive a contradiction.

Exercise 12.18 (p. 437): For the first direction assume that T is Fredholm, let $\mathcal{H}_1$ be $(\ker(T))^{\perp}$, $\mathcal{H}_2$ be $\text{im}(T)$, and use Proposition 4.25 to show that $T|_{\mathcal{H}_1} : \mathcal{H}_1 \to \mathcal{H}_2$ has a bounded inverse. Define $S|_{\mathcal{H}_2}$ to be $\left(T|_{\mathcal{H}_1}\right)^{-1}$ and $S|_{\mathcal{H}_2^{\perp}} = 0$. For the converse apply Exercise 12.17 to the compact operators $ST - I$ and $TS - I$.

Exercise 12.21 (p. 438): Use Fourier series and the isomorphism $\ell^2(\mathbb{Z}) \cong L^2(\mathbb{T})$.

Exercise 12.27 (p. 442): The base cases $n = 0$ and $n = 1$ hold trivially by definition. For $n \geq 1$ consider

$$\frac{1}{\sqrt{p}} S\big(\mathbf{U}_n(f)\big)(v) = \frac{1}{\sqrt{p}} \sum_{v' \sim v} \mathbf{U}_n(f)(v') = \frac{1}{p^{(n+1)/2}} \left(\sum_{v' \sim v} \sum_{\substack{k \leq n, \\ k \equiv n (\text{mod} 2)}} \sum_{w \sim_k v'} f(w) \right)$$

and count how often the term $f(w)$ appears in this sum, distinguishing between the cases $d(w, v) = n + 1$ and $d(w, v) \leq n - 1$.

Exercise 12.29 (p. 442): Use the addition formula for $\sin\big((n + 1)\theta + \theta\big)$.

Exercise 12.30 (p. 442): For (a) use the operator $\mathbf{U}_{2n} - \frac{1}{p}\mathbf{U}_{2n-2}$ and Cauchy–Schwarz on the finite set $\{w \mid w \sim_{2n} v\}$. For (b) define $\tilde{f}(v) = (-1)^{d(v,v_0)} f(v)$ for a fixed vertex v_0. For (c) treat the case $\theta = 0$ first. If $\theta > 0$ recall first that $p \geq 2$ and use Exercise 12.29 and (b) to deduce that it is enough to show that there are infinitely many n with

$$|\sin((2n + 1)\theta)| \geq \tfrac{1}{2} + \varepsilon$$

for some fixed $\varepsilon > 0$. Note that this holds, for example, if $(2n + 1)\theta \in \pi\mathbb{Z} + [\tfrac{\pi}{4}, \tfrac{3\pi}{4}]$. Now consider the following three cases: If $\theta \leq \tfrac{\pi}{4}$, then every closed interval of length $\tfrac{\pi}{2}$ is visited by the rotation on $\mathbb{R}/(\pi\mathbb{Z})$ by 2θ infinitely often. If $\theta = \tfrac{\pi}{2} - \phi$ for some $\phi \in (0, \tfrac{\pi}{4}]$, then

$$(2n + 1)\theta + \mathbb{Z}\pi = \tfrac{\pi}{2} - (2n + 1)\phi + \mathbb{Z}\pi$$

and the same argument applies. In the only remaining case $\theta = \frac{\pi}{2}$ works.

Exercise 12.32 (p. 444): Use multiplication by $\left(\frac{\mathrm{d}\mu}{\mathrm{d}\nu}\right)^{1/2}$.

Exercise 12.44 (p. 451): Show that if $f \in C(\sigma(T))$ then $\lambda \notin f(\sigma(T)) \implies \lambda \notin \sigma(f(T))$ by applying the functional calculus for $g(z) = \frac{1}{f(z)-\lambda}$. For the converse, let λ be an element of $\sigma_{\mathrm{appt}}(T)$ (noting that there is no residual spectrum in this case) and generalize the argument for Theorem 12.37(4) to this case, again using a sequence of polynomials.

Exercise 12.48 (p. 453): Define the positive self-adjoint operators $B = T^*T$ and $A = \sqrt{B}$ using Corollary 12.45. Show that $\|Tv\| = \|Av\|$ for all $v \in \mathcal{H}_1$ and

$$(\mathrm{im}(A))^{\perp} = \ker(A) = \ker(T).$$

Define $Uv = 0$ for $v \in \ker(T)$ and $UAv = Tv$ for $Av \in \mathrm{im}(A)$. Show that U is well-defined and extends to an isometry on $(\ker(T))^{\perp}$ satisfying the claims in the exercise.

Exercise 12.51 (p. 454): Show that

$$\langle f(T)v, v \rangle = \int_X f(g(x))|v(x)|^2 \, \mathrm{d}\mu(x) = \int_{\mathbb{C}} f(y) \, \mathrm{d}\mu_v(y)$$

where μ_v is the push-forward under g of the measure $|v|^2 \, \mathrm{d}\mu$.

Exercise 12.54 (p. 456): Adapt the argument from Section 9.1.2. If $\mathcal{H}$ is not separable, combine these arguments with Zorn's lemma.

Exercise 12.57 (p. 458): Apply the polar decomposition from Exercise 12.48 to find an isometry $U : \mathcal{H}_1 \to \mathcal{H}_2$ and a positive self-adjoint operator $A \in B(\mathcal{H}_1)$ with $T = UA$. Deduce that U and A are bijective and show $A\pi_1(g) = \pi_1(g)A$ and $U\pi_1(g) = \pi_2(g)U$ for all $g \in G$.

Exercise 12.58 (p. 458): For (a), consider the self-adjoint operator $A = B^*B$ with

$$\pi_1(g)A = A\pi_1(g)$$

for all $g \in G$. If $\sigma(A)$ contains more than one point, then there exist two non-zero functions $f_1, f_2 \in C(\sigma(A))$ such that $f_1 f_2 = 0$, which implies that $V = \ker(f_1(A))$ is a closed proper subspace. Then show that V is invariant under $\pi_1(g)$ for all $g \in G$. For (b), apply the same argument to $\frac{B+B^*}{2}$ and $\frac{B-B^*}{2\mathrm{i}}$.

Exercise 12.59 (p. 458): By Exercise 9.55(b) all that remains is to show that irreducibility of the unitary representation implies extremality. Suppose therefore that $\pi_\phi : G \curvearrowright \mathcal{H}_\phi$ and $\phi = \lambda\phi_1 + (1 - \lambda)\phi_2$ for some $\lambda \in (0, 1)$ and $\phi_1, \phi_2 \in \mathcal{P}(G)$. Construct $\pi_1 : G \curvearrowright \mathcal{H}_1$ with generator v_1 and $\pi_2 : G \curvearrowright \mathcal{H}_2$ with generator v_2 using Exercise 9.55(a) so that

$$\phi(g) = \lambda\phi_1(g) + (1-\lambda)\phi_2(g) = \lambda\langle\pi_1(g)v_1, v_1\rangle + (1 - \lambda)\langle\pi_2(g)v_2, v_2\rangle$$
$$= \langle\pi(g)(\lambda^{1/2}v_1 + (1-\lambda)^{1/2}v_2), \lambda^{1/2}v_1 + (1-\lambda)^{1/2}v_2\rangle,$$

where $\pi(g) = \pi_1(g) \times \pi_2(g)$ on $\mathcal{H} = \mathcal{H}_1 \times \mathcal{H}_2$. Thus $v = \lambda^{1/2}v_1 + (1 - \lambda)^{1/2}v_2$ generates a cyclic sub-representation $\mathcal{H}_v$ of $\mathcal{H}$ isomorphic to $\mathcal{H}_\phi$ by Lemma 9.53. Consider the orthogonal projection P from $\mathcal{H}_v \subseteq \mathcal{H}_1 \times \mathcal{H}_2$ onto $\mathcal{H}_1$ and apply Schur's lemma (Exercise 12.58) to deduce that the unitary representations π_ϕ and π_1 are unitarily isomorphic under an isomorphism sending v to v_1. Similarly for π_2, and hence $\phi = \phi_1 = \phi_2$.

Exercise 12.61 (p. 461): For (a) and the first part of (b) notice that $\imath$ is continuous by definition of the weak* topology. For the example in (b) set $T_2 = f(T_1)$ for some function $f \in C(\sigma(T_1))$, or consider a measure μ on $\sigma(T_1) \times \sigma(T_2)$ whose support projects surjectively onto each coordinate and define both operators as multiplication operators, or use, for example, two diagonal 3-by-3 matrices T_1, T_2, each with two different eigenvalues such that $T_1 T_2$ has 3 different eigenvalues.

Exercise 12.65 (p. 462): Show that $\mathrm{d}\mu_{v,w} = \overline{f_0}\,\mathrm{d}\mu_v$ satisfies (12.15) for all $a \in \mathcal{A}$.

Exercise 12.69 (p. 467): For (a) assume that $\mu(U \times \mathbb{N}) = 0$ for some non-empty open set $U \subseteq \sigma(\mathcal{A})$. Use some non-zero $f \in C_c(U) \hookrightarrow C(\sigma(\mathcal{A})) \cong \mathcal{A}$ to derive a contradiction. For (b), notice that continuity of π follows from the definition of the weak* topology. For (c), write $\widehat{a}$ for the Gelfand transform of $a \in \mathcal{A}$ when considered as an element of $\mathcal{A}'$ and show that $\widehat{a} = a^o \circ \pi$ for all $a \in \mathcal{A}$. Now use the characterizing property of spectral measures. For (d) use (c). For (e), note that $\pi(\sigma(\mathcal{A}')) \subseteq \sigma(\mathcal{A})$ is compact and that by (c) we have $\mathrm{Supp}(\mu_{v,w}) \subseteq \pi(\sigma(\mathcal{A}'))$ for all $v, w \in \mathcal{H}$. Now apply (a).

Exercise 12.72 (p. 468): Note that $ST = TS$. By (FC5) this gives $ST^{1/n} = T^{1/n}S$, where $T^{1/n}$ is defined as in Corollary 12.42. Apply Theorem 12.60 to the C^*-algebra generated by I, S and $T^{1/n}$ to realize both as multiplication operators M_g resp. M_h on $L^2(X, \mu)$ for a finite measure space (X, μ) and two positive functions g, h in $L^\infty_\mu(X)$ with $g^n = h^n$ μ-almost everywhere.

Exercise 12.75 (p. 470): Consider first the case $B_1 \subseteq B_2$ and show that in this case $\mathrm{im}\, \Pi_{B_1} \subseteq \mathrm{im}\, \Pi_{B_2}$ using the argument in the proof of Lemma 12.74.

Exercise 12.77 (p. 472): First deal with simple functions using the properties of a projection-valued measure and Exercise 12.75.

Exercise 12.78 (p. 472): It suffices to consider the case $f = 0$. Fix $v \in \mathcal{H}$ and show that $\mu_v(B) = \langle \Pi_B v, v \rangle$ for $B \in \mathcal{B}$ defines a finite measure on X. Then show that $\left\| \int_X f_n(\lambda) \,\mathrm{d}\Pi_\lambda v \right\|^2 = \left\langle \int_X |f_n|^2(\lambda) \,\mathrm{d}\Pi_\lambda v, v \right\rangle = \int_X |f_n|^2 \,\mathrm{d}\mu_v$, and apply dominated convergence.

Exercise 12.82 (p. 477): In both contexts the cyclic subspace is the minimal invariant closed subspace containing a given $v \in \mathcal{H}$ and so it suffices to show that the notions of invariance are equivalent. It is easy to verify using only the definition of convolution that a closed subspace that is invariant under the unitary representation is invariant under convolution. To see the converse, use the same approximation argument as in the proof of Corollary 12.81.

Exercise 12.88 (p. 481): Show that $\widetilde{L^1(G)} \subseteq C_0(\widehat{G})$ is a subalgebra that is closed under conjugation and separates points. Then apply Exercise 2.43.

Exercise 12.90 (p. 483): For (a) suppose that $g_n \to g_0$ and $t_n \to t_0$ as $n \to \infty$. Recall from Proposition 11.43 that the topology on $\widehat{G}$ can be defined by uniform convergence on compact sets, and apply this to $K = \{g_n \mid n \geqslant 1\} \cup \{g_0\}$. For (b), notice first that (a) implies that $\imath(g) : \widehat{G} \to \mathbb{S}^1$ is continuous. Moreover, uniform continuity of $\langle \cdot, \cdot \rangle$ restricted to $K \times L$ for some compact subset $L \subseteq \widehat{G}$ shows that $\imath(g_n)|_L \to \imath(g_0)|_L$ uniformly as $n \to \infty$. By Proposition 11.43 this shows that $\imath(g_n) \to \imath(g_0)$ and so $\imath : G \to \widehat{\widehat{G}}$ is continuous. For (c) approximate $f_1, f_2 \in L^2(G)$ by $f_1', f_2' \in C_c(G)$ and notice that $\langle \lambda_g f_1', f_2' \rangle = 0$ once g is outside a certain compact subset. For (d), apply Theorem 12.85 to see that $M_{g_n} f \to 0$ in the weak topology as $n \to \infty$ for $f \in L^2(\widehat{G})$ and a second time to see that $\widehat{\widehat{\lambda}}_{\imath(g_n)} f \to 0$ weakly as $n \to \infty$ for $f \in L^2(\widehat{\widehat{G}})$. Now use continuity of the unitary representation $\widehat{\widehat{\lambda}}$ of $\widehat{\widehat{G}}$ on $L^2(\widehat{\widehat{G}})$ to conclude that $\imath(g_n) \to \infty$ as $n \to \infty$.

Exercise 12.93 (p. 484): For (a) notice that a character on G/H can be lifted to G using composition with the quotient map $G \to G/H$. For (b) use Theorem 12.84 on G/H to show that $(H^\perp)^\perp \subseteq H$. For (c) apply (a) to $\widehat{G}/H^\perp$.

Exercise 12.94 (p. 484): For (b) suppose first that θ has dense image and conclude from $\widehat{\theta}(\chi_t) = 1$ for some $t \in \widehat{G}$ that $t = 0$. For the converse, use Exercise 12.93 to find a non-trivial character $t \in \widehat{G} \cap (\operatorname{im} \theta)^\perp$ if $\operatorname{im} \theta$ is not dense in G.

Exercise 12.95 (p. 485): For the isomorphism between the dual group of the product and the direct sum of the dual groups show that the elements of the direct sum define characters and that these separate points.

Exercise 12.96 (p. 485): By definition $\varprojlim(G_n, \phi_n)$ is a subgroup of $\prod_{n \geq 1} G_n$. Combine Exercise 12.93 and Exercise 12.95.

Exercise 12.97 (p. 485): Use Exercise 12.96 and Pontryagin duality.

Exercise 13.7 (p. 489): For (c) note that

$$H_0^1((0,1)) \subseteq H^1(\mathbb{T}) \subseteq H^1((0,1))$$

with $\mathbb{1} \in H^1(\mathbb{T}) \setminus H_0^1((0,1))$ and $I \in H^1((0,1)) \setminus H^1(\mathbb{T})$ where $I(x) = x$ for $x \in (0,1)$. Use Fourier series to show that $T_p = -T_p^*$. Use the definition of weak derivatives to show that $T = -T_0^*$ is the weak derivative on $H^1((0,1))$.

Exercise 13.10 (p. 494): Show first that $(\operatorname{im}(B))^\perp = \{0\}$ and deduce that B^{-1} is densely defined. To see that B^{-1} is self-adjoint prove that $\langle B^{-1}u, v \rangle = \langle u, B^{-1}v \rangle$ for $u, v \in \operatorname{im}(B)$ and that $\langle B(B^{-1})^* u, v \rangle = \langle u, v \rangle$ for any $u \in D_{(B^{-1})^*}$ and $v \in \mathcal{H}$.

Exercise 13.11 (p. 494): To see that in (a) and (b) there are no other eigenfunctions than the given ones, use elliptic regularity (Theorem 5.34 and Example 5.20) to conclude that the eigenfunctions satisfy certain differential equations with boundary conditions.

Exercise 13.13 (p. 494): For (a) assume first that there exists a bound N on the number of neighbours and show that $T_{\text{initial}}(f)((v_1, v_2)) = f(v_1)$ and $T_{\text{terminal}}(f)((v_1, v_2)) = f(v_2)$ for any $f \in L^2(\mathcal{V})$ and $(v_1, v_2) \in \vec{\mathcal{E}}$ defines a pair of bounded operators

$$T_{\text{initial}}, T_{\text{terminal}} : L^2(\mathcal{V}) \longrightarrow L^2(\vec{\mathcal{E}})$$

with $T = T_{\text{terminal}} - T_{\text{initial}}$. For the converse consider functions $f = \delta_{v_n}$ so that the vertex $v_n \in \mathcal{V}$ has more than n neighbours. In (b) the operator T^* is defined on a subset of $L^2(\vec{\mathcal{E}})$ and maps $g \in D_{T^*}$ to $T^*(g)(v) = \sum_{w \sim v} \big(g(w, v) - g(v, w) \big)$ for all $v \in \mathcal{V}$.

Exercise 13.14 (p. 494): In (a), show that D_{T^*} is defined by Kirchhoff's law: That is, a function

$$f \in L^2(Q) = \bigoplus_{e \in \vec{\mathcal{E}}} L^2(S_e)$$

is in the domain of T^* if each function $f_e = f|_{S_e}$ of f belongs to $H^1(S_e) \subseteq C(S_e)$ and

$$\sum_{e=(v,w) \in \vec{\mathcal{E}}} f_e(v) = \sum_{e=(w,v) \in \vec{\mathcal{E}}} f_e(v)$$

at every vertex $v \in \mathcal{V}$. For (b) argue as in Section 6.4.2. For (c) show that the eigenfunctions are on each interval defined by an appropriate trigonometric function that vanishes on the three vertices that are not in the centre. Use the Kirchhoff condition in the centre to find the constraint for the eigenvalues. Assume first that the ratios of the lengths are incommensurable and reduce the counting to the counting of poles of another trigonometric function.

Exercise 13.16 (p. 498): Using the natural unitary representation on $\mathcal{H}_1 \times \mathcal{H}_2$ show that $\mathrm{Graph}(T)$ is invariant and that the operators B in the proofs of Theorems 13.9 and 13.15 commute with the unitary representation. Now apply Exercise 12.58(b).

Exercise 13.19 (p. 499): Show that the eigenvalues of H are unbounded, and then use Exercise 4.29.

Exercise 13.21 (p. 500): By the spectral theorem (Theorem 13.15) it is sufficient to consider the multiplication operator M_g for a real-valued function g on a finite measure space.

Exercise 13.22 (p. 500): Use Exercise 13.21.

Exercise 13.23 (p. 501): For (a) take the inner product with $v \in D_S$. For (b) simply calculate $\|Sv \pm iv\|^2$; for (c) and (d) notice that

$$(I - U_S)(Sv + iv) = Sv + iv - (Sv - iv) = 2iv$$

for all $v \in D_S$.

Exercise 13.25 (p. 501): For (a) let $v \in D_S$ and $w = Sv + iv \in D_U$ for $U = U_S$ so that $(I - U)w = 2iv$ and

$$S_U\big((I - U_S)w\big) = i(I + U_S)(w) = i(Sv + iv + Sv - iv) = 2iSv,$$

giving $D_{S_U} = D_S$ and $S_U = S$. For (b), let $w \in D_U$ and $v = w - Uw \in D_S$ for $S = S_U$ so that $Sv + iv = iw + iUw + iw - iUw = 2iw$ and

$$U_S(S_U v + iv) = S_U v - iv = iw + iUw - iw + iUw = 2iUw,$$

giving $D_{U_S} = D_U$ and $U_S = U$.

Exercise 13.26 (p. 501): To show that S_U is self-adjoint apply Theorem 9.2 to see that it is sufficient to consider unitary multiplication operators. Then apply Exercise 13.5(a).

Exercise 13.27 (p. 502): Show that $U = T$ from Example 12.9 is an isometry defined on the whole Hilbert space for which $I - U$ is injective and $\mathrm{im}(I - U)$ is dense, and U cannot be extended to a unitary operator.

Exercise 13.28 (p. 502): By applying the Cayley transform (and its inverse) it is sufficient to consider the associated partial isometries.

Exercise 13.30 (p. 502): Show that if $g \in L^2((0,1))$ satisfies

$$0 = \langle \mathrm{i}f' \pm \mathrm{i}f, g \rangle = \mathrm{i}(\langle f', g \rangle \pm \langle f, g \rangle)$$

for all $f \in H_0^1((0,1))$ it follows that $g \in H^1((0,1))$ satisfies the equation $g' = \pm g$. Conclude from this that $n_+(S) = n_-(S) = 1$.

Exercise 14.16 (p. 523): First take the quotient of $C_c(\mathbb{R})$ by the kernel of $\| \cdot \|_\Lambda$ using Lemma 2.15. Apply Theorem 2.32 to obtain the completion $\mathcal{A}_\Lambda$ of the quotient of $C_c(\mathbb{R})$. Use the Banach algebra inequality in Theorem 14.6 to show that the convolution operation extends to $\mathcal{A}_\Lambda$ and gives it the structure of a Banach algebra. Now use (14.4) and the automatic extension property in Proposition 2.59 to extend the canonical map from $C_c(\mathbb{R})$ to $\mathcal{A}_\Lambda$ to a map from $L^1(\mathbb{R})$ to $\mathcal{A}_\Lambda$.

Exercise 14.20 (p. 526): Argue as in the proof of Lemma 14.2.

Exercise 14.21 (p. 527): For (a) use the fact that the characters on the abelian group $(\mathbb{Z}/q\mathbb{Z})^\times$ form an orthonormal basis of $L_m^2\big((\mathbb{Z}/q\mathbb{Z})^\times\big)$ where m is the counting measure multiplied by $\frac{1}{\phi(q)}$ (notice that this is a special case of Exercise 3.50). For (b) notice that the coefficient of the trivial character in the Fourier expansion of f_a is $\frac{1}{\phi(q)}$. Now combine the assumption and PNT itself in the form (14.1) (see also Exercise 14.3).

Exercise 14.22 (p. 527): Argue along the lines of the proof of Proposition 14.4, but use Lemma 14.12 to control the error term (as we cannot use monotonicity).

Exercise 14.23 (p. 527): Estimate

$$\|f\|_\chi = \limsup_{h \to \infty} \left| \sum \frac{\chi(n)\Lambda(n)}{n} f(\log n - h) \right| \leqslant \limsup_{h \to \infty} \sum \frac{\Lambda(n)}{n} |f|(\log n - h)$$

and apply Lemma 14.12.

Exercise 14.24 (p. 527): For any $b \in \mathbb{Z}$ define $\Lambda_b = \Lambda f_b = \Lambda \mathbb{1}_{\{k \in \mathbb{Z} \mid k \equiv b \pmod{q}\}}$ so that $\Lambda = \sum_{b=0}^{q-1} \Lambda_b$ and $\nu_\Lambda = \sum_{b=0}^{q-1} \nu_{\Lambda_b}$. Now use the convergence in Lemma 14.12 and argue as in the proof of Proposition 14.14 to find a subsequence on which $(\lambda_h \nu_{\Lambda_b})$ converges for all $b = 0, \ldots, q-1$. Finally, note that $\nu_{\chi\Lambda} = \sum_{b=0}^{q-1} \chi(b)\nu_{\Lambda_b}$.

Exercise 14.25 (p. 528): For (a) apply Abel summation (Lemma 14.33) with the choices $a_n = \chi(n)$ and $b_n = \log^2 n$ and use the fact that $|A_n| \leqslant q$ for all $n \geqslant 1$ to see that

$$\left| \sum_{n=1}^m A_n(b_n - b_{n+1}) \right| \leqslant q \sum_{n=1}^{m-1} (b_{n+1} - b_n) + qb_m = 2qb_m \leqslant 2q \log^2(1+y)$$

where $m = \lfloor y \rfloor$. For (b) use (14.26), re-order the summation, and apply (14.11). For (c) argue as in Corollary 14.13 (using Corollary 14.13 to control errors).

Exercise 14.26 (p. 528): The argument is similar to the proof of the algebra inequality for $\| \cdot \|_\Lambda$, but much simpler. Use Exercise 14.25(c) to obtain

$$\|f_1 * f_2\|_\chi = \limsup_{h \to \infty} \left| \int f_1 * f_2 \, \mathrm{d}\lambda_h \rho \right|$$

where ρ is defined by $d\rho = \frac{1}{t}\,d(\nu_{\chi\Lambda} * \nu_{\chi\Lambda})$, and then repeat the argument for (14.22).

Exercise 14.27 (p. 528): Verify that the proof of Theorem 14.17 works if $\|\cdot\|_{\Lambda}$ is simply replaced by $\|\cdot\|_{\chi}$ throughout.

Exercise 14.28 (p. 528): Use the same f_0 and f as in the corresponding case of Theorem 14.18. If now $\|f\|_{\chi} = \left|\int f D_{\chi}\,dm\right| = \|f\|_1 = \|f_0\|_1$ for the density D_{χ} from Exercise 14.24, then $|D_{\chi}(t)| = 1$ almost everywhere for t in $[-2,2]$. However, this forces $D_{\chi} \in \operatorname{im}\chi$ almost everywhere, which leads to a contradiction.

Exercise 14.29 (p. 529): Use (14.27) as a replacement for Mertens' theorem in the proof of the $\xi = 0$ case in Theorem 14.18. Use this, Exercise 14.27 and Exercise 14.28 to conclude that $\|\cdot\|_{\chi} = 0$ for any non-trivial Dirichlet character χ. Conclude by using Exercise 14.22.

Exercise 14.34 (p. 533): If $\Re(s) = \sigma > \sigma_0$ then $\left|\frac{a_n \log n}{n^s}\right| \ll \frac{|a_n|}{n^{(\sigma+\sigma_0)/2}}$, which implies that $g = -\sum_{n\geqslant 1} \frac{a_n \log n}{n^s}$ converges absolutely and uniformly on compact subsets of the half plane $\{s \in \mathbb{C} \mid \Re(s) > \sigma_0\}$. Integrating g term-by-term along line segments shows that $f' = g$.

Exercise B.20 (p. 561): Define $d\mu' = |g|\,d\mu$ and $g'(x) = \begin{cases} 1 & \text{for } g(x) = 0, \\ \arg g(x) & \text{for } g(x) \neq 0. \end{cases}$

Notes

[1] (Page v) This description — natural in light of the fact that there seem to be more than seven hundred books in *Mathematical Reviews* whose title contains the phrase 'Functional Analysis' — appears in the preface to the monograph of Aubin [2] and is doubtless older than that.

[2] (Page 6) The Laplace operator is intimately connected with both geometry and physics. An elegant brief discussion in the notes of Arnold [1, Ch. 4] points out the connection between the Laplace operator applied to a surface $f : \mathbb{R}^2 \to \mathbb{R}$ with $|f|$ small viewed as a perturbation of a flat sheet, the area of the surface defined by f, and the work required to bend the surface into this shape. An aspect we are not able to explore here — essential to the physical meaning of the Laplace operator — is reflected in Arnold's comment *"The enemies of physics define the Laplace operator in their mathematical textbooks by [relation (1.5)], which renders this physical object relativistically meaningless (it depends not only on the function to which the operator is applied, but also on the choice of the coordinate system). On the contrary, the operators [...] and Δ depend only on the Riemannian metric and do not depend on the coordinate system."*

[3] (Page 24) The proof here is taken from a note by Väisälä [107], and the original result is in a paper of Mazur and Ulam [70].

[4] (Page 36) The Dvoretzky–Rogers theorem [24], answering a question of Banach, states that a Banach space is finite-dimensional if and only if every unconditionally convergent series is absolutely convergent. The difficult part of this result is to show that in any infinite-dimensional Banach space there is an unconditionally convergent series that is not absolutely convergent. This is often relatively easy to show for a concretely given Banach space (and in particular, for a Hilbert space) but in general requires analysis of the geometry of convex bodies in Banach spaces.

[5] (Page 42) A more constructive proof can be given using Bernstein polynomials [10] which are defined by $B_{f,n}(x) = \sum_{k=0}^{n} \binom{n}{k} x^k (1-x)^{n-k} f(\frac{k}{n})$ for any function $f \in C([0,1])$ and $n \geqslant 1$. The original proof due to Weierstrass [109] uses convolutions with a Gaussian heat kernel and is much closer in spirit to Exercise 3.68.

[6] (Page 43) The strongest result in this direction is Mergelyan's theorem [71]. This states that if $X \subseteq \mathbb{C}$ is a compact set for which $\mathbb{C} \smallsetminus X$ is connected, then any continuous function $X \to \mathbb{C}$ whose restriction to the interior X° is holomorphic, is a uniform limit of a sequence of polynomials. Without the additional hypothesis that the function be holomorphic on the interior the result is simply false, as indicated. If $\mathbb{C} \smallsetminus X$ is not connected a similar result holds using rational functions instead of polynomials.

© Springer International Publishing AG 2017 589
M. Einsiedler, T. Ward, *Functional Analysis, Spectral Theory, and Applications*,
Graduate Texts in Mathematics 276, DOI 10.1007/978-3-319-58540-6

(7) (Page 59) Bergman [8] introduced the space of holomorphic functions in a complex domain with sufficiently regular behaviour at the boundary to ensure they are absolutely integrable. Part of their importance is that they are Banach spaces; we refer to the monograph of Hedenmalm, Korenblum and Zhu [43] for an accessible treatment.

(8) (Page 77) In fact, the space $L^p_\mu(X)$ (or $\ell^p(\mathbb{N})$) is uniformly convex for any p in $(1, \infty)$, but the proof for p in $(1, 2)$ is more involved; we refer to Clarkson [18] for the details.

(9) (Page 82) The property that all closed subspaces are complemented in fact *characterizes* Hilbert spaces in the following sense. Lindenstrauss and Tzafriri [63] showed that if $(V, \|\cdot\|)$ is a Banach space in which every closed subspace is complemented then the norm is equivalent to one induced by a scalar product.

(10) (Page 92) In fact the existence of a left-invariant Borel measure is closely related to local compactness. Weil [110] showed that if a group has a left-invariant measure for which a convolution can be defined, then there is a topology on the group with the property that the completion of the group in that topology is locally compact, and the left-invariant measure is essentially the Haar measure on the completion. Oxtoby [83], in investigating what invariant measures can be found on groups that are not locally compact, showed that a complete separable metric group possesses a left-invariant Borel measure if and only if the group is locally compact and dense in itself.

(11) (Page 93) In particular, the convergence in L^2 does not imply convergence of the Fourier series at any given point, and *a priori* does not even imply convergence almost everywhere. In the classical setting $G = \mathbb{T}$, these questions have been of central importance. Dirichlet proved that the Fourier series converges at each point if $f \in C^1(\mathbb{T})$, and Paul du Bois-Reymond showed that there is a function $f \in C(\mathbb{T})$ whose Fourier series diverges at one point. Lusin conjectured that the Fourier series converges almost everywhere to the function for $f \in L^2(\mathbb{T})$, and Kolmogorov [56] found a function in $L^1(\mathbb{T})$ whose Fourier series diverges almost everywhere. Carleson [16] proved the convergence almost everywhere for $f \in L^2(\mathbb{T})$, an extremely difficult result later extended to $f \in L^p(\mathbb{T})$ for $p \in (1, \infty)$ by Hunt [48]. We refer to Lacey [58] for a modern, approachable, account. The situation is more complicated for functions on compact abelian groups, in part because there is no canonical way to sum over the group of characters.

(12) (Page 95) This form of uncertainty principle is pointed out for finite cyclic groups as part of a wider investigation by Donoho and Stark [23]. In the case where G is the group $\mathbb{Z}/p\mathbb{Z}$ for a prime p, Tao [104] proved the stronger result that

$$|\operatorname{Supp} f| + |\operatorname{Supp} \widehat{f}| \geqslant |G| + 1,$$

but the proof requires methods in matrix theory beyond our scope.

(13) (Page 126) The Baire category theorem result is a powerful tool across much of topology and analysis. It was shown by Osgood [81] for $\mathbb{R}$, and independently by Baire [3] for $\mathbb{R}^d$. It was later applied in functional analysis by Banach and Steinhaus [4].

(14) (Page 128) This analogy is pursued in a monograph by Oxtoby [82], motivated by work of Sierpiński [98] and Erdős [28], who showed that under the assumption of the continuum hypothesis there is an injective function $f : \mathbb{R} \to \mathbb{R}$ with $f = f^{-1}$ with the property that $f(A)$ is a null set if and only if A is of first category. An approach to constructing sets with prescribed Diophantine approximation properties was given by Schmidt via what we now call *Schmidt games* [93]. The simplest of these takes the following form: Let X be a metric space, $S \subseteq X$ any subset, and fix constants $\alpha, \beta \in (0, 1)$. The game is played as follows: the first player, Bob, chooses any open ball $B_0 \subseteq X$ with radius ρ_0. Then Alice, the second player, chooses a ball $B_1 \subseteq B_0$ with radius $\rho_1 = \alpha\rho_0$. Bob then chooses a ball $B_2 \subseteq B_1$ with radius $\rho_2 = \beta\rho_1$, Alice chooses a ball $B_3 \subseteq B_2$ with radius $\rho_3 = \alpha\rho_2$, and so on. The intersection of all the balls B_n for $n \geqslant 1$ comprises a single point x. If $x \in S$ then Alice wins the game, if not Bob wins. If Alice can force a victory, then the set S is called (α, β)-winning, and S is said to be α-winning if it is (α, β)-winning for all $\beta \in (0, 1)$. Clearly S needs to be dense if it is (α, β)-winning, and it may be shown that there are

some null sets that are also meagre and α-winning. Moreover, any countable intersection of α-winning sets is again α-winning.

[15](Page 143) This is shown by Meyers and Serrin [72]; if the closure is taken of functions that are smooth up to the boundary then the situation is different. We refer to Evans [30] for an accessible account.

[16](Page 182) Horn's conjecture [47], which was proved in two parts, one by Klyachko [53] and the other by Knutson and Tao [54] says the following. If A and B are Hermitian $n \times n$ matrices, then an ordered triple (I, J, K) of subsets of $\{1, \ldots, n\}$ with the same cardinality is called admissible if the inequality

$$\sum_{i \in I} \lambda_i(A + B) \leqslant \sum_{i \in J} \lambda_i(A) + \sum_{i \in K} \lambda_i(B)$$

holds. Horn's conjecture was that all such admissable inequalities together with the trace identity (6.13) characterize the possible eigenvalues of pairs of Hermitian matrices and their sum. We refer to the survey article by Knutson and Tao [55] for the details and references.

[17](Page 182) This is one of a large number of results in matrix analysis and its applications by Weyl [111]. Courant and Hilbert [20, p. 286] give this inequality the following physically intuitive meaning, familiar to anyone who has used a stringed musical instrument: If a dynamical system stiffens, then the frequency of its fundamental tone or resonance, and that of all the overtones, increases.

[18](Page 196) This relation between semi-norms and traces may be found in the work of Bernstein and Reznikov [9].

[19](Page 199) We refer to Courant and Hilbert [20] for a thorough classical treatment of Bessel functions.

[20](Page 202) Weyl's motivation came from a problem in black body radiation, though it was well understood at the time that the mathematical questions also arose in the theory of vibrations. The result was foreshadowed by Lord Rayleigh [100] in 1877, who used a three-dimensional lattice point counting problem to count vibrational modes in a cube, allowing him to asymptotically count the number of 'overtones'. Somerfeld and Lorentz conjectured in 1910 that in fact the quantity was also independent of the *shape*, giving the context in which Weyl proved this remarkable theorem. Weyl also gave error terms in dimensions 2 and 3, and conjectured the form of a second term in terms of the area of the boundary of U in dimension 3.

[21](Page 202) Milnor [73] noted that a remarkable pair of lattices in $\mathbb{R}^{16}$ constructed by Witt [115] gives rise to a pair of 16-dimensional tori that have the same eigenvalues but different shapes. Much later Gordon, Webb, and Wolpert [40] exhibited two non-convex polygons in $\mathbb{R}^2$ with the same eigenvalues but different shapes. In the positive direction, Zelditch [117] showed that the answer to Kac's question is yes for a large class of convex subsets of $\mathbb{R}^2$ with analytic boundary.

[22](Page 215) A sequence (z_n) of complex numbers with $|z_n| < 1$ for all $n \geqslant 1$ is said to satisfy the Blaschke condition [12] if $\sum_{n=1}^{\infty}(1 - |z_n|) < \infty$; in this case the Blaschke product $B(z) = \prod_{n=1}^{\infty} \frac{|z_n|}{z_n} \frac{z_n - z}{1 - \overline{z_n} z}$, where the product is taken over all n with $z_n \neq 0$, and with a factor z if $z_n = 0$, is analytic in the open unit disk and vanishes at each z_n. Finite Blaschke products as used here may be characterized as the analytic functions on the open unit disk with continuous extension to the closed unit disk.

[23](Page 223) This was shown by Banach and Tarski in 1924 [5]; we refer to the monograph of Wagon [108] for more details, other related paradoxical decompositions, and the history of this kind of result.

[24](Page 226) This alludes to the observation that if every room is occupied in a hotel with infinitely many rooms then a new guest can always be accommodated. If there are countably many rooms, this is done by moving each guest to the 'next' room: *"Sobald nun ein neuer Gast hinzukommt, braucht der Wirt nur zu veranlassen, dass jeder der alten Gäste das*

Zimmer mit der um 1 höheren Nummer bezieht, und es wird für den Neuangekommenen das Zimmer 1 frei" (from a lecture of Hilbert in 1924; see [31, p. 730]).

[25] (Page 262) We refer to Parthasarathy [84] for a more detailed treatment of the theory of probability measures on compact metric spaces, and to [27, Ch. 4] for material on equidistribution from a dynamical point of view.

[26] (Page 267) This is the Kryloff–Bogoliouboff Theorem [57], and it means that a continuous transformation on a compact metric space always gives rise to one (and perhaps to many) measure-preserving systems.

[27] (Page 296) The theory of distributions is of central importance in partial differential equations, where it sometimes allows solutions to be found in the sense of distributions when they cannot be readily found in the classical sense (as seen in Chapters 5 and 6). The theory of generalized functions was initiated by Sobolev [99] to provide weak solutions to certain partial differential equations, and then developed systematically by Schwartz [94], [95].

[28] (Page 342) The uncertainty principle has many extensions, generalizations, and applications. We would struggle to do better than to quote Folland and Sitaram [34] both for its extensive bibliography and for its elegant description: *"The uncertainty principle is partly a description of a characteristic feature of quantum mechanical systems, partly a statement about the limitations of one's ability to perform measurements on a system without disturbing it, and partly a meta-theorem in harmonic analysis that can be summed up as follows. A non-zero function and its Fourier transform cannot both be sharply localized."*

[29] (Page 353) The approach developed here is close to that of von Neumann, whose lectures on the original work of Haar are now available in a convenient form [80].

[30] (Page 403) Margulis' argument showed in particular that the quotients $SL_3(\mathbb{Z})/\Lambda$ by finite index subgroups Λ are (via a standard graph structure on them) an expander family. To prove this, we will discuss unitary representations of the group $SL_3(\mathbb{Z})$ (that is, actions of $SL_3(\mathbb{Z})$ by unitary transformations on a Hilbert space). There is also a family of *certain* finite quotients $SL_2(\mathbb{Z})/\Lambda$ that give an expander family, but the proof of this lies deeper and goes beyond what we will be able to cover. We refer to the monographs of Sarnak [92], Lubotzky [66] or the notes [26] for the details.

[31] (Page 419) This is the simplest result in the topic of *automatic continuity*, which asks for algebraic conditions on Banach algebras $\mathcal{A}$ and $\mathcal{B}$ that ensure that any algebra homomorphism $\chi : \mathcal{A} \to \mathcal{B}$ is continuous. We refer to the monograph of Dales [21] for a thorough account.

[32] (Page 439) We refer to the monograph of Lubotzky [66, Sec. 4.5] and the papers of Kesten [52] and Buck [14] for more details (and for generalizations to other Cayley graphs).

[33] (Page 442) The reader should not confuse the word 'classical' with 'outdated'. Apart from playing an important role in approximation theory and differential equations these relations and the resulting polynomials are, in part because of their relation to regular trees, of great importance for number theory and related areas; we refer to the work of Lindenstrauss on arithmetic quantum unique ergodicity [62] for a striking instance of this.

[34] (Page 503) The priority for the elementary proof and for some of the steps toward it is contested; we refer to Goldfeld [39] for a detailed account.

[35] (Page 503) The conventional complex-analytic proof of the PNT involves showing that it is equivalent to the non-vanishing of the Riemann zeta function $s \mapsto \sum_{n \geqslant 1} \frac{1}{n^s}$ on the line $\Re(s) = 1$. The question of error rates in the prime number theorem also involves behaviour of the Riemann zeta function on the critical line $\Re(s) = \frac{1}{2}$, and hence ultimately the Riemann hypothesis itself.

[36] (Page 538) Some of these are well explained in the monograph of Wagon [108]; a particularly striking one is the existence of paradoxical decompositions, which was discussed in Section 7.2.3.

References

1. V. I. Arnold, *Mathematical understanding of nature* (American Mathematical Society, Providence, RI, 2014).
2. J.-P. Aubin, *Applied functional analysis*, in *Pure and Applied Mathematics (New York)* (Wiley-Interscience, New York, second ed., 2000).
3. R. Baire, 'Sur les fonctions de variables réelles', *Annali di Mat.(3)* **III** (1899), 1–123.
4. S. Banach and H. Steinhaus, 'Sur le principe de la condensation de singularités', *Fundamenta* **9** (1927), 50–61.
5. S. Banach and A. Tarski, 'Sur la décomposition des ensembles de points en parties respectivement congruentes', *Fund. Math.* **6** (1924), 244–277.
6. B. Bekka, P. de la Harpe, and A. Valette, *Kazhdan's property (T)*, in *New Mathematical Monographs* **11** (Cambridge University Press, Cambridge, 2008).
7. F. Benford, 'The law of anomalous numbers', *Proc. Amer. Phil. Soc.* **78** (1938), no. 4, 551–572.
8. S. Bergman, *The Kernel Function and Conformal Mapping*, in *Mathematical Surveys, No. 5* (American Mathematical Society, New York, N. Y., 1950).
9. J. Bernstein and A. Reznikov, 'Sobolev norms of automorphic functionals', *Int. Math. Res. Not.* (2002), no. 40, 2155–2174.
10. S. N. Bernstein, 'Démonstration du Théorème de Weierstrass fondée sur le calcul des Probabilités', *Comm. Soc. Math. Kharkov 2* **XIII** (1912), no. 1, 1–2.
11. B. Blackadar, *Operator algebras*, in *Encyclopaedia of Mathematical Sciences* **122** (Springer-Verlag, Berlin, 2006). Theory of C^*-algebras and von Neumann algebras, Operator Algebras and Non-commutative Geometry, III.
12. W. Blaschke, 'Eine Erweiterung des Satzes von Vitali über Folgen analytischer Funktionen', *Berichte über die Verhandlungen der Königlich-Sächsischen Gesellschaft der Wissenschaften zu Leipzig, Mathematisch-Physische Klasse* **67** (1915), 194–200.
13. N. Bourbaki, 'Sur certains espaces vectoriels topologiques', *Ann. Inst. Fourier Grenoble* **2** (1950), 5–16 (1951).
14. M. W. Buck, 'Expanders and diffusers', *SIAM J. Algebraic Discrete Methods* **7** (1986), no. 2, 282–304.
15. C. Carathéodory, *Vorlesungen über reelle Funktionen*, in *Third (corrected) edition* (Chelsea Publishing Co., New York, 1968).
16. L. Carleson, 'On convergence and growth of partial sums of Fourier series', *Acta Math.* **116** (1966), 135–157.
17. A. L. Cauchy, *Sur l'equation á l'aide de laquelle on détermine les inégalités séculaires des mouvements des planétes*, in *Oeuvres Complétes (IInd Série)*, **9** (Gauthier–Villars, 1829).
18. J. A. Clarkson, 'Uniformly convex spaces', *Trans. Amer. Math. Soc.* **40** (1936), no. 3, 396–414.

© Springer International Publishing AG 2017

M. Einsiedler, T. Ward, *Functional Analysis, Spectral Theory, and Applications*,
Graduate Texts in Mathematics 276, DOI 10.1007/978-3-319-58540-6

19. J. B. Conway, *A course in functional analysis*, in *Graduate Texts in Mathematics* **96** (Springer-Verlag, New York, second ed., 1990).

20. R. Courant and D. Hilbert, *Methods of mathematical physics. Vol. I* (Interscience Publishers, Inc., New York, N.Y., 1953).

21. H. G. Dales, *Banach algebras and automatic continuity*, in *London Mathematical Society Monographs. New Series* **24** (The Clarendon Press Oxford University Press, New York, 2000). Oxford Science Publications.

22. F. Diamond and J. Shurman, *A first course in modular forms*, in *Graduate Texts in Mathematics* **228** (Springer-Verlag, New York, 2005).

23. D. L. Donoho and P. B. Stark, 'Uncertainty principles and signal recovery', *SIAM J. Appl. Math.* **49** (1989), no. 3, 906–931.

24. A. Dvoretzky and C. A. Rogers, 'Absolute and unconditional convergence in normed linear spaces', *Proc. Nat. Acad. Sci. U. S. A.* **36** (1950), 192–197.

25. M. Einsiedler and T. Ward, *Homogeneous dynamics and applications.* http://www.personal.leeds.ac.uk/~mattbw. In preparation.

26. M. Einsiedler and T. Ward, *Unitary representations and spectral gap.* http://www.personal.leeds.ac.uk/~mattbw. In preparation.

27. M. Einsiedler and T. Ward, *Ergodic theory with a view towards number theory*, in *Graduate Texts in Mathematics* **259** (Springer-Verlag London Ltd., London, 2011).

28. P. Erdős, 'Some remarks on set theory', *Ann. of Math. (2)* **44** (1943), 643–646.

29. P. Erdős, 'On a new method in elementary number theory which leads to an elementary proof of the prime number theorem', *Proc. Nat. Acad. Sci. U. S. A.* **35** (1949), 374–384.

30. L. C. Evans, *Partial differential equations*, in *Graduate Studies in Mathematics* **19** (American Mathematical Society, Providence, RI, second ed., 2010).

31. W. Ewald and W. Sieg (eds.), *David Hilbert's lectures on the foundations of arithmetic and logic 1917–1933*, in *David Hilbert's Foundational Lectures* **6** (Springer-Verlag, Berlin, 2013).

32. G. B. Folland, *A course in abstract harmonic analysis*, in *Studies in Advanced Mathematics* (CRC Press, Boca Raton, FL, 1995).

33. G. B. Folland, *Real analysis*, in *Pure and Applied Mathematics (New York)* (John Wiley & Sons Inc., New York, second ed., 1999). Modern techniques and their applications, A Wiley-Interscience Publication.

34. G. B. Folland and A. Sitaram, 'The uncertainty principle: a mathematical survey', *J. Fourier Anal. Appl.* **3** (1997), no. 3, 207–238.

35. M. Fornasier and D. Toniolo, 'Fast, robust and efficient 2d pattern recognition for re-assembling fragmented images', *Pattern Recognition* **38** (2005), 2074–2087.

36. H. Furstenberg, 'Strict ergodicity and transformation of the torus', *Amer. J. Math.* **83** (1961), 573–601.

37. H. Furstenberg, 'Disjointness in ergodic theory, minimal sets, and a problem in Diophantine approximation', *Math. Systems Theory* **1** (1967), 1–49.

38. O. Gabber and Z. Galil, 'Explicit constructions of linear-sized superconcentrators', *J. Comput. System Sci.* **22** (1981), no. 3, 407–420.

39. D. Goldfeld, 'The elementary proof of the prime number theorem: an historical perspective', in *Number theory (New York, 2003)*, pp. 179–192 (Springer, New York, 2004).

40. C. Gordon, D. Webb, and S. Wolpert, 'Isospectral plane domains and surfaces via Riemannian orbifolds', *Invent. Math.* **110** (1992), no. 1, 1–22.

41. A. Gorodnik and A. Nevo, *The ergodic theory of lattice subgroups*, in *Annals of Mathematics Studies* **172** (Princeton University Press, Princeton, NJ, 2010).

42. G. Greschonig and K. Schmidt, 'Ergodic decomposition of quasi-invariant probability measures', *Colloq. Math.* **84/85** (2000), no. 2, 495–514.

43. H. Hedenmalm, B. Korenblum, and K. Zhu, *Theory of Bergman spaces*, in *Graduate Texts in Mathematics* **199** (Springer-Verlag, New York, 2000).

44. S. Helgason, *Differential geometry, Lie groups, and symmetric spaces*, in *Pure and Applied Mathematics* **80** (Academic Press, Inc. [Harcourt Brace Jovanovich, Publishers], New York-London, 1978).

45. E. Hewitt and K. A. Ross, *Abstract harmonic analysis. Vol. I*, in *Grundlehren der Mathematischen Wissenschaften* **115** (Springer-Verlag, Berlin, second ed., 1979).

46. D. Hilbert and E. Schmidt, *Integralgleichungen und Gleichungen mit unendlich vielen Unbekannten*, in *Teubner-Archiv zur Mathematik [Teubner Archive on Mathematics]*, *11* (BSB B. G. Teubner Verlagsgesellschaft, Leipzig, 1989). Edited and with a foreword and afterword by A. Pietsch.

47. A. Horn, 'Eigenvalues of sums of Hermitian matrices', *Pacific J. Math.* **12** (1962), 225–241.

48. R. A. Hunt, 'On the convergence of Fourier series', in *Orthogonal Expansions and their Continuous Analogues (Proc. Conf., Edwardsville, Ill., 1967)*, pp. 235–255 (Southern Illinois Univ. Press, Carbondale, Ill., 1968).

49. H. Iwaniec and E. Kowalski, *Analytic number theory*, in *American Mathematical Society Colloquium Publications* **53** (American Mathematical Society, Providence, RI, 2004).

50. M. Kac, 'Can one hear the shape of a drum?', *Amer. Math. Monthly* **73** (1966), no. 4, part II, 1–23.

51. J. L. Kelley, *General topology* (Springer-Verlag, New York-Berlin, 1975). Reprint of the 1955 edition [Van Nostrand, Toronto, Ont.], Graduate Texts in Mathematics, No. 27.

52. H. Kesten, 'Symmetric random walks on groups', *Trans. Amer. Math. Soc.* **92** (1959), 336–354.

53. A. A. Klyachko, 'Stable bundles, representation theory and Hermitian operators', *Selecta Math. (N.S.)* **4** (1998), no. 3, 419–445.

54. A. Knutson and T. Tao, 'The honeycomb model of $GL_n(\mathbb{C})$ tensor products. I. Proof of the saturation conjecture', *J. Amer. Math. Soc.* **12** (1999), no. 4, 1055–1090.

55. A. Knutson and T. Tao, 'Honeycombs and sums of Hermitian matrices', *Notices Amer. Math. Soc.* **48** (2001), no. 2, 175–186.

56. A. Kolmogorov, 'Une série de Fourier-Lebesgue divergente presque partout', *Fundamenta math.* **4** (1923), 324–328.

57. N. Kryloff and N. Bogoliouboff, 'La théorie générale de la mesure dans son application à l'étude des systèmes dynamiques de la mécanique non linéaire', *Ann. of Math. (2)* **38** (1937), no. 1, 65–113.

58. M. T. Lacey, 'Carleson's theorem: proof, complements, variations', *Publ. Mat.* **48** (2004), no. 2, 251–307.

59. P. D. Lax, *Functional analysis*, in *Pure and Applied Mathematics (New York)* (Wiley-Interscience [John Wiley & Sons], New York, 2002).

60. C. G. Lekkerkerker, *Geometry of numbers*, in *Bibliotheca Mathematica, Vol. VIII* (Wolters-Noordhoff Publishing, Groningen; North-Holland Publishing Co., Amsterdam-London, 1969).

61. V. B. Lidskiĭ, 'Non-selfadjoint operators with a trace', *Dokl. Akad. Nauk SSSR* **125** (1959), 485–487.

62. E. Lindenstrauss, 'Invariant measures and arithmetic quantum unique ergodicity', *Ann. of Math. (2)* **163** (2006), no. 1, 165–219.

63. J. Lindenstrauss and L. Tzafriri, 'On the complemented subspaces problem', *Israel J. Math.* **9** (1971), 263–269.

64. M. Loève, *Probability theory. I*, in *Graduate Texts in Mathematics* **45** (Springer-Verlag, New York, fourth ed., 1977).

65. M. Loève, *Probability theory. II*, in *Graduate Texts in Mathematics* **46** (Springer-Verlag, New York, fourth ed., 1978).

66. A. Lubotzky, *Discrete groups, expanding graphs and invariant measures*, in *Modern Birkhäuser Classics* (Birkhäuser Verlag, Basel, 2010). With an appendix by Jonathan D. Rogawski, Reprint of the 1994 edition.

67. G. A. Margulis, 'Explicit constructions of expanders', *Problemy Peredači Informacii* **9** (1973), no. 4, 71–80.

68. G. A. Margulis, 'Explicit constructions of expanders', *Problems of Information Transmission* **9** (1975), no. 4.

69. F. I. Mautner, 'Geodesic flows on symmetric Riemann spaces', *Ann. of Math. (2)* **65** (1957), 416–431.

70. S. Mazur and S. Ulam, 'Sur les transformationes isométriques d'espaces vectoriels normés', *C. R. Math. Acad. Sci. Paris* **194** (1932), 946–948.

71. S. N. Mergelyan, 'Uniform approximations to functions of a complex variable', *Amer. Math. Soc. Translation* **1954** (1954), no. 101, 99.

72. N. G. Meyers and J. Serrin, '$H = W$', *Proc. Nat. Acad. Sci. U.S.A.* **51** (1964), 1055–1056.

73. J. Milnor, 'Eigenvalues of the Laplace operator on certain manifolds', *Proc. Nat. Acad. Sci. U.S.A.* **51** (1964), 542.

74. H. Minkowski, *Geometrie der Zahlen*, in *Bibliotheca Mathematica Teubneriana, Band 40* (Johnson Reprint Corp., New York, 1968).

75. C. C. Moore, 'The Mautner phenomenon for general unitary representations', *Pacific J. Math.* **86** (1980), no. 1, 155–169.

76. C. H. Müntz, 'Über den Approximationssatz von Weierstraß', *Schwarz-Festschr.* (1914), 303–312.

77. M. R. Murty, *Problems in analytic number theory*, in *Graduate Texts in Mathematics* **206** (Springer-Verlag, New York, 2001). Readings in Mathematics.

78. J. v. Neumann, 'Proof of the quasi-ergodic hypothesis', *Proc. Nat. Acad. Sci. U.S.A.* **18** (1932), 70–82.

79. J. v. Neumann, 'On a certain topology for rings of operators', *Ann. of Math. (2)* **37** (1936), no. 1, 111–115.

80. J. v. Neumann, *Invariant measures* (American Mathematical Society, Providence, RI, 1999).

81. W. F. Osgood, 'Non-uniform convergence and the integration of series term by term.', *Amer. J. Math.* **19** (1897), 155–190.

82. J. C. Oxtoby, *Measure and category. A survey of the analogies between topological and measure spaces* (Springer-Verlag, New York, 1971). Graduate Texts in Mathematics, Vol. 2.

83. J. C. Oxtoby, 'Invariant measures in groups which are not locally compact', *Trans. Amer. Math. Soc.* **60** (1946), 215–237.

84. K. R. Parthasarathy, *Probability measures on metric spaces*, in *Probability and Mathematical Statistics, No. 3* (Academic Press Inc., New York, 1967).

85. F. Peter and H. Weyl, 'Die Vollständigkeit der primitiven Darstellungen einer geschlossenen kontinuierlichen Gruppe', *Math. Ann.* **97** (1927), no. 1, 737–755.

86. R. R. Phelps, *Lectures on Choquet's theorem*, in *Lecture Notes in Mathematics* **1757** (Springer-Verlag, Berlin, second ed., 2001).

87. M. S. Pinsker, 'On the complexity of a concentrator', in *Proceedings of the Seventh International Teletraffic Congress (Stockholm, 1973)*, **318** (1973), 318/1–318/4. unpublished.

88. M. S. Pinsker, 'On the complexity of a concentrator', *Problems of Information Transmission* **9** (1975), no. 4, 325–332.

89. M. S. Raghunathan, *Discrete subgroups of Lie groups* (Springer-Verlag, New York, 1972).

90. J. G. Ratcliffe, *Foundations of hyperbolic manifolds*, in *Graduate Texts in Mathematics* **149** (Springer, New York, second ed., 2006).

91. M. Reed and B. Simon, *Methods of modern mathematical physics. II. Fourier analysis, self-adjointness* (Academic Press [Harcourt Brace Jovanovich, Publishers], New York-London, 1975).

92. P. Sarnak, *Some applications of modular forms*, in *Cambridge Tracts in Mathematics* **99** (Cambridge University Press, Cambridge, 1990).

93. W. M. Schmidt, 'On badly approximable numbers and certain games', *Trans. Amer. Math. Soc.* **123** (1966), 178–199.

94. L. Schwartz, *Théorie des distributions. Tome I*, in *Actualités Sci. Ind., no. 1091* = *Publ. Inst. Math. Univ. Strasbourg* **9** (Hermann & Cie., Paris, 1950).

95. L. Schwartz, *Théorie des distributions. Tome II*, in *Actualités Sci. Ind., no. 1122* = *Publ. Inst. Math. Univ. Strasbourg* **10** (Hermann & Cie., Paris, 1951).

96. A. Selberg, 'An elementary proof of the prime-number theorem', *Ann. of Math. (2)* **50** (1949), 305–313.

97. J.-P. Serre, *A course in arithmetic* (Springer-Verlag, New York-Heidelberg, 1973). Translated from the French, Graduate Texts in Mathematics, No. 7.

98. W. Sierpiński, 'Sur les fonctions jouissant de la propriété de Baire de fonctions continues', *Ann. of Math. (2)* **35** (1934), no. 2, 278–283.

99. S. Soboleff, 'Méthode nouvelle à résoudre le problème de Cauchy pour les équations linéaires hyperboliques normales', *Rec. Math. [Mat. Sbornik] N.S.* **1(43)** (1936), no. 1, 39–72.

100. J. W. Strutt, *The theory of sound. Second edition, revised and enlarged. Volume II.* (London: Macmillan. 520 S. 8°, 1896) (English). 3rd Baron Rayleigh.

101. M. Takesaki, *Theory of operator algebras. I* (Springer-Verlag, New York, 1979).

102. M. Talagrand, 'Pettis integral and measure theory', *Mem. Amer. Math. Soc.* **51** (1984), no. 307, ix+224.

103. T. Tao, *A Banach algebra proof of the prime number theorem; Urysohn's lemma; The prime number theorem in arithmetic progressions; Elementary multiplicative number theory* (https://terrytao.wordpress.com). Accessed: 29th October 2015.

104. T. Tao, 'An uncertainty principle for cyclic groups of prime order', *Math. Res. Lett.* **12** (2005), no. 1, 121–127.

105. T. Tao, *An introduction to measure theory*, in *Graduate Studies in Mathematics* **126** (American Mathematical Society, Providence, RI, 2011).

106. F. Trèves, *Topological vector spaces, distributions and kernels* (Academic Press, New York, 1967).

107. J. Väisälä, 'A proof of the Mazur–Ulam theorem', *Amer. Math. Monthly* **110** (2003), no. 7, 633–635.

108. S. Wagon, *The Banach-Tarski paradox*, in *Encyclopedia of Mathematics and its Applications* **24** (Cambridge University Press, Cambridge, 1985). With a foreword by Jan Mycielski.

109. K. Weierstrass, 'Über die analytische Darstellbarkeit sogenannter willkürlicher Functionen einer reellen Veränderlichen', *Verl. d. Kgl. Akad. d. Wiss. Berlin* **2** (1885), 633–639.

110. A. Weil, *L'intégration dans les groupes topologiques et ses applications*, in *Actual. Sci. Ind., no. 869* (Hermann et Cie., Paris, 1940).

111. H. Weyl, 'Das asymptotische Verteilungsgesetz der Eigenwerte linearer partieller Differentialgleichungen (mit einer Anwendung auf die Theorie der Hohlraumstrahlung)', *Math. Ann.* **71** (1911), 441–479.

112. H. Weyl, 'Über die Gleichverteilung von Zahlen mod Eins', *Math. Ann.* **77** (1916), 313–352.

113. E. T. Whittaker and G. N. Watson, *A course of modern analysis*, in *Cambridge Mathematical Library* (Cambridge University Press, Cambridge, 1996).

114. N. Wiener, 'Tauberian theorems', *Ann. of Math. (2)* **33** (1932), no. 1, 1–100.

115. E. Witt, 'Eine Identität zwischen Modulformen zweiten Grades', *Abh. Math. Sem. Hansischen Univ.* **14** (1941), 323–337.

116. D. Witte Morris, *Introduction to arithmetic groups* (Deductive Press, 2015).

117. S. Zelditch, 'Spectral determination of analytic bi-axisymmetric plane domains', *Geom. Funct. Anal.* **10** (2000), no. 3, 628–677.

Notation

$\mathbb{N}$, natural numbers, v

$\mathbb{N}_0$, non-negative integers, v

$\mathbb{Z}$, integers, v

$\mathbb{Q}$, rational numbers, v

$\mathbb{R}$, real numbers, v

$\mathbb{C}$, complex numbers, v

$\Re(\cdot), \Im(\cdot)$, real and imaginary parts, v

$\ll, o, O$, relations between growth in functions, vi

$k(\phi)$, matrix of rotation through ϕ on $\mathbb{R}^2$, 2

$SO_2(\mathbb{R})$, group of rotations of the plane, 3

χ_n, character $\phi \mapsto e^{2\pi i n\phi}$, 3

Δ, Laplace operator, 6

$\mathbb{S}^{d-1}$, $(d-1)$ unit sphere in $\mathbb{R}^d$, 7

$C(X)$, continuous functions on X, 16

$\ell^1(\mathbb{N})$, space of summable sequences, 20

c_c, space of finitely supported sequences, 20

$\mathscr{L}^1_\mu(X)$, space of integrable functions, 21

$B(X)$, space of bounded functions, 27

$C_b(X)$, space of continuous bounded functions, 27

$C_0(X)$, space of continuous functions vanishing at infinity, 27

λ_{count}, counting measure, 28

$\mathscr{L}^\infty(X)$, space of bounded measurable functions, 29

c_0, space of null sequences, 39

$C_{\mathbb{R}}(X), C_{\mathbb{C}}(X)$, real- and complex-valued continuous functions, 42

$\mathbb{1}_A$, indicator function of the set A, 50

$\{\cdot\}$, fractional part of a real number, 51

$B(V, W)$, bounded linear maps V to W, 55

$B(V)$, bounded linear maps V to V, 55

V^*, continuous linear functionals on V, 55

$H^p(D)$, Hardy space, 59

$A^p(D)$, Bergman space, 59

$Y^\perp$, orthogonal complement of Y in a Hilbert space, 79

$\langle S \rangle$, linear hull of S, 82

$\mu_1 \perp \mu_2$, μ_1 and μ_2 are mutually singular, 83

$\mathbb{P}(X)$, set of all subsets of X, 90

$\mathcal{H}_\chi$, weight space associated to character χ, 110

G_δ, countable intersection of open sets, 127

$H^k(\mathbb{T}^d)$, Sobolev space, 135

n^α, shorthand for $(n_1^{\alpha_1}, \ldots, n_d^{\alpha_d})$, 138

Δ, weak Laplace operator, 154

$K(V, W), K(V)$, space of compact operators, 168

A^*, adjoint of operator A, 175

$HS(\mathcal{H})$, space of Hilbert–Schmidt operators on $\mathcal{H}$, 195

ω_d, volume of the unit ball in $\mathbb{R}^d$, 202

$\mathbb{T}^d_R$, scaled torus, 202

F_2, free group on two generators, 225

Σ_X, simple integrable functions on X, 235

$T_*\mu$, push-forward of a measure, 265

$\mathscr{D}(U)$, space of distributions on U, 296

L^1, space of equivalence classes of integrable functions, 297

C_c^∞, space of smooth compactly supported functions, 297

L_{loc}^1, space of equivalence classes of locally integrable functions, 297

$\mathcal{P}(X)$, space of Borel probability measures on X, 304

$\mathscr{D}(U)$, space of test functions, 312

$\mathscr{S}(X)$, space of Schwartz functions on X, 312

$L^2(X,\mu)$, alternative for $L_\mu^2(X)$, space of square-integrable functions, 313

$\delta_{a,b}$, function equal to 1 if $a = b$ and 0 if not, 316

U_T, unitary operator associated to T, 327

x^α, shorthand for the monomial $x_1^{\alpha_1} \cdots x_d^{\alpha_d}$, 340

$\mathscr{S}(\mathbb{R}^d)$, Schwartz space of functions on $\mathbb{R}^d$, 341

$\pi : G \curvearrowright \mathcal{H}$, unitary representation, 344

$\mathcal{P}_1(G)$, positive-definite functions on a group, 345

λ_g, left regular representation, 353

λ_g, shift in domain, 362

AB, product of sets in a group, 371

$\mathcal{H}^G$, subspace of invariant vectors in a unitary representation, 375

$\mathbb{P}^1$, projective line, 382

$A_\mathcal{G}$, adjacency matrix for a graph, 403

$M_\mathcal{G}$, averaging operator for a graph, 404

$\rho(a)$, resolvent set of operator a, 409

$B(\mathcal{H})$, algebra of bounded operators on a Hilbert space, 417

$\beta\mathbb{N}$, Stone–Čech compactification of $\mathbb{N}$, 421

σ_{disc}, discrete spectrum, 433

σ_{appt}, approximate point spectrum, 434

σ_{approx}, approximate spectrum, 434

σ_{cont}, continuous spectrum, 434

σ_{resid}, residual spectrum, 434

σ_{ess}, essential spectrum, 437

Λ, von Mangoldt function, 504

Λ_2, second von Mangoldt function, 509

$B_\varepsilon(\cdot)$, ε ball in a metric space, 539

$\lim \mathcal{F}$, limit of a convergent filter, 540

$\lim_\mathcal{F} f$, convergence along a filter, 540

$\sigma(\mathcal{C})$, σ-algebra generated by $\mathcal{C}$, 551

General Index

Abel summation formula, 529
absolute convergence, 32
absolutely continuous, 83
absorbent, 293
adjacency matrix, 403
adjoint, 12, 175
  densely defined operator, 488
  operator, 175
    densely defined, 488
    star operator, 417
admits Følner sets, 362
affine
  action, 373
  map, 24
  subspace, 66
alert reader, 390, 414, 493
algebra, 48
  Banach, 61
    automatic continuity, 592
    bounded operators, 433
    C^*, 417
    continuous functions, 62
    dual space, 418
    Gelfand dual, 418
    Gelfand transform, 422
    homomorphism, 448
    ideal, 168
    integrable functions, 333
    maximal ideal, 419
    von Neumann, 431
    spectral radius, 410
    spectrum, 409
    unital, 409
  C^*, 417
    normal element, 417
    self-adjoint element, 417
    star operator, 417
  Calkin, 170, 437
  commutative, 61
  homomorphism, 418
  von Neumann, 311, 431
algebraically complemented, 82

almost
  everywhere, 22, 555
  injective, 437
  invariant, 363, 376
    vector, 375
  surjective, 437
amenable, 362
  fixed point of action, 373
  group, 70, 219
    admits Følner sets, 362
    Følner sets, 371
    left-invariant mean, 362
    quotient, subgroup, 220
    Reiter condition, 363
  growth in groups, 374
  radical, 221, 262
analytic, 414
  Blaschke product, 591
  boundary, 591
  methods, 503
  power series, 323
  strongly, 260
  weakly, 260
annihilator
  double, 484
approximate
  eigenvalue, 495
  eigenvector, 434
  identity, 102
  invariant measure, 266
  point spectrum, 434
  spectrum, 434
Arzela–Ascoli theorem, 39
  locally compact, 42
asymptotic, 504
atom, 314
atomic measure, 314
automatic continuity, 592
averaging operator, 404
axiom of choice, 223, 537

Baire category theorem, 126
  topological, 127

balanced, 293
ball, 6, 18
    closed, 77
    metric, 402
    non-compactness, 38
    open, 26
    unit, 17
Banach
    Alaoglu theorem, 256
    algebra, 61
        automatic continuity, 592
        bounded operators, 433
        C^*, 417
        continuous functions, 62
        dual space, 418
        examples, 62
        field, 411
        Gelfand dual, 418
        Gelfand transform, 422
        generated, 410
        homomorphism, 448
        ideal, 168
        integrable functions, 118, 333
        inverse of an element, 411
        maximal ideal, 419
        von Neumann, 431
        resolvent, 411
        spectral radius, 410
        spectrum, 409
        unital, 62, 409
        without a unit, 421
    limit, 217
    space, 4
        compact operator, 174
        dual, 70, 209
        reflexive, 113, 209, 214
        topology, 253
        trace-class norm, 195
        uniformly convex, 77
    Steinhaus theorem, 121
        application to Fourier ana-
            lysis, 123
    Tarski paradox, 223
barycentre, 304
    existence, 305

invariant, 579
    measure, 310
    uniquely determined, 304
base field, 296
Benford's law, 51
Bergman space, 590
Bernstein polynomial, 589
Bessel
    equation, 200
    function, 200
bidual, 213
    isometric embedding, 213
bilinear pairing, 227
binary relation, 541
    reflexive, transitive, filter prop-
        erty, 541
black box, 91, 184, 187
Blaschke product, 591
Bochner
    integral, 111
    theorem, 346, 381
        for $\mathbb{R}^d$, 346
Borel
    measure, 170
    probability measure, 373
    set, 240
    σ-algebra, 28, 51
        Lusin's theorem, 558
boundary
    conditions, 66
        Dirichlet, 199, 494
        Neumann, 494
        periodic, 494
    of a set in a graph, 402
    value problem, 67
        Dirichlet, 8
bounded
    derivative, 28
    function, 16, 29
    functional, 418
    linear operator, 55
        extension, 58
    operator, 55
        compact, 169
    sequence, 517

limit, 70

C^*-algebra, 417
  normal element, 417
    spectral radius formula, 418
  self-adjoint element, 417
  star operator, 417
Calkin algebra, 170
Carathéodory extension theorem,
  242, 552
category
  first, 126, 590
  second, 126
Cauchy
  formula, 330
  inequality, 276
  inequality with an ε, 276
  integral formula, 260, 414, 564
  integration, 411
  interlacing theorem, 183
  Schwarz inequality, 72
  sequence, 27, 545
    convergent subsequence, 33
cautious reader, 204
Cayley transform, 500
Čech, 421
Césaro average, 218
chain, 90
character, 3
  modular, 371
  multiplicative, 527
  separate points, 92
  weight, 110
characteristic
  function, 50
    smooth approximate, 151
  polynomial, 12
cheat, 296, 477
Chebyshev polynomial of the second
  kind, 442
Choquet's theorem, 306
circle rotation, 267
clopen, 240
  set, 538
closable operator, 487
closed

graph theorem, 131
linear hull, 82, 213
operator, 131
set, 538
coarser
  filter, 540
  strictly, 300
  topology, 539
coercive, 81
common refinement, 471
commutative
  algebra, 419
    ideal, 419
    quotient, 419
  Banach algebra, 418
    unital, 419
  C^*-algebra, 423
  diagram, 317
  ring, 420
compact, 545
  integral operator, 170, 177
  intersection property, 545
  operator, 167–169
    Hilbert–Schmidt, 64, 171
    ideal in a Banach algebra,
      168
    preserved by limits, 169
    regularity property, 169
    spectral theorem, 177
  sequentially, 545
  totally bounded, 546
  Tychonoff, 546
compactification
  one-point, vi
  Stone–Čech, 421
complemented, 82
  algebraically, 82
  topologically, 82
complete, 27, 88
  diagonalizability, 178
  metric space
    Baire category theorem, 126
completely multiplicative function,
  529
completeness of characters, 92

completion, 36, 38, 58
   Hardy space, 59
   metric space, 38
   unique extension, 58
compression, 183
concave, 182, 306
conditional
   convergence, 36
   expectation, 85, 86
conjugate
   exponent, 97, 557
   space, 209
connected
   graph, 401
   network, 400
content, 241
continuous
   addition, 18
   extension, 139
   function, 539
   functions dense in L^p, 51
   group action, 107
   scalar multiplication, 18
   sequentially, 259
   uniformly, 40
convergence
   absolute, 32
      Banach space, 33
   almost everywhere, 590
   conditional, 36
   equivalent norm, 30
   Fourier series, 197, 590
   in measure, 295
   measure, 70
   strong, 123
   unconditional, 36
   uniform, 27, 122
convergent
   convex combination, 304
   filter, 539
   sequence, 539
convex, 15
   combination, 15, 304
   hull, 301
   set, 70

absorbent, 293
balanced, 293
   space, 76
   strictly, 24
convolution, 96, 118
   associative, 369
   Dirichlet, 507
   multiplicative, 507
   operator, 115
Courant–Fischer–Weyl theorem, 182
cover, 40
   finite subcover, 545
   open, 545
co-volume, 393
cross-section, 227
cyclic
   Hilbert space, 316
   operator, 458
   representation, 316
      generator, 316
   subspace, 317
   vector, 455, 456, 590

decay
   at infinity, 563
   boundary, 143
   Fourier transform, 340
   super-polynomial, 341
deficiency index, 502
δ-measure, 350
dense, 48
densely defined operator, 487
   adjoint, 488
   self-adjoint, 488
diagonal group, 392
diameter, 401
differentiable vector, 351
differential equation
   fundamental solution, 69
   partial, 6, 8, 9, 200, 592
directed set, 541
   tail field, 541
Dirichlet
   boundary condition, 199, 494
   boundary problem
      wave equation, 10

boundary value problem, 8, 135, 152, 161, 165
character, 527
convolution, 507
kernel, 99, 100
Neumann bracketing, 202
series, 529, 533
theorem, 526
discrete spectrum, 433
distribution, 296
divergence theorem, 162
dual
    Banach algebra, 418
    Gelfand, 418
    space, 209
duality, 473
Dvoretzky–Rogers theorem, 589
dynamical systems, 327

edge, 400
eigenvalue, 12, 167
    cancellation, 157
    variational characterization, 181
eigenvector, 12, 167
    approximate, 434
elliptic
    differential operator, 153
    regularity, 69, 135, 153, 155
        on the torus, 157
entire function, 414
equicontinuous, 39
equidistributed, 48
    probability measure, 263
    sequence of measures, 263
equivalence relation, 225
equivalent norm, 18, 131
ergodic, 265
    circle rotation, 267
    mean theorem, 175, 261
    relation to indecomposable, 265
    theory, 304
essential
    radius, 437
    range, 410, 443
    spectrum, 437
    supremum norm, 29

essentially
    bounded, 337
    self-adjoint, 502
Euler totient function, 526
even function, part, 1
expander
    family, 402
    graph, 375, 400
        logarithmically small diameter, 402
expectation, 342
    conditional, 85, 86
exponential growth, 374
extension, 488
    bounded linear operator, 58
    Carathéodory theorem, 552
    continuous, 139
    natural, 90
    operator, 146
    Tietze theorem, 549
extremal subset, 301
extreme, 266
    point, 70

false hope, 314
Fejér kernel, 100, 102
Fekete's lemma, 415
Fell topology, 376
filter, 539
    compactness, 546
    convergence, 540
    convergence along, 540
    convergent, 254
    finer, 547
    finer, coarser, 540
    neighbourhood, 540
    tail, 540
finer, 540
    filter, 540
    topology, 539
finite intersection property, 302
first category, 126, 590
Følner
    sequence, 222
    sets, 362

Fourier
  analysis, 4
  back transform, 475
  coefficient, 4
  inversion theorem, 335
  series, 4
    convergence, 590
    convergence almost every-
      where, 590
    diverges almost everywhere,
      590
    non-convergent, 125
  transform, 329, 426, 428
    Gaussian, 330
    not an isometry, 428
Fredholm operator, 437
frequency variable, 330
Friedrichs extension, 499
Fréchet
  Riesz theorem, 80
  space, 295, 312, 341
Fubini theorem, 270, 556
function
  continuous, 539
  even, 1
  even, odd part
    uniqueness, 1
  generalized, 296
  odd, 1
  simple, 553
    integral, 553
  test, 296
  weight, 3
functional, 209, 296
  calculus, 324, 454
    continuous, 444
    measurable, 444
fundamental
  domain, 150, 384
  solution, 69
  tone, 591
fundamental frequency, 4

gauge function, 211
Gauss elimination, 378
Gaussian, 479

distribution, 330
G_δ-set, 127
Gelfand
  dual, 418
    and Pontryagin dual, 426
  Pettis integral, 113
  transform, 422
    not an isometry, 428
generalized function, 296
generic point, 328
geometric series, 65
geometry of numbers, 392
Gram–Schmidt procedure, 88
  non-separable, 90
graph, 400, 437
  adjacency matrix, 403
  averaging operator, 407
  boundary of a set, 402
  connected, 401, 404
  diameter, 401
  edge, vertex, 400
  expander, 401–403
    family, 405
  k-regular, 400
  Laplace operator, 404
  linear operator, 131
  metric, 401, 494
  oriented edge, 494
  path, 401
  quantum, 495
  regular tree, 438
  simple, 400
  sparsity, 400
  spectral gap, 405
  undirected, 400, 437
  Weyl law, 495
Green function, 67
group
  action, 2
    affine, 373
    associated unitary operator,
      107
    measure-preserving, 107
  amenable, 70, 219
  character, 3

continuous action, 107
diagonal, 392
left action, 118
orthogonal, 392
topological, 91, 353
unipotent, 392
unitary representation, 107
growth, 374
    polynomial, exponential, inter-
       mediate, 374

Haar measure, 92, 353, 396
Hadamard three-lines theorem, 234
Hahn–Banach
    lemma, 209
    theorem, 211
Hamel basis, 88
Hardy space, 59, 81, 89
harmonic
    function, 152, 161
       mean value principle, 161
    weak, 154
    weakly, 154, 156
harmonic function, 8
Hausdorff space, 539
Hausdorff–Young inequality, 340
heat
    equation, 7, 8, 10
    kernel, 589
Heine–Borel theorem, 20
Hellinger–Toeplitz theorem, 132
Herglotz
    theorem, 252
       positive-definite sequence, 314
       torus, 320
Hermitian, 176
    matrix, 176
Hilbert
    hotel, 226
    Schmidt
       norm, 195
       operator, 64, 171
    space, 71
       infinite-dimensional, 88
       norm is strictly sub-additive,
       76

orthogonal complement, 79
orthogonal projection, 80
real, 74
Hölder
    conjugate, 97, 234
    inequality, 97, 234, 557
Holmgren operator, 174
homogeneous, 211
    ordinary differential equation,
       63
Horn conjecture, 182
Howe–Moore theorem, 311
hull
    closed linear, 82
    linear, 82
hyperplane, 305
hypersurface, 396

ideal, 419
    commutative algebra, 419
    maximal, 419
    proper, 419
impatient reader, 334
inclusion–exclusion principle, 508
induced representation, 385
induction, 384
inequality
    Cauchy–Schwarz, 72
    Hölder, 234, 557
    Jensen, 109
    triangle, 539, 557
infinite
    dimensional, 167
       Hilbert space, 88
    dimensional space, 38
    intersection property, 545
initial
    topology, 542
    values, 63
inner product, 71
    space, 71
       sesqui-linear, 72
       strict positivity, 71
integral
    Bochner, 111

Gelfand–Pettis, 113
Lebesgue, 551
operator, 62
    compact, 170
    kernel, 64
Pettis, 113
simple function, 553
strong, 368
weak, 113
interlacing theorem, 183
intermediate growth, 374
invariant
    ergodic measure, 265
    measure, 265
irreducible
    representation, 458
    unitary representation, 345, 458,
        498
isometry, 24
isomorphic
    unitarily, 313
isomorphism
    isometric, 58
Iwasawa decomposition, 393
    Haar measure, 396

Jensen inequality, 565
joining, 327
Jordan
    block, 167
    measurable, 196
    normal form, 12

kernel, 22, 64
    Dirichlet, 99, 100
    Fejér, 100
    heat, 589
    Landau, 104
    reproducing, 81
    semi-norm, 22
Kirchhoff's law, 585
Krein–Milman theorem, 301

λ-semi-norm, 505
Landau kernel, 104
Laplace

equation, 8
operator, 6
    compact self-adjoint, 198
    eigenfunction, 157
    elliptic regularity, 155
    graph, 404
    infinitesimal, 6
    regular tree, 437
lattice, 384
    Minkowski's first theorem, 393
Lax–Milgram lemma, 81
Lebesgue
    decomposition, 83, 322
    integral, 111, 551
    spectrum, 318, 327
left
    action, 118
    Haar measure, 92
    inverse, 130
left-invariant
    functional, 219
    mean, 362
        topological, 368
    measure, 590
Leibniz' rule, 150
LF space, 312
Lidskiĭ's theorem, 193
limit
    Banach, 217
    topology, 542
Lindenstrauss–Tzafriri theorem, 590
linear
    functional, 55, 209
    hull, 82
        closed, 82
    operator, 55
        extension, 58
    order, 538
Lipschitz
    condition, 58
    constant, 18
locally
    compact group, 473
        bidual, 473
    constant, 153

convex
   extreme point, 301
   topology, 292
   vector space, 293
  finite, 51
   measure, 108
  H^k, 155
  L^p, 155
Lusin's theorem, 60, 558

von Mangoldt function, 504, 528
  second, 509
Mantegna fresco, 11
matrix
  adjacency, 403
  coefficient, 311, 344
  compression, 183
  diagonal, 12
  group, 361
  Hermitian, 176, 181, 182
  permutation, 406
  rotation, 2
  self-adjoint, 176
Mautner phenomenon, 378, 381
maximal
  element, 538
  ideal, 420
  spectral type, 321
maximum modulus theorem, 235
Mazur–Ulam theorem, 24
meagre, 126
mean, 362
  ergodic theorem, 175, 261
  value principle, 161
measurable
  functional calculus, 444, 462,
    464
  Jordan, 202
measure
  absolutely continuous, 83
  atomic, 314, 326
  Borel, 176, 265
  equivalence class, 321
  Fourier coefficient, 314
  Haar, 353
  invariant, 107, 265

  ergodic, 265
  locally finite, 51, 108, 239
  preserving, 107
  preserving system, 265
   disjoint, 328
  probability, 262
  projection-valued, 468
  push-forward, 265
  regular, 240, 558
  σ-finite, 83
  space, 553
   σ-finite, 553
  spectral, 444
Mergelyan's theorem, 589
meromorphic extension, 533
Mertens' theorem, 503
metric, 539
  graph, 494
  pseudo, 539
  space, 539
   completion, 38
   separable, 41
metrizable
  group, 239, 353
  subset, 305
  topology, 257, 304
min-max principle, 182
Minkowski
  first theorem, 393
  theorem
   Carathéodory's form, 306
miracle, 248
mixing, 311
Möbius inversion, 508
modular character, 360, 371
de Moivre formula, 566
multiplication operator, 176
  self-adjoint, 176
  spectrum, 410, 435
multiplicative
  convolution, 507
  function, 529
  inverse, 419
  unit, 409
multiplicity, 323

Müntz' theorem, 214

natural extension, 90
neighbourhood, 538
    filter, 540
net, 541
von Neumann
    algebra, 311, 431
    series, 64
Neumann boundary conditions, 494
non-diagonal spectral measure, 461
norm, 16
    defines a metric, 18
    equivalent, 18
    operator, 55
    pseudo-, 21
    semi-, 21
    trace-class, 195
    uniformly convex, 77
normal
    space, 547
    topological space, 547
normed
    linear space
        dual, 209
    space
        bidual, 213
    vector space
        conjugate, 209
        inner product, 71
nowhere dense, 126
nuclear space, 312
null set, 555
numerical range, 181

odd function, part, 1
one-point compactification, vi
open
    cover, 545
    mapping theorem, 126
    neighbourhood, 539
    set, 538
operator
    averaging, 407, 438
    closable, 487
    compact, 168

spectral theorem, 177
    densely defined, 487
    adjoint, 488
    self-adjoint, 488
    eigenvalues, 68
    equality, 488
    Fredholm, 437
    Hilbert–Schmidt, 64, 171
    integral, 64, 171
        compact, self-adjoint, 177
    Laplace, 438
    multiplication, 176
    norm, 55
    partial differential, 284
    positive, 191, 450
    restriction, 147
    self-adjoint, 132
        spectral theorem, 177
    summing, 438
    symmetric, 498
    trace-class, 183
    unbounded, 487
    uniformly elliptic, 284
    unitary, 175, 313
        spectral theory, 314
    unitary multiplication, 313
order
    linear, 538
    partial, 538
ordinary differential equation, 62
    homogeneous, 63
    initial value, 63
    Sturm–Liouville, 66
    Volterra, 64
oriented
    edge, 494
    path, 414, 416
orthogonal
    complement, 79
    group, 392
    projection, 80, 185
orthonormal, 86
    basis, 88

pairing, 227

paradoxical decomposition, 223
parallelogram identity, 74, 564
    characterizes Hilbert space, 74
Parseval
    formula, 95, 97
    theorem, 93
partial
    differential equation, 6, 8, 9,
        200, 592
        heat equation, 7
        wave equation, 10
    differential operator, 133, 157,
        284
    isometry, 501
    order, 538
        maximal element, 538
        reflexive, 538
        transitive, 538
partition, 112
    finer, 471
    of unity, 161, 549
        approximate, 356
path, 401
permutation matrix, 406
Pettis integral, 113
phase shift, 332
Plancherel formula, 337, 339, 479
PNT, 504
Poisson summation formula, 342
polar decomposition, 192, 453, 561
polarization identity, 74, 325, 564
polynomial
    Bernstein, 589
    characteristic, 12
    growth, 374
        amenable, 374
    trigonometric, 48
Pontryagin duality, 120, 426, 473
positive operator, 191, 450
positive-definite
    function, 344, 346
        boundedness, 345
    sequence, 314
        Herglotz's theorem, 315
power set, 90

pre-dual, 254, 303
pre-Hilbert space, 71
prime number theorem, 503, 525
    error rate, 503
principal matrix coefficient, 344
probability
    distribution, 362
    measure, 70, 262, 265
        equidistribution, 263
        ergodic, 265
        invariant, 265
    space, 553
product topology, 542
projection, 82
    bounded, 82
    orthogonal, 80, 185
    valued measure, 325, 468
projective
    topology, 542
property (T), 375–377
    relative, 375
    spectral gap, 376
pseudo-metric, 21, 539
pseudo-norm, 21
push-forward, 265
Pythagoras' theorem, 79

quantum graph, 495

Radon
    measure, 239, 353, 514
        vague topology, 515
    Nikodym derivative, 83
    σ-finite case, 85
    signed measure, 520
random walk, 437
reduced word, 225
reflection, 25
reflexive, 113, 209, 214, 254
reflexivity, 473
regular
    graph, 400
        connected, 404
    measure, 240, 558
    representation, 376, 478
regularity

elliptic, 69, 153, 155
  Laplace operator, 135
  on the torus, 157
  Sobolev, 156
Reiter condition, 363
relative trace, 196
Rellich's theorem, 198
representation
  induced, 385
  induction, 384
  irreducible, 345, 458, 498
  left regular, 368
  regular, 478
  unitarily isomorphic, 344
  unitary, 107
reproducing kernel, 81
residual spectrum, 434
resolvent, 411, 413
  function, 413, 416
  set, 409, 412, 500
resonance, 11, 591
restriction operator, 147
  trace, 148
Riemann
  hypothesis, 503
  integral, 111, 112, 368
  sum, 112
  zeta function, 532, 592
Riemann–Lebesgue lemma, 334
Riesz representation, 239
  locally compact, 248
  on $C(X)$, 248
Riesz–Thorin interpolation, 233

Schauder basis, 88
Schmidt game, 590
Schur lemma, 458
Schwartz space, 341
second category, 126
Selberg symmetry formula, 503, 507
self-adjoint, 167
  integral operator, 177
  operator, 132
    densely defined, 488
    positive, 191
    spectral theorem, 177

semi-inner product, 499
semi-linearity, 72
semi-norm, 21
  absorbent set, 298
  continuous, 22
  defines a norm, 22
  Fréchet
    space, 295
  kernel, 22
  locally convex topology, 292
  strong operator topology, 290
sequence, 539
  dense, 48
  equidistributed, 48
sequentially
  compact, 545
  continuous, 259
sesqui-linearity, 72
set
  clopen, 538
  closed, 538
  open, 538
  theory, 537
Siegel set, 398
simple
  approximation, 473
  function, 235, 553
simultaneous spectral theorem, 181
smooth
  boundary, 148
  partition of unity, 161
Sobolev
  embedding theorem, 135, 139,
    150
  regularity, 156
  space, 8, 135
space variable, 330
spectral
  gap, 375, 405
  measure, 317, 324, 444, 453
    atomic, 327
    continuity, 381
    non-diagonal, 461
  radius, 410, 418
  resolution, 181

simultaneous theorem, 181
theorem, 176
  compact self-adjoint operator, 177
  proof, 180
theory, 12, 70
  unbounded self-adjoint, 487
  unitary flow, 344
  unitary operator, 314
  unitary operators, 313
spectrum, 410
  approximate, 434
  approximate point, 434
  discrete, point, 433
  essential, 437
  essential range, 443
  Lebesgue, 318
  pure discrete, 327
  residual, 434
stadium, 165, 568
star operator, 417
star-shaped, 144
Stone
  Čech compactification, 421
  theorem, 351
  Weierstrass theorem, 42, 49
    locally compact, 47
strictly convex, 24
strong
  analytic, 260
  convergence, 123, 261
  integral, 111, 112, 114, 368
  operator topology, 290
  topology, 253
stronger topology, 539
Sturm–Liouville
  boundary value problem
    integral operator, 69
  equation, 66, 68, 69, 167
sub-additive
  sequence, 415
  strictly, 24, 76
sub-multiplicative sequence, 415
super-polynomial decay, 341
superposition, 330

support of a function, 549
symmetric
  operator, 498

tail filter, 540, 541
Taylor
  approximation, 6, 43
  coefficient, 89
  expansion, 89
test function, 141, 296
thermal equilibrium, 8
Tietze's extension theorem, 61, 549
Toeplitz's theorem, 255
Tonelli's theorem, 556
topological
  complement, 82
  group, 91, 353
    abelian, 91
    character, 92
  space
    compact, 545
    neighbourhood, 538
    normal, 547
  vector space, 295
topology, 538
  coarser, weaker, 539
  finer, 543
  induced from a subset, 541
  initial, limit, projective, weak, 542
  locally convex, 292
  product, 542
  strong, 253
  strong operator, 290, 469
  stronger, finer, 539
  ultra-strong operator, 312
  ultra-weak operator, 312
  uniform convergence on compact sets, 429
  uniform operator, 290
  vague, 515
  weak, 253
    operator, 292
  weak*, 254
torsion element, 431
total

derivative, 6, 491
weight, 354
totally
    bounded, 169, 245, 546
        compact, 545
    disconnected, 240, 241
        cover, 245
    ordered set, 90
trace, 147, 183
    map, 182
    relative, 196
    restriction operator, 148
trace-class
    compact, 187
    norm, 183, 187, 195
    operator, 183
tree
    averaging operator, 438
    Laplace operator, 438
    regular, 438
    summing operator, 438
triangle inequality, 539
    L^p norm, 557
trick, 396
trigonometric polynomial, 48, 96
Tychonoff's theorem, 41, 546

ultrafilter, 546, 547
    convergent, 547
uncertainty principle, 342, 592
unconditional convergence, 36
uniform
    boundedness, 121
    convergence, 27
    convergence on compact sets,
        294
    operator topology, 290
uniformly
    continuous, 40
    convergent, 122
    convex norm, 77
    convex space, 76
    distributed, 48
    elliptic operator, 284
unimodular, 361
unipotent

group, 392
    subgroup, 378
unique
    best approximation, 77
    ergodicity, 267
    extension, 58
    prime factorization, 508
    solution, 66
unit, 62
    ball, 19
        compact in weak* topology,
            256
    ball is non-compact, 38, 39, 89
    sphere, 259
unital, 62
    Banach algebra, 409
unitary, 167, 175
    flow, 344
    isomorphism, 313
    operator, 107, 175
    representation, 107
        cyclic, 316
        irreducible, 345, 458, 498
upper envelope, 306
upper-semicontinuous, 306, 307
Urysohn's lemma, 547

vague topology, 515
vector space
    inner product, 71
    norm, 16
    topological, 295
vertex, 400
Volterra equation, 64

Walsh system, 95
wave equation, 10
    Dirichlet boundary problem,
        10
    vibrating string, 11
weak
    analytic, 260
        function is analytic, 260
    derivative, 141
    harmonic, 154

integral, 113, 114
operator topology, 292
topology, 252, 253, 542
weak* topology, 252, 254
weaker topology, 539
weakly harmonic, 156
function, 154
Weierstrass
approximation theorem, 42

Landau kernel, 104
weight, 110
of a function, 3
space, 110
Weyl
law, 196, 202
monotonicity principle, 182
Wiener lemma, 428

Zorn's lemma, 538

Printed in the United States
By Bookmasters